International Conference on Aerospace Technology, Communications and Energy Systems (ATCES 2017)

IOP Conference Series: Materials Science and Engineering
Volume 302

Samara, Russia
28 - 30 September 2017

Editors:

Evgeniy V. Shakhmatov
Vladimir Lupandin

ISBN: 978-1-5108-5839-8
ISSN: 1757-8981

Printed from e-media with permission by:

Curran Associates, Inc.
57 Morehouse Lane
Red Hook, NY 12571

Some format issues inherent in the e-media version may also appear in this print version.

Printed by Curran Associates, Inc. (2018)

Institute of Physics
Dirac House, Temple Back
Bristol BS1 6BE UK

Phone: 44 1 17 929 7481
Fax: 44 1 17 920 0979

techtracking@iop.org

Additional copies of this publication are available from:

Curran Associates, Inc.
57 Morehouse Lane
Red Hook, NY 12571 USA
Phone: 845-758-0400
Fax: 845-758-2633
Email: curran@proceedings.com
Web: www.proceedings.com

Table of contents

Volume 302

2017 International Conference on Aerospace Technology, Communications and Energy Systems (ATCES 2017)

28–30 September 2017, Samara, Russian Federation

Accepted papers received: 11 January 2018
Published online: 8 February 2018

Preface

Papers

Power plants and Engines construction

Control systems and Monitoring

IOP Conf. Series: Materials Science and Engineering **302** (2018) 011001 doi:10.1088/1757-899X/302/1/011001

PREFACE

ATCES 2017 was held in Samara University, Russia, from 28-30 September, 2017. ATCES 2017 was Co-organized by Samara University and Asia Pacific Institute of Science and Engineering. The conference provides a useful and wide platform both for display the latest research and for exchange of research results and thoughts in Aerospace Technology, Communications and Energy Systems. The participants of the conference were from almost every part of the world, with background of either academia or industry, even well-known enterprise. The success and prosperity of the conference is reflected high level of the papers received.

The proceedings are a compilation of the accepted papers and represent an interesting outcome of the conference. This book cover 5 chapters: Power plants and Engines construction; Aerospace engines working process; Control systems and Monitoring; Energy Saving & Eficiency; Production and advanced technologies.

We would like to acknowledge all of those who supported ATCES 2017. Each individual and institutional help were very important for the success of this conference. Especially we would like to thank the organizing committee for their valuable advices in the organization and helpful peer review of the papers.

We sincerely hope that ATCES 2017 will be a forum for excellent discussions that will put forward new ideas and promote collaborative researches. We are sure that the proceedings will serve as an important research source of references and the knowledge, which will lead to not only scientific and engineering progress but also other new products and processes.

The event was prepared and conducted with the financial support of the Russian Foundation for Basic Research on the basis of Grant No. 17-08-20535 (g).

Prof. Evgeniy V. Shakhmatov, Samara University, Russia
Dr. Vladimir Lupandin, Magellan Aerospace Corporation, Mississauga (Toronto), Canada

IOP Conf. Series: Materials Science and Engineering **302** (2018) 011002 doi:10.1088/1757-899X/302/1/011002

COMMITTEES

Honorable Chair

Prof. Kasim Mousa Al-Aubidy, Philadelphia University, Jorda

Conference Chairman

Prof. Evgeniy V. Shakhmatov, Samara University, Russia

Conference Co-Chair

Prof.Andrey B. Prokofiev, Samara University, Russia

Prof. Alexander M. Mebel, Florida International University, USA

Program Chair

Prof. Aleksandr I. Ermakov, Samara University, Russia

Dr. Yaser Qudaih, Kyushu Institute of Technology, JAPAN

Financial Chair

Dr. Jinping Lei, The Hong Kong University of Science and Technology, Hong Kong

Local Chairs

Assoc. Prof. Dmitry A. Uglanov, Samara University, Russia

Assoc. Prof. Ilya N. Krupenich, Samara University, Russia

Technical Committee Member

Prof. Vladimir A. Zrelov, Samara University, Russia

Prof. OKUYAMA Keiichi, Kyushu Institute of Technology, Japan

Prof. Yehia Abdel-Aziz, Head of Space Research Lab, NRIAG, Egypt.

Prof. Aleksandr A. Igolkin, Samara University, Russia

Prof. Dmitry K. Novikov, Samara University, Russia

Prof. Tieying Wu, Nanjing University of Aeronautics and Astronautics, China

IOP Conf. Series: Materials Science and Engineering **302** (2018) 011002 doi:10.1088/1757-899X/302/1/011002

Prof. Adel A. Elbaset, Minia University, EGYPT

Assoc. Prof. Viktor Ya. Sverbilov, Samara University, Russia

Assoc. Prof. Aleksandr S.Vinogradov, Samara University, Russia

Dr. Mohammed Mahdi Ali, Philadelphia University, Jordan

Dr. Tarek Hassan Mohamed, Aswan University, Egypt

Dr. Thongchart Kerdphol, Kyushu Institute of Technology, Japan

Dr. Mohamed Yahia Edries, National Authority for Remote Sensing and Space Sciences, NARSS, Egypt

Dr. Ifigeneia Antoniadou, The University of Sheffield, UK

Dr. John Page,The University of New South Wales, Australia

Dr. Kouzou Abdellah, Djelfa University, Algeria

Dr Pooya Sareh, Imperial College London,UK

Dr. Hassan Migdadi, American university of Madaba, Jordan

IOP Conf. Series: Materials Science and Engineering **302** (2018) 011003 doi:10.1088/1757-899X/302/1/011003

Peer review statement

All papers published in this volume of *IOP Conference Series: Materials Science and Engineering* have been peer reviewed through processes administered by the proceedings Editors. Reviews were conducted by expert referees to the professional and scientific standards expected of a proceedings journal published by IOP Publishing.

IOP Conf. Series: Materials Science and Engineering **302** (2018) 012001 doi:10.1088/1757-899X/302/1/012001

The Study the Vibration Condition of the Blade of the Gas Turbine Engine with an All-metal Wire Rope Damper in the Area Mount of the Blade to the Disk

Vladimir S. Melentjev[1] and Alexander S. Gvozdev[1]

[1]Samara National Research University, Russia, 443086, Samara, Moskovskoe shosse, 34.

E-mail: vladamgenja@gmail.com

Abstract. Improving the reliability of modern turbine engines is actual task. This is achieved due to prevent a vibration damage of the operating blades. On the department of structure and design of aircraft engines have accumulated a lot of experimental data on the protection of the blades of the gas turbine engine from a vibration. In this paper we proposed a method for calculating the characteristics of wire rope dampers in the root attachment of blade of a gas turbine engine. The method is based on the use of the finite element method and transient analysis. Contact interaction (Lagrange-Euler method) between the compressor blade and the disc of the rotor has been taken into account. Contribution of contact interaction between details in damping of the system was measured. The proposed method provides a convenient way for the iterative selection of the required parameters the wire rope elastic-damping element. This element is able to provide the necessary protection from the vibration for the blade of a gas turbine engine.

1. Introduction

Improving the reliability of modern turbine engines is actual task. This is achieved due to prevent a vibration damage of the operating blades of turbo machines. The main factor, which limits the level of vibration, is a damping capacity of the blades.

At the Department of construction and design of aircraft engines and in the laboratory "Vibration resistance and reliability of the engines and aircraft systems" for many years studied the problems with damping of the gas turbine engine blades. There many designs of damping elements from tapes, wire ropes and the pressed wire material MR, used in modern gas turbine engine have been developed. This work engaged Ulanov A. [1, 2], Frolov V. [3], Ponomarev Yu. [4], Ermakov A. [5], Belousov A. [6], Chegodaev D., Lazutkin G. [7], Kotov A., Davydov D. and many others [8]. Among the publications of foreign scientists, who are engaged in similar studies, can be distinguished [9-14].

Elastic-damping elements made of wire rope are common in aeronautical engineering and transport engineering. Their work is based on the structural damping. The main advantages of these elements is a good load-bearing capacity, resistance to temperature and aggressive environment, high damping, compact size and long life cycle [1, 4]. The metal wire rope can be laid separate segments or in one piece in the form of a serpentine.

In [6] given the various schemes of fixing of the compressor blades using elastic-damping elements. Schemes can be distributed into three groups: I. The disc of the impeller and the blade of a compressor interact only through the elastic-damping element; II. The design of the root attachment of blade and the retaliatory groove of disk provides active work the elastic-damping element, located under the base of the root attachment. III. The elastic-damping element is established under the base of the root attachment of blade with virtually no change in the retaliatory groove of disk.

The purpose of this work was to create a model and development method of calculation of vibration of the turbine engine blade with a wire rope element installed in the root attachment of blade.

IOP Conf. Series: Materials Science and Engineering **302** (2018) 012001 doi:10.1088/1757-899X/302/1/012001

The proposed method allows to achieve collaboration a system "disk - blade" (is used the contact interaction) and elements from the wire rope in the same model. The finite element method and the method of Euler-Lagrange are used. To calculate is required to create a model that takes into account the joint work a relatively "rigid" attachment (having some flexibility) and more "soft" the elastic-damping elements. At that, "rigid" attachment has little damping in comparison with more "soft" elements. The 3D model of the blade was created in the CAD-package [15]. In this paper we used the software ANSYS. We used a finite element COMBIN14 for modeling the elastic-damping elements. We used contact interaction to simulate the "rigid" attachments. The study of the application of a contact in ANSYS to simulate structural damping is dedicated to a number of foreign works [16, 17].

2. Materials and methods

The method uses the same finite element model of the compressor blade with three modifications that depend on attachment conditions. The diagram (Fig. 1) shows a condition of attachment and the main parameters, that you must to calculate for the simulation.

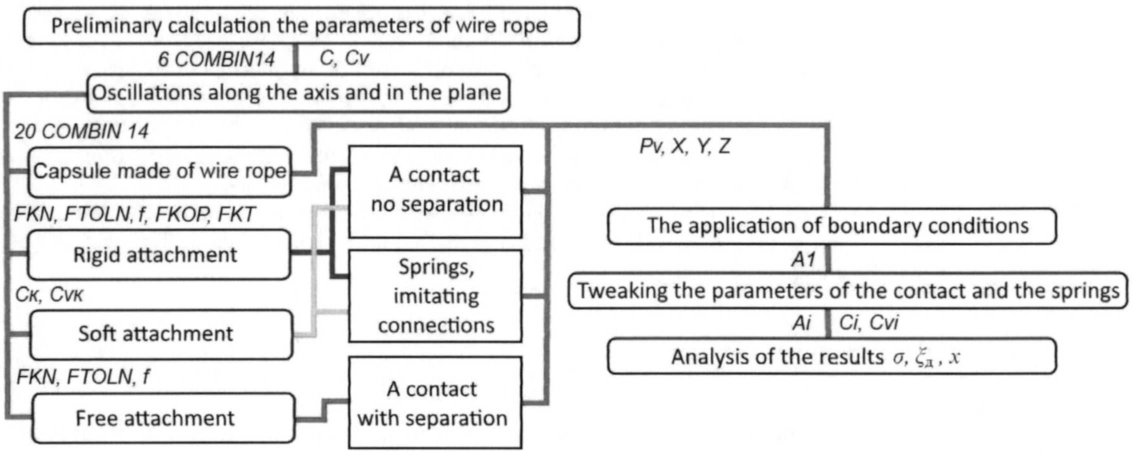

Figure. 1. Block diagram of the algorithm for calculating the compressor blade vibrations with a wire rope element in the root attachment

The desired type of attachment must be selected:

1. The root attachment of blade from all sides surrounded by elements from the wire rope. The blade interacts with the disk of a rotor through "the capsule".

For the following three types of attachments the elastic-damping element is mounted only at the base of the root attachment of blade.

2. If attachment is "rigid", the retaliatory groove in the disk does not change. Stiffness of the connection of the compressor blades with the disk of the impeller is many times higher than the stiffness of the wire rope.

3. If attachment is "soft", stiffness of the connection of the compressor blades with the disk is commensurate with the stiffness of the wire rope. Attachment type 2 and 3 can be modeled in two ways: without the use of contacts (springs, imitating connections), and with contacts.

4. Attachment of the compressor blades in the disk is performed so as to restrict only lateral movement and upward movement (Y-axis). It is "free" attachment. The standard contact can be used to describe such behavior. The contact surface of the disc and blades may drift apart.

The oscillation amplitude A_i and logarithmic damping decrement ξ_i of the oscillation are determined as a result of analysis of the graphs of the oscillation.

In the first step required to determine the dependence of the stiffness coefficient C and the damping coefficient Cv a wire rope when working on the compression from its design parameters and deformation amplitude.

The essence of the experiment is described below. For this study, the standard wire rope (diameter 6 mm) was divided into separate strands. Then the separate strands were re-merged in form of the quasi-continuous ring [18]. Therefore rings have a different radius.

Then the quasi-continuous rings were subjected to cycles of loading and unloading with the help of specially equipped a tensile testing machine. During the experiment, the values of displacement of the plate 1 (Fig. 2) and the corresponding reaction force recorded using a dynamometer in real time.

As can be seen from Fig. 2, equipment includes a fixed base (thickness 10 mm), the movable plate 1. The parts 3, 5 controls a movement of the plate 1. Between the base 1 and the plate member 2 was placed a wire rope. On the outer side of each plate is welded tube 3 having a square cross section and dimensions of 40x40 mm. The thickness of the pipe wall is 3 to 5 mm. The welding is made on the contour 4 at several points (Fig. 2). During compression the steel rod 5 presses directly on the plate 1 due to the through-hole in the pipe 3.

For this work were used the tested rings with the smallest radius. The parameters of the element is the thickness $H = 5.5$ mm, the outer diameter $D_1 = 53.4$ mm, inner diameter $D_2 = 42.4$ mm, material steel $E = 200\ 000$ MPa.

It was assumed that an element to the beginning of the calculation already has deformed. This is necessary to compensate for the shift of the compressor blade by centrifugal force (a static load). The biggest hysteretic loop for the calculation of the shift amount was analyzed (Fig. 3).

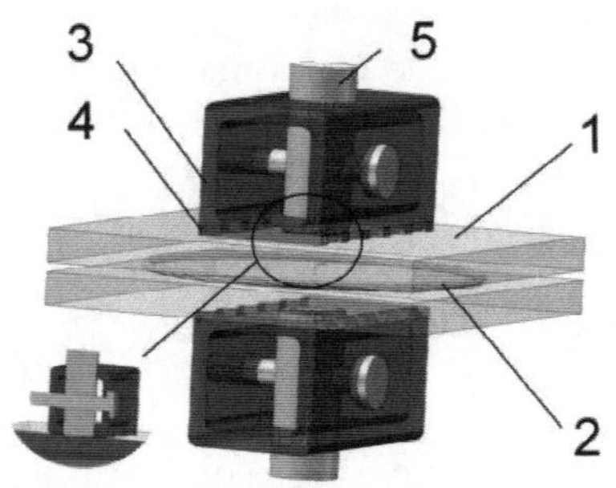

Figure. 2. Equipment for compression testing the wire rope

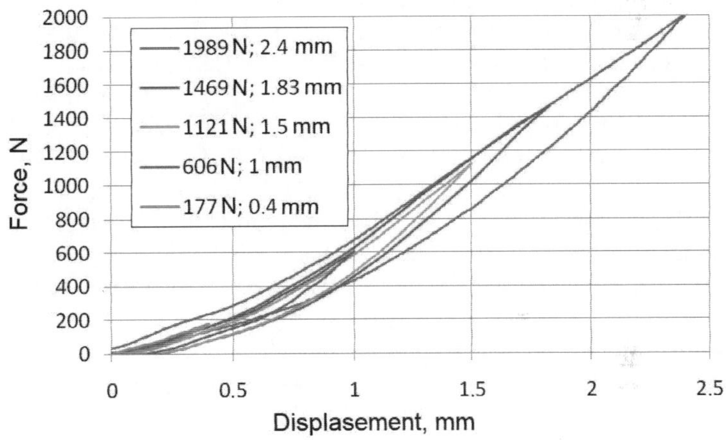

Figure. 3. The hysteretic loops of wire rope at different amplitudes

Several variants of the static characteristics with different maximum load for determination the dynamic rigidity of the elastic-damping element have been received (Fig. 5).

The graphs of the mean-cyclic stiffness C and the energy dissipation coefficient ψ from the deformation amplitude A were approximated by the expressions (Fig. 4, 5):

$$C = -681.6 \cdot A^2 + 1775 \cdot A + 546, \tag{1}$$

$$\psi = -0.34 \cdot \ln A + 0.916. \tag{2}$$

The dimensionless criteria were introduced in this work: the relative amplitude $\varepsilon_A = A/H$; the relative stiffness $\varepsilon_C = C/(l \cdot E)$. Graphs of the characteristics of the elastic-damping element made of wire rope in dimensionless coordinates are shown in Fig. 4 and 5.

Area of the base of a compressor blade $S = 199$ mm^2. Thickness of the wire rope $H = 5.5$ mm. Therefore, the segments of the wire rope with a maximum length $L = 36$ mm can be placed under the compressor blade.

A wire rope as MR material is a material with non-linear properties. Therefore, the calculation of oscillations using elastic-damping element made of a wire rope, has several iterations [1].

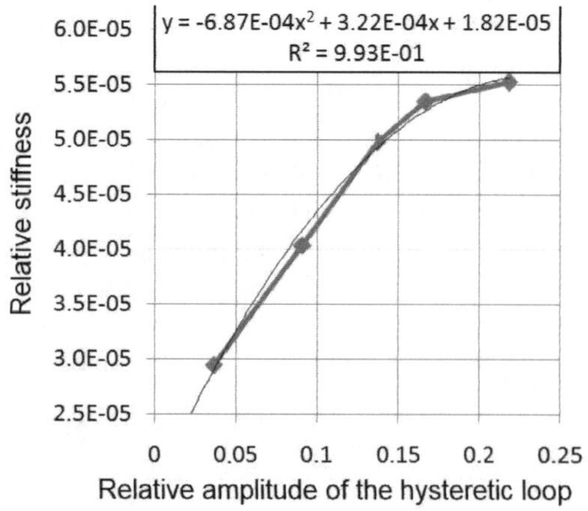

Figure. 4. Relative mean-cyclic stiffness

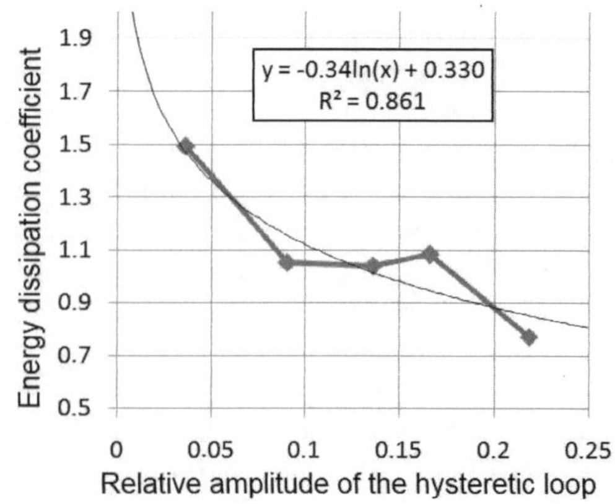

Figure. 5. The dependence of the energy dissipation coefficient ψ from the relative amplitude

Parameters of the first iteration point: A_1 = 1 mm, $\varepsilon_A = 0.182$, $\varepsilon_C = 5.4{\cdot}10^{-5}$, $\psi = 0.917$, ν = 86 Hz, C_1 = 388000 N/m, Cv_1 = 660 N·sec/m. The final amplitude of the deformation is 1.08 mm. The discrepancy is considered acceptable. The parameters are C_i = 388000 N/m, Cv_i = 660 N·sec/m. Auxiliary elements simulating the impeller disc for use a contact in model were created.

Modeling "rigid attachment" was in the first calculation. The values of the real variables of the contact was: FKN = 1, FTOLN = 0.1, *f*(MU) = 0.1, FKOP = 1, TAUMAX = 10^{25}, FKT = 1. Parameters of elastic-damping element (*C* = 172000 N/m, *Cv* = 114 N·sec/m) were left unchanged.

"Soft" contact was modeled in two subsequent calculations. The values of the real variables of contact for the first calculation are FKN = 0.1, FTOLN = 0.1, *f*(MU) = 0.1, FKOP = 1, TAUMAX = 10^{25}, FKT = 1.The average frequency ν = 200 Hz was used to calculate the value of *Cv*.

3. Results

The following equations for the wire rope which running to transverse compression, obtained and approximated. The relative mean-cyclic stiffness is (R^2 = 99.3%):

$$\varepsilon_C = -6.873 \cdot 10^{-4} \varepsilon_A^2 + 3.218 \cdot 10^{-4} \varepsilon_A + 1.82 \cdot 10^{-5}, \tag{3}$$

The energy dissipation coefficient is (R^2 = 86.2%):

$$\psi = -0.344 \ln(\varepsilon_A) + 0.331 \,. \tag{4}$$

The maximum calculated deviation of the compressor blade for scheme "capsule" is 17 mm (ξ = 3.23). Displacement was measured at node of the bottom of the compressor blade airfoil (1/3 the height of the blade airfoil). A strain sensor was located in this place during the experiment.

It was found that the load-bearing capacity of the wire rope is less than that of the material MR investigated in [1]. The maximum displacement was 2.0 mm (the wire rope, *C* = 183 N/mm, *Cv* = 780 N·sec/m) versus 0.6 mm (MR, *C* = 1709 N/mm, *Cv* = 810 N·sec/m).

The maximum deviation of the compressor blade for "rigid" attachment without contacts is 1.07 mm. The maximum amplitude was 0.44 mm. Logarithmic damping decrement of oscillations was ξ = 0.03. The maximum displacement of the blade with the "rigid" attachment and contact amounted to 0.64 mm. The oscillation amplitude amounted to 0.037 mm, ξ = 0.184. The maximum displacement elements COMBIN14 were 0.16 mm. In general, the main energy dissipation comes in the place of contact the blade with the impeller disc groove, when using a contact for communication "the disk - the root attachment of blade". The elastic-damping element made of the wire rope in this case is an auxiliary element for case, when amplitude oscillation is large.

Equivalent stress (von Mises stress) has been determined for this model. Stress was calculated only from the vibration action. Stress from centrifugal force was not taken into account. The maximum value was 100 MPa, which is acceptable.

The results of studies of various parameters of the wire rope for "soft" contact were shown in Fig. 6. As a result of the work, the model with a variety of variants of placement of elastic-damping elements made of the wire rope in the root area of the disc of the impeller was created. A wide range of the logarithmic damping decrement of oscillations from 0.104 to 0.235, in the simulation of the most common attachments was obtained [6].

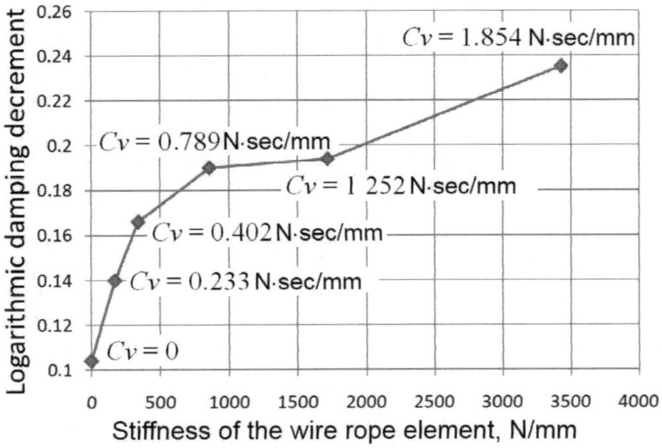

Figure. 6. Graph of logarithmic damping decrement at different stiffness of the wire rope: $P_v = 84$ kPa; FKN = 0.1, FTOLN = 0.1, *f*(MU) = 0.1, FKOP = 0.5

Calculations confirmed the effectiveness of the dampers based on the wire rope for damping vibrations of the compressor blades. The greatest efficiency of vibration isolation is achieved with a modified design of the root attachment of blade. In this case a slight slippage of the bottom of the compressor blade relative to the groove in the plane (normal to axis) of the disc rotor is possible.

4. Discussion and Conclusion

In general, the examined wire rope ($H = 5.5$ mm) was too soft for damping the compressor blades by shceme "capsule". It requires research a wire ropes with more the load-bearing capacity. This will require the tensile testing machine with a more wide range of loads.

The proposed method allows to consider a joint operation of the system "the disk - the blade" with the contact interaction and elastic-damping elements made of the wire rope in a single model. The method is based on the finite element method and the Euler-Lagrange method to calculate the contact. This allows the easy selection of the required parameters (by iterations) of the elastic-damping element made of the wire rope, capable of ensuring the protection from the vibration of the blade of a gas turbine engine.

Acknowledgment
This work was supported by the Ministry of Education and Science of the Russian Federation in the framework of the implementation of the Program ‘‘Research and development on priority directions of scientific-technological complex of Russia for 2014– 2020”

References

[1] Ulanov, A., and S. Bezborodov (2016) Calculation method of pipeline vibration with damping supports made of the MR material. Procedia Engineering, Vol. 150, 2016, pp. 101-106 doi: 10.1016 / j.proeng.2016.06.725

[2] Yan, H., Zhang, L., Jiang, H. -.,& Ulanov, A. M. (2015). Calculation of multi-layer plate

damper under one-axial load. *Chinese Physics B*, 25(2) doi:10.1088/1674-1056/25/2/024306

[3] Frolov, V. (2011). Dempfirujushhie ustrojstva v konstrukcijah lopatok kompressorov dvigatelej semejstva NK [Damping device at the structures of compressor blades an engine family NK]. *Vestnik samarskogo gosudarstvennogo ajerokosmicheskogo universiteta* [Vestnik of the Samara State Aerospace University], no. 3(27): 242-250.

[4] Ponomarev, Yu., A. Ermakov, O. Simakov [et al.] (2013). Metallic counterpart of rubber: A material for vibration and shock protection. *Metal Science and Heat Treatment,* no. 55 (1-2): pp. 8-13.

[5] Ermakov, A. I., Urlapkin, A. V., & Fedorchenko, D. G. (2014). The features of resonance stress scatter in turbine wheels with a weak connectivity of blade vibrations. Research Journal of Applied Sciences, 9(11), 795-799. doi:10.3923/rjasci.2014.795.799

[6] Belousov, A. I., & Badamshin, I. K. (2015). Conceptual approach to design of materials for GTE turbine blades. *Russian Aeronautics*, 58(3), 325-330. doi:10.3103/S1068799815030137

[7] Troynikov, A. A., & Lazutkin, G. V. (2014). Experimental determination of the variation of the elastic damping properties MR material in a continuous operation. ARPN Journal of Engineering and Applied Sciences, 9(12), 2876-2879.

[8] Melentjev, V. S., & Gvozdev, A. S. (2015). Method of static characteristics calculation of adjustable two-ring type full-metal vibroinsulators in a non-linear approach. Modern Applied Science, 9(2), 267-278. doi:10.5539/mas.v9n2p267

[9] Tatzko, S., L. Panning-von Scheidt, J. Wallaschek [et al.] (2013). Investigation of alternate mistuned turbine blades non-linear coupled by under platform dampers. Proceedings of ASME Turbo Expo 2013: Turbine Technical Conference and Exposition, GT2013, San Antonio, Texas, USA

[10] Krack, M. , Panning, L., et al. (2012) "Robust design of friction interfaces of bladed disks with respect to parameter uncertainties", Proceedings of the ASME Turbo Expo, Volume 7, Issue PARTS A AND B, 2012, pp. 1193-1204.

[11] Tamai, R., Tanaka, R., et al. (2013) "Vibration analysis of shrouded turbine blades for a 30 MW gas turbine", ASME 2013 Turbine Blade Tip Symposium, TBTS 2013; Hamburg; Germany; 30 September 2013 through 3 October 2013; Code 103231.

[12] Nan, G. (2013) "An analytical method for vibro-impact between shrouded blades under a multiple-harmonic excitation", Journal of Vibroengineering, Volume 15, Issue 4, 2013, pp. 1947-1960.

[13] Wang, Y., Guo, X. , et al. (2011) "Nonlinear vibratory characteristics and bifurcations of shrouded blades", Lixue Xuebao/Chinese Journal of Theoretical and Applied Mechanics, Volume 43, Issue 4, July 2011, pp. 755-764.

[14] Wadia, A.R. and Szucs, P.N. (2008) "Inner workings of shrouded and unshrouded transonic fan blades", Journal of Turbomachinery, Volume 130, Issue 3, July 2008, Article number 031010.

[15] Ryazanov, A.I. (2016) "Automated 3D modeling of working turbine blades", Russian Engineering Research, Volume 36, Issue 9, 1 September 2016, Pages 751-754

[16] Bo, You, Yi Luo, Xiaodong Wang (2013) Contact algorithm of finite element analysis for prediction of press-fit curve. Journal of Information & Computational Science, 10:9 (2013): pp. 2591–2600

[17] Ender Cigeroglu, Ning An, Chia-Hsiang Menq (2009) Forced response prediction of constrained and unconstrained structures coupled through frictional contacts. Journal of Engineering for Gas Turbines and Power, Vol. 131 / 022505-1

[18] Ponomarev, Yu., V. Arhangelsky, et al. The method of manufacturing the elastic-friction elements of wirerope insulators. RU Patent 2199683, F16F 7/14, publ. October 20, 2002.

IOP Conf. Series: Materials Science and Engineering **302** (2018) 012002 doi:10.1088/1757-899X/302/1/012002

Defining the Field of Existence of Shrouded Blades in High-Speed Gas Turbines

Anatoliy I. Belousov[1]and Sergeiy V. Nazdrachev[1]

[1]Samara National Research University, Russia, 443086, Samara, Moskovskoe shosse, 34.

E-mail: aibelousov@mail.ru

Abstract: This work provides a method for determining the region of existence of banded blades of gas turbines for aircraft engines based on the analytical evaluation of tensile stresses in specific characteristic sections of the blade. This region is determined by the set of values of the parameter, which forms the law of distribution of the cross-sectional area of the cross-sections along the height of the airfoil. When seven independent parameters (gas-dynamic, structural and strength) are changed, the choice of the best option is proposed at the early design stage. As an example, the influence of the dimension of a turbine on the domain of the existence of banded blades is shown.

1. Introduction

The tendency to reduce the number of stages of the turbine determines the need to increase the gas dynamic load on the blade rims. To achieve a high efficiency of the stage, an increase in peripheral rotation speeds is required, which leads to a quadratic increase in the centrifugal load on the rotor parts.

The circumferential speeds of rotation of 500–600 m / s, achieved in existing turbines, predetermine the difficulties in the strength design of blades and disks. This problem is especially important for banded working blades, since the shroud platform additionally loads the pen, the locking connection and the disk. In high-speed turbines, the effect of centrifugal force on the shroud can become critical, which will lead to the impossibility or inexpediency of creating a bandage blade [1–3]. Also the modernization of gas turbine units for ground application requires the using of turbine rotor with hard centrifugal force constrain, and the reducing blade weight is an important problem [4, 5].

2. The influence of the bandage volume on the impeller mass

We determine the influence of the volume of the shroud on the mass of the impeller of the turbine on the following example: a peripheral speed of 480 m / s, a radius of the center of the weight of the shroud of 309.5 mm, a blade height of 45 mm. We take the material of the blade and the shroud platform - ZhS32 (density 8.8 g / cm^3).

Let's consider three variants of blades with volumes of shrouds:
- 0 mm^3 (unshrouded blade);
- 320 mm^3 (corresponds to a lightweight shroud platform);
- 500 mm^3 (corresponds to an unweighted shroud platform).

IOP Conf. Series: Materials Science and Engineering **302** (2018) 012002 doi:10.1088/1757-899X/302/1/012002

The chosen parameters of the shroud platform correspond to the design parameter so the high-pressure turbine blades of NK-110 and NK-93 prop fan engines developed in the ND Kuznetsov construction bureau.

For the given volumes of the shroud platform, the distribution of the cross-sectional areas along the blade's height was selected and the load-carrying sections of the disks 'figure 1' were determined with practically the same strength.

'Table 1' shows the weights of the blades and disks. The weight of the impeller of a high-speed turbine can double when the blades are shrouded, and the volume of the shroud can have a decisive influence on the structural shape of the unit.

For this reason, in the early stages of turbine design, it is important to predict the possibility of bandage blades and to estimate the weight of the working wheel.

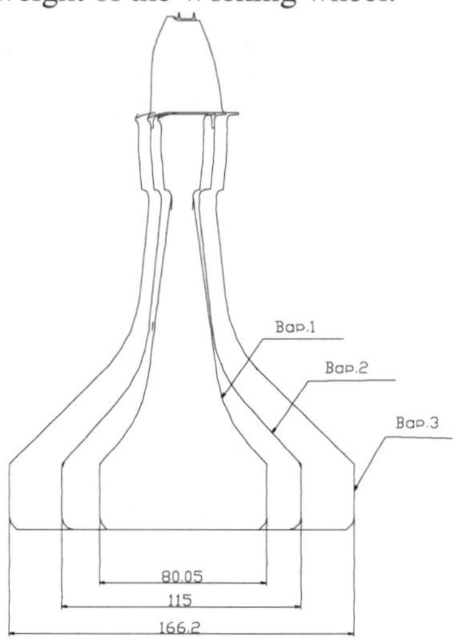

Figure 1. The contours of discs for different blade options

Table 1. Weight characterization of blades and discs

№ Variant	Mass of the blade, g	Mass of the disc, kg	Mass of the rotor, kg
Вар. 1	46	42,1	50
Вар. 2	69	58,3	71
Вар. 3	99	91,8	110

3. The precidion of the field of existence of shrouded blades

For impure blades, an evaluation is made of the strength of the tension stresses in the characteristic sections of the feather [6]. Let us take the power law of distribution of areas over the blade height in the following form:

$$F_i = F_{вт}\left(\frac{R_{вт}}{R_i}\right)^{m}, \tag{1}$$

Where F_i –Area of the i-th cross section on the radius R_i;

m – The exponent characterizing the distribution of the cross-sectional areas along the blade height.

For blades with shrouds, the tensile stresses from centrifugal forces in the bushing cross section $\sigma_{p.вт}$ can be written in the form:

$$\sigma_{p.вт} = 2\rho U_{cp}^{2} \frac{h}{d_{cp}} \Phi + 4\rho V_{n} U_{cp}^{2} \frac{\frac{d_{cp}}{2} + h + \delta}{F_{вт} d_{cp}^{2}} \tag{2}$$

Where, ρ – density of the blade material; U_{cp} – circumferential velocity on the average radius of the blades.; h –blade height; d_{cp} – average diameter of the turbine stator; V_{n} – volume of the blade shroud platform; δ – The difference between the position of the center of gravity of the shroud platform and the peripheral sub-sectional section of the airfoil, which is 2-4mm in size for different sized blades;Φ – coefficient of the airfoil shaping of the blade (is a function of *m*).

Having integrated the centrifugal force from the lower sections by the height of the blade, it is possible to determine the tensile force and tensile stresses in any section of the blade.

Possible configurations of the blade with a shroud are realized in the range of values of m under the following two conditions:

1) the stresses in the bushing and peripheral sections of the pen do not exceed the permissible level;
2) the ratio of the areas of the hub and peripheral sections of the blade, which determines, in the first approximation, the vibrational state of the blade, is in the required range.

Thus, it is possible to introduce the concept of "the domain of existence of shrouded blades". Under the definition we mean the set of values of the parameter m for which the two conditions indicated above are satisfied.

Let us analyze the first condition. In the bushing cross section, the stress level can be taken from theexpected material temperature and the required blade life:

$$\sigma_{p.вт} = \frac{\sigma_{b,\tau}^{t}}{k} \tag{3}$$

Where, $\sigma_{b,\tau}^{t}$ – long-time strength of the blade under temperature *t* and operation time τ;

k – structural load factor.

In the peripheral section, the tensile stresses should not exceed the value $\mu\sigma_{вт}$, где μ =0,3 – 0,7.The physical meaning of the coefficient μ lies in the fact that it determines the law of variation of stresses along the blade's height. The stresses are determined only by the centrifugal load from the shroud, expressed by the second term of formula (2).

The second condition is based on the fact that the parameter χ can vary in the range 1,2 – 7,0 depending on the height of the blade.

Then the domain of existence of shrouded blades can be described by the following system of equations and inequalities:

$$\frac{F_{вт}}{F_{пер}} = \left(\frac{R_{n}}{R_{вт}}\right)^{m,}; \tag{4}$$

$$\sigma_{p.вт} = \rho\omega^2 \left(\frac{V_n \cdot (R_n + \delta)}{F_{вт}} + \frac{R_n^{2-m} \cdot R_{вт}^m - R_{вт}^2}{2-m} \right); \tag{5}$$

$$\sigma_{p.n} = \frac{\rho\omega^2 V_n (R_n + \delta)}{F_{вт} \cdot \left(\frac{R_{вт}}{R_n} \right)^m}; \tag{6}$$

$$\sigma_{p.n} < \mu^* \sigma_{p.вт} < \frac{\sigma^t_{b,\tau}}{k};$$

$$\chi_{\min} < \left(\frac{R_n}{R_{вт}} \right)^{m,} < \chi_{\max} \tag{7}$$

The limiting case of the existence of blades is reached when the stresses in the bushing and peripheral sections are equal to the maximum. Linking the second and third expressions in (7), we obtain:

$$\frac{V_n \cdot (R_n + \delta)}{F_{вт}} + \frac{R_n^{2-m} \cdot R_{вт}^m - R_{вт}^2}{2-m} = \frac{V_n (R_n + \delta)}{\mu F_{вт} \cdot \left(\frac{R_{вт}}{R_n} \right)^m}, \tag{8}$$

From where we get,

$$V_n = F_{вт} \cdot \frac{(R_n^{2-m} \cdot R_{вт}^m - R_{вт}^2)\mu}{(m-2)(R_n + \delta)(\mu - \left(\frac{Rn}{R_{вт}} \right)^m)} \tag{9}.$$

Volume of the blade shroud platform

$$V_n = F_{вт} \cdot \frac{\mu \sigma_{p.вт.}}{(R_n + \delta)\rho\omega^2} \left(\frac{R_n}{R_{вт}} \right)^m \tag{10}$$

Equating expressions (9) and (10), we obtain an equation for determining

$$\bar{\sigma}_{p.вт} = \frac{\sigma_{p.вт}}{R_{вт}^2 \cdot \rho\omega^2} = \frac{\left(\frac{R_n}{R_{вт}} \right)^2 - \left(\frac{R_n}{R_{вт}} \right)^m}{(m-2)(\mu - \left(\frac{R_n}{R_{вт}} \right)^m)} \tag{11}.$$

Equation (11) establishes a relationship between the geometric characteristics of the blade and its stressed state through the parameter m at the limiting stress values. The domain of existence of shrouded blades lies in the range of numbers m, determined by the following conditions:

$m > m_{пред}$,where, $m_{пред}$Is a solution of equation (11)–The condition for ensuring the strength of the blade to tensile stresses;

$\chi_{\min} < \left(\frac{R_n}{R_{вm}}\right)^{m,} < \chi_{\max}$ The condition of the correspondence of a given ratio of cross-sectional areas to ensure a favorable vibration state.

4. Graphical interpretation of the field of existence

The Graphical interpretation of the domain of existence is shown in Fig.2.

The left-hand side of equation (7) is graphically represented by a straight line parallel to the horizontal axis. The right-hand side corresponds to the curve of a monotonically decreasing function.

In 'Figure' 2 shows the results for a turbine with a blade height of 50 mm, an average radius of 0.4 m and a peripheral speed of 440 m / s (the first stage of a turbine turbojet thrust 300-350 kN). For comparison, the area of existence of a blade without a shroud for the same specified conditions is also shown.

We draw lines corresponding to different values of the peripheral section strength μ (Fig. 3). It is convenient to reconstruct the illustration of the field of existence for practical application in terms of the ratio of the areas χ (Fig. 4).

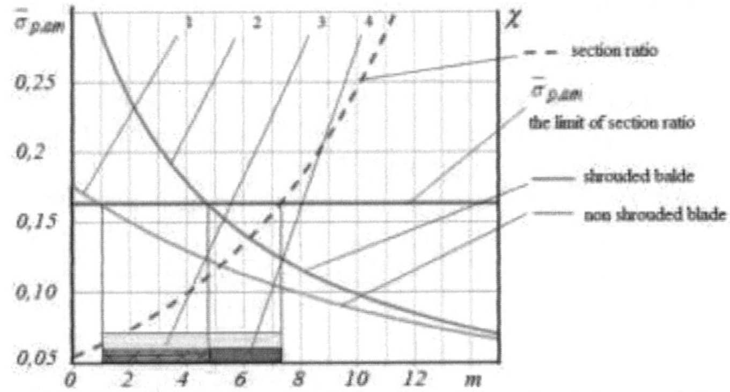

Figure 2. Graphical interpretation of the area of existence of the blades:
1- the line of ultimate stretching stresses of non-shrouded blades;2–line of limiting spreading stretches of shrouded blades;3- area of the existence of the blade;4- area of existence of blades with a shroud

Reducing the volume of the shroud, leading to a decrease in the strength of the peripheral section of the blade, expands the area of possible existence of blades. In the extreme case, when μ =0, the existence region of the shelving and non-platformed blades coincide.

In practice, the blades with numbers $\mu < 0{,}2 - 0{,}3$ Do not apply because of low loading of the peripheral part and increased in connection with this mass of the blade.

For a given dimension, shrouded turbine blades may be in the range of area ratio χ from 3 to 7 depending on the specified strength of the peripheral section.

IOP Conf. Series: Materials Science and Engineering **302** (2018) 012002 doi:10.1088/1757-899X/302/1/012002

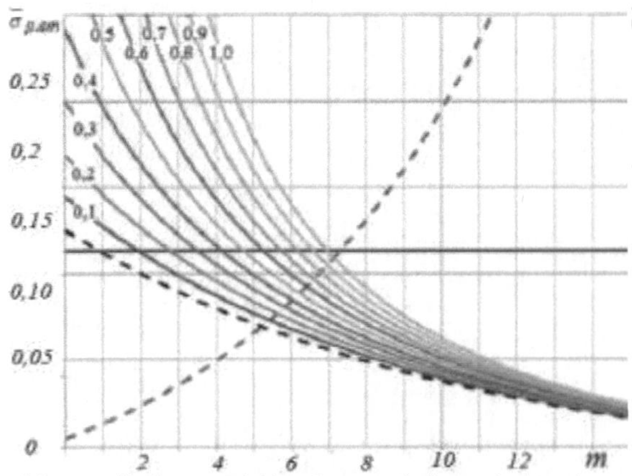

Figure 3. Areas of Existence of shrouded and non-shrouded blades with different level of tension in the peripheral section.

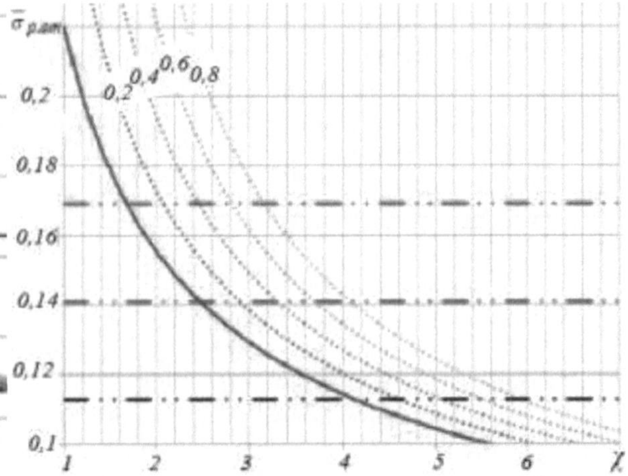

Figure 4. Borders of the areas of existence of the blades for a of smaller dimension turbine

For comparison, we will construct similar graphs for a turbine of smaller dimensions: an average radius of 220 mm and a blade height of 40 mm (corresponding to the first stage of a turbine of a turbojet with thrust of 140-160 kN), for the same circumferential velocity of 440 m / s (see Fig. 4).

For this dimension of the turbine at permissible tensile stresses in the bushing section of 160 MPa, the permissible ratio of the areas is in the range 4.8-6.

5. Conclusion

With an increase in the dimension of the turbine, the banded blades are realized in a wider range of the area ratio. This is due to a quadratic increase in the centrifugal load due to an increase in the rotational speed with a decrease in the average diameter of the turbine.

So, the gas dynamic parameters of the turbine, the specific strength of the material and the design features of the blade itself determine the domain of existence of the blades with the shrouds. Its definition will allow choosing the parameters of the bearing sections and the configuration of the shroud platform that are most suitable for the purpose of the engine being designed.

References

[1] Belousov A.I., Nazdrachev S.V. (2015) *Dialectics of relief of the shrouds of turbine blades*. Pumps. Turbines. Systems. 2015. № 3 (16). P. 7-14.

[2] Belousov A.I., Nazdrachev S.V. (2016) *Principles of facilitating the shroud of the turbine blade.* Izvestiya Vuzov. Aviation equipment, 2016. № 1. P. 48 - 53.

[3] Belousov A.I., Nazdrachev S.V. (2016) *Weight optimization principles for turbine blade shroud platform*. Russian Aeronautics (Iz. VUZ), 2016. Vol. 59. No 1. P. 378–382. DOI 10.3103/S1068799816010086 Print ISSN 1068- 7998; online ISSN 1934-7901.

[4] Belousov A.I., Nazdrachev S.V. (2014) *Methodology of modernizing the serial converted gas turbine unit. Russian Aeronautics*, 2014. Vol.57. № 4. P. 378–382. DOI 10.3103/S1068799814010102.

[5] Belousov A.I., Nazdrachev S.V. (2014) *Methodology of modernizing the serial converted gas turbine unit. Izvestiya Vuzov*. Aviation equipment, 2014. № 4. P. 36 – 38.

[6] Kholshchevnikov K.V. (1970) *Theory and calculation of aircraft scapular machines* / ¬ Moscow: Mechanical Engineering. 356 p.

LEAKAGE ACCOUNT FOR RADIAL FACE CONTACT SEAL IN AIRCRAFT ENGINE SUPPORT

A S Vinogradov* and T V Sergeeva

Department of Design and Engineering of Aircraft Engines, Institute of Aircraft Engines Design and Power Plants, Samara National Research University, Samara, 34 Moskovskoye shosse, Russia

*E-mail: a.s.vinogradov@list.ru

Abstract. The article is dedicated to the development of a methodology for the radial face contact seal design taking into consideration the supporting elements deformations in different aircraft engine operating modes. Radial face contact seals are popular in the aircraft engines bearing support. However, there are no published leakage calculation methodologies of these seals. Radial face contact seal leakage is determined by the gap clearance in the carbon seal ring split. In turn, the size gap clearance depends on the deformation of the seal assembly parts and from the engine operation. The article shows the leakage detection sequence in the intershaft radial face contact seal of the compressor support for take-off and cruising modes. Evaluated calculated leakage values (2.4 g/s at takeoff and 0.75 g/s at cruising) go with experience in designing seals.

1. Introduction

Support seals of the aircraft engine determine the reliable bearing operation and its thermal state [1, 2, 3].

Oil leakage in the air cavity is unacceptable because it leads to irretrievable oil loss and air contamination for the cockpit and cabin pressurization. Apart from that the oil from the support air cavity can get into the engine air path and burn together with fuel in the combustion chambers. This leads to the deterioration of the combustion conditions and the carbon deposition on the chambers walls. Radial face contact seal (further - RFCS) consists of a cover for separating oil and air cavity. Besides, RFCS provides the sealing between the rotor and stator. This seal is insensitive to large axial displacements, so it can be applicable on the turbine supports, where large axial displacements have taken place. RFCS provides an opportunity to have a high temperature in the seal area and it simpler face contact seal in constructional terms [1, 4, 5]. Leakage in RFCS is much less than in the labyrinth seal [5, 6, 7].

2. Finite-element model creation of the support

In this work intershaft RFCS of the engine compressor NK-56 was investigated. The support design with the seal is shown in figure 1. The support front wall is washed by the air from the compressor eighth stage. The same air is used for front support RFCS and intershaft RFCS pressurization. Back RFCS is pressurized with air from the tenth stage.

RFCS is installed on the nut on the medium pressure compressor shaft, and this seal consists of a split carbon ring. It is pressed to the hub by the cylindrical band with air entering from the support air

IOP Conf. Series: Materials Science and Engineering **302** (2018) 012003 doi:10.1088/1757-899X/302/1/012003

cavity. The hub is held down on the high pressure compressor shaft. Carbon ring also clamped by the end face to the other hub. This hub is pressed onto the nut. Hub and seal ring are fixed against axial displacement by a nut. The hub on the high pressure shaft is also fixed by a nut. Both nuts are locked by special elements. They are folded in the joggle at the shafts faces [5, 8].

Intershaft RFCS geometrical model is presented in figure 1. The values shown in figure 1 were used to create the calculation models to determine the value of strain and flow through the seal.

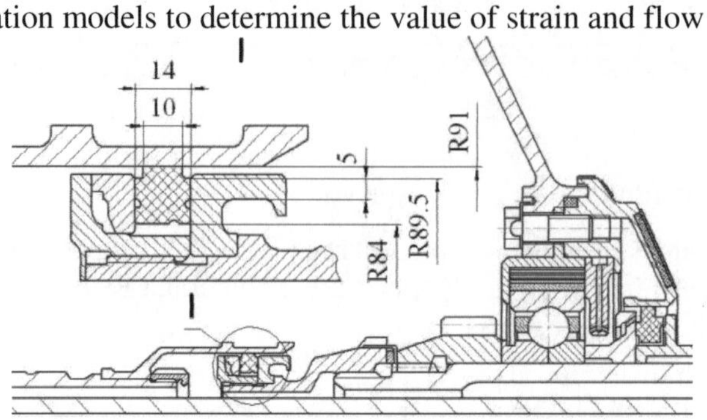

Figure 1. Intershaft RFCS geometrical model.

For that purpose heat, gas-dynamic and hydraulic engine calculations at the maximum (H=0, M=0) and cruising (H=11, M=0.8) mode also were done.

Axial and radial forces, heat generation in the bearing and radial face contact seal have been identified [9, 10]. On the ground of geometrical model flat, axisymmetric and finite-element model was created for subsequent thermal and structural calculations (use model was divided into 16657 elements).

3. Heat calculation

For the displacement and deformation evaluations of the intershaft seal parts the following calculations were performed with the imposition previously identified loads.

Thermal analysis (initial data are the boundary air temperature, the convection heat transfer coefficient and the heat flow released from the intershaft RFCS surfaces due to friction in the bearings) [3, 9].

Structural analysis (initial data are the boundary pressure, axial force, with the imposition of the temperature analysis results, heat generated in contact places). The heat amount generated from friction in the RFCS can have a decisive influence on the seal assembly deformation. This deformation is determined by the formula (1) [3]:

$$Q = q \cdot S \tag{1}$$

q, heat flow rate;

S, contact area.

In the investigated seal S_1 and Q_1 - contact area and heat flow rate released from the friction with the high pressure shaft; values S_2 and Q_2 refer to the average pressure shaft. These parameters are shown in figures 2 and 3.

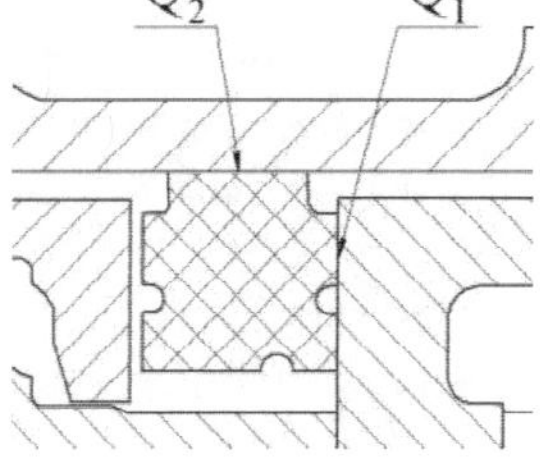

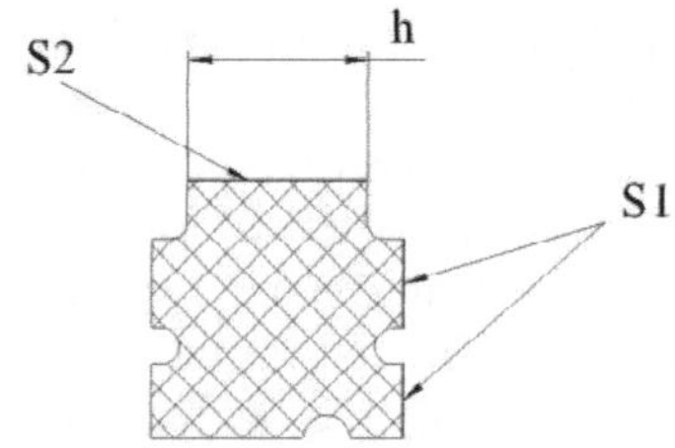

IOP Conf. Series: Materials Science and Engineering **302** (2018) 012003 doi:10.1088/1757-899X/302/1/012003

Figure 2. Heat flow rates locations. **Figure 3.** Contact areas.

Heat flow rate can be defined as in equation (2):

$$q = \Delta p \cdot f \cdot \omega \quad (2)$$

Δp, differential pressure;

f, sliding friction coefficient;

ω, angular ring surface velocity relative to the shaft surface.

Vast amount of troubles caused by the ring rotation speed determination. The angular RFCS ring rotation velocity can be gauged through equation (3). This formula was proposed by the author of [11, 12]:

$$\omega = \left[F_A \left(m_k \cdot R_S \right)^{-1} \cdot \left(\frac{f_T}{f_R} \cdot \frac{R_A + f_R \cdot R_O}{R_O + f_T \cdot R_A} - \varepsilon^{-1} \right) \right]^{1/2} \quad (3)$$

F_A, F_R, unbalanced axial and radial forces from the differential pressure on the sealing ring;

f_T, f_R, friction coefficients in the axial and radial friction pairs (their values can be determined on the basis of the following recommendations [6, 13, 14]);

m_k, carbon seal ring mass;

R_A, resultant action force (from the thrust washer on the RFCS ring) radial location;

R_S, radius of the seal ring mass center;

R_O, RFCS outer radius;

$\varepsilon = F_A \cdot F_R^{-1}$, coefficient.

Based on the foregoing calculation and heat distribution analysis in the model were done.

Previously mentioned findings were used for the structural calculation. Then the counted results of the support and seal assembly deformations were analyzed.

The calculations found the temperature distributions when the engine is running in two modes: cruising and takeoff. As a result, maximum temperature has the following values: 248 °C for cruise mode and 599 °C for takeoff. At cruising carbon ring is heated more evenly. The difference between the maximum and minimum ring temperature is equal to 34 degrees. At takeoff the temperature difference between the ring faces is more sudden and it is equal to 77 degrees. This is two times more compared to cruising mode.

4. Structural calculation

After receiving the calculated temperature distribution and results analysis structural calculations for the researched modes was performed. The calculation results are shown in figures 4 and 5. The maximum deformation in the cruise mode is observed in the upper seal cover. Carbon ring deformations are of the utmost interest.

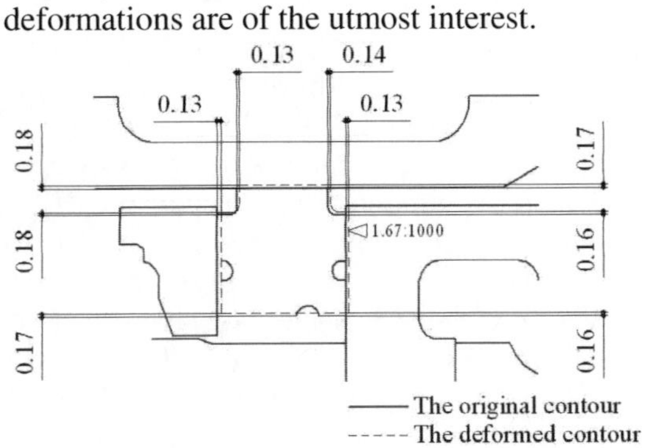

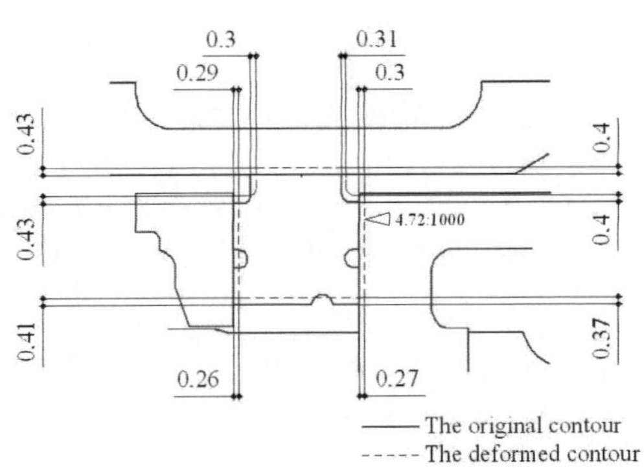

Figure 4. The ring displacement relative to the rest position at cruising.

Figure 5. The ring displacement relative to the rest position at takeoff.

Linear and angular strain corresponding to the cruising regime is shown in figure 4. The figure analysis shows: the ring moves to the right by 0.13 mm and 0.17 mm up from its original position. The angular strain amounted to the following values: about 6' to the face surface, and the same value for the radial. In general, the angular deformation is negligible and both sealing gaps remain plane parallel form. Taper is equal to 1.67 μm.

The maximum deformation for the entire support at takeoff is equal to 2 mm. It corresponds to the same part as on cruise mode and attributed to the same reasons. In the seal assembly (figure 5) maximum strain corresponds also to the upper seal cover and has value of 0.85 mm.

So, the maximum deformation in the seal parts is increased two times more (approximately 110%). The ring is deformed as at cruising: it is shifted to the right by 0.3 mm, and up to 0.17 mm. This conclusion was made after the figure 5 analysis.

The axial deformation increased in 2.3 times, while the radial remained at the same level. The cross section rotation angle has increased significantly. It has grown in 2.7 times (from 6' to 16'). The excessive taper at takeoff (4.72 μm) can has an impact on work of the gas-dynamic grooves. The research have shown the small effect of pressure from the boundary air flow on the deformation, compared to thermal extensions. This value is ~ 1...3% and it is within the calculation error. The heat released from the friction between the contacting surfaces has a decisive influence on the deformation magnitude. It leads to a radial deformation and completely determines its value.

The existence of the seal ring radial deformation causes a gap change in the RFCS and impairs its leakage. The upper seal cover being leads to the same problems. A tangential and slotted gap changing is displayed in figures 6 and 7.

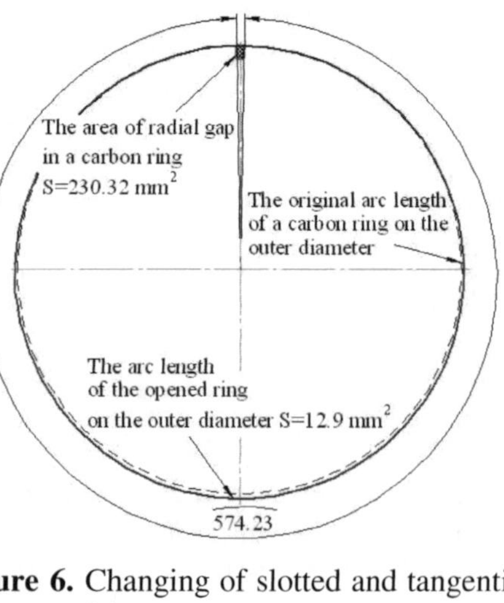

Figure 6. Changing of slotted and tangential gaps at cruising.

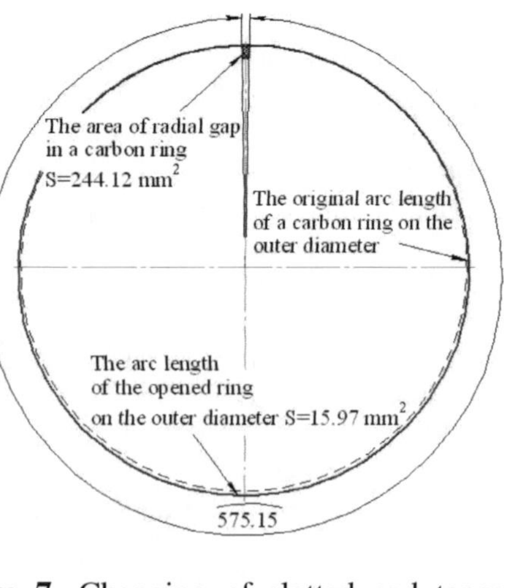

Figure 7. Changing of slotted and tangential gaps at takeoff.

Slotted gap arises in the carbon seal ring split. This can occur because of different seal ring and the upper seal cover thermal expansions.

Obviously, after the transition from the takeoff mode to the cruising, the slotted gap area decreased from 244.12 mm^2 to 230.32 mm^2 (about 6%) and the tangential gap area is also reduced from 15.97 mm^2 to 12.20 mm^2 (about 31%). As one can see, the gaps do not change in proportion; this is due to the predominant influence of the upper seal cover radial deformation. Analysis of changes in the gaps areas clearly shows: the slotted gap has a decisive influence on the seal leakage. For take-off mode conditions it is 94% of the total area, and 95% for cruising. It would appear reasonable, leakage will vary in proportion to the gap size and leaking through the tangential gap will be at the 5% level of

IOP Conf. Series: Materials Science and Engineering **302** (2018) 012003 doi:10.1088/1757-899X/302/1/012003

their total volume. The research have shown that if it is necessary to analyze the seal leakage the flow rate changing through the slotted gap is sufficient to investigate.

5. CFD modeling of the gap

The finite elements tetrahedral mesh was used in this calculation (~490,000 elements in each model). Figure 8 shows the dynamic pressure distribution (ram air pressure) in the air flow path at cruising. The maximum pressure is located in the gap center and it is equal to 291 kPa.

On the calculation results the actuation fluid leakage amount through the gap in the carbon ring on the cruise mode was defined. It has a value of 2.4 g/s. Figure 9 shows the distribution dynamic pressure (ram air pressure) in the air flow path at takeoff. The maximum pressure is located in the gap center and takes the value 75.8 kPa.

On the results of calculation the actuation fluid leakage amount through the gap in the carbon ring on the cruise mode was defined. It is equal to 0.75 g/s. Thus, leakage at takeoff is 3.2 times larger than on cruising (2.4 g/s and 0.75 g/s, respectively).

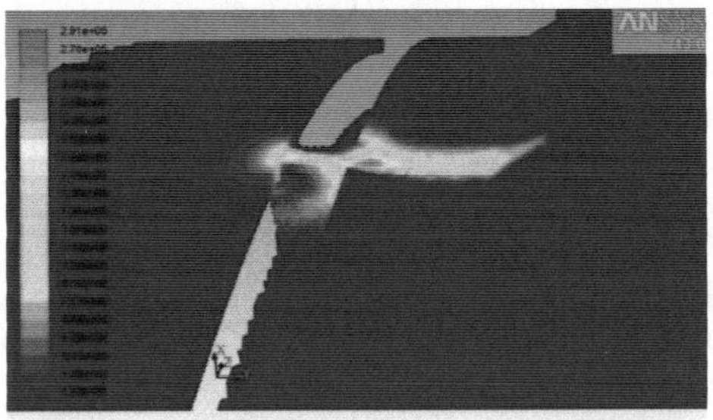

Figure 8. Dynamic pressure field distribution at cruising.

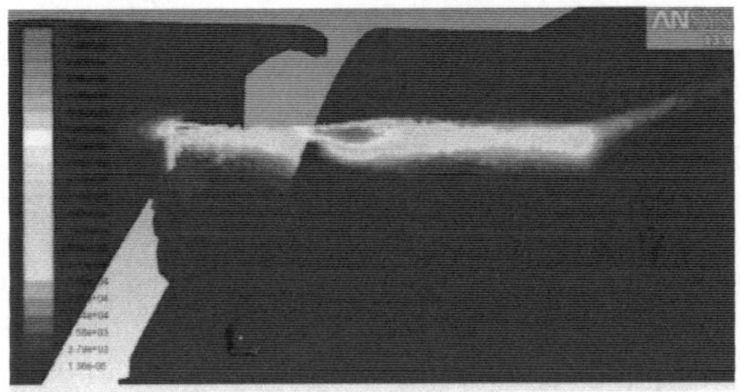

Figure 9. Dynamic pressure field distribution at takeoff.

The correctness of the generated calculation models and designing seals methodologies as a support part were verified by comparison with experimental data (figure 10) [1, 15].

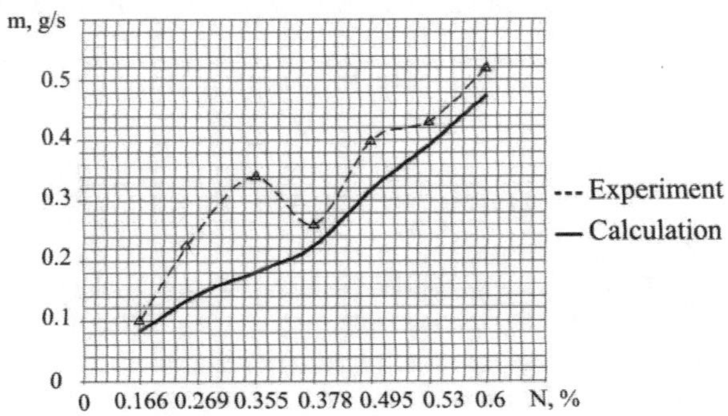

Figure 10. Comparison of the experimental and calculated flow seal characteristics.

The difference between calculation and experimental data is in the range of 8 to 15%. It can be considered as a good result. The used calculation seals models and the seals design method as a support element were correct. It can be approved on the level of the results disarrangement basis.

6. Conclusion about the performed research

The performed research served as a ground for the following conclusions:

Firstly, carbon seal ring deformation behaviour does not change in the transition from the takeoff to cruising mode. Ring deflection in the radial direction also remains constant. It is amounted to 0.17 mm. In axial direction the ring is displaced by 0.3 mm at takeoff and 0.13 mm at cruising (the displacement is reduced by 2.3 times). The cross section rotation angle also varies considerably (from 16' to 6').

Secondly, the heat generated in the contact zone has the greatest influence on the size and shape of the seal gap. The heat generated in the face gap is two times more than in the radial gap. The maximum heat flow value at the face gap is equal to 5147 W at takeoff. Axial force has an effect only on the axial seal ring displacement. It varies between 0.135 mm and 0.029 mm (reduced in 4.7 times) on cruise mode. At takeoff it ranges from 0.3 mm to 0.125 (reduced in 2.4 times). If the pressure is taken into account it leads to the deformation change in the range of 1-3%.

Thirdly, seal assembly parts radial deformation has a decisive influence on the gap magnitude change in the carbon seal ring split. After the transition from the takeoff mode to the cruising, slotted gap area decreases from 244.12 mm^2 to 230.32 mm^2 (about 6%), the tangential gap area is also reduced from 15.97 mm^2 to 12.20 mm^2 (about 31%). Analysis of changes in the gap areas clearly shows: the slotted gap has a decisive influence on the seal leakage. For take-off mode conditions it is 94% of the total area, and 95% for cruising.

Fourthly, leakage calculation through the slotted gap was implemented in FLUENT CFD complex. Calculated leakage values at takeoff is 3.2 times higher than the leakage at cruising, it has a value of 2.4 g/s and 0.75 m/s respectively. These values go with experience in the RFCS designing.

Fifthly, the significant dynamic pressure presence at the seal gap exit (up to 291 kPa at takeoff and up to 75.8 kPa at cruise) should be considered in the RFCS gas-dynamic unloading design process.

On the investigation basis the RFCS leakage calculation methodology was devised. This computing combines: deformation, leakage and thermal analyses. With the aid of the method it is possible to evaluate the leakage on the various engine operation modes.

References

[1] Falaleev S V and Chegodaev D E 1998 *Face non-contact seals of aircraft engines* (Samara: SSAU) p 275

[2] Belousov A I and Zrelov V A 1989 *Design of seals of rotating shafts for turbomachines of aircraft engines* (Samara: SSAU) p 108

[3] Mueller H K and Nau B S 1998 *Fluid sealing technology* (New York: Marcel dekker Inc.) p 485

[4] Marcinkovskij V A and Zagorulko A V 2011 *Vibration reliability and leakage of centrifugal machines* (Sumy: Sumy state university) p 351

[5] Reshetov D N 1989 *Details of machines* (Moscow: Mashinostroenie) p 496

[6] Mikheyev M A и Mikheyeva I M 1977 *Fundamentals of heat transfer* (Moscow: Energija) p 344

[7] Parovay E F and Falaleev S V 2015 Designing of low-flow rate sliding bearings for turbo machinery rotors *Biosciences Biotechnology Research Asia* **12** 731-36

[8] Kuz'michev V S, Tkachenko A Y, Krupenich I N and Rybakov V N 2014 Composing a virtual model of gas turbine engine working process using the CAE system "ASTRA" *Research Journal of Applied Sciences* **9** 635-43

[9] Lebeck A O 1991 *Principles and Design of Mechanical Face Seals* (New York) p 764

[10] Tisarev A, Falaleev S, and Vinogradov A 2014 Calculation of labyrinth seals in the secondary air system of aircraft engine *Open Mechanical Engineering Journal* **8** 424-30

[11] Eskin I D 1984 *Design of dampers and contact seals of the rotor supports of aircraft GTE* (Kuibyshev: KuAI) p 47

[12] Lebeck A O 1980 Mixed Friction Hydrostatic Mechanical Face Seal Model with Thermal Rotation and Wear *ASLE Trans.* **23** 375-87

[13] Lorincevic I V 1991 *Research of RFCS of turbine oil cavity* (Kuibyshev: KuAI) p 84

[14] Novikov D K and Diligensky D S 2015 Calculation algorithm for squeeze film damper *Procedia Engineering* **106** 218-23

[15] Lazutkin G V, Ermakov A I, Davydov D P, Boyarov K V and Bondarchuk P V 2014 Analysis of characteristics of all-metal vibration insulators made of different wire materials *Russian Aeronautics* **57** 327-33

IOP Conf. Series: Materials Science and Engineering **302** (2018) 012004 doi:10.1088/1757-899X/302/1/012004

Development of Advanced Carbon Face Seals for Aircraft Engines

S V Falaleev, P V Bondarchuk and A Yu Tisarev

Samara National Research University,34, Moskovskoye shosse, Samara, 443086, Russia

sergey_falaleev@mail.ru ; bond_ssau@mail.ru; aytisarev@gmail.com

Abstract. Abstract. Modern aircraft gas turbine engines require the development of seals which can operate for a long time with low leakages. The basic type of seals applied for gas turbine engine rotor supports is face seal. To meet the modern requirements of reliability, leak-tightness and weight, low-leakage gas-static and hydrodynamic seals have to be developed. Dry gas seals use both gas-static and hydrodynamic principles. In dry gas seals microgrooves are often used, which ensure the reverse injection of leakages in the sealed cavity. Authors have developed a calculation technique including the concept of coupled hydrodynamic, thermal and structural calculations. This technique allows to calculate the seal performance taking into account the forces of inertia, rupture of the lubricant layer and the real form of the gap. Authors have compared the efficiency of seals with different forms of microgrooves. Results of calculations show that seal with rectangular form of microgrooves has a little gap leading to both the contact of seal surfaces and the wear. Reversible microgrooves have a higher oil mass flow rate, whereas HST micro-grooves have good performance, but they are difficult to produce. Spiral microgrooves have both an acceptable leakages and a high stiffness of liquid layer that is important in terms of ensuring of sealing performance at vibration conditions. Therefore, the spiral grooves were chosen for the developed seal. Based on calculation results, geometric dimensions were chosen to ensure the reliability of the seal operation by creating a guaranteed liquid film, which eliminates the wear of the sealing surfaces. Seals designed were tested both at the test rig and in the engine.

1. Introduction

Engines with advanced parameters require seals operating for a long time with low leakages [1-5]. The basic type of seals applied for gas turbine engine rotor supports is face seal. To meet the modern requirements of reliability, leak-tightness and weight, low-leakage gas-static and hydrodynamic seals have to be developed [6,7]. Dry gas seals use both gas-static and hydrodynamic principles. In dry gas seals microgrooves are often used, which ensure the reverse injection of leakages in the sealed cavity.

The operating principle of such type of seals is based on the balance of axial forces, acting on axially-movable and rotating rings. Complex-shaped microgrooves are produced on the face area [1–2]. At operational conditions pressure rises up in the gap that leads to the displacement of axial-movable ring and minimum positive clearance of 1…3 μm in friction couple is formed. It allows seal to operate with low leakages without the wear for the wide range of operating parameters.

Application of modern materials, complex-shaped microstructures and improvement in manufacturing technology allow to broaden their functional possibilities. Therefore, authors have decided to apply the low leakage seals technology in order to improve existing face seals and to develop advanced face seal constructions applied for aircraft engine.

To develop the seal, the hydrodynamic performance of lubrication layer in the gap with microgrooves has to be determined [8, 9] and the required thermal state has to be achieved [10].

2. Methodology of flow calculation in the seal gap

To design a face seal the following factors have to be taken into account: inertial forces effect on the lubrication layer, possible liquid evaporation in the gap, thermal and force deformation of the seal rings.

The main problem in the calculating of the seal characteristics is to find pressure field on the working surface of seal rings. There are some analytical equations, which allow to calculate the pressure distribution in the ring gap for the seals without microgrooves. These equations were derived from Navier-Stokes equation taking into account averaging the inertial terms for the layer thickness by the Slezkin-Targ method. However, due to the complexity of differential equations these formulas were obtained for particular cases only. To calculate flow parameters of the hydrodynamic seals, the complex forms of microgrooves and deformed gap (figure 1) have to be taken into account (here r_1 and r_2 – inner and outer radius of seal; p_1 and p_2 are pressure inside and outside respectively). In order to calculate these tasks the finite volume method has been used. [8]. The pressure at each point of finite volume was obtained from the condition of flow equality through the control volume in the radial and circumferential directions (figure 2).

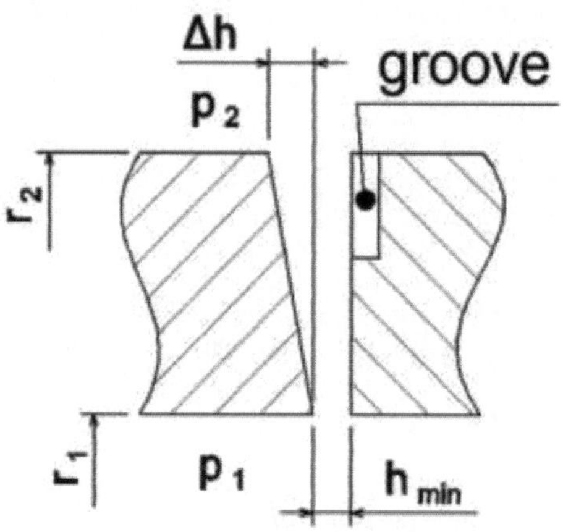

Figure 1. Simplified scheme of complex-shaped clearance

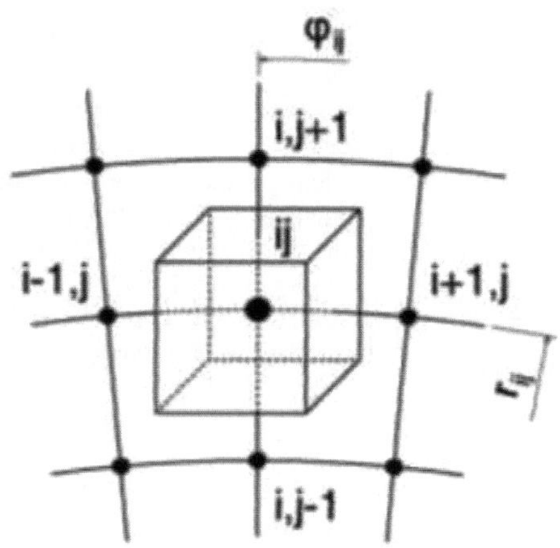

Figure 2. Finite volume

The equation for the pressure calculation at the point *ij* taking into account the inertial forces is following

$$p_{ij} = (B_{ij}p_{i-1,j} + C_{ij}p_{i+1,j} + D_{ij}p_{i,j-1} + E_{ij}p_{i,j+1} + F_{ij} + G_{ij})/A_{i,j}, \quad (1)$$

where coefficients are

$$A_{ij} = \frac{\Delta r}{24\mu\Delta\varphi}\left[2\left(\frac{h^3}{r}\right)_{ij} + \left(\frac{h^3}{r}\right)_{i-1,j} + \left(\frac{h^3}{r}\right)_{i+1,j}\right] + \frac{\Delta\varphi}{24\mu\Delta r}\left[2\left(h^3 r\right)_{ij} + \left(h^3 r\right)_{i,j-1} + \left(h^3 r\right)_{i,j+1}\right] \quad (2)$$

$$B_{ij} = \frac{\Delta r}{24\mu\Delta\varphi}\left[\left(\frac{h^3}{r}\right)_{ij} + \left(\frac{h^3}{r}\right)_{i-1,j}\right] \quad (3)$$

$$C_{ij} = \frac{\Delta r}{24\mu\Delta\varphi}\left[\left(\frac{h^3}{r}\right)_{ij} + \left(\frac{h^3}{r}\right)_{i+1,j}\right] \quad (4)$$

$$D_{ij} = \frac{\Delta\varphi}{24\mu\Delta r}\left[\left(h^3 r\right)_{ij} + \left(h^3 r\right)_{i,j-1}\right] \quad (5)$$

$$E_{ij} = \frac{\Delta\varphi}{24\mu\Delta r}\left[\left(h^3 r\right)_{ij} + \left(h^3 r\right)_{i,j+1}\right] \quad (6)$$

$$F_{ij} = r_{ij}\frac{\omega\Delta r}{4}\left(h_{i-1,j} - h_{i+1,j}\right) \quad (7)$$

$$G_{ij} = \frac{\rho\omega^2\Delta\varphi}{80\mu}\left[\left(h^3 r^2\right)_{i,j+1} - \left(h^3 r^2\right)_{i,j-1}\right] \quad (8)$$

h – a current gap value (in the location of microgrooves location h equals the sum of the gap value and the depth of the groove);
r, φ – point coordinates in the polar coordinate system;
μ – dynamic viscosity;
ρ – density;
ω – rotational speed;
G_{ij} – inertial forces coefficient.

The pressures are found using iterative method. After the calculation of pressure distribution in the gap $p=p(r, \varphi)$, the following parameters has to be determined: carrying capacity of layer W that is a force disclosing the sealing joint; bending moment M; stiffness of lubrication layer C; windage N in the gap:

$$W = \int_0^{2\pi}\int_{r_1}^{r_2} p r dr d\varphi \quad (9)$$

$$M = \int_0^{2\pi}\int_{r_1}^{r_2} p \cdot r^2 dr d\varphi \quad (10)$$

$$C = -dW/dh \quad (11)$$

$$N = \mu\omega^2 \int_0^{2\pi}\int_{r_1}^{r_2} \frac{r^3}{h(r,\varphi)} d\varphi dr \quad (12)$$

Leakage through the seal gap in the radial direction is

$$Q = \frac{1}{12\mu}\int_0^{2\pi}\left[r h^3(r,\varphi)\left.\frac{dp}{dr}\right|_{r=r_1} \pm \frac{3\rho\omega^2 r^2 h^3(r,\varphi)}{10}\right] d\varphi \quad (13)$$

Here the second term in brackets takes into account the influence of inertia forces. The sign – or + is determined by the direction of the flow.

To consider coupled hydrodynamic, thermal and structural processes, iterative conjugate computation has been performed using developed software based on described methodology and software ANSYS Mechanical. In figure 3 an example of temperature distribution at the rings of friction pair is shown. The temperature non-uniformity in the circumferential direction is observed. The reason for that is significant difference in friction power in the seal gap and locations of lubrication rapture. For this reason thermal deformation can cause an undulation that can be determined by iterative calculation.

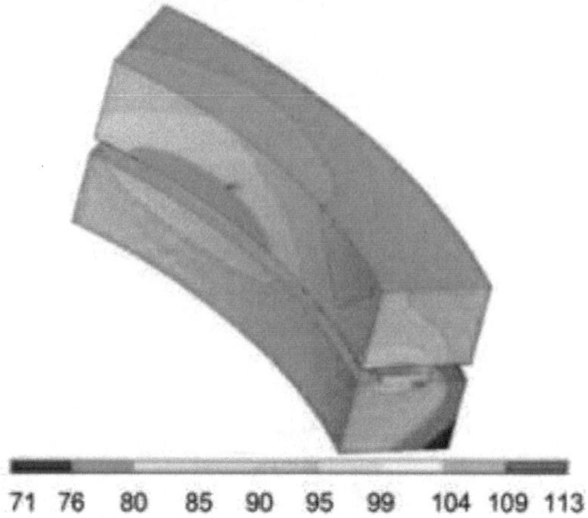

Figure 3. Temperature distribution in the rings of friction couple, °C

3. Performance comparison of seals with microgrooves of different forms

Calculations were performed for 4 types of microgrooves (figure 4) produced on the ring with an outer diameter of 110 mm. The geometry of microgrooves was optimized. Comparison of seal performance is shown in table 1.

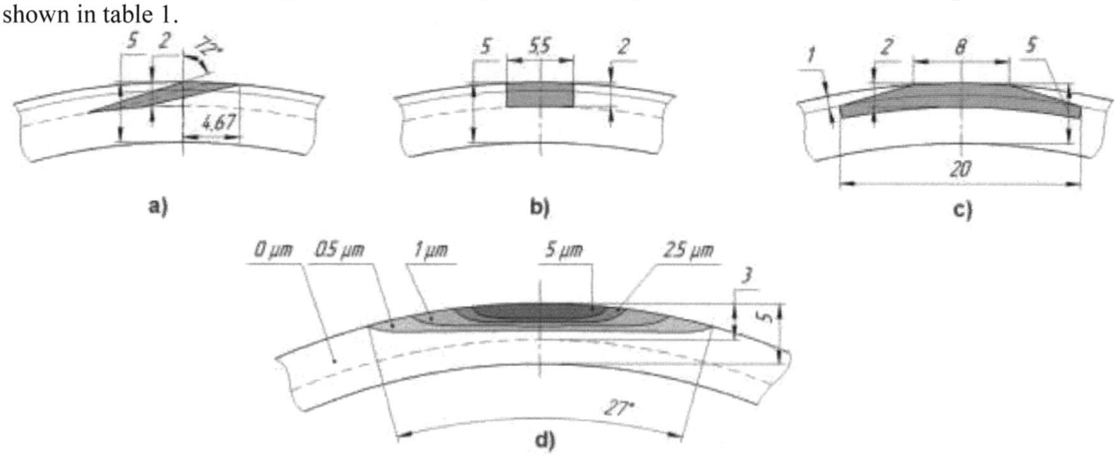

Figure 4. Design of microgrooves:
a) spiral; b) rectangular; c) reversible; d) HST (Hydrodynamic Surface Tension)

Table 1. Performance of seals with different forms of microgrooves

Microgroove type	Gap (µm)	Leakage (g min^{-1})	Frictional power (W)	Liquid layer stiffness (N µm^{-1})
spiral	1,33	1,65	1214	670
rectangular	0,28	0,02	3221	93
microgrooves HST	0,94	0,60	1613	1323
reversible	1,56	2,54	960	186

Results of calculations show that seal with rectangular form of microgrooves has a little gap leading to both the contact of seal surfaces and the wear. Reversible microgrooves have a higher oil mass flow rate, whereas HST micro-grooves have good performance, but they are difficult to produce. Spiral microgrooves have both an acceptable leakages and a high stiffness of liquid layer that is important in terms of ensuring of sealing performance at vibration conditions. Therefore, the spiral grooves were chosen for the developed seal.

4. Seal design and test

Bearing chamber of compressor supports is traditionally sealed by contact seals because this type of seals can ensure the best tightness. Replacement of face-seals with non-contact seal can significantly increase the seal life with slight reduction in tightness. The work on the installation of seals with gas and liquid lubrication in aircraft engines was carried out in cooperation with PJSC "Kuznetsov". Designed and manufactured prototypes of face seals with microgrooves for the compressor support of the aircraft engine are shown in figure 5 and 6.

The seal separates the air cavity of the compressor and the bearing chamber. Several nozzles supply oil to holes in the sleeve, whereby an oil bath is formed due to the centrifugal forces and the oil under pressure enters the gap. As a result, the contact surface A of the seal is lubricated. The piston ring of the seal is made of pyrographite. The contact surface of the steel sleeve is coated with a chromium-molybdenum coating. To ensure a non-contact operating, a structure of microgrooves is applied to the surface A of the sleeve (figure 5).

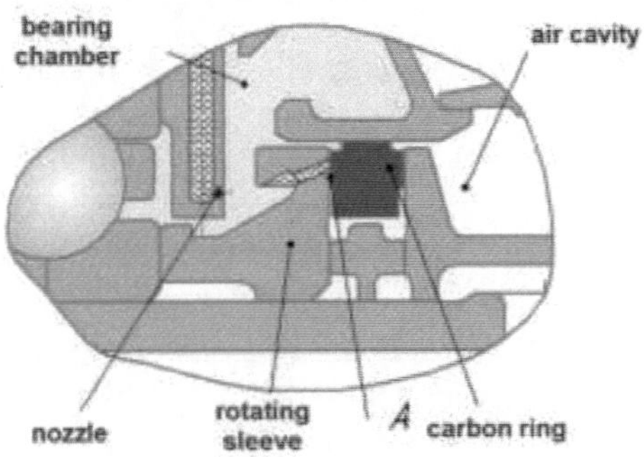

Figure 5. Seal design in compressor support

Figure 6. Sleeve of face seal with microgrooves

The results of calculations have shown that seals operated with the gap of about 2μm in all operating points of the engine. The air leakage through the face gap was less than 0.05 g/s.

Authors have developed a manufacturing technology and have produced a prototype at the Samara University. Tests both at the test rig (3 hours) and in the engine (10 hours with 6 starts and 6 shutdowns) have confirmed the operability of the developed seal. No wear of the operating surfaces was detected.

5. Results and discussions

Authors have developed a calculation technique including the concept of coupled hydrodynamic, thermal and structural calculations. This technique allows to calculate the seal performance taking into account the forces of inertia, rupture of the lubricant layer and the real form of the gap. New hydrodynamic seal was designed for the support of GTE compressor. Based on calculation results, geometric dimensions were chosen to ensure the reliability of the seal operation by creating a guaranteed liquid film, which eliminates the wear of the sealing surfaces. The results of tests both at the test rig and in the engine confirmed the low air leakage and reduced frictional power. As a further research, the authors plan to use ceramic materials with wear-resistant DLC (diamond like carbon) coatings with a low friction coefficient.

Acknowledgment
This work was supported by the Ministry of Education and Science of the Russian Federation in the framework of the implementation of the Program ''Research and development on priority directions of scientific-technological complex of Russia for 2014– 2020.

References

[1] Bondarchuk P and Falaleev S 2014 Direction of improvement of the radial-face seals of rotor supports of the aircraft engines *International Journal of Engineering and Technology* 6(5) pp 2261–2268
[2] Falaleev S V and Bondarchuk P V 2014 Face gas-dynamic seal of high-speed compressor *Research Journal of Applied Sciences* 9(11) pp 707-710
[3] Vinogradov A, Badykov R and Pilla C 2014 Research of the thermal state of aviation engine *supports Research Journal of Applied Sciences* 9(11) pp 789-794
[4] Vinogradov A S 2014 Seal design features for systems and units of aviation engines *Life Science Journal* 11(8) pp 575-580
[5] Vinogradov A S and Badykov R R 2014 Influence of labyrinth seal leakage on the turbine support cooling *Proc. of ASME 2014 Gas Turbine India Conference*
[6] Mueller H K 1990 *Abdichtung bewegter Maschinenteile* (Waiblingen)
[7] Lebeck A O 1991 *Principles and Design of Mechanical Face Seals* (New York: Wiley)
[8] Falaleev S V 2015 Techniques for calculating the hydrodynamic characteristics of mechanical face seals with gaps of complex forms *Journal of Friction and Wear* 36(2) pp 177-183
[9] Novikov D K 2014 Development of squeeze film damper characteristics calculation methods which take into account a liquid inertia forces *Research Journal of Applied Sciences* 9(10) pp 649-653
[10] Balyakin V B and Falaleev S V 2015 Study of temperature state of mechanical gas dynamic sealing *Journal of Friction and Wear* 36(3) pp 213-217

Modal Analysis of Space-rocket Equipment Components

A.A. Igolkin*, A.I. Safin and A.B. Prokofiev

Heat Engineering and Heat Engines Department, Samara National Research University,Samara, Russia

*Corresponding author: igolkin97@gmail.com

Abstract. In order to prevent vibration damage an analysis of natural frequencies and mode shapes of elements of rocket and space technology should be developed. This paper discusses technique of modal analysis on the example of the carrier platform. Modal analysis was performed by using mathematical modeling and laser vibrometer. Experimental data was clarified by using Test.Lab software. As a result of modal analysis amplitude-frequency response of carrier platform was obtained and the parameters of the elasticity was clarified.

1. Introduction

A significant number of space-rocket equipment (SRE) failures is connected to fatigue failures of parts under the alternating stresses caused by vibroacoustic loads. In order to prevent vibration breakdowns, it is necessary to analyze the natural frequencies and vibration modes of the SRE components. Modal tests are mandatory in most countries while in Russia the requirements for such normative documentation (ND) are still being developed nowadays. In this regard, the issue of creating modal analysis technology is of great importance and interest, since it improves the accuracy and efficiency of vibroacoustic testing and reduces the implementation time.

According to the research of the Experimental Mechanics Community [1], the determination of the dynamic characteristics of spaceships and their components can be divided into two categories: strength analysis and analysis of the satellite structural stiffness. The purpose of strength analysis is to confirm the integrity and performance of the components under vibration loads, comparable to those which appear during the flight. The dynamic analysis of the satellite structural stiffness poses the task of determining its modal parameters (natural frequencies and vibration modes, as well as damping at these frequencies). The purposes of determining modal parameters are the following:

1. Determination of resonance frequencies;
2. Checking the accuracy of the mathematical finite element model and its elements;
3. Determination of the value and change of modal parameters of remote elements having a significant spatial extent [2].

Finite element model and experimental modal verification are used for improving pump vibroacoustic performances in research paper [3]. In paper [4] authors describe in detail the contactless laser vibrometry method as applied to the car surface using the three-component laser scanning vibrometer for verify the vehicle finite-element model. Paper [5] presents research on diagnostic parameters using the Doppler Effect laser vibrometer for monitoring conditions of lightweight aircraft engine blades made of composite materials in flight.

In this paper we consider an example of determining the resonance frequencies and refine the mathematical finite element model. The object of the study is the carrier platform (CP), designed for

IOP Conf. Series: Materials Science and Engineering **302** (2018) 012005 doi:10.1088/1757-899X/302/1/012005

fixing the elements of the spaceship apparatus. It should be noted that a laser vibrometer was used in many researches in order to increase the accuracy and speed of modal analysis. The need for using laser vibrometry is determined by the fact that the contact vibration sensors installed on the thin-walled structures of the satellite introduce significant interference due to the additional mass. In addition, the use of a scanning laser vibrometer allows the use of a large number of vibration detection points (and, hence, a more accurate pattern).

The natural frequencies of the component structure are identified by determining and recording the maximum values on the vibration velocity graph from the test report. Modal analysis makes it possible to determine:

-the amplitude-frequency characteristics (by vibration velocity) at points on the surface of the component;

-the amplitude-frequency response averaged over the surface of the component (by vibration speed);

-the vibration form of the component which corresponds to each value of the natural frequency.

2. Methods

Search of the CP for its resonance frequencies is performed using a Polytec PSV-400 laser vibrometer. After the scanning process is completed, the results of vibration measurement of all fragments are connected together. Then transfer functions from the software complex "Polytec Scanning Vibrometer" are imported into "LMS Test.Lab", where calculation of modal parameters is performed. The test procedure for the CP is shown in Figure 1. For the first scheme (Fig. 1a), the CP was suspended on rubber bands without securing to the floor. For the second scheme (Fig. 1b), the measurement object was suspended on an elastic suspension on one side, and its lower part is cantilevered to the base with two hinges on the other side.

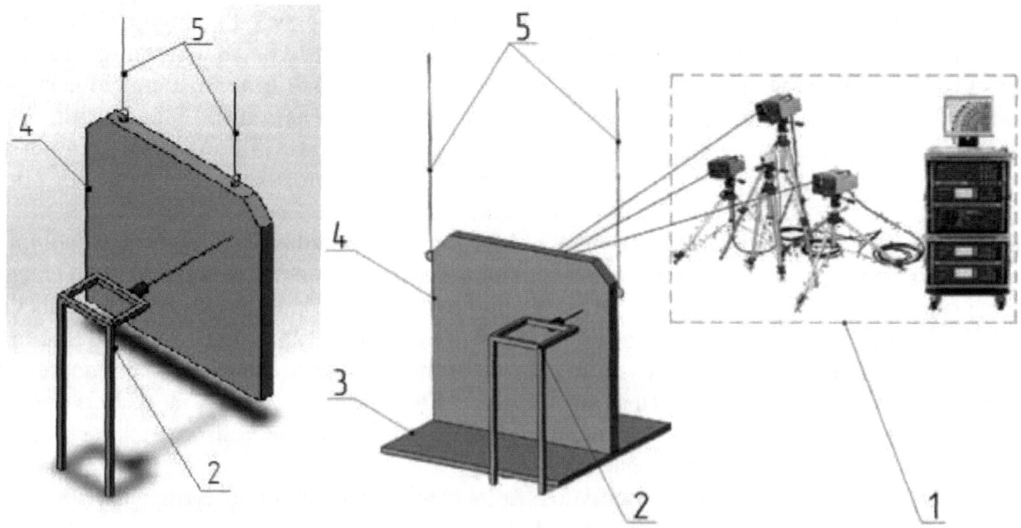

a) b)

Figure 1. Both schemes of conducting experimental studies using a laser scanning vibrometer PSV-400-3D:
a) with no base b) fixed to base
1 –laser scanning vibrometer; 2 – vibration exciter;
3 – base; 4 – measurement object; 5 – elastic suspension.

A vibration exciter is used as a source of excitation, fixed at the point with maximum displacement when generating vibrations. The vibration exciter in both cases of fixing is set equally to minimize the discrepancy under the experimental conditions.

First, the so-called mode overview, during which the object is scanned in a wide frequency range and with a small resolution in frequency, is performed. At this stage, the frequency bands, in which the resonance of the structure may arise, are determined.

However, this analysis in such a wide frequency range gives a general idea of the characteristics, since it is a preparatory stage in the subsequent refinement of the natural frequencies and vibration modes of the object.

Repeated scanning performed in a narrow frequency range with high resolution in frequency allows us to refine the obtained values of the object natural frequencies. In order to reduce the time of experiment, it is advisable to reduce the number of scanning points.

The final step in the calculation is to determine the shape of the vibrations. To do this, the object is scanned at a previously refined frequency with a larger number of scan points.

A review of the received modes using the Test.Lab (Modal Analysis) software allows to identify stable poles, which are analyzed in accordance with the "modal phase collinearity" (MPC) and "maximum phase deviation" (MPD) indicators. Output data of the program are arrays of obtained values of frequency transfer functions and natural frequency values. Figure 2 shows the frequency response of the structure after processing the data in the program. The solid line shows the data without processing in the program; the dotted one shows the data after processing.

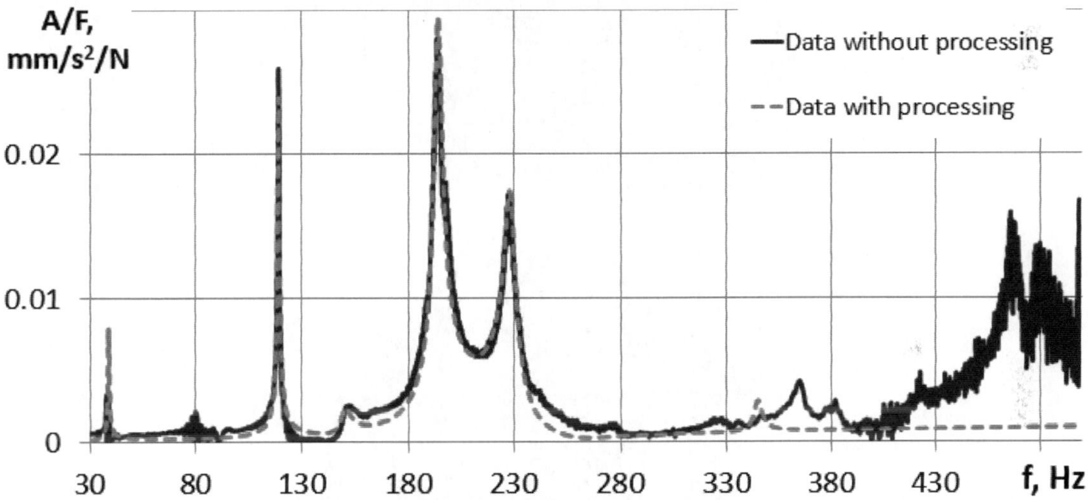

Figure 2. Amplitude-frequency characteristic of CP after using Test.Lab.

The simulation of the CP natural oscillations is carried out by means of the procedure applied for solving the Lanczos eigenvalue problem which is used in the MSC.Nastran finite element system, which is described in detail in [6]. Search of natural oscillations is carried out in the frequency range from 0 to 500 Hz.

3. Results

As a result of scanning the object from 30 to 500 Hz with the first variant of fixing, the first 3 harmonics of the platform have been identified: 74.4; 113.1; 150.0 Hz. Then the values of the object natural frequencies have been refined and the vibration forms at these frequencies have been determined. The simulation of the experiment in MSC.Nastran have also been performed in the frequency range from 0 to 500 Hz.

Figure 3 shows a comparison of the first vibration form obtained from the experiment and the simulation.

As shown in Figure 3 simulation mode was identical to experimental mode. The same results were found for other modes.

Table 1 compares the first three values of the object natural frequencies obtained from the experiment and the simulation.

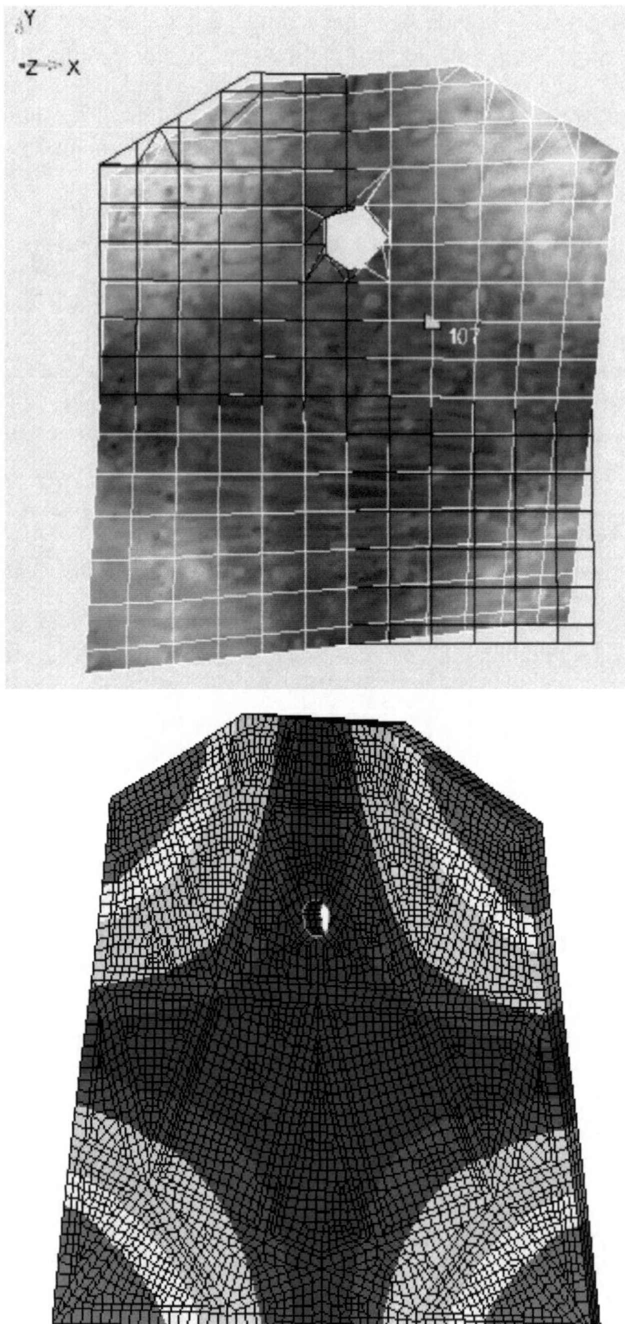

Figure 3. First mode during the experiment (top-74.4 Hz) and during the simulation (bottom-81.5 Hz)

Table 1. Comparison of theoretical and experimental values of natural frequencies

Mode number	The natural frequency in the experiment (Hz)	The natural frequency in the simulation (Hz)	Δ (%)
1	74.4	81.5	9.54
2	113.1	118.8	5.04
3	150.0	158.6	5.73

As a result of scanning the object in the frequency range from 30 to 500 Hz with the second variant of fixing, the first 3 harmonics of the platform are identified: 38.6; 118.5; 194.8 Hz. Results obtained by using second fixing scheme are summarized in Table 2.

Table 2. Comparison of theoretical and experimental values of natural frequencies

Mode number	The natural frequency in the experiment (Hz)	The natural frequency in the simulation (Hz)	Δ (%)
1	38.6	36.1	6.48
2	118.5	118.5	0
3	194.8	191.3	1.8

Figure 4 shows a comparison of the first form of oscillation as a result of experiment and simulation

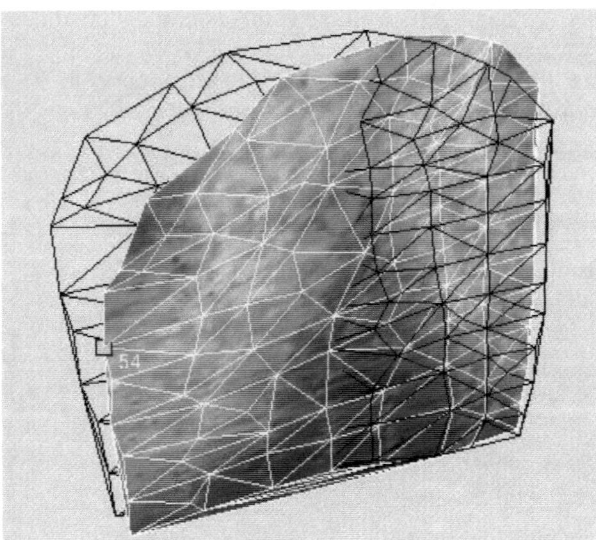

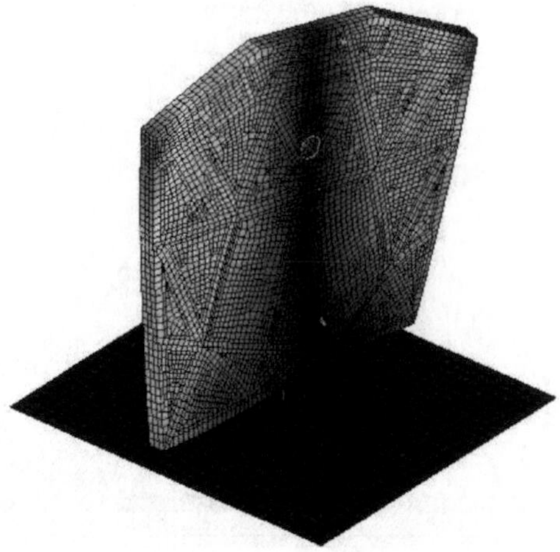

Figure 4. First mode during the experiment (top-38.6 Hz) and during the simulation (bottom- 36.1 Hz)

It was found that simulation modes were identical to experimental modes obtained by using second fixing scheme.

4. Conclusions

In this paper we have investigated that:

1) The proposed method of modal analysis of space-rocket equipment components allows determining the resonance frequencies of the object and verifying the accuracy of the mathematical finite element model and its elements.

2) The results obtained are consistent with the result of finite element modeling of the structure in the considered frequency range.

The results indicate the overall correctness of the finite element models. In comparison with experimental data discrepancy of the simulation results does not exceed 10% which indicates the correctness of finite element models.

Acknowledgments

The results of the work were obtained using the grant of the President of the Russian Federation for state support of young Russian scientists (grant number MD-3082.2017.8).

References

[1] Doupe C., Swenson E., George L. and Black J., 2013. Finite element model tuning with 3D mode from FalconSAT-5J. 50thAIAA/ASME/ASCE/AHS/ASC 17th SDM conference proceedings, 17, 7–12.
[2] Testoedov N., Lysenko E.and Berns V., 2016. Diagnostics of structures of space vehicles based on the results of vibration and acoustic tests: monograph. Siberia State Aerospace University: Standartinform.
[3] Safin A., Rodionov L., Makaryants G. and Prokofiev A., 2016. Gear pump modal analysis. Proceedings of ICSV 2016 - 23rd International Congress on Sound and Vibration: From Ancient to Modern Acoustics.
[4] Makaryants G., Safin A. and Shakhmatov E., 2016. Vehicle modal analysis. Proceedings of 22nd International Congress on Sound and Vibration, ICSV 2015.
[5] Makaryants G., Molchanov D. and Safin A., 2016. Journal of the International Measurement Confederation, 95, 383-388.
[6] Peresypkin K. and Kozhichenkova M., 2013. Formation of the shape of the gain of a size-stable platform by means of numerical optimization. Journal of Samara State Aerospace University, 4 (42), 223-229.

A Gas Lubricant Combined Support-sealing Node

S V Falaleev [1,*], H Nadjari[1] and A S Vinogradov[1]

[1]Aircraft engines design department, Samara national research university, Moscow highway 443086, Samara, Russia

[*] Corresponding author. Email address: sergey_falaleev@mail.ru

Abstract. The purpose of the research provided in this article is to develop a gas-dynamic device capable of performing the functions of support sealing, unloading devices for axial thrust bearings and damping of axial vibrations of the rotor. Some kinds of seals applied in supports of aircraft engines are known. A face gas-dynamic seal is one of the most effective and standard technology solution for compressors. As the basic element of the developed device, a face gas-dynamic seal with spiral grooves is considered. It also includes the fundamental mathematical calculation of such devices and the experimental research outcomes that form the basis of which such devices can be produced and adapted for use.

1. Introduction

With the development of aircraft gas turbine engines, there is a continuous increase in parameters of operating cycle. High-pressure differentials in turbo compressors and considerably high rotating speeds of rotors lead to the formation of over amplifications of forces on radial thrust bearings [1]. Traditionally labyrinth or graphite seals are used to seal the oil cavities of the rotor supports of gas turbine engines [2]. Usually unloading cavities are used to unload the radial thrust bearings, this is performed in the compressor or turbine, where air at high pressure is supplied or air it is dumped into the second contour of the engine. Both of these actions cause a decrease in the axial force applied on the engine rotor. Dry friction or hydrodynamic dampers are used to decrease the radial vibration of rotors [3], dampers are not provided for in axial vibrations of rotors. In some cases in aviation Gas turbine engines with high power parameters, it is impossible to eliminate excitation sources of axial vibration completely [4], therefore it is necessary to use axial vibrations dampers in rotor supports to decrease the vibration to an acceptable level. That is why is necessary to develop highly effective devices for unloading radial thrust bearings of aircraft gas turbine engines.

2. Principle operation of a face gas-dynamic seal

The main element of construction proposed in this paper is a face gas-dynamic seal. As mentioned above, application of such seals is influenced by its ability to self-regulate, which is to maintain the required amount of clearance when changes in external loadings occur.

A face Gas-dynamic seal consists of the rotating hard-alloy ring fixed on the shaft, and an axially-movable carbon-graphite ring placed in the body where preliminary contraction is done by springs. On the rotating ring, there is a pressure head site on which spiral grooves and the sealing ring corbel separating the high-pressure cavity from the cavity of low pressure are executed. Rubber mating rings are usually used as secondary seals. The gas-dynamic face seal works by the principle of equilibration of the gas static and gas-dynamic forces operating on the axially-movable and rotating rings. When at rest, the sealing rings are pressed together by the influence of gas static forces and forces of preliminary compression caused by springs. As the shaft gas comes to the narrowed spiral grooves and encounters resistance of sealing corbel, it compresses, forming zones of elevated pressure. Pressure in the face slot increases therefore the axially movable ring is removed from the rotating ring, forming the guaranteed clearance 0,002mm … 0,005mm. Thus, during operation, the surface seal rings do not contact with each other.

The increase in the loading force leads to a reduction of the clearance in the seal and to a simultaneous increase in the gas-dynamic pressure in the clearance and bearing capacity that leads to recovery of the initial clearance. The reverse process takes place as loading reduces.

There are exists various design options of gas-dynamic chambers [5, 6]. The "Spiral grooves" is most popular and well-studied design form of gas-dynamic chambers. For the first time, the theory of seals with spiral grooves has been studied in the works of E A Muijderman [7].

3. The proposed design of the combined "seals- unloading device damper" node

The additional axial gas-dynamic bearing which serves both for unloading of the rotor, and for damping axial vibrations(that represents axial gas-dynamic damper) will structurally differ from face gas-dynamic seal, in that the non-rotatable seal ring will not be movable in the axial direction, and will rest against the face surface of the housing.

The proposed design of the thrust bearing which will also function as a rotor support seal and damper of axial vibrations is shown on figure1.

The basic elements of the gas-dynamic device: The non-rotatable ring 1 (which can be made from graphite, or silicon carbide with covering) and the rotating ceramic ring 2 on which face surface spiral grooves for creation gas-dynamic pressure in face clearance are executed. The ring 2 is installed in the sleeve 3, sealing is provided by the rubber ring 4, torque transmission by the element 5. The ring 1 is mounted in the housing 6, sealing is provided with the rubber ring 7, torque transfer is provided by the milling 8. The device also provides an encapsulation of support together with the labyrinth seal 13 (that is acts as seals) and additional bearing working together with the ball bearing 9. Ball bearing is mounted in the housing sleeve through the radial damper 11 and the face elastic ring 12.

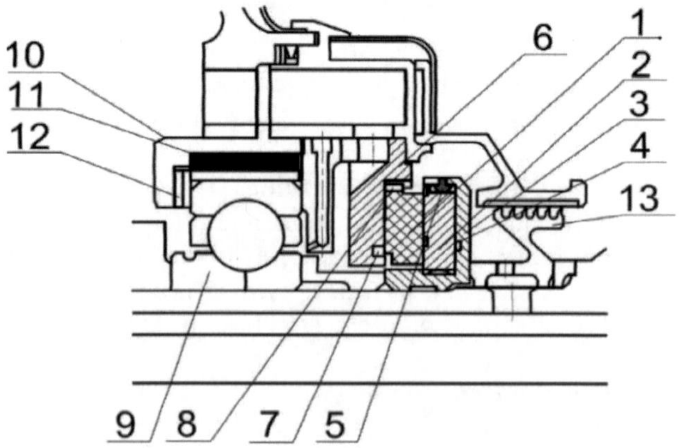

Figure 1. The design of a gas-dynamic device.

In statics, the clearance between rings 1 and 2 is established about 100 microns. In this case, the effect of spiral grooves is low and the device works as a normal slot-hole seal. As the engine begins to operate the axial force increases, the elastic ring 12 is deformed and the rotor is relatively gradually displaced to the left of the stator. The ring 12 should be designed in such a way, when an axial force of 3 tons is applied, the axial shift produced is 90…98 microns. In this case the axial clearance in the gas-dynamic device becomes equal 2 … 10 microns, spiral grooves begins to work more effectively, and the device unloads additional axial forces of rotor and provides an effective sealing for the supports, and damping of axial vibrations.

IOP Conf. Series: Materials Science and Engineering **302** (2018) 012006 doi:10.1088/1757-899X/302/1/012006

4. The fundamental theoretical design of gas-dynamic seals

The structural design of a face seal with spiral grooves is shown in figure2.

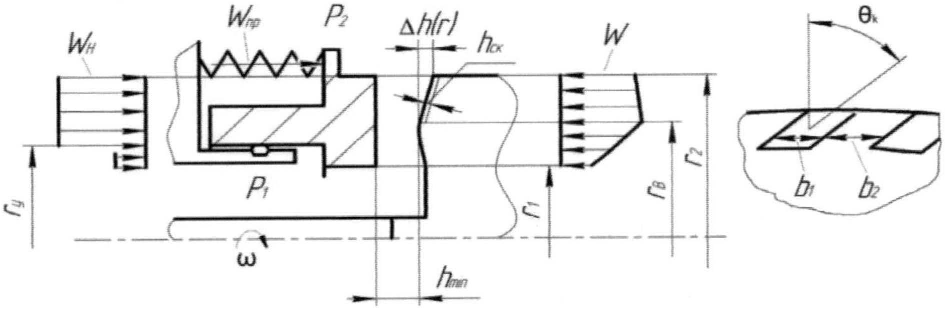

Figure2. The structural design of a face gas-dynamic seal with spiral grooves.

The load bearing capacity (applied on non-rotating ring seal) - W and the loading force - $W_н$ formed from the action of outer pressure - p_2 internal pressure - p_1 and spring forces- W_{np}. From the side of the non-rotatable ring of pressure p_1 and p_2 are also separated by the secondary seal which is on the radius r_y.The sealing clearance is formed by the external radius r_2 and internal radius r_1. The current value of the clearance $h(r)$ is the sum of minimum clearance and the clearance $h_{\min}$ which is formed as a result of deformation $\Delta h(r) h(r) = h_{\min} + \Delta h(r)$. Spiral grooves are carried out on the ring rotating with angular speed ω and characterized by following geometrical parameters: radius of the arrangement of spiral grooves $r_в$, depth of spiral grooves h_{ck}, discharge angle of spiral grooves θ_k, the width of spiral groove and crossing point width between grooves b_1.

Pressure on the border between zones of pressure and forcing - $\bar p_в$:

$$\bar p_в = \left[\frac{\bar p_1^2\int_{\bar r_в}^{\bar r_1}\frac{d\bar r}{\bar r A_0\left(1-B(\bar r-1)\right)^3}+\int_{\bar r_1}^{\bar r_в}\frac{d\bar r}{\bar r H^3}+\frac{2\lambda}{H^2}\int_{\bar r_1}^{\bar r_в}\frac{d\bar r}{\bar r H^3}\cdot\int_{\bar r_в}^{\bar r_2}\frac{B_0\bar r d\bar r}{A_0\left(1-B(\bar r-1)\right)^2}}{\int_{\bar r_в}^{\bar r_2}\frac{d\bar r}{\bar r A_0\left(1-B(\bar r-1)\right)^3}\int_{\bar r_1}^{\bar r_в}\frac{d\bar r}{\bar r H^3}}\right]^{\frac{1}{2}} \qquad (1)$$

Pressure distribution in the leakage zones and pressurization is determined by formulas:

$$\bar p_l = \left[\bar p_1^2 + (\bar p_в^2 - \bar p_1^2)\frac{\int_{\bar r1}^{\bar r}\frac{d\bar r}{\bar r H^3}}{\int_{\bar r_1}^{\bar r_в}\frac{d\bar r}{\bar r H^3}}\right]^{\frac{1}{2}} \qquad (2)$$

$$\bar p_н = \left[1-\left(1-\bar p_в^2\right)\frac{\int_{\bar r_в}^{\bar r}\frac{d\bar r}{\bar r A_0\left(1-B(\bar r-1)\right)^3}}{\int_{\bar r_в}^{\bar r_2}\frac{d\bar r}{\bar r A_0\left(1-B(\bar r-1)\right)^3}}+\frac{2\lambda}{H}\left(\int_{\bar r_в}^{\bar r}\frac{B_0\bar r d\bar r}{A_0\left(1-B(\bar r-1)\right)^2}-\frac{\int_{\bar r_в}^{\bar r_2}\frac{B_0\bar r d\bar r}{A_0\left(1-B(\bar r-1)\right)^2}}{\int_{\bar r_в}^{\bar r_2}\frac{d\bar r}{\bar r A_0\left(1-B(\bar r-1)\right)^3}}\cdot\int_{\bar r_в}^{\bar r}\frac{d\bar r}{\bar r A_0\left(1-B(\bar r-1)\right)^3}\right)\right]^{\frac{1}{2}} \qquad (3)$$

Here $\bar r$ - dimensionless radius $\bar r = r / r_1$; $\bar p$ - dimensionless pressure $\bar p = p / p_2$;

H - Dimensionless clearance $H = h_{\min} / h_{ск}$; λ - compressibility parameter $\lambda = \dfrac{6\mu\omega r^2}{P_2 h_{ск}^2}$;

A_0, B_0 - Constants of the spiral groove; B - conicity parameter $B = \dfrac{\Delta h(r)}{r_2 - r_1} \cdot \dfrac{r_{\min}}{h_{\min}}$.

In pressurization zone, between radiuses r_2 and r_1 of expression for A_0, B_0 are presented in equation [2]:

$$A_0 = \frac{H_1^3 + \bar{b}_k \left(1 - \bar{b}_k\right)\left(H_1^3 - 1\right)^2 \cos^2\theta_k}{\bar{b}_k + H_1^3\left(1 - \bar{b}_k\right)}, B_0 = \frac{\bar{b}_k \left(1 - \bar{b}_k\right)\left(H_1^3 - 1\right)\left(H_1 - 1\right)\sin\theta_k \cos\theta_k}{\bar{b}_k + H_1^3\left(1 - \bar{b}_k\right)} \quad (4)$$

Where H_1 - the parameter considering the amount of clearance on the current radius r $H_1 = 1 + \dfrac{h_{ck}}{h_{\min} + \Delta h(r)}$;

$\bar{b}_k$ - The dimensionless width of the spiral groove $\bar{b}_k = \dfrac{\bar{b}_1}{b_1 + b_2}$.

In leakage zone, between radiuses and the same equations will take the form:

$A_0 = 1 - B(\bar{r} - 1)$, $B_0 = 0$

Knowing pressure distribution in clearance, it is possible to define seals-hermiticity, and also axial thrust perceived by a friction pair.

The gas-dynamic force that opens the clearance is determined by the formula:

$$W = \bar{w} p_2 \pi \left(r_2^2 - r_1^2\right), \bar{w} = \frac{2}{\left(\bar{r}_2^2 - \bar{r}_1^2\right)} \left[\int_{\bar{r}_1}^{\bar{r}_в} \bar{p}_l \bar{r} d\bar{r} + \int_{\bar{r}_в}^{\bar{r}_2} \bar{p}_н \bar{r} d\bar{r} \right] \quad (5)$$

The research on the applicability of face gas-dynamic seals for unloading of the rotor from the excess axial force is based on the theory of 'face gas-dynamic seal'. The geometrical parameters of the face seals are shown in (Table 1).

Table 1. Some geometrical parameters of the friction pair of face gas-dynamic seal.

Parameter	r_1	r_2	r_y	r_b	Quantity of spiral grooves	$h_{ск}$
Value, мм	90	113	193	100	12	0,006…0,008

Figure3, shows the relationship between the extra bearing capacity change of the gas layer in the bearing clearance on the size of the clearance. Experimental values which were taken at the stand are indicated on the graph by x and triangular symbols [8, 9]. Equal to the depth of the spiral groove which a gas bearing can have. The extra load-bearing capacity - is the difference of current bearing capacity and minimum bearing capacity at the size of the clearance. The value of the load-bearing capacity was obtained through calculations taking into account the geometry of the sealing rings, the forces of the springs and the operating pressure, while the experimental clearance was calculated from the leakage of air measured through the end clearance during the experiment. Air differential pressure – 0.4 MPa, rotor rotational speed - 5500 and 10000 RPM.

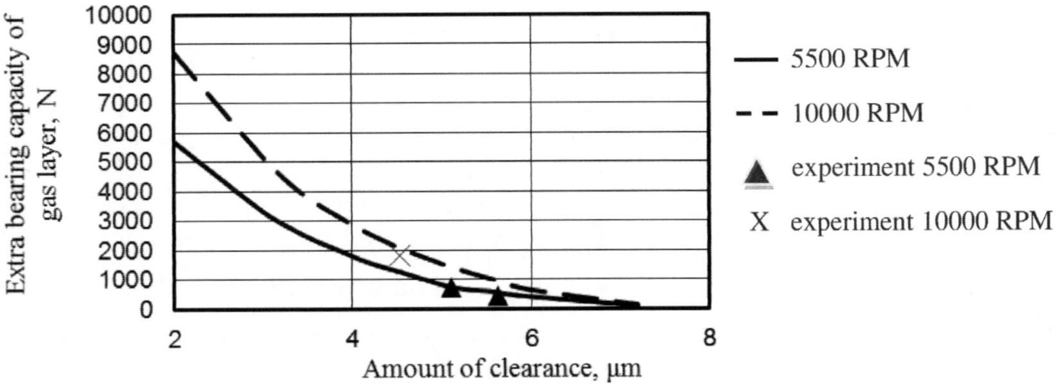

Figure 3. Design and experimental values of bearing capacity of the axial gas-dynamic bearing.

The analysis of 'figure3', shows that at a configuration of the friction pair, where the clearance is more than 2 microns (at smaller values of the clearance, contact of working surfaces and their wear is possible) it is possible "perception" to have the axial force operating on rotor reaching limits of up to 5800 ... 8800N. In order for the bearing to carry the considerably more axial load, it is necessary to increase the area of the friction pair. Therefore, it's suggested to have the width of the friction pair increased from 22.5 mm to 40 mm. The analysis on the amount of clearance in the bearing from the size of the axial load was made at a differential pressure of air of 0.4 MPa and the rotational speed of rotor 10000 RPM. The results of the analysis have shown that at rotational speeds of 10000 RPM, and with an increase in width of the friction pair of up to 40 mm, the bearing is capable of withstanding an axial force equal to 3 tons. Thus, results from experimental analysis of axial gas-dynamic bearing have shown its prospective applicability.

5. Aviation gas turbine engine gas-dynamic axial vibrations rotor damper

Dynamic characteristics analysis of the layer of gas lubricant is used for assessment of the damping force. At the same time, it is possible to use theoretical researches on the face seals [10]. The dynamic reaction of the thin coat lubricant C_d (S), can simply be presented in the form of sums of rigidity and damping, which for incompressible environments are defined by viscosity and geometrical parameters of the node, and for the compressed environments depends also on vibratory frequency.

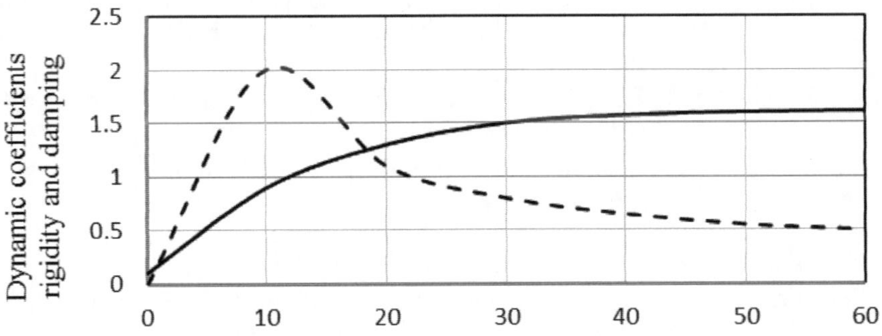

Figure 4. Dependence of dynamic coefficients rigidity $\overline{C}_y$ and damping $\overline{D}$ on compression parameter σ .

At the same time, axial movements damping is defined as.

For plane-parallel clearance: $$b_{zz} = \frac{3\pi\mu}{2h^3}\left[r_2^2 - r_1^2 - \frac{(r_2^2 - r_1^2)^2}{\ln(r_2 / r_1)} \right] \quad (6)$$

This expression is applicable for incompressible environments and can be used as a first approximation for compression. More exact formulas used for the definition of damping gas layer need to be derived by use of special functions [10]. The dynamic reaction of the gas layer on the rotor is defined as the sums of dynamic coefficients of rigidity and damping $\bar{c}_d = c_d h_{\min} /(p_2 \pi (r_2^2 - r_1^2)) = \bar{c}_y + i\bar{D}$. Dependence of dynamic coefficients of rigidity $\bar{c}_y$ (represented by a solid line) and damping $\bar{D}$ (represented by the broken line) against the compression parameter σ for the specified node, $\sigma = 12\mu\omega(r_1 / h_{\min})^2 / p_2$ is presented on 'figure 4'. These results allow us to analyze dynamics of rotor systems.

6. Conclusions

The existing ways of unloading thrust bearings from axial forces in a gas turbine engines are associated with a reduction of their efficiency, or with the considerable complication of design. This article is a proposition for a new, simpler and more effective technical solution, that entails the use of a face gas-dynamic seal as an unloading device. Because seals of this type have the ability to self-regulate, the force-load on the unloading device will change depending on loading conditions, attaining an optimum value for each cycle of operation. The knowledge and experience in design provide for precise prediction of the force-load on the gas-dynamic device. The Discrepancy of experimental data to the theoretical design analysis does not exceed 5%.

A method of analysis of the unloading device characteristics is hereby proposed. These results allow us to conduct research on dynamics of rotor systems.

REFERENCES

[1] Tisarev A, Falaleev S, Koch C, Nagorski M and Staudacher S 2016 *Natural cooling affecting the restart of micro gas turbine Proc. of the ASME Turbo Expo* 8 Paper GT2016-56982

[2] Vinogradov A S and Badykov R R 2014 *Influence of labyrinth seal leakage on the turbine support cooling ASME* 2014 *Gas Turbine India Conf. GTINDIA* 2014 Paper GTINDIA2014-8172

[3] Novikov D K and Diligensky D S 2015 *Calculation Algorithm for Squeeze Film Damper Procedia Engineering*vol 106, 2015, Pages 218-223 2nd *Int. Conference on Dynamics and Vibroacoustics of Machines, DVM* 2014; Samara; Russian Federation; 15 September 2014 through 17 September 2014; Code 119208

[4] Balyakin V B, Novikov D K, Rizinsky I N and Stukalov V P 1990 *Vibrating development of the engine NK-12ST Aviazionnaya Promishlennost* (in Russia) Moscow 8 pp 15–17

[5] Falaleev S V 2014 *Hydrodynamic characteristics of the face seal taking into account lubricant film breakdown, inertial forces and complex clearance form Life Science Journal* 11 (9) 48 pp 337–343

[6] Belousov A I and Falaleev S V 1989 *Gasostatic face seal with elastic working surface Soviet Journal of Friction and Wear* (English translation of Trenie i Iznos) 10 (3) pp 34–38

[7] Muijderman E A 1966 *Spiral groove bearings Springer–Verlag*

[8] Rosseev N I, Medvedev S D, Monakhov A V, Falaleev S V, Balyakin V B and Novikov D K 2001 *Stand for dynamic tests of 'dry' gas seals Gazovaya Promyshlennost* (4) pp 55–58

[9] Balyakin V B and Falaleev S V 2015 *Study of temperature state of mechanical gas dynamic sealing Journal of Friction and Wear* 36 (3) pp 213–217

[10] Chegodaev D E and Falaleev S V 1985 *Dynamic characteristics of gas layer of face seal with elastic surface Soviet Journal of Friction and Wear* (English translation of Trenie i Iznos) vol 6 № 5 pp 136–139

IOP Conf. Series: Materials Science and Engineering **302** (2018) 012007 doi:10.1088/1757-899X/302/1/012007

Simulation of the Vibratory Condition of the Compressor Blade with a Pressed wire Material "MR" Damper Which Located Around the Root Attachment

Alexander S. Gvozdev[1] and Vladimir S. Melentjev[1]

[1]Samara National Research University, Russia, 443086, Samara, Moskovskoe shosse, 34.

E-mail: vladamgenja@gmail.com

Abstract. When you create a modern gas turbine engines urgent task is to improve the reliability by preventing fatigue damages of rotor blades. Such damage is largely determined by the level of vibration stresses. In this paper, using the finite element method and transient analysis of propose a method calculating the damping characteristics of the plates of the pressed wire material "MR" around the root attachment of the compressor blades of a gas turbine engine. Where taken into account contact interaction between the blades and the impeller disk.

1. Introduction

When you create a modern gas turbine engines urgent task is to improve the reliability by preventing fatigue damages of blades. Such damage is largely determined by the level of vibration stresses. The most important factor limiting the vibration is damping capacity of gas turbine engine blades. This characteristic depends on the energy dissipation in a streamlined gas flow, the blades material and structural damping [1-4]. Structural damping it is dissipation of energy in the root attachment and node of contact root platform. In [5], the various schemes of fixing of the compressor blades with elements from the MR material. It can be classified into three groups: A) the impeller disk and blades are in contact only via the plate of the MR material. B) The design of blade and the joggle are provides the active work of plate of the MR material, which is located under the base of the root attachment. C) The plate of the MR material is installed practically free of change in the joggle of the impeller disk under the base of root attachment blade. To calculate of interactions need to create a model that takes into account the joint work blades parts root attachment with small damping and plates of the MR material with a lower hardness, but great damping.

The aim of this work was the construction of the model and the creation of methods of calculation of vibrations blade of compressor with elements of the MR material in the root attachment.

2. Materials and methods

Only one finite element model with three modifications is used in the developed calculation methodology. Each modification depends of fixing conditions. The circuit in Fig. 1 shows the main parameters of a mathematical model that you need to pre-calculate for the successful calculation.

A type of fixing of the compressor blades in the impeller disk has been selected:

1. The root attachment of compressor blades from all sides surrounded by the spacer plates of the MR material. The compressor blade cooperates with the impeller disk through "capsule", which consists of separate plates of the MR material. The plate of the MR material is installed only at the base of the root attachment for the following three types of fixings.

2. The joggle in the impeller disk does not change when the anchorage. Stiffness bracing of the compressor blades of the impeller disk is much higher than the stiffness of the MR material.

3. In soft fixation stiffness bracing of the compressor blades of the impeller disk is comparable with the stiffness of the MR material.

4. "Free" fixation. The joggle design allows the blade to shift lightly to the center of the impeller disk. The plate of the MR material is strongly deformed. The contact surfaces may drift apart.

As a result of analysis of the oscillation curves for each option of fixings are defined amplitude A_i fluctuations and logarithmic damping decrement of oscillations ξ_i. At this stage are selected the ultimate option of fixing and characteristics of MR material.

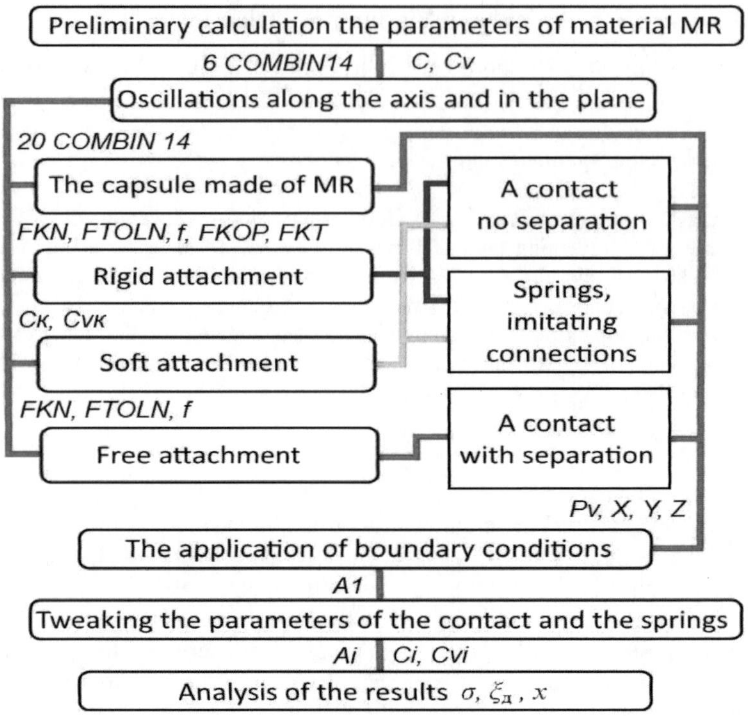

Figure. 1. Block diagram of the algorithm for calculating the compressor blade vibrations

In [6, 7] are given expression that describes the stiffness C and damping Cv of the thin plates of the MR material. In [7, 8] indicates that the MR material has a nonlinear physical properties. The coefficients of stiffness and energy dissipation of the MR material depend on amplitude of its compression and level of the preliminary static strain. In this paper the static and the dynamic characteristics of MR material are assumed to be equal. This hypothesis is supported by the similarity of the dynamic characteristics which calculated based on the static experiment with data of the dynamic experiment [7, 9, 10]. In [7] given the formula of the transition of the energy dissipation coefficient ψ to the equivalent damping coefficient Cv, used in ANSYS.

$$C_V = \psi \cdot C / 2\pi v, \tag{1}$$

where C is the stiffness; v is oscillation frequency.

The expressions for calculating the parameters C and ψ can be represented in the form [7]:

$$C = 14 \cdot \left(1 + 363\varepsilon_A^{3.5}\right)\left(1 + 727\varepsilon_Q^{3.45}\right) \cdot d_w^{1,2} \cdot H^{-0.33} \left(\varepsilon_Q / 0.262\right)^{0.063H} S\left(\bar{\rho}/0.2\right)^{1.1+0.01H/d_w} \tag{2}$$

$$\psi = 1.71 \cdot \left(1 - 7.23H^{0.13}\left(\varepsilon_A - 0.048\right)\right) \cdot \left(1 - 0.1 \cdot \left(10.5\varepsilon_Q - 1\right)^2\right) \cdot d_w^{0.138} H^{0.15} \times \times \left(0.095/\varepsilon_Q\right)^{0.057H} \left(0.2/\bar{\rho}\right)^{0.37+0.015\frac{H}{d_w}} \tag{3}$$

The expressions (2) and (3) comprise the dimensional and the dimensionless parameters. The parameter measurement ranges investigated at [7] given in square brackets. The dimensionless parameters: $\bar{\rho} = m_{MR}/SH\rho_S$ - the relative density of the MR material [0.143...0.26]; $\varepsilon_A = A/H$ - the relative deformation amplitude [0.043...0.183]; $\varepsilon_Q = Q/H$ -the relative preliminary static deformation [0.087...0.262]. The dimensional parameters: A - the deformation amplitude; d_w - the diameter of the wire, [0.09...0.12 mm]; H - the plate thickness, [1.15...4.35 mm]; S – the plate area, m_{MR} - the plate weight; ρ_S - the steel density, 7800 kg/m^3; Q – the preliminary static deformation.

Due to the nonlinearity of the MR material the calculation of oscillation in this case is of iterative nature [7]. Initially, the required parameters of the MR material are unknown. Therefore, the values of the initial parameters were taken close to the average: H = 2.1 mm; d_w = 0.09 mm; Q = 0.5 mm; A = 0; ε_A = 0; $\bar{\rho}$ = 0.2; ε_Q = 0.238. Then, according to equations (2) and (3): C = 227 N/mm, ψ = 1.316. The frequency of oscillation of the system corresponds to the natural frequency.

The model to calculating the vibrations along and the relatively axis the compressor blade is shown on Fig. 2. The area of the blade base is taken as the calculated area of the plate S of the MR material. The fixation of the compressor blades is performed over all axes, except the one about which the oscillations occur.

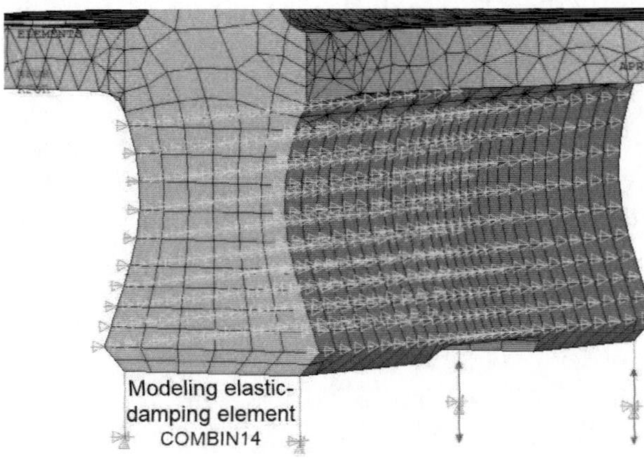

Figure. 2. The modeling by ANSYS the plate of the MR material located at the base of the blade

Firstly, the character of the linear movement the model along the vertical axis has been defined. Secondly, the behavior of the model with plates of the MR material was evaluated in the swing plane. To simulate the spacers on the lateral surfaces of the root attachment needed to add to the model 16 elements COMBIN14. The pressure P_v has been applied to the surface of the pressure side of the airfoil the blade.

Further were carried out the studies of "rigid" and "soft" fixing. In this case, the compressor blade along the lateral surfaces is supported on the impeller disk.

At that the elastic-damping element of MR material is mounted only in the lower part of root attachment and practically does not receives a static load. The "rigid" and the "soft" fixing are different by the method of modeling the interaction of the system "blade - disk." At first the "rigid" fixation was investigated. The slot design of the impeller disk and the blade root is almost unchanged. In this case, the joggle in the impeller disc being done deeper for placement of MR material plate.

Initially the modeling was carried out without the use of contact interaction.

Further were taken the following parameters: $C_к$ = 16820000 N/m, $Cv_к$ = 10 N·sec/m. The angular velocity is not applied. It was assumed that at the beginning of calculation our model is deformed under a static load by the field of centrifugal forces. The amplitude of the external pressure P_v = 84 kPa, duration 0.005 sec.

By simulate the interaction between the impeller disk and the compressor blade instead of COMBIN14 springs can be used contact. The auxiliary elements, which modeling the impeller disk, to be able to use a contact in the model were designed. The surfaces on the impeller disk were chosen as targets. The surface of the compressor blade has been chosen as the contact. The contact of type «No separation» was chosen. The option "Close gap" has been set to the initial moment of calculating the surfaces are brought into contact state even when there is a small gap. The following values of real variables contact were set: FKN =1, FTOLN =0.1, *f*(MU) =0.1, FKOP =1, TAUMAX =10^{25}, FKT =1.

However, the contact interaction can be softer. Constructively this can be achieved by the following method. For example, this can be done by changing the construction of the root attachment or by the use of inserts of wire-rope elements. The deformation of elastic-damping element is increased at such fixation, characterized by a decreasing value of the contact stiffness FKN and/or FKOP and increasing value FTOLN. Therefore, the element more active involved in the work.

The following real constants of contact were used to implement the "free fixation" at the calculation: FKN = 0.1, FTOLN = 0.1, *f*(MU) = 0.1.

Further were performed several calculations at the different properties of the MR material. The following parameters of the contact were taken: FKN = 0.1, FTOLN = 0.1, f(MU) = 0.1, FKOP = 0.5, FKT = 1. The value FKOP has been reduced in order to increase the influence on the result of the parameters MR material.

The frequency *Cv* at which the calculation was made was 178 Hz. It is the average frequency. The first peak frequency is 100 Hz (Fig. 3). The dissipation of energy on the first peak has a significant proportion. Therefore, the assumption of a certain average frequency of oscillation is acceptable.

3. Results

For all subsequent schemes fixing blades, the maximum deviations obtained along the X-axis.

Several calculations have been performed at the simulation of the "capsule" scheme:

1. The initial parameters of the MR material: P_v = 100 kPa, $\xi_д$ = 0.847, *A* = 69 mm. Displacement are measured at the node on the blade airfoil of the compressor blades. This node was located in the lower part of the compressor blade to 1/3 of its height. At this place at the experiment was located the strain sensor. External pressure P_v acts sinusoidally only at the initial time.

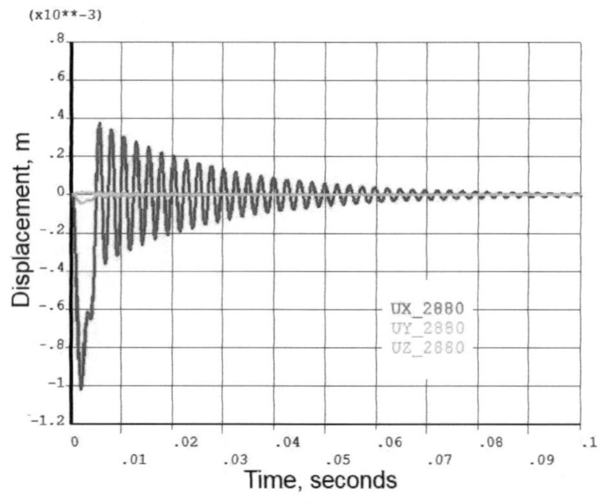

Figure. 3. The graph of vibrations under the loading of the sinusoidal pressure at amplitude P_v = 84 kPa

2. The stiffness of the plate of MR material has been increased to increase loading capacity of the structure. Was obtained the value of the parameter *C*=3115000 N/m, which is close to the maximum possible for the range investigated in [7]. Was obtained: $\xi_д$ =0.011, *A* =0.015 mm. Damping is very small.

3. Parameters have been calculated *C* = 1709000 N/m, *Cv* = 810 N·sec/m. Parameter $\xi_д$ = 1.615, the first peak amplitude of 0.6 mm, the amplitude of the third peak of about 0.1 mm.

In general, the simulation result was considered satisfactory. However, of this type of fixation the MR material should have a stiffness that on one or two orders higher than the stiffness of the MR material, which investigated in [7]. At the same time should not be a significant reduction in the value of ψ. Fig. 3 shows a graph of vibratory displacement at "rigid" fixing without contact. The maximum deviation $A_п$ was about 1 mm, the maximum amplitude A_{max} of about 0.35 mm; parameter $\xi_д$ = 0.123. The maximum stress of the compressor blade was about 95 MPa. This value correlates well with the stresses obtained in the study of the pinned damper, and corresponds to the expected result. The logarithmic damping decrement $\xi_д$ = 0.123 when using plate of the MR material is very close to the logarithmic damping decrement, obtained to the pinned damper by the same compressor blade $\xi_д$ = 0.135 [5,8,11].

For the scheme "rigid" fixation using contacts: the value of displacement was about 0.56 mm, the oscillation amplitude was 0.055 mm (versus 0.35 mm), the energy dissipation coefficient $\xi_д$ = 0.188. It value is greater than the previously obtained value of 0.123. The maximum amount of displacement of the COMBIN14 element was 0.14 mm.

In general, when using the contact at the coupled the compressor blade and the root attachment of the impeller disk, then the main energy dissipation occur into the "root attachment-blade" contact.

The maximum displacement at the "soft contact" was 1.08 mm, the maximum amplitude of 0.22 mm, $\xi_д$ = 0.123. For the case of "free fixation": A = 5.2 mm, $\xi_д$ = 1.223. At such a scheme of fixation you need stiffer elastic-damping elements to normal operation of the construction. The results of studies of the "soft fixation" with different plates of the MR material are shown on Fig. 4.

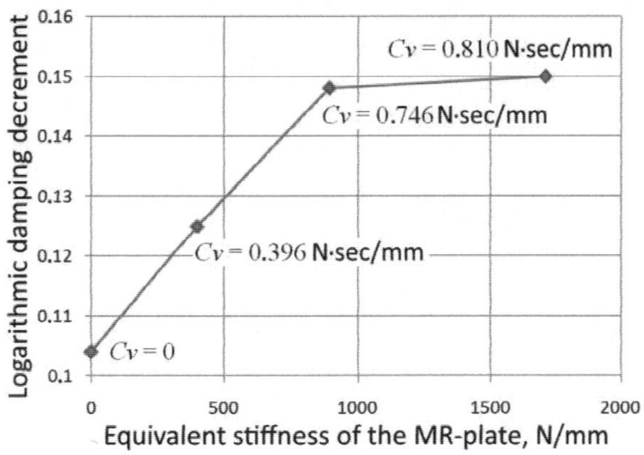

Figure. 4. The logarithmic damping decrement at the different parameters of the MR material

It is evident that due to structural damping at the result of friction the compressor blade and the impeller disk is provided the value of the parameter equal to $\xi_д$ = 0.104. The use of spacers of the MR material can increase this parameter by 44% while maintaining of acceptable amplitudes not exceeding 1.71 mm.

A result of research was designed a model that takes into account the vibrations of compressor blade mounted in the joggle of the impeller disk. The different options for placement of the plate of the MR material into the root attachment of the impeller disk were investigated.

A wide range of the logarithmic damping decrement was obtained: from 1.615 to 0.08 [5]. The calculations confirmed the experimental studies of the ability of MR material to damping of the compressor blades of gas turbine engine [1, 5]. The high-performance of these dampers for vibration damping of the compressor blades. The proposed method is based on the finite element method and the method of the Lagrange-Euler to calculate of contact. The proposed method allows considering a joint operation of the "disk-blade" system and the plates of the MR material implementing contact interaction, into the single model. This model is a convenient tool at the iterative selection of the required parameters of MR material.

4. Discussion and Conclusion

By solve the problem of oscillations the compressor blades with restrictions on the plates of the MR material it is possible without the involvement of the finite element method. You can use the markers (by MSC.ADAMS or ANSYS/Rigid Body) for this method.

By analyzing calculations that are presented above, could be made a conclusion. The interaction of "rigid" and "soft" root attachment fixings of compressor blade at the disk of the impeller can be modeled both by the contact, and by the COMBIN14 elements. The difference between these two methods is only in the values of real constants of these elements (C, C_v, FKN, FKOP etc.).

Acknowledgment

This work was supported by the Ministry of Education and Science of the Russian Federation in the framework of the implementation of the Program ''Research and development on priority directions of scientific-technological complex of Russia for 2014–2020".

References

[1] Belousov, A. I., & Badamshin, I. K. (2015). Conceptual approach to design of materials for GTE turbine blades. Russian Aeronautics, 58(3), 325-330. doi:10.3103/S1068799815030137

[2] Bo, You, Yi Luo, Xiaodong Wang (2013) Contact algorithm of finite element analysis for prediction of press-fit curve. Journal of Information & Computational Science, 10:9 (2013): pp. 2591–2600

[3] Ender Cigeroglu, Ning An, Chia-Hsiang Menq (2009) Forced response prediction of constrained and unconstrained structures coupled through frictional contacts. Journal of Engineering for Gas Turbines and Power, Vol. 131 / 022505-1

[4] Kolmakova, D., Popov, G., Shklovets, A., & Ermakov, A. (2014). Techniques and methods to improve the dynamic strength of gas turbine engines compressor rotor wheels. Paper presented at the ASME 2014 Gas Turbine India Conference, GTINDIA 2014, doi:10.1115/GTINDIA2014-8203

[5] Frolov, V. (2011). Dempfirujushhie ustrojstva v konstrukcijah lopatok kompressorov dvigatelej semejstva NK [Damping device at the structures of compressor blades an engine family NK]. Vestnik samarskogo gosudarstvennogo ajerokosmicheskogo universiteta [Vestnik of the Samara State Aerospace University], no. 3(27): 242-250.

[6] Lazutkin, G. V., & Ulanov, A. M. (2014). Mechanical characteristics of vibration isolators made of various wire materials. Research Journal of Applied Sciences, 9(11), 711-715. doi:10.3923/rjasci.2014.711.715

[7] Ulanov, A., and S. Bezborodov (2016) Calculation method of pipeline vibration with damping supports made of the MR material. Procedia Engineering, Vol. 150, 2016, pp. 101-106 doi: 10.1016 / j.proeng.2016.06.725

[8] Troynikov, A. A., & Lazutkin, G. V. (2014). Experimental determination of the variation of the elastic damping properties MR material in a continuous operation. ARPN Journal of Engineering and Applied Sciences, 9(12), 2876-2879

[9] Ulanov, A. M., & Bezborodov, S. A. (2014). Life-time of vibration insulators made of metal rubber material under random load. Research Journal of Applied Sciences, 9(10), 664-668. doi:10.3923/rjasci.2014.664.668

[10] Yan, H., Zhang, L., Jiang, H. & Ulanov, A. M. (2015). Calculation of multi-layer plate damper under one-axial load. Chinese Physics B, 25(2) doi:10.1088/1674-1056/25/2/024306.

[11] Melentjev, V. S., & Gvozdev, A. S. (2015). Method of static characteristics calculation of adjustable two-ring type full-metal vibroinsulators in a non-linear approach. Modern Applied Science, 9(2), 267-278. doi:10.5539/mas.v9n2p267

IOP Conf. Series: Materials Science and Engineering **302** (2018) 012008 doi:10.1088/1757-899X/302/1/012008

The Development of A Squeeze Film Damper Parametric Model in the Context of a Fluid-structural Interaction Task

Dmitrii K. Novikov[1] and Dmitrii S. Diligenskii[1]

[1]Samara National Research University, Russia, 443086, Samara, Moskovskoe shosse, 34.

E-mail: novikovdk_ssau@mail.ru

Abstract. The article considers the work of some squeeze film damper with elastic rings parts. This type of damper is widely used in gas turbine engines supports. Nevertheless, modern analytical solutions have a number of limitations. The article considers the behavior of simple hydrodynamic damping systems. It describes the analysis of fluid-solid interaction simulation applicability for the defying properties of hydrodynamic damper with elastic rings ("allison ring"). There are some recommendations on the fluid structural interaction analysis of the hydrodynamic damper with elastic rings.

1. Introduction

The one of the most common devices to deal with the vibration in rotary systems is a damper that is mounted inside of the support assembly. The classification of different forms and designs for these devices are represented in [1]. In that particular case will be considered several hydraulic damper operational aspects: structural-fluid simulation fidelity on the basis of the flexible part of the elastic ring (figure 1,4). The function of the flexible element is presented by a thin-wall ring that has equiangularly spaced smooth pedestals. Depending on whether are used a single or a pair of rings pedestals will be found on both sides of the ring in a check-board arrange (figure 1,a,2) or on one side but in the same criss-cross pattern with pedestals from another ring (figure 1,b,3).

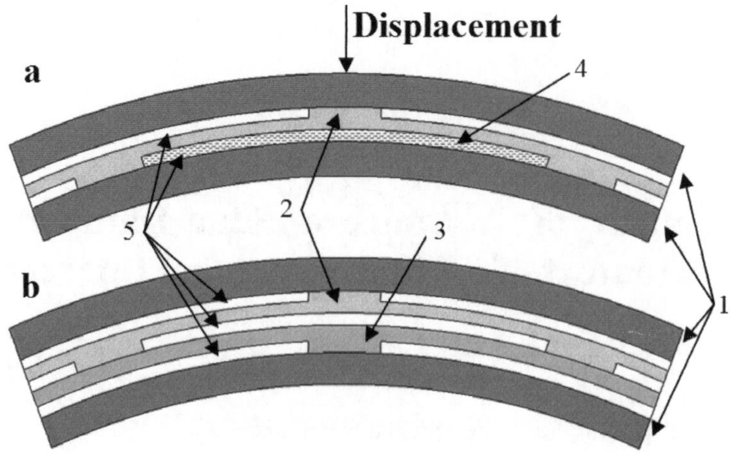

Figure 1. Damping support with the elastic ring: a – single elastic ring, b – doubled elastic element, 1 – rigid case (elasticity modulus > 1*10^15 Pa) , 2 – elastic ring with pedestals on both sides (inside and outside), 3 – elastic ring with pedestals on one side (inside), 4 – flexible part of the damping element, 5 – gap for the oil flow.

Therefor there is always a gap for the oil between ring and between rings and casing (figure 1,a,b,5). The amount of pedestals and other geometry of elements is subjected to [2]. The height of pedestals on two sides could be different. The damping is provided by the squeezing process of the lubricant fluid through gaps. There are many patents (i.e. [3]) related to the design of a this type damper. Both present and further investigations in the field of SFDs are highly influenced by the work of Adolfo Delgado and Luis San Andreas the actual head of the rotordynamics laboratory in Texas A&M University [4].

This type of damper is fairly widespread and has many advantages, including small size and ability to be designed with predetermined rigidness and anisotropy, high damping capacity, possibility for the tuning from the resonance frequency. At the same time there are several major deficiencies: the required high precision of tolerance values, angle taper, etc., adjustment on the working mechanism due to the narrow range of existed industrial standard on the elastic ring geometry and the combined action if there are several.

Apart from this IS there is a couple of papers [5,6] on defying the stiffness of these rings. In the articles rings are presented as set of coupled beams attached from their ends. Moreover there is an assumption that in reality there is only a one half of the ring working. In [7] has been studied the effect from dampers on GTE dynamics.

To estimate hydraulic forces and reactions from them there are also some analytical research, [8,9]. where are mentioned approaches to predict elastic rings flexibility. Thus, in these papers the damping is considered on the basis of theory for fluid oscillations between flat plates that could be found in the S.I. Sergeev monography [9], and the elastic element as a curved beam with the stiffness equals *C*. In reality the damper is a structure where the rigid part with a complex design (ring with pedestals) is mounted between support casing and the rotor and all gaps are filled with a lubricant. During oscillation of the rotor takes place the mutual dynamic effect between oil and elastic parts of the ring between pedestals. In that case the research of that type of the damper requires the solution of so called fluid-structure interaction task.

2. Hydraulic-elastic interlinked task

2.1Analytical model

To solve that kind of the task it is required the special damper model that however should be revised by proven analytical solutions. For this reason the model development should be started from the study of a single elastic ring section. In the figure 2 is presented an aligned section of the elastic ring. In that case the oil flow in the gap could be considered as flow between two parallel plates, oscillating by the harmonic law. To defy the damping value it is possible to use some analytical expressions from the Sergeev monography [9].

The common scheme is represented in the figure 2. Where L - an overall size of the damping element; h - the initial gap between vibrator and stator; ζ- harmonic oscillation speed; b - element width, P(x) - pressure distribution during the vibration, Vx(z) - fluid flow speed distribution over the gap height.

The reaction force on the damper vibrator is defined by the expression (1)

$$\left.\begin{aligned} &P = P_c = B\int_{-l}^{+l} p_э dx = -\mu\Lambda^3 BF\zeta(\alpha,\tau)\dot{\zeta} = -C\dot{\zeta}, \\ &F\zeta(\alpha,\tau) = (1-\alpha\sin\tau)^{-3}, \alpha \equiv \frac{\alpha}{h_c} \end{aligned}\right\} \quad (1),$$

where $\Lambda = \frac{L}{h}$,

By (1) has been performed several calculations the results of which are represented in figure 3. In figure 3 the vertical axis has a logarithmic scale. According to the diagram it is apparent that with the increase of the gap size (figure 1,5) the reaction force value tends toward zero, and if the gap become smaller the force reaction tends toward the infinity. Therefore the analytical approach is reasonable only for the oscillations with the amplitude value within the size of the gap, i.e. without surfaces contact. Whereas in contrast the interactional numerical computation allows to estimate the damping force even in case there is a rigid parts interaction.

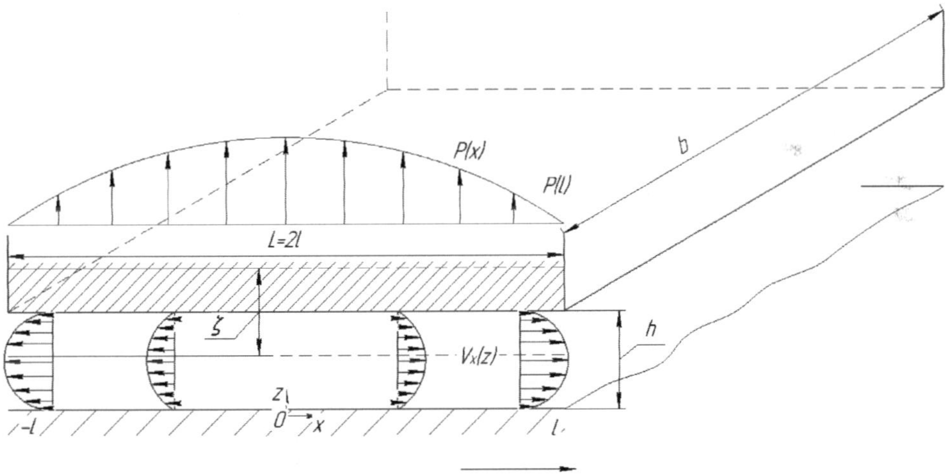

Figure 2. The section of a damper with a plain vibrator making oscillations in the normal direction to the stator surface.

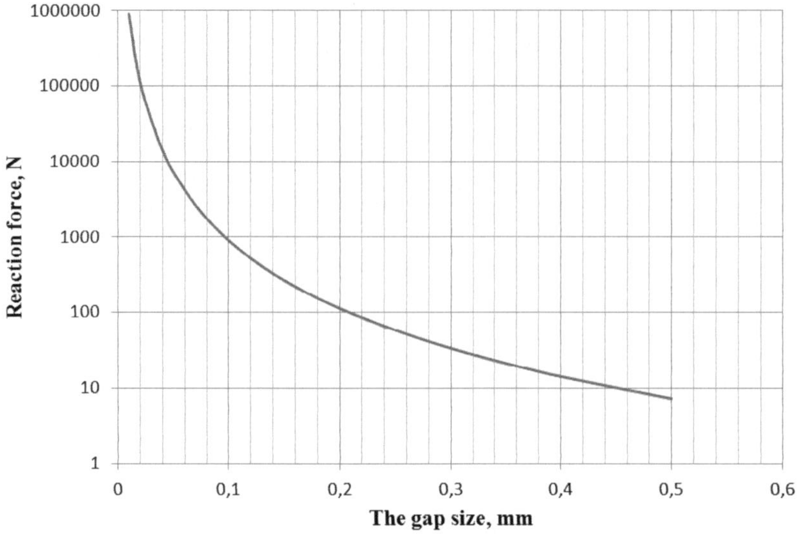

Figure 3. The force reaction on the surface of the plain damper defined according to Sergeev analytics for different gap sizes. Other parameters are taken from the table 1

2.2. Finite-element model

The finite element model is presented by a parametrical 2-way fluid-solid interaction model (figure 4) consists of 4 blocks: solid body part - "Transient Structural"(figure 4,a), fluid body part - "Fluid Flow (Fluent)"(figure 4,b), parameterization block - "Parameter Set" (figure 4,c), solution and results transmission block - "System Coupling" (figure 4,d).

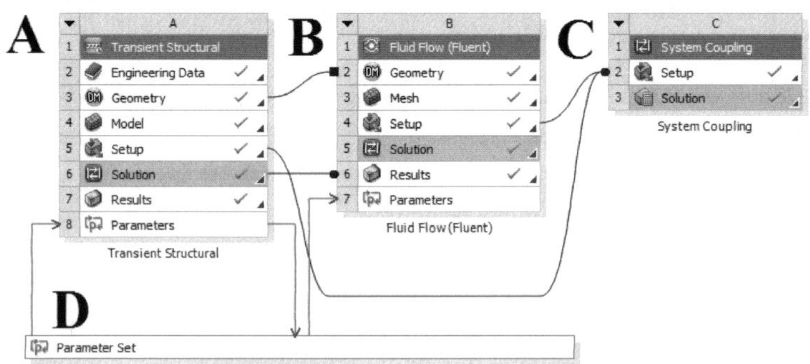

Figure 4. 2way-FSI analysis block scheme: a-structural, b-fluid, c- structural/fluid coupling, d-parameters

2.3. Principle assumptions and boundary conditions

The model represents an interaction model of a hydraulic elastic task. There are two main parts - mechanical and fluid.

For the solid part have been selected following options:

- fixing of the lower body by the bottom surface in all direction
- fixing the upper surface of the top body in one tangential direction to prevent the sliding
- fixing the upper surface of the top body in another tangential direction by periodic symmetry border to prevent the sliding
- applying the oscillation on the top body by the function $y = a \sin \omega t$
- The global mesh size has the maximum of 1 mm

For the fluid part have been selected following options:

- heat transmission off
- flow stream is laminar
- the gap is closed by walls from two sides
- from two other sides are applied two flow supply borders "pressure-inlet" with the pressure value P= 0
- the mesh size has the maximum of 1 mm along the surface and 0,15 mm by height
- the analysis is performed only for several cycles to exclude false results due to the unsteady-state process in the beginning (5 cycles to minimize the processing time)

Both models has borders there they are interact each other. Those are two surfaces from the solid part and two from the fluid (figure 5).

The parameterization is made for all geometry, mesh sizes, fluid body layers and pressure value from both inlets.

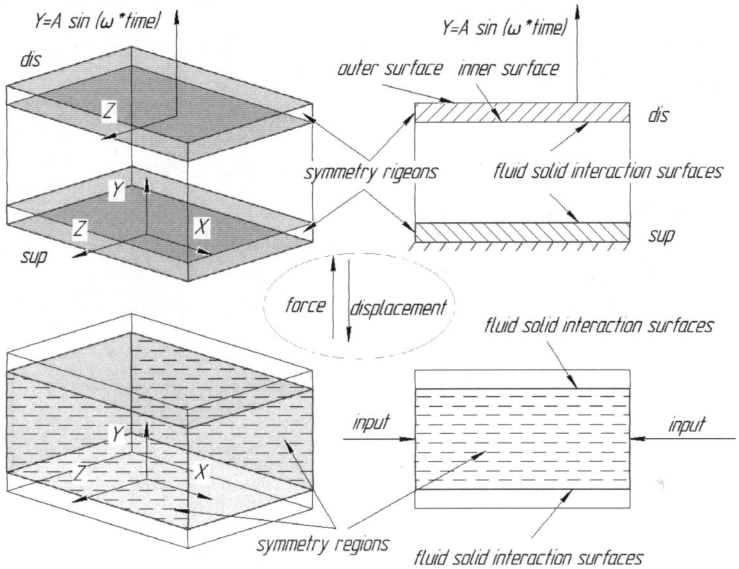

Figure 5. Boundary conditions scheme

2.3.1.Convergence analysis. As long as the model has a lot interacting parts and is in unsteady state the processing time will increase. Has been performed a group of calculations two estimate the acceptable level for mesh size and fluid layers and oscillation cycles number. Initial conditions are listed in the table 1.

Table 1. Parameters for flat damper hydrodynamic analysis

Parameter	**Value**
Overall length L, mm	50
Gap height h, mm	0,4
Oscillations amplitude a, mm	0,03
Oscillations frequency, Hz	50
Oil viscosity μ, kg/ms	0,022
Oil density ρ, kg/m^3	800

IOP Conf. Series: Materials Science and Engineering **302** (2018) 012008 doi:10.1088/1757-899X/302/1/012008

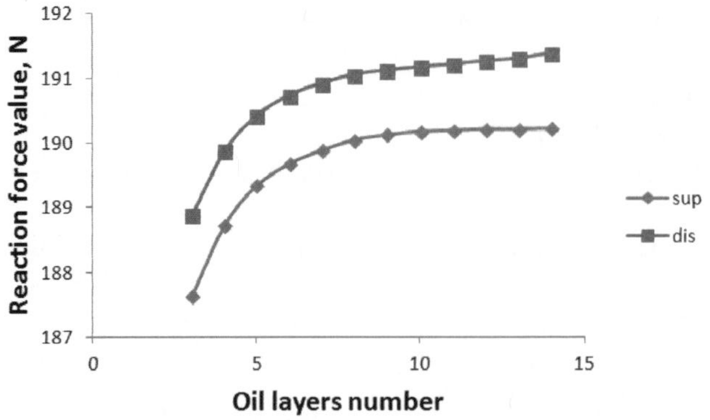

Figure 6. The dependency of the reaction on the flat damper from fluid layers number: "sup" - stationary lower body surface, "dis" - oscillating upper body surface (figure 5)

The increase in the number of fluid body layers leads to the convergence. The difference of the reaction value for 3 and for 13 layers does not exceed 1.5% (figure 6).

In figures 7 is presented the dependency of the reaction force on the mesh size for different amount of layers.

The decreasing of the mesh size leads to convergence. From the figure 7 it is shown that for 3 and for 7 layers the convergence process looks different, so that for 7 layers the optimal mesh size would be 0,4 mm by the edge length and for 3 layers it is element with 0.75-0.8 mm edge length. During the further mesh size reduction process appears the divergence. The relative difference of the control values in any case is not above 1%. As a result, for the time processing economy was taken a decision to use a 3 fluid layers and 1mm size mesh for this analysis.

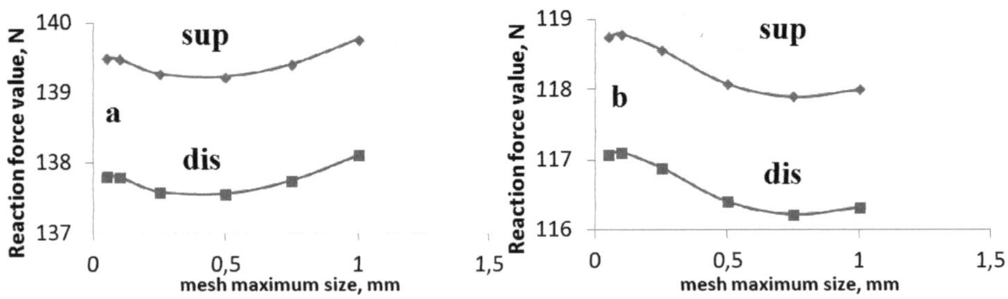

Figure 7. The dependency of the flat damper reaction on the mesh size for the case of: a - with 7 mesh layers, b - with 3 mesh layers.

In the figure 8 is shown the dependency of the force reaction on the time. It is clear that during a short period of time (before 0,025 s) the process is unsteady. Therefore for the meaningful comparison with analytical calculations should be considered results from the relatively steady part that is appeared only after 0,03 s. It is also should be mentioned that the total value of the reaction is not steady and cyclically drops to the zero. It means that the comparison should be made for maximum values.

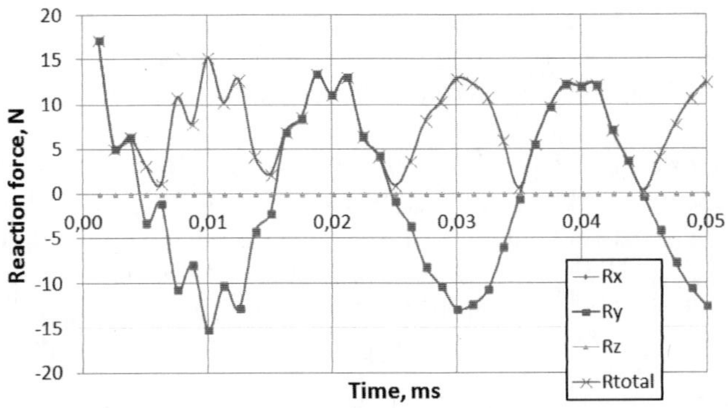

Figure 8. The dependency of the damper reaction force on the time: Rx, Ry, Rz - reaction by x, y and z axis direction relatively, Rtotal - the total reaction on the surface.

2.3.2. Analysis. With applying parameters that have been received previously is defined the effect of the gap sizes and oscillations amplitude over the force reaction in the damper. For both dependencies processing results have been compare with analytical solution by Sergeev [9]. All of the received data is represented in figure 9.

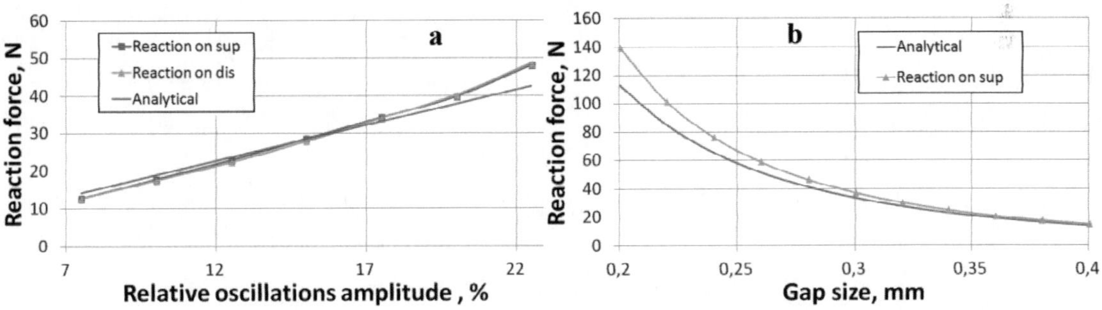

Figure 9. Dependency of the reaction force on: a - the gap size, b - the oscillations amplitude value

3. Discussion and Conclusion

The processing results are in a good correspondence with the analytical data. On the basis of this solution it is expected to perform the further analysis including full damper elastic ring models in the context of the 2way fluid-solid interaction task. That is required further preparation by analysis of cylindrical dampers behavior in single-axial and double-axial oscillation, and prism-type.

- Has been developed the flat damper model in the context of interconnected elasticity-hydrodynamic cross-simulation.
- Has been defined optimum mesh parameters for better convergence
- Has been conducted the validation of the developed fluid-solid interaction model by comparison with analytical solution
- Has been defined further steps to study squeeze film damper in the context of a fluid-solid interaction task.

References

[1] Diligenskii, D.S., Novikov, D.K. Analiz konstrukcii gidrodinamicheskih dempferov // International science-technology forum dedicated to 100 years of "Kuznetsov" JSC and 70 years of SSAU. - 2012, Vol.2, p. 47-48.
[2] «GTE rotors elastic rings»// Industry-specific standard// ISS 1 10778-72, ISS 1 10779-72, 1972. - 11p.
[3] Mohsiul Alam, Doug K. Spencer, Compact resilient anisotropic support for bearing // US

2006/0204153, Int. Cl. F16C 27/00., Pub. date: Sep.14, 2006
[4] Delgado, A., and San Andrés, L., 2010, "Identification of Force Coefficients in a Squeeze Film Damper with a Mechanical Seal: Large Contact Force," ASME Journal of Tribology, Vol. 132(July), p. (1-7)
[5] Leontiev, M.K. Konstrukcija i proektirovanie uprugodempfernyh opor rotorov VRD / Hronin, D.V., Leontiev, M.K., Borozdyko, E.V. - Moscow Aviation Institute. - Moscow, 1988. - 37p.
[6] Artemov, E.A. Jeksperimental'noe i raschetnoe opredelenie podatlivosti uprugih opor turbomashin // «News of Higher Educational Institutions. Aviation technology». - 1965, №2 - 8p.
[7] Falaleev, S. V., & Balyakin, V. B. (2014). Application of a hydrogasdynamic axial vibration damper for reducing GTE vibration. Russian Aeronautics, 57(3), 314-318. doi:10.3103/S1068799814030167.
[8] Lobanov, V.K., Hrustalev, A.B. The study of damping properties in particular type of GTE elastic support [Text]//. «Vibration strength and reliability of the engine and other aircraft systems» - Kuibyshev aviation institute, - Kuibyshev, 1977, №4 - 3p.
[9] Sergeev S.I., Dempfirovanie mehanicheskih kolebanij [Text] Book / edited by Maingard S.A. - State publishing House of physical and mathematical literature, Moscow, 1959 - 408p.

Developing Parametric Models for the Assembly of Machine Fixtures for Virtual Multiaxial CNC Machining Centers

A.V. Balaykin, K.A. Bezsonov, M.V. Nekhoroshev and A.P. Shulepov

Samara University, 32 Moskovskoye Shosse, Samara, 443086, Russia

E-mail: balaykinav@ssau.ru

Abstract. This paper dwells upon a variance parameterization method. Variance or dimensional parameterization is based on sketching, with various parametric links superimposed on the sketch objects and user-imposed constraints in the form of an equation system that determines the parametric dependencies. This method is fully integrated in a top-down design methodology to enable the creation of multi-variant and flexible fixture assembly models, as all the modeling operations are hierarchically linked in the built tree. In this research the authors consider a parameterization method of machine tooling used for manufacturing parts using multiaxial CNC machining centers in the real manufacturing process. The developed method allows to significantly reduce tooling design time when making changes of a part's geometric parameters. The method can also reduce time for designing and engineering preproduction, in particular, for development of control programs for CNC equipment and control and measuring machines, automate the release of design and engineering documentation. Variance parameterization helps to optimize construction of parts as well as machine tooling using integrated CAE systems. In the framework of this study, the authors demonstrate a comprehensive approach to parametric modeling of machine tooling in the CAD package used in the real manufacturing process of aircraft engines.

1. Introduction

Tooling is a part of technological complements to the machinery that is designed to perform specific tasks in the process. The main group of tooling consists of fixtures, i.e. machine accessories that help fix the workpiece in a specified position in accordance to operation specifications.

Use of fixtures helps:

- get rid of pre-machining workpiece sketching and make machining more accurate;
- improve the performance of operations;
- reduce the cost of production;
- make operations easier and safer;
- improve the technological capabilities of equipment;
- organize multi-machine servicing;
- use technologically plausible timings and reduce the number of workers involved.

To facilitate and speed up the design and production process in state-of-the-art systems, CAD/CAM systems are used. These systems enable one to create a parametric model, which is an applied program that controls the build of a new design object version upon any alteration or modification of its parameters. Fixtures as design objects have a number of features that make it feasible to develop their parametric models as user applications.

Siemens NX has a lot of parameterization tools, see Figure 1. The parameterization methods we propose have certain pros and cons; they are optimal for their specific modeling tasks. In its turn, each method provides for various parameterization techniques.

Our research was to integrate variance parameterization into a top-down design methodology to create a multi-variant and flexible fixture assembly model.

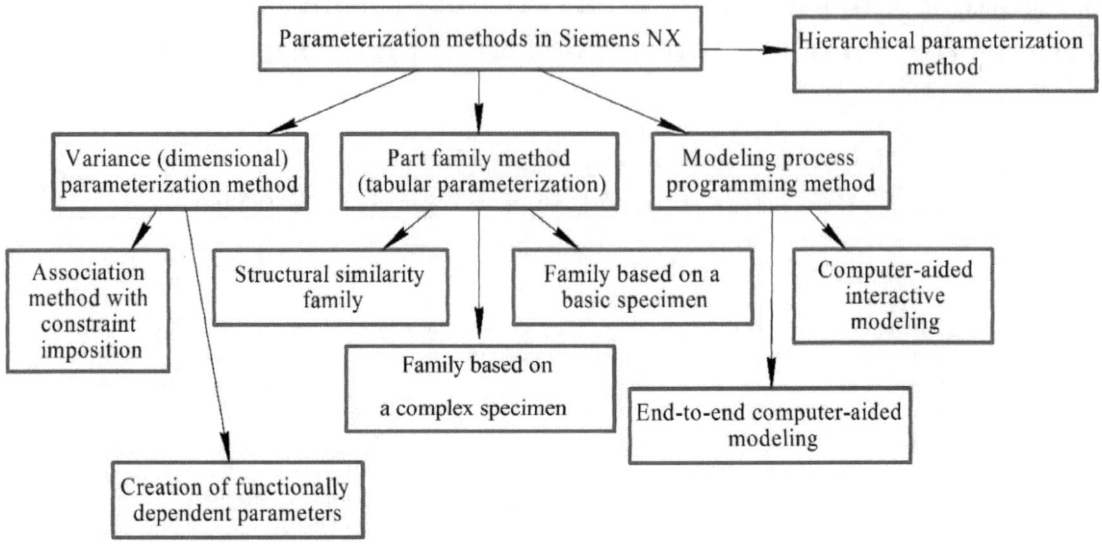

Figure 1. Siemens NX parameterization methods

Variance (dimensional) parameterization is the most common parameterization method implemented in Siemens NX. This method enables derivation of association models with geometric constraints and dimensions imposed during the building process. This method also helps specify the dependencies of different model parameters. Functional relations can be specified for the active sketch dimensions and parameters of 3D object operations.

Variance (dimensional) parameterization links the 3D model of a fixture to e-tables to contain geometric dimensions and calculation ouput.

2. Materials and methods

Variance or dimensional parameterization is based on sketching, with various parametric links superimposed on the sketch objects and user-imposed constraints in the form of an equation system that determines the parametric dependencies.

[1] summarizes the results obtained when solving the problems of CAD-enabled part and assembly modeling where possible 3D deviations were factored in. It therefore presents a new part and assembly modeling theory that factors in their spatial tolerances and enables one to analyze part assemblability at the stage of their geometric development in CAD systems.

[2] uses one of the CAD/CAE systems that helps quickly generate alternative designs at various levels of building modeling.

[3] dwells upon geometric modeling of spatial deviations in parts and assemblies. For each type of position and shape tolerances, it describes a method for building a configuration space, a point wherein helps obtain various positions of the marked frame and/or the modified coefficients of the quadratic forms of this surface within the tolerance; i.e. a full description of the 3D position of this surface.

[4] dwells upon a top-down 3D design and life-cycle management technology. This approach enables use of parallel business processes and organize efficient management of changes in a product. It works not only for 3D models at various stages of design, but also for requirement structures as well as functional and logical structures of the product. A 3D model developed by top-down design is an

integral part of the enterprise data and process model.

[5] dwells upon a 3D-based interface for modeling new and intuitive assemblies to support conceptual exploration works. In their modelign interfaces, unmarked segmented object components are assembled to create new 3D models.

In the future, 3D models will fully replace 2D drawings of parts and assemblies. 3D models built contain source data for CAE analysis based on the computational solution of differential equations [6,7].

For engineering and scientific computations, 3D models are divided into finite elements, whereupon boundary conditions are imposed. Computations helps refine the geometric parameters of part and make necessary adjustments. A modified 3D model of the part is then transferred to the CAM system to develop the manufacturing process [8-13].

Computing CAD models in CAE systems helps optimize the materials used in design [14,15].

To sum up, it is safe to conclude that in this focus area, researchers actively use various parameterization methods to link 3D models of fixtures to e-tables that will contain geometric dimensions and calculation output.

3. Developing top-down parametric models for the assembly of fixtures using virtual multiaxial CNC machining centers

In our research, variance parameterization is fully integrated in a top-down method. This method enables the creation of multi-variant and flexible fixture assembly models as all the modeling operations are hierarchically linked in the build tree.

The top-down method (see Figure 2) implies that all parts are created jointly using a combined build tree.

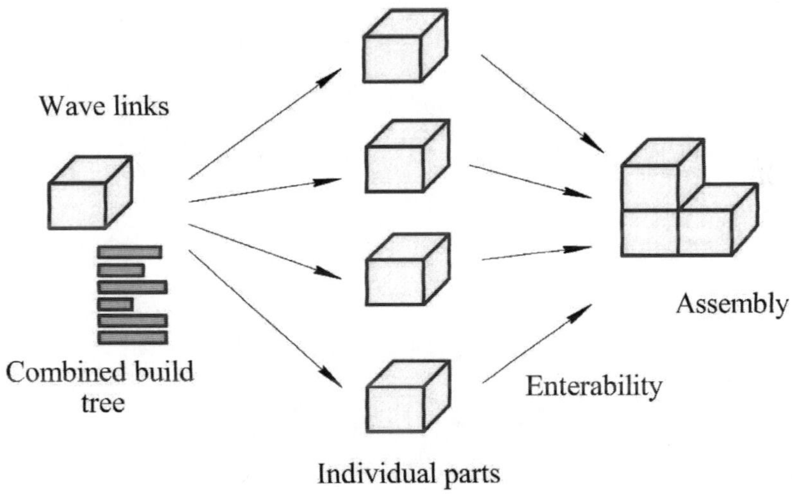

Figure 2. Top-down design

Thus, the resulting parts in the product layout only contain geometric copes derived from the service part that contains the combined build tree. This method has the following advantages:

- all parts are designed interrelatedly, no need to manually harmonize them;
- relations are physically included in the model, i.e. changes made to one of the parts will be reflected in all other parts it is connected to;
- the conceptual design is included in the model, i.e. top-level parameters and forms are connected to the parts and control them for consistency with the concept;
- the design process is easy to divide into the required number of interrelated stages (sketch project, work project, etc.);

- the process becomes considerably less prone to error when buing based on the links relations between parts rather than engineers' imagination and memory.

Let's take a 3D machine fixture model as an example, see Figure 3. The parent of the product layout is an assembly model that combines the following components:

- source;
- work;
- result.

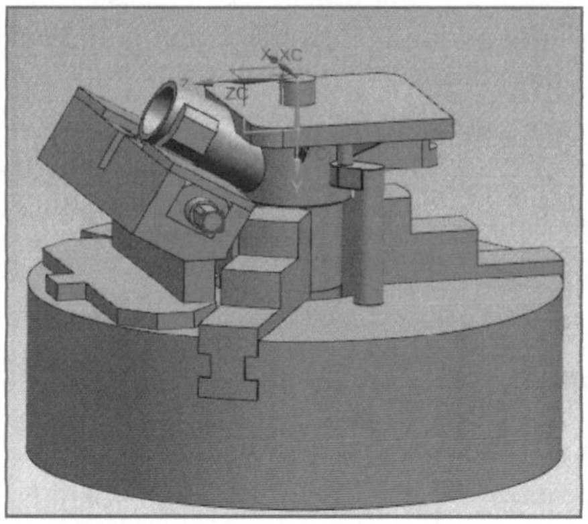

Figure 3. Fixture assembly model

For the source, we created a 3D Journal model, the sketches of which were controlled by a certain number of conceptual parameters, see Figure 4:

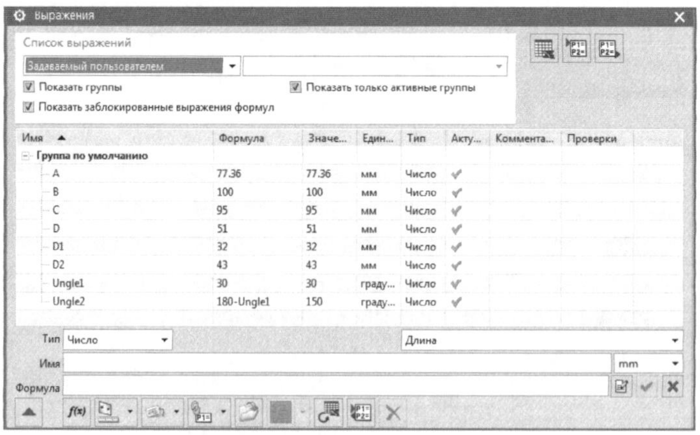

Figure 4: Concepual Parameters Window

A is the length of the sponge-fixed nozzle;

B is the width of the platform (machined surface);

C is the length of the platform (machined surface);

D, D1, D2 are the diameters of the surfaces of the part;

Ungle1 is the inclination angle of the sponge-fixed nozzle;

Ungle2 is the angle at which the column is inclined relative to the chuck.

We further made the source sketches of the fixture (see Figure 5) that are associatively copied to the work by means of Wave. The work uses expression relations and Wave links to sketches to take data from the source; thus, all changes in the source will be automatically traced in the work.

IOP Conf. Series: Materials Science and Engineering **302** (2018) 012009 doi:10.1088/1757-899X/302/1/012009

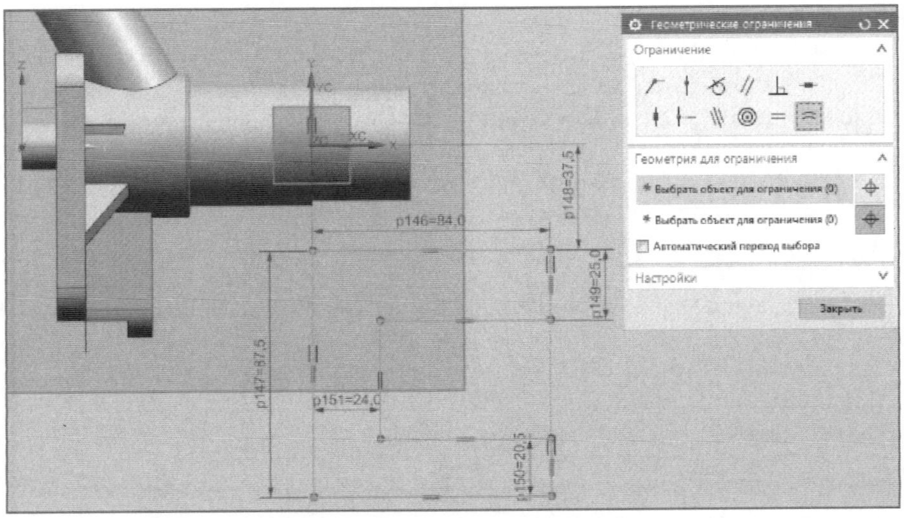

Figure 5. Source chuck sketch

Source sketches are the basis for all subsequently made sketches, which in their turn are the basis for building the solids representing the geometry of components.

This method can be seen in action when altering a conceptual parameter in the source. Let's change Ungle1 from 30 degrees to 45 degrees. Changing these parameters alters the inclination angle of the sponge-fixed nozzle as well as the inclination angle of the column. The initial form of the fixtures and the result of its parametric change are shown in figures 6 and 7.

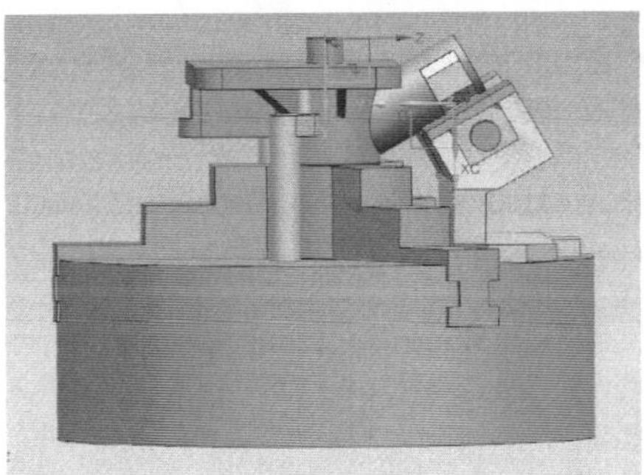

Figure 6. Unaltered column

Figure 7. Altered column

Comparing the results obtained, we conclude that appropriate sketching and rigid fixation of fixture sketches and components relative to the part will cause geometric changes in the part to alter the geometry of such components.

The top-down method has an undeniable advantage which is that all the geometric properties of a product are harmonized, and the entire modeling process is fully contextual.

Stated above the top-down design methodology enables to create multi-variant and flexible model of fixture assembly models in which all operations of the modeling are connected among themselves hierarchically. This parametrization technique is proposed for use in a real production process in the manufacture of parts on multi-axis machining centers with CNC, which will significantly shorten the time for designing the tooling, as changing the geometry of the models will occur automatically and interrelated with the original part. In the Siemens NX CAD system, machine tooling files contain associative 2D documentation and processing strategies on CNC equipment, and when the 3D model is rebuilt, this data is automatically changed.In the future it is planned to integrate engineering calculations into the model of adaptation which helps to optimize the design of both parts and all equipment.

4. Conclusion

This paper gives an analytical review of papers on parametric modeling to prove the relevance of this focus area. Progressive top-down design technology is persented herein. The paper describes the advantages of this method as well as how it can be used when designing a machine fixture assembly model.

Top-down methodology comabined with NX CAE and NX CAM can considerably reduce product design time and make the design and production stages less prone to error.

References

[1] Gaer M A, Zhuravlev D A 2011 Bulletin of the Irkutsk state technical university vol 4(51) pp 24-26
[2] Sacks R , Eastman C, Lee G 2004 Automation in Construction vol 13 pp 291-312
[3] Gaer M A, Zhuravlev D A, Yatsenko O V 2011 Bulletin of the Irkutsk state technical university vol 10(57) pp 32-36
[4] Lihachev M V 2015 Vestnik SibGAU vol 16 pp 423–429
[5] Jaiswal P, Huang J, Rai R 2016 Computer-Aided Design vol 74 pp 45–54
[6] Ermakov, A.I. and Urlapkin, A.V., The inf luence of a blade Vibrations connectivity on a degree of disturbance of turbine wheels rotation symmetry, Res. J. Appl. Sci, 2014, vol. 9, no. 11, pp. 800–805.
[7] Ryazanov A.I. Mathematical model and numerical solution of the process of heating and melting of a traveling cylinder fed into a rocket chamber, ARPN J. Eng. Appl. Sci., 2014, vol. 9, no. 10, pp. 1859–1865.
[8] Popov, G.M., Baturin, O.V., Kolmakova, D.A., and Krivtsov, A.V., Improvement results of TK-32 turbocompressor turbine with gas-dynamics and strength CAE-systems, Int. J. Eng. Technol., 2014, vol. 6, no. 5, pp. 2297-2303.
[9] Ermakov A.I., Shklovets A.O., Popov G.M., and Kolmakova D.A. Investigation of the effect of the gas turbine compressor supports on gas flow circumferential nonuniformity, Res. J. Appl. Sci., 2014, vol. 9, no. 10, pp. 684-690.
[10] Khaimovich, I.N. and Khaimovich, A.I., Analytical modeling of the microstructure evolution of titanium alloys during high-speed stamping of the blades of gasturbine engines, Russ. J. Non-Ferrous Met., 2015, vol. 56, no. 2, pp. 181-189.
[11] Nekhoroshev, M.V., Pronichev, N.D., Smirnov, G.V. Computer simulation of high-speed anodic dissolution processes of geometrically-complex surfaces of GTE details Open Mechanical Engineering Journal, 8, pp. 436-440.
[12] Melentev, V.S. and Gvozdev, A.S., Methods of building a parametric CAD-model of a piston micromotor with the systems, Int. J. Eng. Technol., 2014, vol. 6, no. 5, pp. 2331–2338.
[13] Pechenin, V.A., Bolotov, M.A., and Ruzanov, N.V., Development of a method of ICP algorithm accuracy improvement during shaped profiles and surfaces control, Int. J. Eng. Technol., 2014, vol. 6, no. 5, pp. 2229-2235.
[14] Balaykin, A.V., Nosova, E.A., Galkina, N.V Study of the Ti-5Al-5Mo-5V-1Cr-1Fe titanium alloy grain structure uniformity after bending and annealing. IOP Conference Series: Materials Science and Engineering, Vol. 177, I. 1. 2017.
[15] Alexeev, V.P., Balyakin, A.V., Khaimovich, A.I. Influence of the direction of selective laser sintering on machinability of parts from 316L steel. IOP Conference Series: Materials Science and Engineering, Vol. 177, I.1. 2017.

IOP Conf. Series: Materials Science and Engineering **302** (2018) 012010 doi:10.1088/1757-899X/302/1/012010

CFD Analysis of A Starved Four-Pad Tilting-Pad Journal Bearing with An Elastic Support of Pads

E F Parovay[1] **and S V Falaleev**[1]

[1]Aircraft Engines Design Department, Samara National Research University, 34, Moskovskoye shosse, Samara, 443086, Russia

Email: selena_pa@mail.ru

Abstract. Tilting-pad journal bearings are widely used in technics. Oil starvation operation regime is not common for hydrodynamic bearings. However, correctly designed low-flow journal bearing have to operate efficiently and consistently on high rotor speeds. An elastic support of bearing pads is a set of elastic pins made of steel. Elastic support allows pads to self-align and achieve an optimal operational mode. The article presents the thermohydrodynamic performance of an axial journal bearing. The study deals with 60 mm diameter four-pad tilting-pad journal bearing, submitted to a static load varying from 1000 to 30000 N with a rotating speed varying from 1000 to 10000 rpm. The investigation focuses on numerical studying the characteristics of low-flow tilting-pad journal bearings under oil starvation conditions. Dependencies of the bearing performance on the load, rotational speed of the shaft, and the size of the radial clearance are presented.

1. Introduction

Tilting-pad journal bearings (TPJB) are widely used in high-speed rotating machines due to their high dynamic stability, damping characteristics, ability to operate at high rotational speeds, and long service life [1]. TPJB can be classified by the type of mounting: bearings with pads mounted on hinges; rolling-pad bearings with an elastic support of pads. The lower surface of the smaller radius of rolling pads leans on the internal surface of the housing. An example of such a bearing is H. Hashimoto's rolling-pad journal bearing with six pads [2]. These bearings are more rigid, compared to bearings with hinged pads. Hydrodynamic bearings with elastic pads increase the load bearing capacity by up to 30%, compared to conventional thrust bearings with tilting (pivoted) pads [3]. However, there is a tendency to improve the efficiency of gas turbine engines. Therefore, designing low-flow tilting-pad journal bearings for gas turbine engines with high rotational speeds is relevant.

The boring of bearing pads into the shaft radius with a close radial clearance of the bearing speeds up the onset of fluid friction. Forced load of all pads can prevent "pad fluttering" that is typical for tilting-pad journal bearings [4]. Such a bearing can operate without forced supply of oil to the oil grooves ("oil starvation" mode). These features significantly reduce the engine weight and oil flow, thus increasing the efficiency of the engine overall.

Elastic support of bearing pads is used to even out the load among pads (thrust bearings), and for force closure of the running clearance (journal bearings). The main requirements to the elastic support of pads are maintenance of set stiffness properties and high damping properties. Elastic support of bearing pads can be realized with elastic pins (Figure 1), "squirrel cage", Belleville washers, elastic plates or elastic porous materials like the metal analogue of rubber (Metal analogue of Rubber, designed by the Samara University).

TPJBs with an elastic support of pads can utilize such lubricant properties as continuity and wettability to maintain their lifting properties and improve their flow characteristics [5].

In order to utilize these properties (under "oil starvation"), it is necessary to make it so that the running clearance in passive pads that are opposite to the carrying ones is not closed completely, while the elastic support is capable of both closing and opening the running clearance. A bearing with elastic pins and Metal analogue of Rubber is capable of this, unlike "squirrel cage" supports, Belleville washers, elastic plates, and the "Allison" ring. A combined support with two elastic pins, maximally spread across the length of the pad, compensates for this flaw to a certain extent. The advantage of such a support is the technological simplicity of the boring of pads into the shaft radius with subsequent closure of the clearance into the shaft diameter by boring on technological fixing pins. Furthermore, the presence of elastic pins in the support solves the problem of circumferential fixation of pads in constructions with a support, for instance, in "Allison" rings.

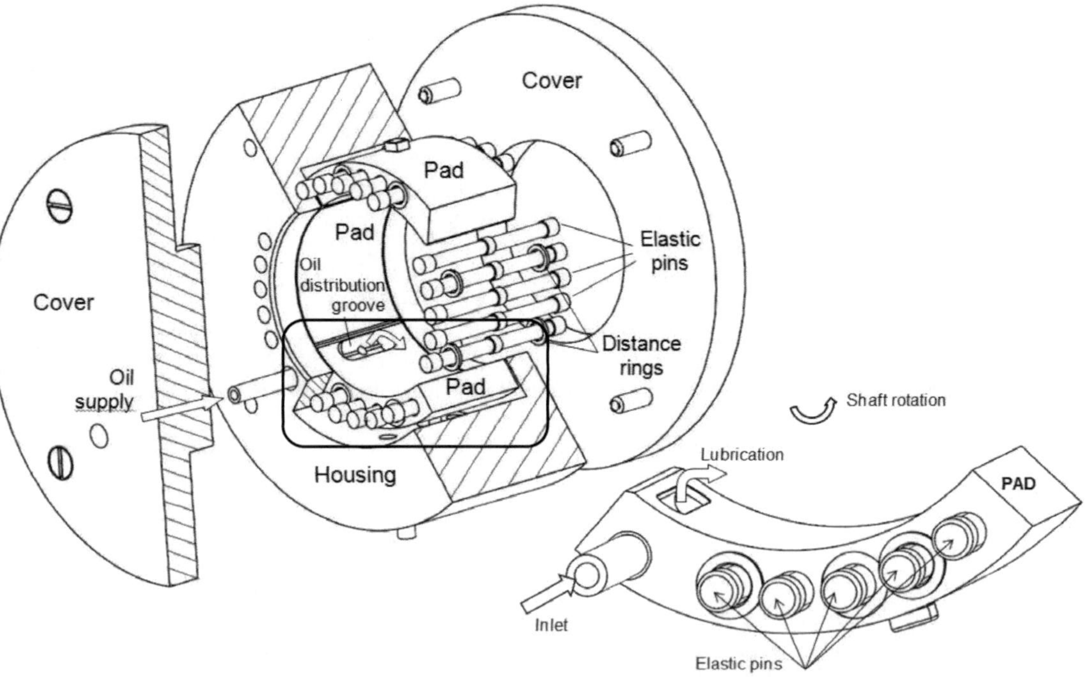

Figure 1. Tilting-pad journal bearing with elastic support of pads (elastic pins configuration)

The bearing operates in the hydrodynamic lubrication mode. The lubricant is supplied to the running clearance through an oil groove, after which it is set in motion by the surface of the rotating shaft.

At present, an adequate technique of design of the bearing under consideration that would take into account the peculiarities of its design and utilize modern tools of computer-aided engineering is lacking. Furthermore, the experimental development of studied bearings is also complicated due to the numerous factors that affect their characteristics. Thus, the design of reliable computational techniques is a crucial component when creating and developing the theory of design of the bearings under consideration. The purpose of this research is to study numerically the characteristics of low-flow tilting-pad journal bearings under oil starvation conditions.

2. Literature survey

Many studies of tilting-pad journal bearings are experimental.

The conventional (analytical) design method of journal bearings is based on Reynold's equations and continuity [6]. Nowadays, these models are modified to include components that take into account the effect of roughness, heat effect, cavitation, turbulence, multiphase nature of lubricant flow, use of non-Newtonian fluids, etc. [7]. Thermoelastohydrodynamic models are used to study the characteristics of tilting-pad journal bearings [8].

The literature survey showed that the development of mathematical models of TPJB must be confirmed by the results of numerous experiments.

Thus, empirical and semi-empirical models are only applicable to a certain structure of the bearing and even to a certain range of their standard size [9, 10, 11]. Furthermore, most existing techniques of thermoelastohydrodynamic analysis are unsuitable for starved bearings. V.A. Voskresensky offered a purely analytical method of designing tilting-pad journal and thrust bearings [9]. This method was adapted to the conditions of oil starvation in the part of heat balance problems and estimation of oil flow through the bearing and is currently used to design roughly the low-flow journal bearings at such Russian companies as "Kuznetsov" and "Salyut", as well as at the Samara University.

The analysis of pressure distribution in the clearance of bearings with an axial boring of pads and boring of pads into the shaft radius showed the following. The type of dependencies matched the theoretical ones for tilting-pad journal bearings [9]. The nature of pressure distribution with the boring of pads into the shaft radius is smoother ("filled" distribution); with axial boring, the distribution is "peaked" and the maximum pressure is almost twice as high as with boring into the shaft radius. Identical results for conventional bearings were obtained in studies [12]. A smoother distribution of pressure in the lubricant film is indicative of an extended service life of such bearings [6, 9].

The above methods have obvious restrictions when it comes to bearings with new geometry (inaccurate description of flow recirculation, mixing, mass balance, etc.). These restrictions can be eliminated by means of computational fluid dynamics (CFD) [13]. CFD provides for accurate calculations [14].

3. CFD solution of a two-phase problem

The multiphase nature of the working fluid significantly affects the accuracy and physics of processes in the designed model of the bearing. A finite element model of internal cavities of the bearing was generated in ICEM CFD for the CFD analysis (Figure 2, A). Here 1 is a mapped grid of the running clearance; 2 – mapped grid of the inlet; 3 – mapped grid of the discharge area.

The feature of the finite element model is the mapped grid of finite elements in the running clearance and inlets – lubricant discharge and face clearances of pads.

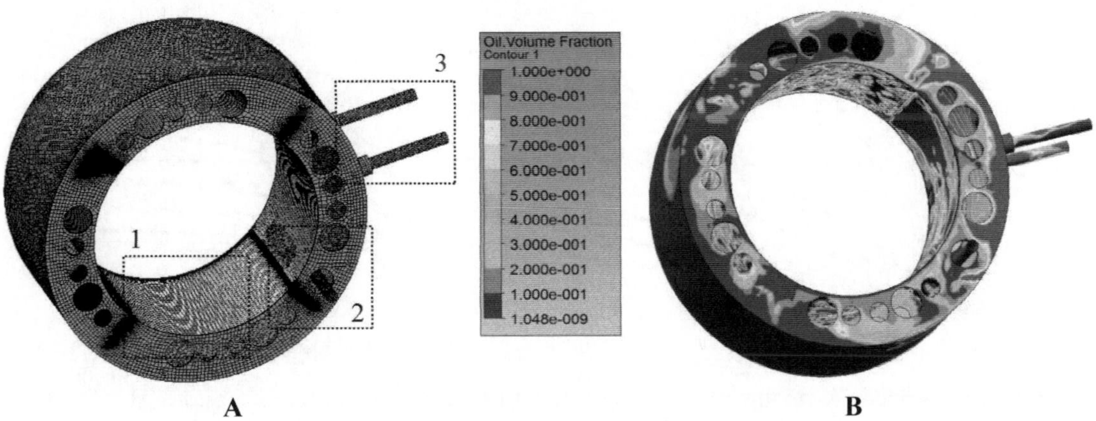

Figure 2. Bearing model

The modelling of a two-phase operating environment (oil + air) was performed with the Ansys CFX package for gas-hydrodynamic analysis with regard to surface tension, adhesion, and modelling of a free boundary between the phases. A surface tension coefficient with the continuum surface force option and the initial liquid phase was set to model the free boundary between the liquid and gaseous phase.

4. Results

A series of computations in accordance with the modified method showed the dependencies of the characteristics of bearings on the clearance size and the rotational speed of the shaft. The maximal oil temperature in the bearing rises with the reduction of the running clearance and with the increase of the rotational speed of the shaft (Figure 3, A). The load factor reduced with an increase in the rotational speed of the running clearance and a reduction of the clearance (Figure 3, B). The lubricant flow through the bearing increased linearly with the increase of the clearance and the rotational speed (Figure 3, C).

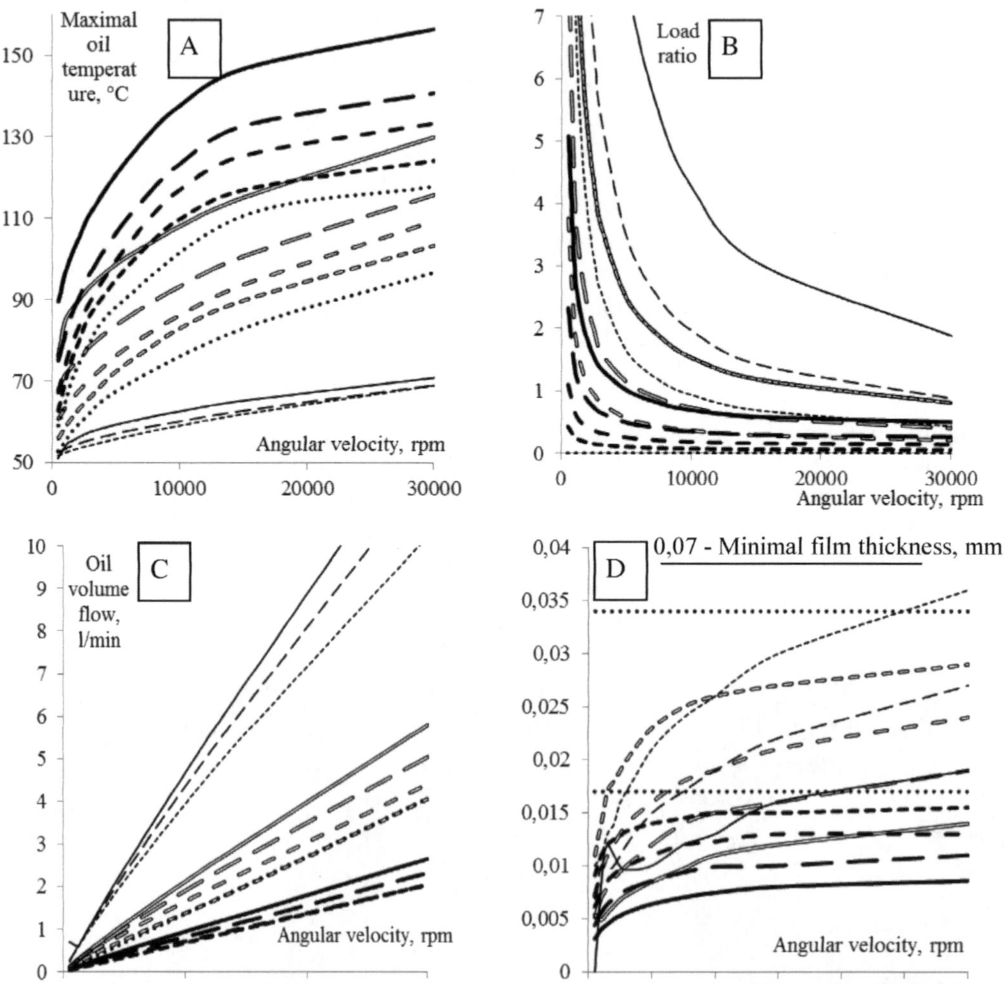

Figure 3. Dependencies on the rotational speed: A - of the maximal oil temperature; B -of the load factor on the rotational speed; C - of the maximal oil temperature; D - of the minimal clearance thickness on the rotational speed

The minimal thickness of the clearance increases with an increase in the rotational speed of the rotor (shaft floating) and with the reduction of the running clearance (Figure 3, D). The dependency chart has an asymptote (for unloaded bearings). The hydrodynamic design of a two-phase model of a TPJB showed the distribution of model phases and the flow parameters: total and for each modelled phase.

The main problem of rotary bearings is that they consume too much oil in order to maintain reliability and their bearing capacity (heat sink, reduction of friction). High oil consumption is the main reason behind the low efficiency and environmental unfriendliness of high-speed machinery.

Journal bearings are commonly used in Russia in ground gas-pumping plants. Tests with the industrial compressor station that first used the prototype of a low-flow TPJB showed good results. After the oil flow was dropped from the design 12 l/min (oil bath) to 0.2 l/min (oil starvation), the bearing operated for more than an hour according to the plan of the test. When the unit was disassembled, it was found that the bearing was in ideal state. At that, the lower carrying pad had a surface polished to a mirror finish – a result of functional bedding in the rotating shaft during the spin-up and stop of the rotor.

Foreign experience. Scientists laid an extensive theoretical and experimental foundation for the development of techniques of designing TPJB. The hydrodynamic theory of lubrication that scientists use in most theoretical studies has proven to be efficient in solving problems that are related to the description of the operation of journal bearings. The elaboration of the hydrodynamic theory of lubrication – M. Fillon's, M. Khonsari's [12, 15] and other researchers' thermohydrodynamic and thermoelastohydrodynamic theories for journal bearings, are interesting and helpful. The results of experimental studies of various TPJB [16], including the results that prove the workability of tilting-pad journal bearings with an elastic-pivot pad [17], as well as the outstanding results of D. Childs [18,19] and results of computer simulation [20] show the following trends in the study of TPJB:

1) engineering study of different TPJB designs, including elastic-pivot pad and elastic-pad TPJBs;
2) development of advanced technological computational methods, including CAE;
3) start of experimental studies of oil starvation in journal bearings.

In order to increase the service life of future high-speed bearings, it is necessary to refer again to the hydrodynamic theory of lubrication, which, in turn, requires high oil flow, but can guarantee high bearing capacity. In a hydrodynamic mode, the flow can be reduced by using new bearing designs and new unconventional materials. At present, unconventional materials with a low coefficient of friction are being developed. At that, these materials lack a specific area of application, including TPJB.

In Russia, several scientific schools have worked successfully for decades in the area of hydrodynamics of journal bearings. However, their main inventions concern tilting-pad journal bearings (operating in an oil bath). Thus, the experimental studies of low-flow TPJB should be substantiated theoretically. Uniting the studies of low-flow TPJB into a single complex – a simultaneous solution of the hydrodynamic problem, the thermal problem, and the problem of the wear of the friction pair surface – provides for a comprehensive solution of the outlined problem.

In future, the developed technique of designing starved TPJB with an elastic support of pads can help cut the time required to design and tune the bearing and cut the production costs of experimental models. Thus, low-flow TPJB with an elastic support of pads can solve the problem of cutting lubricant consumption in high-speed machinery.

5. Discussion

Obtained dependencies of the characteristics of bearings on the size of the clearance and the rotational speed of the shaft are in line with the TPJB characteristics that were obtained by other authors [9, 14]. At that, the diagrams of dependencies of the minimal clearance thickness on the rotational speed for different dimensions of the radial clearance of the bearing show a hopped peak at a load of 10000 N and 5000 N and a clearance Δ_1 = 25 μm. This peak is caused by "shaft floating". High loads, small radial clearance, and low rotational speed fail to generate force that is sufficient to support the shaft on the hydrodynamic wedge. By default, V.A. Voskresensky's method [9] uses tabulated results of hydrodynamic computations, which results in a limited applicability of this method. However, hydrodynamic computations, the results of which are presented in tables, are performed in boundary conditions that correspond to an oil bath – unlimited oil supply to pads and zero excess pressure across the pads, while low-flow journal bearings requires computations at numerically specific lubricant flow through the bearings and their pads.

The graphic presentation of a non-stationary design of a two-phase bearing model with oil spillage during the shaft spin-up in the bearing (Figure 2, B) shows that the lubricant spreads in the form of an elongated stain. The lubricant is sucked from the oil groove into the running clearance of the bearing. The rough surface of the shaft bearing helps small parts of the lubricant to attach to it and moistens the impact of rubbing surfaces. A similar flow pattern of a two-phase liquid in the face clearance was obtained by Lebeck in his experimental study of face seals [21]. He found liquid traces, elongated in the direction of shaft rotation, at the breakage of the lubricant layer in the gaseous phase. This proves the reliability of results that were obtained during the present research. The herein presented thermal design technique for radial journal bearings could also be used to design face bearings and seals [22].

The CFD method is applicable to all journal bearings, including smooth bearings that work on oil that requires multiphase hydrodynamic analysis: water journal bearings; bearings that work at high rotational speeds and (or) on low-viscous lubricants; low-flow bearings. The CFD method can serve as the basis for linked unilateral and bilateral CAE analysis of smooth and tilting-pad journal bearings and for direct optimization. A comprehensive technique of designing TPJB is lacking at present. Herein, a comprehensive technique implies a design technique that takes into account the peculiarities of the TPJB structure and actual load, and utilizes to the extent possible the modern computational means (CFD). Thus, developing reliable computational techniques is a crucial element when creating and developing the theory of design of new TPJB structures.

The present paper is based on the existing concept of hydrodynamic lubrication. However, the case under consideration lies at the very edge of the applicability of this concept (small clearances).

Computations with this technique for bearings with conventional clearance sizes yielded similar results, which prove its validity. Open literature lacks computational results with small clearances; however, examples of positive tests of bearings under these conditions prove the quality of the developed technique. Further researches will involve specialized experimental studies of low-flow small-clearance bearings. The herein considered issues of designing low-flow journal bearings with a ceramic friction pair with a low coefficient of friction enable making bearings with improved and efficient performance characteristics. This will enable compressor and pump designers to create turbomachines with advanced parameters.

6. Conclusion

The scientific novelty of this research consists in the discovered regularities of lubricant flow in small clearance of TPJBs, as well as in the regularities of their main characteristics.

Researchers of journal bearings may also be interested in the herein developed two-phase CAE model of a low-flow journal bearing with a small radial clearance (20 μm), which takes into account the effect of lubricant adhesion, surface roughness, inertia, and the presence of a free boundary between the liquid and gaseous phase. The analysis of results showed that the lubricant spreads in the two-phase low-flow bearing in the form of a stain. The rough surface of the shaft bearing helps small parts of the lubricant to attach to it and moistens the impact of rubbing surfaces. Thus, it was shown, that even with small clearances under oil starvation, the lubrication of the surfaces of the bearing is performed, besides the velocity head of the lubricant, by a continuous layer of fluid particles. Adhesion and microscopic roughness facilitates the presence and preservation of these particles.

The shortcoming of the designed bearing model is that it does not consider the effect of pad deformation on the form of the clearance and the characteristics of the bearing.

References

[1] Chaadaev K N and Novikov D K 2009 *Russian Aeronautics* **52 (4)** 426-31
[2] Mikami M, Kumagai M, Uno S and Hashimoto H 1988 J. of Trib. 110 73-9

[3] Wang N, Meng Q, Wang P, Geng T and Xiaoyang Yu 2013 ASME J. of Fluids Engineering. 135
[4] Parovay E F and Falaleev S V 2015 Biosciences Biotechnology Research Asia 12(1) 731-6
[5] Parovay E F and Falaleev S V 2017 Procedia Engineering 176 253-63
[6] Bruce R W 2012 Handbook of Lubrication and Tribology: Theory and Design (CRC Press Taylor & Francis Group) p 121
[7] Dimond T, Younan A and Allaire P 2011 A review of tilting pad bearing theory Int. J. of Rot. Machinery 10.1155(2011)908469
[8] Fillon M, Bligoud J-C, Frene J 1992 J. of Trib.114 579-88.
[9] Voskresensky V A 1980 Computing and design of sliding bearings (Moskow: Mashinostroenie) p 224
[10] Tipei N I 1964 Journal bearings: computing, design, lubrication (Bucharest) p 456
[11] Chernavsky S A 1963 Journal bearings (Moskow: Mashinostroenie) p 344
[12] Khonsari M M, Jang J Y and Fillon M 1996 J. of Trib. 118 571-9
[13] Uhkoetter S, Der Wiesche S A and Joos F 2013 Proceedings of the ASME Turbo Expo 7 B 1115(GT2013)94364
[14] Guo Z, Hirano T. and Kirk R G 2005 ASME J. Eng. Gas Turb. Power 127 445-51
[15] Fillon M, Bligoud J-C and Frene J 1992 J. of Trib. 114 579-87
[16] Andres L S, Tao Y and Li Y 2015 J. Eng. Gas Turb. Power 137 022505-1
[17] Zhang F, Ouyang W, Hong H, Guan Y, Yuan X and Dong G 2015 Trib. International 88 228-35
[18] Childs D and Harris J 2009 Static J. Eng. Gas Turb. Power 131 062502
[19] Wilkes J C and Childs D W 2013 J. Eng. Gas Turb. Power 135 012503
[20] Suh J and Palazzolo A 2015 J. of Trib. 137 041704
[21] Lebeck A O 1991 Principles and Design of Mechanical Face Seals (New York) p 746
[22] Falaleev S V 2015 J. of Friction and Wear 36(2) 177-83

Experimental Research into Noise Emission of A Gear Micropump with Plastic Rotor

L V Rodionov[1] **and P D Rekadze**[1]

[1] Institute of Engine and Power Plant Engineering, Samara National Research University, 34, Moskovskoye shosse, Samara, 443086, Russia

Email: leonid@rodionoff.net

Abstract. The previous researches show that it's possible to replace several parts of gear pump to plastic ones. This substitution leads to cost and noise reduction of the pump. Therefore, the series of acoustic experiments on a test bench were carry-out. Sound pressure levels were recorded with microphone, located in a pipe made of a vacuum rubber. Conducted experiment shows that acoustic characteristics of the micropump depend on the different material of driven rotor. Experimental result indicates that the proposed measures for replacing metal rotor to plastic one reduce micropump noise on the studied modes. The maximum achieved acoustic efficiency on equivalent level is 11 dB.

1. Introduction

Engineers pay particular attention to vibroacoustic loads in different technical objects [1, 2, 3, 4, 5], caused by constant wear. Moreover, there are exist strict requirements on noise levels [6].

The vital part of any hydraulic systems is flow sources, pumps. Positive displacement machines are most common in those systems. Among these types of machines, gear pumps are most popular. Gear pumps have a widespread use because of the simplicity of their design, low sensitivity to mechanical impurities, ease of manufacturing and low cost. Along with the advantages, there are disadvantages: a high noise level [7], high-quality requirements for the production of gears, high-pressure ripples causes the pipeline instability, cavitation processes [8]. These disadvantages are caused by different phenomena. For example, the authors [9] describe measures by pressure pulsations and vibrations loading reduction, which lead to high noise level. Kollek [10] shows that the fluid trapped between gears is the main reason of gear pump noise. Other papers were devoted to the instability of pipeline caused by high-pressure ripples [11, 12]. These articles provide numerical models of vibrations in pipeline due to pressure ripples.

Nomenclature

b	face width (mm)
E	Young's modulus (MPa)
d_{in}	initial diameter (mm)
f	operator
f_{sh}	shaft frequency (Hz)
$f_{teeth1,2}$	first and second teeth frequency (first and second harmonic) (Hz)
k	harmonic number
m	module of the wheel (mm)

IOP Conf. Series: Materials Science and Engineering **302** (2018) 012011 doi:10.1088/1757-899X/302/1/012011

n	shaft speed (rpm)
P_{out}	output pressure (bar)
$Q_{act/theor}$	actual and theoretical flow (l/min)
z	number of teeth
ΔL	sound pressure level efficiency (dB)
ΔLeq	equivalent sound pressure level efficiency (dB)
Greek characters	
α	coefficient of linear thermal expansion (1/K at 20 °C)
η	loss factor
η_v	volumetric efficiency
λ	thermal conductivity (W/(m ·K))
μ	friction coefficient
σ	strength limit (MPa)

Polymeric materials (PM) are used in mechanical engineering more and more due to higher than steel and aluminum loss factor η [13]. A strength limit σ of several existed PM is over 500 MPa while Young's modulus E, in comparison with steel, is considerably lower [14]. These properties have been causing inevitable substitution of traditional materials (TM) of mechanical engineering from 70teens [15] till nowadays [16].

The papers show possibility PM instead TM implementation with appropriate efficiency for hydraulics [17, 18] as well as for different parts of external and internal gear pumps [18, 19]. Moreover, PM application leads to pumps noise and vibration reduction [20, 21].

The main PM advantages for mechanical engineering are increased chemical resistance [14, 22], the ability to work under lubrication lack [23], high enough loss factor η [15, 24].

In this paper authors assessed acoustic efficiency from plastic rotor implementation in a gear micropump. Only the driven rotor was substituted.

2. Experimental study

The PEEK was chosen as the material for driven rotor [14], on the basis of it enormous properties in contrast to other PM's. The comparison of PEEK (driven gear) and steel 42Cr4 (driver gear) shown in [13].

Sound pressure levels emitted by the micropump with plastic driven rotor were recorded by microphone, a signal processing module NI usb-4431 and PC with installed LabView software. In addition, a flow rate, inlet and outlet pressure of the micropump were collected with the flowmeter, pressure transducer, vacuum gauge and portable registration module.

The acoustic characteristics of the micropump had measured for the following configurations of drive and driven rotors: "steel-steel" and "steel-PEEK".

These rotors packages were studied on a gear micropump. A number of modes were formed as a combination of outlet pressure and shaft speed. Selected modes are presented in Table 1.

Table 1. Test modes.

Characteristics	Unit	Value			
Shaft speed, n	(rpm)	1000	1500	2000	2500
Outlet pressure, P_{out}	(bar)	3	6	9	12

The algorithm of the experiment is:

- Set the shaft speed by frequency drive and outlet micropump pressure by a throttle.
- Register a pressure and flow rate.
- Record an acoustic signal.

The experiments were carry-out on a hydraulic test bench. This test bench was based on previous one, which was designed earlier [25]. The hydraulic system scheme is illustrated in Figure. 1.

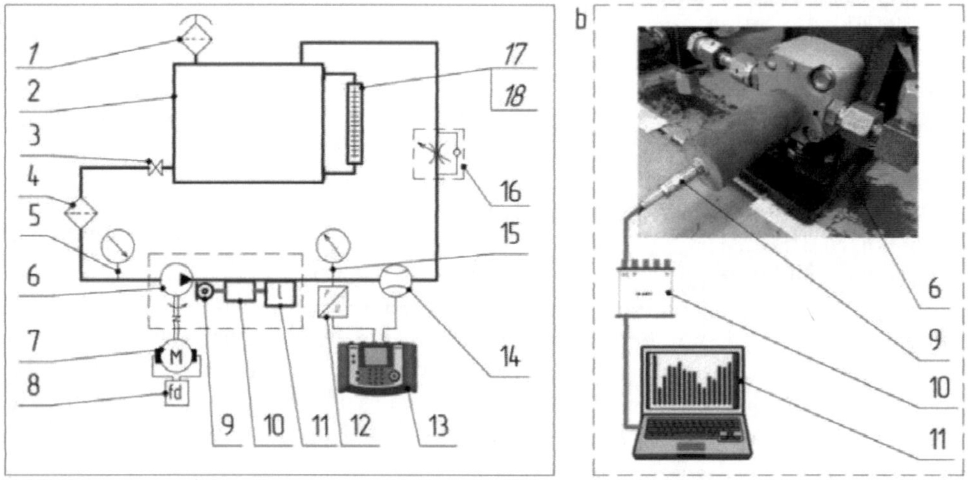

Figure 1. (a) scheme of the test bench: 1 – air filter, 2 – tank, 3 – valve, 4 – preliminary filter, 5 – vacuum gauge, 6 – gear micropump, 7 – motor, 8 – frequency drive, 9 – microphone, 10 – measuring system, 11 – laptop, 12 – pressure transducer, 13 – portable registration module, 14 – flowmeter, 15 – pressure gauge, 16 – throttle, 17 – fluid level gauge, 18 – temperature gauge; (b) scheme of the microphone arrangement.

This system provides an opportunity to measure the flow under different operating conditions, a shaft speed and output pressure, while the oil temperature is kept constant. Sound pressure levels were recorded with microphone, located in a pipe made of a vacuum rubber. A thickness of the pipe is 9 mm (Figure 1).

3. Results and discussion

The collected data were carefully processed. Then an acoustic efficiency caused by the implementation of plastic rotor instead the steel one were calculated, thoroughly analyzed and presented:

- For shaft and teeth frequencies depended on the micropump outlet pressure (Figure 2).
- As equivalent efficiency depended on the shaft speed (Figure 3(a)).
- As equivalent efficiency depended on the micropump outlet pressure (Figure 3(b)).

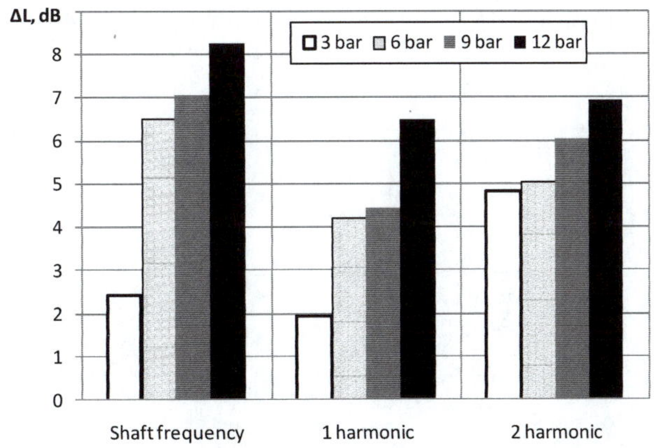

Figure 2. Acoustic efficiency $\Delta L = f(f_{sh}, f_{teeth1}, f_{teeth2})$ on main working process frequencies at n=2000 rpm.

The teeth frequency is determined as (1):

$$f_{teeth} = kzf_{sh} = k\frac{n \cdot z}{60} \tag{1}$$

The Figure 2 shows that acoustic efficiency on shaft frequency is slightly higher. The efficiency strictly depends on outlet pressure of the micropump. This tendency observed on other shaft speeds.

Figure 3 depicts the results of equivalent acoustic efficiency like shaft speed ΔLeq= f(n) and output pressure ΔLeq=f(Pout) function as a difference between a noise emitted from "steel-steel" and "steel-PEEK" packages:

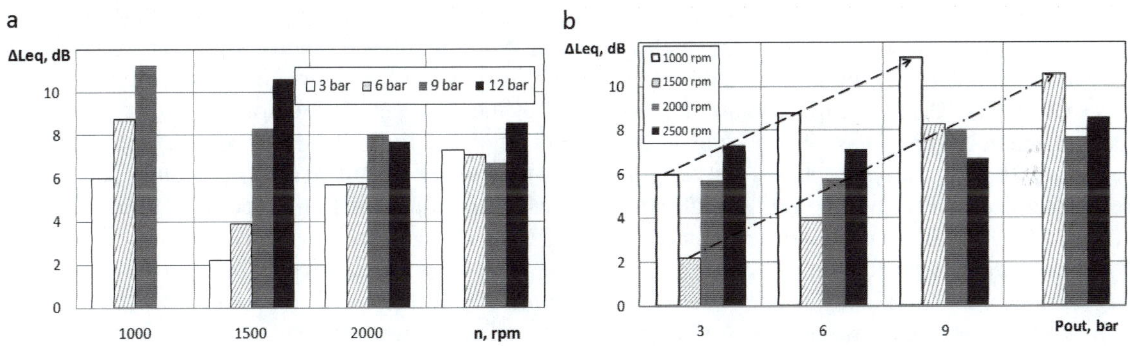

Figure 3. (a) An acoustic efficiency $\Delta Leq = f(n)$; (b) An acoustic efficiency $\Delta Leq = f(P_{out})$.

The equivalent acoustic efficiency at low shaft speeds (1000 and 1500 rev/min) rise sharply (Figure. 3(a)). At 2000 and 2500 rev/min the efficiency increased slightly. As was closely observed (Fig. 3(b)), there is a strong correlation between acoustic efficiency and outlet pressure of the micropump on the 1000 rpm and 1500 rpm. Apart from that, there is no strict dependence between acoustic efficiency and outlet pressure of the micropump on the 2000 rpm and 2500 rpm.

The sound pressure level did not register on working mode n=1000 rpm and Pout =12 bar due to low micropump delivery.

The volumetric efficiency and the theoretical delivery [26] for pumps with teeth number z = 8…16 should be determined as (2):

$$\eta_v = \frac{Q_{act}}{Q_{theor}} = \frac{Q_{act}}{6{,}5 d_{in} mbn} \tag{2}$$

Figure 4, shows the volumetric efficiency of the gear micropump versus the output pressure, for two configurations.

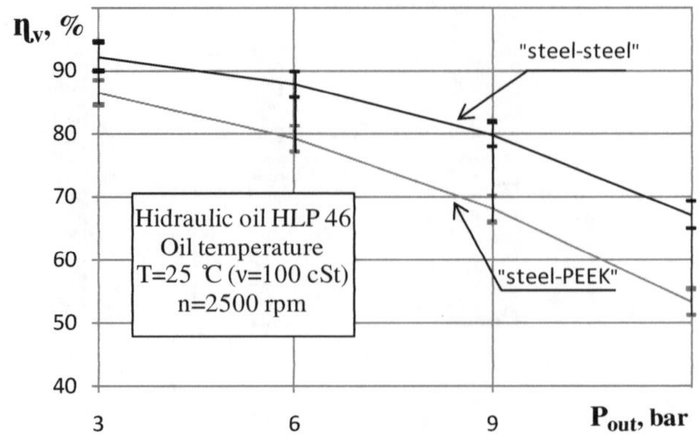

Figure 4. Volumetric efficiency of the gear micropump vs the output pressure at n=2500 rpm.

The inlet micropump pressure was registered by a vacuum gauge. The maximum value was 0,8 bar at n=2500 rpm and P_{out} =3 bar. This value was not taken into account in the equation (2) as additional leakage to a suction area [27]. Thus, theoretical volumetric efficiency, presented in figure 4 is higher than it is.

It is necessary to emphasize the fact that volumetric efficiency because of plastic rotor implementation, as expected, are lower in comparison with steel rotor package [28].

The volumetric efficiency η_v of gear pumps often expected in a range from 0,8 till 0,95 [29]. In current research volumetric efficiency were under value 0, 8 at 6 bars. Nevertheless, these results are suitable for pumps with elements made of plastics [22, 28]. For example, the Tuthill company produces gear pumps with rotors made of PEEK and maximum outlet pressure Pout=16 bar. Therefore it clearly demonstrates an importance of outlet pressure increasing for pumps made of plastics.

According to Bashta [26] a volumetric coefficient of the pump is affected on its total length of the gaps. The declining of the ratio of gaps length to theoretical delivery leads to volumetric coefficient increase. Hence, the low pump delivery leads to the lower volumetric coefficient. Thus, it is recommended to run the pump at a high shaft speeds. This hypothesis is supported by the obtained experimental results: volumetric efficiency η_v =0, 53...0, 67 were achieved at outlet pressure Pout =12 bar, instead expected (η_v =0, 8…0, 95) [27]. The maximum volumetric efficiency η_v =0, 67…0, 92 was recorded at highest shaft speed (n=2500 rpm) for "steel-steel" package. The tendency to volumetric efficiency growing is observed in all studied range of shaft speeds. Starting from n=1000 rpm the volumetric efficiency rising up on 0,075 for each 500 rpm.

4. Conclusion

Fluid power machines are commonly used in the design of the hydraulic driving and systems of machines and appliances. Development prospects of the machines are still open. Together with the improvement of their efficiency and outlet pressure limits, it is possible to achieve a noise emission declining.

The carried-out experiment provides the results, which firmly establishes the relationship of acoustic efficiency to rotor materials in gear pumps. This is specifically referred to plastic materials. The achieved acoustic efficiency is 2...8 dB depends on outlet pressure on the main working process frequencies of the gear micropump at n=2000 rpm, and 2...11 dB on equivalent level depends on outlet pressure and shaft speeds.

The volumetric efficiency of the gear micropump depended on outlet pressure was η_v=0,53…0,87 for "steel-PEEK" package. The principal cause of this wide range value was random side gaps in the micropump during it assembling.

Acknowledgements
This work was supported by the Ministry of Education and Science of the Russian Federation (Grant № 9.1299.2017/4.6) and partly by the Russian Federation President's grants (project code МК-1098.2017.8, МК-6480.2016.8).

References
[1] Makaryants G M, Prokofiev A B and Shakhmatov E V 2015 Vibroacoustics analysis of punching machine hydraulic piping Proc. Eng. 106 17-26
[2] Makaryants G M, Gafurov S A, Zubrilin I A, Kruchkov A N, Prokofiev A B and Shakhmatov E V 2013 Design methodology of hydrodynamic noise silencer Proc. of the 20th ICSV vol 3 (Thailand) pp 2531-2536
[3] Shakhmatov E V, Prokofiev A B, Kruchkov A N and Makaryants G M 2012 The instability of the pipeline due to transporting fluid's pressure ripples Proc. of the 28th ICAS vol 3 (Australia) pp 2101-2108
[4] Makaryants G M, Sverbilov V Ya, Prokofev A B, Makaryants M V and Shakhmatov E V 2012 The tonal noise reduction of the proportional pilotoperated pneumatic valve Proc. of the 19th ICSV vol 1 (Lithuania) pp 689-696
[5] Carletti E, Miccoli G, Pedrielli F and Parise G 2016 Vibroacoustic measurements and simulations applied to external gear pumps. An integrated simplified approach Arch. of Acoustics 41 (2) 285-296
[6] SN 2.2. 4/2.1.8.562-96 Sanitarnye normy. Shum na rabochikh mestakh, v pomeshheniyakh zhilykh, obshhestvennykh zdaniy i na territorii zhiloy zastroyki. Available from: http://gostbank.metaltorg.ru/data/ norms_new/sanpin/1.pdf. (In Russ.)
[7] Gafurov S A, Rodionov L V and Makaryants G M 2016 Simulation of gear pump noise generation Proc. of the ASME 2016 9th FPNI Ph.D Symposium on Fluid Power (Brazil)
[8] Gafurov S A and Rodionov L V 2014 Acoustic visualization of cavitation in fuel combination pump Proc. of the 19th ICSV (Lithuania) pp 3916-3923
[9] Gafurov S A, Prokofiev A B and Shakhmatov E V 2014 Reduction of vibroacoustic loads in aviation combined pumps, Proc. of the 29th ICAS (Russia) pp 1-10
[10] Kollek W and Stryczek J 1980 Influence of trapping fluid in cuts between gears on pumps noise Sterowanie and naped hydraul 5 3-7
[11] Shakhmatov E V, Prokofiev A B, Kuchkov A N and Makaryants G M 2012 The instability of the pipeline due to transporting fluid's pressure ripples Proc. of the 28th ICAS vol 3 (Australia) pp 2101-2108
[12] Prokofiev A B, Makariyants G M and Shakhmatov E V 2010 Modeling of pipeline vibration under the pressure ripples in the working fluid, Proc. of the 17th ICSV vol 2 (Egypt) pp 1142-1149
[13] Rekadze P, Rodionov L and Riman O 2016 Gear material physical properties effects on vibroacoustic characteristics of the pump, Proc. of the 23rd ICSV (Greece)
[14] Zagotovki iz tekhnicheskikh polimerov - sterzhni, listy, vtulki. Available from: http://polimer1.ru/assets/files/downloads/catalog/katalog-tehnicheskie-plastiki.-rukovodstvo.pdf.
[15] Heckl M and Müller M A 1980 Spravochnik po tekhnicheskoy akustike (Leningrad: Sudostroyeniye)
[16] Mancini S, Neto A, Cioffi M and Carlos B 2016 Replacement of metallic parts for polymer composite materials in motorcycle oil pumps Journ. of reinforced plastics and composites 36 (2) 149-160
[17] Stryczek J, Biernacki K, Krawczyk J and Wołodźko J 2014 Application of plastics in the building of fluid power elements Vestnik NTU "KPI" Eng. series 2 (71) 5-10.
[18] Krawczyk J and Stryczek J 2016 Designing of the gerotor pump body made of plastics Proc. of the ASME 2016 9th FPNI Ph.D Symposium on Fluid Power (Brazil)

[19] Stryczek J, Bednarczyk S and Biernacki K 2014 Gerotor pump with POM gears: design, production technology, research Arch. of Civil and Mechanical Eng. 14 391–397
[20] Rodionov L, Pomatilov F and Rekadze P 2015 Exploration of acoustic characteristics of gear pumps with polymeric pinion shafts Proc. Eng. 106 36–45
[21] Rekadze P and Rodionov L 2015 A gear pump with plastic gears noise and vibration research Proc. of the 22rd ICSV vol 2015 (Italy) pp 1–8
[22] Magnetic drive plastic pump TypE FNP M (Pumpen). Available from: http://www.tuthillpump.com/dam/2708.pdf.
[23] Raman A and Gladson F 2014 Dry lubrication of polymer gears using epoxy and poly urethane paints, Int. Jour. for Techn. Research in Eng. 1(5) 260-265.
[24] Bogolepov I I 1986 Promyshlennaya zvukoizolyatsiya (Leningrad:Sudostroyeniye)
[25] Rekadze P, Rodionov L, Riman O and Almurzin P 2016 Designing a compact test bench for gear pump vibroacoustic performances research Proc. of the 23rd ICSV (Greece)
[26] Bashta T M 1967 Gidravlicheskiye privody letatelnykh apparatov (Moscow: Mashinostroyeniye)
[27] Yudin E M 1964 Shesterennyye nasosy (Moscow: Mashinostroyeniye)
[28] Metalsi Tani F and Bourdim A 2012 Study of feasibility of plastic gear to reduce noise in a gear pump Acad. Journ. 7(2) 143
[29] Durnov P I 1985 Nasosy, ventilyatory, kompressory (Kiev:Vyshcha shkola)

IOP Conf. Series: Materials Science and Engineering **302** (2018) 012012 doi:10.1088/1757-899X/302/1/012012

Unsteady RANS Simulation of Acoustic Wave Propagation in Pulsation Damper

K A Romanov[1] and G M Makaryants[1]

[1]Department of Automatic Systems of Power Plants, Institute of Engine and Power Plant Engineering, Samara National Research University, 34 Moskovskoe shosse st., Samara, 443086. Russia

Email: romanov.kirill.94@mail.ru

Abstract. A research on dynamic characteristics of a pipeline pulsation damper is presented in this paper. The study covers an investigation of acoustic wave propagation in a pipeline numerical model. For this purpose, the numerical model which was previously obtained for estimation of hydrodynamic noise after the damper diffuser has been verified. The verification is conducted using URANS model. The simulation results illustrate the adequacy of the present model in calculating acoustic wave propagation. Hence, the acoustic characteristics obtained in further calculations can be regarded as reliable. The numerical results demonstrate that an initially plane acoustic wave does not change its form and the overall amplitude level of acoustic oscillations varies in accordance with changes in the geometry flow cross-section.

1. Introduction

The study of oscillations in piping has become an important aspect of designing water supply and treatment systems. Acoustic waves induced by water pressure pulsations which are caused by periodic flow occurring downstream the pump, vortex shedding downstream the bends and changes in pipe cross-section are of particular interest and complexity. The pressure pulsations of working fluid may also result in unacceptable level of pipeline vibration [1, 2] as well as in high acoustic noise [3-7]. One way to solve this problem is to use pressure pulsation dampers [8-14] in the piping systems.

The acoustic noise induced by discontinuities of the damper structure is critical for damper effectiveness and efficiency in case of high flow rates. A diffuser connecting a damper with a pipeline can be considered as one of such discontinuities. In order to solve the problem of damper sound generation in a proper manner the process of vortex pulsation generation in the diffuser must be simulated. Therefore, the proposed model must have the necessary characteristics to calculate the propagation of acoustic waves.

The features of the flow passing through the diffuser and the hydrodynamic processes inside the diffuser have been widely discussed in the literature. McDonald & Fox [15] and then Kwong & Dowling [16] have demonstrated that a conical diffuser is affected by transitory stall. Dequand et al. [2] concluded that separation flow in the diffuser induces duct acoustic pulsations caused by vortex shedding.

In the observed papers great attention is paid to common phenomenon of the fluid passing through the diffuser. However, the methods and techniques of calculating the sound generation in the pipeline system components are also of particular interest. Several studies have been conducted on this issue. One part of the research papers is devoted to predict the vortex-born noise by means of semi-empirical techniques. For example, Karekull et al. [5] searched for a scaling law for the sound power of the orifices, bends and dampers. The research in the other part of the papers is focused on applying Unsteady Computational Fluid Dynamics (CFD) methods in order to directly simulate the vortex sound generation, see, for example Lam et al. [17], Gloerfelt & Lafon [18], Singh & Rubini [13]. This approach allows us to obtain more detailed information on vortex noise.

Although considerable research has been devoted to direct computation of the noise radiated by vortex flows, rather less attention has been paid to CFD calculation of turbulent flows in a diffuser. The third-octave spectrum obtained from the numerical simulation of the damper agrees well in the low-frequency region, however, at higher frequencies there is a strong discrepancy with the experimental data in the newly released research paper [19]. Thus, it was necessary to check the impact of mesh resolution on direct hydrodynamic noise computation.

The aim of this paper is to verify the proposed numerical model of the pulsation damper in order to simulate pressure wave passed through the damper duct obstructed by the diffuser.

2. Operation principle of a piping noise damper

The piping noise damper consists of a non-flowing vessel placed parallel to a constricted duct (figure 1). In high frequency range the vessel has significant acoustic conductivity while the constricted duct poses high acoustic resistance. That is why in high frequency range the flow oscillation energy is isolated inside the non-flowing vessel and it does not pass through the damper. Isolated oscillation energy is attenuated in hydraulic throttles connected to the constricted duct and the non-flowing vessel. The vessel and the duct act as reactive wave resistances and the hydraulic throttle is the active one. While the wave impedance of the pipeline is frequency independent and active the hydraulic throttle must provide the similar character of wave resistant in the inlet and outlet of the dampener. Therefore, this type of dampener is called the dampener with constant active resistance [12].

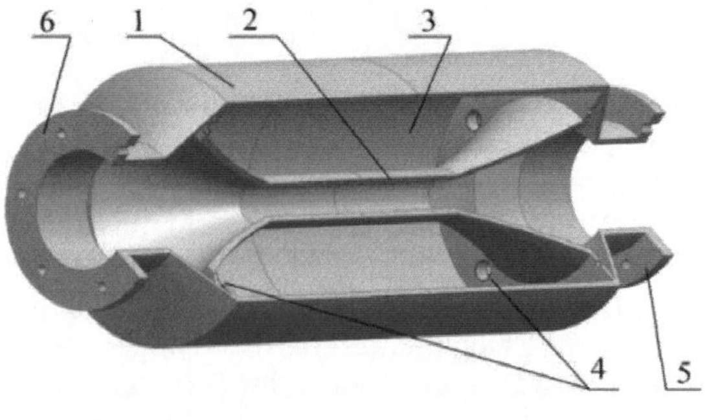

Figure 1. Schematic representation of the damper: (1) – dampener housing assembly, (2) – constricted duct, (3) – non-flowing vessel, (4) – hydraulic throttles, (5, 6) – connecting flanges.

3. Methods

The process of the modeling of the acoustic wave propagation was devoted to the flow analysis using the unsteady Reynolds-averaged Navier-Stokes (URANS) model. Calculating the acoustic wave propagation with the flow rate assumed to be zero involves the user-defined functions for specifying the pressure pulsations at the inlet. During the calculation the low-Reynolds $k-\omega$ SST turbulence model was used to closure governing equations. The choice of such turbulence model was made due to its best ability to predict the flow separation from the wall [20]. Figure 2 depicts the geometry of flow domain region.

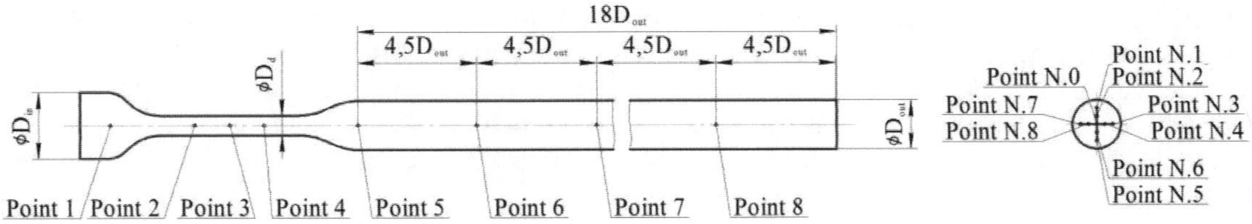

Figure 2. Geometry of computational domain.

3.1 Governing equations

The prediction of pressure pulsations in the diffuser as well as their propagation requires the calculation of the flow parameters time dependency. For this purpose the three-dimensional complete system of hydrodynamics equations which consists of the continuity equation:

$$\frac{\partial \rho}{\partial t} + \frac{\partial \rho u_j}{\partial x_j} = 0 \, , \tag{1}$$

and the momentum equation

$$\frac{\partial \rho u_i}{\partial t} + \frac{\partial \rho u_j u_i}{\partial x_j} = -\frac{\partial p}{\partial x_i} + \frac{\partial \tau_{ij}}{\partial x_i} \, , \tag{2}$$

where ρ is a density, u_j and x_j ($j = 1,2,3$) are components of velocity and displacement on the axes of a Cartesian system, p is a pressure, τ_{ij} is a component of stress tensor, t is a time, is solved. The complete system of hydrodynamics equations includes the state equation. The assumption of the adiabatic wave processes allows one to eliminate the temperature from the state equation. Density and pressure are correlated using Tait equation of state:

$$\rho = \rho_0 \sqrt[n]{\frac{K_0 + n \Delta p}{K_0}} \, , \tag{3}$$

where ρ0 is a reference fluid density, K0 is a reference bulk modulus, n is a density exponent, $\Delta p = p - p_0$, p is an absolute fluid pressure, p0 is an absolute reference fluid pressure.

3.2 Boundary conditions and time discretization

The evaluation of acoustic oscillations has been conducted using URANS analysis. Harmonic pressure of the fluid was set as a boundary condition at the inlet. Flow rate was set to zero at the outlet, thus ensuring the propagation of acoustic waves.

The time step value which was selected in the preceding research based on the required maximum frequency in the spectrum of vortex pulsations has been considered as insufficient. In the present paper the Courant number has been regarded as a key factor when choosing the time step value:

$$\Delta t = CFL \cdot \frac{\Delta x_{min}}{c_\infty} \tag{4}$$

where Δt is the time step value, CFL is the Courant number which describes how fluid is moving through computational cells, Δxmin is the minimal cell size, c∞ is the speed of sound in a medium.

4. Results and discussions

4.1. Comparison of different time step values

In what follows, the results of the numerical calculation of the acoustic wave propagation are described. The acoustic wave propagation was calculated in order to determine whether the generated mesh model can adequately calculate acoustics, i.e. whether the wave changes during its downstream propagation as the wave should remain plane at the outlet and how the mesh size affects the results of the further calculation. Figure 3 depicts the results of pressure pulsation calculation for different time step values selected based on the Courant number for $CFL = 10$, $\Delta x_{min} = 10^{-3}$ mm and $c_{\infty} = 1200$ m/s. The data has been obtained from the center points of various cross-sections (figure 2, points 3.0 and 6.0).

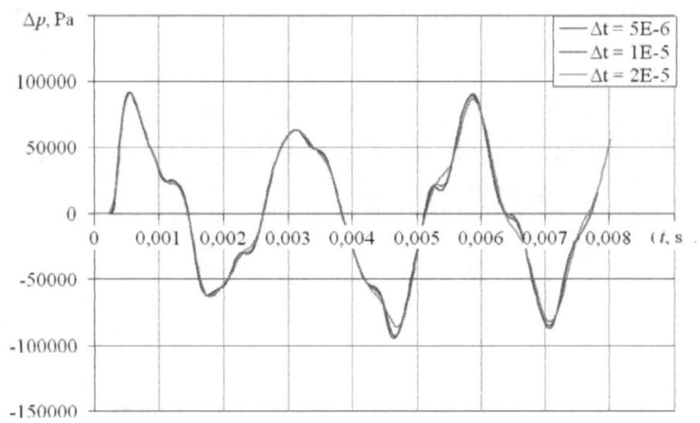

(a) Point 3.0

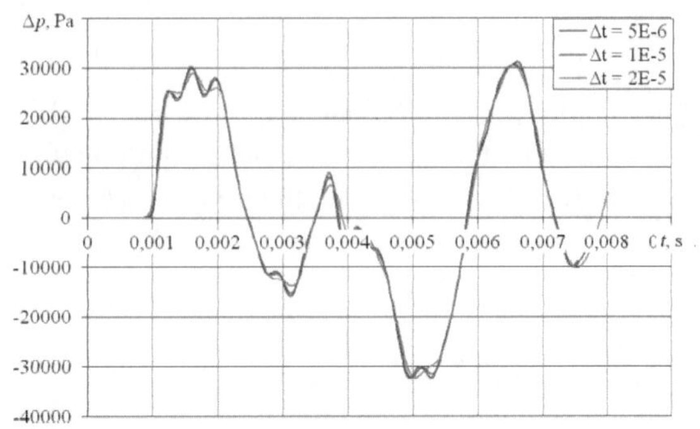

(b) Point 6.0

Figure 3. Pressure pulsations for different time step values

As can be seen from the graphs on figure 3, the time step $\Delta t = 10^{-5}$ s provides the necessary accuracy, but it does not require much calculation time as for $\Delta t = 5 \cdot 10^{-6}$ s. On the other hand, the time step $\Delta t = 2 \cdot 10^{-5}$ s does not provide adequate results of the processes which occur in the pipeline.

Figure 4 shows the static pressure distribution field for the case of Δt = 10-5 s and t = 0.0075 s which illustrates the wave propagation.

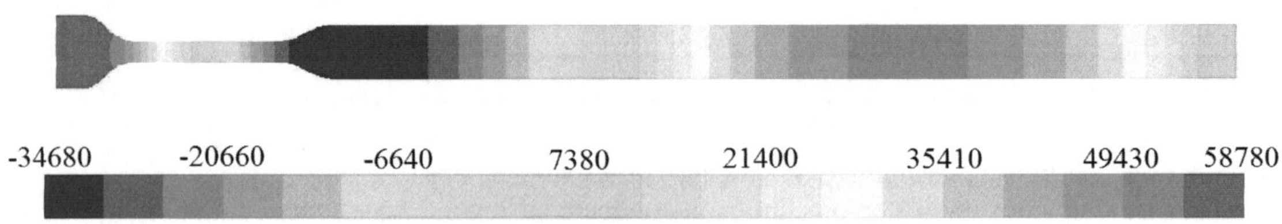

Figure 4. Static pressure distribution

4.2 Check of the wave planeness

To check the planeness of the propagating acoustic wave, measurements have been taken at 5 points in 8 different cross-section planes for the time step $\Delta t = 10^{-5}$ s. In the case of a plane wave, all values of the pressure oscillations should not differ by more than 5% in each cross-section, which can be observed in figure 5. Figure 5 depicts the results of pressure pulsation calculation for different cross-sections. The data has been obtained from cross-sections 2 and 7 (figure 2, points 2.0, 2.1, 2.3, 2.5, 2.7 and 7.0, 7.1, 7.3, 7.5, 7.7).

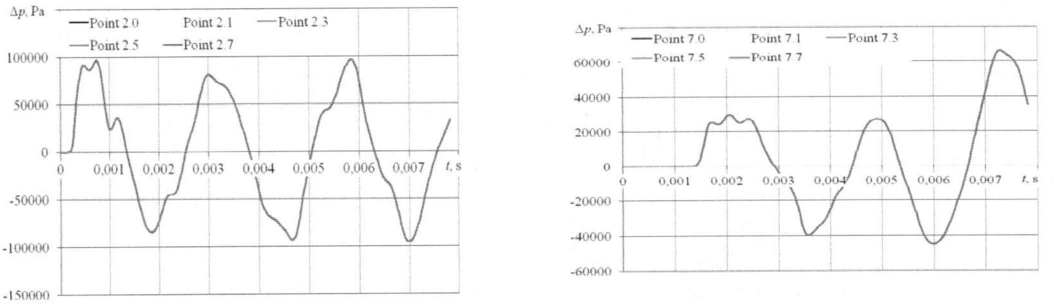

(a) Cross-section no. 2 (b) Cross-section no. 7

Figure 5. Pressure pulsations for different cross-sections

The graph from figure 6 which shows the overall level of pressure oscillations for the time step $\Delta t = 10^{-5}$ s also demonstrates that the proposed numerical model allows one to adequately calculate the acoustics and that the discrepancy of the calculation data with the experimental one from [20] is not connected with the characteristics of the mesh model. The data has been obtained from the center points of every cross-section.

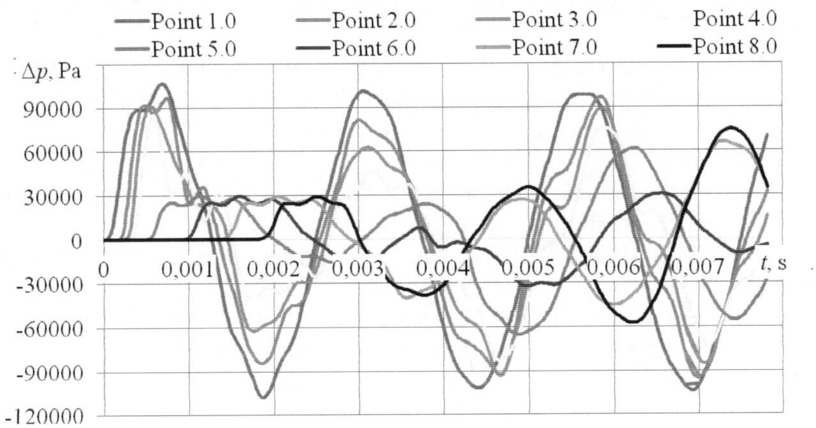

Figure 6. Overall level of pressure pulsations

From all of the above, we can conclude that the calculated model allows us to adequately calculate the propagation of acoustic waves, while one should pay attention to the calculation parameters, for example, the time step value.

5. Conclusion

A numerical simulation of acoustic wave propagation was performed. It was noted that the accepted time step value calculated from the Courant number is sufficient to obtain adequate data. An acoustic wave passing through the damper model is a plane wave. The proposed design model makes it possible to further calculate vortex pulsations, for more accurate calculation of which further investigations are required.

Acknowledgements

This work was supported by the Ministry of Education and Science of the Russian Federation.

References

[1] Tian J, Yuan Ch, Yang L, Wu Ch, Liu G and Videa Zh 2016 The vibration analysis model of pipeline under the action of gas pressure pulsation coupling Eng. Failure Analysis 66 328–40
[2] Dequand S, van Lier L, Hirschberg A and Huijnen J 2002 Aeroacoustic response of diffusers and bends: comparison of experiments with quasi-steady incompressible flow models J. Fluids and Structures 16 957–69
[3] Musaakhunova L F, Igolkin A A and Shabanov K Y 2015 The vibroacoustic characteristics research of the gas pipeline Proc. Eng. 106 316–24
[4] Igolkin A A, Musaakhunova L F and Shabanov K Y 2015 Method development of the vibroacoustic characteristics calculation of the gas distribution stations elements Proc. Eng. 106 309–15
[5] Kårekull O, Efraimsson G and Åbom M 2014 Prediction model of flow duct constriction noise Appl. Acoustics 82 45–52
[6] Gafurov S, Rodionov L and Makaryants G 2016 Simulation of gear pump noise generation Proc. of the 9th FPNI Ph.D. Symp. on Fluid Power, FPNI 2016, Florianópolis, Brazil, 26–28 October, doi:10.1115/FPNI2016-1531
[7] Rodionov L and Rekadze P 2015 Exploration of acoustic characteristics of gear pumps with polymeric pinion shafts Proc. Eng. 106 36–45
[8] Shorin V P and Sanchugov V I 1978 On estimating the operating efficiency of suppressors of liquid pulsations, which contain resonant loops in their structure Power Eng. (New York) 16 113–20
[9] Ermilov M A, Kryuchkov A N, Balyaba M V and Shabanov K Y 2015 Development of a pressure pulsation damper for gas pressure regulators with account of operation parameters Proc. Eng. 106 277–83
[10] Ermilov M A, Balyaba M V, Kryuchkov A N and Shabanov K Y 2015 The experimental development of the pulsation damper in a gas reduction line Proc. of the 22nd Int. Congress on Sound and Vibr., Florence, Italy, 12–16 July
[11] Golovin A N and Shorin V P 1982 Designing fluid-oscillation dampers Power Eng. (New York) 20 132–38
[12] Singh N K and Rubini P A 2015 Large eddy simulation of acoustic pulse propagation and turbulent flow interaction in expansion mufflers Appl. Acoustics 98 6–19
[13] Xiwen Dai 2016 Vortex convection in the flow-excited Helmholtz resonator J. Sound and Vibr. 370 82–93
[14] Makaryants G M, Gafurov S A, Zubrilin I A, Kruchkov A N, Prokofiev A B and Shakhmatov E V 2013 Design methodology of hydrodynamic noise silencer Proc. of the 20th Int. Congress on Sound and Vibr., Bangkok, Tailand, 7–11 July
[15] McDonald A T and Fox R W 1966 An experimental investigation of incompressible flow in conical diffusers Int. J. Mech. Sc. 8 125–30
[16] Kwong A H M and Dowling A P 1994 Unsteady flow in diffusers J. Fluids Eng. 116 842–47
[17] Lam G C Y, Leung R C K and Tang S K 2014 Aeroacoustics of duct junction flows merging at different angles J. Sound and Vibr. 333 4187–202
[18] Gloerfelt X and Lafon P 2008 Direct computation of the noise induced by a turbulent flow through a diaphragm in a duct at low mach number Computers and Fluids 37 388–401
[19] Makaryants G 2017 Vortex pulsation simulation of a piping noise dampener Proc. of the 24th Int. Congress on Sound and Vibr., London, UK, 23–27 July
[20] Menter F R and Sarkar S 1994 Two-equation eddy-viscosity turbulence models for engineering applications AIAA Journal 32 1598–605

IOP Conf. Series: Materials Science and Engineering **302** (2018) 012013 doi:10.1088/1757-899X/302/1/012013

Self-Ignition in Porous Media: Critical Phenomena

E Shchepakina[1]

[1]Scientific and Educational Center for Fluid Dynamic Research, Samara National Research University, 34, Moskovskoye shosse, Samara 443086, Russia

E-mail: shchepakina@ssau.ru

Abstract. The self-ignition of flammable liquid in an inert porous medium is studied. We obtained the complete classification of the possible scenarios of the process using asymptotic and geometrical techniques. This approach allows us to reveal a critical regime which plays a role of a watershed between the safe processes and self-accelerating regimes that lead to the explosion. The realizability conditions for the critical regime are obtained as the explicit asymptotic expression for the control parameter.

1. Introduction

A wide spectrum of thermophysical and chemical processes occurring during combustion, especially in multiphase media, is characterized by a significant difference in time scales (for example, temperature changes vs. concentration changes). Therefore, the singularly perturbed systems of differential equations are used to model combustion processes. For studying such systems different asymptotic methods are usually applied. The asymptotic geometric method used in this paper is based on the theory of integral manifolds (see, for instance, [1, 2] and the references therein). This method allows us to replaces the original system by another system on an integral manifold of lower dimension. The lowering of the dimension occurs due to the decomposition of the original system in the vicinity of the integral surface into the independent "slow" subsystem and the "fast" subsystem. If the slow integral manifold is attractive, then the analysis of the original system can be replaced by the analysis of the slow subsystem retaining a desired degree of accuracy. The use of such approach makes it possible not only to substantially simplify the initial system by reducing its dimension and by eliminating the computational rigidity, but it also gives us a full picture of the dynamics of the physical processes and provides an effective tool for modelling the critical phenomena in these processes. The critical regime corresponds to a chemical reaction separating the domains of self-accelerating reactions and domains of safe reactions. This regime is modelled by a unique trajectory that contains an unstable integral manifold.

This paper deals with the investigation of the critical conditions for self-ignition of combustible fluids in porous insulation materials. This phenomenon is usually caused by a leaking of a combustible liquid into lagging material surrounding a hot pipework. Due to highly insulation environment heat losses are remarkable low and self-ignition may occur as a result of exothermic oxidation reaction. To determine the conditions under which the critical regime is realized, the canard technique [3-6] is used.

We shall study a process which may be defined as the self-ignition in two-phase medium (combustible liquid and inert porous matrix). The possible depletion of oxygen, its diffusion into porous structure, and transport of the liquid or its vapor within the insulation are all ignored. We focus our attention on the competitive effects of the reactive term of the dispersed liquid and evaporative heat loss. The dimensionless model in this case has the following form [7]:

$$\dot{u} = QK_1xe^{-1/u} - (u - u_a) - Q_cK_2xe^{-\beta_e/u}, \dot{x} = -K_2xe^{-\beta_e/u} - K_1xe^{-1/u}. \quad (1)$$

Here, u is the dimensionless temperature of the reactant phase; the dimensionless concentration x represents the mass fraction of combustible liquid present in the porous material; the dimensionless parameters Q and K_1 characterize the heat of reaction and the reaction frequency, respectively, for the exothermic oxidation reaction, while Q_c and K_2 are the similar terms for the endothermic evaporation reaction; β_e is the ratio of the enthalpy of vaporization to the activation energy of the oxidation reaction; u_a is ambient temperature.

Let us introduce the new variables θ and τ by $u = \theta + \beta^2$, $t = \tau\exp(1/\beta)$, $\beta = u(0)$, and parameters

$$\varepsilon = \exp\left(-\frac{1}{\beta}\right), a = K_2\exp\left(\frac{1-\beta_e}{\beta}\right), \theta_a = \frac{u_a - \beta}{\beta^2}, \mu = \frac{QK_1}{\beta^2}\exp\left(-\frac{1}{\beta}\right), \nu = \frac{Q_cK_2}{\beta^2}\exp\left(\frac{-\beta_e}{\beta}\right).$$

That leads (1) to the system

$$\varepsilon\dot{\theta} = \mu x\exp\left(\frac{\theta}{1+\beta\theta}\right) - (\theta - \theta_a) - \nu x\exp\left(\frac{\beta_e\theta}{1+\beta\theta}\right) = F(x,\theta), \quad (2)$$

$$\dot{x} = -ax\exp\left(\frac{\beta_e\theta}{1+\beta\theta}\right) - K_1x\exp\left(\frac{\theta}{1+\beta\theta}\right) = G(x,\theta), \quad (3)$$

which is the singularly perturbed [1] due to the smallness of the parameter ε for typical combustible liquids. The chemically relevant phase space Ω of system (3), (4) is defined by $\Omega = \{x \geq 0, \theta \geq -1/\beta\}$.

In [8] the system (1) was investigated numerically under quasi-steady-state assumption that corresponds to the assumption $\varepsilon = 0$ for the system (2), (3). This approach allows determining the main types of chemical regimes of the investigated process. The quasi-steady-state assumption is widely used in the theory of combustion and gives good results to draw conclusions about the qualitative behavior of the full system for sufficiently small ε. However, the critical phenomena are highly sensitive with respect to the parameters. Hence, this fact implies the considerable difficulties under the numerical calculations. Thus, detailed study of this mathematical object is possible with taking into account the small perturbations and use of asymptotic methods, for example, methods of the integral manifolds theory for singularly perturbed systems.

2. Analysis

The trivial solution is the final steady state of the system. The degenerate equation $F(x,\theta)=0$ describes *the slow curve S* of (2), (3) [1]. The subset $S^s(S^u)$ of S with $\partial F(x,\theta)/\partial\theta < 0\ (>0)$ is called the *stable* or *attractive* (*unstable* or *repulsive*) part of *S*. A point *A* on *S* in which $\partial F/\partial\theta=0$ is called the *jump* or *turning point*. Stable and unstable parts of the slow curve are zeroth order approximations of corresponding stable and unstable slow invariant manifolds. The invariant manifolds lie in an ε-neighborhood of the slow curve, except near jump or turning points [1]. The slow curve intersects the axis $O\theta$ at the point $\theta = \theta_0$ and has an asymptote $\theta = \theta_\infty = (\beta_e - 1 - \beta\ln(\mu/\nu))^{-1}\ln(\mu/\nu)$. The shape of the curve *S* varies with the relation between values of the parameters, which leads to a change in qualitative behavior of the system. So, if we change the value of one parameter, with fixed values of the other parameters, we can change the type of chemical reaction. Following [8], we consider β_e as a control parameter. For $\theta_\infty = \theta_0$ we have $\beta_e = b_0$, where $b_0 = 1 + (1+\beta\theta)\ln(\mu/\nu)/\theta_a$.

Consider the case $\beta > u_a$ and $b_0 < 1$. For $\beta_e < b^T$ the lower branch of S can consist of two stable parts (s_1^s) and one unstable part (s_1^u), see Figure 1(a). These parts are divided by two turning points, which merge with one another and disappear at a value $\beta_e = b^T$ [8], see Figure 1(b). In both cases the trajectories of system (2), (3) move along the stable part of the slow curve to the final steady state. These trajectories correspond to the slow regimes, which are safe, see Figure 2. For $\beta_e > b_0$ the upper and lower branches of S consist of stable (s_1^s and s_2^s) and unstable (s_1^u and s_2^u) parts, which are divided by the turning points A_1 and A_2, see Figure 1(d). The system's trajectories starting at any point of the basin of attraction of s_2^s correspond to the slow regimes.

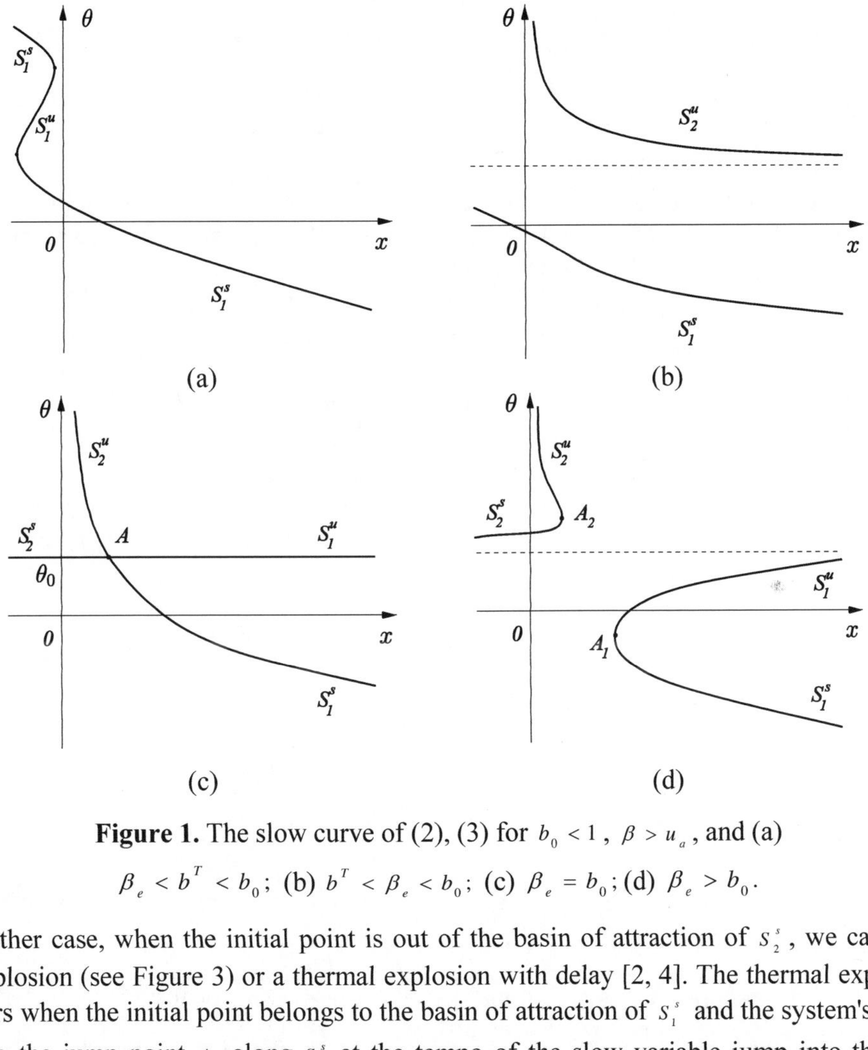

Figure 1. The slow curve of (2), (3) for $b_0 < 1$, $\beta > u_a$, and (a) $\beta_e < b^T < b_0$; (b) $b^T < \beta_e < b_0$; (c) $\beta_e = b_0$; (d) $\beta_e > b_0$.

In the other case, when the initial point is out of the basin of attraction of s_2^s, we can observe a thermal explosion (see Figure 3) or a thermal explosion with delay [2, 4]. The thermal explosion with delay occurs when the initial point belongs to the basin of attraction of s_1^s and the system's trajectories by reaching the jump point A_1 along s_1^s at the tempo of the slow variable jump into the explosive regime.

For $\beta_e = b_0$ the point A_1 merges with A_2 to give one self–intersection point A of the slow curve, see Figure 1(c). As it was noted above, in ε-neighborhood of the subset s_1^s (s_2^u) there exists a stable (unstable) slow invariant manifold $s_{1,\varepsilon}^s$ ($s_{2,\varepsilon}^u$). For some value $\beta_e = b^* = b_0 + O(\varepsilon)$, $(\varepsilon \to 0)$, the stable and unstable slow invariant manifolds $s_{1,\varepsilon}^s$ and $s_{2,\varepsilon}^u$ are glued at the point A. As a result, for $\beta_e = b^*$

system (2), (3) has a *canard* trajectory [1, 2, 9] which, at first, follows an attractive invariant manifold and then a repulsive one. In both cases the distances are travelled with $O(1)$ as $\varepsilon \to 0$, see Figure 4. This canard simulates the critical regime, separating slow chemical regimes from regimes with a self–acceleration in the case $b_0 < 1$.

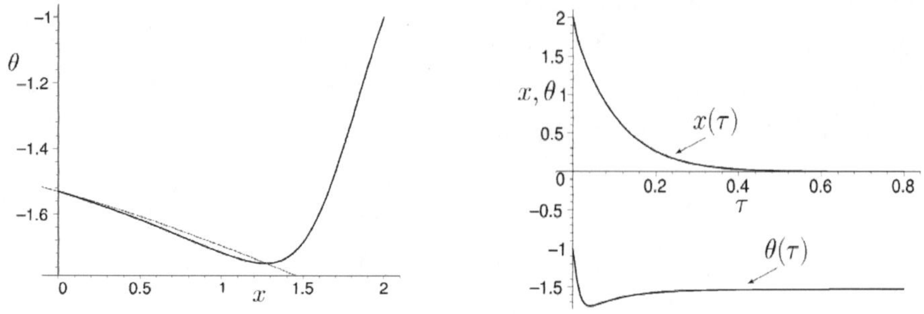

Figure 2. (left) The trajectory (the thick line) and the slow curve (the thin line) of (2), (3), and (right) the x– and θ –components of the solution in the case of the slow regime: $b = 0.3685$.

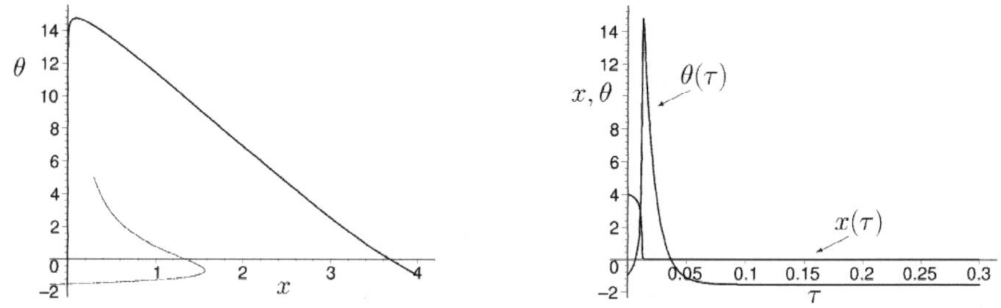

Figure 3. The case of the thermal explosion: $b = 1$.

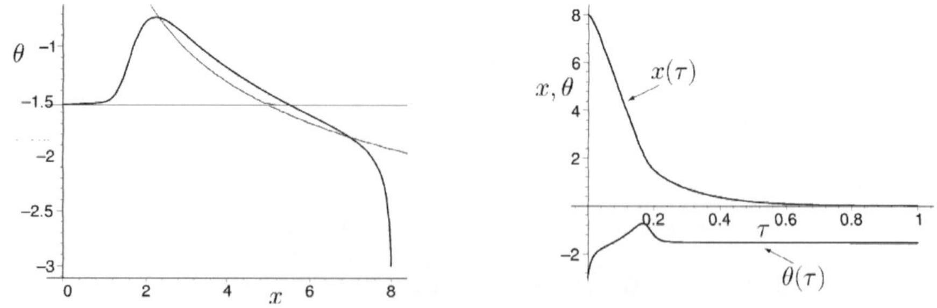

Figure 4. The case of the critical regime: $b = b^* = 0.56827$.

If $\beta > u_a$, one can observe the similar transformation of the slow curve (and the qualitative behavior of the system) as shown in Figure 1 but with a decreasing value of the parameter β_e ($b^T > b_0$ in this case).

For $b_0 > 1$ plots of the slow curve are mirror images of the graphs shown in Figure 1 with respect to the vertical axis. The thermal behavior of the chemical system is safe for nonsignificant values of the initial concentration of a combustible liquid. Otherwise, the value $\beta_e = b^T$ determines the boundary of the safe region [8].

Our goal is to reveal the sufficient conditions for realization of the critical regime for the case $b_0 < 1$. As it has been noted above, the main feature is that during the critical regime the temperature attains a high value without explosion. The interest in critical phenomena is occasioned by not only for reasons of safety, but for the fact that in many cases the critical regime is the most effective in technological processes [2-6].

3. Critical phenomena: realizability conditions

Using the method of integral manifolds and the canard techniques [1, 2] it is possible to find the critical value of the parameter $\beta_e = b^*$ and corresponding trajectory in the form of the asymptotic representations

$$\theta = \varphi(x,\varepsilon) = \varphi_0(x) + \varepsilon\varphi_1(x) + o(\varepsilon),\ \ \beta_e = b^* = b_0 + \varepsilon b_1 + o(\varepsilon). \tag{4}$$

We write (2), (3) as

$$\varepsilon\theta'\left[ax\exp\left(\frac{\beta_e\theta}{1+\beta\theta}\right) + K_1 x\exp\left(\frac{\theta}{1+\beta\theta}\right)\right] = \theta - \theta_a + \nu x\exp\left(\frac{\beta_e\theta}{1+\beta\theta}\right) - \mu x\exp\left(\frac{\theta}{1+\beta\theta}\right),$$

or, taking into account (4),

$$x\left(\varepsilon\varphi_0' + \varepsilon^2\varphi_1'\right)\left[K_1\exp\left(\frac{\varphi_0}{1+\beta\varphi_0}\right)\left(1+\varepsilon\frac{\varphi_1}{(1+\beta\varphi_0)^2}\right) + a\exp\left(\frac{b_0\varphi_0}{1+\beta\varphi_0}\right)\left(1+\varepsilon\frac{(1+\beta\varphi_0)b_1\varphi_0 + b_0\varphi_1}{(1+\beta\varphi_0)^2}\right)\right]$$

$$= \varphi_0 + \varepsilon\varphi_1 - \theta_a + \nu x\exp\left(\frac{b_0\varphi_0}{1+\beta\varphi_0}\right)\left(1+\varepsilon\frac{(1+\beta\varphi_0)b_1\varphi_0 + b_0\varphi_1}{(1+\beta\varphi_0)^2}\right) - \mu x\exp\left(\frac{\varphi_0}{1+\beta\varphi_0}\right)\left(1+\varepsilon\frac{\varphi_1}{(1+\beta\varphi_0)^2}\right) \tag{5}$$

By setting $\varepsilon = 0$ in (5) we obtain the slow curve' equation:

$$F(x,\varphi_0) = \mu x\exp\left(\frac{\varphi_0}{1+\beta\varphi_0}\right) - \varphi_0 + \theta_a - \nu x\exp\left(\frac{b_0\varphi_0}{1+\beta\varphi_0}\right) = 0. \tag{6}$$

Conditions for self-intersection of the slow curve in point $A(x_s, \varphi_0(x_s))$

$$\partial F(x_s, \varphi_0(x_s))/\partial x = \partial F(x_s, \varphi_0(x_s))/\partial\varphi_0 = 0 \tag{7}$$

give us the coordinates of the self-intersection point and the zeroth-order approximations for critical value b^*. Indeed, from (6), (7) we obtain

$$x_s = \frac{(1+\beta\theta_a)\theta_a}{\mu\exp\left[\theta_a/(1+\beta\theta_a)\right]\ln(\nu/\mu)},\quad \varphi_0(x_s) = \theta_a, \tag{8}$$

$$b_0 = 1 + (1+\beta\theta_a)\theta_a^{-1}\ln(\mu/\nu). \tag{9}$$

Equating the coefficients with ε^1 in (5) we get

$$x\varphi_0'\left[a\exp\left(\frac{b_0\varphi_0}{1+\beta\varphi_0}\right) + K_1\exp\left(\frac{\varphi_0}{1+\beta\varphi_0}\right)\right] = \varphi_1\left\{1 + \frac{x}{(1+\beta\varphi_0)^2}\left[b_0\nu\exp\left(\frac{b_0\varphi_0}{1+\beta\varphi_0}\right) - \mu\exp\left(\frac{\varphi_0}{1+\beta\varphi_0}\right)\right]\right\}$$

$$+\nu x\exp\left(\frac{b_0\varphi_0}{1+\beta\varphi_0}\right)\frac{b_1\varphi_0}{(1+\beta\varphi_0)}. \tag{10}$$

From (7) we note that the expression in brackets in the right-hand side of (10) is equal to zero at point A. To avoid a discontinuity in function $\varphi_1(x)$ at x_s, taking into account (8), we put

$$b_1 = \frac{\varphi_0'(x_s)(1+\beta\theta_a)}{\nu\theta_a}\left[K_1 \exp\left(\frac{(1-b_0)\theta_a}{1+\beta\theta_a}\right) + a\right],$$

or, taking into account (6) and (7),

$$b_1 = \frac{2\mu^3(1+\beta\theta_a)^4}{\theta_a^4}\ln^3\frac{\nu}{\mu}\exp\left(\frac{3\theta_a}{1+\beta\theta_a}\right)\left(\frac{K_1}{\mu}+\frac{a}{\nu}\right)\left[2\beta(1+\beta\theta_a) - 2 + \frac{(1+\beta\theta_a)}{\theta_a}\ln\frac{\nu}{\mu}\right]. \quad (11)$$

Thus, the expressions (6), (9)–(11) determine the first-order approximation for the canard and the corresponding critical value $\beta_e = b^*$. It should be noted that it is not possible to explicitly solve equation (6) with respect to φ_0, while the critical value b^* has been found in the explicit form. However, one can use the implicit or parametric representation for slow invariant manifold [1] to obtain an approximation of the canard.

4. Conclusion

In this paper the model of self-ignition of combustible fluids in an inert porous medium has been studied. The realizability conditions for the critical regime have been obtained as the explicit asymptotic expression for the control parameter. It was shown that the critical regime is modelled by the canard. This regime plays the role of a watershed between the safe processes and regimes with self-acceleration that leads to an explosion. It should be noted that the critical regime is not a slow regime, since the temperature may attain a high value, nor an explosive regime, as the temperature increases at the tempo of the slow variable. Thus, for the examined model the new type of the safe regime has been revealed.

Acknowledgments

The paper is supported by the Ministry of Education and Science of the Russian Federation (project RFMEFI58716X0033).

References

[1] Shchepakina E, Sobolev V and Mortell M P 2014 *Singular Perturbations. Introduction to System Order Reduction Methods with Applications (Lecture Notes in Math.* vol 2114*)* (Cham: Springer)
[2] Shchepakina E and Sobolev V 2005 Black Swans and Canards in Laser and Combustion Models Singular Perturbation and Hysteresis ed M P Mortell, R E O'Malley, A Pokrovskii and V Sobolev (Philadelphia: SIAM) chapter 8 pp 207–255
[3] Sobolev V A and Shchepakina E A 1996 Differential Equations 32 1177
[4] Gol'dshtein V, Zinoviev A, Sobolev V and Shchepakina E 1996 Proc. R. Soc. Lond. A 452 2103
[5] Shchepakina E and Sobolev V 2001 Nonlinear Analysis 44 897
[6] Shchepakina E 2003 Nonlinear Analysis: Real Word Applications 4 45
[7] McIntosh A C, Bains M, Crocombe W and Griffiths J F 1994 Combust. Flame 99 541
[8] McIntosh A C and Griffiths J F 1996 IMA Journal of Applied Mathematics 55 83
[9] Benoit E, Callot J L, Diener F and Diener M 1982 Collect. Math. 32 37

IOP Conf. Series: Materials Science and Engineering **302** (2018) 012014 doi:10.1088/1757-899X/302/1/012014

Signaling Device for the Pre-Emergency State of the Elements of the Rotating Assembly of Steam Compressor of Desalination Plant

A I Danilin[1], A Zh Chernyavskiy[1], S A Danilin[1] and E V Blagin[1]

[1]Samara University, 34, Moskovskoye shosse, Samara, 443086, Russia

E-mail: aidan@ssau.ru

Abstract This article deals with non-contact exploitation control method based on the treatment of the radio wave signal reflected from controlled gear teeth and its advantages in comparison with traditional methods of gear teeth control. Justification of necessity to use such control method during multiplier gears condition determination during its exploitation is given. Also this article deals with influence of different types of gear wear on typical information parameters of analyzed signals. Disadvantages of the method which are the impossibility of determination of certain types of wear are also taken into account. Certain stages of the development of mathematical model for interaction of first converter with controlled surface. Suggested mathematical model uses only the laws of geometric optics without taking wave processes into account but considering first converter direction diagram influence during its interaction with controlled surface. Structural scheme of developed experimental system for gears teeth condition control for steam compressor. Operation of the experimental system of gear control is given on the base of structural scheme. Core of the developed device is microcontroller STM32 which treat the information received from the sensors as well as connection with computer. Certain elements of the experimental control system as well as its components are described separately. Photos of experimental unit for control for control method development in laboratory conditions are presented. Design of the first converter is given in short.

1. Introduction

Modern desalination plants widely use steam compressors to achieve economical operation modes [1]. A steam compressor is a complicated mechanical block, which includes the rotor with blade assembly in the form of impeller (Figure 1). Operating rotational rotor speed is 32000 rpm. The axial and radial play of the elements of such a rotating assembly are subjected to strict requirements, which are provided by support bearings and thrust bearings, to the shaft vibration parameters and to the blades of the compressor wheel. Support and thrust elements of such dynamical system are subjected of increased requirements both on manufacturing quality and monitoring of their operational status. Many papers are published currently aimed to optimization of operating modes of different compressors. Those include investigations of the scientists of Samara University [2, 3, 4], which display the features of compressor workflow optimization.

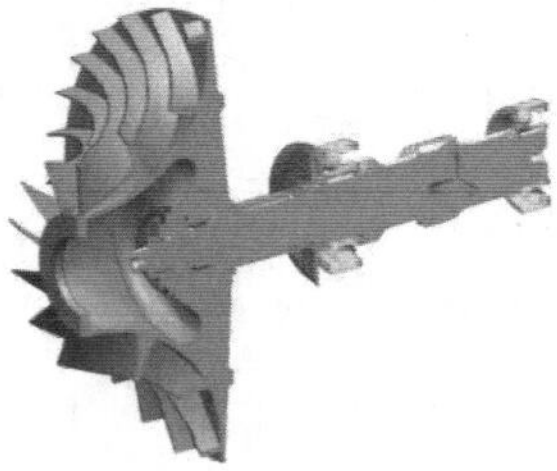

Figure 1. General view of the elements of the rotating assembly of steam compressor rotor.

2. Problem formulation

At the present time, the problem of condition monitoring of the blades of the impeller and steam compressor shaft in a static state on stopped compressor is solved by visual inspection of the blade row with the help of technical endoscopes, eddy current and magnetic defectoscopes. This is very painstaking work, requiring professional skills and responsibility from technical personnel. Despite all measures taken, according to statistics, there are emergencies associated with the destruction of the impeller blades and shaft (Figure 2). There are known publications that discuss the dynamic diagnostic system based on spectrum analysis of vibration sensors [5, 6]. But the spectral diagnostic method has some serious drawbacks. Diagnosis on the spectrum of the envelope signal is quite complicated for physical interpretation of the defects of the blade apparatus, because it usually requires a good theoretical training of vibro-diagnostician or presence of the specialized computer expert system in his technical arsenal. Also personal practical experience of vibro-diagnostician and his knowledge of the internal structure of the controlled equipment have great influence on the accuracy of estimate of the current technical condition and reliability of diagnostics of defects of the state of the gear pairs.

Figure 2. Photo of the destroyed impeller blades.

In fact, the deviation of the shape of the blades of compressor wheel (scoring, blades folding) from the design variant leads to decreasing the efficiency of steam compressor, and finally leads to the destruction of the steam compressor rotor, to recover of which, at best, significant material and time expenses are required. Thus, the diagnosis of the defective condition of the elements of the steam compressor rotor of desalination plant is important at present.

As the analysis of scientific literature shows, among various methods and means of diagnostics and control elements of the rotating machinery on operating steam generators, there is prospective contactless discrete-phase method (DPM) allowing to determine the individual deformation condition of each blade of impeller and change in the gaps between the inner surface of the steam compressor body and blade tips in the blade row [7]. The obtained information is compared with the geometric position of a specific blade in the wheel at certain times and the corresponding interpretation of the obtained values in the field of mechanical stresses and deformations.

Based on the theoretical prerequisites of the DPM, a block diagram of the signaling device for the pre-emergency state of the blade assembly of the steam compressor [8] is presented, shown in Figure 3. The block diagram is based on the principle of measuring the current time intervals (steps) between the ends of the rotating blades, determining the maximum and minimum values of each step, finding the average value of each step and comparing them with the average pitch of the blade wheel at each turn of the rotor of the steam compressor.

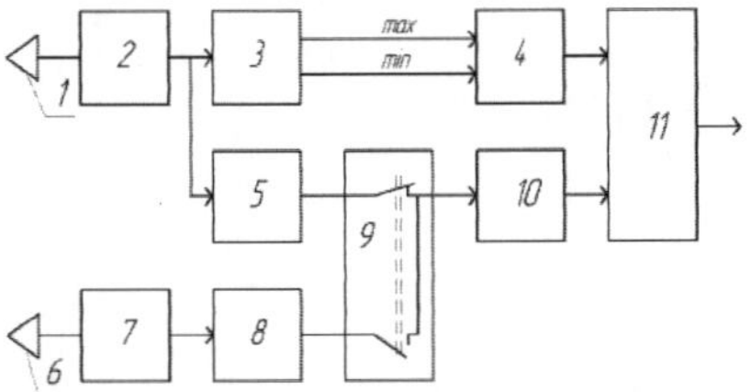

Figure 3. The block diagram of the signaling device for the pre-emergency state of the blade assembly of the steam compressor.

3. Analytical expressions

In accordance with Figure 3 block diagram works as follows. Peripheral sensor 1 mounted in the housing of steam compressor over the trajectory of the ends of the blades generates an electrical signal which by the shaper 2 is converted into a rectangular pulse. The time intervals τ_i between the rectangular pulses corresponding to a circumferential step between the blades are converted into digital codes in the block 3. In the same block, the values of the maximum and minimum values for each step are determined on N present rotor revolutions. Obtained extreme values of each step are received by block 4, which determines the average value τ_{si} of each step accordingly with expression:

$\tau_{si} = (\tau_{i\max} + \tau_{i\min})/2$.

In addition, the rectangular pulses from shaper 2 are received by block 5, where the time intervals τ_i are summed within N rotor revolutions, and then the average rotation period for N rotor revolutions is determined according to expression (1):

$$T_s = \frac{1}{N}\sum_{j=1}^{N}\sum_{i=1}^{K}\tau_i \quad (1)$$

If the design implementation of the steam compressor allows the sensor 6 of the reference revolution mark to be installed, then its electrical signals are converted into rectangular pulses by means of the shaper 7, which are fed to the block 8, in which then the average rotation period of the rotor is determined for N turns according to expression (2):

$$T_s = \frac{1}{N}\sum_{j=1}^{N} T_j \quad (2)$$

The switch block 9 allows to provide the necessary mode of operation of the device, both with the sensor of the reference revolution mark, and without it. The values T_s obtained in block 10 are divided by the number of blades K in the wheel, thereby determining the average pitch along the impeller for N rotor revolutions: $\tau_s = T_s / K$.

Further, the so determined τ_{si} and τ_s are compared among themselves in the comparison unit 11. If it τ_s differs from τ_{si} on a certain threshold value p, then a signal about the occurrence of a defect in one or more blades is generated at the output of the block 11.

When using amplitude criterion in signaling device for the pre-emergency state of the blade assembly of the steam compressor [9], accordingly with Figure 4, for intact, defect-free blades one can write: $A_s = \dfrac{y_{ssmax} - y_{ssmin}}{4}$.

In accordance with this formula, for all characteristic operating modes of the compressor, it is possible to determine the average amplitude of the blade vibrations, and after statistical analysis to select the maximum and minimum values of the vibration amplitudes of the blades, which, when compared with the average amplitude of the oscillations, give an algebraic difference

$\Delta = A_{max} - A_s = \left| A_{\min} - A_s \right|$, characterizing the structural and technological deviations that occur in the manufacturing of blades of a specific wheel.

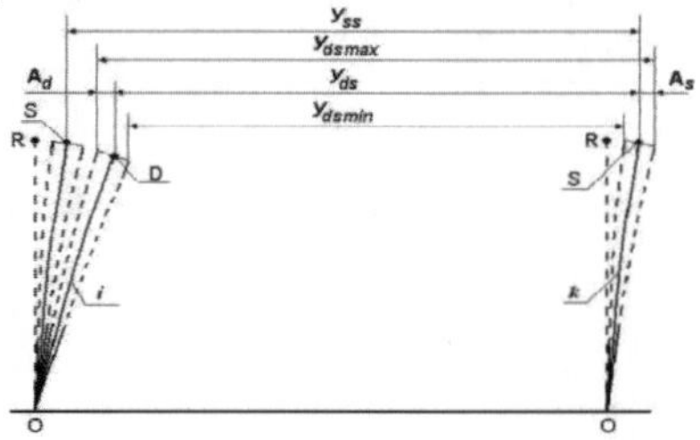

Figure 4. Element of circumferential development of the blade row with two adjacent blades on the compressor operating mode.

Further, taking into account the average vibration amplitude of the blades on the operating mode of the compressor the current average value of the oscillation amplitudes of the blades is continuously determined and compared with the average value.

As soon as $\left| A_{cur} - A_s \right| > \Delta$, it is stated the change in the amplitude of oscillations of the i-th blade, i.e. registering the appearance of the defect in it. For identified interblade interval, the amplitude of vibrations of defective blades is determined as it is stated in equation (3):

$$\begin{cases} y_{dsmax} = y_{ds} + A_d + A_s \\ y_{dsmin} = y_{ds} - A_d - A_s \end{cases}, \qquad (3)$$

therefore $A_d = \dfrac{y_{dsmax} - y_{dsmin}}{2} - A_s$.

4. Results and discussions

Based on the proposed algorithm and the block diagram, a signaling device for the pre-emergency state (SDPES) of the blades assemblies of steam compressors operated at desalination plants was developed. The appearance and functional composition of the SDPES are shown in Figure 5. It includes: pulse sensor installed in the housing of steam compressor; measuring unit with USB interface to connect to computer; noise-proof communication cables; special software.

A signal about the occurrence of a defect of one or more blades in the form of warning light and sound signals is issued to the operator control panel. If the defect of the blade continues to develop, the SDPES generates a signal for the automatic control system of the equipment for normal (or emergency) shutdown of the steam compressor.

To monitor the current state of the blades, express analysis of their operability, to memorize their operational status, to record the inspection results, to create a technical passport of the blades, it is possible to communicate SDPES with a personal computer via a serial USB interface. To implement this possibility, special software has been developed that allows real-time field operational diagnostics of blades.

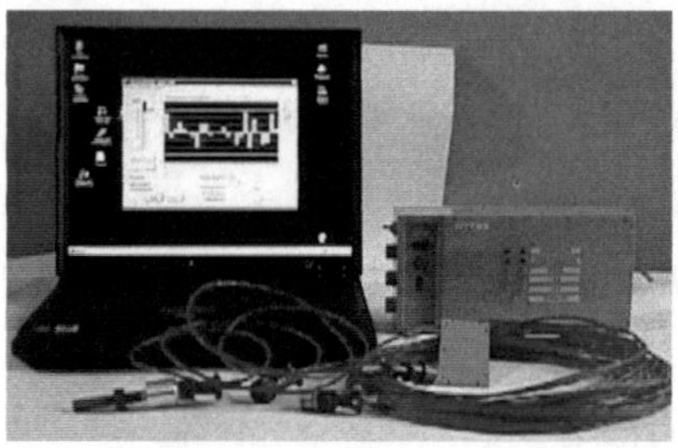

Figure 5. The appearance and functional composition of the SDPES.

In addition, the SDPES provides an autonomous internal storage device, which is activated after the warning signal is issued and fixes the dynamics of the development of the defect of the blade. In this case, the date, time and deformation state of all impeller blades are recorded in the internal nonvolatile memory. SDPES was developed and implemented on a modern electronic component base, which allowed to receive a compact and reliable device with a two-fold redundancy with an operational adaptation to any kind of steam compressors and impellers with a different number of blades.

SDPES has the following technical specifications:

1. Operating rotational speed range of the steam compressor rotor: 100 ... 33000 rpm.
2. Operating temperature range of pulse sensors:
   - option 1 -40 ... +150°C;
   - option 2 -40 ... +1000 °C.
3. The operating temperature range of the measuring unit -40 ... +50 °C.
4. Power supply of SDPES 27 VDC.

Laboratory tests of the SDPES were carried out, in which a defect was introduced in one of the compressor blades in the form of deflection of the end of the blade. Figure 6 shows the screenshot of the operator's program of the SDPES, which shows the deviations of all blades, including simulation of the defect, reaching 6 mm. The natural technological spread of the deviations of the other blade ends, the so-called misplacement, lies within ± 0.5 mm.

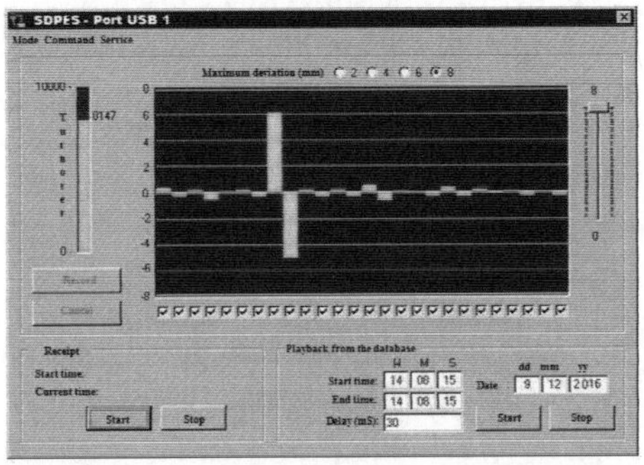

Figure 6. The screenshot of the operator's program of the SDPES with a defective blade.

On the basis of the experimental studies, recommendations have been made on the choice of threshold levels corresponding to the pre-emergency (2 mm) and emergency (2.5 mm) deviations of the ends of the blades for the steam compressor blades.

5. Conclusions

The experience gained in the testing of various steam compressors showed that the initial dynamic position of the blades in the wheel is such that the spacing of the blades (misplacement) can be very different from the average pitch for the wheel. The difference sometimes reaches 1.5...2 %, so comparing the current operational measurements with the average step, which may be greater than some inter-blade interval, leads to an incorrect interpretation of the deformation state of the blades.

Thus, when comparing the current deformation states of the blades with the threshold values (set points), the question arises of the reference value relative to which the comparison will be made.

To solve this problem, usually at the beginning of operation, the time intervals between pulses of the peripheral sensor are measured for the responsible operating conditions of the turbine-unit for each blade [10]. The measurements are carried out for a certain number of revolutions of the compressor rotor, then the measured values of the time intervals are averaged and the average base time intervals, as a rule, referred to the rotation period of the rotor, which are taken as basic or reference positions of the peripheral parts (ends) of the intact blades. Thus, a kind of passport of deformation state of the blades of the controlled stage of the compressor is created.

In the subsequent operational period, for each blade, the current values of the time intervals between the sensors are measured and averaged for each blade. Then the obtained results are compared with the corresponding averaged basic (passport) time intervals, after which the magnitude and speed of the discrepancy between the results of the comparison are judged about the occurrence and development of damages or defects of the corresponding blades.

This approach to determining the set-points takes into account the vibrational motion of the blades and, therefore, can serve as an objective justification for finding the basic support positions of the peripheral parts (ends) of the blades by averaging for each blade the required number of measurements in the beginning of the operational regime.

Acknowledgments

Work was supported by the Ministry of Education and Science of the Russian Federation in the framework of the implementation of the Program "Research and development on priority directions of the scientific-technological complex of Russia for 2014–2020 years".

References

[1] Patent No. 2280011 Russian Federation, IPC C02F 1/04. A desalination plant for salt water and a method for desalination of salt water using a plant [In Russian] / Luzhkov Yu.M., Solomonov Yu.S., Karaygin M.V., et al. Publ. 20.07.2006, Bul. No. 20.

[2] Matveev V.N., Baturin O.V., Popov G.M. Seven-stage axial compressor optimization // Engineering Optimization 2014. Edited by Aurelio Araujo. CRC Press 2014. Pages 821–826.

[3] Goryachkin E., Popov G., Baturin O., Kolmakova D. Three-Stage Low Pressure Compressor Modernization by Means of Optimization Methods // ASME Turbo Expo 2015: Turbine Technical Conference and Exposition. Vol. 2C: Turbomachinery. Montreal, Quebec, Canada, June 15–19, 2015.

[4] Popov G., Baturin O., Novikova Y., Goriachkin E., Novikov A. Using of low-pressure compressor to design the concept of the air brake system // 55th AIAA Aerospace Sciences Meeting. AIAA SciTech Forum. AIAA 2017-0111.

[5] Hongkun Li, Qingkai Han, Xiaowen Zhang Acoustic modes simulation analysis and experimental verification for centrifugal compressor // The 21st International Congress on Sound and Vibration. 13-17 July, 2014, Beijing, China.

[6] Petry N., Benra F.K., König S. Experimental study of acoustic resonances in the side cavities of a high-pressure centrifugal compressor excited by rotor / Stator Interaction, In Proceedings of ASME Turbo Expo, Glasgow, United kingdom, 2339-2351, 2010.

[7] Danilin A.I. Beskontaktnye izmereniya deformacionnykh parametrov lopatok v sistemakh kontrolya i upravleniya turboagregatami [Contactless measurement of the deformation parameters of the blades in the systems of monitoring and control of turbines, in Russian]. Samara, 2008. 217 p.

[8] Patent No. 2177145 Russian Federation, IPC G01H1/08. Signalizator predavariynykh deformatsiy lopatok turbomashin [Signaling device for the pre-emergency deformations of blades of turbomachines] / Danilin A.I., Chernyavskiy A.Zh., Publ. 20.12.2001, Bul. No. 35.

[9] Patent No. 2341781 Russian Federation, IPC G01H11/06. Sposob izmereniya amplitudy kolebaniya lopatok turbomashiny i ustroystvo dlya ego osuschestvleniya [Method for measuring the oscillation amplitude of turbomachine blades and the device for its implementation] / Danilin A.I., Adamov S.I., Chernyavskiy A.Zh., Danilin S.A., Arefieva O.V., Publ. 20.12.2008. Bul. No.35.

[10] Danilin A.I., Chernyavskiy A.Zh. Criteria for discrete-phase diagnostics of the working conditions of the turbine blades and their feasibility in the automatic control systems of turbo-units [in Russian] // Vestnik of Samara University. Aerospace and Mechanical Engineering. Samara. Vol. 1 (17). 2009. Pages 107-115.

The Monitoring System of the Operating State of the Gear Wheels of the Torque Multiplier of the Desalination Plant Steam Generator

A I Danilin[1], V V Neverov[1], S A Danilin[1], A A Shimanov[1] and A B Tsapkova[1]

[1]Samara National Research University, 34, Moskovskoye shosse, Samara, 443086, Russia

E-mail: aidan@ssau.ru

Abstract. The article describes a noncontact operational control method based on the processing of a microwave signal reflected from the controlled teeth of the wheel. In this paper describes the influence of wear patterns on the characteristic information parameters of the analyzed signals. The block diagram in section 3 shows the experimental system for monitoring the operating state of the gear wheels of the steam compressor torque multiplier. The design of the primary converter is briefly described.

1. Introduction

Gear transmissions are widely used in the mechanical equipment parameters for converting rotary motion. Their service life determines the indicator of failure-free operation of the mechanism. What is one of the critical links in the tree of failures? Failures of mechanical equipment due to wear or destruction of the cog wheels result in long downtime. It is costly to restore functionality.

To the torque multipliers that are used in the desalination plant vapor compressors, high demands are placed on the reliability, precision of the production of gears and the economy in operation. These torque multipliers provide an increase in the number of revolutions up to 32,000 per minute. Gears used in this mechanism are under heavy loads. This factor leads to their accelerated wear and tear and frequent replacement. Therefore, control and diagnostics of the gears in the torque multipliers during their work is an actual problem.

At this time, the diagnosis of defects, wear and integrity of the gear wheels is mainly in the static state. In the dynamic mode, a wide application was made of the spectral method of analysing vibrational signals obtained from a controlled mechanism. But the spectral method of diagnosis has many serious drawbacks. Diagnosis by the envelope spectrum of the signal is complex for the physical interpretation of the defects in the gears. It usually requires good theoretical skills of the engineer or the presence in his technical arsenal of a specialized computer expert system. A very great influence on the accuracy of the assessment of the current technical condition and on the reliability of the diagnosis of defects in the state of gear pairs is provided by the personal practical experience of the engineer and his knowledge of the internal arrangement of the monitored equipment [1].

2. Our method

For operational monitoring and diagnostics of the operating state of the gear of the torque multiplier of the steam compressor, a non-contact radio-wave method [2, 3, 4]. It is based on real-time processing of signals obtained after detecting the reflected from the teeth of the probe microwave radiation. The parameters of the information signals are compared with the reference signals that were obtained at the start of the operation of the gear wheel. This method allows you to obtain information directly about the degree of wear of each tooth and the appearance of defects in it, in a dynamic mode. The parameters of the probing signal are determined by the geometrical dimensions of the teeth and the technological configuration of the control zone. The microwave radiation can exist in the oil environment in which the torque multiplier gears work. The advantages of this method include the absence of a large number of sensors and the need for their accurate alignment. Also, there is no need for frequent maintenance of the sensor. Who works in a fairly aggressive environment?

The following variants of tooth fracture are distinguished: tooth breakage, tooth chipping, and damage to the ends of the teeth, abrasive wear, and the appearance of defects in the form of cracks, flaking or deep contact destruction of the material [1]. During the diagnostics, the reflected probing flow is converted into an electrical signal. From which several information parameters are distinguished. Of all the above types of damage using the proposed method, it is not possible to diagnose only the appearance of cracks in the teeth if they are not located on the surface under investigation.

The need to develop a mathematical model for the interaction of the probing flow and the object of investigation appeared during the development of the method. The process of developing a mathematical model can be divided into several stages, characterizing its complication in order to improve the metrological characteristics of the results obtained.The first stage is the model in the plane, the second stage is the spatial model [5].

The first step of building a model in the two-dimensional coordinate system. The basis is the laws of geometrical optics.

A simplified tooth model is a trapezoid. A simplified model of the radiator is a segment, each point of which is both a radiator and a receiver. By varying the position of the emitter relative to the gear can observe the change in the output signal and thereby select an optimum sensor arrangement, a relatively controlled gear. The mechanism of interaction between the sensor and the surface is illustrated in Figure 1. From each radiation point of the sensor, within a linear angle determined by the direction diagram, rays emerge with a certain angular pitch specified in advance. The weighting factor for each beam is used to take into account the angular position in the radiation pattern of the sensor. The weight coefficient decreases with increasing angle of incidence relative to the sensor radiator normal. In figure 1, this is shown in the example of ray 1.

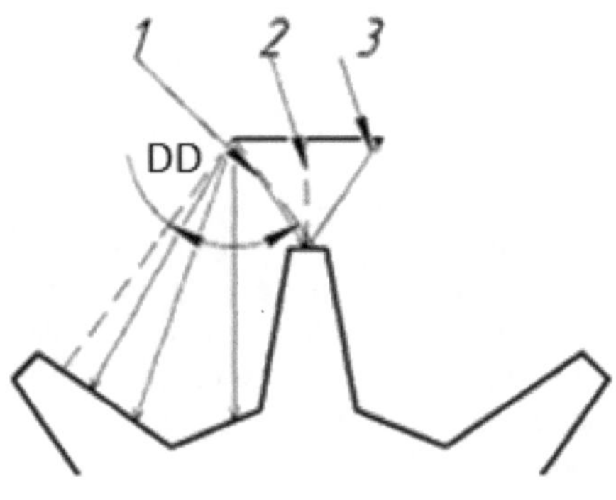

Figure 1. The mechanism of interaction.

Then investigated and determined by the intersection points of the rays with the surface. From the points of intersection, straight lines perpendicular to the surface under investigation are restored. In the figure 1, this is the perpendicular 2 for ray 1. After this, reflected rays with reflection angles equal to the angles of incidence with respect to the normal restored to the surface are constructed. The specified weight coefficients are also used for reflected rays. After that, the intersection points of the reflected rays and the sensor surface are calculated. For each reflected ray intersecting the emitter segment, a certain weighting factor is assigned according to its intersection point.

The resulting weight coefficients from the reflected rays are summed and stored. Further, there is a rotation of the functions describing the cogwheel, relative to its center by a certain step set in advance and everything is repeated anew. Based on the number of intersections of the reflected rays with the sensor surface, a reflected signal is constructed. The construction of a mathematical model in a two-dimensional coordinate system is a necessary step of simplification for the subsequent construction of a three-dimensional model in space (figure 2) [6].

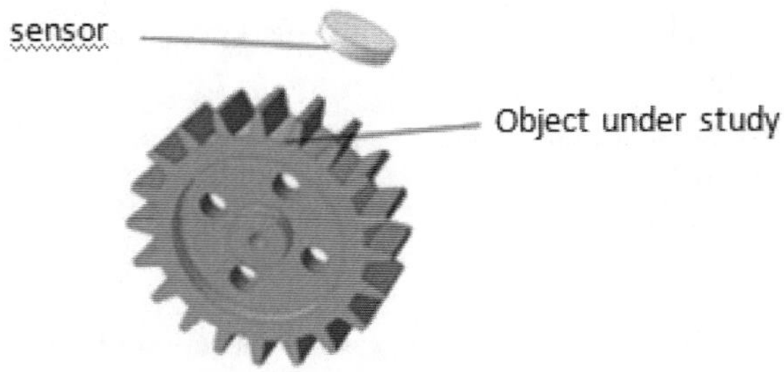

Figure 2. General view of the three-dimensional model.

Here the sensor from the segment is transformed into a circle. The tooth of the wheel is a trapezoidal parallelepiped. The radiation area of the sensor is divided by a rectangular grid into fragments with a certain step specified in advance. From each fragment rays are built with a certain angular pitch To take into account the direction diagram of the sensor, a weighting factor is used for a group of rays having the same solid angle. The weight coefficient decreases inversely proportional to the increase in the solid angle relative to the normal. To construct the interaction of beams with a cogwheel in space, the results obtained for the model in the plane are used, with appropriate changes for the possibility of their application in the spatial model.

Mathematical model allows: to create a theoretical basis for the planning of practical experiment, to justify the optimal relative position of the sensor and the test gear, consider the impact properties of the medium between the sensor and the object under study.

For the experiments, was developed an experimental stand, presented in figure 3, and manufactured by exemplary gear with the desired geometry of the tooth.

Figure 3. Appearance of the experimental stand.

The results of theoretical and experimental studies have shown that the received reflected signal after detection has the form of a quasi-bell-shaped pulse [5, 6, 7]. Information parameters on the basis of which the evaluation of the degree of wear of the gear wheel is made: signal amplitude, normalized signal edge lengths, and normalized signal duration, absence of signal.

Each of the types of wear affects certain information parameter, extracted from the reflected flux. So the absence of a signal indicates a broken tooth. Chipping of teeth affects the following information parameters:

- signal amplitude;
- normalized signal duration.

Abrasive wear of teeth affects the following information parameters:

- amplitude of the signal;
- the normalized duration of signal edges.

Thus, each of the types of wear of the teeth of the controlled wheel affects several measured information parameters.

3. Our device

The block diagram of the control system of the gears of the torque multiplier is shown in figure 4.

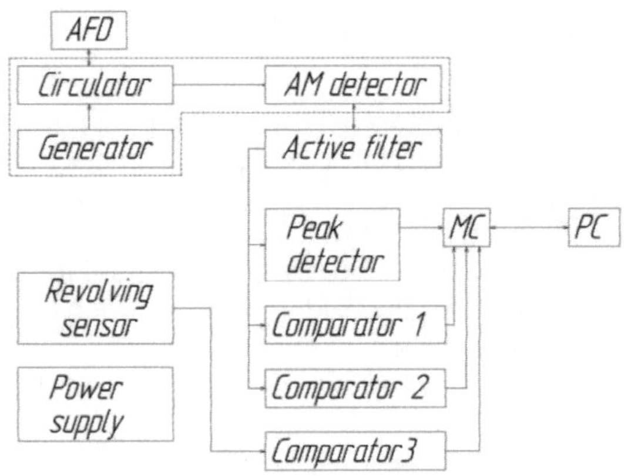

Figure 4. Structural diagram of the experimental device.

The signal reflected from the surface of the teeth is received by the antenna-feeder device (AFD). It is channeled through the circulator into an amplitude detector. The detected signal enters the active filter. In the active filter, high-frequency noise is separated and amplified. Next, the peak detector (PD), fixes the maximum value of the pulse. With the help of ADC, this value is read and stored in the memory of the microcontroller (MC). After filtering and amplification, the information signals fall into the comparators 1 and 2, which act as the drivers of the rectangular pulses.

In the MC, the duration of the pulses and their fronts is determined. With the help of a revolving sensor there is a period of rotation of the cogwheel, which is used to normalize the time parameters of the information signal. In addition, the presence of a reverse pulse and its time position allows each tooth of the controlled gear to be identified.

AFD is a coaxial transmission line, one end of which is connected to the generator module, and the other to a modified waveguide radiator. The modified waveguide radiator consists of two parts: a waveguide-coaxial junction and a waveguide with a dielectric filling, for example, quartz glass. The modified waveguide radiator is screwed directly into the gearbox case, perpendicular to the gear ring of the controlled wheel. The use of dielectric filling makes it possible to reduce the dimensions of the waveguide while maintaining the wavelength. Dielectric filling prevents splashing of oil into the waveguide and formation of carbon deposits there. The design of such a primary converter is shown in figure 5.

IOP Conf. Series: Materials Science and Engineering **302** (2018) 012015 doi:10.1088/1757-899X/302/1/012015

Figure 5. The design of the primary microwave converter.

The signal from the detector output has an amplitude of 0.2 V. To use the ADC effectively, it is necessary that the signal level aspire to the source of its reference voltage. The reference voltage for the ADC is 2.5 V. Therefore, to amplify the signal and to filter it from high-frequency noise, this monitoring system uses a third-order Butterworth active filter. The active filter is designed on two operational amplifiers. The monitoring system uses a Butterworth filter because it has the most smooth amplitude-frequency response within the bandwidth. Analog Devices ADA4062 are used as operational amplifiers.

Comparators are used to form the boundaries of measured intervals. The information signal arriving at the input of the first and second comparator from the output of the active filter has the form shown in figure 6. The comparator 1 is set to operate at a level of 0.1 from the amplitude of the reference information signal. Comparator 2 is set to operate at a level of 0.7 from the amplitude of the reference input signal (figure 6).

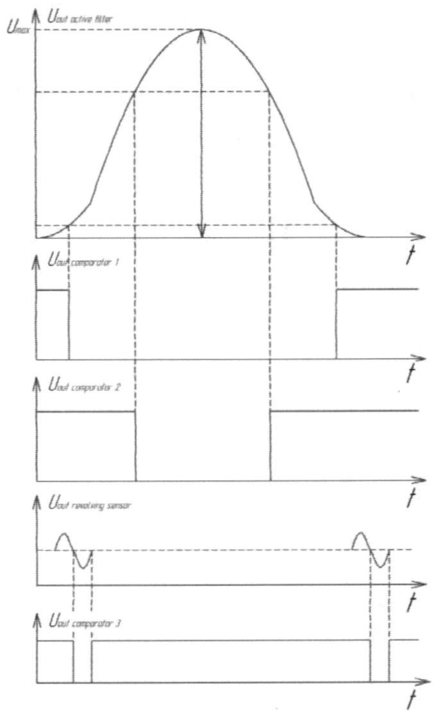

Figure 6. Diagrams explaining the operation of time-slice units.

IOP Conf. Series: Materials Science and Engineering **302** (2018) 012015 doi:10.1088/1757-899X/302/1/012015

4. Conclusion

The proposed system for monitoring the working state of the gear wheels of energy-loaded multipliers can be used for diagnostics in operation in any dynamic modes.

Ultimately, the proposed system allows automating the control process, objectively evaluating and recording its current operating state, and reducing the number of device preparations. The monitoring system generates a pre-emergency signal of the torque multiplier and thereby significantly reduces the probability of failure of a complex and responsible mechanical system and ensures operation of the torque multiplier of the steam compressor based on its actual technical state.

5. Acknowledgments

This work was carried out by lead performer of research and development effort with financial support of Russian Federation Ministry of Science and Education with the realization of Government decree №218 from 09.04.2010 according to contract about providing and using subsidy № 02.G25.31.0150 from 01.12.2015.

References

[1] Rusov V I 2012 Diagnostics of defects of rotating equipment on vibration signals (Perm) p 200

[2] Danilin A.I. e.a., Method for contact-free determination of technical state of gear wheels and device for its implementation. Patent RF, no. 2602488, 2016.

[3] Danilin A I and Neverov V V 2016 Methods of evaluation of the working condition of cogwheels of powerful gear systems Proceedings of the SSC RAS (Samara: SSC Press) vol 18 no.4(6) pp. 1265-1267

[4] Danilin A I and Neverov V V 2014 Methods of evaluation of the working condition of cogwheels of powerful gear systems Proc. All-Russian Conf. Actual problems radioelectronics and telecommunications (Samara: SSAU Press) pp. 138

[5] Danilin A I and Neverov V V 2016 Building the mathematical model of interaction microwave sensor with controllable elements cogwheels Russia Proceedings of the SSC RAS (Samara: SSC Press) vol 18 no. 4(6) pp. 1261-1264

[6] Danilin A I, Neverov V V and Zhukov S V 2015 Construction of a spatial model of the interaction of the microwave converter with the elements of the crown of the cogwheel Proc. Int. Conf. XIII Korolyov Readings (Samara: SSAU Press) pp. 138

[7] Danilin A I and Neverov V V 2016 Experimental device for monitoring the operating state of cogwheels. Proc. 3rd Int. Conf. Dynamics and vibroacoustics of machines (Samara: Samara University Press) pp. 222-224

IOP Conf. Series: Materials Science and Engineering **302** (2018) 012016 doi:10.1088/1757-899X/302/1/012016

Experimental Study of under-platform Damper Kinematics in Presence of Blade Dynamics

D Botto, C Gastaldi, M M Gola and M Umer[1]

DIMEAS Politecnico di Torino, Italy

[1]Corresponding author: muhammad.umer@polito.it

Abstract. Among the different devices used in the aerospace industries under-platform dampers are widely used in turbo engines to mitigate the blade vibration. Nevertheless, the damper behaviour is not easy to simulate and engineers have been working in order to improve the accuracy with which theoretical contact models predict the damper behaviour. Majority of the experimental setups collect experimental data in terms of blade amplitude reduction which do not increase the knowledge about the damper dynamics and therefore the uncertainty on the damper behaviour remains a big issue. In this paper, a novel test rig has been purposely designed to accommodate a single blade and two under-platform dampers to deeply investigate the damper-blade interactions. In this test bench, a contact force measuring system was designed to extensively measure the damper contact forces. Damper kinematics is rebuilt by using the relative displacement measured between damper and blade. This paper describes the concept behind the new approach, shows the details of new test rig and discusses experimental results by comparing with previously measured results on an old experimental setup.

1. Introduction

Some test rigs [1- 4] were developed in the past to mainly study the overall effect of the dampers on the blade in terms of vibration amplitude reduction and resonant frequency shift. These test rigs do not provide facility to analyse in depth the dynamics of the damper and its kinematics in terms damper/under-platform relative displacement. The effect of the dampers on the blade dynamics can be studied by introducing the two contact parameters known as normal and tangential contact stiffness. In [5] a method has been proposed to determine the contact stiffnesses of complex contact geometries. Test rigs were developed in [6 , 7] and used to measure the friction coefficient and the tangential contact stiffness, in controlled laboratory conditions. A Damper Rig [8] was built in 2008 to deeply investigate the behaviour of several dampers in terms of kinematics and force transmission characteristics [9, 10]. A numerical model of the damper/test-rig system was presented in [11], together with the first version of the contact parameters estimation procedure, subsequently improved in [12]. The Damper+Blade Rig presented here in this paper share with the Damper Rig the same ultimate goal: measuring the hysteresis cycle produced by the damper between the platforms. However, their structure and working principles are different. In the first rig (Damper Rig) the trajectory and amplitude of the input motion can be finely controlled by means of two piezo actuators connected to a dummy platform (no blades are present). However, the frequency range, the rig can explore is limited (lower than 160 Hz). In the second rig, (see section 2) the motion is achieved by exciting a blade at resonance; therefore, the variety of platform kinematics the user can investigate are limited. However, the rig is excited in the frequency range actually encountered by the blades during working. Furthermore, in this second rig it is possible to directly measure the effect of the damper hysteresis cycle on the blade amplitude of vibration. These two test rigs have complementary capabilities. Furthermore, they offer the possibility to test the same dampers, compare the resulting performance in terms of hysteresis cycles, and contact parameters. Section 4 contains such a comparison for the pre-optimized damper geometry presented in [12].

2. Test Rig Description

Figure 1 shows the top view, a section and details of the rig. The test rig is composed of three main sub-assemblies, namely a central block and two lateral blocks. The central block is made of two symmetric parts, (1A) and (1B). The two symmetric parts are fixed to an optical table through a base plate (2) with 32 vertical bolts. The two parts house a clamping mechanism. The clamping mechanism applies the centrifugal load on the blade. This clamping mechanism is made of two wedge blocks (3) and (4), with a slope of 1:10. These wedges convert a force F_B applied by main bolt (5) perpendicular to the longitudinal axis of the blade into a pushing force F_P acting along the longitudinal axis. Due to the 1:10 slope between the two wedges, the nominal force amplification A_{nom} between the bolt force F_B and the pushing force F_P is 10. A strain gages based load cell (6) measures the applied bolt force F_B. A thrust ball bearing (7) is placed between the main bolt and the load cell to allow their relative rotation. The subassembly composed of the main bolt, the thrust ball bearing and the load cell is enclosed in a casing (8) tightened to the central block with screws. Two rails of linear flat roller bearings (9) are introduced between the moving contact surfaces of wedge blocks to minimize the friction losses. The pushing force F_P is transmitted to the blade (11) through the pushing block (12). Between the pushing block and the upper wedge an aligning pin (13) allows small rotation between the upper wedge and the pushing block. The aligning pin allows to apply a uniform pressure at the blade root even if a small misalignment is present between the blade and the upper wedge. To avoid damage due to the high contact pressure, the self-aligning pin and its counterparts on the pushing block and the upper wedge have been designed with conforming contact surfaces. Pressure distribution on conforming contact surfaces has been deduced by the graphs found in [13]. The blade is inserted in the blade adapter (14) in which a bucket groove has been machined to match the blade root geometry. The blade adapter is pushed against the shoulders on the central block. As the blade adapter is a replaceable part, blades with a different root can be tested by simply changing the adapter according to any blade root geometry. Eight long stud bolts are inserted across the two parts of the central block, by means of through-holes (15), to hold the complete assembly altogether and stiffen the structure. These stud bolts co-operate with the vertical bolts to increase the rigidity of the overall structure. A complete description of this test rig is present in [14].

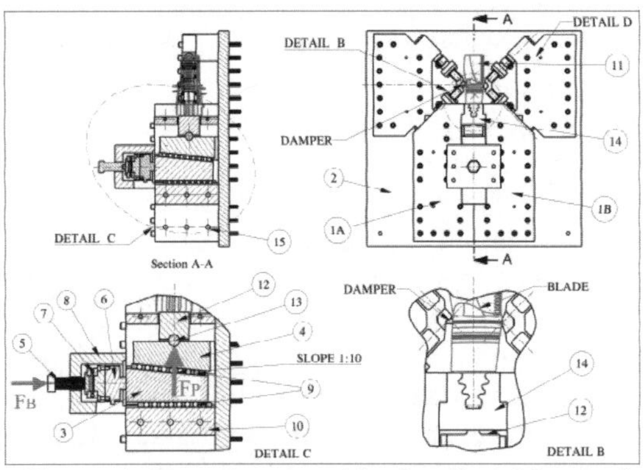

Figure 1. Test rig assembly details with a real turbine blade and two dampers.

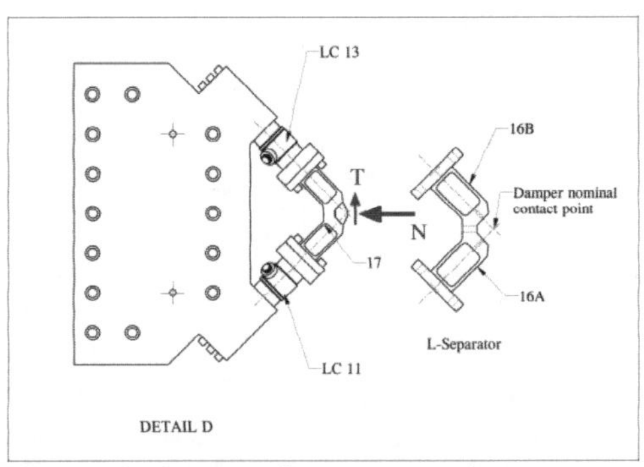

Figure 2. Contact force measuring system used to contact forces on one side of the damper.

Measuring the contact forces acting on the both dampers is a significant feature of this novel test rig. Two contact force measuring devices, one for each damper, are available to measure their respective force components. A device is shown in figure2 that has a two arms structure (16A and 16B) named L-Separator. The axes of the two arms are oriented at 90° to each other which intersect at the point of nominal contact between the damper and the pad. Two piezoelectric load cells LC (LC_{11}, LC_{13}) are coaxially mounted at the other end of both arms. The ratio between the longitudinal and transverse stiffness is around 100 to neglect any transverse force component. The contact forces N and T are deduced by the signals measured by each load cell. The selected load cells are able to measure both static and dynamic force component due to very low time drift factor (i.e. lower than 5 mN/s).

3. Experimental Procedure:

Experimental tests were performed on a dummy blade with a fir-tree root clamped with a constant pushing force of 50kN, a realistic value for similar blades in real turbine. Tests were carried out with different blade excitations at several centrifugal loads applied to the dampers in form of dead weights. For each testing condition a standard force-controlled Frequency Response Function (FRF) was obtained in which acceleration at the tip of the blade was measured. As FRFs alone do not give an insight into the damper behaviour, the present test rig is capable of linking a given point on a FRF to the corresponding damper-platform contact forces and relative displacements. In the following paragraph, all measured and reconstructed quantities are described. This set of quantities will then be used in section 4 to analyse the dampers and the blade behaviour together with their mutual influence.

The readings of the contact force measuring load cells mentioned in section 2 and shown in figure 2 give the complete in-plane force components at the cylindrical side of the dampers. As already mentioned the measurement chain composed of load cell and a charge amplifier allows measuring the static component of forces as well as their dynamic variation. The static components are necessary to reconstruct the damper equilibrium and to estimate friction coefficients. The tangential and normal to the contact force components, here termed T_{cyl} and N_{cyl} are computed from measured forces (F12 and F14, F11 and F13). Furthermore, the damper static equilibrium is reconstructed by neglecting damper mass inertia forces (lower than 0.1 N in the frequency range of interest) and therefore assuming contact and centrifugal forces to pass through one point, as described in [10] and shown in figure 3. The relative displacements at the damper platforms are measured by using a differential laser head. A complete set up of the test rig to measure the damper contact forces and relative displacement is shown in figure 4. These displacements are plotted against the tangential component of corresponding contact force to obtain the hysteresis loops. Hysteresis loops are commonly used to describe the characteristics of the frictional contacts. An example of these hysteresis with corresponding FRFs can be seen in figure 5.

4. Results and discussion

As shown in figure 6a the damper shifts the first bending mode resonant frequency from 410 Hz to 567.5 Hz in full stick condition ($|F_E|$=1N) and lowers the force-normalized blade amplitude of motion of one order of magnitude. This effect is produced by the dampers, which start to slip against the platforms and dissipate energy, therefore reducing the amplitude of the FRF peak and reduces the stiffness introduced by the contact in full stick. Details of all these experiments and methodology is discussed in [15].

4.1. Measured contact hysteresis loops explain puzzling measured FRFs

A general understanding of the damper effect on the blades is nowadays common knowledge in the turbine dynamics field. However, there are still a series of open questions which turbine designers and experimenters face. The first question is related to the experimental characterization of damper-blades systems. FRFs sometimes produce unexpected results which are hard to understand and justify. For instance, with reference to figure 6a and figure 7, it can be noticed how the amplitude and the frequency of the peak decreases for $|F_E|$ ranging from 1 N to 50 N, as expected. However, for |FE | ≥80 N the peak amplitude and frequency start increasing again. This behaviour is repeatable, and is observed at different loads on the damper. Without an insight on the damper behaviour this

IOP Conf. Series: Materials Science and Engineering **302** (2018) 012016 doi:10.1088/1757-899X/302/1/012016

phenomenon would remain unexplained. Thanks to the test rig capabilities it is now possible to relate the FRF shape to the corresponding damper behaviour. Figure 6b shows the hysteresis cycle at the cylindrical contact for two different values of $|F_E|$ (50 N and 100 N respectively). It can be seen that at $|F_E|$=50N the cylindrical interface is in micro-slip/at the onset of gross slip: this behaviour is easily simulated with the numerical code by considering the blade platform as rigid body. At $|F_E|$ = 100 N the contact is slipping and therefore dissipating energy, however during part of the cycle the damper appears to be "glued" to the platform. Specifically the tangential force changes its value but no relative movement between damper and platform is recorded. This behaviour introduces an additional stiffness to the system, thus producing an increase of the peak frequency.

Another experimental observation which puzzles designers during testing is the marked lack of repeatability for the same test conditions. The results reported in figure 6a and figure 7 are obtained starting from |FE|=1N and going up to $|F_E|$=100N. If the same set of experiments is repeated, without unloading the dampers, in the opposite order (from $|F_E|$=100N to 1N) the FRFs obtained in the micro-slip region ($|F_E|\leq$20N) will be markedly different from the first set. An example can be found in figure 5. The results of several test campaigns are reported in figure 5a to show that the results are not randomly distributed. They are in-fact quite repeatable, provided that the initial conditions are the same. In the micro-slip regime, the steady state solution is non-unique; it depends on the initial conditions, namely the value of the static component of contact forces as pointed out also by the Yang and Menq in [16]. More recently these authors showed in [17] that using the damper static balance equations to compute the normal preloads acting on the damper sides does not allow calculating a unique solution if the contact is partially stuck. These authors numerically demonstrated the effect of this under-determinacy on under-platform dampers in [10]. Using the test rig presented here, the numerical observations described above have found an experimental counterpart. By reconstructing the static equilibrium on the damper as explained in section 3, N-flat and T-flat and their point of application on the damper are determined. Figure 5c shows that the static component of the normal contact forces undergo a 50 % increase if the damper enters the gross-slip regime (i.e. if the blade is excited with $|F_E|$=100N). This produces different steady state hysteresis cycles (figure 5b): specifically, if the static normal components of contact forces are low, the damper is in micro-slip (blue). If the normal components increase the steady state cycle will be larger (black dashed line) but the damper will also remain glued to the platform for a portion of the cycle, similarly to what was happening at $|F_E|$=100N. The damper behaviour at the contact is perfectly compatible with the FRFs: the one corresponding to higher contact forces displays a higher peak frequency (due to the additional stiffness) and a larger amplitude.

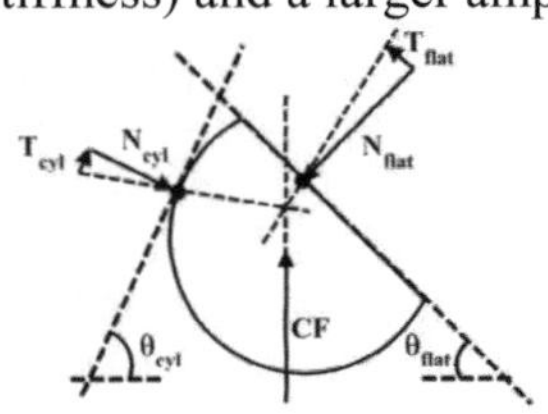

Figure 3. Damper force equilibrium.

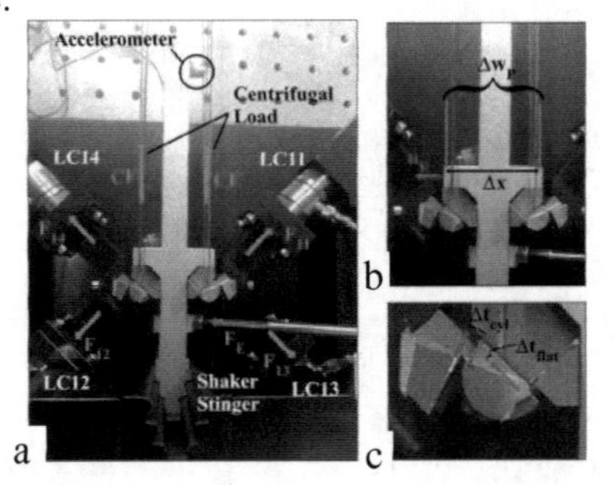

Figure 4 (a) Test rig setup, measured and applied force. (b) Laser setup to record the blade platform in-plane motion. (c) Laser setup to measure tangential relative displacement at the flat-on-flat and cylinder-on flat contacts.

4.2. Damper performance simulation

The numerical model used to simulate the damper behaviour, fully described and validated in [10], is shown in figure 8a. It represents the damper, modelled as a rigid body, between two platforms, with assigned motion. The contact is modelled through standard macro-slip elements. Inputs to the numerical model are: the platform motion signal and the friction contact parameters estimated in [18] and confirmed by the measurements in section 4.3. The resulting force equilibrium is reported in figure 8c. It compares well with the measured counterpart (see figure 8b), thus confirming the reliability of the measurements and the interpretation of experimental results.

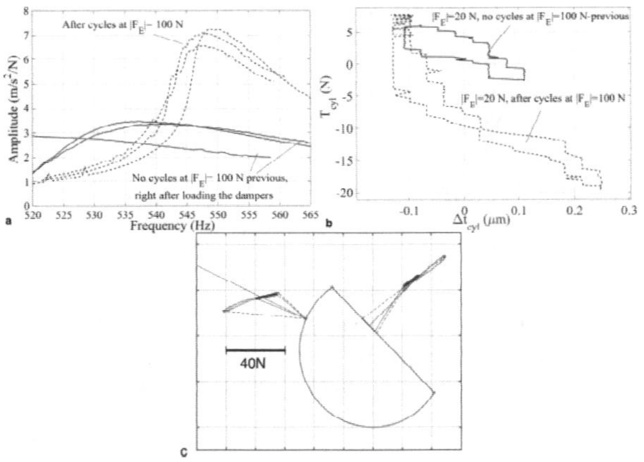

Figure 5 (a) FRFs (b) Hysteresis cycles at damper cylindrical contact(c) Equilibrium at CF=4.6kg F_E=20N initial condition comparison.

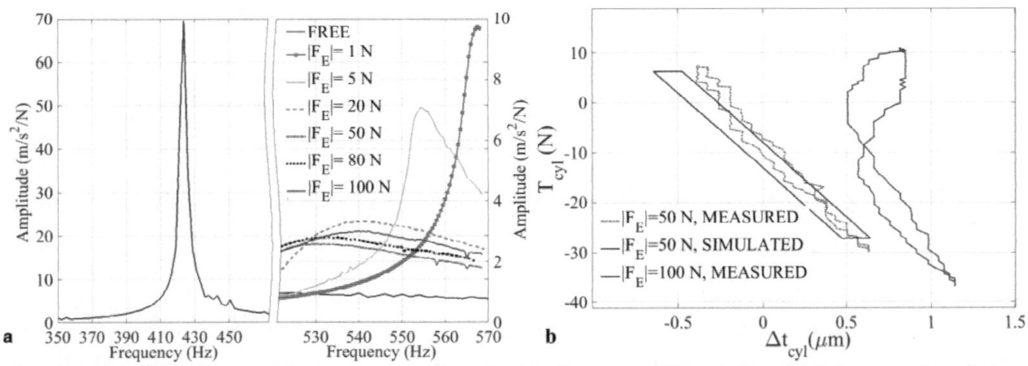

Figure 6(a) FRF free (without dampers) and with dampers CF= 4.6kg (b) Measured and simulated hysteresis at cylindrical side of the damper for working conditions shown in figure 5a.

4.3. Comparison of damper parameters and performance on two independent test rigs

To numerically validate the experimental results, these authors developed a contact parameter estimation technique in [10], further improved in [18], and based on the experimental observation of the piezo damper only test rig, specifically it was observed that; friction coefficient and contact stiffness are obtained from the ratio of tangential/normal force and corresponding slope of the hysteresis cycle respectively. The similar technique was adapted on the obtained experimental data on this novel test rig and computed results were compared with the results presented in [12] and [18] where exactly same dampers were tested. A case with $|F_E|$=100N (similar to the one in figure 6b) has been selected in order to ensure the cylindrical contact to reach gross-slip. Specific points on the hysteresis diagram have been marked by a symbol and a number to allow a cross-comparison. From the combined analysis of the T/N diagram and of the hysteresis at the cylindrical contact in figure 9 it holds:

- from markers 1 to 2 the cylindrical contact is sliding since the T_{cyl} /N_{cyl} is constant and equal to a maximum, μ_{cyl}=0.6, exactly the same value measured on the damper-only test rig [12];
- from markers 2-3 and 5-1 presents a sharp increase of tangential force with no movement and from markers 2-5 the state is stick, the measured tangential stiffness kt_{cyl} ≈35N/μm±9 N/μm, perfectly compatible with the 30 N/μm ±7 N/μm measured on the damper only test rig [18].

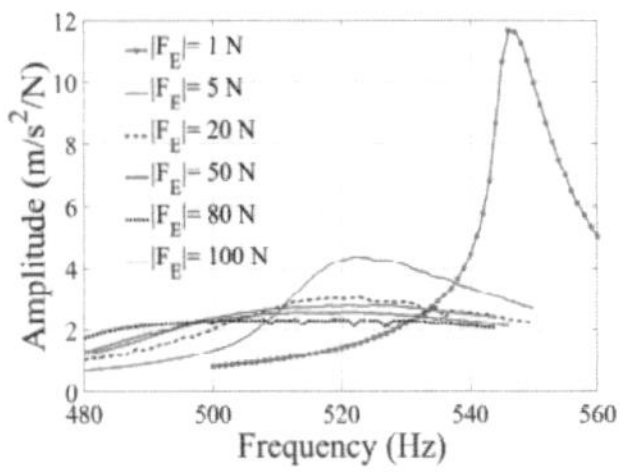

Figure 7 FRFs CF=2.6kg.

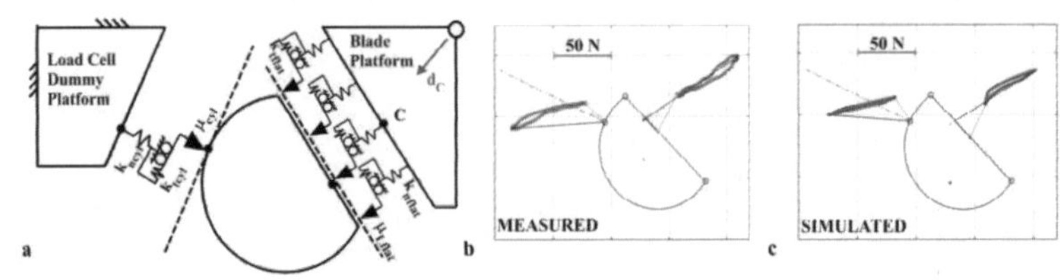

Figure 8(a) Damper numerical model (b, c) Measured and simulated forced equilibrium at F_E=100N, CF=4.6kg.

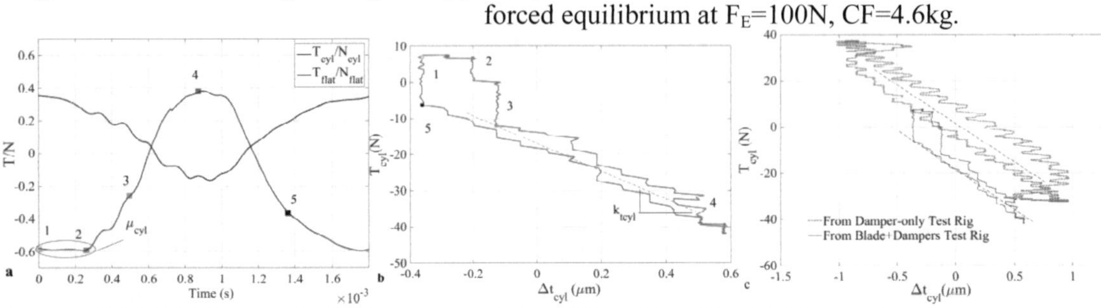

Figure 9 (a) T/N force ratio (b) Hysteresis at F_E=100N, CF=4.6kg (c) Hysteresis slope comparison of cylindrical to flat contact obtained on two independent and different test rigs.

5. Conclusion

The capabilities of the novel test rig allow an insight investigation into the damper behaviour. Namely for each point on a FRF the corresponding hysteresis at the contacts, force equilibrium and platform kinematics can be produced. This set of diagrams is particularly helpful in explaining unexpected phenomena (e.g. FRFs lack of repeatability) which often makes experimental characterization of damper-blade systems difficult. The test rig can be used to estimate friction contact parameters, which compare extremely well with the values found for the same damper on an independent test rig (damper-only test rig). These findings speak for the soundness of both test rigs and for the estimation procedure itself. The obtained experimental results are compared with the results obtained on a previous reference damper-only test rig. These results comparison highlight the importance of considering the blade mode shape to achieve a full understanding of the system dynamics and as a mean to achieve a better and faster damper shape optimization.

References

[1] Panning L, Sextro W and Popp K 2000 Optimization of inter-blade friction damper design. *In Proc.of ASME Gas Turbine & Aero engine congress and exhibition (Munich)*

[2] Petrov E P and Ewins D J 2002 Analytical formulation of friction interface elements of nonlinear multi-harmonic vibrations of bladed discs. *In Proc. of ASME Turbo Expo (Amsterdam)*

[3] Firrone C M 2009 Measurement of the kinematics of two under- platform dampers with different geometry and comparison with numerical simulation *J. of sound and vibration* **323** 313-33

[4] Bessone A, Toso F and Berruti T 2015 Investigation on the dynamic response of blades with asymmetric under platform dampers. *In proc. of ASME Turbo Expo Montreal (Quebec)*

[5] Botto D, Lavella M 2015 A numerical method to solve the normal and tangential contact problem of elastic bodies *Wear* **330-331** 629-35

[6] Filippi S, Akay A and Gola M M 2004 Measurements of tangential contact hysteresis during micro-slip *J. of Tribology* **126** 482–89

[7] Lavella M, Botto D and Gola M M 2013 Design of a high- precision, flat-on-flat fretting test apparatus with high temperature capability *Wear* **302** 1073–81

[8] Gola M M, Braga dos Santos M and Liu T 2010 Design of a new test rig to evaluate under-platform damper performance *In Proc. of ESDA (Istanbul)*

[9] Gola M M, Bragas Dos Santos M and Liu T 2012 Measurement of the scatter of under-platform damper hysteresis cycle: experimental approach. *In Proc. of ASME IDETC (Chicago)* pp 359–69

[10] Gola M M and Gastaldi C 2014 Understanding complexities in under-platform damper mechanics *In Proc. ASME Turbo Expo (Düsseldorf)* vol.7A

[11] Gola M M and Liu T 2014 A direct experimental-numerical method for investigations of a laboratory under-platform damper behaviour *Int. J. of Solids and Structures.* **51** 4245–59

[12] Gastaldi C and Gola M 2016 Pre-optimization of asymmetrical under-platform dampers *J. Eng. Gas Turbines Power* **139 (1)**

[13] Johnson K L 1985 *Contact Mechanics* (Cambridge: Cambridge University Press) pp114–18

[14] Botto D, and Umer M 2018 A novel test rig to investigate under-platform damper dynamics *Mech. Syst. Signal Proc.* **100** 344-59

[15] Botto D, Gastaldi C, Gola M M and Umer M 2017 An experimental investigation of the dynamic of a blade with two under platform damper *J. Eng. Gas Turbines Power* **140 (3)** 032504-9

[16] Yang B D and Menq C H 1998 Characterization of contact kinematics and application to the design of wedge dampers in turbomachinery blading: Part1-stick-slip contact kinematics *J. Eng. Gas Turbines Power* **120 (1)** 410–17

[17] Zucca S, Botto D and Gola M M 2008 Range of variability in the dynamics of semi-cylindrical friction dampers for turbine blades *In Proc. of ASME Turbo Expo (Berlin)* pp 519-29

[18] Gastaldi C and Gola M M 2016 Testing, simulating and understanding under-platform damper dynamics *In VII European Congress on Computational Methods in Applied Sciences and Engineering (Crete)*

IOP Conf. Series: Materials Science and Engineering **302** (2018) 012017 doi:10.1088/1757-899X/302/1/012017

The Present State of "Solid" under-Platform Damper Mechanics at AERMEC - POLITO

M M Gola[1], C Gastaldi

DIMEAS Politecnico di Torino, Italy
AERMEC LABORATORY - http://www.aermec-dimec.polito.it

[1]Corresponding author: muzio.gola@polito.it

Abstract. The numerical simulation of friction-damped blades in turbomachinery applications requires knowledge of contact parameters to be introduced in contact models. The equilibrium and kinematics of dampers are dominated by phenomena occurring at the two interfaces between the damper and the corresponding blade platforms. A precise knowledge of parameter values is vital to ensure trustworthy predictions of blades vibration frequency and of the available amount of amplitude damping. Test rigs developed at the AERMEC lab. during the last decade are presented, and reasons for their improvements are explained. These test rigs have the primary purpose to allow direct observations of contact forces and displacements, and the ensuing estimate of contact parameters. The reliability of the measurements and of their processing is demonstrated through a worked out example.

1. Introduction

As openly stated in [1] development of models for dynamic analysis of bladed disks with friction dampers are going on since over thirty years, yet, under-platform damper (UPD) models have been developed (at 2008) "*only for dampers of simple geometric shape for the so-called wedge or cottage-roof dampers*" Such cottage roof dampers have however been modified in a number of ways, witnessed by the many patents available, not all of them investigated in depth, at least openly, in the scientific literature. A deviation from the basic cottage-roof shape has been introduced in order to allow a safe contact to be obtained not only in ideal "In-"Phase" (IP) relative motion between blade platforms, but also in the "Out-of-Phase" (OOP) motion. This problem is evident, for instance, back in 1998 [2], when, while using a cottage roof damper, Šanliturk et al. introduce a correction factor in the contact stiffness to reflect the damper rolling effect due to edge contact. However, the obvious correction is to use a flat-curved solid damper having one of the two contact surfaces in cylindrical form, thus allowing a perfect mating with the under-platforms.

There are several papers on the subject, passing through the comparative study of wedge-shaped dampers compared with others having a couple of curved contacts [3, 4], through a comparison between cottage-roof and fully cylindrical [5, 6]. And through others which is not possible to properly mention in this short paper. Worth noticing that a mixed geometry damper having one surface flat and the opposite cylindrical finally appears. A notable work is [7], where the following useful information is found (quoted) "*The idea of combining the advantages of both mentioned damper types results in an asymmetrical damper with a flat contact surface on one side and a curved contact surface on the other. ... ... This damper type is already used in Siemens PG gas turbines since 1985 ...* ". A laboratory damper which, although apparently different but in fact having the same mechanics is studied in [8, 9, 10, 11]. Followed by the combined experimental-theoretical studies on proper flat-curved dampers performed in the AERMEC laboratory [12, 13, 14, 15, 16].

Paper [1] compares flexible-thin dampers with a cottage-roof solid damper having flat surfaces on both sides, and - on the same line - probably the latest work at the time of writing this paper [17] studies both theoretically and experimentally a test rig on which a flat-flat is again inserted. In all cases of this nature, two arrays of a large number of contact points are modelled so that they can take into account stick, slip, and separation during the vibration cycle. The last feature allows lift-off to be modelled, which entails, if lift-off does not occur, satisfaction of equilibrium to in-plane rotation of the damper and consequent correct position of the resultant forces on the opposite contact surfaces.

As a matter of fact, the damper can undergo two types of rolling / lift-off with concentration of contact at the edges of the surface: one due to intentional angle mismatch which can occur in OOP, as described above, the second due to the position of the resultant contact force which reaches the contact surface edge, as it may happen in IP platform motion if measures have not been taken to avoid it [2, 11, 15]. Such measures are clear and easily governed ex-ante, and independently of the amount of platform displacement, if a flat-curved damper is adopted [15]. Not so in the case the flat-flat damper.

Although theoretically necessary on the ground of contact indeterminacy, and although at AERMEC lab. we maintain that it should be adopted on the grounds of sound design and mechanical predictability, we must honestly acknowledge that flat-curved contact surface, usually but not exclusively in the cylindrical shape having a line contact with the platform, is deemed really necessary. Patent EP1249576 of 2002 [17] claims that a wedge shaped damper having contact faces subtending an angle lower - about 10 degrees - than the platforms' angle allows a further vibration damping mechanism (called "*percussive alternate contact*") to arise from the oscillations of the damper, added to the main sliding mechanism. It is all too evident, however, that this should entail the introduction of a percussive dissipation model in the simulation software. However, already in [2], but also in [11], it was seen on experimental grounds that the overall damping capability is greatly reduced by damper rolling, a fatal blow to its desirability.

This problem was implicitly known to the inventor of a 1992 US Patent [18], where the claim concerns a damper, wedge shaped in cross section and provided with a pad on one side and two pads on another side. The pads are located such that they do not lift off the platform surfaces for conditions up to the maximum coefficient of friction even in the most severe IP motion [18].

What differentiates the AERMEC approach from all others in studies of damper mechanics and optimisation is the adoption of an original and innovative test rig named "Piezo+Damper Rig I". Devices simulating bevelled platform surfaces are arranged in such a way that one is put in OOP or IP motion (or their combinations) by a couple of orthogonal piezo actuators, while the opposite one is supported by an orthogonal couple of load cells. This design allows platform displacements and forces to be directly measured; a laser arrangement is added that measures damper displacements and rotations, relating them to platform displacements and contact forces [9, 10, 11]. It was, and still is, believed that this is the most accurate way to determine contact parameters (normal and tangential contact stiffness, friction coefficient) and check, at the same time, the observed damper kinematics against the predictions of any numerical model.

All other approaches rely on measuring the frequency response of blades, where the FRF of the blade is the object of experimental observation, and the results are compared with those from a model: this is the case from the early pioneering 1980 studies [19] through the decades down to the latest 2017 achievements [20]. Contact parameters are typically tuned so that the experimental FRF of a damped blade matches its numerical counterpart. This procedure is very debatable, as it is well known that there exist multiple combinations of the four normal and tangential contact stiffness values which produce a satisfactory match between a given experimental FRF and its numerical counterpart [21].

It is interesting to remark a growing interest in using damper kinematics as a means to better understand contact conditions. In [20] a laser measurement of damper kinematics is also introduced much in the way already adopted by AERMEC in 2010 [10] and used throughout its following papers [11, 12, 13, 16]. Allowing for the differences in approach, what nonetheless makes this Imperial

College and Rolls Royce initiative [20] akin to the AERMEC line on damper testing are the twin objective, i.e., to improve the fidelity of damper modelling and to rigorously assess processes needed for reliable predictions.

2. Test Rigs available at AERMEC

Table 1 summarises the main features of the test rigs in use at the AERMEC lab.

Figure 1 shows the "Piezo+Damper Rig I" in the initial 2009 version, where the load cells on the right dummy platform are loaded through a "tripod" arranged parallel to the direction of the piezo actuators The load cells are pre-loaded by means of a wire pulled through a pulley and dead-weights (not visible) arrangement [11]. The piezo actuators, hardly visible on the left and up, realize the platform motion through decoupling parallelograms. Figure 2 shows a detail of load cells in the later 2013 version, stiffened for higher operating frequency and tripod rotated by 45°. In "Piezo+Damper Rig I" the load measuring load cells are Dytran 1051V2, subject to leakage and thus unable to measure the static force component, thus requiring lengthy and annoying load removal procedure [11] in order to get the total force value which is necessary for equilibrium calculations. In the later rigs these load cells were replaced by KISTLER 9323AA load cells with charge amplifiers which do not filter out the static component of the measured force.

Table 1. Capabilities of AERMEC test rigs for solid dampers

	Piezo+Damper Rig I (designed 2009)	**Damper+Blade Rig (designed 2014)**	**Piezo+Damper Rig II (designed 2016)**
Scope	Stand-alone damper behaviour	Damper coupled with a blade (real or dummy)	Stand-alone damper behaviour
Input motion	All plane motions, 2 piezo actuators	Constrained by blade-platform kinematics, shaker excitation	One-directional motion, adjustable direction and amplitude, 1 piezo act.
FRF of blade	no	yes	no
Frequency range	up to 150 Hz (100 Hz practical)	Blade FRF up to 4500 Hz, damper mechanics tested up to 500 Hz	up to 500 Hz (target, to be tested)
Contact pressure (flat-on-flat)	up to 2 MPa	up to 6 MPa	up to 6 MPa
Temperature	Room temperature	Room temperature	Room temperature

Figure 1. View of the “Piezo+Damper Rig I” in the initial 2009 version

Figure 2. Tripod rotated by 45° during damper rotation observation

In 2014 AERMEC started the design of new test rig mechanically assembled the following year [22], presently used for investigations on real turbine blades [23], with capabilities and main results described also in [24]. In short, this test rig named “Damper+Blade Rig” allows a combination of FRF measurement on a blade, in principle not a dummy but a real customer’s blade, with the direct measurement of forces transmitted between damper and platforms, gaining the advantages of both methods.

A third rig at an advanced commissioning stage, named “Piezo+Damper Rig II”, is substantially an improved version of the first one, however incorporating the latest improvements. All the rigs described in this paper are intended for solid dampers, while strip damper experimental investigations require a completely different approach.

3. Experimental capabilities

Table 2 summarises the parameters which are either directly measured or derived through equilibrium and kinematic calculations, together with the standard deviations of their estimates. Within the narrow limits of this paper it will suffice to prove tangibly that measurements can be very accurate and reliable, thus making the ensuing determination of contact parameters reliable. As described in [13], by pointing two laser beams on the bottom of the damper, Figure 3a, its rotation and its displacements in a direction along the beams can be determined, while the orthogonal component comes from the constraint condition requiring the contact point to move along the platform surface direction.

Table 2. parameters either observed or derived on the basis of damper equilibrium and kinematics

	Estimated parameters	Standard deviation values	
OBSERVED	• Right Contact Forces	• 3% (≈0.5 N)	forces
OBSERVED	• Platforms relative displ. • Damper rotation • Damper vertical displ. • Damper tangential relative displ. at the contact	• 0.08 μm • 5% (≈0.6·10-4 rad) • 0.08 μm • 0.08 μm	kinematics
DERIVED	• Left Contact Forces + force application point • T/N force ratios	• 3-5% (≈0.7-0.9 N), 0.5 mm • <4% (right), <9% (left)	forces

IOP Conf. Series: Materials Science and Engineering **302** (2018) 012017 doi:10.1088/1757-899X/302/1/012017

	• Damper tangential relative displ. at the contact	• 5% (≈0.5 μm)	kinematics

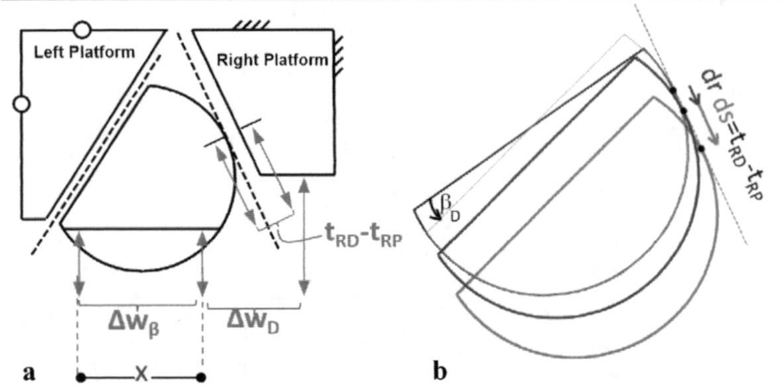

Figure 3.

a) laser measurement points for damper motion, absolute (**Δw**) and relative to the right platform (**t_{RD}-t_{RP}**)

b) sketch of contact displ. due to rolling (dr) and damper tangential displ. due to spring/sliding (ds= **t_{RD}-t_{RP}**)

From damper rotation and displacement kinematics the motion at the contact is derived [13], as explained by Figure 3b. An example of such results is shown in Figure 4a, where the red line represents the tangential relative damper-platform displacement. The same **t_{RD}-t_{RP}** displacement is directly observed by pointing to laser beams at the contact location, Figure 3a. As shown in Figure 4b the observed and the derived displacement do match extremely well. Thus the tangential displacement vs. the corresponding variation of tangential force, observed as well, gives a very reliable estimate for the tangential contact stiffness (k_{tR}=30±7 N/μm) for a cylindrical contact length of 8 mm. The same values are obtained for this damper when coupled to a blade on the "Damper+Blade Rig" [23, 24].

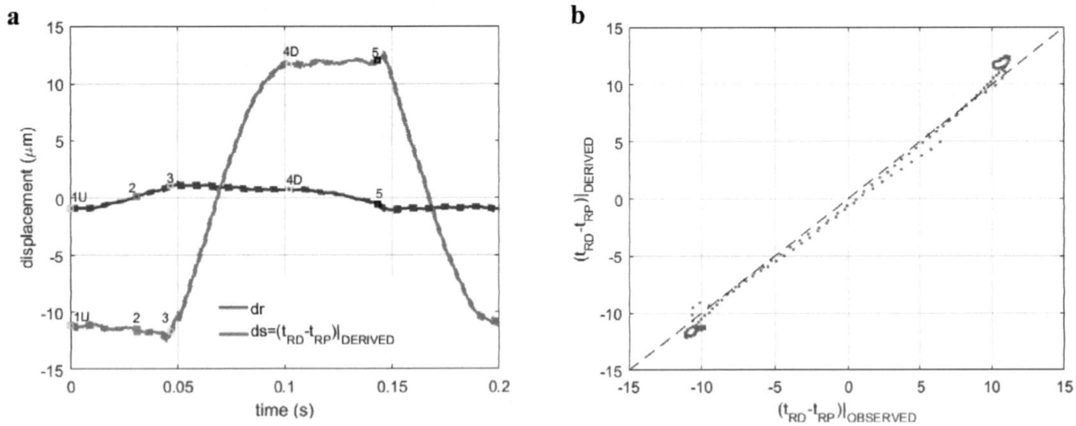

Figure 4. a: dr and ds= **t_{RD}-t_{RP}** as derived from laser measurements of damper kinematics
b: comparison of kinematically derived ds= **t_{RD}-t_{RP}** vs. its observed counterpart

4. Conclusion

The test rigs developed so far for the simultaneous and direct experimental observation of forces exchanged between a solid damper and the contacting blade platforms and damper kinematics, have been presented. At the same time, the main reasons for their improvements have been highlighted. Methods currently in use for the determination of contact parameters have been concisely reviewed and commented. It has once more been stressed that contact parameter tuning through FRFs is an undetermined problem, since the same result may depend on parameter combinations. One example for all has been put forward, showing how direct experimental observation can be devised to the purpose of determining tangential contact stiffness. The tangential force directly observed by means of load cells is related to the corresponding tangential damper-platform relative displacement which is either observe directly or derived from an independent reconstruction of damper kinematics. Both methods produce results with a striking degree of similarity. This cross-confirmation between two independent determinations of the same parameter makes the whole process very trustworthy. References are provided which allow the reader to explore how similar methods can be applied to the determination of all other contact parameters.

References

[1] Petrov E P 2008 Explicit finite element models of friction dampers forced response analysis of bladed discs *J. Eng. Gas Turbines Power* 130 022502-11

[2] Şanlitürk K Y, Ewins D J and Stanbridge A B 1999 Under-platform dampers for turbine blades: theoretical modelling, analysis and comparison with experimental data *J. Eng. Gas Turbines Power* 123 919-929

[3] Jareland M H 2001 A parametric study of a cottage roof damper and comparison with experimental results. *Proc. ASME. Turbo Expo: Power for Land, Sea, and Air June 4–7 2001 New Orleans 4* V004T03A038

[4] Jareland M H 2001 Experimental investigation of a platform damper with curved contact areas *Proc. ASME Design Eng. Technical Conference September 2001 Pittsburgh* DETC2001/VIB-21391

[5] Panning L, Sextro W and Popp K 2000 Optimization of interblade friction damper design *Proc. ASME Turbo Expo: Power for Land, Sea, and Air, May 8-11 2000 Munic* 4 V004T03A068-8

[6] Panning L, Sextro W and Popp K 2003 Spatial dynamics of tuned and mistuned bladed disks with cylindrical and wedge-shaped friction dampers *Int. Jour. Rotating Machinery* 9 219–228

[7] Panning L, Popp K, Sextro W, Götting F, Kayser A and Wolter I 2004 Asymmetrical under -platform dampers in gas turbine bladings: theory and application *Proc. ASME. Turbo Expo : Power for Land, Sea, and Air June 14-17 2004 Vienna* 6 269-280

[8] Firrone C M 2009 Measurement of the kinematics of two under- platform dampers with different geometry and comparison with numerical simulation *J. of sound and vibration* 323 313-333

[9] Gola M M, Braga dos Santos M and Liu T 2010 Design of a New Test Rig to Evaluate Under-Platform Damper Performance *Proc. 10th Biennial Conf. ASME Engineering Systems Design and Analysis July 12-14 2010 Istanbul* 5 85-94

[10] Gola M M, Braga dos Santos M and Liu T 2012 Measurement of the scatter of under-platform damper hysteresis cycle: experimental approach. *Proc. ASME Int. Design Engineering Technical Conf. and Computers and Information in Engineering Conf. August 12-15 2012 Chicago* 1 359-369

[11] Gola M M and Liu T 2014 A direct experimental–numerical method for investigations of a laboratory under-platform damper behaviour *Int. Jour. Solids and Structures* 51 4245–59

[12] Gola M M and Gastaldi C 2014 Understanding complexities in under-platform damper mechanics *Proc. ASME Turbo Expo Power for Land, Sea, and Air June 16-20 2014 Dusseldorf* 7A V07AT34A002-12

[13] Gola M M and Gastaldi C 2014 Latest investigations on under-platform damper inner mechanics *Int. Conf. on Science and Technology Problems and Prospects of the Propulsion Engineering Development June 25-27 2014 Samara* 238-249

[14] Gastaldi C and Gola M M 2016 On the relevance of a microslip contact model for under-platform dampers *Int. Jour. Mechanical Sciences* 115-116 145-56

[15] Gastaldi C and Gola M M 2016 Pre-optimization of asymmetrical under-platform dampers *Jour. Eng. Gas Turbines and Power* 139 012504-9

[16] Gastaldi C and Gola M M 2017 Estimation accuracy vs. engineering significance of contact parameters for solid dampers *Global Power and Propulsion Society Forum January 16-18 2017 Zurich* 1-7

[17] Stuart Y and Goodman P J 2002 Vibration damper for a gas turbine *Patent EP1249576*

[18] Bobo M 1992 Vibration damping of gas turbine engine buckets *US Patent* 5,156,528
[19] Griffin J H 1980 Friction damping of resonant stresses in gas turbine engine airfoils *J. Eng. Power* 102 329-333
[20] Pesaresi L, Salles L, Jones A, Green J S, Schwingshackl C W 2017 Modelling the nonlinear behaviour of an under-platform damper test rig for turbine applications *Mech. Syst. Signal Pr.* 85 662–679
[21] Gastaldi C and Gola M M 2015 A random sampling strategy for tuning contact parameters of under-platform dampers *Proc. ASME. Turbo Expo: Power for Land, Sea, and Air June 15-19 2015 Montreal* 7B V07BT33A004-12
[22] Gola M M 2015 Overview of test rigs at AERMEC Lab *4th Workshop on Joints Modelling October 2015 Dartington UK*
[23] Botto D, Umer M, Gastaldi C and Gola M M 2017 An experimental investigation of the dynamic of a blade with two under platform dampers *Proc. of ASME Turbo Expo June 26-30 2017 Charlotte NC USA*
[24] Botto D, Umer M, Gastaldi C and Gola MM 2017 Experimental study of under-platform damper kinematics in presence of blade dynamics *Int. Conf. on Aerospace Technology, Communications and Energy Systems September 28-30 2017 Samara*

IOP Conf. Series: Materials Science and Engineering **302** (2018) 012018 doi:10.1088/1757-899X/302/1/012018

System Statement of Tasks of Calculating and Providing the Reliability of Heating Cogeneration Plants in Power Systems

V V Biryuk[1], A B Tsapkova[1], E A Larin[2], M Y Livshiz[3] and L P Sheludko[3]

[1]Samara National Research University, Russian
[2]Saratov State Technical University, Russian
[3]Samara State Technical University, Russian

Email: aidan@ssau.ru

Abstract. A set of mathematical models for calculating the reliability indexes of structurally complex multifunctional combined installations in heat and power supply systems was developed. Reliability of energy supply is considered as required condition for the creation and operation of heat and power supply systems. The optimal value of the power supply system coefficient F is based on an economic assessment of the consumers' loss caused by the under-supply of electric power and additional system expences for the creation and operation of an emergency capacity reserve. Rationing of RI of the industrial heat supply is based on the use of concept of technological margin of safety of technological processes. The definition of rationed RI values of heat supply of communal consumers is based on the air temperature level iside the heated premises. The complex allows solving a number of practical tasks for providing reliability of heat supply for consumers. A probabilistic model is developed for calculating the reliability indexes of combined multipurpose heat and power plants in heat-and-power supply systems. The complex of models and calculation programs can be used to solve a wide range of specific tasks of optimization of schemes and parameters of combined heat and power plants and systems, as well as determining the efficiency of various redundance methods to ensure specified reliability of power supply.

1. General provisions

One of the required conditions for the creation and operation of heat and power supply systems is to ensure a specified (optimal) reliability of energy supply to consumers. Reliability is -understood as ensuring the quality of electricity and electricity supply to consumers in accordance with energy consumption schedules over a certain period of time and avoiding dangerous situations.

CCP and heat-and-power supply systems (STE) on their basis are attributed to the class of structurally complex multi-functional reconstructed systems

The system statement of the tasks of calculating and providing reliability includes:

- Selection and justification of unit, complex and integral reliability indicators (RI) in accordance to the functional purpose of the researched thermal power plants;
- Development of probabilistic methods and models for the calculation of the RI elements of power equipment, heat power plants in general, and heat-and-power supply systems based on them;
- Justification of the rationed values of RI of heat and power supply systems;

IOP Conf. Series: Materials Science and Engineering **302** (2018) 012018 doi:10.1088/1757-899X/302/1/012018

- Formulation and solution of the task of optimal distribution of RI between the elements of the heat energy installations;
- Development of probabilistic methods and models for calculating the required emergency power reserve in the system to provide a set (rationed) RI value;
- Development and justification of methods for increasing (optimizing) the RI of thermal power plants in general;
- Formulation and solution of tasks of economically most advantageous methods for increasing the reliability of heat power installations.

Let us consider some aspects of solving systems of these problems.

2. Selection and justification of RI of installations and systems

The known complex and integral indicators of reliability often do not allow to correctly solve problems of maintaining the reliability of power supply of consumers. For refinement of them proposed a probabilistic system index is suggested – the efficiency coefficient of the functioning of Kef (t), which takes into account both operational failures, including partial failures, and functional failures at time t.

$$K_{эф}(t) = P\{t, N_p \geq N_s\} \cdot P\{t, N_s \geq N_{mp}\} \tag{1}$$

Where $P\{t, N_p \geq N_s\}$ is the probability that at the moment of time t the level of working capacity of N_p will not be below some given level Ns; $P\{t, N_s \geq N_{mp}\}$ - the probability that the power level N_s will not be below the required level. The first cofactor of the expression (1) is defined by the structure of the STE, the RI of its individual elements, as well as commonly used reservation methods, etc. The second cofactor is defined by the schedules of electrical and thermal energy consumption. The application of this index determines the possibility and expedienc of decomposition methods application when calculating the reliability of structurally complex STEs, allows analyzing STE both from the position of its structure and from the point of view of providing energy consumption schedule.

3. Rationing of RI of power supply systems

The optimal value of the power supply system coefficient F is based on an economic assessment of the consumers' loss caused by the under-supply of electric power and additional system expences for the creation and operation of an emergency capacity reserve. Rationing of RI of the industrial heat supply is based on the use of concept of technological margin of safety of technological processes. The definition of rationed RI values of heat supply of communal consumers is based on the air temperature level iside the heated premises. Herewith, failures of heat supply systems are ranged in accordance to the disturbance of heat supply during the coldest period of the heating season at air temperature inside the premises: 1 rank of failure – $t_n \leq 0°$ C ; 2 rank - $t_в < 10°$ C ; 3 rank - $t_в < 15°$ C ; 4 rank - $t_в < 20°$ C . The rationed values of RI of the heat supply system for the first and second failure ranks are the probabilities of reaching the corresponding temperatures and are equal to $P_1 = 0.97$ and $P_2 = 0.86$ [1]. The third and fourth failure ranks do not lead to large material damage, but identify the extent of comfort living. Consequently, the third and fourth failure ranks are rationed by the values of the system efficiency coefficients, which are, respectively, $k_{E3} = 0.97$ and $k_{E4} = 0.89$ [1].

4. Probabilistic model for the calculation of the RI elements of power equipment

Deterministic methods for calculating the RI of power equipment elements, based on the calculation of relative or absolute strength and durability reserves, have been widely used, but do not take into account a number of real operational factors. In actual operating conditions of power equipment the calculation of the RI of power equipment elements should be based on probabilistic methods that allow for statistical dispersion of the strength and loading characteristics using the theory of random variables and random functions [2, 3].

The basis of the mathematical model for calculating the reliability of WHRB and its failure-free operation is:

- WHRB is divided into zones, each one contributes to the probability of failure: economizing, evaporating and overheating zones;
- Effective voltages are random variables, the spread of which with respect to the expected mathematical value is determined by many factors [4];
- Operational characteristics of the materials used (long-term strength limit and endurance limit) are random variables distributed according to normal or log-normal laws;
- The assessment of failure-free operation is carried out from the condition of determining the probability of not exceeding the effective voltages over the permissible values in each design section.

The failure-free performance should be understood as the probability of not exceeding the loading value $X(t)=\{x_1,...,x_i,...,x_{\hat{i}},t\}$ strength limit value $Y(t)=\{y_1,...,y_i,...,y_{\hat{i}},t\}$, that is, the random working capacity function (WCF) Z (t) = X (t) -Y (t) hits a range of negative values. If there are $m(m = I,\hat{m})$ zones in WHRB with the same reliability, and the failure-free performance of each is characterized by $n(n = I,\hat{n})$ WCF, then the condition of a failure-free operation will be written as

$$Z_{n,m}(t)=\left\{\min\left[Y_{nk}(t)-X_{nk}(t)\right]\right\}_m>0,\ m\in\hat{m};\ n\in\hat{n} \tag{2}$$

Where k is the number of analysed WHRB zones.

Therefore, the working capacity function $z_{n,m}(t)$ is a random function, fluctuating around some random stationary level $z^{cm}_{n,m}(t)$, the position of which is determined by time-constant deviations of the arguments Δx_i and Δy_i from mathematical expectation, $M(Z_{n,m}(t))$ calculated with nominal values x_i^H and y_j^H.

In a general case when several WFC determine failure-free performance, that is, when $\hat{n}>1$, the probability of the failure-free operation will be written as

$$P(t)=P\{Z_{n,m}(t)>0\}_m,\ m\in\hat{m} \tag{3}$$

If WCF $Z_{n,m}(t)$ are independent, then $P(t)=\prod_{m=1}^{\hat{m}}\prod_{n=1}^{\hat{n}}P_{nm}\{Z_{n,m}(t)>0\}$ (4)

The value $P_{nm}\{Z_{n,m}(t)>0\}$ is found with formula

$$P_{nm}\{Z_{n,m}(t)>0\}=P\{Z^{cm}_{n,m}(t)>0\}P_{nm}(t)=p\{Z^{cm}_{n,m}(t)>0\}\exp\{-\nu_{n,m}t\} \tag{5}$$

Where the value $P\{Z^{cm}_{n,m}(t)>0\}$ describes the probability that the random value of the effective voltage will not exceed permissible values at time t, and value $P_{nm}=\exp\{-\nu_{n,m}t\}$ corresponds to the probability that the random WCF $Z_{n,m}(t)$ will not be ejected into the negative range of values during a period of time t.

The probability that for the mth section of the WCF (omitting index m) the effective voltages $\sigma(t)$ will exceed the permissible $\bar{\sigma}(t)$, that is $Z(t)\geq 0$ will be defined as

$$F[\sigma(t)>\bar{\sigma}(t)]=\iint f(\sigma,\bar{\sigma},t)\,d\sigma\,d\bar{\sigma} \tag{6}$$

Where $f(\sigma,\bar{\sigma},t)$ is joint differential probability-distribution function of reduced voltage and long-term strength.

Since the condition for operability of the WHRB parts has the form of $\xi(t)=\sigma(t)-\bar{\sigma}(t)<0$, then, using it to define integration ranges (10), we get

$$P\{\sigma(t) < \overline{\sigma}(t)\} = 1 - F[\xi(t)] = \int_{-\infty}^{\infty}\int_{-\infty}^{\xi+\overline{\sigma}} f(\sigma,\overline{\sigma},t)\,d\sigma\,d\overline{\sigma} \tag{7}$$

In the event of values σ and $\overline{\sigma}$ being distributed according to the normal laws, the probability of failure-free operation will be written as

$$P\{\sigma < \overline{\sigma}, t\} = 0,5 - \Phi\left[\frac{M(\sigma(t)) - M(\overline{\sigma(t)})}{\sqrt{S_{\sigma}^{2}(t) + S_{\overline{\sigma}}^{2}(t)}}\right] \tag{8}$$

Where Φ[β] is Laplace's function; $M(\sigma(t)), M(\overline{\sigma(t)}), S_{\sigma}^{2}(t), S_{\overline{\sigma}}^{2}(t)$ are, correspondingly, expected values and dispersions of operating stationary and ultimate stress at the moment of time t [5].

5. Method of calculating the structural reliability of heating CCPs

For combined thermal-clamping gas-vapor CHPP, generally designed for electric power, pocess steam and hot water generation, the correlation of produced energy types is inherent. This leads to the necessity of calculating the interdependent RIs for the release of each type of generated energy.

This method is based on the Markov model of the evolution of system's conditions, which is based on a description of its functioning by means of a Markov process with a discrete set of conditions and continuous time. The basis for the use of the Markov model is the adoption of exponential laws for the distribution of operative and recovery time.

Real thermal-clamping CCTs are structurally complex systems. Therefore, to calculate the RI of CCT, its structural scheme is represented as n (i = 1, n) consistently connected aggregate blocks, each of which contains Li = Mi + Pi elements, where Mi and Pi are respectively the number of working and reserve elements of the i-th Block. The matrix of possible plant conditions is represented as a set of vertices and arcs. The condition vertices are described by (n + k) -dimensional graph of the form

$$\{X_{1}^{z}, X_{2}^{z}, \ldots, X_{n}^{z}, X_{p}^{z}, \ldots, Q_{p}^{z}, N_{p}^{z}, B_{m}^{z}\} = \{X_{n}^{z}, N_{n}^{z}, Q_{n}^{z}, B_{m}^{z}\}$$
$$z = 1, Z; i = 1, n; r = 1, R \tag{9}$$

Where Z is the number of possible plant's conditions; n – number of selected aggregate blocks; R-number of heat carriers produced by the plant; X_{i}^{z} - number of failed elements of the i-th block in state z; $Q_{P_r}^{z}, N_{P}^{z}$ - respectively, the output capacity of the r type of heat carrier in condition Z and electric power; B_{m}^{z} - firing rate in CCT for producing electric and heat power.

Procedures for the formation of a set of conditions of a power plant, the matrix of transition intensities, and solutions of a system of differential (algebraic) equations for the change in the probability of conditions are formalized and presented as a set of calculated programs [5,6]. To each possible condition of the plant correspond to certain the levels of working capacity of the power plant for the release of electric N_{p}^{z} and thermal $Q_{P_r}^{z}$ energy correspond. Presenting the required conditions for the operation of the CCT in the form of a deterministic piecewise constant function $N_{sj}(j = 1, G)$ and $Q_{sRj}(j = 1, G)$, the set of CCT's conditions can be divided into subsets. One of them $(z \in S^{+})$ can be described by operability levels $N_{p}^{z} \geq N_{s}$ and $Q_{P_r}^{z} < Q_{P_s}$.

Probability of finding a plant in the aggregate condition S+, i.e. the availability coefficient for producing electrical power and thermal output in relation to a fixed working capacity level for a period of time t is determined by the formulas

$$p_{H}(t) = K_{r}^{э}(N_{s}, t) = \sum_{z = S^{+}} P_{z}\left(N_{p}^{z} > N_{s}, t\right) = 1 - \sum_{Z = S} P_{z}\left(N_{p}^{z} < N_{s}, t\right) \tag{10}$$

$$p_H(t) = K^{Э}_{z_R}(Q_s, t) = \sum_{z=S^+} P_z\left(Q^z_p > Q_s, t\right) = 1 - \sum_{Z=S} P_z\left(Q^z_p < Q_s, t\right) \tag{11}$$

Mean integral values of availability factors for the period T are

$$K^Э_Г(N_s) = \frac{1}{T}\int_0^T K^Э_Д(N_S, t)dt, \quad K^Q_Г(Q_s) = \frac{1}{T}\int_0^T K^Q_Д(Q_S, t)dt \tag{12}$$

The average value of the CCT failure flow parameter is determined as the sum of the multiplied probabilities of failures of the operability to the intensity of the system transitions from the range of the corresponding operable conditions S^+ to the inoperative condition S^-, i.e.

$$\omega^Э(t) = \sum_{z=S^+}\left[P_z(t)\sum_{z=S^-_N}\lambda_{ij}\right], \quad \omega^Q(t) = \sum_{z=S^+}\left[P_z(t)\sum_{z=S^-_Q}\lambda_{ij}\right] \tag{13}$$

The subset of states S^+ contains conditions with varied available power (N_s, N_p, N_H) and productivity (Q_s, Q_p, Q_H). To account for partial failures of the power plant, it is possible to calculate the availability rate by formulas

$$K^Э_Г(t) = \sum_{z=S^+} P_z(t)\overline{N}^z; \quad K^Q_Г(t) = \sum_{z=S^+} P_z(t)\overline{Q}^z \tag{14}$$

Where $\overline{N}^z, \overline{Q}^z$ is the relative electrical and thermal power capacity of the power plant in the z-th condition.

The developed method for calculating reliability indexes of CCT and power supply systems based on them, was used to solve the problems of providing a specified (optimal) level of system reliability by using the optimal redundancy method.

6. Conclusion

- A probabilistic model is developed for calculating the reliability indexes of combined multipurpose heat and power plants in heat-and-power supply systems. The complex of models and calculation programs can be used to solve a wide range of specific tasks of optimization of schemes and parameters of combined heat and power plants and systems, as well as determining the efficiency of various redundance methods to ensure specified reliability of power supply.
- It was established that reliability indexes of multipurpose combined plants are interdependent. Calculation of reliability indexes of thermal-clamping plants should be made for each type of energy produces to consumers.

References

[1] http://minenergo.gov.ru/aboutminen/energostrategy
[2] Reliability of energy systems and their equipment. Reference book /G.N. Antonov, etc. // under the general. ed. Y.N. Rudenko. Moscow: Energoatomizdat, 1994. 480 p.
[3] Biryuk V V, Gorshkalev A A, Kayukov S S, Uglanov D A Open Mechanical Engineering Journal 2014 8 441-444
[4] Monakhov G V, Krasovsky B M District heating systems 1985, 151-166
[5] Kiselev N A, Burtsev S A, Strongin M M Measurement Techniques 2015 58 1016-1022
[6] Ankudinova M S, Larin E A, Sandalova L A Izvestia Vuzov. Problems of energy 2016 3-4 50-61

IOP Conf. Series: Materials Science and Engineering **302** (2018) 012019 doi:10.1088/1757-899X/302/1/012019

Selecting the Parameters of the Orientation Engine for a Technological Spacecraft

A.I. Belousov and A.V. Sedelnikov

Aircraft engines design department, Samara national research university, Russia

aibelousov@mail.ru; axe_backdtaft@inbox.ru

Abstract. This work provides a solution to the issues of providing favorable conditions for carrying out gravitationally sensitive technological processes on board a spacecraft. It is noted that an important role is played by the optimal choice of the orientation system of the spacecraft and the main parameters of the propulsion system as the most important executive organ of the system of orientation and control of the orbital motion of the spacecraft. Advantages and disadvantages of two different orientation systems are considered. One of them assumes the periodic impulsive inclusion of a low thrust liquid rocket engines, the other is based on the continuous operation of the executing elements. A conclusion is drawn on the need to take into account the composition of gravitationally sensitive processes when choosing the orientation system of the spacecraft.

1. Introduction

Low-thrust liquid-fuel rocket engines (LT LRE) are one of the output links of a complex motion control system (MCS) of many modern spacecraft (SC).

The MCS carries out the following functions:

- navigation – control of the movement of the center of mass of the spacecraft;
- stabilization – parrying possible turns of the spacecraft relative to the center of mass, keeping its body in the position set by the orientation system;
- orientation –rotation the spacecraft at specified angles to perform assigned tasks, including the implementation of space technologies in orbit.

Space technologies have led to the emergence of a number of unique materials and products. Thus, monocrystals are used in the manufacture of high-temperature turbines blades of aircraft engines. However, growing such monocrystals is a gravitationally sensitive process: the impact of even a small force field in the process of crystallization is capable of deflecting the axis of the grown monocrystal. This can give rise to a significant heterogeneity and anisotropy of the mechanical properties of the monocrystal [1]. For example, the anisotropy of the modulus of elasticity of grown single crystals will contribute to the dispersion of the natural oscillation frequencies of such blades, which leads to a decrease in the resource of the turbine. It was shown in [2] that such anisotropy is capable of reducing the limit of the long-term strength of monocrystalic blades to 40%.

Having a porous structure and good mechanical strength characteristics metallic foam is another perspective material . When implementing the process, it is necessary that the gaseous component of the foam metal be thoroughly mixed with the melt. Otherwise, during crystallization, inhomogeneities arise, which substantially worsen the strength properties.

Most gravitationally sensitive processes are energy-intensive. To successfully conduct them in space, it is necessary to make serious demands on the orientation of the spacecraft (SC) at the stage of its operation. Effective orbital stabilization of spacecraft relative to the Sun allows to obtain the required electric power from solar panels (SP).

The most common is the orientation scheme, which involves the use of active-passive orientation devices (APOD). Such devices include power gyroscopes and gyrodynes. With the perturbing effect of internal and external factors on the orbital motion of the spacecraft, the angular velocity of rotation of the APOD flywheels increases. This prevents disorientation of the spacecraft itself [1]. However, such an increase cannot be unlimited. At a certain value of the angular velocity taken as critical, in view of the asymmetry of the flywheels, radial runout occurs. It consists in the fact that the alternating centrifugal force of inertia, caused by asymmetry, causes an alternating reaction in the flywheel shaft supports. This, in turn, provokes a significant vibration disturbance. Such a situation is unacceptable in the realization of gravitationally sensitive processes, since favorable conditions for their flow are disrupted. There is a need for the orientation engine to unload the angular momentum of the APOD[3].

2. The role of LT LPRE in the MCS of a technological SC

As an orientation engine for a medium-class spacecraft, as a rule, LT LPRE is used. When the critical angular velocity of the flywheel rotation is attained, the APOD turns on the liquid-propellant engine in the pulsed mode, discharging the kinetic moment of the flywheels. The change of the pulsed thrust of the engine with time is shown in 'Figure 1' [4].

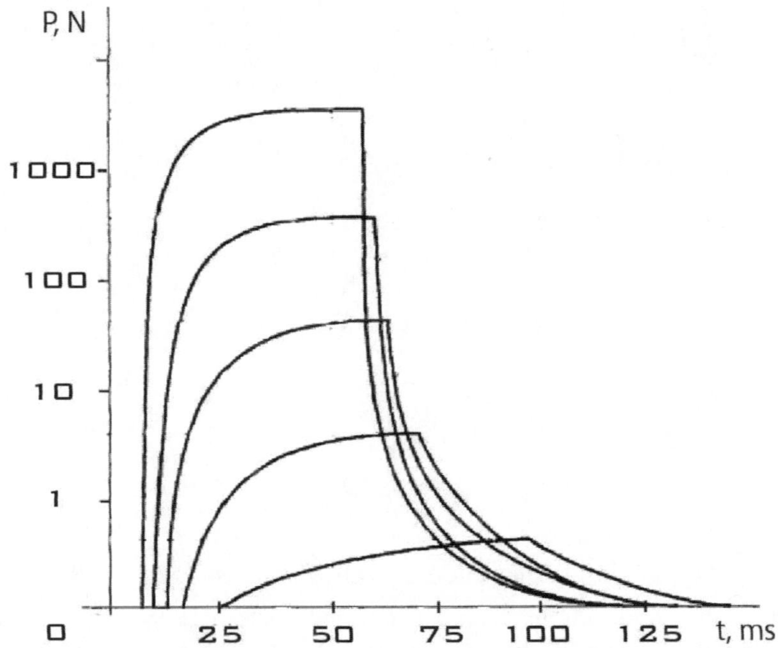

Figure 1. The rate of change of thrust of the LT LRE with time

The field of micro acceleration causes the appearance of anisotropy of the properties of the obtained materials and can significantly worsen their quality. Figure 2 shows the dependence of the level of micro acceleration on time in the location of the technological equipment when using liquid rocket engines in impulse mode [5].

IOP Conf. Series: Materials Science and Engineering **302** (2018) 012019 doi:10.1088/1757-899X/302/1/012019

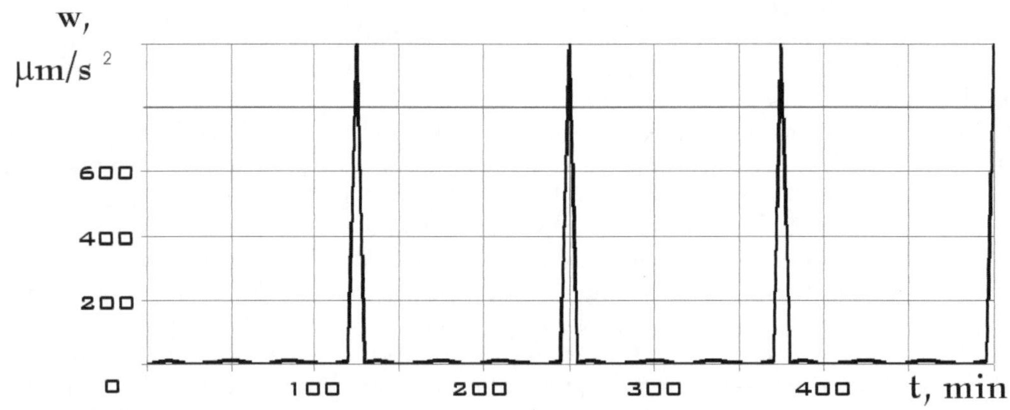

Figure 2. The rate change of micro acceleration with time when using a LT LPRE

The peak value of microaccelerations, as well as the waiting time of favorable conditions for the realization of gravitational-sensitive processes, directly depend on the thrust of the orientation engine. The equations of controlled motion of the spacecraft in the decomposition of elastic elements into individual rigid sections have the form [6]:

$$\begin{cases} \dot{\omega}_x = \dfrac{M_x - \sum_{i=1}^{N} I_{ix}\dot{\omega}_{ix} - m(y_0\ddot{z}_0 - \ddot{y}_0 z_0) - \sum_{i=1}^{N} m_i\left[(y_i - y_0)\cdot(\ddot{z}_i - \ddot{z}_0) - (\ddot{y}_i - \ddot{y}_0)\cdot(z_i - z_0)\right]}{I_x + \sum_{i=1}^{N} I_{ix}}; \\ \dot{\omega}_y = \dfrac{M_y - \sum_{i=1}^{N} I_{iy}\dot{\omega}_{iy} - m(\ddot{x}_0 z_0 - x_0\ddot{z}_0) - \sum_{i=1}^{N} m_i\left[(\ddot{x}_i - \ddot{x}_0)\cdot(z_i - z_0) - (x_i - x_0)\cdot(\ddot{z}_i - \ddot{z}_0)\right]}{I_y + \sum_{i=1}^{N} I_{iz}}; \\ \dot{\omega}_z = \dfrac{M_z - \sum_{i=1}^{N} I_{iz}\dot{\omega}_{iz} - m(x_0\ddot{y}_0 - \ddot{x}_0 y_0) - \sum_{i=1}^{N} m_i\left[(x_i - x_0)\cdot(\ddot{y}_i - \ddot{y}_0) - (\ddot{x}_i - \ddot{x}_0)\cdot(y_i - y_0)\right]}{I_z + \sum_{i=1}^{N} I_{iz}}, \end{cases} \quad (1)$$

Where, $\vec{M}(M_x, M_y, M_z)$– the moment created by the thrust vector of the motor relative to the center of mass of the spacecraft in the main coupled coordinate system (Figure 3). The torque generated by the thrust vector of the motor in the main coupled coordinate system (Figure 3); I_{ix}, I_{iy}, I_{iz} – the diagonal components of the inertia tensor of the *i*-th section of the elastic element in the coordinate system, whose origin is located at the center of the mass of the section, and the axes are parallel to the axes of the principal coupled coordinate system (Figure 3); I_x, I_y, I_z – the diagonal components of the inertia tensor of the hull in the principal coupled coordinate system; $\vec{r}_0(x_0, y_0, z_0)$ – the radius vector of the center of mass of the spacecraft in the main coupled coordinate system (Figure 3); $\dot{\vec{\omega}}(\dot{\omega}_x, \dot{\omega}_y, \dot{\omega}_z)$ – the vector of angular acceleration of the spacecraft body in the main coupled coordinate system; $\vec{\omega}_i$ – the vector of relative angular velocity (relative to the central body) in the main coupled coordinate system (Figure 3); m_i – mass of i-th section of a part of an elastic element; $\vec{r}_i(x_i, y_i, z_i)$ – the radius vector of the center of mass of the i-th section of the elastic element in the principal coupled coordinate system (Figure 3); N – the number of sections into which the elastic elements of the spacecraft are divided.

$$\begin{cases} w_x = \dot{\omega}_y z_M - \dot{\omega}_z y_M + \omega_x(\omega_x x_M + \omega_y y_M + \omega_z z_M) - (\omega_x^2 + \omega_y^2 + \omega_z^2)x_M; \\ w_y = \dot{\omega}_z x_M - \dot{\omega}_x z_M + \omega_y(\omega_x x_M + \omega_y y_M + \omega_z z_M) - (\omega_x^2 + \omega_y^2 + \omega_z^2)y_M; \\ w_z = \dot{\omega}_x y_M - \dot{\omega}_y x_M + \omega_z(\omega_x x_M + \omega_y y_M + \omega_z z_M) - (\omega_x^2 + \omega_y^2 + \omega_z^2)z_M, \end{cases} \quad (2)$$

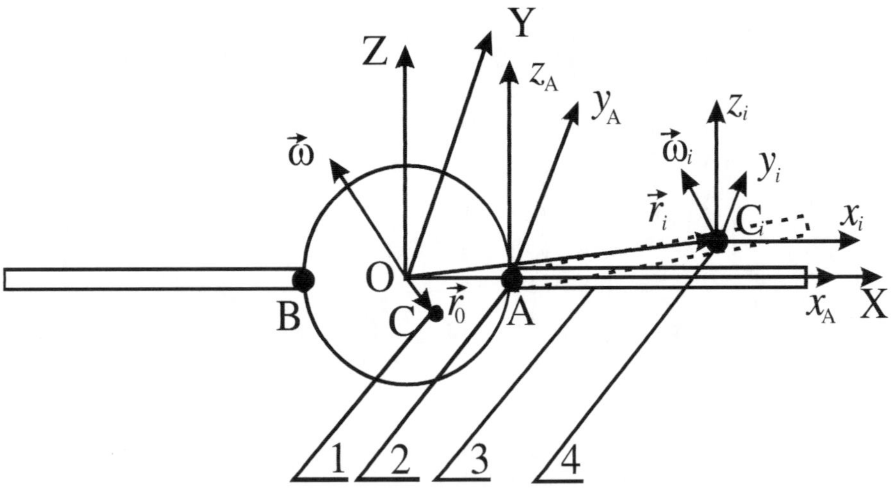

Figure 3. Schematic diagram of the SC

Then the components of micro accelerations can be defined as follows:
1 – is the center of mass of the spacecraft;
2 – the point of attachment of the elastic element to the body of the spacecraft;
3 – undeformed position of the elastic element;
4 – center of mass of the i-th part of the elastic element;
OXYZ is the main connected coordinate system;
$A\ x_A\ y_A\ z_A$ – is the local coordinate system associated
With the point of attachment of the elastic element to the body of the spacecraft;
$C_i\ x_i\ y_i\ z_i$ – is the local coordinate system associated
with the mass center of the i-th part of the elastic element

Analysis in expressions (1) and (2) shows that for the technological assignment, low thrust values should be selected. So, for SC "Spot-4" this value is 4 *N* [7], for the spacecraft "Foton" - up to 6 *N* [8]. Such a value of thrust increases the duration of the engine operation and the time for the release of the flywheel angular momentum of the APOD. For example, for a space remote sensing satellite, this shortcoming is significant. However, from the point of view of the realization of gravitationally sensitive processes, large values of thrust will cause significant oscillations of large elastic elements. Therefore, in spite of the fact that the time for unloading flywheels of APOD will be reduced, the waiting time of favorable conditions for microacceleration will significantly increase. Thus, for the successful implementation of gravitational-sensitive processes, it is necessary to choose the thrust of the orientation engine to 10 *N*, depending on the inertial-mass characteristics of the spacecraft.

The wide application of small spacecraft (MCS) for the solution of technological problems dictates the necessity of using the orientation of micromotors for performing tasks [9]. Thus, the electrothermal micromotor, tested for the dilution of five different MCSs along the orbits of functioning, produced a thrust of 30 *mN* and was also effective for solving orientation problems [10]. Due to minimal impact, the use of such engines does not create such significant microaccelerations, as in the case of the LT LRE. For the orientation of MCA, both pulsed motors similar to the LT LRE can be used, as well as continuously operating, for example, electric rocket engines (ERE) [11] or a set of flywheel control motors (SFCM) [12, 13] In the second case, there will be no peaks of microacceleration caused by impulsive switching of the engines, however, the constant operation of the ERD or SFCM will lead to the appearance of other microaccelerations (Figure 4).

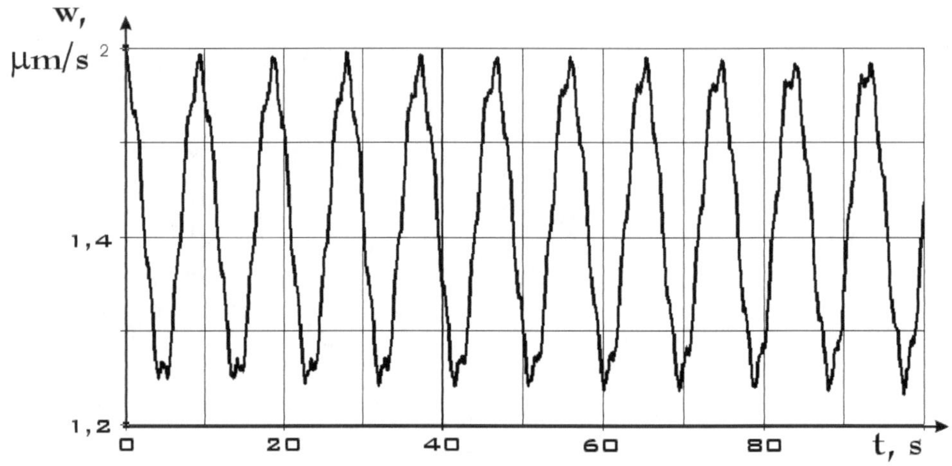

Figure 4. The rate of change of microacceleration, caused by the continuous operation of thrust 30 *mN*

3. Conclusion

Analysis of Figures 2 and 4 shows that for a short-term gravitational-sensitive processes requiring a micro-acceleration level below 1 $\mu m/s^2$, a pulse control scheme for orbital orientation is more appropriate. Since at a certain time interval between the peaks of the microacceleration (Figure 2), the level can drop to the required values. For a constantly operating ERE, this effect is not observed (Figure 4).

Thus, when choosing the scheme and parameters of the orientation engine, it is important to take into account not only the class of spacecraft and the requirements for its orientation, but also the composition of the gravitationally sensitive processes realized on its board.

References

[1] L.A. Maharramova and B.E. Vasiliev; The influence of the orientation of a monocrystal on the stress-Deformed condition and strength of gas turbine blades; Bulletin of the Moscow Aviation Institute, 19 5 (2012) 89–97.

[2] K.B. Alekseev and G.G. Bebenin; Control of space vehicles; Moscow, Mashinostroenie, (1974) 340.

[3] A.I. Belousov, A.V. Sedelnikov and K.I. Potienko; Study of Effective Application of Electric Jet Engine as a Mean to Reduce Microacceleration Level; International Review of Aerospace Engineering, 8 4 (2015) 157–160.

[4] B.A. Titov and A.L. Sirant; The study of the dynamics of a spacecraft with an orientation system based on two-component low thrust liquid-propellant rocket engines; SSAU Bulletin, 1 (2007) 98–105.

[5] A.V. Sedelnikov and A.A. Kireeva; Alternative solution to increase the duration of microgravity calm period on board the space laboratory; Acta Astronautica, 69 6-7 (2011) 480–484.

[6] A.I. Belousov and A.V. Sedelnikov Probabilistic estimation of fulfilling favorable conditions to realize the gravity-sensitive processes aboard a space laboratory; Russian Aeronautics, 56 3 (2013) 297–302.

[7] A.V. Sedelnikov and K.I. Potienko How to estimate microaccelerations for spacecraft with elliptical orbit; Microgravity Sciences and Technology, 28 1 (2016) 41–48.

[8] A.N. Kirilin, R.N. Akhmetov, G.P. Anshakov, A.D. Storozh and N.R. Stratilatov; A New Step Towards Unique Technologies In Space: The Foton-M4 Spacecraft; All-Russian Scientific-Technical Journal Polyot, 2 (2015) P. 3–9.

[9] V.N. Blinov, N.N. Ivanov, Yu. N. Setshenov and V.V. Shalay; Small spacecraft. Book 3. Minisatellites.The unified space platforms for small spacecrafts; Publishing house of OMGTU, Omsk (2010) 348.

[10] V.N. Blinov, V.V. Shalay and E.B. Charushina; Mathematical modeling of ammoniac electrothermal micromotors based on the results of full-scale tests of corrective propulsion systems; Information and space, 2 (2016) 104–112.

[11] A.I. Belousov and A.V. Sedelnikov Problems in formation and control of a required microacceleration level at spacecraft design, tests, and operation; Russian Aeronautics, 57 2 111–117.

[12] V.I. Arbashkin, K.E. Voronov, I.V. Piyakov et al.; Rotational motion of the satellites «Foton-M» № 4; Cosmic Research, 54 4 (2016) 315–322.

[13] A.V. Sedelnikov; Fractal Assessment of Microaccelerations at Weak Damping of Natural Oscillation in Spacecraft Elastic Elements. II; Russian Aeronautics, 57 2 111–117.

IOP Conf. Series: Materials Science and Engineering **302** (2018) 012020 doi:10.1088/1757-899X/302/1/012020

Numerical Investigation of the Impact of the Compressor Operation Mode on Working Process of the Combustion Chamber

M Yu Orlov, S V Lukachev and V M Anisimov

Samara National Research University, Scientific Educational Center of Fluid Dynamics Research, 34, Moskovskoye Shosse, Samara, Russian Federation, 443086

Email: adler65@mai.ru, teplotex_ssau@bk.ru, vradik@mail.ru

Abstract. The method of integrated compressor/combustor simulation was used to investigate the impact of flow distortion, appeared due to compressor blades, during the combustion chamber workflow. The method was improved in terms of generating a common grid and of principles of the boundary conditions settings. The geometric model includes four geometric volume bodies: guide vanes of the penultimate stage of high-pressure compressor, the impeller and guide vanes of the last stage and the flow path of combustion chamber. The calculation was carried out for some operation mode of the engine (nominal, 0.7 of nominal and 0.5 of nominal regimes) with and without compressor. The results were compared with the results of combustion chamber simulation without the compressor. Simulations showed that blade wakes extend up to the flame tube head. These wakes influence on the flame tongue, pressure field, temperature and velocity in the recirculation-mixing zone. It can influence on combustion efficiency, ecological performance and on temperature field at the combustor outlet. Thus, the simulations, which take into account combustion chamber and compressor, are more fully represent the characteristics of the working process of the combustion chamber and increase the efficiency of the design of new products.

1. Introduction

Currently the performance improving of modern gas turbine engines is related to improvement of workflow in a combustion chamber. Operation of combustion chamber is a part of engine workflow and associated with working process of a compressor and a turbine. Previously, a design and further development of compressor, combustion chamber and turbine were carried out separately. It does not allow reaching required characteristics of their joint operation at the engine. Problems identified at the stage of the test of engine-demonstrator are quite complicated to solve. Therefore, it is necessary to know the relationships between compressor workflow and parameters variations inside the combustion chamber to improve workflow of the engine more effective. It will reduce the financial and time costs for its creation and further development at the phase of design and improve product competitiveness.

The use of the CFD-methods at the design and further development phases can help to solve this problem. In this case, simulations can determine not only the integrated, but also the local values of the flow parameters. It significantly increases the opportunities in the case of optimization of working process.

Airflow through the axial compressor stage may be regarded as the flow through the diffuser channel system. Also, it is important to establish the flow distortion, appeared due to compressor blades, and its impact on the combustion chamber workflow.

Currently, development of the method of integrated simulation is very urgent. There are many papers related to this study [1-9]. The aim of these papers is development of integrated simulation of compressor and diffuser of combustion chamber. In addition, it includes the investigation of the compressor influence on the workflow at the diffuser. All these studies are the steps of achievement the goal – create virtual gas turbine engines (GTE). The aim of this paper is the improvement of the method of integrated compressor/combustor simulation and investigation of the compressor influence on the workflow in the flame tube of the combustion chamber.

2. Models and methods

The method of integrated compressor/combustor simulation was developed by creating a common mesh model of compressor and combustion chamber (CC). The object of research is combustion chamber of aero-derivative power plant working on gaseous fuel.

The geometrical model, which was the basis of the grid model, is shown in figure 1. It includes four different geometric volume bodies: the flow path of combustion chamber, the impeller and the guide vanes of the last stage and guide vanes of the penultimate stage of high-pressure compressor [10-11].

The sector of the geometrical model was used instead of full model to decrease time calculation. The angle of the combustion chamber and blade rows sectors was the same. Fuel rate depends on the thrust rating and for the even and odd burners it may differ. For example, the decrease of operation mode leads to decrease of fuel rate through even burners but fuel rate through the odd keeps constant. Hence, the sector includes two burners. The number of blades might differ from the actual number by no more than 10% to ensure the periodicity condition, agreement of compressor and combustion chamber sectors, and the accuracy of calculation. The periodical sector is the geometrical volume of compressor and CC. The whole circumferential model can be obtained by repeating the sector.

Finite-volume model was generated for each geometric model by advanced size function Proximity and Curvature in Ansys Meshing (see figure 2).

Figure 1. Geometrical model of compressor and combustion chamber.

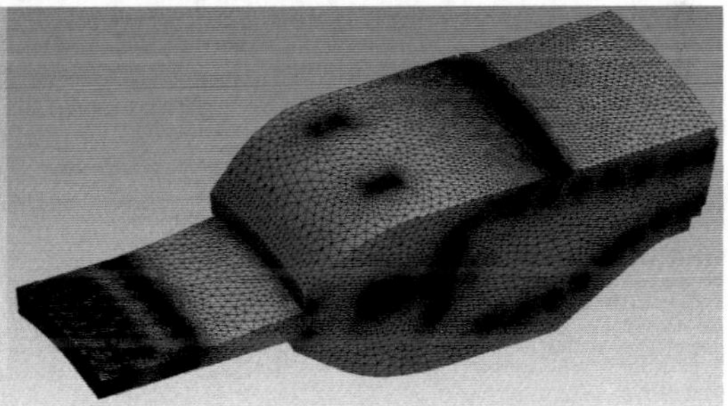

Figure 2. Finite-volume model of compressor and CC.

One of the main conditions for the periodicity is the mesh identity on each side of the geometry. This condition can be achieved by using the grid function Match Control. Unstructured grid model contains less than 12 million elements. Skewness is an important quality parameter of the grid, which should not exceed 0.96. As a result, this parameter became equal 0.85, which proves that the grid is acceptable. Inlet and outlet of air and fuel were set as well as paired "interfaces" on contiguous faces blade rows and the combustion chamber.

Three-dimensional simulation package Ansys Fluent 16.2 was used for simulation of joint action of compressor and combustion chamber, because the simulation methods were well-mastered in previous research works [12-17]. The methods of simulation of compressor and CC workflows were created and tested.

After mesh generation, mathematical model and boundary conditions were formulated. Mathematical model includes calculation models, which are necessary for workflow simulation in combustion chamber. Reynolds Stress Model (RSM), selected as a turbulence model, is one of the most complex turbulence models proposed by Fluent [18-20]. The calculations took into account the effect of the radiation, as the heat flux from the radiation is comparable to the heat flux from convection or conduction in combustion chamber. This is due to high level of temperature. Radiation model Discrete Ordinates (DO) was chosen for the simulation [21]. Adiabatic boundary conditions were used for walls. Model wsggm-domain-based (weighted-sum-of-gray-gases model) was used to calculate the absorption coefficient of the mixture [22]. Flamelet Generated Manifold was selected to simulate combustion processes in combustion chamber. Kinetic scheme GRI 3.0 was used to simulate mechanism of methane combustion [23]. Model of Turbulent flame speed was used with the FGM. The boundary conditions are specified by input parameters of pressure and temperature of the airflow coming from the impeller of the penultimate stage of the compressor, and fuel supplying to the combustion chamber.

The simulation on the grid model, which included only the combustion chamber without the compressor, were also performed for a more complete investigation of the compressor effect on the combustion chamber workflow. Averaged parameters were set on the inlet of the CC for the simulation without compressor. The parameters values were received by thermogasdynamic calculation. The compressor determined the parameters at the inlet of the diffuser for the case of CC with compressor.

Direction vectors were set in cylindrical coordinates, because the airflow runs onto the guide vanes at a certain angle: in the radial direction – 0, in the circumferential direction – 0.74, in the axial direction – 0.67.

Periodic conditions on the lateral faces of the combustion chamber and blade rings sectors were set by text commands. The interaction between certain previously created "interfaces" was defined to transfer values of the flow parameters from one body to another (see figure 3).

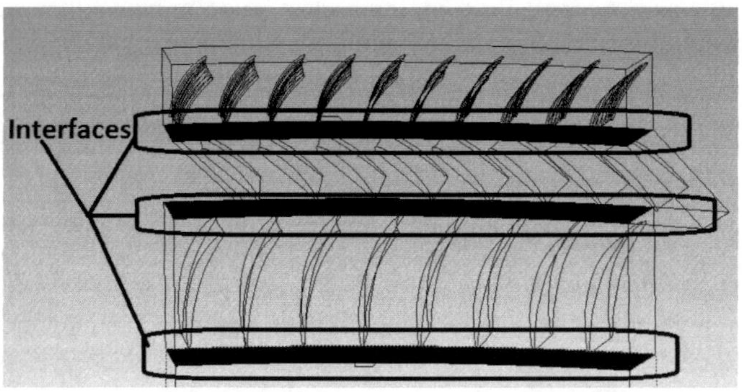

Figure 3. Interfaces on contact sides of the bodies.

The calculation was carried out in a transient case. It is necessary to specify certain time step and number of time steps required for the convergence. According to the recommendations proposed by software package ANSYS, the value of time step was calculated depending on the number of impeller blades. Number of time steps was set so that the impeller has three committed full rotations. [21].

3. Results and discussion

The integrated compressor/combustor simulation was carried out in a transient case for several operation modes of the engine (nominal, 0.7 of nominal and 0.5 of nominal regimes) with and without compressor. Three surfaces were used for results visualization (see figure 4). The URANS simulations showed that blade wakes extend up to the flame tube head (see figure 5). These wakes were obtained by specific time step. These wakes influence on the flame contour (see figures 6-8) and on the temperature field at the exit of combustion chamber (see figures 9-11).

The figure 12 shows the radial diagram of outlet temperature obtained by simulation of three operation modes with and without compressor. The radial distortion was calculated by:

$$\Theta_{rad} = \frac{T_{Gi} - T_C}{T_G - T_C}. \quad (1)$$

A comparison of radial diagrams shows that compressor rotation influences on the diagram view. As well as in the case of temperature circumferential distribution, the smallest distribution takes place at the nominal operation mode.

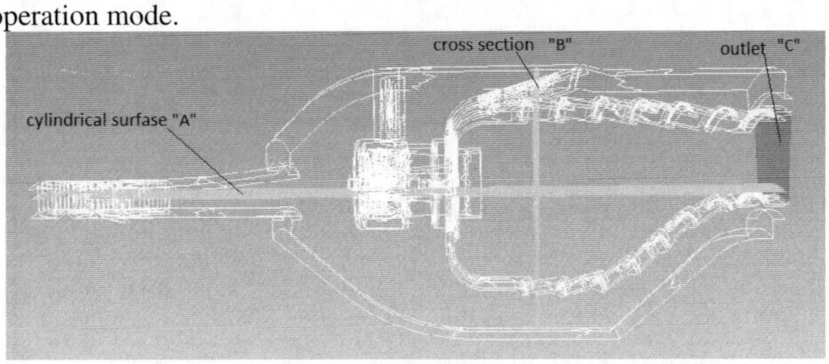

Figure 4. Surfaces mapping the calculation results.

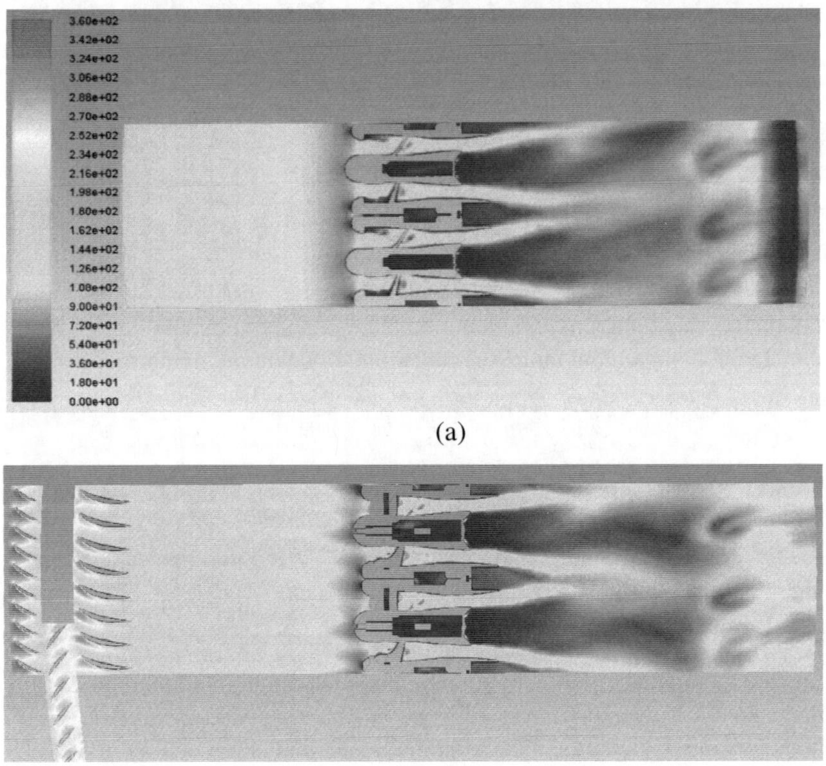

(a)

(b)

Figure 5. Velocity distribution on cylindrical surface “a” of CC without (a) and with (b) compressor.

IOP Conf. Series: Materials Science and Engineering **302** (2018) 012020 doi:10.1088/1757-899X/302/1/012020

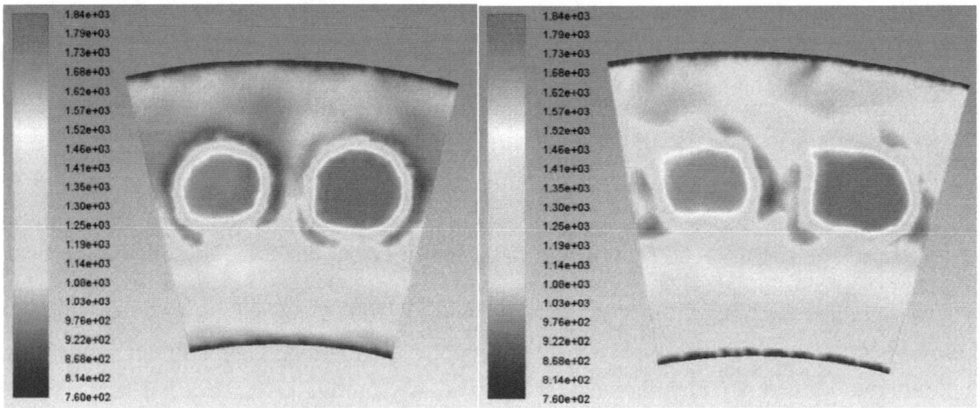

(a) (b)

Figure 6. Temperature distribution at cross-section “b” without (a) and with (b) compressor at nominal operation mode.

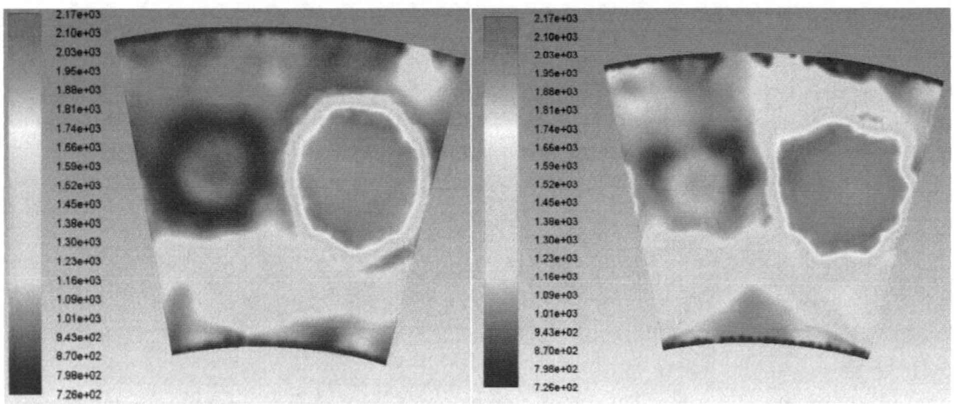

(a) (b)

Figure 7. Temperature distribution at cross-section “b” without (a) and with (b) compressor at 0.7 of nominal operation mode.

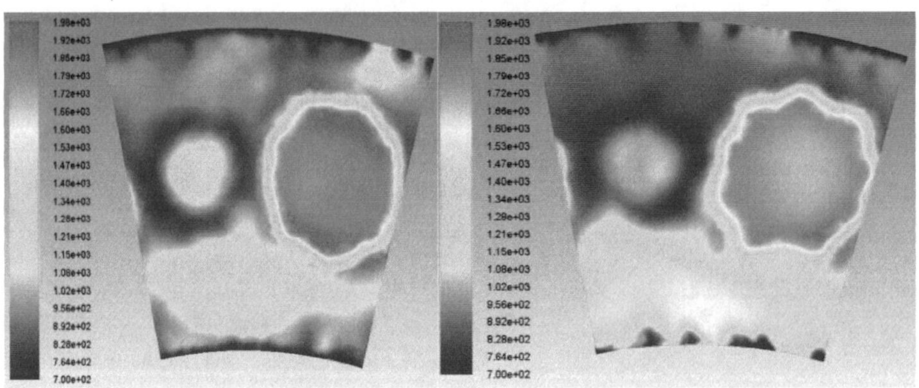

(a) (b)

Figure 8. Temperature distribution at cross-section “b” without (a) and with (b) compressor at 0.5 of nominal operation mode.

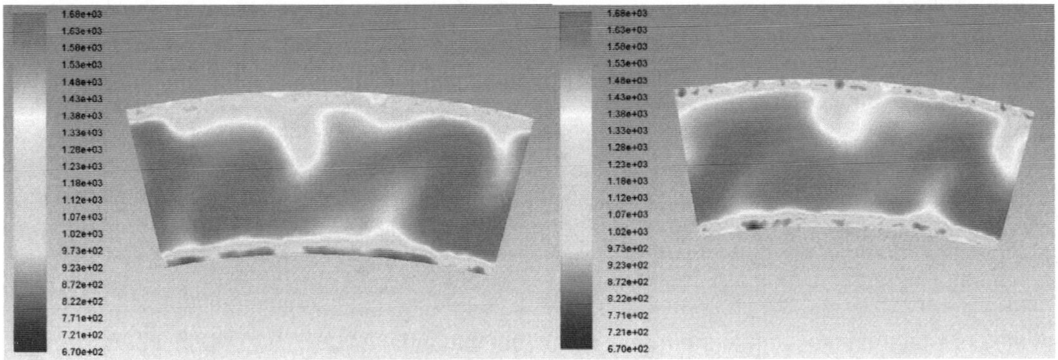

(a) (b)

Figure 9. Temperature distribution at the outlet "c" of the CC without (a) and with (b) compressor at nominal operation mode.

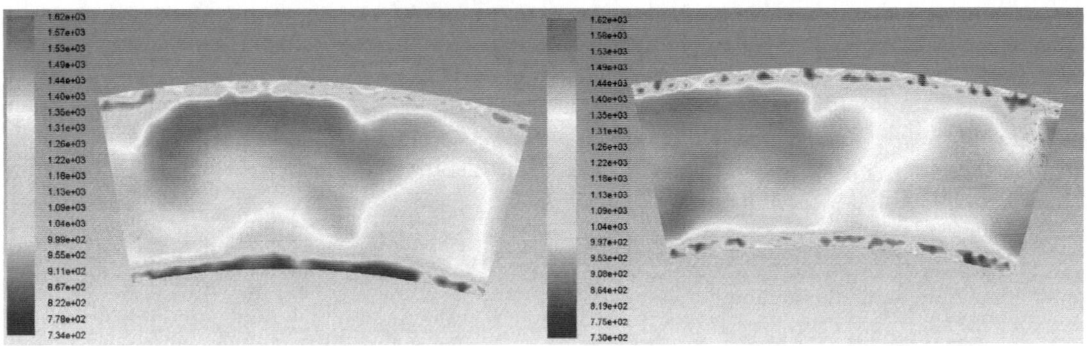

(a) (b)

Figure 10. Temperature distribution at the outlet "c" of the cc without (a) and with (b) compressor at 0.7 of nominal operation mode

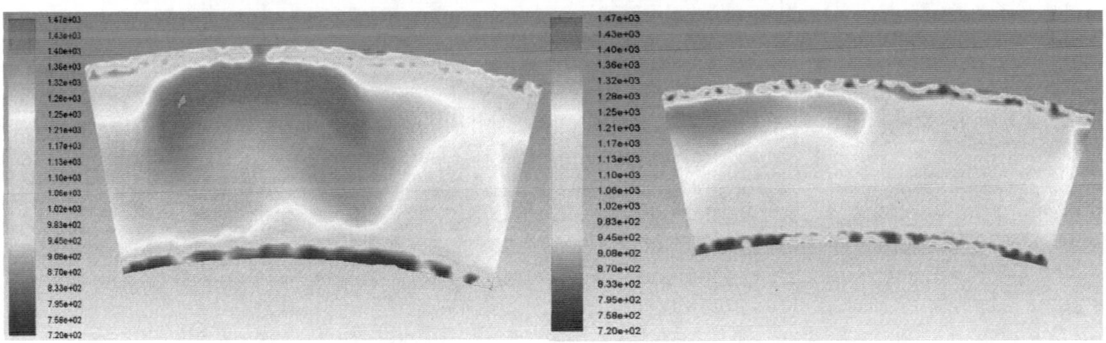

(a) (b)

Figure 11. Temperature distribution at the outlet "c" of the cc without (a) and with (b) compressor at 0.5 of nominal operation mode.

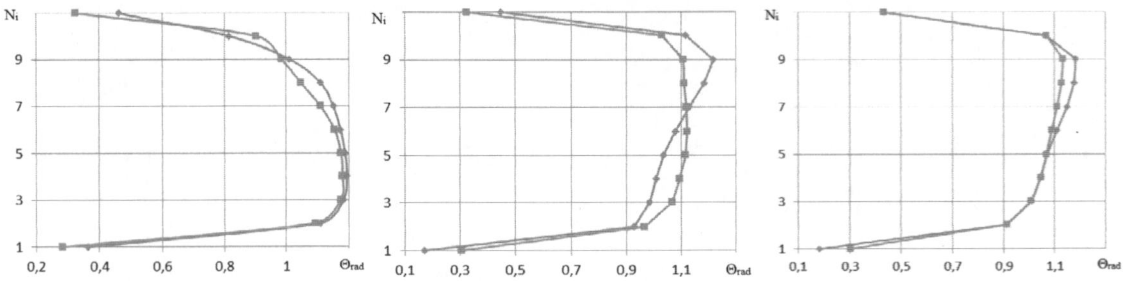

(a) (b) (c)

Figure 12. Radial diagram of temperature distortion at the outlet of the CC at nominal (a), 0.7 (b) and 0.5 (c) of nominal operation modes (orange with and blue without compressor).

Also, the fields of the rate of thermal NO distribution were carried out at the cross section behind the burner (see figure 13) and at the cross section on a distance of 70 mm behind the burner (see figure 14). It can be seen that the inclusion of the compressor leads to changes in the shape and size of the area of NOx formation zone and, consequently, to changes in its concentration in the combustion products.

The influence of the compressor on the concentration of NOx in the combustion products can be explained as follows. Taking the compressor into account in the simulation changes the velocity field in front of the burner swirler (see figure 15).These flow disturbances impact on the fuel distribution in the burner and behind it (see figure 16).

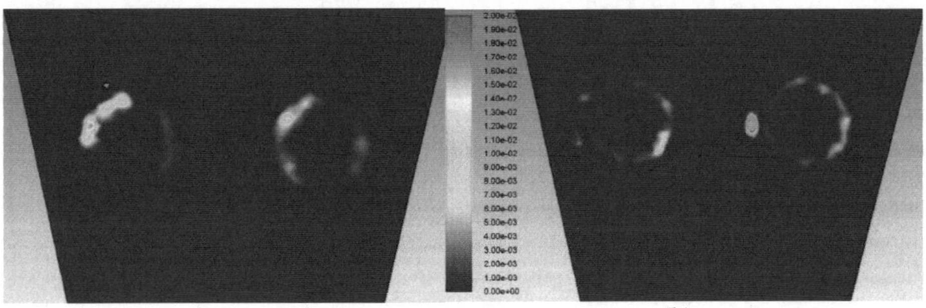

(a) (b)

Figure 13. Rate of thermal no at the cross-section behind the burner of the CC without (a) and with (b) compressor at nominal operation mode.

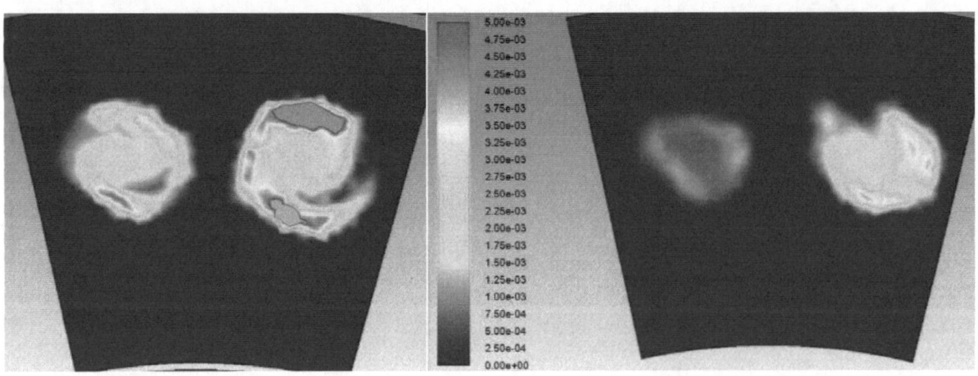

(a) (b)

Figure 14. Rate of thermal no at the cross-section on a distance of 70 mm from the burner of the CC without (a) and with (b) compressor at nominal operation mode.

IOP Conf. Series: Materials Science and Engineering **302** (2018) 012020 doi:10.1088/1757-899X/302/1/012020

At the nominal operation mode of the engine for even and odd burners, the equivalence ratio is equal. However, as shown in figure 16, local fuel distribution is different for the CC simulation with and without compressor. Hence, the disposition of fuel rich local zones is varies, as well as high temperature zones. It leads to differences in concentrations of NOx at the outlet of the combustion chamber. Thus, the calculation data of NOx shows significant differences in the level of its formation for the both cases.

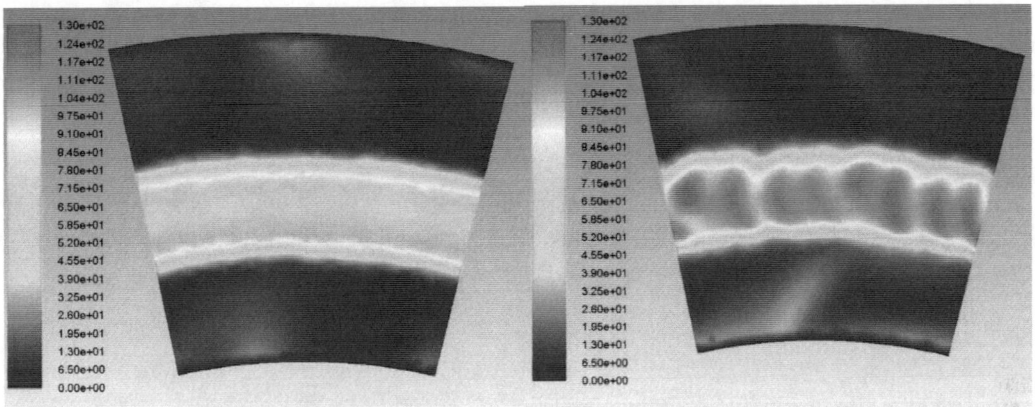

(a) (b)

Figure 15. Velocity field at the cross-section in front of the swirler of the cc without (a) and with (b) compressor at nominal operation mode.

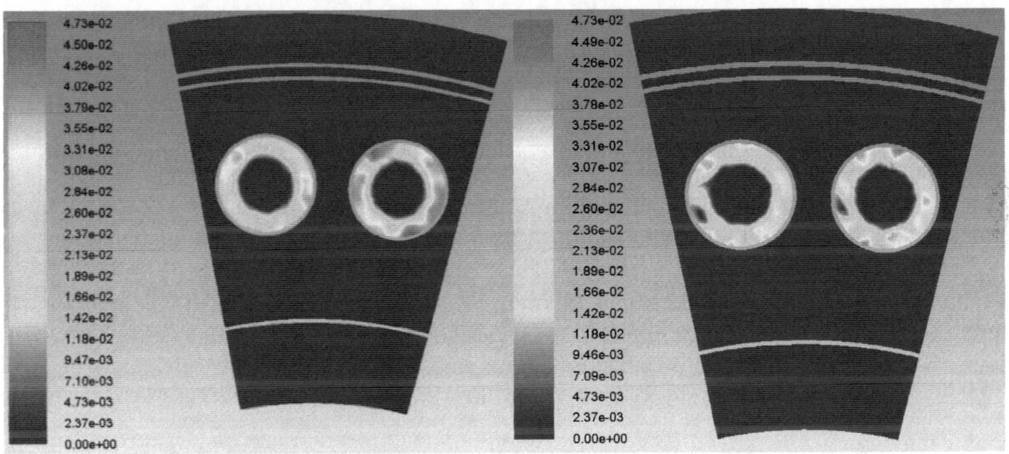

(a) (b)

Figure 16. Fuel distribution behind the swirler for the cc without (a) and with (b) compressor at nominal operation mode

4. Conclusion

Summarizing all the above it can be noted that:

- The simulation taking the compressor into account allows obtaining the qualitatively and quantitatively different data from the results of simulation without compressor.
- The influence of the compressor in the simulation depends on the design of flame tube head, burner and engine operation mode. The results showed that the least impact of the compressor is at the nominal operation mode and the biggest – at the 0.7 of nominal regime.

- Blade wakes extend up to the flame tube head and influence on the flame contour, pressure field, temperature field and velocity at the recirculation-mixing zone. These changes can effect on combustion efficiency, ecological characteristics and outlet temperature.
- The simulation of combustion chamber with compressor can better reflect the special features of combustion chamber workflow in a greater extent. It increases the efficiency of design new products.

Developed method can be used for optimization of combustion chamber workflow with interaction of other GTE components, and during its development. In addition, it can be used for analyzing the effect of various factors on the combustion chamber workflow. For example, the location of aircraft in space, changes of GTE operation mode, foreign objective in the engine flow path etc.

Acknowledgments

This work was supported by the Ministry of education and science of the Russian Federation in the framework of the implementation of the Program "Research and development on priority directions of scientific-technological complex of Russia for 2014-2020" (RFMEFI58716X0033).

References

[1] Kannan K V and Page G J 2016 Automated multi-code URANS simulation of compressor-combustor components Proceedings of the ASME Turbo Expo 2016 GT2016-56904
[2] Kannan K V and Page G J 2014 Coupling of Compressible Turbomachinery and Incompressible Combustor Flow Solvers for Aerothermal Applications ASME Turbo Expo 2014: Turbine Technical Conference and Exposition
[3] Walker A D, Carrotte J F, Peacock G L, Spencer A and McGuirk J J 2013 Experimental Study of the Unsteady Aerodynamics of the Compressor-Combustor Interface of a Lean Burn Combustion System San Jose, 49th AIAA/ASME/SAE/ASEE Joint Propulsion Conference
[4] Schlüter J U, Wu S K, Alonso J J and Pitsch H 2003 Integrated RANS-LES Computations Of Turbomachinery Components: Generic Compressor/Diffuser Center for Turbulence Research Annual Research Briefs ADA420749
[5] Boudier G, Gicquel L Y M, Poinsot T, Bissières D, and Bérat C 2007 Comparison of LES, RANS and experiments in an aeronautical gas turbine combustion chamber Proceedings of the Combustion Institute 31(2) 3075-3082
[6] Schluter J U, Wu X, Kim S, Alonso J J and Pitsch H 2004 Coupled RANS-LES computation of a compressor and combustor in a gas turbine engine In 40 th AIAA/ASME/SAE/ASEE Joint Propulsion Conference and Exhibit
[7] Sehra A K 2006 The numerical propulsion system simulation: A vision for virtual engine testing In Proceedings of the ASME Turbo Expo 2006: Power for Land, Sea and Air ASME New York, USA.
[8] Matveev S S, Zubrilin I A, Orlov M Y and Matveev S G 2015 Numerical investigation of the influence of flow parameters nonuniformity at the diffuser inlet on characteristics of the GTE annular combustion chamber. Source of the Document Proceedings of the ASME Turbo Expo 4A GT2015-42676
[9] Orlov M Y and Matveev S S 2014 Numerical simulation of an influence of a compressor and a turbine on characteristics of a combustion chamber of a small-sizes gas turbine engine Life Science Journal 11 (11) 119 pp 650-654
[10] Popov G, Kolmakova D, Shklovets A and Ermakov A 2015 Optimization of the axial compressor flow passage to reduce the circumferential distortion IOP Conference Series: Materials Science and Engineering 9th International Conference on Compressors and Their Systems London United Kingdom, Article number 012053
[11] Matveev V N, Baturin O V, Popov G M and Egorov I N 2014 Seven-stage axial compressor optimization Engineering Optimization IV - Proceedings of the 4th International Conference on Engineering Optimization ENGOPT 2014 Lisbon Portugal pp 821-826

[12] Orlov M Y, Matveev S S, Makarov N S and Zubrilin I A 2014 Numerical modeling problems of operating process of combustion chambers of GTE and solution approaches Source of the Document ARPN Journal of Engineering and Applied Sciences 9 (12) pp 2894-2899
[13] Anisimov V, Zubrilin I and Orlov M 2016 Investigation of thermal and stress states of the annular combustion chamber flame tube walls Proceedings of the ASME Turbo Expo GT2016-57479
[14] Orlov M Y and Matveev S S 2014 Numerical simulation application for the design and fine-tuning of small-sized gas turbine engine combustor. Source of the Document Open Mechanical Engineering Journal 8 pp 450-456
[15] Zubrilin I A, Dmitriev D N, Matveev S S and Matveev S G 2015 Numerical investigation of the nonreacting swirling flow structure downstream of industrial gas turbine burner with the central body Proceedings of the ASME Turbo Expo 4A GT2015-42181
[16] Matveev S S, Zubrilin I A, Orlov M Y, Matveev S G and Chechet I V Investigation of fuel distribution in partially premixed swirled burner with pilot flame Proceedings of the ASME Turbo Expo 4B-2016 GT2016-57478
[17] Zubrilin R A, Matveev S S, Zubrilin I A and Matveev S G 2016 Gaseous fuel flame stabilization in a modular swirled burner Proceedings of the ASME Turbo Expo 4B-2016 GT2016-57441
[18] Gibson M M and Launder B E 1978 Ground Effects on Pressure Fluctuations in the Atmospheric Boundary Layer J. Fluid Mech 86 491–511
[19] Launder B E 1989 Second-Moment Closure: Present... and Future? Inter. J. Heat Fluid Flow vol. 10 no. 4 pp. 282–300
[20] Launder B E, Reece G J and Rodi W 1975 Progress in the Development of a Reynolds-Stress Turbulence Closure J. Fluid Mechanics vol. 68, no. 3, pp. 537–566
[21] ANSYS 16.2 User's Guide.
[22] Coppalle A and Vervisch P 1983 The Total Emissivities of High-Temperature Flames". Combustion and Flame. 49. 101–108.
[23] Anderson W and Bonhus D L1994 An Implicit Upwind Algorithm for Computing Turbulent Flows on Unstructured Grids". Computers Fluids. 23(1) Pp. 1–21.

IOP Conf. Series: Materials Science and Engineering **302** (2018) 012021 doi:10.1088/1757-899X/302/1/012021

Investigation of the Effect of the Non-uniform Flow Distribution After Compressor of Gas Turbine Engine on Inlet Parameters of the Turbine

M Yu Orlov, S V Lukachev and V M Anisimov

Samara National Research University, Scientific Educational Center of Fluid Dynamics Research, 34, Moskovskoye Shosse, Samara, Russian Federation, 443086

Email: adler65@mai.ru, teplotex_ssau@bk.ru, vradik@mail.ru

Abstract. The position of combustion chamber between compressor and turbine and combined action of these elements imply that the working processes of all these elements are interconnected. One of the main requirements of the combustion chamber is the formation of the desirable temperature field at the turbine inlet, which can realize necessary durability of nozzle assembly and blade wheel of the first stage of high-pressure turbine. The method of integrated simulation of combustion chamber and neighboring nodes (compressor and turbine) was developed. On the first stage of the study, this method was used to investigate the influence of non-uniformity of flow distribution, occurred after compressor blades on combustion chamber workflow. The goal of the study is to assess the impact of non-uniformity of flow distribution after the compressor on the parameters before the turbine. The calculation was carried out in a transient case for some operation mode of the engine. The simulation showed that the inclusion of compressor has an effect on combustion chamber workflow and allows us to determine temperature field at the turbine inlet and assesses its durability more accurately. In addition, the simulation with turbine showed the changes in flow velocity distribution and pressure in combustion chamber.

1. Introduction

The performance improvement of modern gas turbine engine and power plants involves not only the optimization of parameters averaged over the cross section, but also the formation of the cross section distribution. Combustion chamber is one of the main elements of gas turbine engines and power plants, which directly produce an effect on almost all characteristics of GTE. For example, formation of the desirable temperature field at the turbine inlet determines the necessary durability of nozzle assembly and blade wheel of the first stage of high-pressure turbine.

Joint operation of compressor, combustion chamber and turbine implies that workflows of these elements are linked between each other. Compressor parameters change may affect the workflow in the combustion chamber, and then in the turbine. Thus, during the design of the combustion chamber it is important to determine not only the temperature distribution, but also to obtain an understanding of how it is influenced by various factors, including the change of the rotor revolutions and the corresponding change in the compressor.

Existing studies related to integrated simulation of combustion chamber and neighboring nodes [1-7] consider mainly the interaction of the compressor and the combustion chamber and in a less degree consider processes throughout the whole engine. Meanwhile computer simulation can make the engine design processes more effective. Therefore, the aim of this study is to investigate the influence of the compressor on the processes in the combustion chamber and through them on the parameters of the gas prior to the turbine.

2. Models and methods

The object of the investigation is core engine of power plant. Ansys Fluent three-dimensional simulation package was used as a calculation tool. Earlier, using this package, the method of combustion process simulation was tested and verified with experiments [8-15]. Geometrical model includes: the impeller and guide vanes of the last stage and guide vanes of the penultimate stage of high-pressure compressor, combustion chamber, nozzle assembly and blade wheel of high-pressure turbine [16-18].

Periodical sector model of engine core was used instead of full-sized model to decrease necessary computer memory and calculation time. The sector of combustion chamber includes two burners, because fuel rate depends on the thrust rating and for the even and odd burners it may differ. The blade number at sectors of compressor and turbine might differ from the real, by no more than 10%, to ensure the sector periodicity. Unstructured finite element model is presented in figure 1.

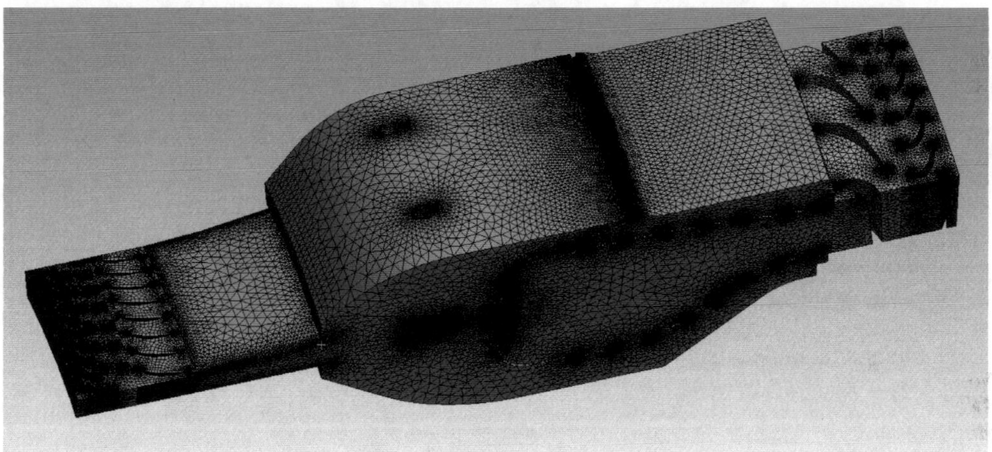

Figure 1. Finite element model of engine core.

Number and size of finite elements on contact surfaces of the compressor, combustion chamber and turbine should be approximately equal. Otherwise, such a discrepancy may cause an error during data transfer from one body to another. The mesh tools Match Control was used to set the periodicity on the lateral faces of the model. The number of finite elements is about 14 million; the skewness is less than 0.96. Inlet and outlet boundaries were set for air and fuel, also paired "interfaces" on contiguous faces of blade rows and the combustion chamber were set. After mesh generation, boundary conditions were set and mathematical model was specified. Mathematical model includes computational models, which are necessary for simulation workflow in combustion chamber [19-24]. The interaction between certain pre-established “interfaces” was set for transferring parameter values from compressor to combustor and from combustor to turbine. Turbine rotation was set equal to compressor rotation, because they are on the same engine shaft [25]. The calculation was carried out in transient case for two operation modes: nominal and 0.5 of nominal.

3. Results and discussion

Velocity field shown on figure 2 was calculated at nominal operation mode. Temperature fields at the cross-section of recirculation mixing zone (see figures 3 and 4) and at the outlet (see figure 5 and 6) were calculated on nominal and on 0.5 of nominal operation mode. In addition, radial diagram of temperature distortion at the outlet of CC was obtained by simulation (see figure 7). The radial distortion was calculated by:

$$\Theta_{rad} = \frac{T_{Gi} - T_C}{T_G - T_C}. \quad (1)$$

The fields and diagram are presented by the comparison with similar fields obtained by CC simulation (without compressor and turbine) and by integrated compressor/combustor simulation.

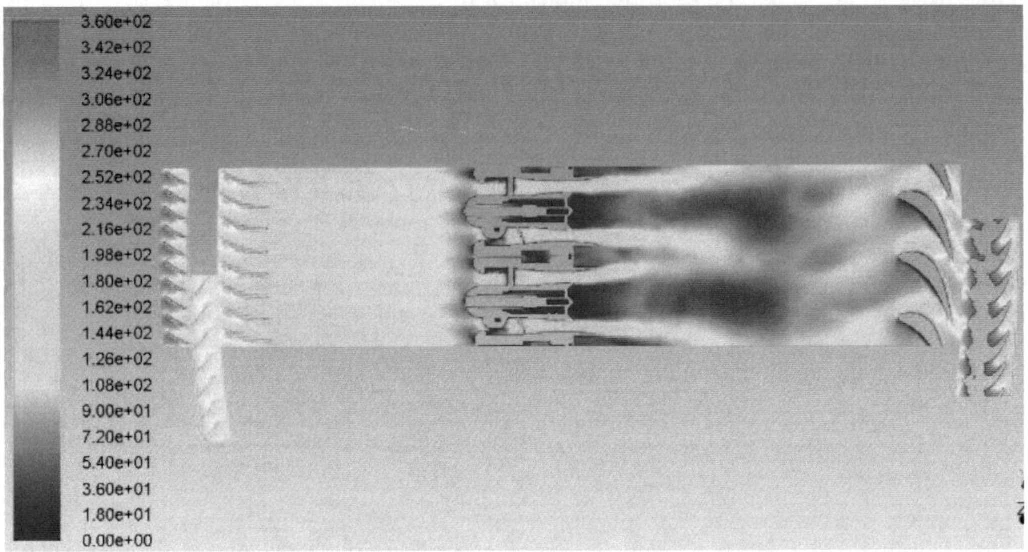

Figure 2. Velocity field at the radial section of engine core at nominal operation mode.

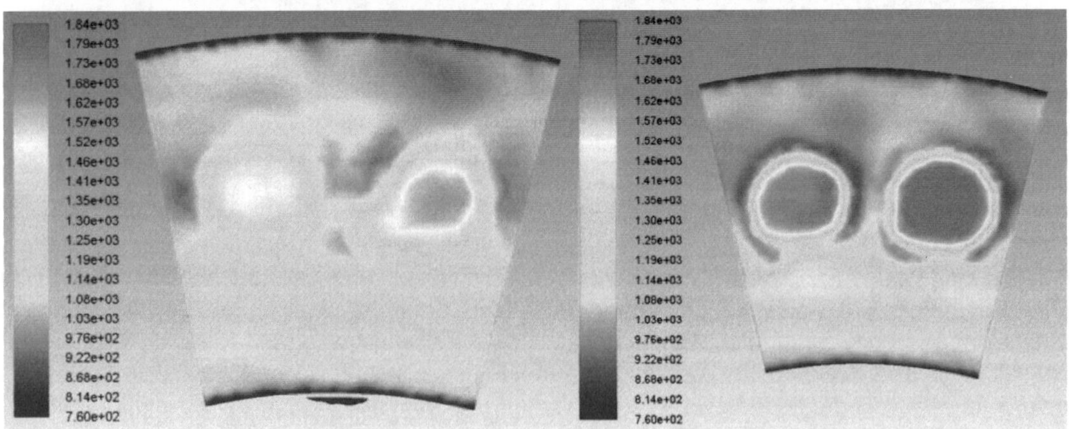

(a) (b)

Figure 3. Temperature field at the cross-section of recirculation mixing zone at nominal operation mode (a – cc with turbo compressor, b – combustor).

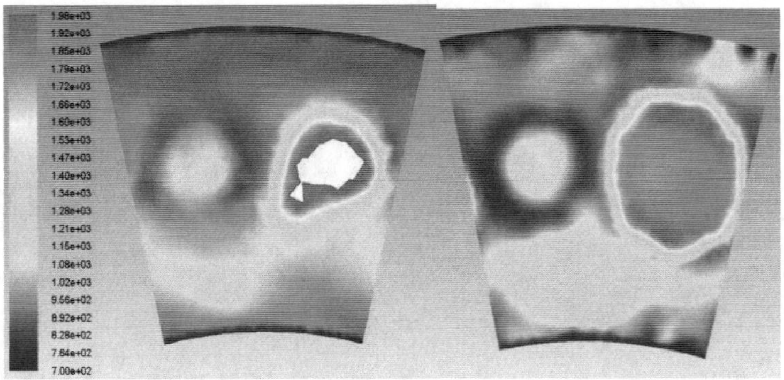

(a) (b)

Figure 4. Temperature field at the cross-section of recirculation mixing zone at 0.5 of nominal operation mode (a –cc with turbocompressor, b – combustor).

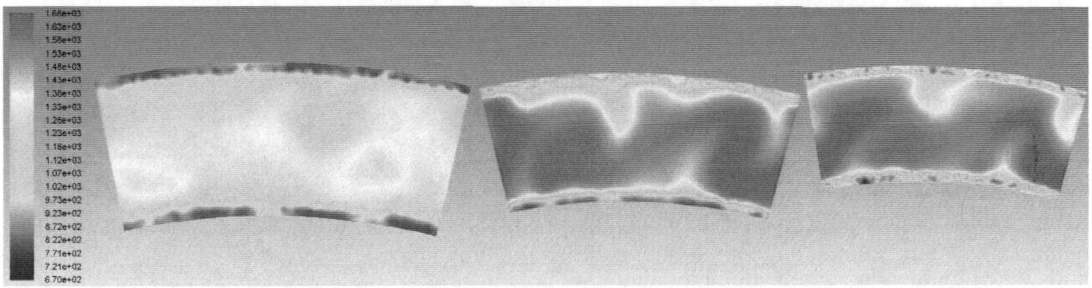

(a) (b) (c)

Figure 5. Temperature field at the cc outlet at nominal operation mode (a – cc with turbo compressor, b – combustor, c – cc with compressor).

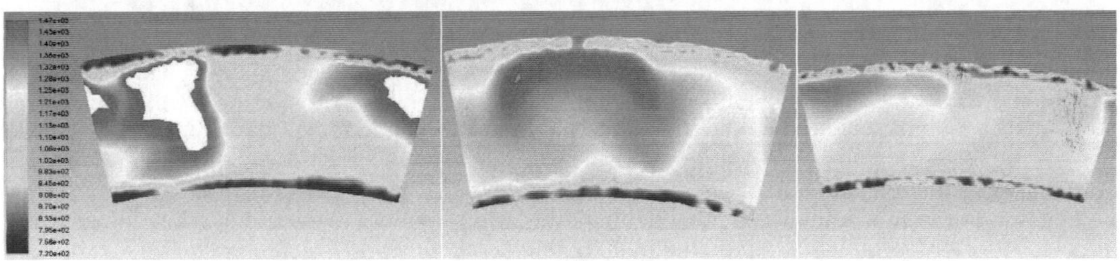

(a) (b) (c)

Figure 6. Temperature field at the outlet of combustion chamber at 0.5 of nominal operation mode (a – cc with turbo compressor, b – combustor, c – cc with compressor).

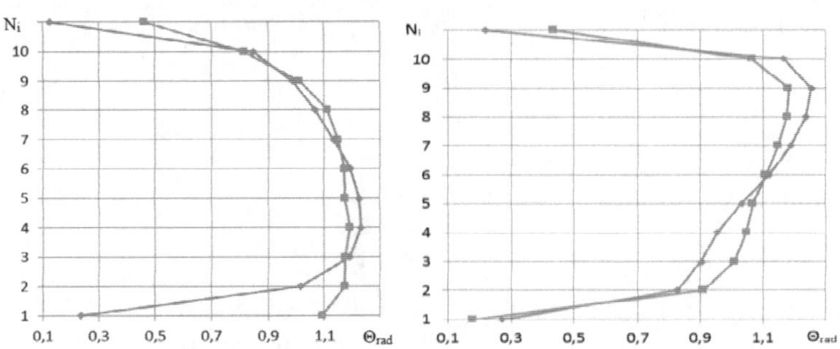

(a) (b)

Figure 7. Radial diagram of temperature distortion at the outlet of cc at nominal (a) and 0.5 (b) of nominal operation mode (orange – combustor, blue – cc with turbo compressor).

IOP Conf. Series: Materials Science and Engineering **302** (2018) 012021 doi:10.1088/1757-899X/302/1/012021

Mass fraction of NOx was obtained to estimate the influence of different simulation cases (CC with or without turbo compressor) on performance characteristics at 0.5 of nominal operation modes (see figure 8-9).

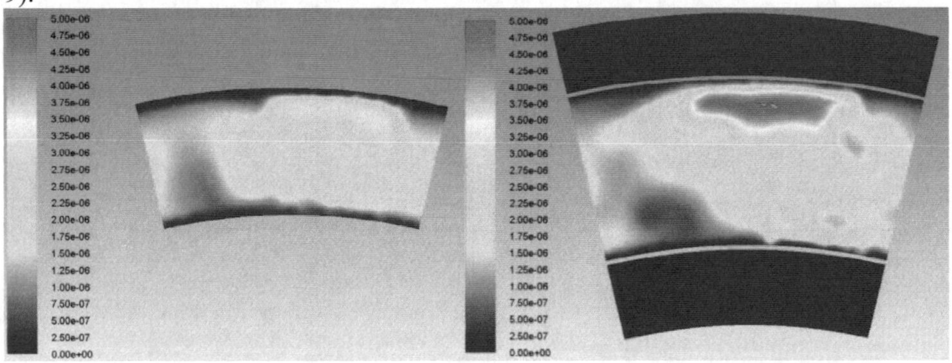

(a) (b)

Figure 8. Mass fraction of NOx at the outlet of cc (a) and at the cross-section after dilution holes (b) for the cc simulation without turbo compressor at 0.5 of nominal operation mode.

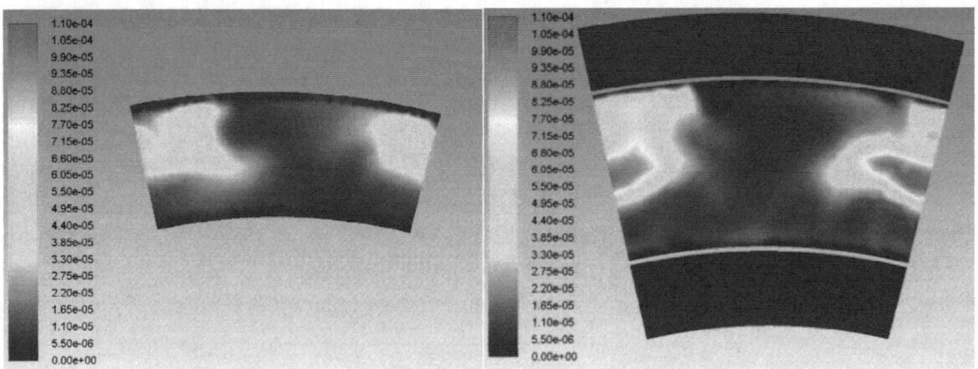

(a) (b)

Figure 9. Mass fraction of NOx at the outlet of cc (a) and at the cross-section after dilution holes (b) for the cc simulation with turbo compressor at 0.5 of nominal operation mode.

Velocity and pressure fields were obtained by simulation for both operation modes. Figures 10 and 11 show the velocity changing at the CC outlet.

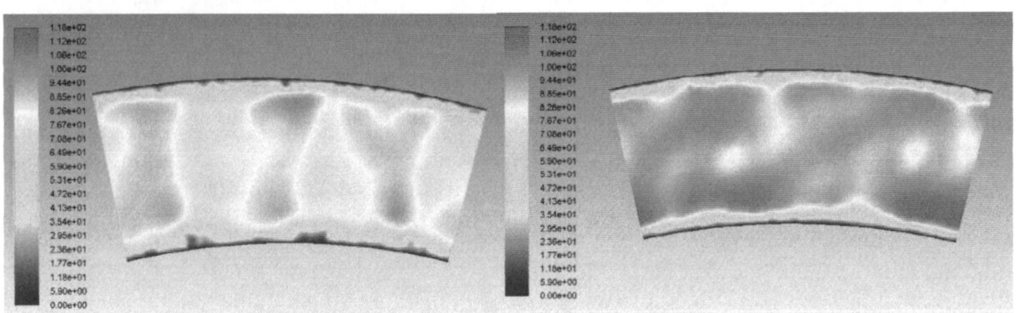

(a) (b)

Figure 10. Velocity field at the cc outlet for the cc simulation with turbo compressor (a) and without turbo compressor (b) at nominal operation mode.

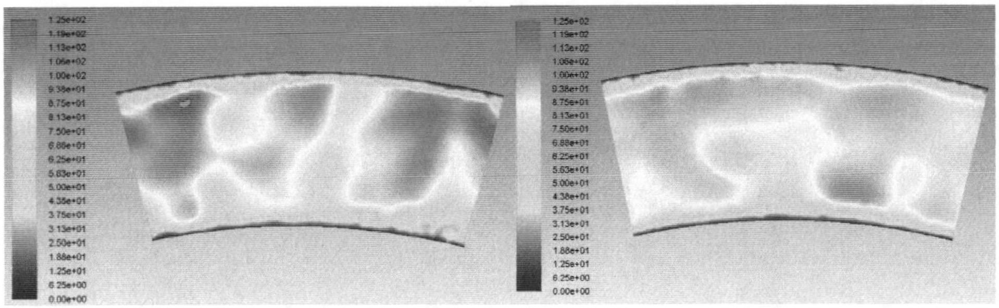

(a) (b)

Figure 11. Velocity field at the cc outlet for the cc simulation with turbo compressor (a) and without turbo compressor (b) at 0.5 of nominal operation mode.

The comparison of simulation results show significant differences in the cases of combustion chamber simulation, integrated compressor/combustor simulation and CC with turbo compressor. Besides, differences occur in quantitative (see Fig. 3) and qualitative ratio (see Fig. 7). At both operation modes in the case of CC simulation with turbo compressor the flame contour, its size and temperature are different from similar parameters in the case of combustion chamber simulation. For example, the temperature difference is about 300 K. Since these parameters influence on organization of the CC workflow, it can be assumed that the received changes will affect the mixing zone. For example, mass fraction of NOx at the outlet of combustion chamber differs by two orders of magnitude, and velocity differs by 20 m/s. This is confirmed by the results presented in Figures 5,6–11.

Differences in obtained results can not be explained only by compressor influence. In this case calculation results of combustor simulation with the turbo compressor and combustor with only compressor will be equal. As a comparison shows the differences (see Fig. 5, 6) it is evident that the presence of turbine in the calculation affects the workflow in the combustion chamber. Thus, the influence of such elements as the compressor, the combustor and the turbine on each other, takes place in forward and reverse direction.

Calculation of total pressure loss for the combustion chamber with turbo compressor and without showed that the obtained values may differ from each other by up to 30%.

4. Conclusion

Summarizing all the above it can be noted that:

- The flow distortion after the compressor influences on the processes occurring in the combustion area of the combustion chamber.
- Changes in processes occurring in the combustion zone of the combustion chamber, lead to gas flow parameters change at the turbine inlet.
- Calculation of NOx emission values and a total pressure loss in the simulations of combustion chamber with and without turbo compressor gives different results.
- Operation of the turbine influences on the formation of the temperature field at the turbine inlet, particularly making it more distorted in circumferential direction.

The research of this field is ongoing in order to inspect the chemical processes more clearly. Reactor models are used. Course of chemical processes of the harmful substances formation after combustion chamber in a turbine is under estimation.

Acknowledgments

This work was supported by the Ministry of education and science of the Russian Federation in the framework of the implementation of the Program "Research and development on priority directions of scientific-technological complex of Russia for 2014-2020" (RFMEFI58716X0033).

IOP Conf. Series: Materials Science and Engineering **302** (2018) 012021 doi:10.1088/1757-899X/302/1/012021

References

[1] Kannan K V and Page G J 2016 Automated multi-code URANS simulation of compressor-combustor components. Proceedings of the ASME Turbo Expo 2016, GT2016-56904

[2] Kannan K V and Page G J 2014 Coupling of Compressible Turbomachinery and Incompressible Combustor Flow Solvers for Aerothermal Applications. ASME Turbo Expo 2014: Turbine Technical Conference and Exposition. ASME.

[3] Walker A D, Carrotte J F, Peacock G L, Spencer A and McGuirk J J 2013 Experimental Study of the Unsteady Aerodynamics of the Compressor-Combustor Interface of a Lean Burn Combustion System San Jose, 49th AIAA/ASME/SAE/ASEE Joint Propulsion Conference.

[4] Schlüter J U, Wu S K, Alonso J J and Pitsch H 2003 "Integrated RANS-LES Computations Of Turbomachinery Components: Generic Compressor/Diffuser", Center for Turbulence Research Annual Research Briefs ADA420749.

[5] Boudier G, Gicquel L Y M, Poinsot T, Bissières D and Bérat C 2007 "Comparison of LES, RANS and experiments in an aeronautical gas turbine combustion chamber," Proceedings of the Combustion Institute, 31(2), 3075-3082.

[6] Schluter J U, Wu X, Kim S, Alonso J J and Pitsch H 2004 "Coupled RANS-LES computation of a compressor and combustor in a gas turbine engine." In 40 th AIAA/ASME/SAE/ASEE Joint Propulsion Conference and Exhibit.

[7] Sehra A K 2006 "The numerical propulsion system simulation: A vision for virtual engine testing". In Proceedings of the ASME Turbo Expo, 2006: Power for Land, Sea, and Air, ASME, New York, USA.

[8] Matveev S S, Zubrilin I A, Orlov M Y and Matveev S G 2015 Numerical investigation of the influence of flow parameters nonuniformity at the diffuser inlet on characteristics of the GTE annular combustion chamber. Source of the Document Proceedings of the ASME Turbo Expo 4A, GT2015-42676

[9] Orlov M Y and Matveev S S 2014 Numerical simulation of an influence of a compressor and a turbine on characteristics of a combustion chamber of a small-sizes gas turbine engine. Life Science Journal 11 (11), 119, pp. 650-654

[10] Orlov M Y, Matveev S S, Makarov N S and Zubrilin I A 2014 Numerical modeling problems of operating process of combustion chambers of GTE and solution approaches. Source of the Document ARPN Journal of Engineering and Applied Sciences 9 (12), pp. 2894-2899

[11] Anisimov V, Zubrilin I and Orlov M 2016 Investigation of thermal and stress states of the annular combustion chamber flame tube walls. Proceedings of the ASME Turbo Expo, 2016. GT2016-57479.

[12] Orlov M Y and Matveev S S 2014 Numerical simulation application for the design and fine-tuning of small-sized gas turbine engine combustor. Source of the Document Open Mechanical Engineering Journal 8, A450, pp. 450-456

[13] Zubrilin I A, Dmitriev D N, Matveev S S and Matveev S G 2015 Numerical investigation of the nonreacting swirling flow structure downstream of industrial gas turbine burner with the central body. Proceedings of the ASME Turbo Expo 4A, GT2015-42181

[14] Matveev S S, Zubrilin I A, Orlov M Y, Matveev S G and Chechet I V 2016 Investigation of fuel distribution in partially premixed swirled burner with pilot flame. Proceedings of the ASME Turbo Expo 4B-2016, GT2016-57478

[15] Zubrilin R A, Matveev S S, Zubrilin I A and Matveev S G 2106 Gaseous fuel flame stabilization in a modular swirled burner. Proceedings of the ASME Turbo Expo 4B-2016, GT2016-57441

[16] Popov G M, Baturin O V, Kolmakova D A and Krivcov A V 2014. «Improvement results of TK-32 turbocompressor turbine with gas-dynamics and strength CAE-systems». International Journal of Engineering and Technology, 6(5), pp. 2297-2303.

[17] Kolmakova D A, Baturin O V and Popov G M 2014. «Knowledge lack impact assessment of the source date on numerical simulation results of operational process in axial flow turbine blade row». ARPN Journal of Engineering and Applied Sciences 9(12), pp. 2880-2889.

[18] Matveev V N, Popov G M, Baturin O V, Goryachkin E S and Kolmakova D A 2015. « Workflow optimization of multistage axial turbine». 51st AIAA/SAE/ASEE Joint Propulsion Conference 2015, Orlando, United States, p. 13
[19] Gibson M M and Launder B. E 1978, "Ground Effects on Pressure Fluctuations in the Atmospheric Boundary Layer," J. Fluid Mech., 86 491–511.
[20] Launder B E 1989 "Second-Moment Closure: Present... and Future?", Inter. J. Heat Fluid Flow, vol. 10, no. 4. pp. 282–300.
[21] Launder B E and Reece G J 1975 "Progress in the Development of a Reynolds-Stress Turbulence Closure," J. Fluid Mechanics, vol. 68, no. 3, pp. 537–566.
[22] Coppalle A and Vervisch P 1983 "The Total Emissivities of High-Temperature Flames". Combustion and Flame. 49. 101–108.
[23] Anderson W and Bonhus D L 1994 "An Implicit Upwind Algorithm for Computing Turbulent Flows on Unstructured Grids". Computers Fluids. 23(1). Pp. 1–21.
[24] Oijen A and Goey L P H 2000 "Modelling of Premixed Laminar Flames Using Flamelet-Generated Manifolds". Combust. Sci. Tech. 161. Pp. 113–137.
[25] ANSYS 15.0 User's Guide.

IOP Conf. Series: Materials Science and Engineering **302** (2018) 012022 doi:10.1088/1757-899X/302/1/012022

Influence of Thrust Level on the Architecture and Optimal Working Process Parameters of a Small-scale Turbojet for UAV

V S Kuz`michev, E P Filinov and Ya A Ostapyuk

Department of Aircraft Engine Theory, Institute of Engine and Power Plant Engineering, Samara National Research University, 34, Moskovskoye shosse, Samara, 443086, Russia

Email: kuzm@ssau.ru, filinov.evg@gmail.com, oya98@mail.ru

Abstract. This article describes how the thrust level influences the turbojet architecture (types of turbomachines that provide the maximum efficiency) and its working process parameters (turbine inlet temperature (TIT) and overall pressure ratio (OPR)). Functional gasdynamic and strength constraints were included, total mass of fuel and the engine required for mission and the specific fuel consumption (SFC) were considered optimization criteria. Radial and axial turbines and compressors were considered. The results show that as the engine thrust decreases, optimal values of working process parameters decrease too, and the regions of compromise shrink. Optimal engine architecture and values of working process parameters are suggested for turbojets with thrust varying from 100N to 100kN. The results show that for the thrust below 25kN the engine scale factor should be taken into the account, as the low flow rates begin to influence the efficiency of engine elements substantially.

1. Introduction

Small-scale gas turbine engines are widely used as a source of mechanical energy or propulsion for electric generators, power plants, UAVs (cruise missiles and target aircraft), auxiliary power units, etc [1, 2].

Design process of a gas turbine engine sometimes spreads over 8-10 years and begins with conceptual design stage: setting the engine architecture and values of working process parameters. This problem is usually solved using multi-criteria optimization.

As the size (thrust level) of small-scale engines decreases, so does the size of engine blades (vanes) and other characteristic sizes. Thus, the Reynolds number decreases, the relative axial/radial clearances and thicknesses increase and the relative boundary layer thickness increases as well. This impairs the efficiency of turbomachinery and finally the efficiency of engine [3]. Therefore, the first task of this investigation is to determine the value of engine thrust to begin taking the sizing factor into the account.

Radial turbomachinery is affected by the scale factor to a much lower extent, but generally have less efficiency. Finding the trade-off point (in terms of thrust) between these factors for both compressors and turbines is the second task.

2. Problem setting

This investigation is based on the results of multi-criteria optimization of the working process parameters (turbine inlet temperature T_4^* and overall pressure ratio of the compressor $\pi_{c\Sigma}^*$) of turbojet (as a propulsion unit of UAV) with thrust varying from 100 N to 100 kN.

Axial/axicentrifugal/centrifugal compressors and axial/radial turbines were examined for each value of thrust, thus providing the following engine architectures: centrifugal compressor + radial turbine (CC+RT), centrifugal compressor + axial turbine (CC+AT), axicentrifugal compressor + axial turbine (ACC+AT), axial compressor + axial turbine (AC+AT).

As engine thrust decreases, correction to the compressor and turbine efficiency was made depending on the type of turbomachine and the mass flow rate as described in [1, 4, 5].

As the turbojet is an element of higher-level system, aircraft efficiency criteria should be used as optimization criteria. Total mass of fuel required for mission and engine M_{e+f} and the fuel consumption per ton of payload per km of distance C_{t*km} are usually used describe the efficiency of aircraft in terms of mass and fuel efficiency. Both C_{t*km} and SFC provide similar results of optimization for a long-distance missions, thus the SFC was considered an optimization criterion for this investigation.

The following gasdynamic and strength constraints were taken into account: height of blades at the exit of compressor (Hc), height of blades at the inlet of turbine (Ht), circumferential velocity at the compressor blade tips (Uc), turbine pressure ratio (π_t^*), turbine inlet temperature at take-off mode ($T_4^*{}_0$). Impossibility to satisfy the constraints was the indicator for changing the architecture of engine. Principal design data of the turbojet and aircraft is shown in table 1.

Table 1. Principal design data

Parameter	Description	Value
F{cr}/F{t.o}	Cruise to take-off thrust ratio	0,5
H	Altitude, km	0,8
M	Mach number	0,85
L{f}	Flight range, km	2000
η*{pol.c.b}	Baseline value of compressor polytrophic efficiency	0,86
η*{t.b}	Baseline value of turbine efficiency	0,86
H{cf}	Minimal height of the centrifugal compressor blade, m	0,005
H{c.ax}	Minimal height of the axial compressor blade, m	0,01
H{t.ax}	Minimal height of the axial turbine blade, m	0,01
U{c}	Maximum circumferential velocity of the compressor, m/s	600
π{t *max*}	Maximum value of turbine pressure ratio	3

3. Results of optimization

Results of optimization were obtained using the computer-aided system ASTRA, developed at the Samara national research university and are listed in table 2. The computer-aided system ASTRA [6, 7] has built-in optimization module, which was used to obtain the results described in this article. If the optimization problem includes more than 10 variables, use of specialized design optimization software (e.g. IOSO [8, 9]) would be more effective.

For each value of thrust the locally optimal regions were determined. The value of deviation was set to 2%, ensuring the intersection of locally optimal regions while satisfying the constraints. The optimal point was determined at the trade-off area as equally spaced from the locally optimal points and having minimal value of $T_4^*{}_0$ (as providing higher TIT is more complicated task than providing higher OPR values). The results for thrust of 1kN are shown in figure 1 (the optimal point is marked with "x").

The figure 2, 3 shows the influence of thrust value on the optimal regions for the M_{e+f} and SFC (C_{sp}) criterion. The results show that for a given aircraft parameters and mission the optimal regions shrink with the decrease of thrust.

The optimal engine architectures for a set of thrust ranges were identified:

- 0.1-0.7 kN – centrifugal compressor and radial turbine;
- 0.7-1.3 kN – centrifugal compressor and axial turbine;
- 1.3-7 kN – axicentrifugal compressor and axial turbine;
- 7-100 kN – axial compressor and axial turbine.

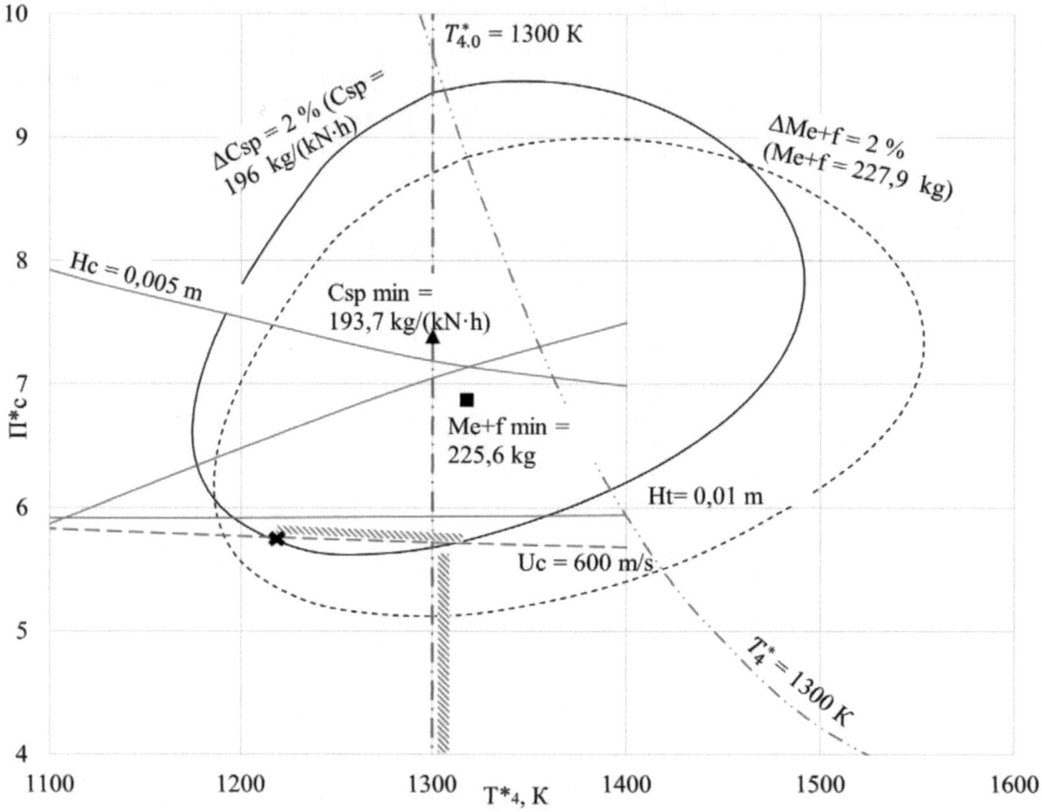

Figure 1. Locally optimal regions, constraints and optimal point for the turbojet with 1 kN of thrust (CC+AT architecture).

The optimal values (using both criteria) of overall pressure ratio and turbine inlet temperature against thrust values are shown in figure 4 and 5.

Table 2. Results of optimization for the thrust values from 0.1 kN to 100 kN

F	$\pi^*_{c\Sigma}$	$T^*_{4.0}$	$G_{a\ sp.0}$	M_{e+f}	C_{sp}	$C_{sp.0}$	Compressor	Turbine
kN		K	kg/s	kg	kg/(kN*hr)	kg/(kN*hr)		
0,1	2,7	1160	0,24	35,5	295	173	centrifugal	radial
1	5,4	1225	1,64	230	199	118	centrifugal	axial
2	7,5	1275	3,03	418	181	108	axicentrifugal	axial
10	13,5	1325	13,8	1882	154	93,9	axial	axial
25	15,8	1335	34,3	4613	149	90,9	axial	axial

50	16,7	1345	68,2	9183	148	90,3	axial	axial
75	16,8	1347	102	13761	148	90,2	axial	axial
100	16,9	1350	136	18338	148	90,2	axial	axial

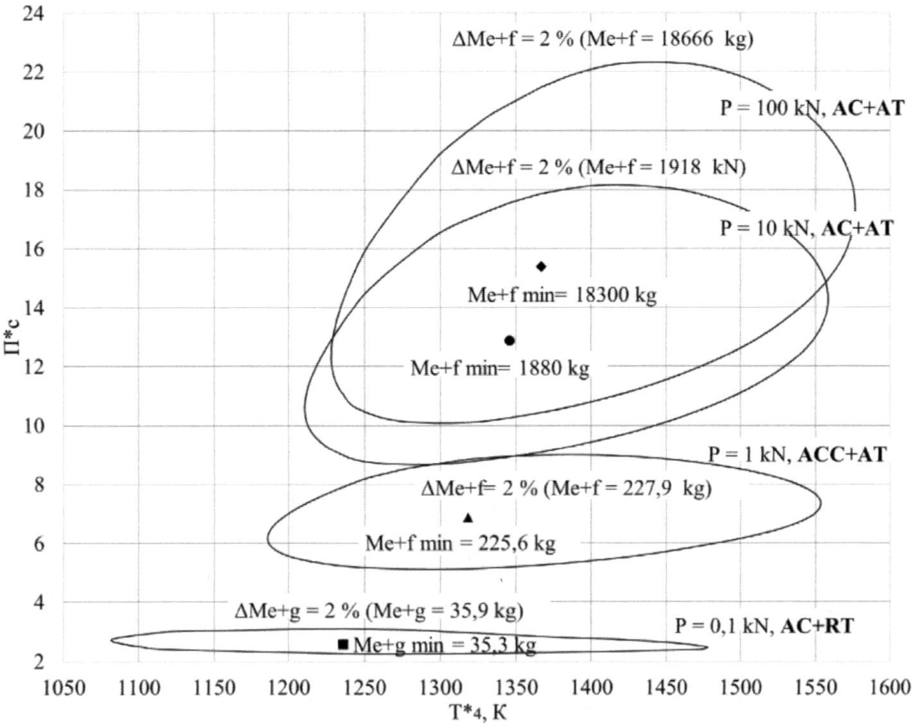

Figure 2. Influence of turbojet thrust on the locally optimal regions for the M_{e+f} criterion.

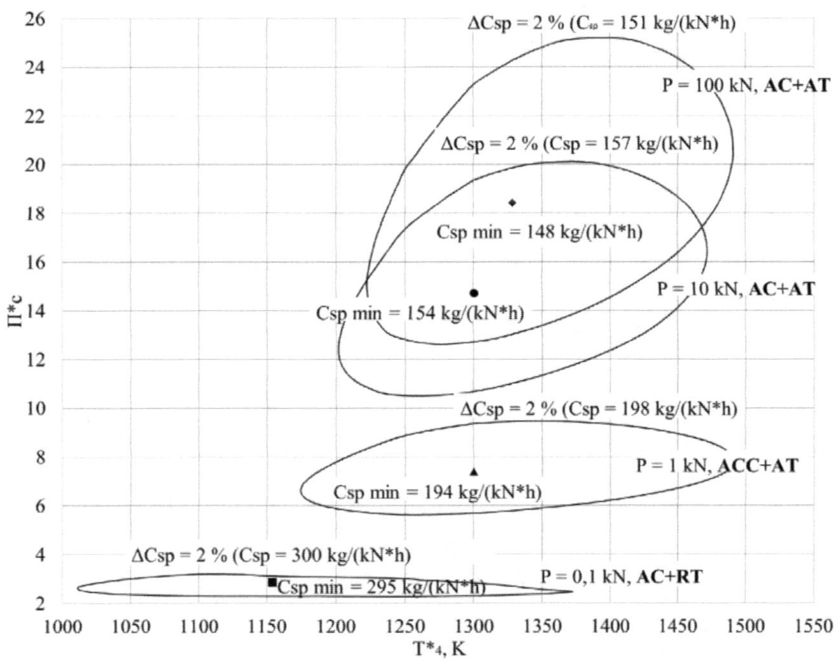

Figure 3. Influence of turbojet thrust on the locally optimal regions for the C_{sp} criterion.

4. Conclusions

As the thrust of turbojet at the take-off conditions decreases from 100kN to 0.1kN, the optimal value of overall pressure ratio becomes 6 times less and the turbine inlet temperature becomes 10-15% less and the locally optimal regions shrink as well.

The results of optimization with and without taking the correction into the account show that the engine scaling factor should be taken into account for the thrust values of 25 kN and lower.

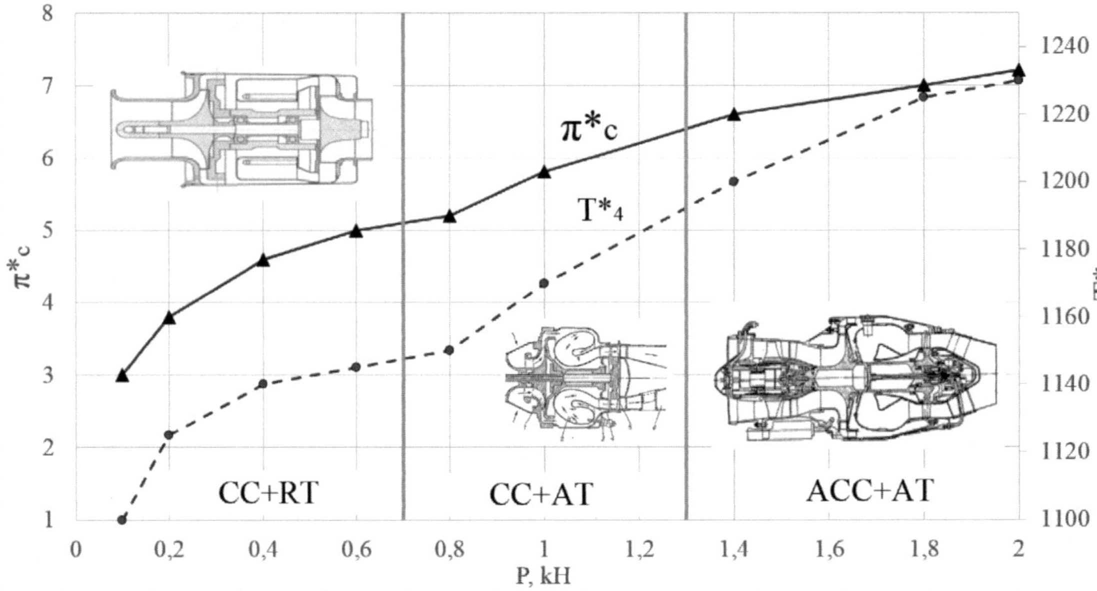

Figure 4. Optimal values (multi-criteria) of OPR and TIT and engine architectures of turbojets, thrust from 0,1 to 2 kN

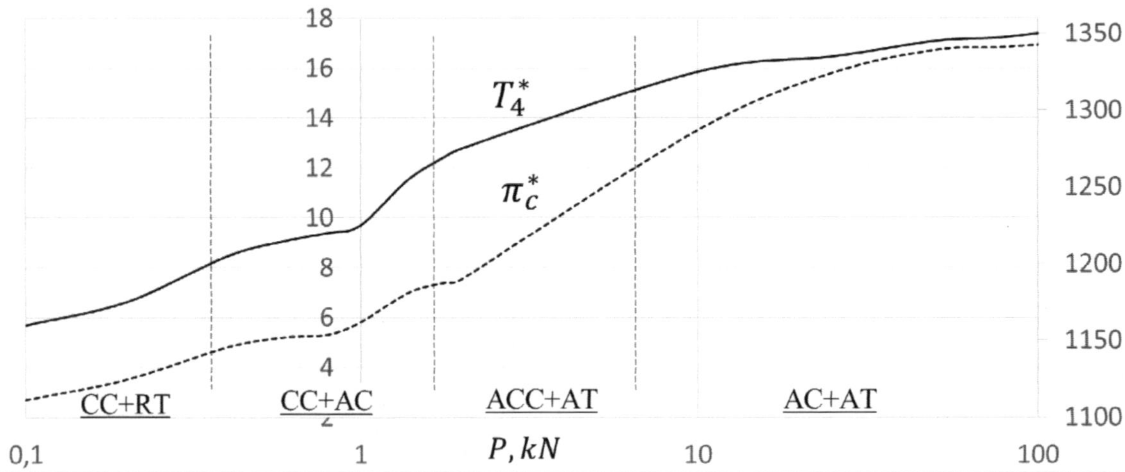

Figure 5. Optimal values (multi-criteria) of OPR and TIT and engine architectures of turbojets, thrust from 0,1 to 100 kN

The results of this investigation may be used to identify the optimal engine architecture for a given value of thrust. Axial compressor is optimal for thrust of 7 kN and higher, and as the thrust decreases, axicentrifugal compressor becomes more appropriate down to the 1.3 kN of thrust. For the thrust lower than 1.3 kN centrifugal compressor provides higher efficiency. Axial turbine is effective down to the 0.7 kN of thrust, and radial turbine – for smaller engines. It should be noted that these results correspond to the turbojet as a propulsion unit of cruise missile.

Acknowledgments
This work was supported by the Ministry of education and science of the Russian Federation in the framework of the implementation of the Program of increasing the competitiveness of Samara University among the world's leading scientific and educational centers for 2013-2020 years.

References

[1] Grigor'ev V A, Zrelov V A, Ignatkin Yu M, Kuz'michev V S, Ponomarev B A, Shakhmatov E V 2007 *Helicopter gas turbine engines* (Moscow: Machine Industry Press) p 491
[2] Sehra A K, Whitlow W Jr Propulsion and power for 21st century aviation *Progress in Aerospace Sciences* **40** (4-5) 199-235
[3] Akhmetov Yu M, Akhmetzyanov D A, Mikhaylova A B, Mikhaylov A E Features of functional design of the unmanned aircraft engines *UGATU Herald* **17** 78-86
[4] Grigor'ev V A, Zhdanovskiy A V, Kuz'michev V S, Osipov I V, Ponomarev B A 2009 *Parameter selection and thermogasdynamic calculations of aircraft gas turbine engines* (Samara: Samara St. Aerosp. Univ. Press) p 202
[5] Epstein A H 2003 Millimeter-scale, MEMS gas turbine engines *Proc. of ASME Turbo Expo 2003* vol 4 p. 669-696
[6] Kuz'michev V S, Ostapyuk Y A, Tkachenko A Y, Krupenich I N, Filinov E P Comparative analysis of the computer-aided systems of gas turbine engine designing *IJMERR* **6** (1) 28-35
[7] Kuz'michev V S, Tkachenko A Y, Krupenich I N, Rybakov V N Computer-aided system of virtual testing of gas turbine engines *MATEC Web of Conferences* **77**
[8] Popov G M, Baturin O V, Kolmakova D A, Krivcov A V Improvement results of TK-32 turbocompressor turbine with gas-dynamics and strength CAE-systems *International Journal of Engineering and Technology* **6**(5) 2297-2303
[9] Matveev V N, Popov G M, Baturin O V, Goryachkin E S, Kolmakova D A Workflow optimization of multistage axial turbine *51st AIAA/SAE/ASEE Joint Propulsion Conference* 132015

IOP Conf. Series: Materials Science and Engineering **302** (2018) 012023 doi:10.1088/1757-899X/302/1/012023

Technology for Transient Simulation of Vibration during Combustion Process in Rocket Thruster

V.M. Zubanov[1], D.V. Stepanov[1] and L.S. Shabliy[1]

[1] Samara National Research University, 34, Moskovskoe shosse st., Samara, Russia

E-mail: waskes91@gmail.com

Abstract. The article describes the technology for simulation of transient combustion processes in the rocket thruster for determination of vibration frequency occurs during combustion. The engine operates on gaseous propellant: oxygen and hydrogen. Combustion simulation was performed using the ANSYS CFX software. Three reaction mechanisms for the stationary mode were considered and described in detail. The way for obtaining quick CFD-results with intermediate combustion components using an EDM model was found. The way to generate the Flamelet library with CFX-RIF was described. A technique for modeling transient combustion processes in the rocket thruster was proposed based on the Flamelet library. A cyclic irregularity of the temperature field like vortex core precession was detected in the chamber. Frequency of flame precession was obtained with the proposed simulation technique.

1. Introduction

The combustion process in the rocket engine chamber largely determines operating parameters of rocket engine. Transient combustion processes are the most interesting for rocket engine engineers because of vibration induced by burning [1]. Also rocket thrusters often operate exclusively in pulsed mode [2]. The parameters in the chamber at this case are extremely unsteady during the entire period of the engine's operation time; the methods of research (both simulation and experimental) of stationary processes are not applicable for studying of such processes. But the CFD-simulation tool can be allowed for modeling of combustion processes in a transient mode with formally any degree of time discretization [3, 4]. The purpose of this work is to create a technology of the CFD-tools using, which will allow visualizing the unsteady workflow of any rocket thruster operating on gaseous components of oxygen and hydrogen to obtain vibration frequency. This technique was developed based on the previously published method for modeling stationary combustion [5, 6], therefore only its differences from the previous one will be indicated.

2. Features of CFD-model for transient simulation

2.1. Timestep selection

The time step should be assigned based on the average velocity of the flow along the mesh elements. The time step size usually amounts to microseconds, considering the high flow rate in the chamber. For example, with an axial velocity V_z = 5,000 m/s, a camera length L = 0.1 m and with averaged number of cells along the chamber length N = 200, the actual order of the time step Δt (1):

$$\Delta t = \frac{L/N}{V_Z} = \frac{0{,}1/200}{5000} = 0{,}1 \text{ microsecond} \quad (1)$$

It should be noted that only an approximate order of the time step can be obtained by such calculations. The exact value is selected during trial calculations by estimating the rate of development of transient processes in the calculated zone, when the solution is stable (more small steps over time make the decision more stable).

The total simulation time is chosen to "cover" the simulation time of the process of interest with a small range. Usually it includes 200 ... 300 time steps.

2.2. Meshing

The mesh model creation for transient calculation mode does not fundamentally differ from the mesh model creation for stationary simulation [6], however, there are some features. For example, if the ignition process is modeled by setting a high-temperature boundary wall on the discharge surface of a candle, then the boundary must be represented in the grid model (Figure 1).

The size of the grid cells must be that the flow, moving with its speed, "moved" by the distance between neighboring cells in one time step Δt (with the same cell sizes, it is equal to the size of the cells themselves). Thus, the grid must have a smooth change in the cells dimensions. The *Expansion Ratio* parameter, which reflects the "growth rate" of the cells, should not exceed 10 ... 50 value. And it is necessary to correlate the rate of change in the cells dimensions with the flow velocity change in the calculated zone.

2.3. Boundary conditions

The boundary conditions can be set as constant (in this case their description does not differ from [6]) or varying with time by a given dependencies. Constant boundary conditions can be used, for example, to study the combustion process of a steady flow. However, the real process of ignition of many thrusters occurs with significantly variable fuel components flow parameters that can be specified in the form of explicit (*CEL-expression*) or tabular (*User-Function*) dependencies.

The ignition initiation can be simulated by specifying the energy transfer (*Heat transfer*) or by the high temperature (*Temperature*) in the prechamber ignition zone. This zone can be conditionally modeled by a point (*Source Point*) or an extended area of the design area (*Subdomain*) which simulated a spark discharge zone, or by heating the boundary wall (*Wall*), which is the surface of the spark plug (figure 1).

3. Setting the properties of the working fluid

Hydrogen is strictly promising rocket fuel due to its high energy and ecology [7] the properties of the working fluid must be specified as *Ideal Variable Composition Mixture* through the properties of simple components: the initial reagents and reaction products including final and intermediate ones. The component composition of the simulated mixture is determined by a specific set of reactions that are considered in the calculation and which may include several dozen reactions.

A more actual «oxygen-hydrogen» reaction system, consisting of 26 equations, was found in work [8]; the equations describe the direct and inverse interaction of nine components (H, H_2, O, O_2, O_3, HO_2, OH, H_2O_2, H_2O), considering their relative catalytic effect. This system has been adapted to the combustion process in the thruster by excluding catalytic reactions involving nitrogen and argon.

The properties of all components except ozone (O_3) were taken from the «Gas Phase Combustion» ANSYS CFX library. Ozone properties were taken from the open Chemistry WebBook (http://webbook.nist.gov/chemistry) of the National Institute of Standards and Technology (USA) [9], where they are given by the coefficients of the Shomate equation and reformatted in coefficients of NASA-polynomial by method described in details in [10].

4. Initial solution for transient calculation

Setting the initial conditions for transient calculations should be done more carefully: for transient calculation, as a rule, the initial conditions are given as the results of a preliminary stationary calculation, in contrast to the stationary calculation, while the initial conditions of the stationary calculation can be arbitrary values. It is very important that the reaction mixture components set of the preliminary stationary calculation should be equal to the components set of transient calculation.

The initial values can be obtained using the *Eddy Dissipation Model (EDM)* of combustion. EDM does not consider the rate of chemical transformation of components in an explicit form [11]. The calculations with this model agree well and give an acceptable accuracy for specifying the initial distribution field of the parameters. Nevertheless, the EDM model, as a rule [6], overestimates the flow temperature (Figure 2).

It should be noted that the stability of the calculation with the complex multi-stage reaction system caused by the feature, which consists in the reaction rate, is neglected, and calculated from the reaction products in the EDM; parameter «B» of model EDM in ANSYS CFX has default value "-1". Otherwise, combustion model EDM can only be used for the system consisting of only two reactions (if option B is equal to "0.5" as the default in the ANSYS Fluent [12]).

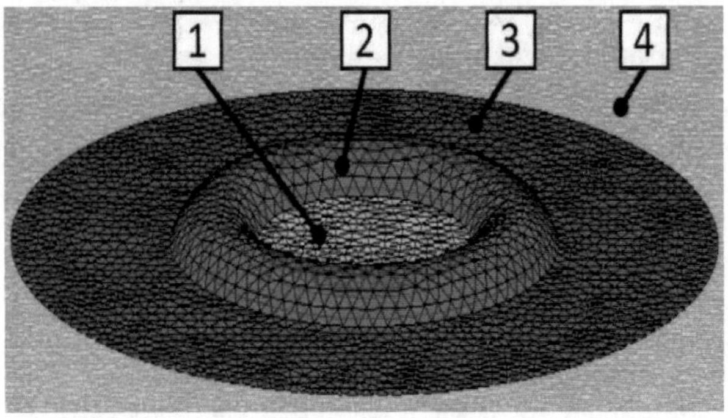

Figure 1. The discharge surface of a surface discharge plug in the prechamber mesh model: 1 - central electrode, 2 - flow zone (subdomain) around appearance of the spark, 3 - side electrode, 4 - prechamber wall

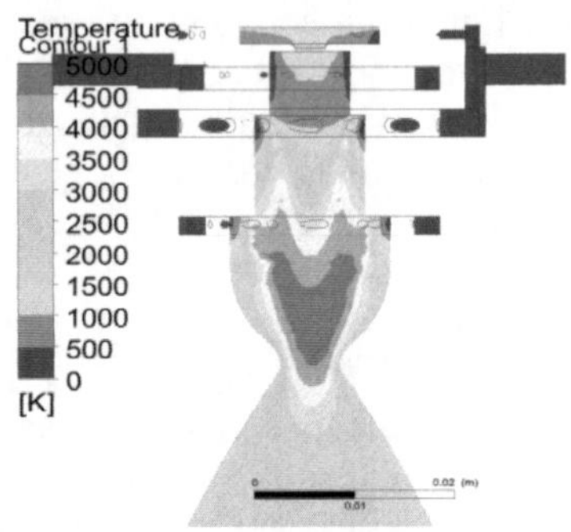

Figure 2. Static temperature distribution along the thruster chamber using EDM combustion model with chemical scheme of 24 reactions [13]

5. Combustion model and chemical reaction scheme

The choice of the combustion model and the system of chemical equations is a rather complicated task as in the case of stationary calculation [6]. An assessment of the different combustion models applicability for the transient processes modeling in thrusters in combination with different sets of reactions was the main objective of this study. To solve it, the combustion model *Finite Rate Chemistry* (*FRC*) [11] and *Flamelet* model of micro-flame combustion were considered. FRC model considers the rate of chemical reactions, calculated by the Arrhenius equation. The EDM combustion model was immediately excluded from consideration, since its application is justified only for combustion processes in which the rate of chemical reactions strongly prevails over the rate of their mixing (Damköhler number $D_a >> 1$); the latter cannot be guaranteed for combustion in a rapidly moving flow in the thruster chamber.

Combustion in the transient processes simulation can be described both by a single gross reaction and by a set of chemical reactions of varying detail degrees (as in the case of stationary processes [6]). A smaller number of reactions makes possible to obtain a faster and more stable solution by ignoring the intermediate chemical processes; the combustion temperature is usually overestimated, since the losses from underburning of the intermediate components are not considered. The most branched reaction scheme with the most detailed component description is necessary for modeling the transient combustion processes, including ignition; since the ignition process is largely determined by the interaction of unstable rapidly decomposing components present in the reacting mixture at low concentrations.

The following combinations of the "combustion model" - "set of reactions" were considered:

- FRC model with a reaction system [6], composed of 24 equations taken from [13], involving 8 components;
- FRC model with a reaction system [8], composed of 26 equations, involving 9 components (with addition of ozone O3);
- Flamelet-model, which uses Flamelet-library of 8 components to describe the combustion scheme.

The first two systems are a traditional representation of chemical interactions in the form of chemical equations, and the third is a tabulated information on the combustion products concentration at different fuel / oxidizer ratios in the flame front. The Flamelet library must be created before the calculation with the CFX RIF combustion library generator [11]. The Built In algorithm of the Flamelet library creation was used based on the combustion model of hydrogen fuel in air (Fuel Module - H2) with Kinetic Scheme – without NOx. The replacement of air with pure oxygen was accomplished by changing the component composition of the oxidant.

Attempts to use the FRC model did not allow one to obtain a solution with any of the investigated reaction schemes. The calculation with the reaction system [6] had no converged solution. The calculation with the system [8] converged, but the simulated combustion was not self-sustaining: the flame quickly faded during the first few time steps [10].

Calculation using the Flamelet-model stably converges and gives an adequate picture (Figure 3): For transient modeling, the flame front slightly fluctuates relative to its "average" position, determined for stationary simulation, the parameters of the flow (velocity, components concentration) practically do not differ from the values obtained for stationary simulation, except for a serious drop in temperature to values only slightly exceeding the maximum temperature of hydrogen combustion in pure oxygen (3100-3200 K).Comparison of the results of stationary modeling (Figure 2) and transient (Figure 3) simulation shows that combustion with the Flamelet model occurs at the mixing boundaries of the fuel and oxidizer not so rapidly that is associated with a finite rate of chemical reactions not considered earlier in the EDM model. This is confirmed by a certain amount existence of unreacted initial fuel components and intermediate combustion products in the critical section.

Thus, Flamelet-combustion model allows obtaining a stable and adequate solution, it can be recommended for use as a basic combustion model in the calculation of transient burning processes in thrusters.

6. Results analysis of transient simulation

Figure 4 shows the change in the parameter static temperature in time. A periodically rotating circle of flame unevenness was unexpectedly found in the cylindrical section of the subcritical tapered nozzle on the conical section of the thruster chamber during analysis of obtained data when testing the transient methods. It is like a vortex core precession. Figure 4 is given as an illustration of the fact that the use of transient simulation gives much more opportunities for investigating inter-chamber workflows. For example, based on the results obtained, the frequency and amplitude of the vortex precession can be calculated. From figure 4 is evident that during simulation time (0, 6 ms from 0,4 ms to 1,0 ms) flame moving was ~240 degrees on clockwise direction. So flame moving frequency can be calculated as (2):

$$f = \frac{\sim 240/360}{0.0006} \approx 1{,}1 \text{ kHz} \quad (2)$$

The simulation time can be increased for more accurate calculation. The application of the described technique should be accompanied by its validation under the conditions of a specific application, as in [14] and experimental verification as in [15].

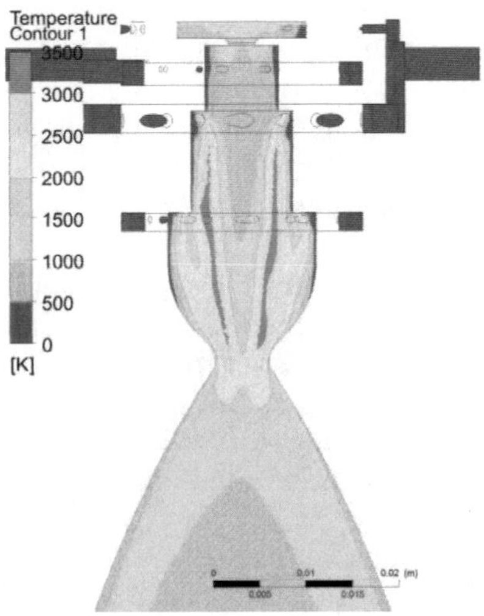

Figure 3. Static temperature distribution along the thruster chamber using *Flamelet*-combustion model

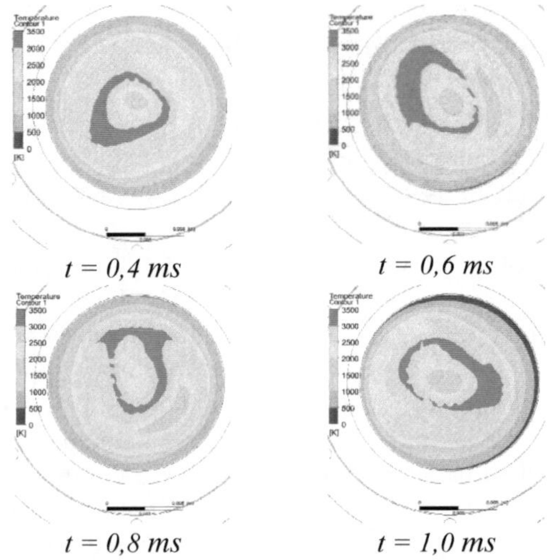

Figure 4. Static temperature distribution in the cylindrical section of the thruster chamber

7. Conclusion

Technology for transient simulation of combustion processes of rocket thruster on gashouses oxygen and hydrogen were created. Several modeling methods of the hydrogen-oxygen transient combustion in the chamber thruster were investigated using detailed chemical kinetics and the mechanisms of the different combustion models in ANSYS CFX were analyzed. From the considered combustion mechanisms only the *Flamelet*-combustion model allowed obtaining the qualitatively adequate results. Using the FRC combustion model is unreasonably difficult due to the large variation of rate constants input data of the Arrhenius equation for a set of elementary chemical reactions. The EDM combustion model can be used for quick obtainment of reference solutions for initialization, despite the overestimated temperatures. The *Flamelet*-combustion model was recognized as a qualitatively adequate model; it can be used in the future to study instabilities of combustion processes and transient processes like precession of vortex core (PVC). PVC frequency has been determined.

Acknowledgments

The work was completed with the financial support of the Ministry of Education and Science of Russia and with Grant Council of the President of the Russian Federation in the framework of the SP-3790.2015.1 scholarship.

References

[1] Dobrovolskiy M V 2005 *Liquid propellant rocket engines. Basics of designing.* (Moscow: MGTU Publ.) p 488

[2] Ivanov V K, Kashkarov A M, Romasenko E N and Tolstikov L A 2006 LRE turbopump unit of NPO Energomash Conversion in engineering Vol 1 pp 15-21

[3] Kozlov V E, Chechet I V, Matveev S G, Titova N S and Starik A M 2015 Modeling study of combustion and pollutant formation in HCCI engine operating on hydrogen rich fuel blends International Journal of Hydrogen Energy 41(5) pp 3689–3700 doi:10.1016/j.ijhydene.2015.12.078

[4] Biryuk V, Kayukov S, Zvyagintsev V and Lysenko U 2014 Ways of speed increase for internal combustion engine fuel injectors Research Journal of Applied Sciences 9(11) pp 721–724 doi:10.3923/rjasci.2014.721.724

[5] Egorychev V S, Shabliy L S and Zubanov V M 2016 Modeling intracameral workflow of small

thrust rocket engine operating on gaseous oxygen and hydrogen in ANSYS CFX. (Samara: Samara University Publishing) p 140

[6] Zubanov V, Egorychev V and Shabliy L 2015 Design of rocket engine for spacecraft using CFD-modeling Procedia engineering 104 pp 29–35 doi: 10.1016/j.proeng.2015.04.093

[7] Sulinov A V Shabliy L S and Zubanov V M 2017 Simulation methodology of the screw-centrifugal pump for liquid hydrogen Journal of Physics: Conf. Series 803 012161 doi:10.1088/1742-6596/803/1/012161

[8] Gerasimov G Ya and Shatalov O P 2013 Kinetic mechanism of combustion of hydrogen-oxygen mixtures Journal of Engineering Physics and Thermophysics (New York: Springer) 87(5) pp 1063-1070 doi: 10.1007/s10891-014-1108-z

[9] Chase M W Jr 1998 NIST-JANAF themochemical tables, fourth edition J. Phys. Chem. 9 pp 1–1951

[10] Zubanov V M, Stepanov D V and Shabliy L S 2017 The technique for simulation of transient combustion processes in the rocket engine operating with gaseous fuel "hydrogen and oxygen" Journal of Physics: Conf. Series 803 012187 doi:10.1088/1742-6596/803/1/012187

[11] Shabliy L, Krivcov A and Baturin O 2014 Separated computation of the whole jet engine workflow Proceedings of the 4th International Conference on Simulation and Modeling Methodologies, Technologies and Applications SIMULTECH 2014 pp 274–279 doi: 10.5220/0005108602740279

[12] Matveev S S, Zubrilin I A, Orlov M Y and Matveev S G 2015 Numerical investigation of the influence of flow parameters nonuniformity at the diffuser inlet on characteristics of the GTE annular combustion chamber Proceedings of the ASME Turbo Expo 4A V04AT04A051 doi:10.1115/GT2015-42676

[13] Gardiner W C 1984 Combustion Chemistry (New York: Springer-Verlag) p 351

[14] Zubanov V M, Shabliy L S and Krivcov A V 2015 Rational technique for multistage centrifugal pump CFD-modeling Proceedings of the ASME Turbo Expo 2B doi:10.1115/GT201542070

[15] Zubanov V, Egorychev V and Shabliy L 2014 Hydrogen-oxygen rocket engine design using CFD-modeling Research Journal of Applied Sciences 9(10) pp 660–663 doi:10.3923/rjasci.2014.660.663

The Technique for CFD-Simulation of Fuel Valve from Pneumatic-Hydraulic System of Liquid-Propellant Rocket Engine

L S Shabliy[1], D V Malov[1] and D S Bratchinin[1, 2]

[1] Samara National Research University, 34, Moskovskoe shosse st., Samara, Russia
[2] PSC Kuznetsov, 29, Zavodskoye shosse, Samara, Russia

E-mail: shelbi-gt500@mail.ru

Abstract. In the article the description of technique for simulation of valves for pneumatic-hydraulic system of liquid-propellant rocket engine (LPRE) is given. Technique is based on approach of computational hydrodynamics (Computational Fluid Dynamics – CFD). The simulation of a differential valve used in closed circuit LPRE supply pipes of fuel components is performed to show technique abilities. A schematic and operation algorithm of this valve type is described in detail. Also assumptions made in the construction of the geometric model of the hydraulic path of the valve are described in detail. The calculation procedure for determining valve hydraulic characteristics is given. Based on these calculations certain hydraulic characteristics of the valve are given. Some ways of usage of the described simulation technique for research the static and dynamic characteristics of the elements of the pneumatic-hydraulic system of LPRE are proposed.

1. Introduction

The hydraulic processes occurring in the liquid-propellant rocket engine (LPRE) are the subject of close attention of the research engineers, since it largely determines the efficiency and reliability of the engine. The calculated prediction of the nature and characteristics of the fluxes in the engine elements allows to identify possible design errors even before the product is embodied in the metal, and find ways to improve existing engines which much lower costs in comparison with the experimental study. However, using of popular in recent years CFD-simulation technologies of hydrodynamic processes of liquid rocket engines [1, 2], allows to perform even optimization researches on high-level models (in fact, virtual prototypes), but in the domestic rocket-engines industry it is still not widespread. Therefore, at the present time, young specialists are faced with the task of developing and mastering CFD-simulation technologies, which in the conditions of modern re-equipment of computational tools (the emergence of cluster computing technologies, supercomputers), will provide a fast pace of designing, manufacturing and testing LPRE aggregates [3, 4, 5]. This work shows clearly modern simulation tools applicable to space rocket industry.

2. Pneumatic-hydraulic system of LPRE

The work of the LPRE chamber is provided by the pneumatic hydraulic system (PGS) of the propulsion system. The term "PGS" is understood to mean a set of pneumatic-hydraulic devices and pipes, providing storage of fuel components and gases on board aircraft, their feeding during engine operation under a certain pressure and with a certain flow rate in the combustion chamber and engine gas generator, starting and stopping the engine, as well as performing some other operations determined by the purpose and specificity of the operation of the aircraft [6, 7]. Modeling the flow in pipelines, connectors and other elements, which does not change its shape of flowpath when the engine is running, usually does not represent complexity. Another situation occurs with valves - necessary elements of any PGS.

The article describes the approaches to the modeling of valves of the pneumohydraulic LPRE system by computational fluid dynamics (CFD) on the example of differential valve simulation. Such valves are used, for example, in the main trunking of the components of the LPRE NK-33 (AJ-26 in NASA classification). The NK-33 engine is a multiple-launch engine, those allows providing multiple start-up (acceptance test and operation as part of carrier rocket) without bulkheading of the main units and blocks [8, 9]. Despite the fact that after each start-up, it is necessary to replace certain single-acting units, the usage of differential valves allows not to remount the engine after the test.

2.1. Design of differential valve

The differential valve under consideration has a straight-flow construction (Figure 1). The body 1 consists of an outer shell and a central body mounted on pylons. In the central body is installed the stem of the movable valve 3, pressed by a spring 4 to the seat of the cover 2, connected to the body. To ensure the tightness of the valve in the closed state, the valve and cover contact is provided by the seal 5, which can be constructed from rubber or, in the case of a cryogenic component, a fluoroplastic.

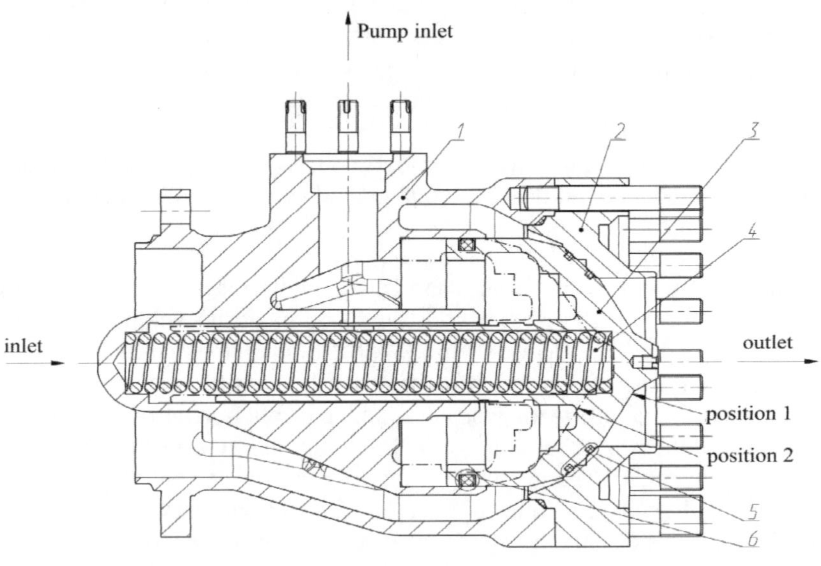

Figure 1. Longitudinal section of the differential valve under consideration: 1 – body; 2 – cover; 3 – valve; 4 – spring; 5, 6 – seals

2.2. Principle of operation of the differential valve

The valve is designed to control the timing of the fuel supply to the gas generator or the engine chamber, and also for stopping its feeding when the engine is switched off. Before starting the engine, the valve is closed by the action of a spring. Fuel is piped from the pump to the valve inlet, filling the cavity A before the seals. Also filled cavity B, connected through the flange to the pump inlet. While the pressure in cavities A and B is the same, the valve remains closed under the action of the spring, and the fuel does not flow past the valve to the consumer. After the turbo-pump unit (TPU) is unleashed by the pyro-turbine, the pressure in the supply line and the cavity A connected to it increases. The valve 3 (figure 1) is shifted to the left due to the fuel pressure in the gap between the valve 3 and the cover 2 on the tapered surface of the valve. When the valve opens, the fuel begins to act on the entire surface of the valve, ensuring its full opening. At the same time, on the side of the cavity B, the pressure remains equal to the pressure at the pump inlet. Thus, the valve remains in the open state as long as the pressure in the cavity A exceeds the pressure in the cavity B, which corresponds to the operation of the pump. The pressure difference at which the valve starts to open is determined by the initial spring force and the pressure at the pump inlet. For closed-circuit LPRE, when starting with a safe feed sequence, the components are the one at which an oxidizer first enters the consumer. At given pressures at the pump inlet, determined by their cavitation characteristics, and the dynamic characteristics of the pumps, the opening moment of the valves in the engine system is given by the initial spring compression force.

When the engine exits to the main thrust stage, the rotational speed of the TPU rotor increases, and the increasing pressure in the supply line increases the force that presses the valve 3 to the body 1. In this case, the valve remains in the open position. After operation the engine is switching off by stopping the supply of fuel to the gas generator from the control. In this case, the combustion in the gas generator ceases, the rotational speed of the rotor TPU drops and the pressure at the outlet from the pumps drops too. As a result, the pressure between the cavities A and B is equalizing, the valve 3 (see figure 1) under the action of the spring 4 returns to its original state - it is pressed against the cover 2, ensuring the closing of the valve.

3. Technique of CFD-simulation for differential valve

The above principle of operation of the differential valve shows the complexity of the valve operating process due to the presence of a differential pressure at the pump inlet and outlet. In this study, flow was simulated through the valve cavity A at different positions of the movable element: from fully open to minimally open. Also, the forces acting on the moving part of the valve were evaluated. The flow in the cavity B was not simulated.

As with any CFD simulation, the calculation process consisted of five stages: geometric modelling, meshing, imposition of boundary conditions, calculation and analysis of the obtained results.

3.1. Geometry

The geometrical model of the flow part was obtained by modifying the geometric model (Figure 2) created by the authors [10, 11] during the work on the project "Development of technology and methodological support for the creation of a virtual LPRE", implemented as part of the activities of block 2 "Development And increasing the effectiveness of scientific and innovative activities "and block 3 "Development of the information science and education environment and infrastructure" Development programs of the SSAU for 2009 - 2018. It is limited by the internal surfaces of the valve body, as well as the supply and discharge pipeline, the surfaces of the central body holding its pylons, and the movable part of the valve. Geometry reflects the main surfaces that affect the flow. At the same time, elements that do not have a significant effect on the flow (connecting gaps, small blind holes for connecting elements, etc.) were not reflected in the model, so as not to complicate the further construction of the mesh.

Since the task of this study was to simulate the flow at different valve positions, the geometric model was supplemented by another geometric body modelling the volume occupied by the flow part valve at its intermediate positions (shown in figure 2 in a darker color). Thus, the modified geometric model contained two geometric elements: a solid body representing the entire flowing part of the fully open valve (see figure 2) and the second solid body representing the moving element of the valve. This division allowed the relative displacement of two bodies to be modelled without the need to rebuild the computational mesh, since during the simulation of the "opening" of the valve, the reduction in the cavity B (see figure 1) was not modelled (see figure 2), and the displacement occurred unobstructed penetration of the moving body into the stationary one.

This division allowed the relative displacement of two bodies to be modelled without the need to rebuild the computational grid, since during the simulation of the "opening" of the valve, the reduction in the cavity b (see figure 1) was not modelled (see figure 2), and the displacement occurred unobstructed penetration of the moving body into the stationary one.

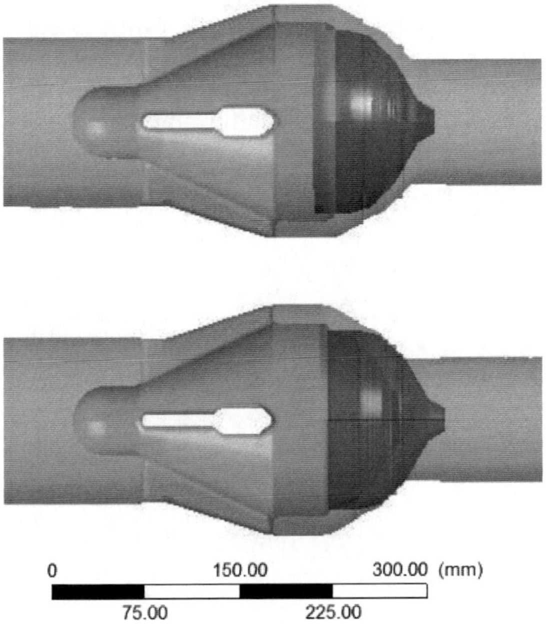

Figure 2. Geometrical model of the valve flowpath. The position of an additional geometric body that fills the volume of the movable part of the valve when its position corresponds to the fully closed state (below) and the semi-open (upper) valve mode

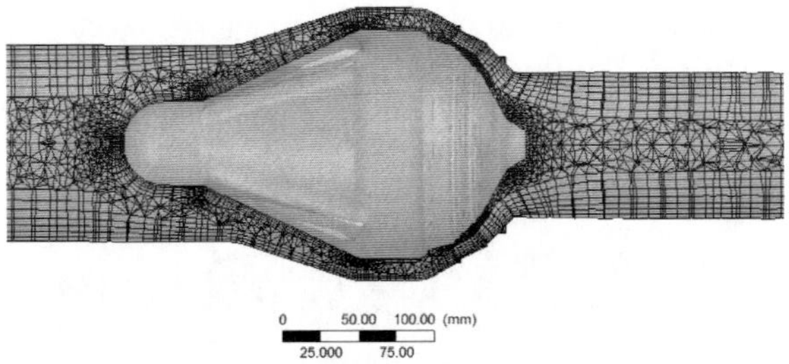

Figure 3. Mesh visualization for the model of a fully open valve in a plane at an angle of 45 ° to the plane of symmetry of the pylons of the central body

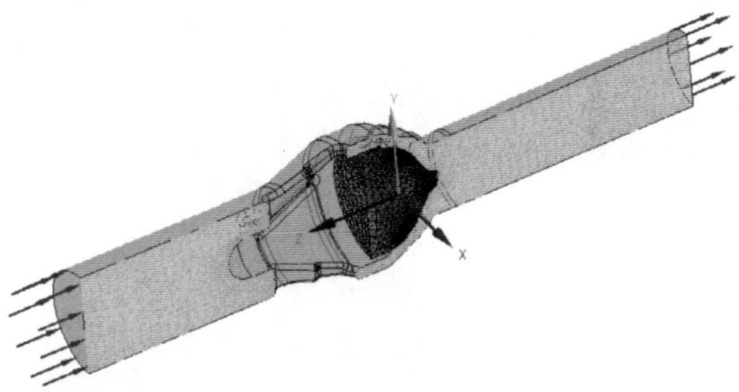

Figure 4. Visualization of the calculation model of the flowpath with immersed solid body represented via the mesh model

3.2. Meshing

The mesh model was created in the ANSYS Meshing program. The following settings were used:

- association of the dimensional function of grid elements - to the curvature (Use Advanced Size Function → On: Curvature);
- the limiting sizes of elements: (Min Size) – 0.1 mm, maximum (Max Size) – 23 mm, maximum size of the side of the element (Max Face Size) – 12 mm.
- coefficient of ratio of the sizes of neighbouring cells (Growth Rate) – 1.2
- smooth transition to thickening of the boundary layer (Inflation Option → Smooth transition)

Fully generated mesh (Figure 3) contained 2.3 million elements (tetrahedral - pyramids with triangular base, triangular prisms and pyramids with a quadrangular base)

3.3. Moving parts simulation

As already mentioned, the simulation of overlapping by the moving part of the valve of the flowing part was carried out without rebuilding of mesh, using the model of "solid body immersed in liquid" (Immersed Solid) [12]. A mesh of the flow part corresponding to the fully open valve is used (Figure 3), but a part of the cells marked by the program as cells occupied by a solid body (Figure 4) become impermeable to the flow. This is clearly seen from the current lines in different modes (Figure 5). In this case, the displacement of the Immersed Solid body was specified at the stage of the solution start as a parameter of the model "displacement along the local Z-axis" (Figure 4).

3.4. Stationary simulation

The ones shown in Figure. 5 flow patterns were obtained by calculations in a stationary setting. The data obtained from such calculations can characterize the throttling properties of the valve (Figure. 6). However, specific throttling characteristics can be obtained only for the previously known operating conditions of the valve determined by the PGS: a constant pressure difference between the valve and the variable flow or vice versa, a constant flow at a varying pressure at one or two boundaries. The ones shown in Figure. 5 and 6, the results were obtained by simulating a constant flow rate.

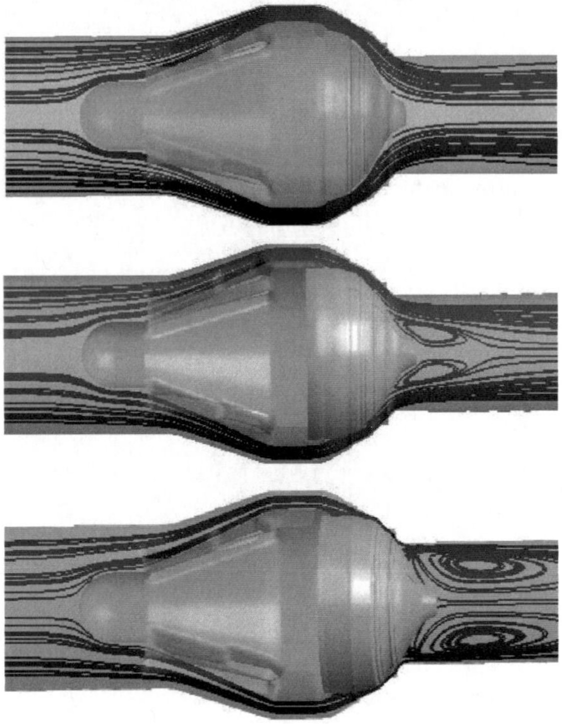

Figure 5. Lines of current in the valve: fully open (top), partially open (in the middle), almost closed (bottom) with 3 mm displacement above the fully closed position

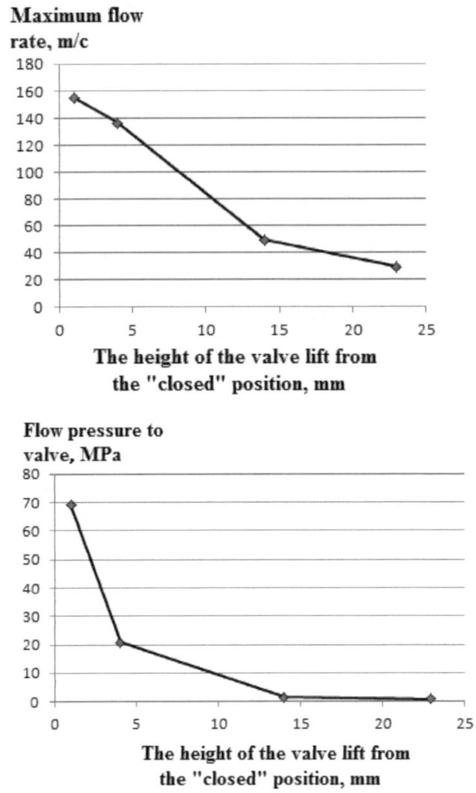

Figure 6. Valve characteristics depending on the lift height (displacement) of the valve

3.5. Transient simulation

At the second stage, the flow through the valve was modeled, taking into account the movement of its moving parts in an unsteady setting. In this case, the motion of the locking element can be set according to the laws of mechanics of solids (Rigid Body [12]). In this case, the ANSYS CFX solver takes into account the superposition of inertial forces, spring elasticity and fluid pressure based on the specified mechanical properties of the moving element (orientation, degrees of freedom, center of mass, mass, moments of inertia, initial velocity and acceleration, etc.) and calculates the displacement in Given instants of time. The visual result of this simulation is the video of the valve working process under given conditions, which can be used to study the work on virtual prototypes.

4. Conclusions

Thus, the described set of software tools, models and sequence of their application can be considered as an experimented technique that allows modeling of valves, as well as other elements of the fluid hydraulic system of LPRE, and determination of their characteristics with the framework of optimization studies [13] or gas-dynamic tests [14].

Acknowledgments
The authors wishing to acknowledge assistance or encouragement from colleagues [10, 11] for the high quality of the geometric model of the flow-through part of the valve that they created in the framework of the project "Development of technology and methodological support for the creation of a virtual liquid-propellant rocket engine (LPRE)", which was taken as a basis for creating a modified geometric model used in this study.

The work was completed with the financial support of the Ministry of Education and Science of Russia within the basic part framework of the state assignment and with Grant Council of the President of the Russian Federation in the framework of the SP-3790.2015.1 scholarship.

References

[1] Zubanov V, Egorychev V and Shabliy L 2015 Design of rocket engine for spacecraft using CFD-modeling *Procedia engineering* **104** pp 29–35 doi: 10.1016/j.proeng.2015.04.093
[2] Zubanov V, Egorychev V and Shabliy L 2014 Hydrogen-Oxygen rocket engine design using CFD-modeling Research Journal of Applied Sciences 9(10) pp 660–663 doi: 10.3923/rjasci.2014.660.663
[3] Egorychev V S and Ryazanov A I 2016 A method of design calculation for the two-component jet-jet gas nozzle with outer mixing Russian Aeronautics 59(4) pp 619–622 doi: 10.3103/S1068799816040280
[4] Zubanov V M, Shabliy L S and Krivcov A V 2015 Rational technique for multistage centrifugal pump CFD-modeling Proceedings of the ASME Turbo Expo 2B doi:10.1115/GT201542070.
[5] Shabliy L and Cherniaev A 2014 Optimization of gas turbine compressor blade parameters for gas-dynamic efficiency under strength constraints Proceedings of the 4th International Conference on Simulation and Modeling Methodologies, Technologies and Applications SIMULTECH 2014 pp 523-528 doi: 10.5220/0004994905230528
[6] Belousov A I and Sedel'nikov A V 2014 Problems in formation and control of a required microacceleration level at spacecraft design, tests, and operation Russian Aeronautics 57(2) pp 111–117 doi:10.3103/S1068799814020019
[7] Vaulin S D, Vaulina O S, Salich V L and Schipicin A G 2005 The research of dynamics processes in the plant of the liquid industrial wastes utilization on the basis of the rocket engine on gas propellants Izvestiya Akademii Nauk. Energetika 1 pp 101–109
[8] Borisov V A, Zhizhkin A M and Melent'ev V S 2011 Liquid-propellant rocket engine NK-33 (Samara: Samara University Publ.) p 27
[9] Parovay F V, Borisov V A and Parovay E F 2014 Research of temperature influence on the contact pressure in radial seals Research Journal of Applied Sciences (Medwell Journals) 9(11) pp 761–766 doi: 10.3923/rjasci.2014.761.766
[10] Gimadiev A G, Bystrov N D, Gafurov S A and Bratchinin D S 2013 Design of operating process model of main fuel valves of LPRE (Samara: Samara University Publ.) p 73
[11] Gimadiev A G, Stadnik D M and Bratchinin D S 2014 Analysis of the flow force in the fuel components supply valves of the aircraft engines ARPN Journal of Engineering and Applied Sciences 9(12) pp 2864-2866
[12] ANSYS Inc. 2014 CFX Modeling Guide
[13] Shablii L S and Dmitrieva I B 2014 Blade geometry transformation in optimization problems from the point cloud to the parametric form Russian Aeronautics (Allerton Press Incorporation) 57(3) pp 276–282 doi: 10.3103/S1068799814030106
[14] Matveev V N, Shabliy L S and Krivcov A V 2017Application of stereolithography prototypes for gas dynamic tests and visualization. Journal of Physics: Conf. Series 803 012097 doi:10.1088/1742-6596/803/1/012097

Gas-Dynamic Methods to Reduce Gas Flow Nonuniformity from the Annular Frames of Gas Turbine Engines

D Kolmakova[1] and G Popov[1]

[1] Researcher, Samara National Research University, Samara, Russia

E-mail: daria.kolmakova@ssau.ru

Abstract. Gas flow nonuniformity is one of the main sources of rotor blade vibrations in the gas turbine engines. Usually, the flow circumferential nonuniformity occurs near the annular frames, located in the flow channel of the engine. This leads to the increased dynamic stresses in blades and consequently to the blade damage. The goal of the research was to find an acceptable method of reducing the level of gas flow nonuniformity. Two different methods were investigated during this research. Thus, this study gives the ideas about methods of improving the flow structure in gas turbine engine. Based on existing conditions (under development or existing engine) it allows the selection of the most suitable method for reducing gas flow nonuniformity.

1. Introduction

Circumferential nonuniformity of gas flow is one of the main sources of turbomachinery rotor blades destruction [1-6]. Several approaches are implemented in practice to reduce the level of gas flow nonuniformity. The main idea of these methods is to eliminate the negative effect of nonuniformity. Firstly, it is possible to modify the design of turbomachinery by changing the axial clearance between the RW and GV or by changing the stagger angles and pitch of GV blades to non-uniform [7-9]. Secondly, the modifications can be implemented for the annular frame by changing the number of struts or using the inclined struts. However, it is difficult to use them for existing engines.

This article is dedicated to reducing the circumferential nonuniformity of gas flow appeared form the annular frame of axial compressor. The goal of this research was to find the ways of the flow non-uniformity reduction, which would lead to a decrease in the level of dynamic stresses in RW blades located in front of the annular frame, but would not require a fundamental engine design changing.

Both ways of reduction of flow nonuniformity including variations in design of compressor and frame were considered in the paper. The main advantage of this research is detailed consideration of the reasons of negative effects of flow nonuniformity and possible ways to reduce the nonuniformity.

2. Motivation of the research. Description of the compressor.

The motivation for this study was the real-life problem of the destruction of rotor blades of the fifth rotor wheel in intermediate pressure compressor (RW5 in the Figure 1). This compressor is part of a gas turbine engine NK-36ST, which has been designed and manufactured at JSC "Kuznetsov" [10].

The cause for the blade destruction is circumferential flow nonuniformity from the struts of the annular frame. The struts have different thickness and uninform distribution (Figure 2). The struts cause the local areas of high pressure (Figure 3).

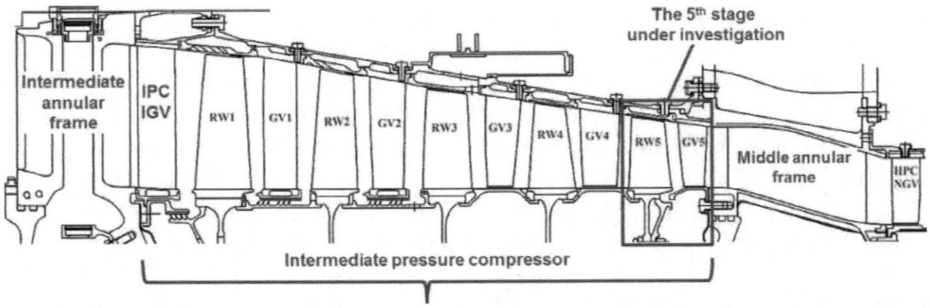

Figure 1. Scheme of the flow section of the test compressor.

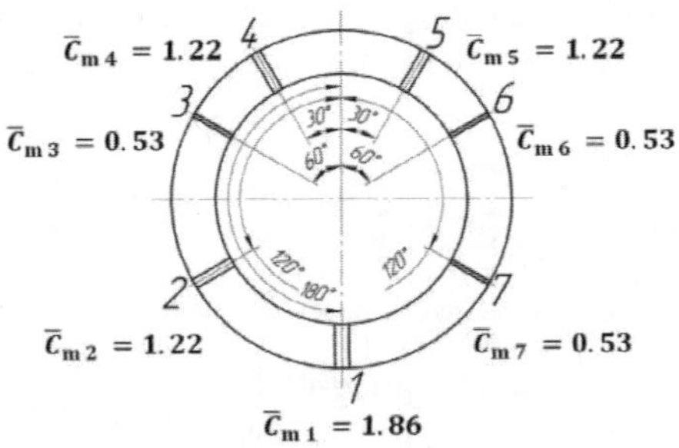

Figure 2. The appearance of the middle engine annular frame.

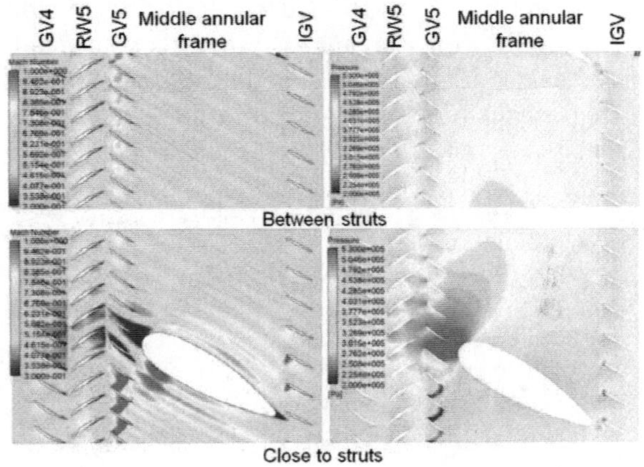

a) b)

Figure 3. Mach number (a) and pressure (b) fields.

3. Method of calculation of the level of gas flow nonuniformity

Step 1. CFD simulation of the sector IPC model at the main GTE modes which consists of intermediate annular frame, all blade rows of IPC, middle annular frame, and HPC IGV domains. All blades were simulated as one blade passage with periodical boundary conditions; frames were simulated as a sector of one strut. To transfer flow parameters between the domains, the *Mixing Plane* interface averaging the flow parameters in the circumferential direction was used [11]. The aim was to determine the distribution of the flow parameters along the height of flow section (total pressure, total temperature and flow angle) behind the RW4. These parameters then were used as boundary conditions for subsequent calculations.

Step 2. Calculation of the "full circle" model. Full circle model is used to detect the flow circumferential unevenness, which arises from the annular frame struts. Full circle model consists of GV4, RW5 and GV5, the middle annular frame and HPC IGV. Gas dynamic load on the all blades and static pressure distribution along the circumference of the RW5 are determined in this calculation.

Finite volume mesh of the full circle model was created in *AutoGrid5* [11]. Parameters of the mesh and configuration of computational models were chosen according to the recommendations from [12, 13]. Computational models have been verified [7, 14].

Total pressure, total temperature, flow direction at the inlet obtained from sector model and static pressure at the outlet of were set as boundary conditions for the full circle model. To transfer parameters between rotating and stationary domains of the full circle model, the interface *Frozen Rotor* was used. This interface transmits the flow parameters between domains without their averaging in the circumferential direction [15]. Thus, it can be used to calculate the circumferential flow nonuniformity.

4. Gas-dynamic methods to reduce gas flow nonuniformity

One of the methods of circumferential nonuniformity reduction is to use guide vanes with different circle pitch and different stagger angles in front of annular frame struts [8, 9]. This allows to redistribute the flow between the blade passages and to adjust the position of high pressure zones [16].

To implement the proposed method, a parametric model of IPC GV5 was created. Guide vanes were divided into 7 groups according to the location of 7 struts of the annular frame. Changes in the stagger angles and pitch were performed within each group. The minimum number of variable vanes was a key factor, since the production of many blades with different geometry greatly increases the charges.

The stagger angle of has not been changed for GV5 vanes located in the symmetry plane of the strut, the first and the last vane in the group. Changing the stagger angles was carried out according to the linear law. Vanes located on the opposite sides of symmetry plane of the strut were rotated in opposite directions (Figure 4). When the vanes are located closer to the strut, they were rotated by a larger angle. If the stagger angle increases, there is sign "+" before the angle If the stagger angle decreases, there is sign "-" before the angle. Stagger angels were measured from the leading edge plane.

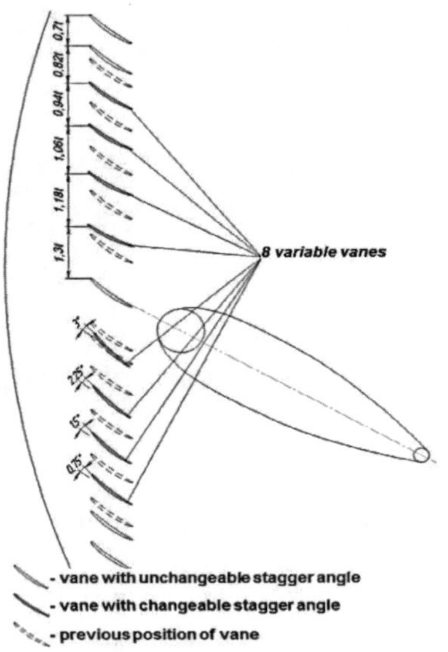

Figure 4. Scheme of GV vanes position change.

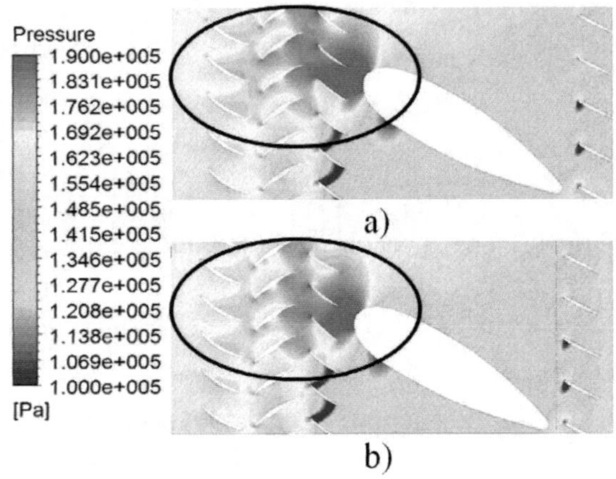

Figure 5. Static pressure distribution of basic variant (a) and GU variant No. 9 (b).

In case of the pitch change, the number of variable vanes was not restricted. Different pitch was set within -0.35...+0.35 of the initial pitch. The sign + (-) indicates that the pitch increase (decrease). The number indicates the maximum increase (decrease) in the pitch between the vanes in the group in relative values from the initial pitch with evenly spaced vanes. The position of extreme vanes in the groups has not been changed. The law of the pitch changing was also linear.

There were investigated 11 different configurations of GV5 with circumferentially different stagger angles and pitch. The variants with the greatest decrease of dynamic stresses are shown in the Table 1.

Ununiform stagger angles and pitch allow to achieve a significant reduction in gas flow nonuniformity. Flow structure becomes more uniform due to the redistribution of the flow (Figure 5). In addition, the values of pressure peaks behind the RW5 were reduced (Figure 6). These improvements lead to the decrease in dynamic stresses in rotor blades. That was confirmed in [17] with the help of calculation method proposed by researches A.I. Ermakov and A.O. Shklovets.

Implementation of this method leads to a complication of manufacturing technology, since it is necessary to produce the guide vanes using a new technology in this case.

Another method for reducing the circumferential nonuniformity is based on the results of [16]. It is in distancing the leading edge of frame struts from the trailing edge of GV blades located upstream. An important requirement is to maintain the internal cavities of frame struts the engine system.

Struts can be divided into three groups according to the maximum thickness of profile Cm: thick (strut No. 1), medium (struts No. 2,4,5) and thin (struts No. 3,6,7) (Figure 2).

To implement the method of the nonuniformity reduction, the parametric models were developed for each type of the strut. The models allow to adjust the position of the leading edge, maintaining the thickness and the rest of the profile by specifying the desired value of shifting (Δb) (Figure 7). Parametric models consider the limitations related to the location of the engine systems in the struts.

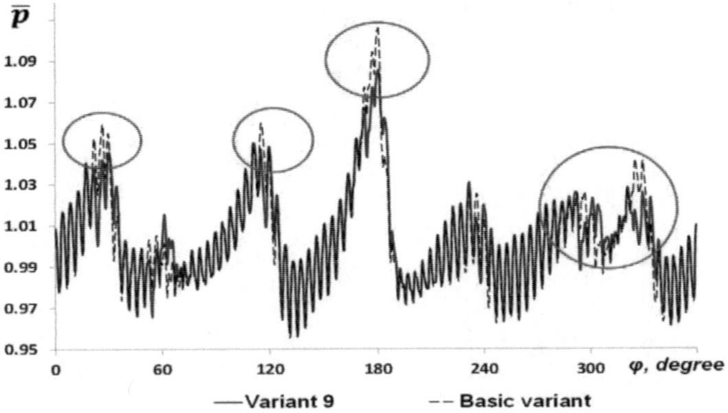

Figure 6. Graphs of the relative static pressure behind RW5 of the basic (a) and GV variant No. 9 (b).

Table 1. The results of the parametric IPC model calculation.

No	Parameter of stagger angles, maximum stagger angles (No. of blades) for groups:			Parameter of alternating blade pitch for the groups:					Number of variable blades	Dynamic stresses MPa
	2, 5, 6	1, 3, 7	4 (3)	1, 7	2, 6	3	4	5		
4	3 (6)	3 (6)	3 (8)	0.3	0.3	0.3	0.3	0.3	42	31.877
9	6 (2)	6 (2)	6 (2)	0	0	0	0	0	14	46.737

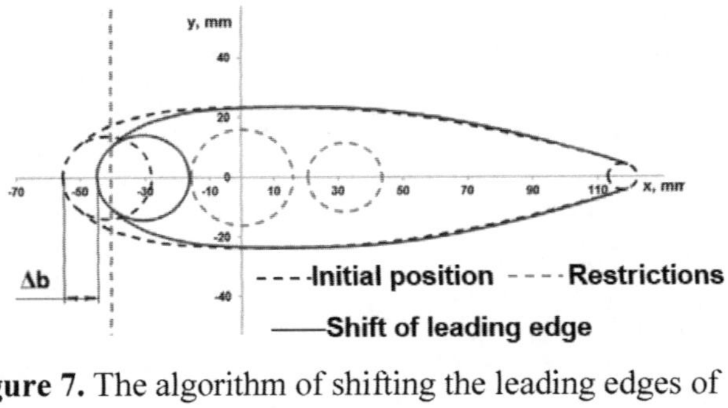

Figure 7. The algorithm of shifting the leading edges of struts

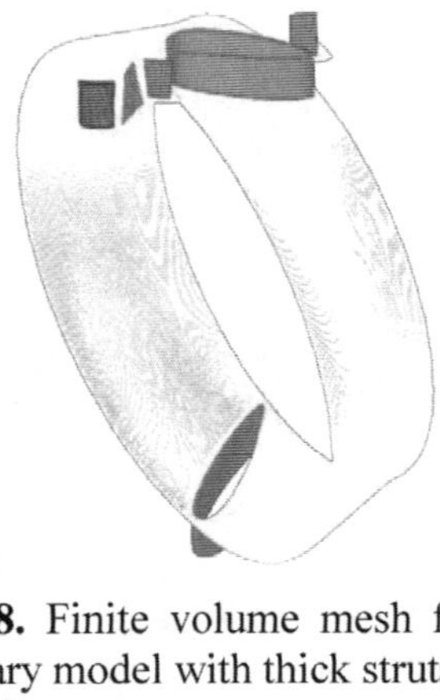

Figure 8. Finite volume mesh for the elementary model with thick struts.

To analyze the effect of leading edge shifting on the circumferential nonuniformity of the gas flow, the set of computational models of the compressor was created using the parametric models of the struts. All of them were full circle models and consisted of GV4, RW5 and GV5, middle annular frame, and IGV of HPC. The annular frame contained only two struts opposite each other with the same thickness in each model (Figure 8). These elementary model were created to assess the contribution of each type of strut in the total circumferential flow non-uniformity. In other words, they were used to calculate the pressure peaks occurring opposite each type of strut.

Moreover, leading edge shifting of struts with $\overline{\Delta b} = 0$, $\overline{\Delta b} = 0.2$, $\overline{\Delta b} = 0.4$, $\overline{\Delta b} = 0.6$, were set for each type of struts in different models. Thus, the total number of computational models was 12.

The calculation and analysis of the results were carried out in *Ansys CFX* using the supercomputer of SSAU "Sergei Korolev".

The circumferential distributions of relative static pressure in the section behind the RW for each type of strut with all values of leading edge shifting were obtained. It should be noted that obtaining of the pressure distribution only for one model took a long time, both in terms of creation model, and the calculation process. The total time for testing of just one variant is approximately 12 hours. Computational models were verified in [7, 17] while assessing the level of vibrations in rotor blades. These distributions were averaged to obtain the so-called elementary peaks (Figure 9).

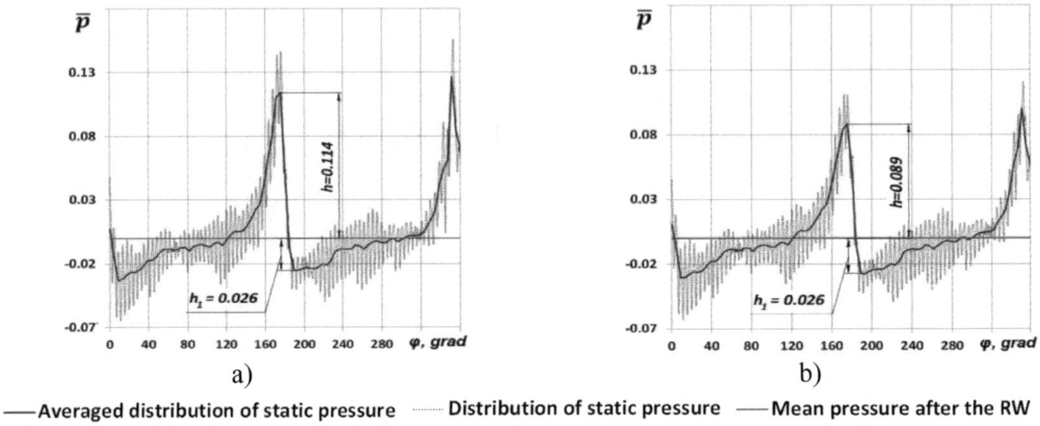

Figure 9. The distribution of the relative static pressure for thick strut with $\overline{\Delta b} = 0$ (a), $\overline{\Delta b} = 0.6$ (b).

Annular frame struts lead to the complex pressure peaks after the RW. The peaks have not only the maximum jump *h* (positive upper part of the graph), but also it goes into the decline *h1* (Figure 9).

Maximal jump of pressure peaks *h* decreases while the leading edge shift is increasing. Maximum drop of pressure peaks *h1* does not depend on the shift of the leading edge. Similar results were obtained for the other types of struts. To verify the adequacy of the dependences, the graph from the individual elementary peaks of each type of the struts was assembled and compared with the graph for the reference design of annular frame (Figure 10). In general, graphics derived from the elementary peak, describe the pressure distribution after the RW well.

Quadratic equations were derived for each type of strut using the regression analysis that describe the effect of the leading edge shift (Δb) on the pressure peak height (h). Equation 1 corresponds to the thick strut, Equation 2 to the medium struts and Equation 3 to the thin struts.

$$h = 0.001\Delta b^2 - 0.135\ \Delta b + 7.5 \quad (1)$$

$$h = 0.001\Delta b^2 - 0.095\ \Delta b + 4 \quad (2)$$

$$h = 2.64 \cdot 10^{-}18\Delta b2 - 0.02\ \Delta b + 2 \quad (3)$$

Application of leading edge shift of annular frame struts allows for not only quantitative changes in the values of pressure peak (Figure 11). The proposed method also provides better quality of flow structure: field becomes more uniform, high-pressure zone reduces.

The received regression equations were used for optimization of the location and shift of annular frame struts to reduce the dynamic stresses on the rotor blades [18].

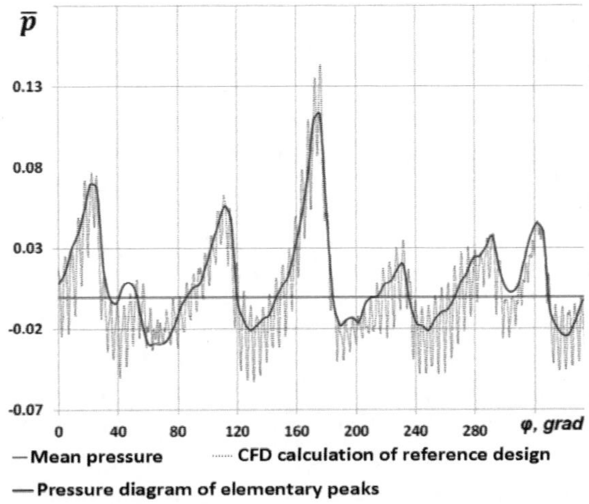

Figure 10. Graph of relative static pressure after RW obtained by different methods.

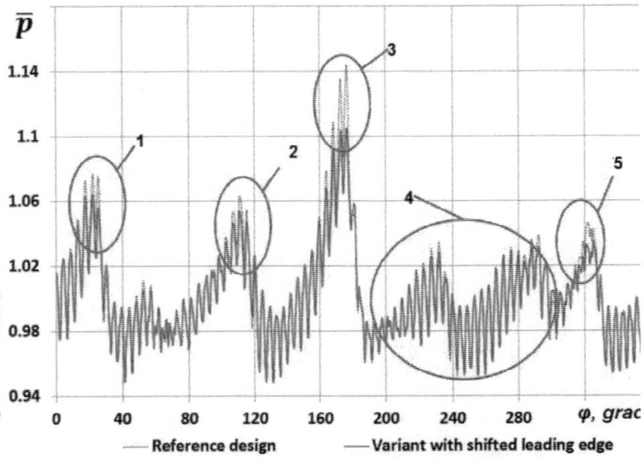

Figure 11. Circumferential distribution of the relative static pressure for the reference design and frame with shifted leading edges.

5. Conclusion.

Several approaches to decrease the gas flow nonuniformity in the axial compressor because of the annular frame have been developed because of conducted research. Proposed approaches allows to reduce the level of pressure peaks after the RW of the last compressor stage. Firstly, it was suggested to modify the guide vanes of the last compressor stage. It was found that the different stagger angles and alternating pitch flatten the circumferential variation after the RW5. In addition, flow field becomes more uniform and periodic. Secondly, the circumferential unevenness of gas flow in the compressor can be reduced by distancing its source (the support rack) from the elements, experiencing the negative impact of circumferential unevenness of gas flow (rotor blades located upstream). This is accomplished by shifting the leading edges of the support racks. The advantage of the second method is that it makes possible to predict the pressure peak level behind the RW without the long process of preparation of the computation model, calculation and post-processing of the results by using surrogate models. This can accelerate the design process of new compressors and design refinement of existing units.

Acknowledgments

The work was financially supported by the Ministry of education and science of Russia in the framework of the implementation of the Program of increasing the competitiveness of SSAU among the world's leading scientific and educational centers for 2013-2020 years.

References

[1] Ivanov V 1983 *Kolebaniya rabochego kolesa turbomashin (Vibrations of turbomachinery impeller)* (Moscow: Mashinostroenie) p 224
[2] Vorob'ev Yu 1988 Kolebaniya lopatochnogo apparata turbomashin (Vibrations turbomachinery blading) (Kiev, Nauk. Dumka) p 224
[3] Cohen H, Rogers G and Saravanamuttoo H 1996 Gas Turbine Theory, 4-th edition (Longman Group Limited) p 455
[4] Logan E. 2003 Handbook of Turbomachinery (CRC Press) p 880
[5] Birger I, Shorr B and Iosilevich G 1993 Raschet na prochnost' detalei mashin (Strength analysis machinery parts) (Moscow: Mashinostroenie) p 640
[6] Cumpsty N A 2004 Compressor Aerodynamics (Krieger Publishing Compan, Edition 2) p 552
[7] Kolmakova D, Popov G, Shklovets A and Ermakov A 2014 Proc. of the ASME 2014 Gas Turbine India Conference GTINDIA2014 Paper No. GTINDIA2014-8203

IOP Conf. Series: Materials Science and Engineering **302** (2018) 012025 doi:10.1088/1757-899X/302/1/012025

[8] Sladojević I, Sayma A and Imregun M 2007 Proc. of the ASME Turbo Expo 5 683–700
[9] Yang Y, Yang A, Dong R, Chen E and Dai R 2012 Hangkong Xuebao/Acta Aeronautica et Astronautica Sinica 33(4) 588-96
[10] JSC "Kuznetsov", Accessed July 20, 2015. http://www.kuznetsov-motors.ru/en
[11] NUMECA User Manual AutoGrid5 Release 8.4 (NUMECA.inc., Belgium) January 2008
[12] Matveev V, Popov G, Goryachkin E and Smirnova Y 2014 Research J. of Applied Sciences 9(11) 784-8
[13] Popov G, Goryachkin E, Baturin O and Kolmakova D 2014 Int. J. of Engineering and Technology 6(5) 2236-43
[14] Ermakov A, Shklovets A, Popov G and Kolmakova D 2014 Research J. of Applied Sciences 9(10) 684-90
[15] ANSYS® Help System, Ansys CFX-Solver Modeling Guide (Release 12.0, ANSYS, Inc.)
[16] Saren V 1984 Flow around irregular lattice of plates placed in front of the cylinder Technical Report (Moscow: Central Institute of Aviation Motors) p 36
[17] Popov G, Kolmakova D, Shklovets A and Ermakov A 2015 IOP Conf.Series: Materials Science and Engineering 90(1) Article number 01205
[18] Kolmakova D, Popov G, Shklovets A and Ermakov A Proc. of the ASME Turbo Expo GT2016 Paper No. GT2016-56991

IOP Conf. Series: Materials Science and Engineering **302** (2018) 012026 doi:10.1088/1757-899X/302/1/012026

Development of A New Equations Describing Profile Losses in Axial Turbine Blade Row

O Baturin[1] and D Kolmakova[2]

[1] Associate professor, Samara National Research University, Samara, Russia
[2] Researcher, Samara National Research University, Samara, Russia

E-mail: oleg.v.baturin@gmail.com

Abstract. Based on the previous investigations, regression equation for calculating the profile losses was found by the least squares based on the analysis of the profile losses nature and using mathematical optimization techniques. This equation makes possible to define the profile loss of axial turbine more accurate than the existing models and equations. It allows the calculation of profile loss values in the axial turbine that differ from the actual value by 10±61% with a probability of 95%. The obtained equation is not only superior to other loss models in statistical criteria, but it also considers the increasing number of factors affecting the profile losses.

1. Introduction

Axial turbine is the most common device for production of mechanical work. The number of turbines operating in various industries amounts to tens of thousands. For this reason, the problem of improving the turbine efficiency is relevant and its solving promises the great economic effect.

Most of the researchers have focused on finding the ways to improve the structure of the flow in the turbine blade rows. A large part of these studies was conducted using the methods of computational fluid dynamics (CFD). The major disadvantage of CFD is that CFD is only confirmatory method. It is an expert system that allows drawing a conclusion about quality of the specific variant of design.

Design of turbomachinery channels and formation of geometry of the blades is still carried based on 1D and 2D computations. CFD calculations only allow identifying and correcting design errors, as well as considering the features of the flow, which are not accounted in 1D and 2D calculations.

The better the design calculation was made, the smaller iterations with computationally expensive CFD models will be required, and the sooner the best variant will be found. For these reasons, the improvement of methods of 1D and 2D design of turbines is a promising task.

The most important aspect affecting the accuracy of prediction of turbine characteristics in 1D and 2D calculations is a reliable prediction of the energy losses in its flow part. Currently, more than ten complex models are known, allowing the calculation of the losses in axial turbines, as well as dozens of equations that allow calculating the individual loss components [1-5].

In the research described in [6], authors compared existing loss models. Report [7] of the Central Institute of Aviation Motors (CIAM) (Russian) [8] was used for there purpose. This report contains the results of experimental determination of the profile losses in more than 170 nonswirling cascades of axial turbines with constant height section. The study showed that although it is possible to establish the best loss model, these models can give significant errors during the calculation. For this reason, the decision was made to create a new equation for calculating the profile losses based on the available experimental data. This was the goal of current research described in the paper.

2. Selection of the variables that affect losses. Justification of the selection of the equation forms

The first task towards this goal was to choose the variables affecting the value of the loss and type of the equation. For this purpose, the analysis of scientific literature was carried out [9, 10, 11, 12], which showed that the energy losses on the surface of the streamlined blades meant by the profile losses excluding the effect of the end surfaces (the hub and shroud endwalls). It is considered that the profile losses consist of four components: friction, edge, wave and losses of flow separation.

In the analysis of the components of profile losses the following fact was highlighted. All components of the profile losses except for the wave losses vary within the entire range of flow velocities. Moreover, friction and edge losses will decrease and asymptotically tend to a certain value as velocity increases. There are no wave losses at low flow rates and they occurred only at $\lambda_{w2s} = 0.75 \ldots 0.85$. With the velocity growth, this type of loss increases (Figure 1). The opposite effect of these two factors determines the form of the dependences $\xi = f(\lambda_{w2s})$.

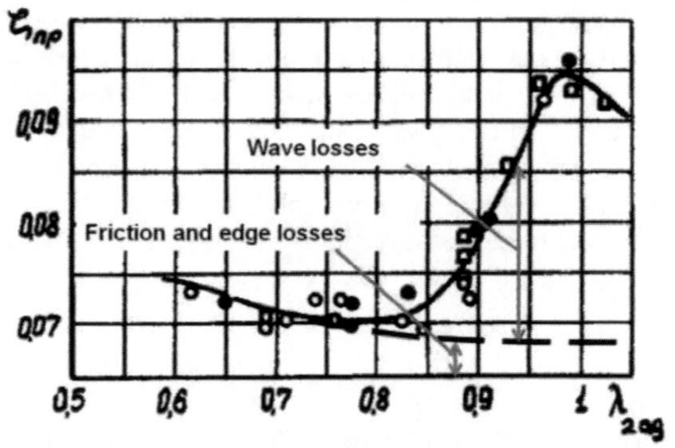

Figure 1. For an explanation of the ratio between friction, edge and wave losses.

Based on the analysis, the geometrical and operational parameters of the profile have been chosen that have a significant impact on the value of profile losses. According to the authors, these parameters must be considered when deriving the equation for the calculation of profile losses. Since derivable equation should be universal, most of the selected parameters were normalized. Table 1 provides a list of selected variables.

Table 1. Variables affecting the value of profile losses.

Designation	Nomination	Loss type on which it affects
$\frac{\Delta\beta}{100}$	Relative value of flow deflection angle in turbine cascade	Friction losses
$\bar{X}_c = \frac{X_c}{b}$	Relative position of maximum thickness	Friction losses
$\frac{C_m}{b}$	Relative maximum blade thickness	Friction losses
λ_{w2s}	Isentropic specific velocity at the cascade outlet	Friction losses, wave losses
$\frac{t}{b}$	Relative spacing	Friction losses, edge losses
$\frac{i}{\beta_1}$	Relative blade angle	Separation losses
Re	Reynolds number	Friction losses
β_2	Flow angle at the cascade outlet	Edge losses

δ	Bending angle of profile	Edge losses
r_2	Trailing edge radius	Edge losses

Based on the analysis of the form of the equations describing the profile losses in the existing models of losses [1-5,9,13], the expected form of the equation has been selected:

$$\xi_{prof} = (K_i \cdot K_{re} \cdot \xi_{fr} + \xi_{eg}) + \xi_\lambda \quad (1)$$

where $\xi_{fr} = A_2\left(\frac{\Delta\beta}{100}\right)^2 + A_3\overline{X}_c + A_4\left(\frac{C_m}{b}\right) + A_8\lambda_{w2s} + A_7\left(\frac{t}{b}\right) + A_9$ - a term, considering the profile losses of friction in the boundary layer in the reference conditions;

$K_i = 1 + A_1\left(\frac{i}{\beta_1}\right)^2$ – adjustment coefficient considering the losses of flow separation;

$K_{re} = A_6\left(\frac{2\cdot 10^5}{Re}\right)^{0.25}$ – adjustment coefficient considering the effect of Reynolds number;

$\xi_{eg} = A_5\frac{r_2}{t\cdot\sin(\beta_2+\delta)}$ – a term considering the edge losses.

$\xi_\lambda = A_{10}\sin(A_{14}\lambda_{w2s} + A_{11}) + A_{12}\lambda_{w2s} + A_{13}$ – a term considering the wave losses.

A1, A2, A3, A4, A5, A6, A7, A8, A9, A10, A11, A12, A13, A14 – coefficients of equations.

The desire to ensure that all terms of the equation are of the same degree, as well as preliminary study of equations of another type by algorithm described above, influenced the choice of the form of equation.

Thus, the problem of the derivation of a new equation of profile losses reduced to finding the numerical values of the coefficients.

3. Search algorithm of equation coefficients

The problem of finding the coefficients of the equation is a search of function extremum. Indeed, such a combination of coefficients must be found that provides a minimum difference between the results of calculation and experimental data. For this reason, the optimization program IOSO was used to find the values of coefficients [14-16].

The core of IOSO software is optimization method based on the construction of response surface, which is being refined and evolves at each iteration. Each iteration of IOSO contains two steps. At the first step, the response function in the form of multi-level graph is constructed based on the earlier accesses to the models with different combinations of varied parameters. The next step is finding of the extremum of a given function. This approach allows users to constantly adjust the response surface in the optimization process. As a result, an unusually small number of starting points are required to its construction and the first results obtaining [15, 17, 18].

IOSO algorithm has a low sensitivity to the topology of the objective functions and allows to accomplish the tasks for smooth, unimodal, multiextremal and non-differentiable functions.

The unknown coefficients of the equation A1-A14 have been selected as variable parameters while setting the optimization task. Their values varied in the range of ±150.

An average integral deviation of calculated values from experimental data in the whole considered range was specified as the optimization criterion:

$$\delta\xi_\Sigma = \frac{\sqrt{\sum_{i=1}^{n}(\delta\xi_i)^2}}{n} \quad (2)$$

where $\delta\xi_i$ - relative deviation from the expected experimental value found for the specific values of conventional specific velocity and the specific cascade.

n - total number of calculated values of the relative deviation for various combinations of conventional specific velocity and definite cascade.

Apparently, the smaller this value is, the better equation describes an experimental data array. Thus, the task was set to IOSO program in a process optimization to find such a combination of factors in which $\delta\xi_\Sigma$ would be minimal.

IOSO formed some combination of varied parameters in the performance of its internal algorithms. They were loaded into a file of Microsoft Excel, which calculates the value of the relative deviation from the expected experimental values $\delta\xi_i$ for different isentropic specific velocities λ_{w2s} and for all 170 cascades. The value of λ_{w2s} for each cascade was varied in the range 0.5-1.4 with a step of 0.05. In total, more than 27.5 thousand values of $\delta\xi_i$ were calculated for a varied combination of variables.

Ultimately, the Excel file calculated the average integral deviation of calculated values from the experimental data in the whole range of $\delta\xi_\Sigma$, as well as the mean values of the whole range of mathematical expectation $\mu_{\Delta\xi}$ and standard deviation $\sigma_{\Delta\xi}$.

Considering that the wave component of the profile losses is not appears at all values of velocity, optimization was carried out in two stages. At the first stage, the coefficients (A1, A2, A3, A4, A5, A6, A7, A8, A9), which consider the friction losses, edge and flow separation losses, were found based on the values of deviations for λ_{w2s} in the range of 0.5...0.75. Then, the coefficients of the term considering the wave losses (A10, A11, A12, A13, A14) were calculated at fixed values of found coefficients, based on the values of the deviations in the whole range of $\lambda_{w2s} = 0.5 ... 1.4$.

4. Final equation and its comparison with other loss models

As a result of optimization, the minimum value of $\delta\xi_\Sigma$, which is equal to 33, was reached. At the same time, mathematical expectation $\mu_{\Delta\xi}$ and standard deviation $\sigma_{\Delta\xi}$. were -10 and 31 respectively. The coefficients of the equation that give such result are shown in Table 2.

Table 2. Equation coefficients

Coefficient	Value	Coefficient	Value
A1	5,52661	A8	-37,16360
A2	1,16463	A9	-53,81240
A3	99,99734	A10	19,82116
A4	69,11952	A11	-53,727
A5	0,68967	A12	1,650231
A6	0,00036	A13	16,11946
A7	99,96786	A14	-9,93227

Thus, the derived equation to calculate the profile losses has the following form:

$$\xi_{prof} = (K_i \cdot K_{re} \cdot \xi_{fr} + \xi_{eg}) + \xi_\lambda \quad (3)$$

where $\xi_{fr} = 1,16463\left(\frac{\Delta\beta}{100}\right)^2 + 99,99734 \cdot \overline{X}_c + 69,11952 \cdot \left(\frac{C_m}{b}\right) - 37,16360\lambda_{w2s} + 99,96786\left(\frac{t}{b}\right) - 53,81240$ – a term, considering the profile losses of friction in the boundary layer in the reference conditions;

$K_i = 1 + 5,52661 \cdot \left(\frac{i}{\beta_1}\right)^2$ - adjustment coefficient considering the separation losses;

$K_{re} = 0,00036 \cdot \left(\frac{2 \cdot 10^5}{Re}\right)^{0.25}$ - adjustment coefficient considering the effect of Reynolds number;

$\xi_{eg} = 0,68967 \frac{r_2}{t \cdot \sin(\beta_2 + \delta)}$ – a term considering the edge losses.

$\xi_\lambda = 19,82116 \cdot \sin(-9,93227 \cdot \lambda_{w2s} - 53,727) + 1,650231 \cdot \lambda_{w2s} + 16,11946$ – a term considering the wave losses.

It was interesting to compare the statistical parameters of the new equation with similar data for other models of profile losses obtained earlier. The obtained equation shows better statistical data, compared with all the loss models discussed above. The derived equation shows the value of the mathematical expectation like CIAM model (both models underestimate the magnitude of profile losses by about 10%), but has better standard deviation by 25% (rel.).

These conclusions are illustrated by Figure 2, where the dependences of the profile loss coefficient ξ from the isentropic specific velocity at the cascade outlet λ_{w2s} are constructed as an example for the cascades No. 42, 55, 119, 135.

Figure 3 shows how the most probable value of the deviation of calculated data from experimental $\Delta\xi$ change depending on the values of isentropic specific velocity λ_{w2s} for all considered loss models.

Obtained maximum and minimum possible values of deviations $\Delta\xi$ with a probability of 95% indicated in the same figure. Data for CIAM loss model, the best of previously considered, are shown for comparison in Figures 2 and 3.

Apparently from Figure 2, the curves obtained with the new equation describe well the experimental data for most of cascades. Often, calculated dependence of $\xi = f(\lambda_{w2s})$ using the new equation is closer to the experimental points than the curve obtained by CIAM equations. It should be noted that dependences $\xi = f(\lambda_{w2s})$ have similar character.

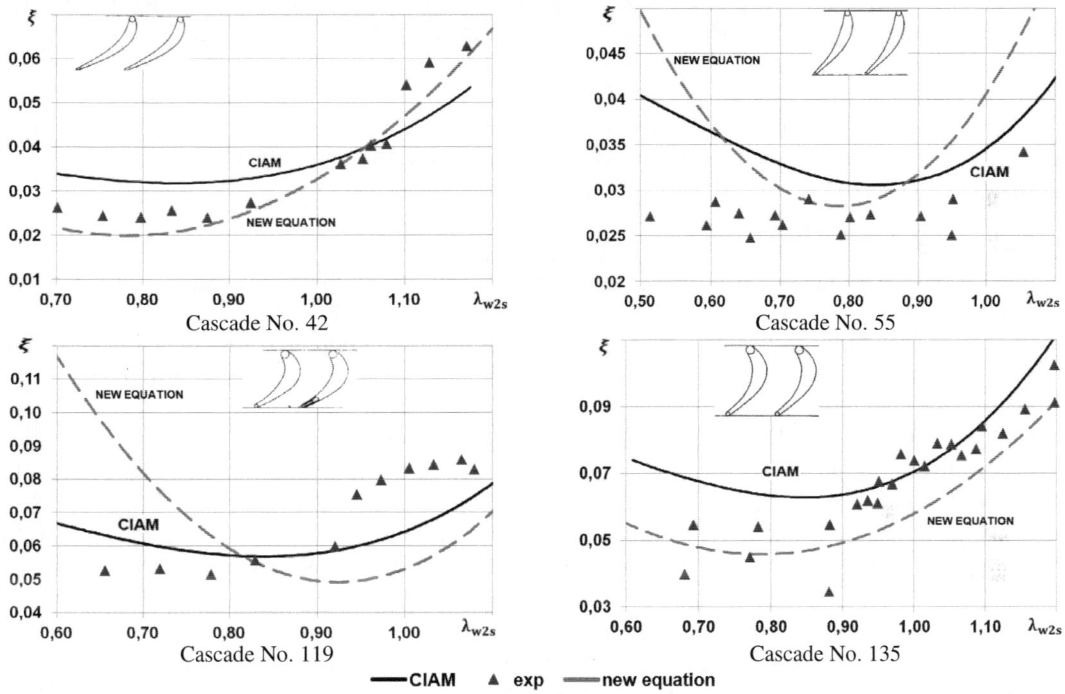

Figure 2. Comparison of the calculation results of profile losses obtained by a new loss model with experimental data and calculations with CIAM model.

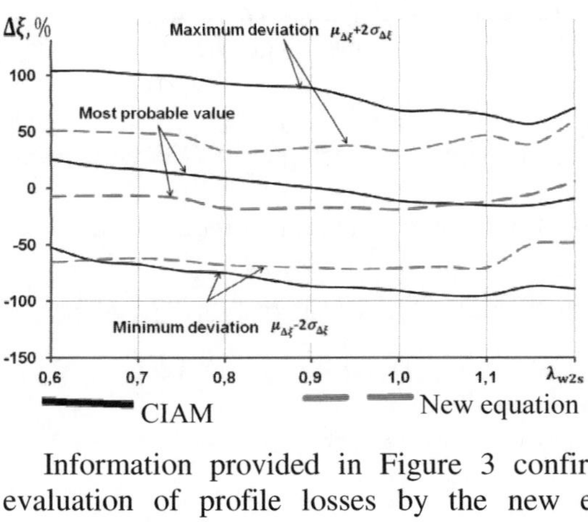

Figure 3. Comparison of changes in the most probable value of the estimated deviation of calculated profile loss values from real value and boundaries of scattering with the probability of 95%, depending on λ_{w2s} for derived equation and CIAM model.

Information provided in Figure 3 confirms the conclusions of the qualitative results in the evaluation of profile losses by the new equation and CIAM model. However, mathematical expectation of the new equation varies slight and is close to zero, indicating the proximity of the calculated and experimental data. New equation provides a smaller dispersion of the experimental and calculated data.

5. Conclusion

Based on the analysis of the profile losses nature and using mathematical optimization techniques, a new equation was proposed, which allows defining the profile loss of axial turbine more accurate than the investigated models. It allows the calculation of profile loss values in the axial turbine that differ from the actual value by 10±61% with a probability of 95%. The proposed new equation allows considering more geometric and operational factors affecting the value of losses.

This work is the first step to finding the best loss model in axial turbine. It considers only profile losses. In the future, it is planned to analyze other types of losses occurring in the blade passage of axial turbine in a similar way. As a result, it is planned to reliably determine which model is the best, or to develop a new one, which is better in accuracy than existing.

Acknowledgments

The work was financially supported by the Ministry of education and science of Russia in the framework of the implementation of the Program of increasing the competitiveness of SSAU among the world's leading scientific and educational centers for 2013-2020 years.

References

[1] Wei N 2000 Significance of Loss Models in Aerothermodynamics Simulation for Axial Turbines *Doctoral Thesis KTH ISBN 91-7170-540-6*
[2] Dahlquist A N 2008 Investigation of Losses Prediction Methods in 1D for Axial Gas Turbines Thesis for the Degree of Master of Science Division of Thermal Power Engineering Lund Institute of Technology (Lund University, Sweden)
[3] Horlock J H 1965 Axial Flow Turbines (Butterworths, UK)
[4] Ainley D G, Mathieson G C R 1955 An Examination of the Flow and Pressure Losses in Blade Rows of Axial-Flow Turbines Reports & memoranda; no.2891 (British ARC)
[5] Dunham J, Came P M 1970 Improvements to the Ainley/Mathieson Method of Turbine Performance Prediction ASME Journal of Engineering for Power 92(3) pp 252-256
[6] Baturin O, Kolmakova D, Gorshkov A and Popov G 2015 Selection of models to assess the profile losses in blade rows using the methods of mathematical statistics Proc. of ASME 2015 Gas Turbine India Conference, GTINDIA 2015 Paper No. GTINDIA2015-1245
[7] Venediktov V D, Granovsky A V 1990 Atlas eksperimental'nykh kharakteristik ploskikh reshetok okhlazhdaemykh gazovykh turbin (The Atlas of Experimental Performances of Cooled Gas Turbine Blade Cascades) p 393
[8] Central Institute of Aviation Motors, Accessed November 01, 2015. http://www.ciam.ru/.
[9] Abiants V Kh 1979 Teoriya gazovykh turbin reaktivnykh dvigatelei (The theory of the gas turbine jet engines) (Mashinostroenie, Moskow) p 246
[10] Belousov A N, Musatkin N F and Rad'ko V M 2003 Teoriya i raschet aviatsionnykh lopatochnykh mashin (Theory and calculation of aircraft turbomachinery), (Samara State Aerospace University, Samara) p 344
[11] Denton J D 1993 The 1993 IGTI Scholar Lecture: Loss Mechanisms in Turbomachines J. of Turbomachinery 115(4) pp 621-656
[12] Lewis R I 1996 Turbomachinery performance analysis, Elsevier Science & Technology Books Pub, p 329
[13] Kacker S C, Okapuu U 1982 A Mean Line Prediction Method for Axial Flow Turbine Efficiency ASME J.l of Engineering for Power 104(2) pp 111-119
[14] Sigma Technology, Accessed October 10, 2014 http://www.iosotech.com

[15] Egorov I N, Kretinin G V, Leshchenko I A, Kuptzov S. V 2002 IOSO Optimisation Toolkit - Novel Software to Create Better Design 9th AIAA/ISSMO Symposium on Multidisciplinary Analysis and Optimisation, (Atlanta, Georgia)
[16] Dennis B H, Egorov I N, Sobieczky H, Dulikravich G S, Yoshimura S 2003 Parallel Thermoelasticity Optimization of 3-D Serpentine Cooling Passages in Turbine Blades American Society of Mechanical Engineers, International Gas Turbine Institute, Turbo Expo (Publication) IGTI GT2003-38051
[17] Komarov O V, Sedunin V A, Blinov V L 2014 Application of Optimisation Techniques for New High-Turning Axial Compressor Profile Topology Design Proc. of ASME Turbo Expo Paper No. GT2014-25379.
[18] Rybakov V N, Kuz'Michev V S, Tkachenko A Y and Krupenich I N 2016 Thermodynamic multi-criteria optimization of the unified engine core for the line of turbofan engines Proc. of ASME Turbo Expo Paper No. GT2016-57854.

Gas Dynamic Modernization of Axial Uncooled Turbine by Means of CFD and Optimization Software

E Yu Marchukov[1] and I N Egorov[2]

[1] Chief designer, Lyulka Design Bureau, Moscow, Russia
[2] Deputy chief designer, Lyulka Design Bureau, Moscow, Russia

E-mail: egorov300657@yandex.ru

Abstract. The results of multicriteria optimization of three-stage low-pressure turbine are described in the paper. The aim of the optimization is to improve turbine operation process by three criteria: turbine outlet flow angle, value of residual swirl at the turbine outlet, and turbine efficiency. Full reprofiling of all blade rows is carried out while solving optimization problem. Reprofiling includes a change in both shape of flat blade sections (profiles) and three-dimensional shape of the blades. The study is carried out with 3D numerical models of turbines.

1. Introduction.

In the design process of turbine, the designer must ensure the number of requirements. Firstly, the turbine must have specified operation parameters (efficiency, mass flow rate, required power, etc). Secondly, turbine elements should meet the requirements of static and dynamic strength, have a predetermined life cycle and the lowest possible weight. Thirdly, parameters of LPT operation must be consistent with the parameters of the engine and parameters of other components. This means that the improvement of LPT parameters must not lead to a deterioration of the parameters of another engine components. Fourthly, the turbine must be manufacturable and have a minimum cost [1].

The development of computer technology and numerical simulation software allow to use optimization techniques in the design of turbomachinery [2-6]. Their application allows the automatic search for the optimal combination of the parameters describing the turbine design.

The goal of this study was to perform a multicriteria optimization of the three-stage LPT (Figure 1) of turboshaft engine. The following problems were addressed in course of multicriteria optimization:

1. Gas-dynamic development of LPT (operation parameters were improved) was carried out;
2. Matching the LPT with output device was improved (residual swirl after LPT was eliminated).

2. Solution algorithm for LPT optimization.

Scheme of the optimization algorithm of LPT used in the research is shown in Figure 2. At each step optimizer *IOSO PM* [3,7] generates Vector of variable parameters *x1…xn.* This vector is a set of variables that describe the geometry of turbine blade in a parametric form. The Vector is transferred to specialized programs *Profiler* [8, 9] and *Profiler 3D* convert LPT blades into geometry files (GF) in a format **.geomturbo* based on the Vector. This format is suitable for importing blade geometry in *Numeca FineTurbo*. Then, computational mesh in *Numeca AutoBlade 5* is generated using geometry files of reprofiled LPT blades (GF1...GFn). Then, computational model is created in *Numeca FineTurbo,* and its calculation is performed. Processing of CFD results is performed in *Numeca CFView*. As a result, several output files are cre containing operation parameters of LPT (constraint vector and optimization criteria *y1…, yn*) in the text format. These parameters are transferred to *IOSO*, which is used for the analysis of optimization results, the selection of best results, and creation of a new vector of variable. This cycle is repeated until the desired result.

Figure 1. Three-stage LPT under optimization

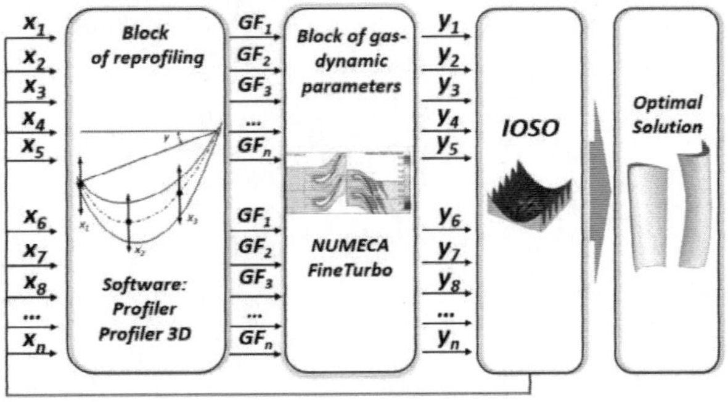

Figure 2. Optimization algorithm used in work.

Optimization procedure in *IOSO* is based on the response surface methodology [10]. Response surfaces are constructed for the objective functions and constraints. Then they are optimized at each iteration in a current search region. The objective function and constraints are then evaluated at the optimal point using the mathematical model of the system. Algorithms of *IOSO PM* have good invariant features, high level of calculation stability while optimizing the complex systems [4, 11, 12]. They also ensure the search for extremum with a presence of incompatibility areas [13, 14].

3. Development of computational model

Development of fast and accurate computational model is the most important part of optimization process (Figure 2). The computational model of turbine was created in *Numeca Auto Grid 5*. The model contained domains of inlet and outlet areas, nozzle vanes and rotor wheels. Total pressure p^*, total temperature T^*, flow angle α and turbulent viscosity were used as inlet boundary conditions. Static pressure was set at the outlet boundary [15].

Five LPT numerical models were created with different number of elements in 2D mesh of blade passage and along the height of the flow passage to perform mesh independence study [15-17] and to select the LPT mesh parameters. The first element height y+ in all created meshes was equal to 1. Parameters of created meshes are shown in Table 1.

Table 1. Basic parameters of created computational meshes of LPT.

Mesh	Number of elements in spanwise direction	Number of elements along the suction side	Number of elements along the height of blade passage	Total number of elements in 2D mesh	Total number of elements
Mesh1	43	113	57	7343	2511201
Mesh2	43	113	85	7125	3633669
Mesh3	51	141	57	10806	3695825
Mesh4	59	177	57	15780	5396545
Mesh5	75	221	57	2330	7978929

Using created computational models, the LPT characteristics were calculated: dependencies of efficiency and mass flow rate from the total pressure ration in turbine (Figure 3). From a comparison of the dependences presented in Figure 3, it was concluded:

1. Increase in the number of elements of computational model along the height (Mesh1, Mesh2) does not lead to a qualitative or quantitative change in the calculated LPT characteristics. For example, calculated values of the mass flow rate through the turbine obtained using computational meshes Mesh1 and Mesh2 differ by 0.004% (rel.), and calculated efficiency values differ by 0.001% (rel.);

2. Increase in the number of 2D mesh elements (Mesh1, Mesh4, Mesh5) lead to some quantitative change in the calculated LPT characteristics, but the qualitative nature of the dependencies does not change. As the number of elements increases, the changes in the values of calculated parameters decrease. Moreover, the mesh convergence is achieved using Mesh4. Calculated values of mass flow rate using Mesh3 and Mesh4 differ by 0.02% and by 0.008% using Mesh4 and Mesh5. Calculated values of efficiency using Mesh3 and Mesh4 differ by 0.06% and by 0.03% using Mesh4 and Mesh5.

At the next step, calculation of turbine characteristics was performed using turbulence models SA, k-ω, k-ε and computational model Mesh4 that ensured mesh convergence (Figure 4).

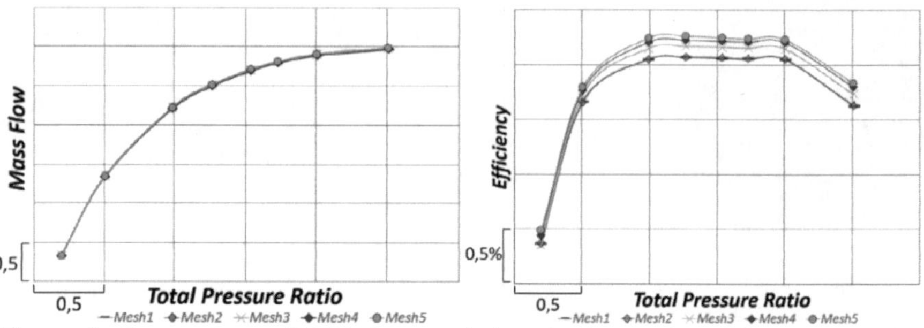

Figure 3. Comparison of LPT characteristics obtained using different models.

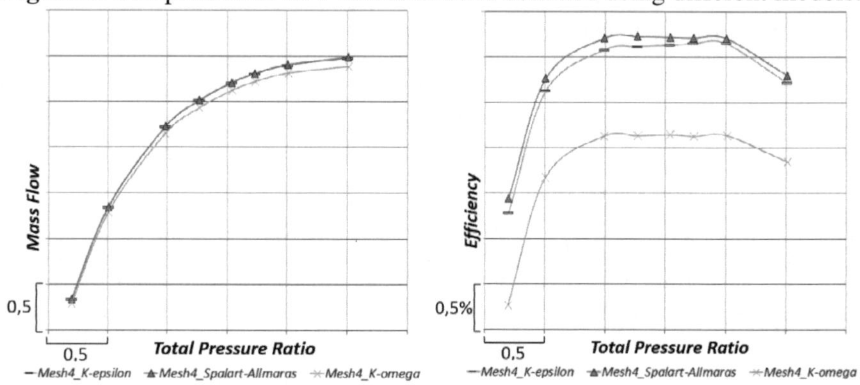

Figure 4. LPC characteristics calculated using different turbulence models.

It is clear from Figure 5 that the LPT characteristics obtained using different turbulence models qualitatively repeat each other, although there is some quantitative difference. Noteworthy is the fact that calculation results obtained using turbulence model *k-ω (Wilcox)* are significantly different from the calculation results obtained with turbulence models *Spalart-Allmaras* and *k-ε (Low Re Yang-Shih).*

Time required for the calculation was considered while selecting the turbulence model for the optimization CFD model of LPT. Calculation time with the Mesh1 on the computational node with 16 processors was 29 minutes using turbulence model *Spalart-Allmaras*, 37 minutes using turbulence model *k-ω(Wilcox)*, and 35 minutes using turbulence model *k-ε (Low Re Yang-Shih).*

Given the fact that numerical model for the optimization study should allow to capture trends in turbine operating parameters and have a minimum time of calculation in the first place, it is decided to conduct the optimization using the computational mesh with Mesh1 (2511201 elements) with *Spalart-Allmaras* turbulence model (the fastest turbulence model). Nevertheless, it is decided during the checking optimization results to use computational Mesh4 (5396545 elements), at which mesh convergence is achieved (increasing the number of elements in a mesh greater than in Mesh3 does not lead to significant differences in the value of the calculated parameters of the turbine).

4. Optimization problem statement and solution.

Gas-dynamic optimization problem was solved in three-criteria formulation with constraints.

The optimization criteria were:

1. Efficiency increase;
2. Decrease of integral flow angle at the LPT outlet α_{out} (Figure 5). It must not be less than 15°;
3. Decrease of the residual flow swirl $\Delta\alpha_{out}$ at the outlet of LPT (Figure 5). It must be not greater than 10°.

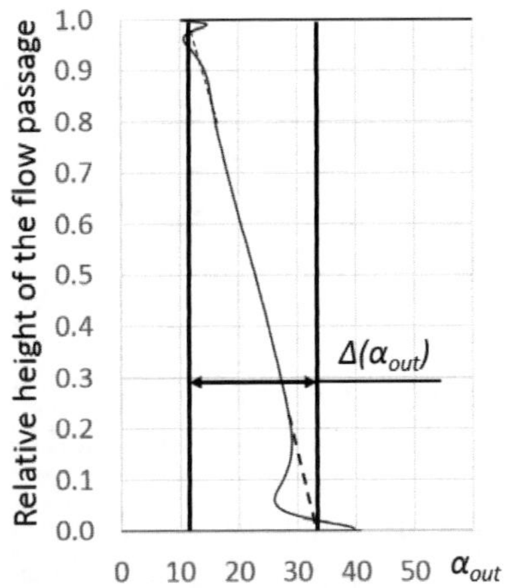

Figure 5. Parameters $\boldsymbol{\alpha_{out}}$ and $\boldsymbol{\Delta\alpha_{out}}$.

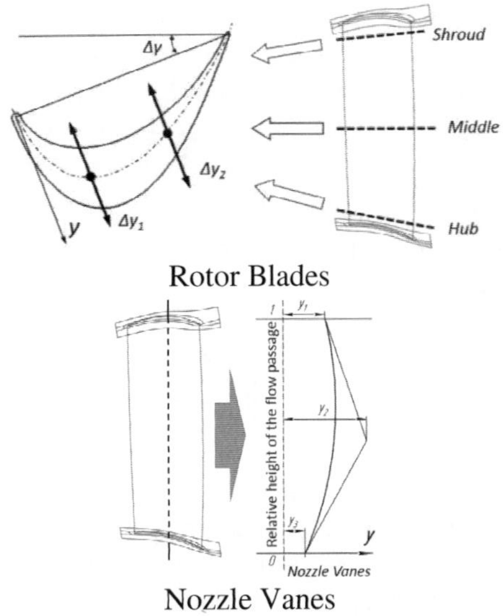

Figure 6. Scheme of variables of gas dynamic optimization problem of LPT.

Constraints for gas dynamic optimization were: mass flow rate *G*, expansion ratio π_T^*, power *N*. These parameters must not be changed during optimization by more than 0.5% relatively to the values of base LPT.

Geometric parameters of rotor blades and nozzle guide vanes were used as variables in gas-dynamic optimization. 2D and 3D reprofiling was performed for all nozzle vanes of LPT. Only 2D reprofiling was done for all rotor blades.

2D reprofiling of nozzle vanes and rotor blades was carried out by changing stagger angles of profile γ and ordinates of midpoints of spline Δy1 and Δy2 in three sections (hub, middle and shroud, Figure 6). The total number of variables for 2D reprofiling was 9 per blade.

3D reprofiling of nozzle vanes was performed by circumferential and axial shifts of the sections, and by scaling the chord of sections. Variation of these parameters along the height of the flow passage was described by Bezier curve with 3 control points (Figure 6). The values of ordinates of 3 control points were changed during optimization. The number of variables used for 3D reprofiling of nozzle vanes was 9 per blade. Total number of variables in the gas-dynamic optimization was 81. Solution of the optimization problem took nearly 1,000 iterations with computational model.

Pareto front was obtained during the optimization for the value of integral flow angle at the outlet of LPT α_{out}, residual swirl $\Delta\alpha_{out}$ and efficiency (Figure 7). The efficiency in the figure is presented not as an absolute value, but as an efficiency drop ΔEff_i relatively the efficiency of the base LPT variant.

$$\Delta Eff_i = Eff_{Base} - Eff_i \qquad (1)$$

where Eff_{Base} – the value of efficiency of base LPT variant; Eff_i – the value of efficiency of i-variant of LPT from Pareto set.

The value of ΔEff_i is shown in Figure 7 as circles. The greater the diameter of the circle is, the greater the ΔEff_i relatively the efficiency of the base LPT variant.

It should be noted that unique geometry of LPT corresponds to every point of Pareto.

Analysis of Pareto front showed that decrease in flow angle α_{out} and residual swirl $\Delta\alpha_{out}$ is followed by efficiency drop. At the same time, minimal efficiency drop is 0.12% and it is achieved with virtually extreme (in terms of optimization constraints) values of α_{out} and $\Delta\alpha_{out}$ (Table 2).

Four points of Pareto set were selected for the further analysis:

1. Point with the minimal value of efficiency drop relatively to the base variant (Variant 1);
2. Point with the minimal value of flow angle at the LPT outlet α_{out} (Variant 2);
3. Point with the minimal value of flow swirl at the LPT outlet $\Delta\alpha_{out}$ (Variant 3);
4. Compromise point (Variant 4).

Values of LPT parameters in selected points of Pareto front are shown in Table 2. Verification of the results of LPT gas-dynamic optimization was performed with Mesh4 as the finest mesh.

Table 2. The values of LPT parameters in the selected points of the Pareto front.

LPT variant	Efficiency drop, ΔEff_i, %	Outlet flow angle α_{out}, degree	$\Delta\alpha_{out}$, degree
Base	0	22.73	12.79
Variant 1	0.12	14.92	9.89
Variant 2	0.386	9.82	9.92
Variant 3	0.292	14.95	3.06
Variant 4	0.262	13.53	4.83

Figure 8 shows the graphs of distribution of the flow angle in absolute motion at the outlet of LPT along the height of flow passage for 4 selected LPT variants compared to the base case. It should be noted that the parameters α_{out} and $\Delta\alpha_{out}$ were significantly decreased.

Figure 9 shows a comparison of Mach numbers, averaged in the circumferential direction for some LPT variants compared to the base one. The main changes in the flow structure occurred in the latest stage. Comparison of the blade shapes of the base and the final LPT variants is shown in Figure 10.

Thus, the outlet angle and the residual swirl value of the flow in the outlet section of the turbine were set to the required range, although this was achieved by some drop in efficiency (Table 2).

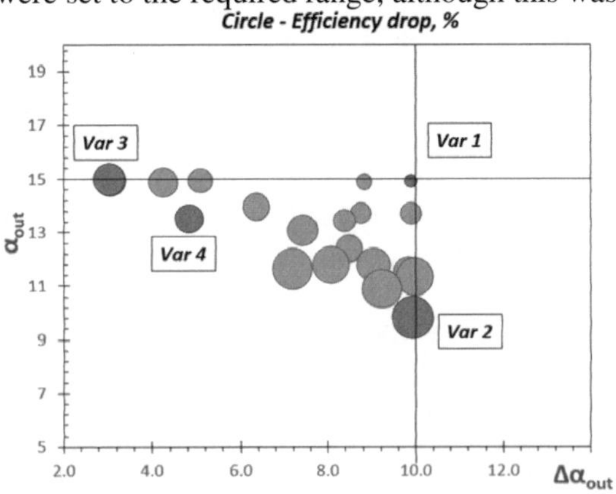

Figure 7. The Pareto Set.

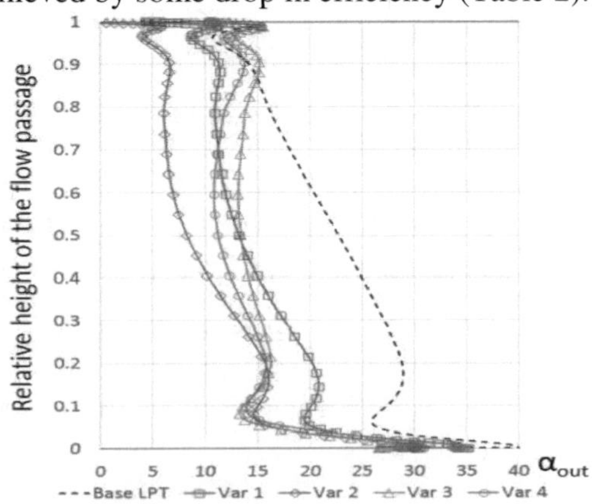

Figure 8. Outlet flow angle in absolute motion along the height of flow passage.

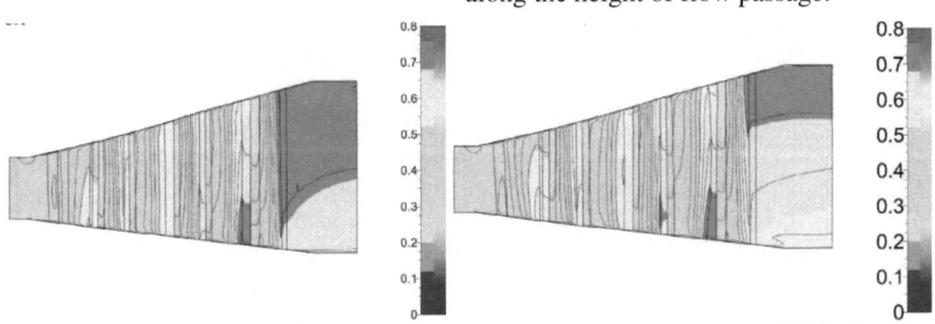

Base variant of LPT *LPT Variant 1*

Figure 9. Comparison of the Mach number in the flow passage of the LPT for different variants.

5. Conclusion

Five numerical models of LPT with different density of computational mesh were created. All these models allow predicting the trend of the LPT characteristics, but they give a quantitative difference in the calculated parameters.

The model Mesh1 with turbulence model Spalart-Allmaras provides the shortest calculation time of one point of LPT characteristics. At the same time, it allows predicting the LPT characteristics and can be used for optimization studies. Mesh4 was used for the final check of the obtained results as the most accurate mesh.

During the optimization, new LPT variants were found. They provide a decrease in the average value of the outlet angle by 12.9 degrees, as well as a decrease in the residual flow swirl at the turbine outlet by 9.8 degrees. These changes are possible due to some drop of efficiency (0.12%).

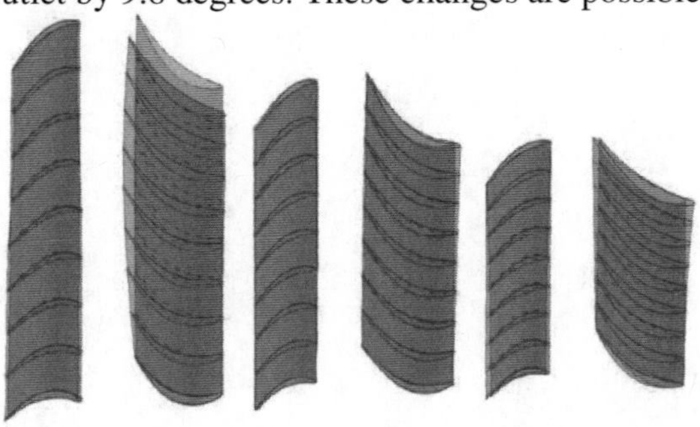

Figure 10. Comparison of the blade geometry of base and optimized LPT variants (blue blades - basic LPT, red blades - optimized LPT)

References

[1] Von Karman Institute for Fluid Dynamics, Aero-engine Design: From State of the Art Turbofans Towards Innovative Architectures, Lecture Series 2008-03.

[2] Inozemcev A A 2008 *Osnovy Konstruirovaniya Aviacionnyh Dvigatelej i Ehnergeticheskih Ustanovok (Fundamentals of Aircraft Engines and Power Plants Designing)* (Mashinostroenie, Moskow) p 366

[3] Egorov I N, Kretinin G V, Leshchenko I A, Kuptzov S V 2002 *9th AIAA/ISSMO Symposium on Multidisciplinary Analysis and Optimization* Code 103018

[4] Matveev V N, Popov G M, Baturin O V, Goryachkin E S and Kolmakova D A 2015 *51st AIAA/SAE/ASEE Joint Propulsion Conf.* Code 130539

[5] Khairuddin U, Costall A W and Martinez-Botas R F 2015 *Proc. of the ASME Turbo Expo* Paper No. GT2015-42053.

[6] Buske, C., Krumme, A., Schmidt, T., Dresbach, C., Zur, S. ,Tiefers, R, 2016 *Proc. of the ASME Turbo Expo* Paper No.GT2016-56079

[7] IOSO Optimization Technology. Access mode: http://www.iosotech.com.

[8] Shabliy L S and Dmitrieva I B 2014 *ARPN J. of Engineering and Applied Sciences* 9 (10) 1849-53

[9] Shabliy L S and Dmitrieva I B *Russian Aeronautics* pp 276-82

[10] Dennis B H, Egorov I N, Dulikravich G S and Yoshimura S 2003 *American Society of Mechanical Engineers, International Gas Turbine Institute, Turbo Expo (Publication) IGTI* GT2003-38051

[11] Popov G, Goriachkin E, Kolmakova D and Novikova Y 2016 *Proc. of the ASME Turbo Expo* Paper No. GT2016-57856

[12] Baturin O V, Popov G M, Goryachkin E S and Novikova Y D 2015 *5th Int. Conf. on Simulation and Modeling Methodologies, Technologies and Applications, Proc.* Pp 227-32

[13] Ivachnenko A G 1982 *Inductive method of selforganization for complex systems models* (Kiev, Naukova Dumka)

[14] Egorov I N and Kretinin G V 1996 *World Publishing Corporation, Aerothermo-dynamics of internal flows III* (Beijing, China) pp 112-20
[15] Popov G, Goryachkin E, Baturin O and Kolmakova D 2014 *Int. J. of Engineering and Technology* 6(5) 2236-43
[16] Abraham S, Panchal K, Ekkad S V, (...), Brown B J and Malandra A 2011 *Proc. of the ASME Turbo Expo* Paper No. GT2011-45188
[17] Macisaac G D, Sjolander S A, Praisner T J, Grover E A and Jurek R 2013 *Proc. of the ASME Turbo Expo* Paper No. GT2013-95670

IOP Conf. Series: Materials Science and Engineering **302** (2018) 012028 doi:10.1088/1757-899X/302/1/012028

Numerical Investigation of the Influence of the Input Air Irregularity on the Performance of Turbofan Jet Engine

Y Novikova[1] and V Zubanov[1]

[1]Samara National Research University, 34, Moskovskoe shosse st., Samara, Russia

E-mail: zubanov.vm@ssau.ru

Abstract. The article describes the numerical investigation of the input air irregularity influence of turbofan engine on its characteristics. The investigated fan has a wide-blade, an inlet diameter about 2 meters, a pressure ratio about 1.6 and the bypass ratio about 4.8. The flow irregularity was simulated by the flap input in the fan inlet channel. Input of flap was carried out by an amount of 10 to 22,5% of the input channel diameter with increments of 2,5%. A nonlinear harmonic analysis (NLH-analysis) of NUMECA Fine/Turbo software was used to study the flow irregularity. The behavior of the calculated LPC characteristics repeats the experiment behavior, but there is a quantitative difference: the calculated efficiency and pressure ratio of booster consistent with the experimental data within 3% and 2% respectively, the calculated efficiency and pressure ratio of fan duct – within 4% and 2.5% respectively. An increasing the level of air irregularity in the input stage of the fan reduces the calculated mass flow, maximum pressure ratio and efficiency. With the value of flap input 12.5%, reducing the maximum air flow is 1.44%, lowering the maximum pressure ratio is 2.6%, efficiency decreasing is 3.1%.

1. Introduction

During operation, on the input to turbofan engine receives air having a non-uniform field of velocity and pressure [1, 2]. This is due to many factors: wind, scythe blowing at work on the ground and at the evolution of the aircraft in air, design and technological projections in the pathflow of the input device. The stable operation of low pressure compressor (LPC) is only possible to a certain degree of irregularity of air in the input device [3]. For this reason, it is necessary to predict the air irregularity influence on the engine workflow. One of the numerical methods that can make such a forecast is modern CFD software package such as NUMECA Fine/Turbo. In such software systems, an approach of using a nonlinear harmonic analysis (NLH-analysis) was implemented to calculate the unsteady effects in turbomachines. NLH techniques allow to significantly reducing the calculation time in comparison with traditional methods of non-stationary 3D modeling [4]. This is achieved by presenting the flow in the form of quasi-three-dimensional flows and thus as a set of two-dimensional tasks, and it makes possible to calculate multistage turbomachinery within a reasonable time.

In the present work, the numerical modeling results of the air flow irregularity impact at inlet turbofan gas turbine engine on the characteristics of its fan were shown.

2. CFD-model description

The test fan (Figure 1) has a wide-blade, inlet diameter of about 2 m and a pressure ratio of about 1.6, the bypass ratio of about 4.8.

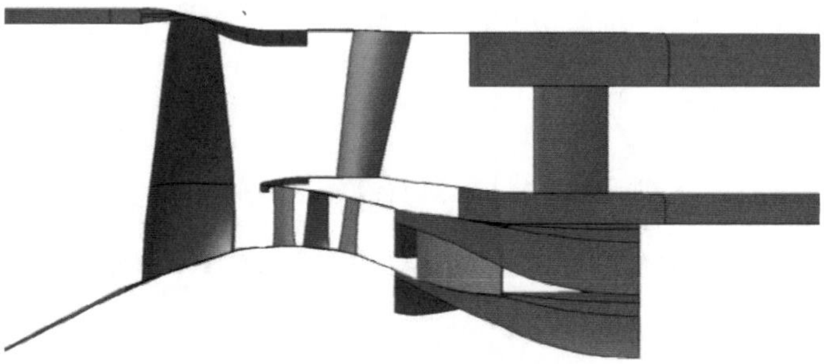

Figure 1. Investigated fan

Numerical study was performed in two stages. In the first stage, the LPC numerical model was created in NUMECA Fine/Turbo CFD software and a comparison of the experimental data and the calculated characteristics in a stationary mode was carried out. In the second stage, the numerical investigation of the air irregularity on fan performance was conducted using NLH method in NUMECA Fine/Turbo [5]. The air irregularity was simulated by input of the flap in the inlet channel of fan. The flap input was carried by an amount of 10 to 22.5% of the diameter of input channel with increments of 2.5%. For each flap position the fan performance was calculated using NLH method in NUMECA Fine/Turbo.

Numerical model of the LPC was created in the NUMECA AutoGrid 5 software. The initial geometric model of the computational domain was constructed based on technical documentation. The geometric model contained the following elements: fan blade, booster and bearing racks (Figure 2). Radial clearance under rotor blades was considered [6].

The LPC numerical model was created in the stationary axisymmetric statement. The blades deformation under the applied forces influence was considered. The shape of the deformed blades was found as a result of the strength calculation, it was loaded into a 3D model of the fan and booster blades.

The created model was divided into finite elements of the structural mesh using program Numeca Autogrid 5 (Figure 3). The total number of elements in a model of about 7 million. The fan domain contains about 2 million elements, each of the other blade domains contains about 0.5 million elements [7].

The regions around the rotor wheels (RW) and the guide vanes (GV) were allocated in the computational domain. The region around GV was calculated in the stationary coordinate system. The region around RW was calculated in the rotating frame, its speed coincides with the rotor speed.

Calculations were performed in a stationary mode using the Mixing Plane interface for data transfer between rotating and stationary domains. The Mixing Plane interface averages the flow parameters in the circumferential direction in the upstream and transmits as a boundary condition in the region located downstream. The turbulence model Spalart-Allmaras was used.

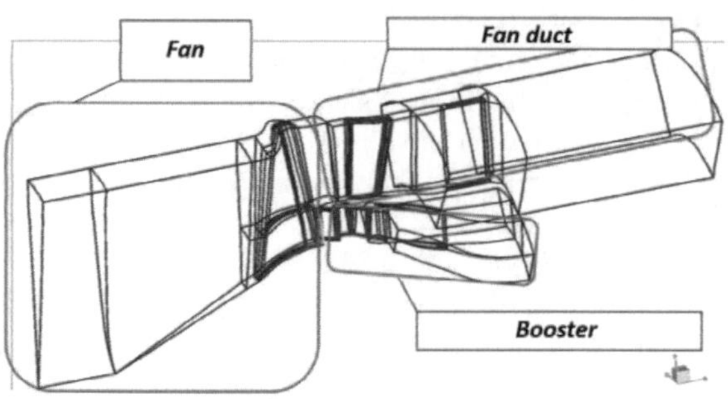

Figure 2. Domains of numerical model of investigated compressor

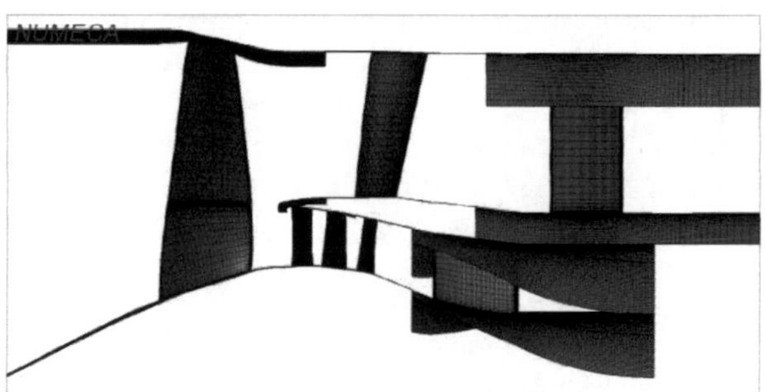

Figure 3. Finite element model of investigated LPC

The values of total pressure $p^{*}=101{,}325\ kPa$ and total temperature $T^{*}=288{,}15\ K$ were set as boundary conditions at the LPC inlet. The flow direction at the inlet region was set as axial. The mass flow values at the outlet of fan duct and booster were set at the outlet boundary conditions. Mass flow ratio determines the value of bypass ratio.

3. Result Analyses

The pressure and efficiency characteristics calculations of booster channel and fun duct was carried out using created numerical model for three modes equal to 0.95, 1, 1.05 of relative rotor speed (Figure 4).

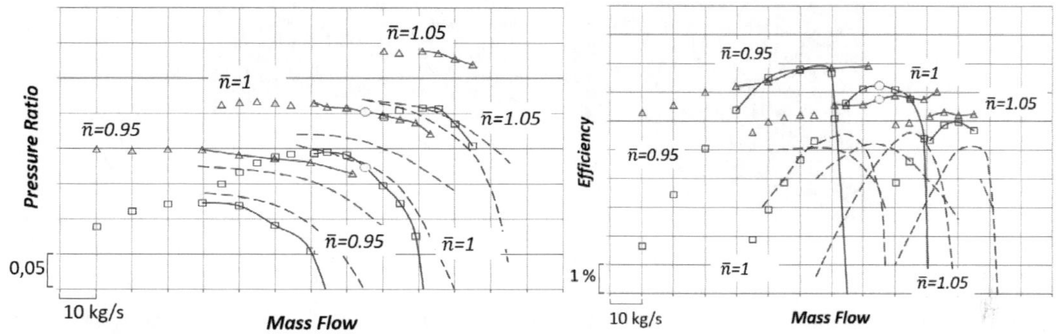

Figure 4. Comparison of experimental and calculated relative characteristics of the LPC of NK-36ST obtained using different computational models

Figure 4 shows the comparison of the calculated LPC characteristics (markers in the form of triangles and squares) and LPC testing experimental data (dashed lines). Markers that are not connected by a line - the calculation at this mode remains stable, but the calculated efficiency and pressure ratio values continuously decrease with the iteration passage, i.e., these modes are between sustainable and unsustainable solution.

From the analysis of dependencies presented in Figure 4 shows that the behavior of the characteristics of the calculated LPC repeats the behavior of the experiment, but there is a quantitative difference. Thus, the calculated values of the efficiency of booster different with experimental by from 1 to 3% and efficiency of fan duct different with experimental from 1.5 to 4%. The calculated values of the degree of pressure ratio in the booster different with the experimental values by up to 2%, the fan duct differ with experimental values by up to 2.5%. Figure 5 shows the calculated field of Mach number in LPC in the meridional plane as averaging in the circumferential direction.

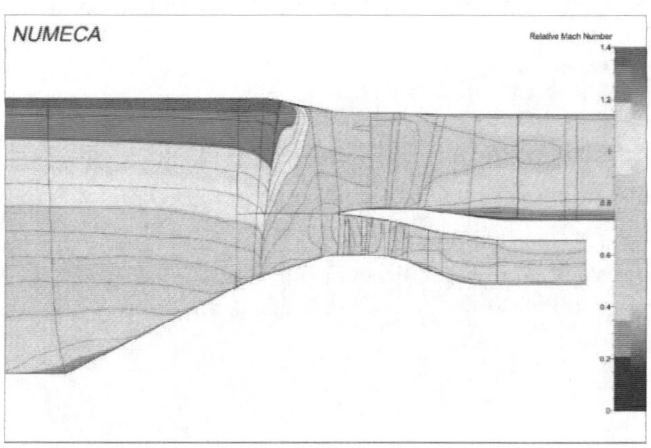

Figure 5. Calculated field of Mach number in LPC

IOP Conf. Series: Materials Science and Engineering **302** (2018) 012028 doi:10.1088/1757-899X/302/1/012028

Calculation of the fan characteristics with simulated input air irregularity was performed using the method of nonlinear harmonic analysis. Since the method limitation does not allow the study of the irregularity effect within the double-duct statement, it was decided to carry out research in the formulation of a single-stage fan. Investigated fan stage (Figure 6) consisted of a fan blade and the guide vane of booster which was "elongated" on the distance from the hub of the inner loop to the peripheral contour of the fan duct. It is assumed that this model will allow qualitative assessment of the air irregularity impact on the fan performance.

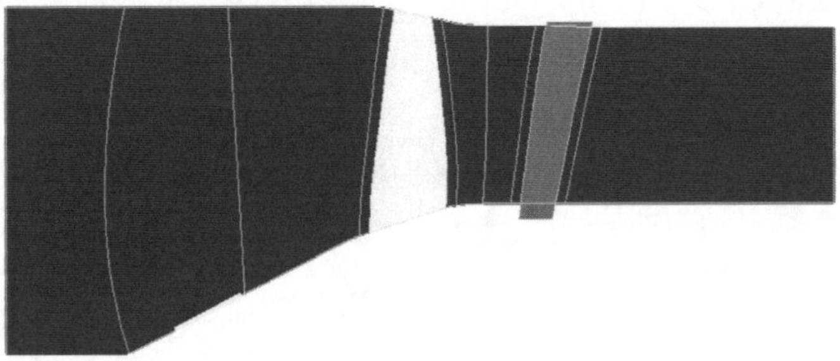

Figure 6. Investigated fan stage

A family of discrete models with the input device and simulator of air irregularity was created to perform the study (Figure 7). The models differ from each other only by the size of the flap extension into the flow of the input device. Input flap carried by an amount of 10 to 22.5% of the diameter of input channel with increments of 2.5%. The input device was modeled entirely (full model), the fan blades and the guide vanes were modeled by sectors corresponding to one inter-blade channel (Figure 8).

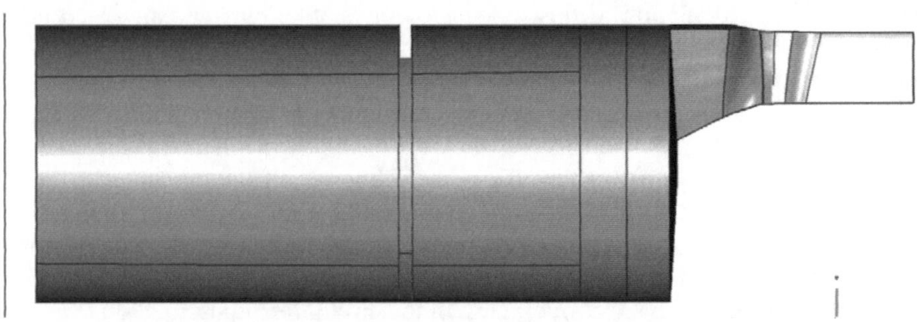

Figure 7. Numerical model of the fan with the simulator of air irregularity

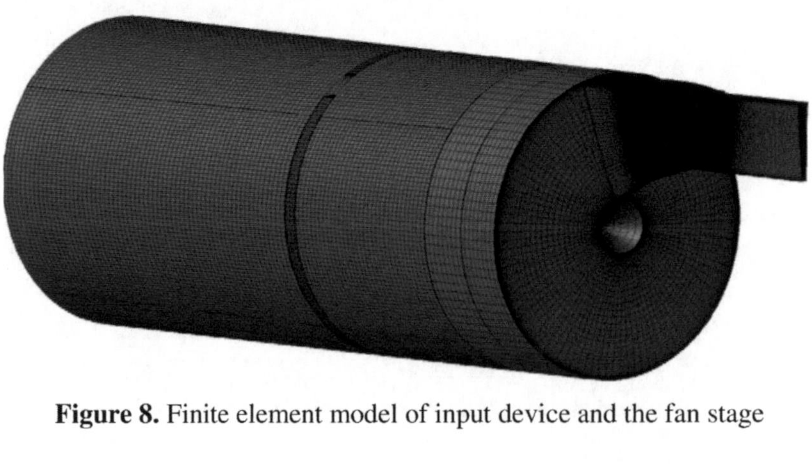

Figure 8. Finite element model of input device and the fan stage

The adiabatic efficiency and pressure ratio characteristics of the fan stage with the air irregularity simulator were calculated with the created numerical model (Figure 9). The calculations were carried out in a non-stationary formulation using the method of nonlinear harmonic analysis. The calculations were performed in the software package NUMECA Fine/Turbo using the following settings of numerical model: *Number of frequencies 3, Max number of perturbations 2.*

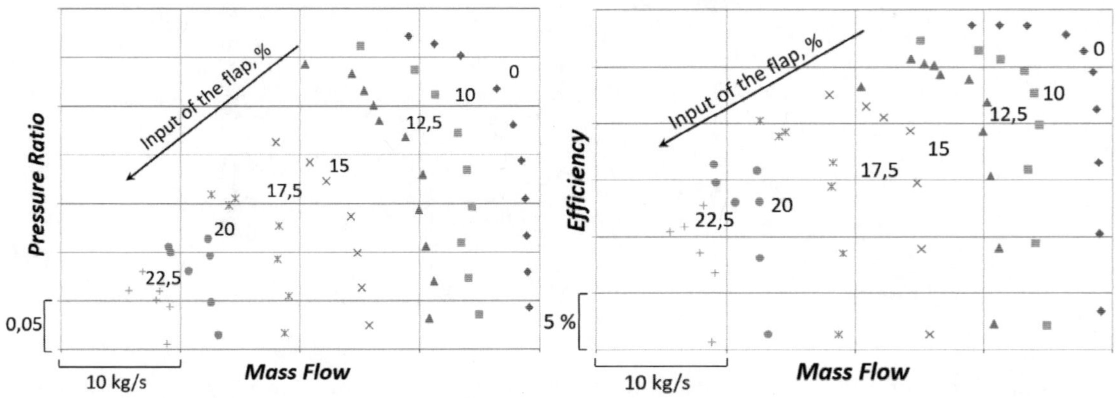

Figure 9. Calculated performance map of fan stage considering the simulator of air irregularity

Analysis of the stage fan characteristics shows that increasing the air irregularity level in the input stage of the fan reduces the calculated value of mass flow, the calculated value of maximum pressure ratio and calculated value of efficiency. With the value of flap input 12.5%, reducing the maximum air flow (absolute, not the given) is 1.44%, lowering the maximum pressure ratio (from the entrance to the inlet device to exit) is 2.6%, decrease in efficiency (including losses in the input device with the flap) is 3.1%.

Figure 10, 11 shows the total pressure field and Mach number field in front of the fan section corresponding to the flap input in 12.5% of diameter of input channel.

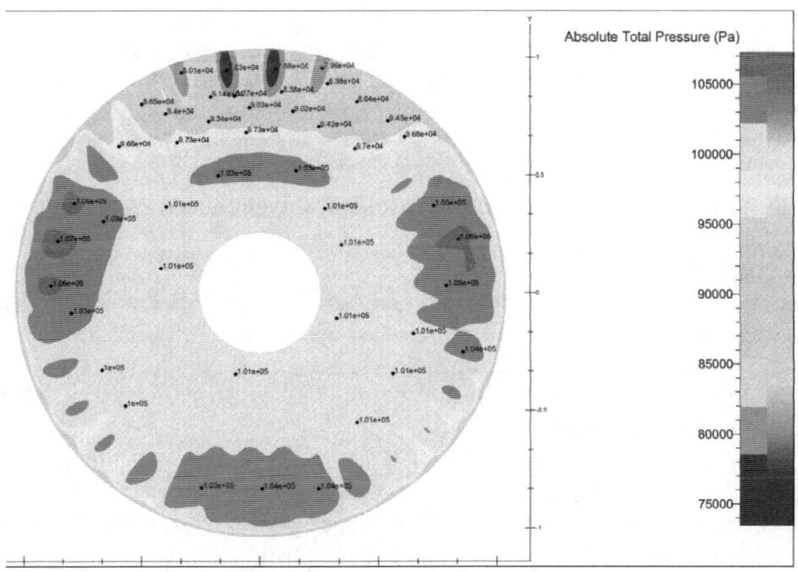

Figure 10. Total pressure field in front of the fan section corresponding to the input of the flap in 12.5%

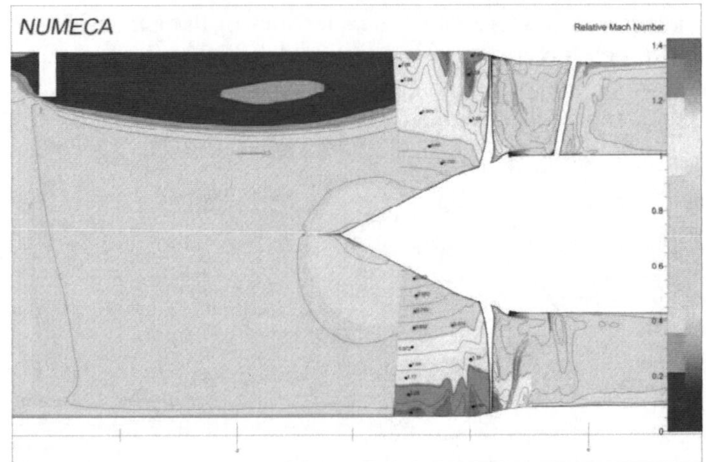

Figure 11. Field of the relative Mach number of the fan channel

4. Conclusion

Calculation model of LPC was created, which allows calculating the characteristics of LPC to a sufficient degree of accuracy. The model consists of fan stages, the fan duct and booster.

The numerical investigation of the air irregularity influence on the fan stage performance was performed. It is shown that the air irregularity increase leads to reduction of the maximum air flow and maximum pressure ratio, to decrease efficiency. It can cause LPC and engine failure.

Acknowledgments

This work was supported by the Ministry of education and science of the Russian Federation in the framework of the implementation of the Program of increasing the competitiveness of Samara National Research University among the world's leading scientific and educational centres for 2013-2020 years.

References

[1] Kulagin, V. V.. 2002, "Teoria, raschet i proektirovanie aviacionnyh dvigateley i energeticheskih ustanovok. Uchebnik. Osnovy teorii GTD. Rabochiy process I termodinamicheskiy analiz", (The theory, calculation and design of aircraft engines and power plants: a Textbook. Fundamentals of the theory of the CCD. Workflow and thermodynamic analysis), 616 p.

[2] Frohnapfel, D.J., Ferrar, A.M., Bailey, J.M., O'Brien, W.F., Lowe, K.T. "Measurements of fan response to inlet total pressure and swirl distortions produced by boundary layer ingesting aircraft configurations", 54th AIAA Aerospace Sciences Meeting, AIAA SciTech Forum, Paper No. AIAA 2016-0533.

[3] Von Karman Institute for Fluid Dynamics, Aero-engine Design: From State of the Art Turbofans Towards Innovative Architectures, Lecture Series 2008-03.

[4] Kolmakova, D., Popov, G. "Methods of improving the axial compressor flow passage to reduce the flow circumferential nonuniformity", Proc. of ASME 2015 Gas Turbine India Conference, Paper No. GTINDIA2015-1276.

[5] NUMECA, Theoretical Manual, FINE/Turbo v8.9 Flow Integrated Environment, NUMECA.inc., Belgium, 2011.

[6] Shablii L. S. and Dmitrieva I. B., 2014, "Blade geometry transformation in optimization problems from the point cloud to the parametric form," Russian Aeronautics, 57(3), 2014, pp. 276-282.

[7] Kuz'michev, V.S., Tkachenko, A.Y., Krupenich, I.N. and Rybakov, V.N., 2014, "Composing a virtual model of gas turbine engine working process using the CAE system "ASTRA"," Research Journal of Applied Sciences, 9(10), pp. 635-643.

IOP Conf. Series: Materials Science and Engineering **302** (2018) 012029 doi:10.1088/1757-899X/302/1/012029

Gas-Dynamic Designing of the Exhaust System for the Air Brake

Yu Novikova[1], E Goriachkin[1]and A Volkov[2]

[1] Associate Professor, Samara National Research University, Samara, Russia
[2] Researcher, Samara National Research University, Samara, Russia

E-mail: y.d.novikova@gmail.com

Abstract. Each gas turbine engine is tested some times during the life-cycle. The test equipment includes the air brake that utilizes the power produced by the gas turbine engine. In actual conditions, the outlet pressure of the air brake does not change and is equal to atmospheric pressure. For this reason, for the air brake work it is necessary to design the special exhaust system. Mission of the exhaust system is to provide the required level of backpressure at the outlet of the air brake. The backpressure is required for the required power utilization by the air brake (the air brake operation in the required points on the performance curves). The paper is described the development of the gas dynamic canal, designing outlet guide vane and the creation of a unified exhaust system for the air brake. Using a unified exhaust system involves moving the operating point to the performance curve further away from the calculated point. However, the applying of one exhaust system instead of two will significantly reduce the cash and time costs.

1. Introduction

Testing of gas-turbine engines (GTE) is an essential part of their life cycle. Tests are carried for the newly created engines, and for the serially produced engines before their delivery to the customer.

During testing the industrial GTE, designed to drive electric generators, compressors pumps and etc. there is a problem of utilization and measurement of generated power [1]. Therefore, it was suggested to use the compressor as the air brake. The air brake utilizes the power generated by GTE by compressing the air. Creation and design of compressors is a difficult and expensive task. This explains the high cost of the air brake for GTE. However, if existing compressor is use as the air brake, it can significantly reduce the cost of developing of the air brake, and reduce the time of its creation. Therefore, it was decided to use a low-pressure compressor of a mass-producing gas turbine engine (JSC "Kuznetsov") to create the air brake [2]. At the previous stages of the research, a compressor was designed that meets all the requirements of the test equipment. It can utilize the required power (25 MW and 32 MW) and has a sufficient stall margin [3]. However, compressor has a residual twist angle of the flow beyond the last guide vanes.

2. Designing of the outlet guide vane

The angle of the flow core in the section at fourth GV outlet at both operation modes is about 31 degrees. Residual twist beyond the fourth GV is undesirable because it can lead to the instability work of the exhaust system of the air brake and test equipment as a whole [4]. It was decided to design the Outlet Guide Vane (OGV) for elimination of residual flow swirling.

IOP Conf. Series: Materials Science and Engineering **302** (2018) 012029 doi:10.1088/1757-899X/302/1/012029

To reduce OGV production costs it was decided to make it with constant cross-section height. Upon request, GV was made of 4 mm thick plate by stamping and subsequent processing of edges. In order to prevent the flow breakdown from the GV back not only in the flow core, but also in the hub and the shroud areas, it was decided to make OGV with an inlet vane angle of 36 degrees. Output vane angle was determined by the condition that the residual flow twist beyond OGV should not exceed 3 degrees. Considering the flow lagging it was decided to make the OGV output guide vanes angle equal to -2 degrees. Based on above requirements, the design of OGV was performed. Flow straightener cross section shown on Figure 1. The number of blades was chosen by the condition of optimum blade profiles density of mesh [5] and it was equal to 56.

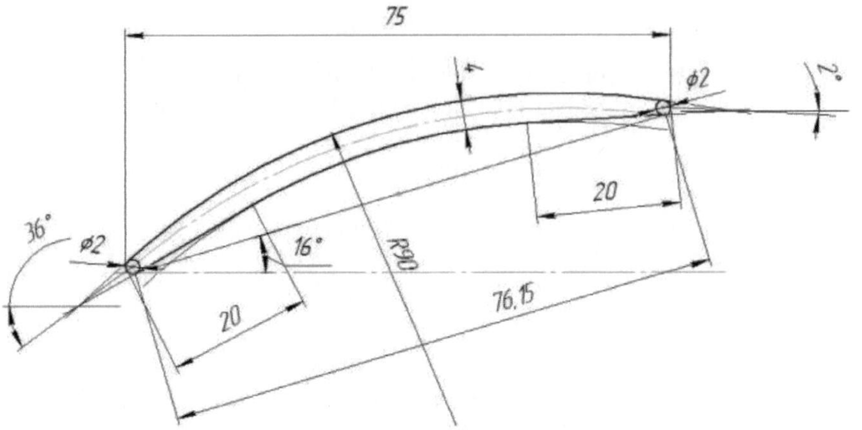

Figure 1. Geometry of the designed OGV.

The meridional profiles of the OGV were determined by design considerations. Thus, the inclination angle of the hub was equal to 7 degrees to provide placing underneath it the required number of unloading disks and cavities. Unloading disks and cavities require to place beyond the 4th blade wheel of the air brake in order to reduce the axial force acting on the ball bearing. The angle of the shroud was chosen to be 10 degrees for the narrowing of flow channel. Created air brake configuration is shown in Figure 2.

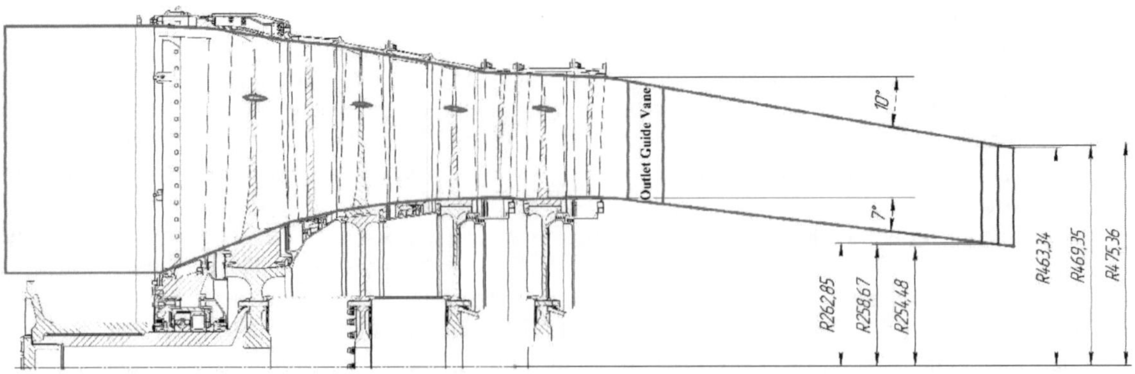

Figure 2. The air brake variant with OGV and with exhaust system.

In order to evaluate the gas-dynamic parameters of the air brake with OGV the calculation of its performance was made (Figure 3, 4). Analysis of the performance showed that this air brake configuration has stall margin 10.66 at 5000 rev / min and 9.01 at 5500 rev / min. However, the angle of the flow beyond OGV was 3 degrees.

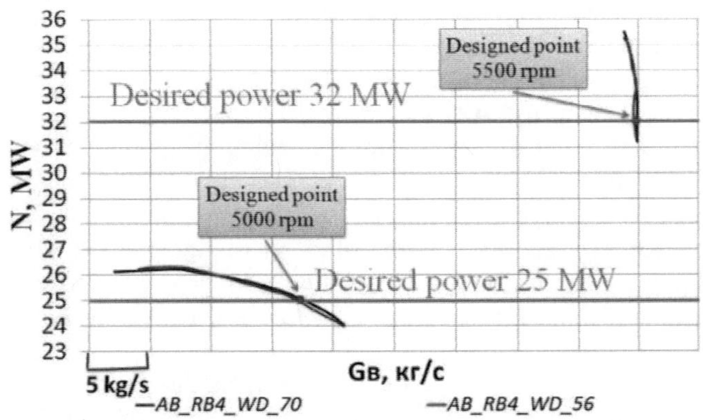

Figure 3. Characteristics of the four stage air brake with different numbers of OGV blades.

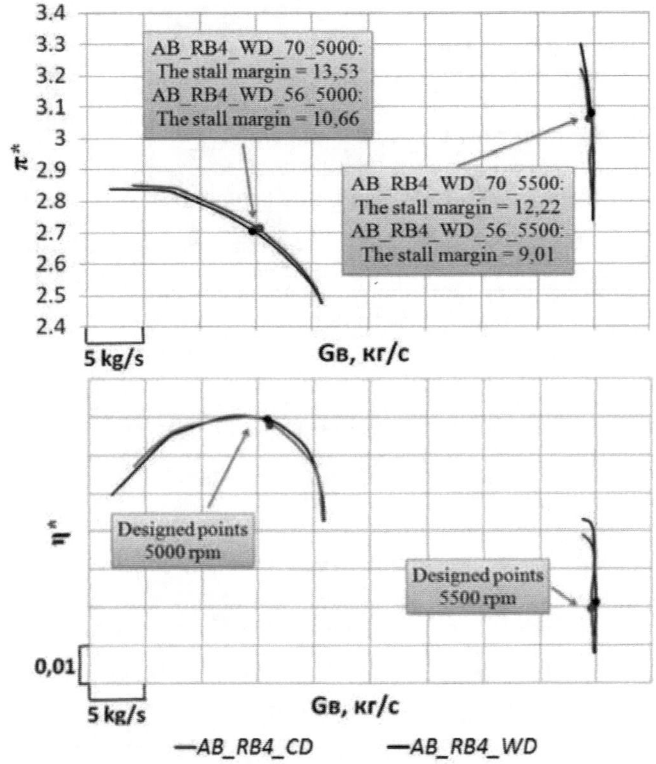

Figure 4. Head characteristics and efficiency of the air brake with different numbers of OGV blades.

Analysis of the flow structure in the air brake showed that the air brake arrangement configuration AB_RB4_WD_56 has flow separation on the hub of the OGV (Figure 2). It has been suggested that this flow separation reduces the stall margin of the air brake. It was decided to increase blade number of OGV to 70 (the air brake configuration AB_RB4_WD_70.) for reducing flow separation. Actually, this decision led to the increase in losses in the straightener (generally, profile losses), but in terms of the air brake it is not critical, because its mission is energy utilization.

Analysis of the air brake AB_RB4_WD_70 characteristics showed that the increase in number of OGV allows increasing stall margin up to 13.53% and 12.22% on operation modes 5000 rev/vin and 5500 rev/min. It also and reduce the flow separation breakdown in the OGV. Moreover, increasing the number of OGV blades has reduced cross-sectional flow angle at the output from the air brake. The output flow angle of the air brake was equal to 0 degrees.

3. Designing of the air brake exhaust system

The variable static pressure at outlet of the air brake calculation model was set during calculating of the characteristics of the air brake earlier. In real operating conditions the ambient pressure will not change and will be equal to atmospheric pressure [6]. For this reason, for the air brake work it is necessary to design the special exhaust system. Mission of the exhaust system is to provide the required level of backpressure at the outlet of the air brake. The backpressure is required for the required power utilization by the air brake (the air brake operation in the required points on the characteristics).

In the case of considered configuration of the air brake the simplest exhaust system is a conical converging nozzle with a central body (Figure 2). When the nozzle works at the supercritical drop (the pressure ratio is more than 1.86) at the nozzle exit the flow speed becomes equal to the speed of sound, and the air flow rate through the nozzle reaches a maximum possible value and the increase of the drop at the nozzle the flow rate does not increase. Thus, the flow rate is set by the constant flow rate through the entire system as a whole: and through the nozzle and through the air brake.

For the design of the nozzle it was necessary to determine the its area of the narrowest cross section by condition of the supercritical flow of air from nozzles at each of the two operation modes (5000 rev / min, 5500 rev / min). So, the total pressure, total temperature, mass flow rate was measured at the outlet from the air brake at the operating points at rotational speeds of 5000 rev / min and 5500 rev / min. The values of the measured parameters are shown in Table 1.

Table 1. The values of the measured parameters at the outlet from the air brake.

n, rev/min	T_{S1}*, K	p_{S1}*, Pa	G_a, kg/sec	N, MW
5000	395.3	268597	246.85	25
5500	411.1	291132	274.97	32

The area of the narrow section at each operation mode was determined by the formula (1):

$$F_C = G_{S1} \cdot T_C^{*1/2} \cdot \left(m_g \cdot p_C^* \cdot q(\lambda_C) \cdot \mu_C\right)^{-1}. \quad (1)$$

Outer and inner diameters of the nozzle with a central body considering the meridional profiles designed in the previous step were determined after determining the areas. The calculated areas of the nozzles for each operation mode and outer and inner diameters are given in Table 2. The figure 2 shows the arrangement of the calculated nozzles together with the air brake.

Table 2. Throat areas and diameters of the air brake nozzle

n, rev/min	F_C, m^2	D_h, m	D_w, m
5000	0.471	0.509	0.927
5500	0.493	0.526	0.951

Since the cross-sectional areas for operation modes of 5000 rev / min and 5500 rev/ min are not significantly different (less than 5%), it was decided to design a one, unified nozzle (Figure 2). Use of the one unified nozzle though move working points of the air brake further from originally design points on performance map, but it can reduce the manufacturing cost (it is necessary to produce not two nozzles, but one), and it simplifies the process of the engine tests with the various powers (not need to change the exhaust nozzle during the engine replacement or the switching the operation mode at the testing engine).

The combined CFD - model of the air brake with a nozzle (Figure 5) was created to check the correctness of the decision about the using of the one, unified exhaust nozzle. Designed model included the air brake, the nozzle and attached volume for correct flow modeling in the nozzle. At the inlet of the calculation model (the inlet of the air brake) the total pressure of 101325 Pa and total temperature 288.15 K were set. At the outlet of the model static pressure is equal to atmospheric pressure was set. (Figure 5). Calculation was made for two operation modes of 5000 rev / min and 5500 rev / min.

As a result of the calculations the integral parameters of the air brake at the two operational modes, considering combined its work with the nozzle were determined. Obtained operating points of the air brake with the nozzle at operation modes of 5000 rev / min and 5500 rev / min were plotted to the corresponding characteristics (Figure 7). From the analysis of the figure it can be concluded that:

1) The air brake power at operation mode 5000 rev/min is slightly less than design power and is equal to 24.8 MW;
2) The air brake power at operation mode 5500 rev/min is slightly more than design power and is equal to 32.3 MW.

The resulting deviation of the working capacity from the designed is feasible and it can be taken into account during testing of engines.

The figure 6 shows the averaged the field of Mach numbers in the circumferential direction in the coupled air brake system with nozzle at the operation mode of 5500 rev / min. From the analysis of the fields of Mach numbers it can be seen that at the nozzle outlet the flow speed is equal to the speed of sound M = 1.

The unified nozzle was designed. This nozzle allows providing air brake operation on required modes.

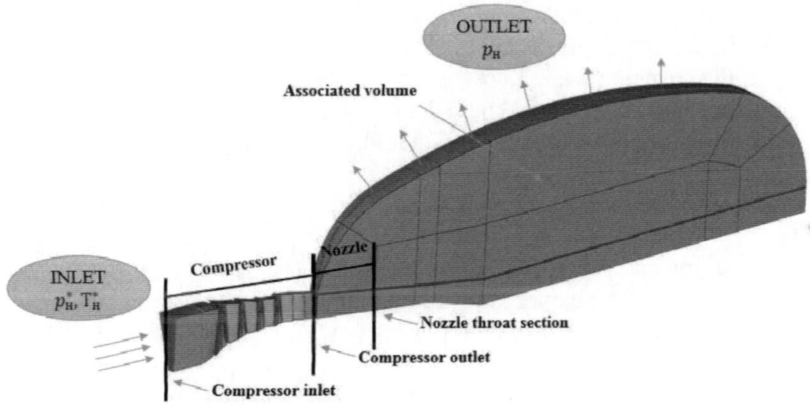

Figure 5. The combined CFD - model of the air brake.

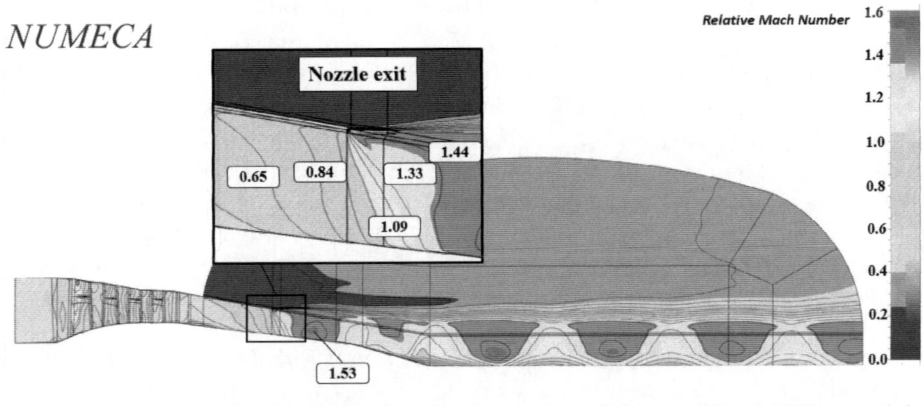

Figure 6. The field of the calculated relative Mach number of the combined CFD - model of the air brake.

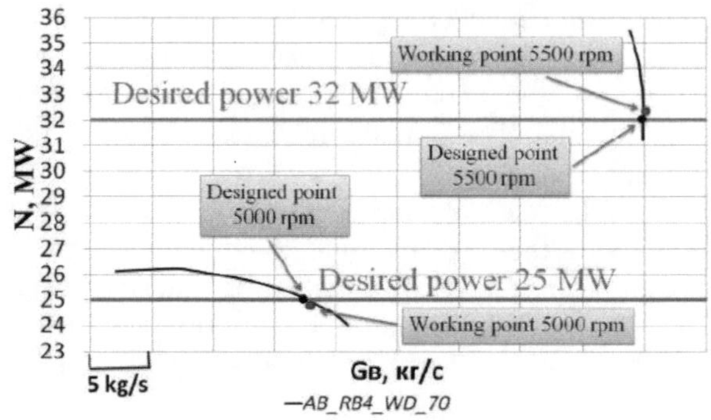

Figure 7. The power characteristics for the combined CFD - model of the air brake

4. Conclusion

The design of the air brake was created for industrial GTE tests with the power of 25MW and 32 MW. The development of the air brake had a number of features:

1) The development was carried out using CFD - program Numeca FineTurbo [7]. However, there was not any preliminary calculation of the turbomachinery using 1D / 2D – methods;
2) The development was carried out, mainly, using the existing parts and assemblies.

These features helped to reduce the development time, and reduce the cost of development and production of the air brake.

During the designing of the air brake the following problems have been successfully solved:

1) The workflow model of the three-stage LPC was created and its validation was made;
2) Stall margin of the air brake were increased by waisting of the meridional flow channel;
3) OGV was profiled which eliminates the twist beyond the air brake;
4) Unified nozzle was designed which provides the air brake work at necessary points on the characteristics.

Acknowledgments

This work was carried out by lead performer of research and development effort with financial support of Russian Federation Ministry of Science and Education with the realization of Goverment decree №218 from 09.04.2010 according to contract about providing and using subsidy № 02.G25.31.0150 from 01.12.2015

References

[1] Korotov, M. V., "Method for testing gas-turbine engine and device for its realization", Patent of Russia Federation, RU 2318195, Date of publication 20.06.2003.

[2] Kuzmenko M.L., Egorov I.N., Shmotin Yu. N., Kretinin G.V., Fedechkin K.S., 2010, "Optimization Studies of the Aircraft High-Bypass ratio Turbo Engine Fan,", Conference ASME Turbo Expo 2010, Glasgow, Scotland, UK, June 14-18.

[3] Kuzmichev, V S., Rybakov, V N., Tkachenko, A. Y, Krupenich, I. N., 2014, "Optimization of working process parameters of gas turbine engines line on the basis of unified engine core", ARPN Journal of Engineering and Applied Sciences, 9(10), pp. 1873-1878.

[4] Belousov, A. N., 2003, "Teorija i raschet aviacionnyh lopatochnyh mashin", Samara, Samarskii Dom Pechati p. 336c.

[5] Lewis, R. I., 1996, "Turbomachinery performance analysis" Publisher: Elsevier Science & Technology Books Pub, p. 329

[6] Kuz'michev, VS., Tkachenko, A.Y, Krupenich, I.N. and Rybakov, VN., 2014, "Composing a virtual model of gas turbine engine working process using the CAE system "ASTRA"," Research Journal of Applied Sciences, 9(10), pp. 635-643.

[7] NUMECA, "User Manual AutoGrid5" Release 8.4, NUMECA.inc., Belgium, January 2008.

IOP Conf. Series: Materials Science and Engineering **302** (2018) 012030 doi:10.1088/1757-899X/302/1/012030

Increase of Gas-Turbine Plant Efficiency by Optimizing Operation of Compressors

V Matveev[1], E Goriachkin[2] and A Volkov[3]

[1] Professor, Samara National Research University, Samara, Russia
[2] Associate Professor, Samara National Research University, Samara, Russia
[3] Researcher, Samara National Research University, Samara, Russia

E-Mail: valeriym2008@rambler.ru; goriachkin.e.s@gmail.com; a44rey@gmail.com

Abstract. The article presents optimization method for improving of the working process of axial compressors of gas turbine engines. Developed method allows to perform search for the best geometry of compressor blades automatically by using optimization software IOSO and CFD software NUMECA Fine/Turbo. The calculation of the compressor parameters was performed for work and stall point of its performance map on each optimization step. Study was carried out for seven-stage high-pressure compressor and three-stage low-pressure compressors. As a result of optimization, improvement of efficiency was achieved for all investigated compressors.

1. Introduction

Using CFD during the axial compressor design can reduce cash costs, but the search for the best version of the flow channel geometry takes a lot of time. The main reason is that the flow channel geometry of axial compressor stages is described by large number of variables. At least two sections, each of which is determined by dozens of independent variables, must be determined to describe the shape of one blade. Furthermore, it is necessary to describe the relative position of the sections relative to each other, shape of meridional section of the flow part, etc. As a result, the number of independent variables for a complete description of one stage reaches several tens which affect contradictory compressor characteristics. Man is not capable physiologically to analyse the problem of such dimension. The problem becomes proportionally larger if it is necessary to investigate the operation of multistage compressor.

In this case, the optimization program can help the engineer. Their use allows the automatization of the search for the optimal combination of independent variables by their automatic variation and analysis of the results based on optimization algorithms.

This article shows the technique for improving of the working process of axial compressor using optimization software IOSO and CFD software NUMECA Fine/Turbo and shows examples of the application of this technique on real seven-stage high-pressure compressor and three-stage low-pressure compressors of NK36-ST gas turbine engine (figure 1). The aim of optimization was to investigate which maximum efficiency is possible to obtain in each compressor [1]. The criteria and constraints for the optimization problems of each compressor have been selected on the basis of thermodynamic calculations of the engine NK36-ST [2].

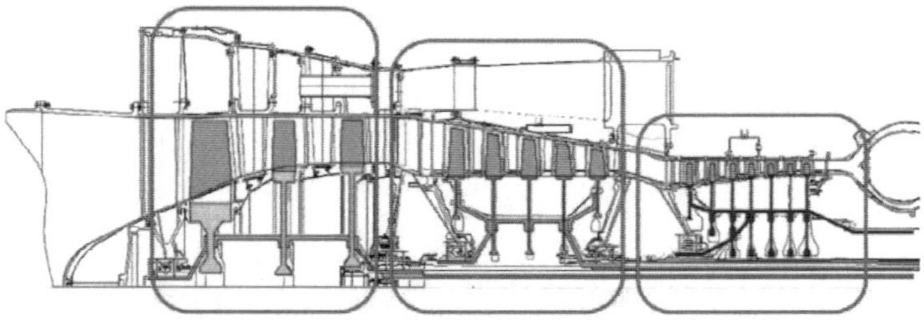

Figure 1. Investigated three-spool compressor.

2. Numerical model of compressors used for optimization

Numerical LPC and HPC models were created for the study based on the project documentation submitted by JSC "Kuznetsov" [3] in the Numeca software (figure 2).

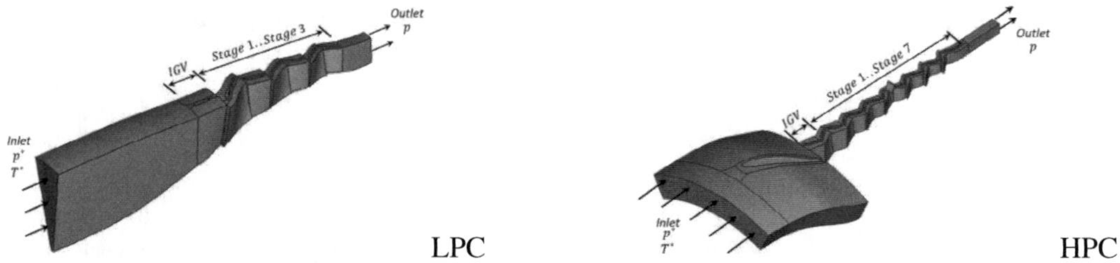

Figure 2. Geometrical models of investigated compressors.

The created model was divided into finite volumes to block-structured grid by internal tools of the program NUMECA. Two computational models ("light" - Model № 1 and "heavy" - Model № 2) were created for each compressor. Light model contained 300 thousand elements per blade row on an average; the value of y + was 3-5. Heavy model contained 500 thousand elements per blade row on an average; the value of y + was 1-2.

As the boundary conditions at the inlet of the HPC and LPC the value of the total pressure was set equal to p*=101,325 kPa and total temperature was equal to T * = 288,15K. Parameters of turbulence at the inlet boundary are k=5m2/c2, ε=30000 m2/c3. In the calculation of characteristics the stall borderline was defined as the point with the minimum flow rate of the working fluid, in which it managed to get the coincided decision [4].

To prove the validity of the numerical models LPC (figure 3) and HPC (figure 4) characteristics were calculated at several operational modes. Then they were compared with the available experimental data for the compressors.

The calculated parameters of the compressor are presented in relative form:

$$\bar{n}_{cor} = n_{cor}/n_{corBASE} \cdot 100\% \quad (1)$$

where $n_{corBASE}$ – *t*he rotor speed at the primary operational mode of the base LPC;

n_{cor}- corrected rotor speed;

$$n_{cor} = n\sqrt{288{,}15/T_H^*} \quad (2)$$

where n - specific rotor speed;

n – physical rotor speed;

T_H^* - air temperature at engine inlet.

Normalized Efficiency – efficiency referred to efficiency at the primary operational mode of the base LPC.

Normalized Pressure Ratio – pressure ratio referred to efficiency at the primary operational mode of the base LPC.

Normalized Mass Flow – air flow rate referred to efficiency at the primary operational mode of the base LPC.

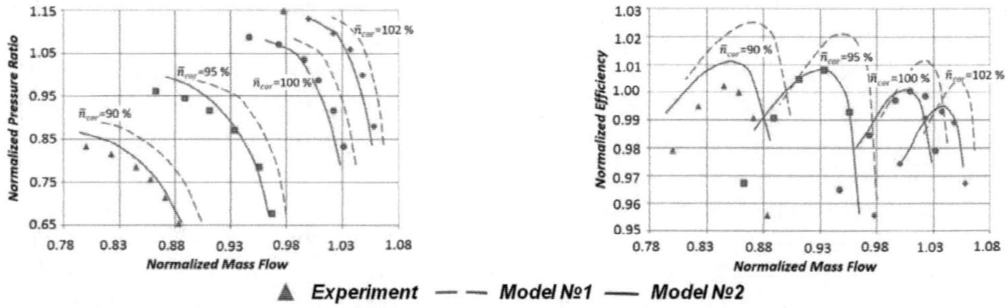

Figure 3. Comparison of calculated and experimental parameters of HPC.

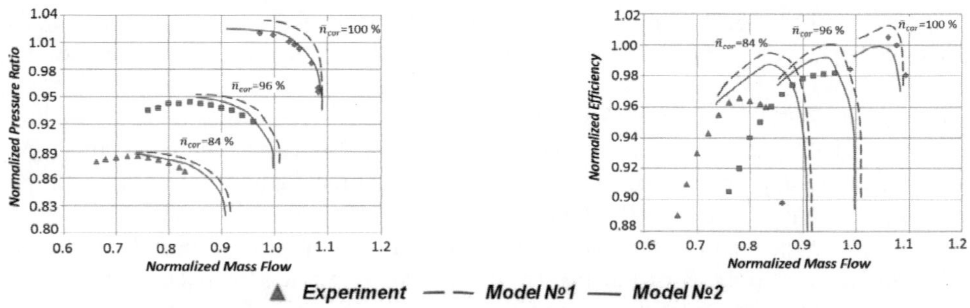

Figure 4. Comparison of calculated and experimental parameters of LPC.

As can be seen from figure 3, both created numerical models show good qualitative coincidence with the experimental results. However, the model №2 shows significantly better quantitative coincidence with experimental results. The difference of values for efficiency and the compression ratio is not more than 1%.

Apparently from figure 4, calculated pressure characteristics of the LPC qualitatively repeat the experimental ones. The maximum deviation of calculated values of the pressure ratio from the experimental value is 1% (abs) at the mode $\overline{n}_{cor} = 84$ %. Also, figure 4 shows that calculated curves of efficiency characteristic of the base LPC qualitatively repeat the experimental ones at all modes. The maximum difference between calculated and measured efficiency is 2.9% (abs.) at the mode $\overline{n}_{cor} = 84$ %. At the mode $\overline{n}_{cor} = 100$ %, the maximum deviation is 0.7% (abs.).

Based on the above, it was concluded that, despite some quantitative discrepancies with the available experimental data, created numerical model can adequately describe the operation of the LPC and can be used for search of configuration with maximum efficiency. For this reason, the model №2 was used for further studies.

3. Optimization of HPC of the engine NK36-ST

The criteria for this optimization problem:
- increase in HPC efficiency at the mode corresponding to the rotational speed of 95%;
- increase in HPC efficiency at the mode corresponding to the rotational speed of 100%.

In order to prevent the shift of characteristics of the compressor, in agreement with JSC "Kuznetsov" [4], the following restrictions were set in the optimization:

Flow rate of the working fluid through the HPC at relative frequency of rotation of 95% was not supposed to be different from the respective flow rate of the base compressor more than ±1,3%;

Flow rate of the working fluid through the HPC at relative frequency of rotation of 100% was not supposed to be different from the respective flow rate of the base compressor more than ±0,6%;

value change of HPC pressure ratio compared with the basic compressor at points of the maximum efficiency at relative rotational speeds of 95% and 100% allowed within ±1,5%.

As varied variables the stagger angles of all rotor blades, guide vanes and IGV of the HPC were selected. The range of change of stagger angles of the vanes of each blade row has been selected so that during the blades rotation their profiles fits into existing blade locks. The number of blades in the row has not changed. This solution allows to find the variant to increase the efficiency of the HPC, which would not require modification of the disk and the body parts of the compressor. The total number of changed variables was 15.

The whole optimization process was completely automated.

As a result, a lot of unimprovable solutions (Pareto set) were obtained, which is a compromise between increase of efficiency at the relative rotational speed of 95% and increase of efficiency on the relative rotational speed of 100%. Each point of Pareto set corresponds to the unique geometry of the HPC represented as an array of stagger angles of all blade rows of HPC.

Analysis of the extreme points of Pareto set shown that at relative rotation frequency of 95% the highest increase of maximum efficiency is 1.8% (abs.) at a substantially constant maximum efficiency at relative rotational speed of 100% (point 1 of Pareto). When the relative rotational speed of 100% the highest increase of maximum efficiency is 0.6% (abs.) at the increasing of maximum efficiency at relative rotational speed of 80% to 1% (point 2 Pareto).

However, for further research one of midpoints of the set Pareto has been selected (point 3 of Pareto), providing increasing of the efficiency as at relative rotation frequency of 100% (0.5% (abs.)) and at the relative frequency rotation of 95 % (1.2% (abs.)).

To analyse the results of the optimization the numerical model of HPC variant was created, which corresponds to the selected point 3 of the Pareto set. By this numerical model the characteristics of optimized version of HPC at the relative rotational speeds of 95% and 100% were obtained as well as their comparison with the characteristics of HPC base version (figure 5) and with the searching results of the optimal combination of stagger angles at the first three stages, described above, performed.

The result of comparison revealed the following characteristics:

- stall margin of operation of optimized HPC compared to the base case at the investigated frequencies of rotation have changed slightly;
- changing the values of the air flow rate and compressor pressure ratio of the optimized HPC at points of maximum efficiency at the investigated frequencies of rotation is within the accepted limits;
- HPC efficiency at the relative rotation frequency of 95% has increased by 1.2% (abs.) and at relative rotation frequency of 100% increase of efficiency was 0.5% (abs.).

4. Optimization of LPC of the engine NK36-ST

On conditions of the problem, it was necessary to ensure the growth of the LPC efficiency by 1.5% (absolute value) with 4% increase of pressure ratio, 2% of the rotor speed and decrease of the working fluid flow rate by 8% with respect to the parameters of the original engine NK-36ST. The goal is to be achieved with unchanged diametrical and axial dimensions of the compressor flow, while maintaining acceptable safety margins of all the rotor elements.

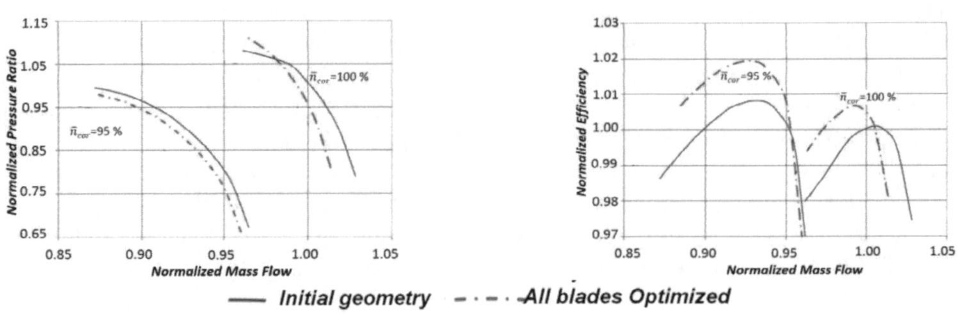

Figure 5. Comparison of the characteristics of optimized versions of the compressor and basic HPC.

Optimization problem was solved in a two-criteria formulation. As optimization criteria were selected:

- increase of LPC efficiency at rotor speed n = 102% (relative to rotor speed of the base compressor);
- decrease of the relative flow rate of air the through the compressor.

The constraints that determine the position of the working point on the LPC characteristic specified for the optimization problem:

- minimum value of the air flow rate in the design point is limited by the range: 0.91 ... 0.96·base;
- the total pressure ratio was maintained at the design point a predetermined range: 1.009 ... 1.046·base;
- the flow angle changing at the LPC outlet was limited by the range: ±5 degrees.

Varying variables were selected as follow. Slightly modified approach from [5] was used to describe the shape of the camber line. Changing the camber line shape of the rotor blades and the relative position of the cross sections relative to each other was carried out by moving the spline middle control points in the circumferential direction in the global coordinate system, as well by varying the stagger angle. This solution helped to keep the value of the blade chord in control sections, which is important in terms of maintaining the stress strain state. The guide vane profile was changed by moving the middle point of the spline in the circumferential direction and moving the point of the trailing edge along both coordinates.

Reforming algorithm of the blade profile along the height with the approaches described above has been implemented in the «in-house» program, developed at the Department of Aircraft Engines Theory of SSAU. This program allows to convert a table of coordinates, which describes the blade shape in design drawing into text files of source data.

To solve the formulated problem of optimization the software package IOSO had 446 calls to the numerical model of the HPC [6]. Each call to the numerical model is calculation of two points on the characteristic of the HPC (points of maximum efficiency on the branches corresponding to the relative frequency of rotation of 95% and 100%) in the software package NUMECA FineTurbo. As a result, the Pareto set of best possible criteria according to two criteria - the relative efficiency and the relative flow rate of the working fluid through the compressor were obtained. Three points were selected for the optimization results validation and analysis.

The pressure characteristics were calculated for each point at the rotor speed of 102% of the base LPC speed and compared with the initial characteristic of the compressor (figure 6).

Comparing the characteristics of the different variants, it can be seen that all the variants have approximately the same efficiency at the desired air flow rate through the compressor. Moreover, this value is greater than efficiency of the base compressor by 1.3% (absolute). Comparison of relative characteristics of the selected LPC variants with characteristic of original compressor Variant №2 was adopted as a final embodiment of the compressor, because it allows to obtain the desired pressure ratio (variant №3 has less value), has a higher efficiency than the variant №1 and stall margin for this variant does not differ from the original compressor stall margin.

Variant №2 was adopted as a final embodiment of the compressor, because it allows to obtain the desired pressure ratio (variant №3 has less value), has a higher efficiency than the variant №1 and stall margin for this variant does not differ from the original compressor stall margin.

5. Conclusion

Summing up the work done it can be concluded that the goal has been reached. The developed algorithm showed its effectiveness for the reprofiling multistage turbomachinery.

HPC variant has been found that provides an increase in the efficiency at the operation mode corresponding to 95% of the rotational speed by 1.2% (absolute) and at the mode corresponding to 100% of the rotational speed by 0.5% (absolute).

IOP Conf. Series: Materials Science and Engineering **302** (2018) 012030 doi:10.1088/1757-899X/302/1/012030

Also a variant of the shape of low pressure compressor blades has been found that provides an increase in efficiency by 1.3% (abs.), while increasing the pressure ratio by 4%, the rotational speed by 2% and decrease in the mass flow rate of working fluid flow by 8% relative to compressor of base engine.

During the search for improved shapes of blades, measures for the conservation of their stress strain state in the former borders were taken. It would be difficult to obtain a similar result with traditional methods of designing compressor as the base variants of the compressor already had a high level of efficiency, and the requirements were contradictory.

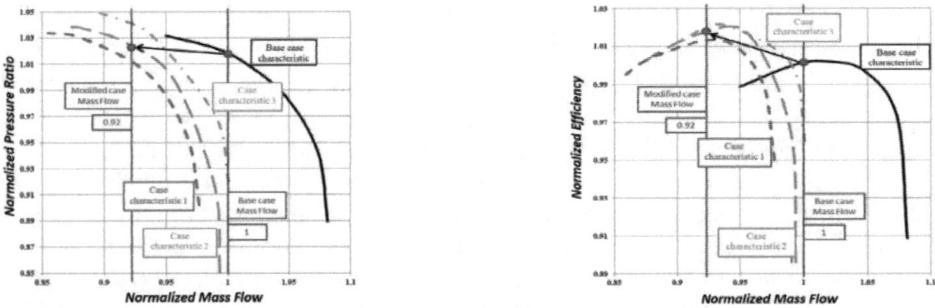

Figure 6. Comparison of characteristics of the selected LPC variants with characteristic of original compressor.

Acknowledgments

This work was carried out by lead performer of research and development effort with financial support of Russian Federation Ministry of Science and Education with the realization of Goverment decree №218 from 09.04.2010 according to contract about providing and using subsidy № 02.G25.31.0150 from 01.12.2015

References

[1] Gritsenki E, Danilchenko V, Lukachev S 2004 *Converting GTE in gas turbine plants for ground applications* p 266

[2] Kuzmichev V, Rybakov V, Tkacheno Y, Krupenich I 2014 Optimization of working process parameters of gas turbine engines line on the basis of unified engine core ARPN Journal of Engineering and Applied Sciences ed J Tarakzai Vol 9 pp 1873-1878

[3] http://www.kuznetsov-motors.ru/en

[4] Kulagin V 2002 The theory, calculation and design of aircraft engines and power plants Fundamentals of the theory of the CCD. Workflow and thermodynamic analysis p 616

[5] Shablii L, Dmitrieva I 2014 Blade geometry transformation in optimization problems from the point cloud to the parametric form Russian Aeronautics Vol 57 ed v Firsov No 3 pp 276–282

[6] Egorov I, Kretinin G, Leshchenko I, Kuptzov S 2002 Optimisation Toolkit - Novel Software to Create Better Design 9th AIAA/ISSMO Symposium on Multidisciplinary Analysis and Optimisation

Study of the Effect of Heat Supply on the Hydrodynamics of the Flow and Heat Transfer in Capillary Elements of Mixing Heads Jet Thrusters

V E Nigodjuk[1] and A V Sulinov[1]

[1]Institute of engines and power plants, Samara University, Samara, Russia

Email: abcsamara@yandex.ru

Abstract. The article presents the results of experimental studies of hydrodynamics and those of loobman single-phase and two-phase flows in capillary nozzle elements propellant thrusters and the proposed method of their calculation. An experimental study was performed in capillaries with a sharp entrance edge of the internal diameter of 0.16 and 0.33 mm and a relative length 188 and 161, respectively, in pouring distilled water and acetone in the following range of parameters Reynolds number $Re = (0,3 ... 10) \cdot 10^{3}$, Prandtl number $Pr = (2 ... 10)$, pressure $p = (0,1 ... 0,3)$ MPa, the heat flux $q = (0 ... 2) \times 10^{6}$ W/m^2, the difference of temperature underheating of liquid $\Delta t_n = (5 ... 80) K$. The dependences for calculation of single phase boundaries, the undeveloped and the developed surface of the bubble and film key singing of subcooled liquid. It is shown theoretically and experimentally confirmed the virtual absence of areas of undeveloped nucleate boiling in laminar flow. The dependence for calculation of hydraulic resistance and heat transfer in the investigated areas of current. It is shown that in the region of nucleate boiling surface in the flow in capillary tubes, influence of the formed vapor phase on the hydrodynamics and heat transfer substantially higher than in larger diameter pipes.

1. Introduction

The need to consider the hydrodynamics of the flow in capillary elements propellant thrusters with supply of heat due to their characteristic working conditions – heating of the capillaries in the process of their work [1]. It is therefore important to know not only the hydraulic characteristics of the capillaries for flow of liquid from supply of heat, but the values of densities of heat flows to the capillaries that lead to the implementation of the possible flow regimes of the liquid with the supply of heat to either account for them or avoid these modes.

Currently there is no General model of fluid flow in tubes with supply of heat, however, a large number of theoretical and especially experimental work performed on pipes of large diameter and is devoted to the study of individual regions of the considered flow.

The aim of this work was research of features of hydrodynamics flow and heat transfer in capillaries with supply of heat, the detection realized with flow regimes and to develop a method of calculation of the boundaries of these modes of heat transfer coefficients and hydraulic resistance.

2. Facilities and experimental conditions

The hydrodynamic study was performed on capillaries made of stainless steel. The experimental study was carried out in a pouring plant, which used the displacement supply of working fluid to the capillary by means of compressed air. The supply pressure of the liquid was monitored by a pressure sensor, the flow rate of the working fluid was measured volumetrically by means of a level gauge, and the time of fills was recorded with an electric stopwatch. The supply of heat flux to the liquid was ensured by passing through the capillary the electric current of constant voltage, the value of which was determined by the voltage drop across the shunt. The voltage drop across the shunt and the capillary was measured with voltmeters. Fluid temperature at the inlet and outlet of the capillary and the temperature of the capillary wall in five sections with flow with supply of heat was measured by a chromel-kopelove thermocouples. An experimental study was performed in capillaries with a sharp inlet edge, an inner diameter of 0.16 and 0.33 mm and a relative length 188 and 161, respectively, in prolivka water and acetone in the following range of parameters the Reynolds number $Re = (0,3...10)\cdot 10^3$, Prandtl number $Pr = 2...10$, fluid pressure $p = 0,1...0,3$ MPa, the density supplied to the capillary heat flux $q = (0...2)\cdot 10^6$ Вт/м^2, subcooled to the temperature of boiling liquid $\Delta t_n = (t_l - t_s) = 5...80$ K. In the capillary phase heat-was preceded by a phase isothermal flow, where they provide hydrodynamic stabilization of the flow of fluid.

3. The physical picture of the fluid flow in the capillaries for supply thereto of heat

The physical picture of the fluid flow in the capillary nozzle elements of the thrusters, taking into account all possible stages of the implementation flow with milk.-the house of heat can be represented as follows (Figure 1): the inlet portion AB is implemented isothermal fluid flow without heat supply; in the subsequent parts of the flow with heat supply BL, which can be liquid-phase flow BC, flow boiling of subcooled liquid CL, consisting of undeveloped plots of CD and DK developed nucleate boiling and film boiling area KL.

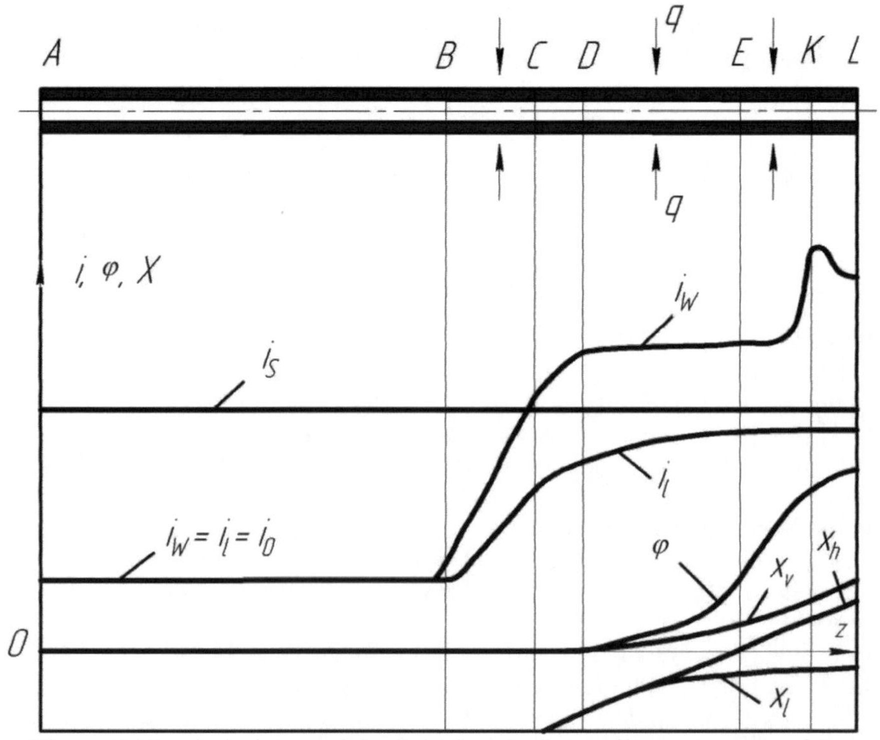

Figure 1. Physical picture of the fluid flow in the capillary elements of the mixing head with supply of heat

In Figure 1 shows the changes of the parameters of the liquid along the channel axis of the capillary, where i_0, i_l, i_w, i_s - the enthalpy of the fluid respectively at the inlet of the capillary, the length of the channel, the wall temperature t_w and temperature of the saturated vapor t_s; φ - is the true volume steam content; x_h – relative enthalpy, determined by heat balance; x_l - is the relative enthalpy of the working fluid, the average over the cross section; x_v – the actual mass vapor content; q – heat flux density; z - is the coordinate along the channel axis of the capillary. Describes the field of real case can have a different length depending on the length of the heat supply, the heat flux, fluid flow, etc.

Studied the field of single-phase current, undeveloped and developed nucleate boiling of subcooled liquid.

4. A study of the boundaries of the regions of fluid flow in capillaries with supply of heat

The transition to undeveloped nucleate boiling occurs at a certain overheating of the capillary wall $t_w > t_s$ and higher density of the heat flow $q_b > q_s$, than that at which $t_w = t_s$:

$$q_s = (i_s - i)\frac{\alpha}{c_p} . \tag{1}$$

The expression to calculate the heat flux at the beginning of the boil q_b can be found under the assumption that the temperature change $(t_w - t_s)$ occurs within the layer as the characteristic dimension which is the diameter of the gas bubble at the moment of conception [2], then:

$$q_b = \left\{\left[\frac{\alpha\sigma}{r\rho_c\lambda_h(1 - T/T_s)}\right]^{1/2} + \left[\frac{\alpha\sigma}{r\rho_c\lambda_h(1 - T/T_s)} + 1\right]^{1/2}\right\}^2 \cdot q_s . \tag{2}$$

The last equation is obtained theoretically and considers fluid properties, and the mode of its flow, and to determine the upper boundary region of single-phase flow.

Under steady laminar flow and heat transfer profile of the enthalpy of the flow is parabolic. This allows proindeksirovat variation i in the section and based on the assumption [3] about the proportionality (ratio about 1) the true volumetric vapor content of the flow area at i > is, given φ = 1 – rs2, to obtain [4]

$$q_d = \frac{q_s}{1 - 2\varphi} , \tag{3}$$

that is, when φ = 0,01, which can be taken abroad developed a boil, qd / qs = 1,02. This indicates an extremely small extent on q region undeveloped boiling in laminar flow.

Experimental determination of the boundaries of two-phase flow regimes showed (Figure 2) that the lower border of the area is undeveloped boiling satisfactorily described (2), and top with turbulent flow – the equation is derived [5]:

$$q_d = \left\{\left[\frac{\alpha\sigma}{r\rho_c\lambda_h(1 - T/T_s)}\right]^{1/2} + \left[\frac{\alpha\sigma}{r\rho_c\lambda_h(1 - T/T_s)} + 2\right]^{1/2}\right\}^2 \cdot q_s . \tag{4}$$

The upper border region of developed nucleate boiling (heat-transfer crisis) in laminar flow is satisfactorily described by the dependence [6]:

$$q_{cr} = \left\{K_\infty r\left[\rho_c\sigma g(\rho_s - \rho_c)\right]^{1/2} + K_0 r u\left(\frac{\rho_s}{\rho_c}\right)^{1/2}\right\}\left[1 + 0{,}0065\left(\frac{\rho_s}{\rho_c}\right)^{4/5}\left(\frac{c_p\Delta t_n}{r}\right)\right], \tag{5}$$

where K∞ = 0.13 and K0 = 0,0012 – empirical coefficients. In turbulent flow the experimental data can also be described by the dependence (5), putting K0 = 0,007.

5. The study of the linear coefficient of hydraulic resistance

In the field of single-phase flow experimentally shown the validity of the ratios to calculate the relative linear coefficient of hydraulic resistance [7]:

$$\bar{\lambda}_l = \frac{\lambda_l}{\lambda_0} = \left(\frac{\eta_w}{\eta_0}\right)^n, \tag{6}$$

moreover, for turbulent flow $n = 0.25$ for laminar $n = 2{,}3\,\bar{z}^{0,3}(\eta_w/\eta_0)^{-0.062}$ if $\bar{z} = (0{,}7...12)\cdot 10^3$.

Experimentally determined values of the Nusselt criterion for stable current, the supply of heat Nu∞ are different from the values of Nu∞=is 4.36 for laminar flow regime at Re > 1,5·103 [4], which proves that the single-phase flow in the flow in capillary tubes, turbulent fluctuations and corresponds to the boundary of the transition determined at isothermal pouring.

The study area of undeveloped nucleate boiling surface in turbulent flow showed that the influence of the vapor phase on the hydrodynamics of the flow is negligible, and the linear coefficient of hydraulic resistance can be described by equation (6), which η_w is determined by the wall temperature.

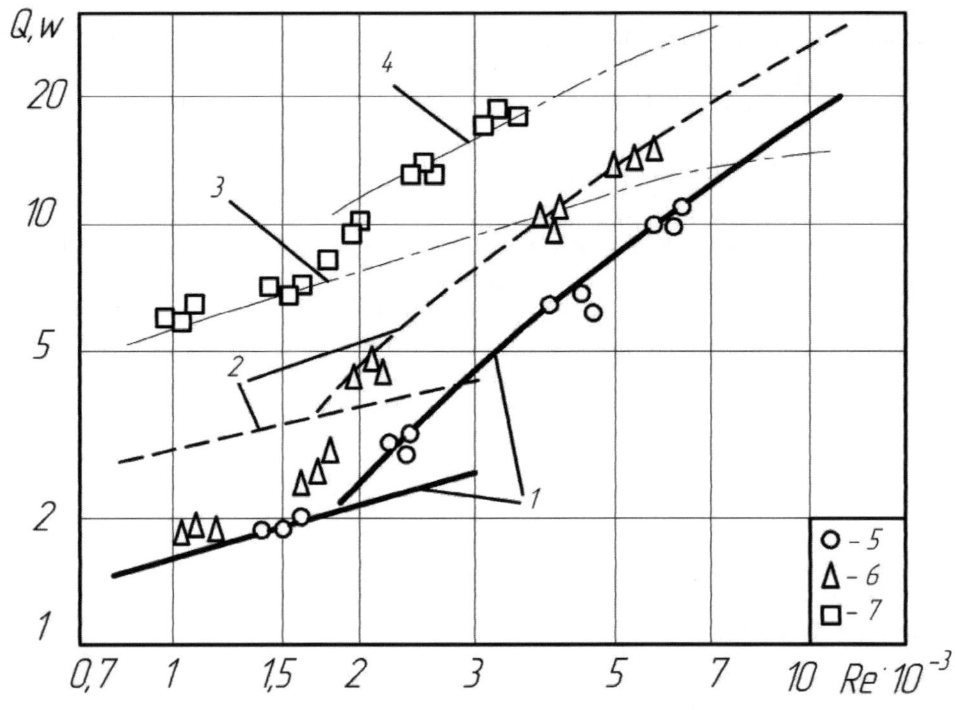

Figure 2. Comparison of calculated and experimental data on determination of boundaries of the regions of the flow with heat supply: 1 – calculation according to (2); 2 – calculation according to (4); 3 – calculation according to (5) with K_0=0,0012; 4 – calculation of (5) with K_0=0,007; experimental results: 5 – the border is boiling; 6 - boundary of the beginning of developed bubble boiling surface; 7 – the upper boundary region of developed nucleate boiling

Generalization of experimental values of relative linear coefficient of hydraulic resistance in the region of developed nucleate surface boiling of subcooled liquid was performed using the complex [8]

$$A = 20\left(\frac{q}{r\rho_c u}\right)^{7/10}\left(\frac{\rho_s}{\rho_c}\right)^{8/100}\left\{\left(\frac{1{,}32\,(i_s - i_b)}{(i - i_b)}\right)\left\langle \ln\left[\frac{1-(i-i_b)}{1{,}32\,(i_s - i_b)}\right]\right\rangle^{-1}\right\} \tag{7}$$

The results are shown in Figure 3a in dependence of the relative (compared with those calculated according to the formula of Poiseuille [9] in laminar flow and the Blasius formula [9] - turbulent) linear coefficient of hydraulic resistance, indicating a more significant effect of the vapor phase on the hydrodynamics of the flow in capillary tubes (line 2) in comparison with the flow in larger diameter pipes (d > 1 mm - 1 line [10]).

The obtained experimental data can be described by a linear dependence in the range $Re = (1...5)\cdot 10^3$, $\Delta t_n = (5...25)K$

$$\overline{\lambda}_d = 1 + 3A\,, \tag{8}$$

allow the calculation of hydraulic resistance of two-phase flow in capillary tubes with a diameter of (0,15...0,35) mm in the region of developed nucleate boiling.

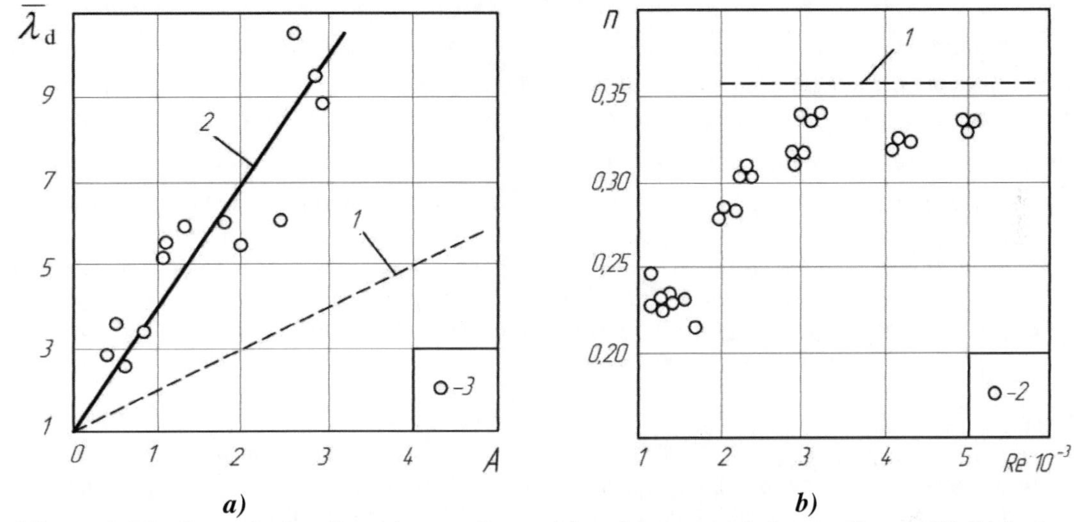

Figure 3. The linear hydraulic resistance factor (a) and exponent "n" according to (9) (b) in the field of developed nucleate surface boiling: a) 1 – calculation according to [10], 2 – calculation according to (8), 3 – experimental data; b) 1 –according to [8], 2 – experimental data

For calculation of heat transfer in this region it is necessary to know the value of the exponent in the generalized dependence [11]

$$t_w^d - t_s = c\,(q - q_s)^n,\;\; c = 28\,T_{cr}^{41/50}\left(p_{cr}\cdot 10^5\right)^{-9/25} M^{9/50}\exp\left(-5{,}6\frac{T_s}{T_{cr}}\right), \tag{9}$$

where Tcr, pcr – critical parameters of the liquid, M – is the molar mass.

The experimental values of the exponent "n" in the expression (9) depend on the Reynolds number and is shown in Figure 3b, which shows the influence of flow regime on the exponent "n": in turbulent flow n=0,33, which is close to the results of [11], the transition to laminar flow of "n" decreases and is 0.23 at $Re < 1500$. This can be explained by the influence of void fraction on the heat transfer.

6. Conclusion

Obtained on the basis of wide experimental investigations of the mathematical model integral describing one - and two-phase flow in the capillary nozzle elements of the thrusters is only true at low differential pressure, when the pressure and hence saturation temperature along the length of the capillary is almost unchanged. When a significant pressure gradient and variable heat-flux density to extend the range of applicability of the models considered are infinitely small in length section of the capillary when the pressure gradient can be neglected. This allows to determine at each integration step for a given heat flux constant hydrodynamic and heat transfer characteristics of flow and total pressure drop on the section and then hold a sum of the length of the capillaries and clarify the flow.

Based on the authors experimental results on the effect of heat input on hydrocortisone and heat transfer under single phase and two phase flow in capillaries has been proposed a method of calculating the hydrodynamics of fluid flow with heat supply. The authors studied single-phase flow field and the undeveloped and developed nucleate boiling of subcooled liquid, while the basic parameters were changed in the following range: $\mathrm{Re} = (0,3 ... 10) \cdot 10^{3}$, $\mathrm{Pr} = (2 ... 10)$, $p = (0,1 ... 0,3)$ МПа, $q = (0 ... 2) \times 10^{6}$ Вт/м2, $\Delta t_n = (5 ... 80) K$.

7. Symbols used in the above ratios

p - pressure; ρ *- density;* $\dot{m}$ *- the average mass flow rate; d - the diameter of the capillary;* λ *- the linear hydraulic resistance coefficient; z – the current coordinate, length;* $\bar{z} = z/(d\,\mathrm{Re}\,\mathrm{Pr})$ *- dimensionless length; Re, Pr, Nu - the criteria of Reynolds, Prandtl, Nusselt; T, t - temperature;* Δt *the temperature difference between the; q - heat flux density;* η *- dynamic viscosity coefficient; r - heat of vaporization;* σ *- surface tension;* λ_h *- thermal conductivity coefficient; i - is the enthalpy;* α *- heat transfer coefficient;* c_p*- - specific heat;* φ *- true volumetric vapor content; M - is the molar mass.*

Indexes: *0 - parameters at the entrance of the channel; the s - parameters of the liquid on the saturation line; w - parameters at the temperature of the wall; l - liquid phase flow region; cr - critical; n - underheating; b - initial boiling point; d - developed nucleate surface boiling; c - steam parameters on the saturation line;* ∞ *- parameter stable current.*

References

[1] Nigodjuk V E and Sulinov A V 2013 The effect of thermal factors on instability of the hydrauliccharacteristics of the capillary nozzle elements LTRE *Proceedings of the Samara scientific center, Russian Academy of Sciences* (Samara) T. 15, No 6 (4), pp 901-904.
[2] Chernobay V A 1970 Determination of the conditions of occurrence of nucleate boiling in forced motion of subcooled liquid Heat Power Engineering (Moscow) No 6, pp 65-67.
[3] Zakharova E A, Kolchugin B A, Labuntsov D A 1970 Calculation of the true volume promotergene in heated channels Heat Power Engineering (Moscow) No 6, pp 58-60.
[4] Godlewski V E, Nigodjuk V E and Sulinov A V 1983 Features of calculation of hydraulic resistance and heat transfer for single-phase and two-phase flows in capillary tubes Engineering-physics journal of (Minsk) 45, No 2, pp 327. (Manuscript, 16 p, ill., bibliographer. 17 the name. DEP. 1.04.83 in VINITI, No 1670).
[5] Bergelson B R and Gerasimov A S 1979 Hydraulic characteristics and the steam content in the channel during boiling of the liquid subcooled to the saturation temperature Engineering-physics journal of(Minsk) T. 37, No 23, pp 784-792.
[6] Leonteva A N 1979 Heat and mass transfer theory (Moscow Higher school) p 495.
[7] Petukhov B S, Genin L G, Kovalev S A 1974 Heat transfer in nuclear power plants (Moscow Atomizdat) p 407.
[8] Tarasova N B, Khlopushin V I, Boronina L V 1967 Local hydraulic resistance at superficial boiling of water in the pipe Thermophysics of high temperatures (Moscow) T. 5, No 1, pp 130-136.
[9] Stoczek N P and Shapiro A C 1978 Hydraulics LRE (Moscow Mashinostroenie) p 118.
[10] Ornatsky A P 1965 Generalization of experimental data on hydraulic resistance at superficial boiling Journal of applied mechanics and technical physics (Moscow) T. 3, No 3, pp 444-451.
[11] Bahvalov J E, Cronin I B, Kurganov I V 1966 Generalization of data on heat transfer during boiling of liquids netegrity Heat Power Engineering (Moscow) No 5, pp 63-68.

IOP Conf. Series: Materials Science and Engineering **302** (2018) 012032 doi:10.1088/1757-899X/302/1/012032

Computer Design Technology of the Small Thrust Rocket Engines Using CAE / CAD Systems

VRyzhkov[1] and ELapshin[1]

[1]Researcher, Samara National Research University, Samara, Russia

E-mail:ke_src@ssau.ru

Abstract. The paper presents an algorithm for designing liquid small thrust rocket engine, the process of which consists of five aggregated stages with feedback. Three stages of the algorithm provide engineering support for design, and two stages - the actual engine design. A distinctive feature of the proposed approach is a deep study of the main technical solutions at the stage of engineering analysis and interaction with the created knowledge (data) base, which accelerates the process and provides enhanced design quality. The using multifunctional graphic package Siemens NX allows to obtain the final product –rocket engine and a set of design documentation in a fairly short time; the engine design does not require a long experimental development.

1. Introduction

Modern graphic packages are used to solve design tasks, production of design documentation, preparation of complex products production of rocket and space technology at the present time. The graphic packages are designed not only to automate the traditionally existing technological links, but also to fundamentally change the approach to designing and producing products.

At the heart of the computer technology for the design of technical devices, the creation of a complete electronic product layout lies in. The creation of three-dimensional models of adequately designed construction, opens opportunities for creating high-quality products (especially complex and science-intensive) in a short time. Ideally, the process of designing and manufacturing multicomponent products requires the creation of spatial electronic models of parts, assemblies, units, systems and the entire product

At the same time, it is necessary to simultaneously solve the tasks of conceptual design, various types of computational and analytical research, modeling of possible dangerous operating modes, as well as the layout of the product and the formation of its design appearance. The data obtained during the design must be used for technological preparation and for production.

In addition, it is required to manage all the created elements of the electronic model (product structure) and of the product creation process.

To implement computer technology design and production the systems of computer-aided design, engineering analysis and technological preparation of high-level production (CAE / CAD / CAM), as well as management systems (PDM) are currently used.

At the same time, the system provides the entire cycle of product creation from the conceptual idea to implementation, also a project-technological environment is created for simultaneous work of project executors with a single virtual electronic product model.

IOP Conf. Series: Materials Science and Engineering **302** (2018) 012032 doi:10.1088/1757-899X/302/1/012032

The system should be structured in a certain way and have several special opportunities to implement this approach to development. The main features are full parameterization, communication both within the model and with its applications, as well as the ability to manage the assembly (build components).

The concept of low-thrust liquid-fueled rocket engines development presupposes the availability of a developed knowledge base in the field of small thrust rocket engines and its wide use in the design (at the initial stages), for solving the main problematic issues at this stage of creating the engine.

2. Process of creating a new small thrust LRE

Conceptually, the process of creating a new small thrust LRE can be represented in the form of the following algorithm (Figure 1), which shows the sequence (technology) of the formation of the design image of thesmall thrust LRE.

This process is presented here in the form of five stages with feedback

The first stage is the preparation of the initial data for the engine being designed.

The second stage is the estimation of the main parameters based on known (classical) ideal methods using thermodynamic, gas dynamic calculations and calculation of heat transfer with access to the main geometric and integral parameters of the engine [1-5]; determination of the design appearance of the engine based on ideal workflow models.

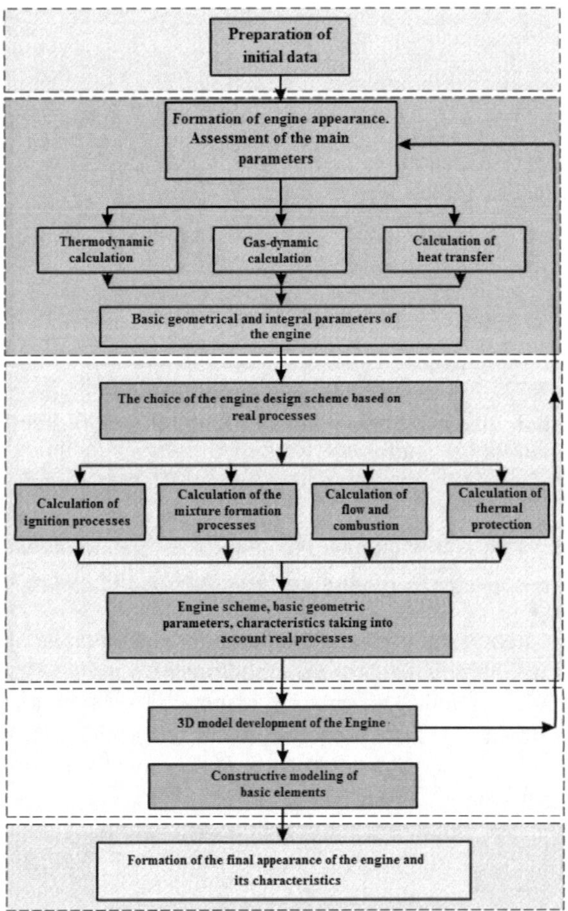

Figure 1. Algorithm for the design image formation of small thrust LRE

The third stage is the choice of the engine design scheme based on real processes, including the formation, considering knowledge bases for circuits, ignition devices, mixing elements, means of providing thermal protection for the engine. Moreover, knowledge bases created in electronic form have an independent value and in the process of selecting a constructive scheme of the engine, information exchange must be established with them not only in the form of enumeration of the base elements, but also in the form of searching for constructive execution in accordance with a certain algorithm.

The choice of the design of the engine, considering the geometry of the gas dynamic path, determined at the previous stage, allows us to form the calculation area for the computational procedures of the mathematical model of the high-level working process of the high-level liquid-fuel rocket engine, which allows to consider the actual processes of mixture formation, combustion, product flow, combustion and thermal protection. At the output of the third stage, the following is to be obtained: the motor circuit, the main geometric parameters and the characteristics of the small thrust LRE.

At the fourth stage, a 3D engine model is being developed. As a computer development environment, one of the graphic packages is used, which makes it possible to obtain a spatial electronic model based on the basic technologies and techniques of 3D electronic modeling of the complex technical devices, the element base of structural blocks, engine components. 3D model can be used as a basis for the release of design documentation for the development, and as part of the PLM-system, for example, propulsion system.

At the fifth stage of the technology of forming the design image of the engine, its final version is determined, verification calculations are carried out, the output parameters and characteristics of the small thrust LRE are determined.

As a computer environment for the development of the small thrust LRE, after the analysis of existing high-level systems, the graphic package Siemens NX (formerly UNIGRAPHICS) [6] was chosen, to the greatest extent, according to the authors, meeting the requirements of the computer technology for designing and manufacturing high-tech and science-intensive products of rocket and space technology.

Among the merits of the package we mention only a few:

- modules of the engineering analysis, based on the built-in solvers, allow to evaluate various scenarios of the working process of the structures being developed;
- the open programming interface allows you to develop your own software, which is integrated into Siemens NX;
- unification of the design processes and elements of the knowledge base accumulated in various fields in the unified system (the technology of design using the knowledge base).

According to the proposed algorithm, before starting the computer design of the small thrust RE (on fuel oxygen-hydrogen), it is necessary to specify the initial data for the development, which must correspond to the calculation results for models that take into account the actual processes in the combustion chamber and nozzle, the parameters of the chosen engine design, Taking into account the provision of the permissible thermal state of the structure and the construction material intended for use. These issues are solved at the initial stages of design - stages 2 and 3 of the algorithm for creating the small thrust LRE (Figure 1).

After receiving all the initial information about the engine, you can proceed directly to the process of developing a three-dimensional electronic model of the small thrust RE. The model is created in the environment of the Siemens NX graphic package.

3. Development of electronic models

Modeling should begin with the development of electronic models of the main components of RDMT with their subsequent virtual assembly at the final stage.

The plane of the exhausing of the outer and inner contours of the mixing head of the engine, as the most crucial element, is punished in Figure 2). The inner contour of the head is a part of the gas dynamic path of the product, and the outer contour is formed taking into account the length of the fuel supply channels to the mixing head. Having set the sizes of the inner and outer contours of the head, by the "rotation" operation, turn the sketch around the central axis, getting the body of rotation - the workpiece of the solid model of the mixing head of the engine. The blank cut is shown in Figure 3.

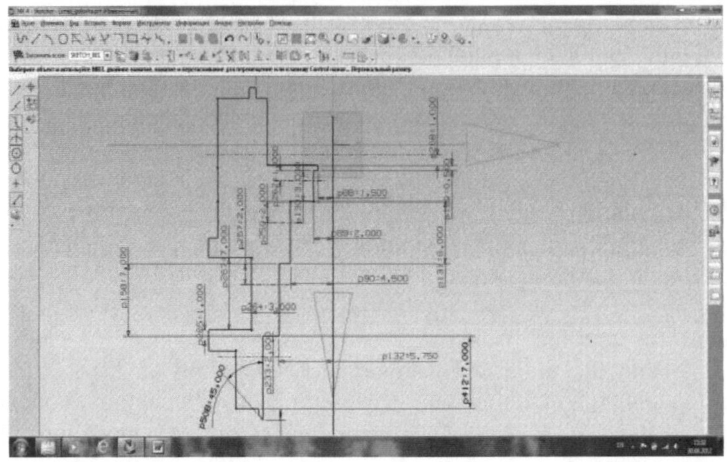

Figure 2. The plane for sketching the contours of the mixing head of the engine

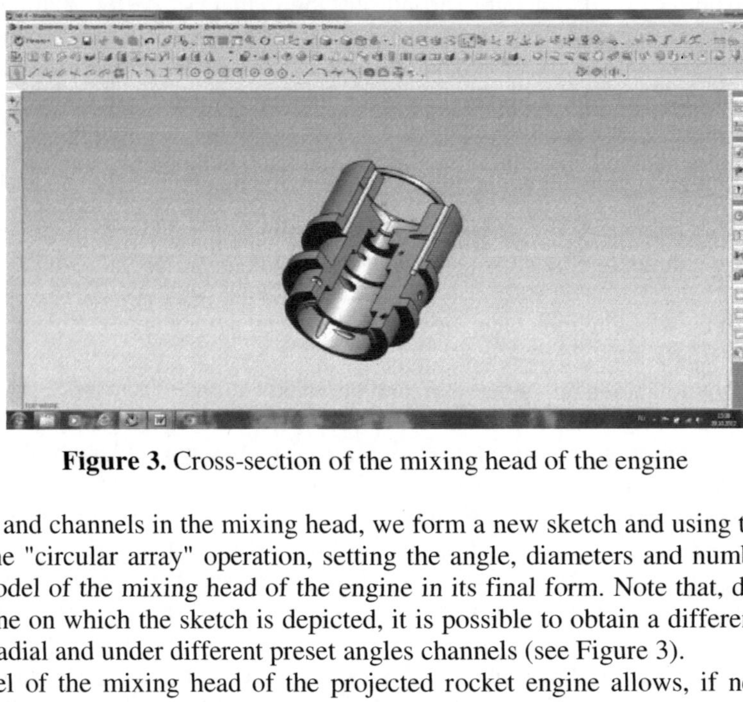

Figure 3. Cross-section of the mixing head of the engine

To create holes and channels in the mixing head, we form a new sketch and using the "subtraction" operation, using the "circular array" operation, setting the angle, diameters and number of holes, we obtain the solid model of the mixing head of the engine in its final form. Note that, depending on the location of the plane on which the sketch is depicted, it is possible to obtain a different version of the holes: tangential, radial and under different preset angles channels (see Figure 3).

The solid model of the mixing head of the projected rocket engine allows, if necessary, on the monitor screen to obtain the types and cross-sections of the future design.

In addition to the control functions that the developer of the electronic model pursues (checking the number, relative position in space, size, coordinates, etc.), visualization of the butt elements with other parts and nodes is possible at this development stage.

The main advantages of the generated virtual spatial solid model, in this case the mixing head of the engine, is the possibility to start preparing its production, after the model is transferred to other participants in the engine making process, meaning that subsequent possible changes will not be fundamental, and the fact that the model can with success to be used in its production on machines with numerical program control.

Similarly, a combustion chamber and a rocket engine nozzle are simulated, which constitute a single unit and other elements of the product.

To create a design flare of the developed rocket engine, a part of the components is taken from the data bank of the LRE components.

The development of electronic models of the main engine parts allows us to proceed to the virtual assembly of the engine. Elements are assembled by conjugating the corresponding faces and axes.

After joining all the components and elements of the rocket engine, we obtain a complete solid-state electronic model of the product, which can be viewed from all the views of interest, including longitudinal sections (Figure 4) and use this model for further work. For example, it is possible to complete an electronic model of a propulsion system, if one is being developed.

4. Create design documentation

The presence of solid models of the components of the rocket engine and subsequent virtual assembly of all components allows to go to the drawing module and create design documentation for both the assembly (assembly drawing) and each part separately, preparing their subsequent manufacturing.

In the Siemens NX environment, parametric models are created and then used, if necessary, at the stages of improvement and modernization.

Figure 5 shows an assembly drawing of the developed rocket engine operating on oxygen-hydrogen fuel.

It is obvious that with the help of the spatial electronic model of the rocket engine, technical issues are simply solved simply technical issues, and technological issues, including selection of tolerances and landings, in accordance with the current GOST. Technical conditions, including leakproofness and strength tests, quality control of welded joints, reconciliation of alignment requirements and a few others, must be included in the design documentation additionally in accordance with the normative acts adopted in the manufacturer of the product.

A set of design documentation released with the help of a computer, responds (can be brought to the state) with all the requirements for design documentation that operate in the relevant industries in the Russian Federation.

Thus, in accordance with the proposed sequence, it is possible to organize the design of thesmall thrust LRE using modern computer technologies, software products and knowledge bases.

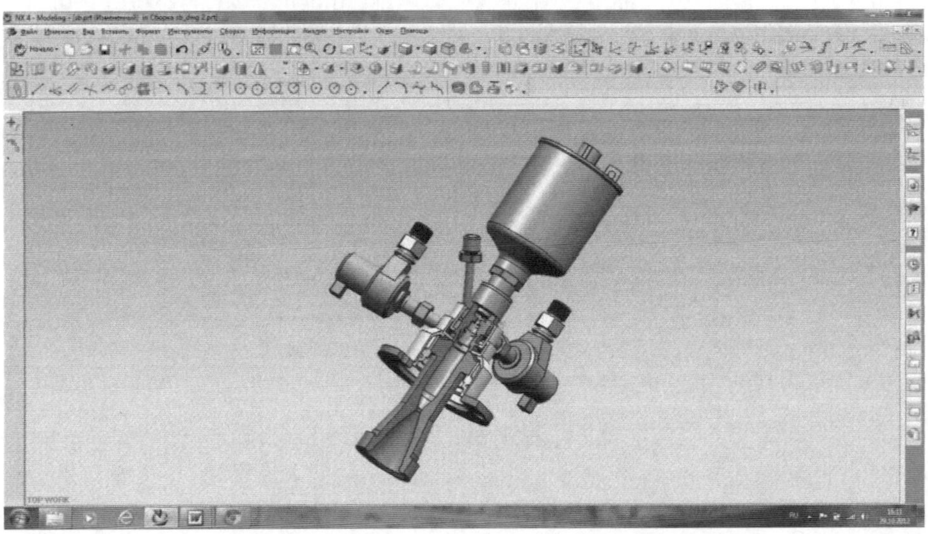

Figure 4. Engine section using a solid model

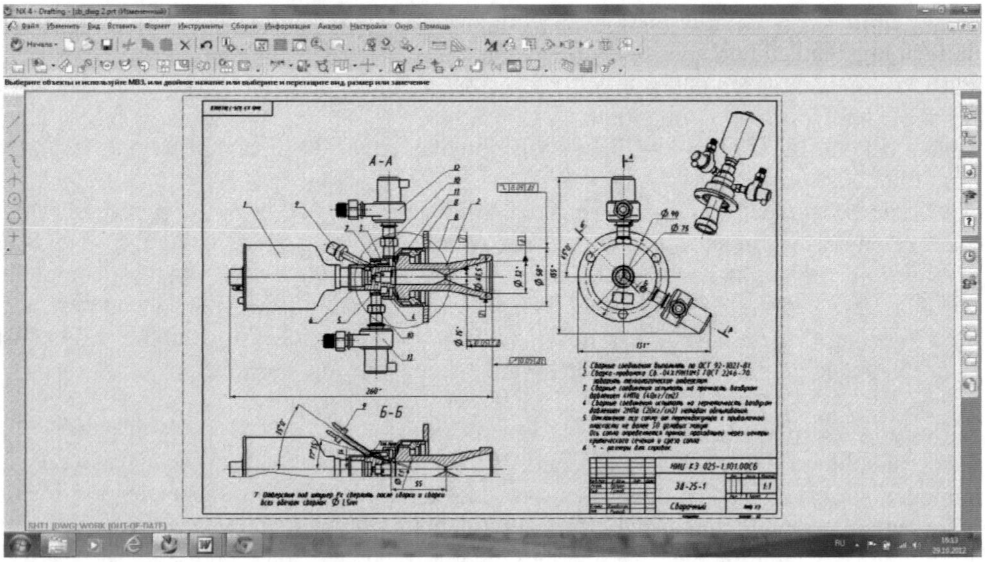

Figure 5. Engine assembly drawing executed using the solid model

References

[1] Alemasov, V., Dergalin, A. and Tishin, A., 1989, The theory of rocket engines (Teorijaraketnyhdvigatelej)// Textbook for university students, Moscow: Mechanical Engineering, p. 464.

[2] Dobrovolsky, M., 2005, Liquid rocket engines. Fundamentals of Design (Zhidkostnyeraketnyedvigateli. Osnovyproektirovanija) // Textbook for Universities, Moscow: Izdatelstvo MGTU named after Bauman, p. 448.

[3] Kudryavtsev, V., 1983, Fundamentals of the theory and calculation of liquid rocket engines (Osnovyteoriiiraschjotazhidkostnyhraketnyhdvigatelej) // Moscow: Higher School, p. 704

[4] Koshkin, V., 1975, Fundamentals of heat transfer in aviation and rocket engineering (Osnovyteploperedachi v aviacionnojiraketnojtehnike) // Moscow: Mechanical Engineering, p. 624.

[5] Glushko, V., 1971-1979, Thermodynamic and Thermophysical Properties of Combustion Products (Termodinamicheskieiteplofizicheskiesvojstvaproduktovsgoranija) // Handbook in 10 volumes, Moscow: VINITI AS USSR.

[6] Krasnov, M., Chigishev, Yu., 2004 Unigraphics for professionals // Moscow: Lori,p. 319.

Pulsations Induced by Vibrations in Aircraft Engine Two-Stage Pump

S A Gafurov[1,2], V A Salmina[1] and H Handroos[2]

[1] Department of Automatic Systems of Power Plants, Samara National Research University, Moskovskoe Shosse 34, 443086, Samara, Russian Federation
[2] Lappeenranta University of Technology, Skinnarilankatu 34, 53850, Lappeenranta, Finland

E-mail: gafurov@ssau.ru

Abstract. This paper describes a phenomenon of induced pressure pulsations inside a two-stage aircraft engine pump. A considered pumps consists of a screw-centrifugal and gear stages. The paper describes the cause of two-stage pump elements loading. A number of hypothesis of pressure pulsations generation inside a pump were considered. The main focus in this consideration is made on phenomena that are not related to pump mode of operation. Provided analysis has shown that pump vibrations as well as pump elements self-oscillations are the main causes that lead to trailing vortices generation. Analysis was conducted by means FEM and CFD simulations as well by means of experimental investigations to obtain natural frequencies and flow structure inside a screw-centrifugal stage. To perform accurate simulations adequate boundary conditions were considered. Cavitation and turbulence phenomena have been also taken into account. Obtained results have shown generated trailing vortices lead to high-frequency loading of the impeller of screw-centrifugal stage and can be a cause of the bearing damage.

1.Introduction

A model of flow induced vibrations in the two-stage pump was developed in paper [1]. Essence lies in trailing vortex stalling with a certain frequency (figure 1). This trailing vortex forms in a radial gap between screw blades and pump case. At that, screw blades are loaded by high-frequency pressure oscillations spreading from a gear stage to screw-centrifugal stage inlet (figure 2). These oscillations can coincide and lead to vibration loading of the pump rotor. Experimental results obtained in paper [2] does not confirm this model as high-frequency oscillation from the gear stage were not registered at the screw-centrifugal stage entrance.

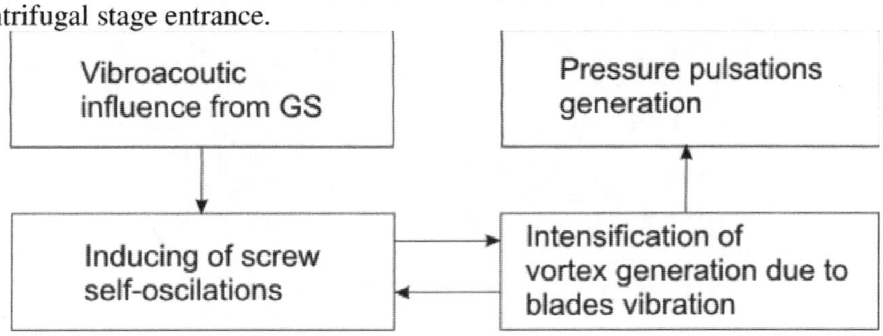

Figure 1. A model of flow induced vibrations in the two-stage pump

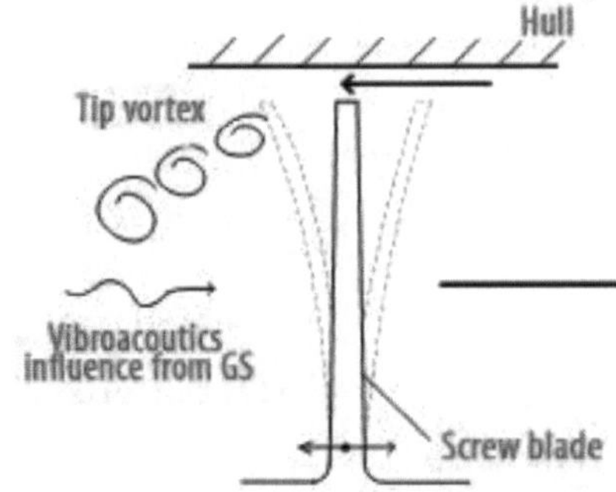

Figure 2. Scheme of pressure pulsations induced vibrations due to the trailing vortex stalling and gear stage influence

A number of factors such as boundary conditions, cavitation model, turbulence model, mesh quality and others influence on accuracy of numerical simulation. CFD approach was used in paper [3] for investigation of the sources of dynamic loading of the radial and thrust bearing inside pumps ND-25 and ND-32. Obtained results showed that the main cause was the so-called “acoustic-vortex resonance”. Acoustic-vortex resonance is the phenomenon when natural frequencies of a screw, pressure pulsations caused by a gear pump and eddy frequency coincide with each other. Star-CD commercial software was used for CFD analysis. However, low quality of the computational mesh, using of the standard $k-\varepsilon$ turbulence model for near-walls flows did not allow to get accurate results and to make final conclusions about the real causes of bearing damages. It is worth to note that some construction elements were not taken into account. Additionally, there is a lack of verification of obtained results. At that, experimental results obtained in paper [3] did not confirm high-frequency pulsations at screw-centrifugal stage inlet. It discards the assumptions made by authors.

2. Two-stage pump modelling

In this work, another cause of high-frequency loading of the two-stage pump elements was accepted. Not only screw blades were considered but also blades of the impeller were taken into account. The schematic design of the pump shows that leakages from the gear stage are poured out to the chamber between stages (figure 3). This champed is also connected with the screw-centrifugal stage flow channel by means of the orifices in the bearing case (figure 4a) as well as by means of discharge holes in the impeller (figure 4b). Such a leakages propagation path results in high-frequency oscillations of blades of the screw and centrifugal wheel.

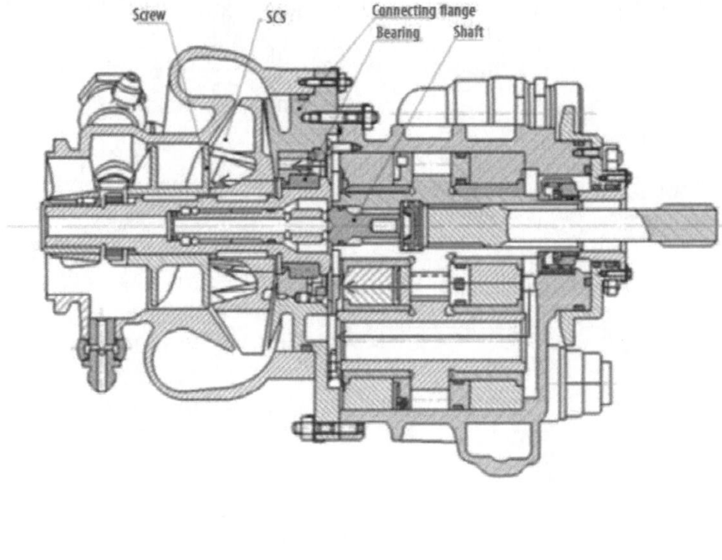

Figure 3. Leakages propagation path from the gear stage to the screw-centrifugal stage

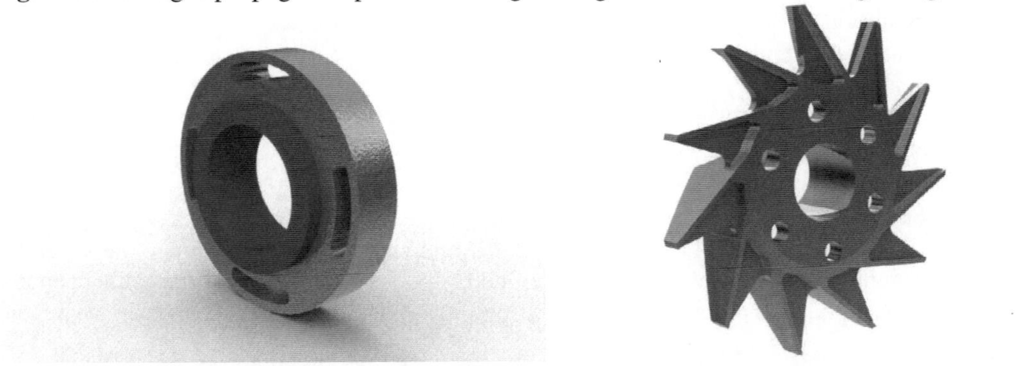

(a) (b)

Figure 4. Construction elements on the screw-centrifugal stage. a) bearing; b) centrifugal

Results obtained in papers [4-6] show that the most important vibration sources of screw and centrifugal wheel blades are trailing vortices. Fluid flow passing the radial clearance between pump case and blades, interacts with the main flow and radial flow from root sections to peripheral. This interaction occurs at blades suction surfaces and results in rolling up of this flow into a vortex. Its radius and circulation depend on pump operating regime, its geometrical parameters and location of vortex appearance. Trailing vortex moves in the circumferential plane of the channel in a direction from the periphery of blades suction sides to their pressure sides.

Near-wall stream, occurring due to the radial flowing, is an unsteady [6]. Disturbances evolution leads to flow separation. These disturbances interact with vibroacoutic influence from the gear stage. This interaction results in oscillations of screw and impeller blades. Oscillations are significantly increased due to the coincidence of their frequencies and natural frequencies of the screw and impeller. Increasing of blades vibrations leads to intensification of pressure pulsations, caused by the hydrodynamic reasons [7,8], and to increasing of vibration loading of the end face of the bearing.

The block diagram of the interaction of vortex disturbances in the screw-centrifugal stage and screw and impeller blades, oscillated due to the vibroacoutic impact from the gear stage, is presented in figure 5.

The general model consists of the sub-models:

- Hydrodynamic model of the screw-centrifugal stage. This model allows to estimate the structure of the flow inside the pump.
- Finite-elements models of the screw and impeller to estimate their natural frequencies.
- Model of leakages from the gear stage. They are represented as sources of oscillations. Their parameters can be obtained theoretically or experimentally.

ANSYS CFX	ANSYS Mechanical		Experimental data
3D geometric model of SCS	**3D geometric model of screw**	**3D geometric model of centrifugal wheel**	**Leakages from GS**
CFD model of SCS	**FEM of screw**	**FEM of centrifugal wheel**	
MODEL OF INTERACTION BETWEEN SCS IMPELLERS AND TIP VORTEX TAKING INTO ACCOUNT GS			

Figure 5. The block diagram of the interaction of vortex disturbances in the screw-centrifugal stage and oscillated screw and impeller blades

IOP Conf. Series: Materials Science and Engineering **302** (2018) 012033 doi:10.1088/1757-899X/302/1/012033

2.1 Boundary conditions, pressure and flow simulation

Boundary conditions for the general model are:

- The eddy frequency and its parameters such as its structure, pressure and velocity distribution inside the eddy.
- Amplitudes of pressure pulsations obtained by the numerical simulation of the screw-centrifugal stage.
- Natural frequencies of the screw and impeller.
- Vibroacoutic performances of the gear stage.

Developed numerical model of the screw-centrifugal stage allows to estimate the separation of the flow inside a pump which results in forming of eddy zone. An accuracy of these turbulence flow can be estimated by yplus parameter distribution along the solid walls. This distribution is shown in figure 6.

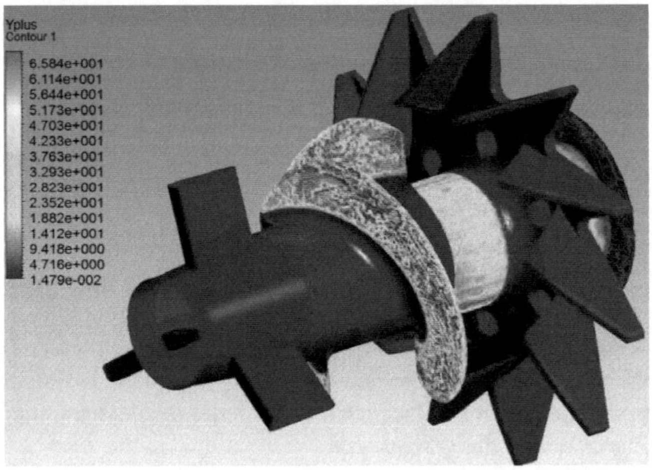

Figure 6. The distribution of yplus parameter along the rotor walls of the screw-centrifugal stage

Figure 7a represents the pressure distribution along the solid walls of the screw-centrifugal stage rotor. There is a separated zone on the suction side of the screw. The flow structure inside the screw is represented in figure 7b.

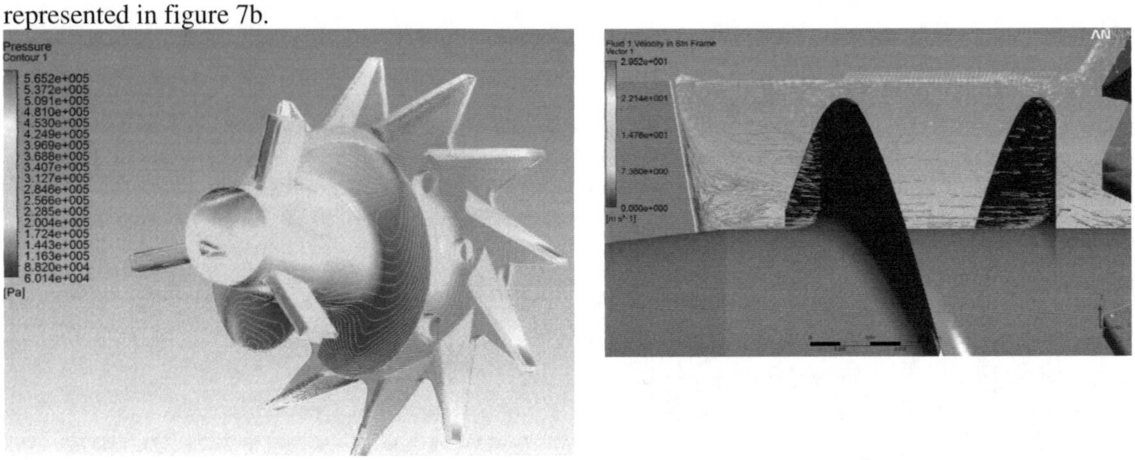

a b

Figure 7. a) Pressure distribution along the solid walls of the screw-centrifugal stage rotor N=4325 rpm, G_{SCS}=8682 kg/h, P_{in}=1.9 atm, T=45 ^{0}C; b) Flow structure on the operating regime N=4325 rpm, G_{SCS}=8682 kg/h, P_{in}=1.9 atm, T=45 ^{0}C

2.2. Separation frequency determining. FEM-analysis

Numerical simulation allows to estimate the parameters of the eddies and to determine its separation frequency by mean of Strouhal number:

$$f = \frac{\upsilon \cdot Sh}{l_{cavity}} = \frac{Sh\sqrt{2g \cdot \Delta H}}{l_{cavity}} \quad (1)$$

where ΔH is pressure difference on the peripheral section of the blade; l_{cavity} is the maximum length of the cavity before the stall break.

Conducted modal FEM-analysis has shown that natural frequencies of the screw and impeller are in the range of 3700-5000 Hz (Table 1).

Table 1. Natural frequencies of the screw and impeller

Natural mode shape	Natural frequency of the screw, Hz	Natural frequency of the impeller, Hz.
f_1	3310.6	3769.6
f_2	3812.7	3846.6
f_3	4373.3	3901
f_4	4887.8	3964.9
f_5	5420.9	4155.6

Modal analysis was performed by ANSYS commercial code. Finite-elements meshes for the screw and impeller are shown in figure 8. Size of the meshes does not exceed 3 mm.

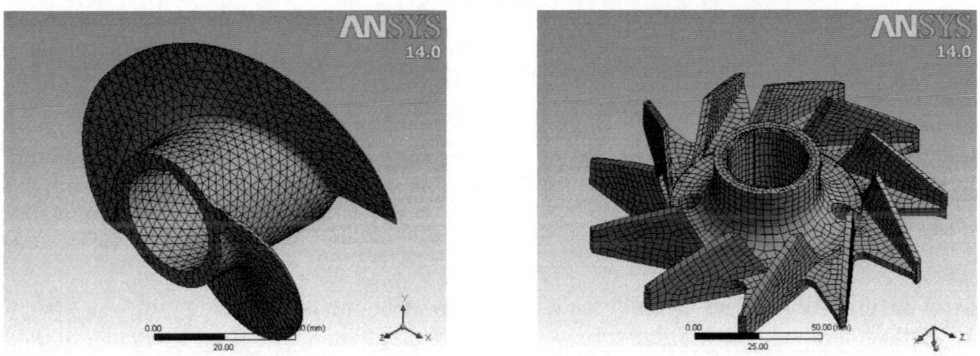

(a) (b)

Figure 8. a) Finite-element meshes of the screw; b) impeller

Natural frequencies of the screw and impeller were also obtained experimentally by means of knocking method. Wheels were hung up on a thread and excited by means of a hummer with a wooden tip. The changing of the sound pressure was being registered on 30 cm distance. Theoretical and experimental spectrums of pressure pulsations are shown in Figure 9. There are coincidences of the eddy frequencies separated from the wheels and their natural frequencies on several operating regimes. It leads to increasing of vibration of the screw and impeller and, as a result, to intensification of the trailer cavitation.

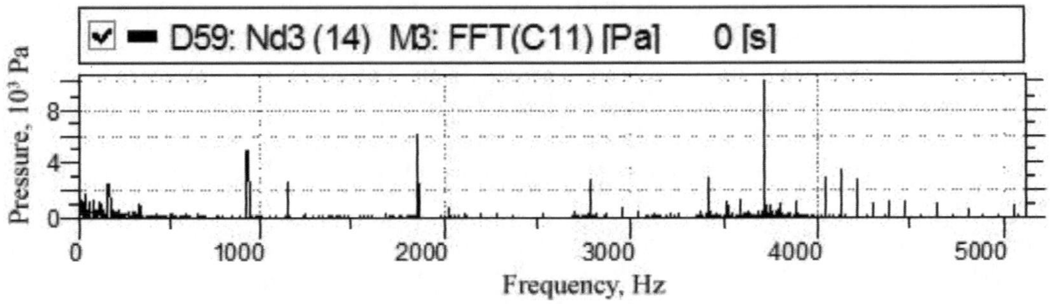

a)

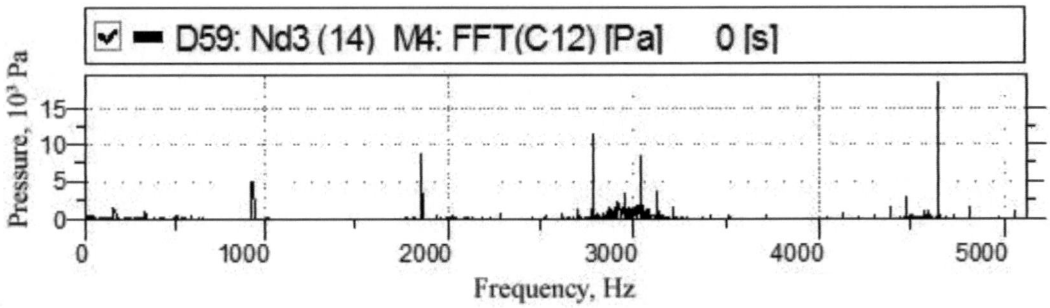

b)

Figure 9. Spectrums of pressure pulsations measured on the pump operating regime - N=4325 rpm, G_{SCS} =8682 kg/h, P_{in} = 1,9 kgf/cm^2, T=45 ^{0}C;
a) outlet of the screw-centrifugal stage;
b) in the chamber between the pump stages

Additionally, operating characteristics of the pumps ND-25 and ND-32 were investigated in paper [9]. This paper shows that the main contribution in total level of vibration makes the high-frequency oscillations above 3 kHz. This paper did not determine the loading source. These investigations confirm the hypothesis assumed here.

3. Conclusion

FEM analysis of natural frequencies of the screw and impeller of the two-stage pump was provided. CFD analysis of flow structure inside the screw-centrifugal stage also was developed. During CFD analysis the main focus was made on pressure distribution and investigation of vortices flow inside the screw-centrifugal stage. FEM-analysis has shown that natural frequencies of the screw and impeller lie inside a range of 3700-5000 Hz. Obtained results were compared with experimental. The comparison has shown a good accuracy of the numerical results. Natural frequencies of screw and impeller and frequencies of trailing vortices were turned out to coincide on pump operating regime. Such coincidence leads to increasing of screw and impeller blades vibration, pressure pulsations, and, as a result, to intensification of the trailer cavitation.

4. Acknowledgement

This work was supported by the Ministry of Education and Science of the Russian Federation in the framework of the implementation of the Program ‘‘Research and development on priority directions of scientific-technological complex of Russia for 2014– 2020 and Grant # 01v-R200-201.

References

[1] Gasparov M K, Pressure pulsations at stream and separate flows, *Samara State Aerospace University*, Samara, (2009) pp. 77

[2] Korotkov F, Ocenka vlijanija jekspluatacionnyh faktorov na rabotosposobnost' agregata ND-25 [Estimation of exploitation factors influence on combined pump ND-25 durability], *OJSC "Scintific and production enterprice "Temp" named after F.Korotkov*, Moscow, (1990)
[3] Wang G H, Unsteady Dynamics of Cloud Cavitating Flows around a Hydrofoil, *Proceedings of the 7th International Symposium on Cavitation CAV2009*, (2009)
[4] Kinelev, V G, Fizicheskaja model' kavitirujushhego shneko-centrobezhnogo nasosa, rabotajushhego v shirokom diapazone rezhimov po rashodu Kiev [Physical model of cavitating screw-centrifugal pump operating in a wide range of mass flow], *Kavitatsionnyye avtokolebaniya v nasosnykh sistemakh,* Naukova Dumka, (1976), pp. 100-107
[5] Karelin Y, Cavitation phenomena in axial and centrifugal pumps, *Moscow:Mashinostroenie,* (1975)
[6] Shapiro S, Structure of real-scale flow in cenrifugal and axial flow, *Moscow: MGIU,* (2004)
[7] Antonov A K, Pressure pulsations at stream and separate flows, *Moscow: Mashinostroyeniye*, (1990)
[8] Rimskiy-Korsakov A B, Phisical principles of noise formation in blowing machines, *Moscow: Nauka*, (1988)
[9] OJSC, Samara construction bureau of engines design, Technical report, *Samara*, (1998) pp.25

IOP Conf. Series: Materials Science and Engineering **302** (2018) 012034 doi:10.1088/1757-899X/302/1/012034

Aircraft Flight Modeling During the Optimization of Gas Turbine Engine Working Process

A Yu Tkachenko, V S Kuz'michev and I N Krupenich

Department of Aircraft Engine Theory, Institute of Engine and Power Plant Engineering, Samara National Research University, 34, Moskovskoye shosse, Samara, 443086, Russia

tau@ssau.ru, kuzm@ssau.ru, uugn@mail.ru

Abstract. The article describes a method for simulating the flight of the aircraft along a predetermined path, establishing a functional connection between the parameters of the working process of gas turbine engine and the efficiency criteria of the aircraft. This connection is necessary for solving the optimization tasks of the conceptual design stage of the engine according to the systems approach. Engine thrust level, in turn, influences the operation of aircraft, thus making accurate simulation of the aircraft behavior during flight necessary for obtaining the correct solution. The described mathematical model of aircraft flight provides the functional connection between the airframe characteristics, working process of gas turbine engines (propulsion system), ambient and flight conditions and flight profile features. This model provides accurate results of flight simulation and the resulting aircraft efficiency criteria, required for optimization of working process and control function of a gas turbine engine.

1. Introduction

Economic and aircraft performance characteristics are usually used as optimization criteria during the selection of working process parameters of gas turbine engine. Mathematical models of these characteristics of various detail levels are described in [1, 2].

Following criteria are most widely used: payload mass, take-off mass of the aircraft, total weight of the propulsion system and required fuel, specific fuel consumption of aircraft, cost of aircraft operation per hour, cost of transportation, net profit per operation hour, reduced cost of transportation, flight range. All these criteria are influenced (directly or indirectly) by the fuel mass onboard.

Most approaches to optimization of the working process parameters of gas turbine engine use the statistic correlation between the SFC at the rated conditions, SFC at off-design modes of operation and the fuel consumption for a given flight range [1]. The purpose of this paper is to provide a method of aircraft flight simulation, providing the interconnection between the propulsion system control program, working process parameters of engine and the aircraft efficiency criteria.

2. Problem setting

Flight conditions influence the working process parameters of gas turbine engine according to the general laws of joint operation of engine components. The value of thrust of the propulsion system to the external drag of the aircraft ratio determines the change in airspeed or flight trajectory. The external drag of an aircraft depends on its aerodynamic characteristics and control parameters and can be calculated for a given climb/descent program or for a specified flight trajectory. Control program of propulsion system determines the aircraft movement parameters along the flight trajectory, flight time, amount of fuel required for the flight and, ultimately, the values of the criteria characterizing the efficiency of the aircraft operation during the given mission. Therefore, the aircraft flight model, taking the interconnection between the engine working process parameters and aircraft behavior into the account is required to provide the correct results of engine working process optimization.

3. Aircraft flight modeling

This aircraft flight model may be developed using the system of differential equations describing the dynamics of the aircraft during the flight. The initial data for this model include take-off mass of the aircraft, aerodynamic and geometric characteristics of the airframe; programs of climbing and descending; altitude of cruising; initial airspeed and coordinates of the aircraft, areas of the characteristic sections of gas turbine engine, performance maps of engine components, control program of engine.

The results of flight simulation include the aircraft parameters variation along the flight trajectory, flight time, fuel consumption per flight, the values of the aircraft performance criteria.

It is necessary to take into account the fact that the flight of aircraft is often a multi-stage process. For example, the scheme of the flight of a subsonic transport (or passenger) aircraft includes the following main stages (figure 1): takeoff; climb; cruising; descending; waiting for landing; pre-landing maneuvering; landing.

Engine operation during the take-off and landing, waiting and pre-landing maneuvering is determined by safety requirements and operational requirements, usually these modes are extreme for engine.

Engine operation during climbing and descending is also regulated, but the engine parameters are not constant and change along the altitude due to variation of ambient conditions.

Engine mode of operation during the cruising changes in a wide range and is determined by the thrust value required for the horizontal flight at a given airspeed.

As the aircraft flight is described in a variety of ways according to the flight stage, modeling of these stages is performed separately. The results of modeling are sequentially transferred as the initial data to the next stages of flight.

Climbing. Airspeed change along the altitude is usually considered initial data. The mode of engine operation is supposed to be constant (nominal mode). This flight stage ends as the aircraft reaches the given altitude of cruising. Flight trajectory is determined by the engine control function (thrust changes according to the ambient conditions).

Cruising. Cruising begins as the aircraft reaches the corresponding altitude and finishes as it reaches the descending point. The engine mode of operation may either change in accordance to the aircraft weight, or be estimated to provide constant airspeed. Horizontal flight conditions provide the means for calculating the angle of attack at each point of cruising. The flight profile may include additional climbing and cruising stages as the weight of aircraft decreases.

Descending. This stage of flight is identical to the climbing stage, except for the idle mode of engine operation.

4. Mathematical model of flight

Aircraft flight is described by three groups of equations: dynamic equations of aircraft motion, kinematic equations and weight change as a result of fuel consumption.

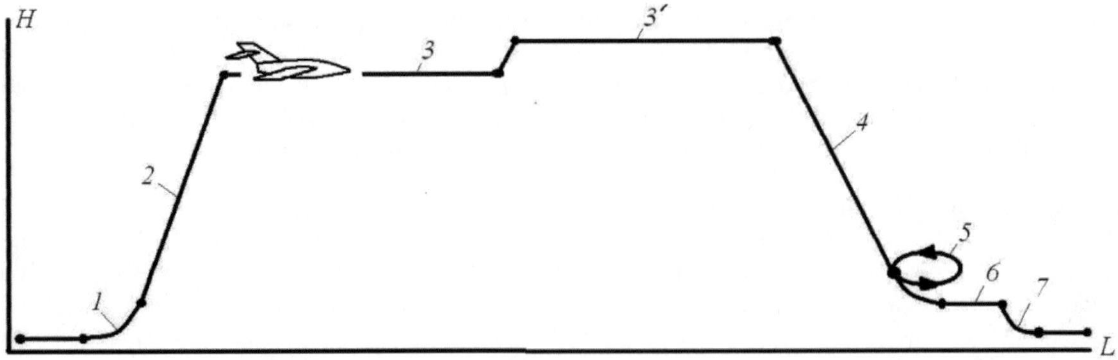

Figure 1. Typical flight profile of a transport aviation aircraft: 1 – takeoff; 2 – climb; 3, 3' – cruising; 4 – descending; 5 – waiting for landing; 6 – pre-landing maneuvering; 7 – landing.

Dynamic equations of aircraft motion relative to the ground (neglecting the wind velocity) include [3, 4, 5]:

$$\left.\begin{aligned}
&M_A\dot{V}_f = P_{\text{ef.ps}}\cos(\alpha+\varphi_{th})\cos\beta - X_a - M_A g\sin\Theta;\\
&M_A V_f\dot{\Theta} = P_{\text{ef.ps}}\left[\sin(\alpha+\varphi_{th})\cos\gamma_a + \cos(\alpha+\varphi_{th})\sin\beta\sin\gamma_a\right] + Y_a\cos\gamma_a -\\
&-Z_a\sin\gamma_a - M_A g\cos\Theta + 2M_A\varpi_z V_f\cos\varphi\sin\Psi + M_A V_f^2\frac{\cos\Theta}{R_{earth}+H};\\
&-M_A V_f\cos\Theta\dot{\Psi} = P_{\text{ef.ps}}\left[\sin(\alpha+\varphi_{th})\times\sin\gamma_a - \cos(\alpha+\varphi_{th})\sin\beta\cos\gamma_a\right]+\\
&+Y_a\sin\gamma_a + Z_a\cos\gamma_a + 2M_A\varpi_z V_f\times(\sin\varphi\cos\Theta - \cos\varphi\cos\Psi\sin\Theta) - M_A V_f^2\cos^2\Theta\sin\Psi\frac{\operatorname{tg}\varphi}{R_{earth}+H}.
\end{aligned}\right\} \quad (1)$$

Coriolis and centrifugal force (associated with ground curvilinearity) for airspeeds less than 1000 m/s are usually not taken into the account.

Flight without aircraft roll and sideslip is described by:

$$\gamma_a = 0; \quad (2)$$

$$\beta = 0; \quad (3)$$

$$Z_a = 0. \quad (4)$$

With the above mentioned assumptions, the equations system (1) takes the following form:

$$\left.\begin{aligned}
&M_A\dot{V}_f = P_{\text{ef.ps}}\cos(\alpha+\varphi_{th}) - X_a - M_A g\sin\Theta;\\
&M_A V_f\dot{\Theta} = P_{\text{ef.ps}}\sin(\alpha+\varphi_{th}) + Y_a - M_A g\cos\Theta.
\end{aligned}\right\} \quad (5)$$

Following equations describe the kinematics of motion and aircraft weight change as a result of fuel consumption:

$$\left.\begin{aligned}
&\dot{H} = V_f\sin\Theta;\\
&\dot{L} = V_f\cos\Theta;\\
&\dot{M}_A = -G_{\text{fuel.ps}}.
\end{aligned}\right\} \quad (6)$$

As the flight stages are usually described with their corresponding distances, use of flight range L as a phase variable is reasonable (figure 1). Thus, the differential equations system of the aircraft flight may be described as follows:

$$\left.\begin{array}{l}\dfrac{dV_f}{dL}=\dfrac{1}{M_{\mathrm{A}}V_f\cos\Theta}\left[P_{\mathrm{ef.ps}}\times\cos\left(\alpha+\varphi_{th}\right)-X_a-M_{\mathrm{A}}g\sin\Theta\right];\\ \dfrac{d\Theta}{dL}=\dfrac{1}{M_{\mathrm{A}}V_f\cos\Theta}\left[P_{\mathrm{ef.ps}}\times\sin\left(\alpha+\varphi_{th}\right)+Y_a-M_{\mathrm{A}}g\cos\Theta\right];\\ \dfrac{dH}{dL}=\mathrm{tg}\,\Theta;\\ \dfrac{dt}{dL}=\dfrac{1}{V_f\cos\Theta};\\ \dfrac{dM_{\mathrm{A}}}{dL}=\dfrac{-G_{\mathrm{fuel.ps}}}{V_f\cos\Theta}.\end{array}\right\} \qquad (7)$$

The resulting system of differential equations include the airframe parameters (external drag X_a, lifting force Y_a, angle of attack alpha) and propulsion system parameters (effective thrust $P_{\mathrm{ef.ps}}$, fuel consumption $C_{\mathrm{fuel.ps}}$). These parameters depend on the control functions of aircraft and engine and may be calculated using the individual sub-models, having their own hierarchy (figure 2).

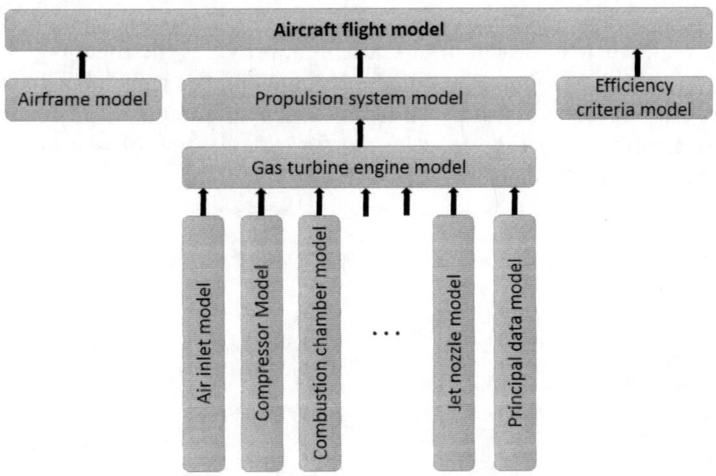

Figure 2. Structure of the mathematical model of aircraft flight.

The mathematical model of gas turbine engine is described in detail in [4] and the mathematical model of criteria.

5. Mathematical model of aircraft

The mathematical model of the aircraft provides the values of lifting force Y_a and external drag X_a as a function of geometrical parameters, aerodynamic characteristics and flight conditions [6, 7, 8]:

$$Y_a=c_{ya}\frac{k}{2}p_H M_f^2 S_w; \qquad (8)$$

$$X_a = c_{xa} \frac{k}{2} p_H M_f^2 S_w. \quad (9)$$

Aerodynamic characteristics of the airframe are usually described by the following functional relations [5]: a) $c_{ya} = f\left(\alpha, M_f\right)$, b) $c_{xa} = f\left(c_{ya}, M_f\right)$: (figure 3).

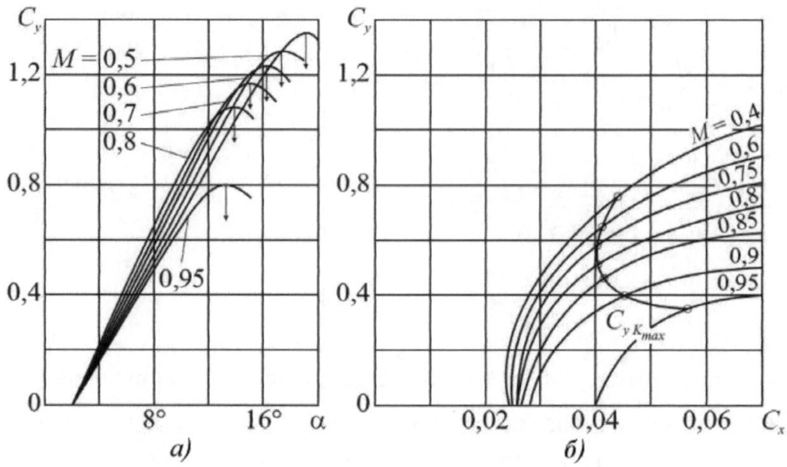

Figure 3. Aerodynamic characteristics of airframe.

6. Conclusions

The described above mathematical model of aircraft flight provides the functional connection between the airframe characteristics, working process of gas turbine engines (propulsion system), ambient and flight conditions and flight profile features. This model provides accurate results of flight simulation and the resulting aircraft efficiency criteria, required for optimization of working process and control function of a gas turbine engine [9, 10].

7. Nomenclature

SFC – specific fuel consumption;
M_A – mass of aircraft;
V_f – airspeed;
$P_{ef.ps}$ – overall effective thrust of propulsion unit;
α – attack angle;
β – sliding angle;
γ_a – rolling angle;
φ_{th} – angle between the thrust vector and aircraft axis;
X_a – drag force;
Y_a – lifting force;
Z_a – side force;
c_{xa} – lift coefficient;
c_{ya} – drag coefficient;
g – free fall acceleration;
Θ – угол наклона траектории;
ϖ_z – угловая скорость вращения земли;
Ψ – угол пути;

R_{earth} – radius of Earth;
H – flight altitude;
$G_{\text{fuel.ps}}$ – fuel consumption rate;
L – flight range;
t – flight time;
k – adiabatic index of ambient air;
p_H – ambient pressure;
M_f – ambient Mach number;
S_w – wing area.

Acknowledgments
This work was supported by the Ministry of education and science of the Russian Federation in the framework of the implementation of the program of increasing the competitiveness of Samara University among the world's leading scientific and educational centers for 2013-2020 years.

References

[1] Maslov V G, Kuzmichev V S, Kovartsev A N, Grigoriev V A 1996 *Theory and methods of the initial design stages of gas turbine engines* (Samara: Samara State Aerospace University Publishing House) p 147
[2] Grigoryev V A, Zhdanovsky A V, Kuzmichev V S 2009 *Selection of the parameters and thermogasdynamic calculations of gas turbine engines* (Samara: Samara State Aerospace University Publishing House) p 202
[3] Bochkarev A F 1985 *Aircraft aeromechanics: Flight dynamics* (Moscow: Mashinostroenie publishing house) p 360
[4] Kulagin V V 2003 *Theory, calculations and design of aircraft engines and power plants. Fundamentals of gas turbine engine theory. Working process and thermogasdynamic analysis. (Book 1). Joint operation of engine components and its characteristics (Book 2)* (Moscow: Mashinostroenie publishing house) p 616
[5] Zhukov O M, Kuz'michev V S, Kovartsev A N Exhaust system configuration evaluation based on engine transport effectiveness in the flight vehicle system 1985 *Soviet Aeronautics* (English translation of Izvestiya VUZ, Aviatsionnaya Tekhnika) **28**(1) 105-109
[6] Bekhtir V P, Rzhevsky V M, Tsipenko V G 1997 *Practical aerodynamics of Tu-154M aircraft* (Moscow: Air Transport publishing house) p 288
[7] Zhang H, Zhang J, Yang G-H, Luo Y 2015 Leader-based optimal coordination control for the consensus problem of multiagent differential games via fuzzy adaptive dynamic programming *IEEE Transactions on Fuzzy Systems* **23**(1) 152-163
[8] Chipperfield A J, Bica B, Fleming P J 2002 Fuzzy scheduling control of a gas turbine aero-engine: A multiobjective approach *IEEE Transactions on Industrial Electronics* **49**(3) 536-548
[9] Kuz'michev V S, Ostapyuk Y A, Krupenich I N, Filinov E.P., Tkachenko A Yu Optimization of working process parameters of small-scale turbojet for unmanned aircraft. *2017 International Conference on Mechanical, System and Control Engineering, ICMSC* 2017 26 June 2017, 7959456, Pages 125-129.
[10] Kuz'michev V S, Krupenich I N, Filinov E.P., Tkachenko A Yu, Novikova Yu D The issues of selecting the working process parameters of gas turbine engine with an element of initial data uncertainty. *53rd AIAA/SAE/ASEE Joint Propulsion Conference*, 2017; Atlanta; Georgia; 10 July 2017 - 12 July 2017; 195549.

Optimizing Parameters of Axial Pressure-Compounded Ultra-Low Power Impulse Turbines at Preliminary Design

D S Kalabukhov[1], V M Radko[1] and V A Grigoriev[1]

[1]Theory of Aircraft Engines Department, Samara University, Samara, 443086, Russia

E-mail: dskalabukhov@ssau.ru

Abstract: Ultra-low power turbine drives are used as energy sources in auxiliary power systems, energy units, terrestrial, marine, air and space transport within the confines of shaft power N_{td} = 0.01…10 kW. In this paper we propose a new approach to the development of surrogate models for evaluating the integrated efficiency of multistage ultra-low power impulse turbine with pressure stages. This method is based on the use of existing mathematical models of ultra-low power turbine stage efficiency and mass. It has been used in a method for selecting the rational parameters of two-stage axial ultra-low power turbine. The article describes the basic features of an algorithm for two-stage turbine parameters optimization and for efficiency criteria evaluating. Pledged mathematical models are intended for use at the preliminary design of turbine drive. The optimization method was tested at preliminary design of an air starter turbine. Validation was carried out by comparing the results of optimization calculations and numerical gas-dynamic simulation in the Ansys CFX package. The results indicate a sufficient accuracy of used surrogate models for axial two-stage turbine parameters selection

1. Introduction

Ultra-low power turbines (ULPT) also known as "microturbines" and they are widely used in the industry, energy and transport sets [1]. Their shaft power usually is up to 10 kW. ULPT have extremely low gas consumption up to 0.1 kg/s, miniature dimensions (average diameter of turbine wheel is typically less than 100 mm and the height of nozzle diaphragm blade is about 5 mm) and a supercritical pressure drop ratio in turbine π_t = 2...100. Rotation speed n depends on the type of power consumer and can vary from a few thousand to (300…500)·10^3 rpm [1, 2]. A turbine wheel is driven by pressurized working fluid (it is usually an air) that stored in a receiver. There are number of features of this type of turbines in comparison with full-size turbines. These include: a thick boundary layer on blades surfaces, large leak through gaps, significant secondary wave losses and other kind of specific losses that cause low energy efficiency of ULPT. Typically, such turbines are made impulse type to reduce leakage through radial gaps.

Ultra-low power turbines usually work on a multi-mode operation and have a variety of scheme designs. Impulse axial and radial centripetal turbines are the most widely distributed among this class of turbomachinery. Studies [3] have shown that in the most cases, an axial small-sized turbine is more effective over a wide range of operating conditions than other types of turbines. Single-stage axial ULPT are most widely used due to the requirements of technological simplicity and severe restrictions on the layout of turbine unit. However, high values $\pi_t \geq 4$ [1] cause increased working loading of turbine stage, which determines the intensive reduction of power efficiency coefficient η_t. In addition, single-stage ULPT are typically operate on the left branch of $\eta_t = f(y_t)$ characteristic (Figure. 1), where

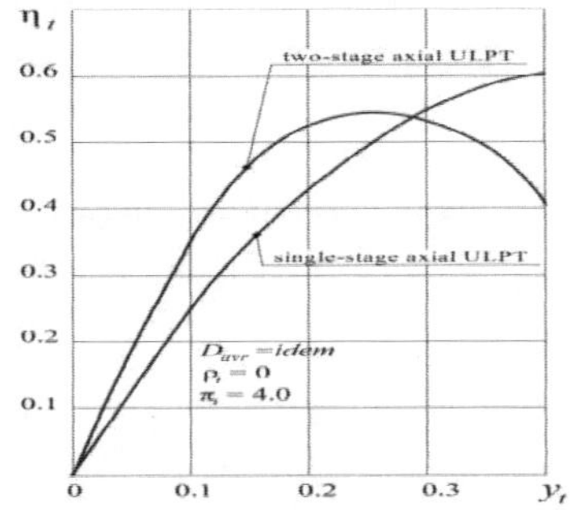

Figure 1. Typical power efficiency characteristics of impulse axial ultra-low power turbines

$$y_t = \frac{u}{c_s} = \frac{\pi D_{avr} n}{60\left[2\frac{k}{k-1}RT_0^*\left(1-\pi_t^{\frac{1-k}{k}}\right)\right]^{1/2}}. \tag{1}$$

An operation at low values $y_t \leq 0.4$ in full-admission turbines leads to high losses of output flow velocity. It is known from theory and practice of impulse turbines [4] that the blade-jet speed ratio in multistage pressure-compounded turbines increases by $\sqrt{z}$ times ($y_{t\,p} = u\sqrt{z}/c_s$, where z – number of stages). During operation on the left branch ($y_{t\,p} < y_{t\,p\,\text{opt}}$), this leads to an increased efficiency with respect to a single-stage turbine. This is explained by using an output velocity in the next stage. Also the wave losses reduce in pressure stages due to reducing pressure ratio π_{st} in each stage (provided that π_t = idem). Furthermore, the higher multistage turbine efficiency is provided by the heat recovery effect [4]. With the proviso that $y_{t\,p} >> y_{t\,\text{opt}}$ a reducing effect of output velocity losses is generally offset by the significantly increase in secondary losses, gaps losses, the friction on turbine wheels, the uneven flow around turbine wheel blades. The optimum ratio $(u/c_s)_{\text{opt}}$ shifts towards lower values with increase of z number. These values, that typical for ULPT, are usually range between 0.1…0.4 [5].

A typical scheme of two-stage pressure-compounded impulse turbine is shown in Figure. 2.

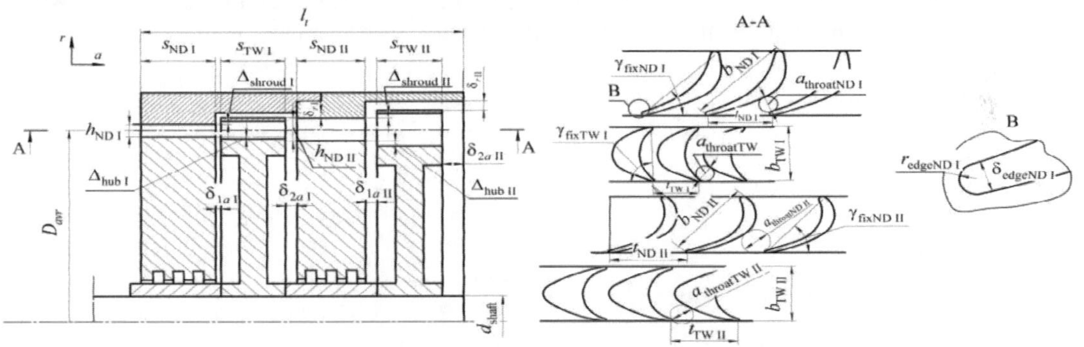

Figure 2. Meridional section of a two-stage pressure-compounded impulse ultra-low power turbine

Thus, the prospect of wider use of two-stage pressure-compounded axial ULPT is justified by the possibility of obtaining the greater efficiency than in single-stage turbines due to the same initial conditions (see Figure. 1). Therefore, the actual problem is the development of rational choice of two-stage pressure-compounded ULPT parameter values. Authors of this paper have already developed a method for single-stage ULPT parameters selecting which allows meet the complex of contradictory requirements. It is shown to be effective at the preliminary design [6]. This work follows, and extends our previous one. We propose a method for selecting the rational parameter values of impulse two-stage pressure-compounded ULPT.

2. Particulars of surrogate models in the method of selecting parameters of two-stage axial pressure-compounded ULPT

The developed method of selecting parameters of a two-stage axial ULPT is based on the principles that was set out in [6] for a single-stage ULPT. The method is implemented in the mathematical package Matlab Optimization Toolbox. Subroutine for the optimization calculation is based on the well-known algorithm of one-dimensional gas-dynamic calculation of a turbine stage [7, 8]. It is based on empirical relations and the amendments on influence of a small size to turbine efficiency.

For example, the regression empirical relation of power efficiency from the above-mentioned operating and geometric parameters has been used in this algorithm. It was obtained by the author of paper [9].

The main feature of the proposed approach to determining the efficiency of a two-stage axial pressure-compounded ULPT is a formation of surrogate models of criteria for evaluating the two-stage turbine efficiency on the basis of mathematical models for a single-stage turbine. The main relationships and dependencies for creating such models are described below.

Gas-dynamic loading of a multi-stage turbine with pressure stages is characterized by the blade-jet speed ratio yt = $\sqrt{\Sigma u^2}$ /cs, but in the special case of rotation speeds equality $y_t = u\sqrt{z}/c_s$ [4]. We propose to derive the blade-jet speed ratio of whole turbine through ratios of it stages y_{stI} and y_{stII}. To do this, it is necessary to perform the following transformations.

The kinetic energy of entire turbine is the sum of the each stage kinetic energy

$$\frac{c_{1s}^2}{2(1-\rho_t)} = \frac{c_{1sI}^2}{2(1-\rho_{stI})} + \frac{c_{1sII}^2}{2(1-\rho_{stII})};\ c_s^2 = c_{sI}^2 + c_{sII}^2, \tag{2}$$

where ρ_{stI}, ρ_{stII} and ρ_t – degree of the first turbine stage, the second stage reaction and the average level of turbine reaction, respectively.

Then

$$y_t = \frac{\left(\Sigma u^2\right)^{1/2}}{c_s} = \left(\frac{u_I^2}{c_{sI}^2 + c_{sII}^2} + \frac{u_{II}^2}{c_{sI}^2 + c_{sII}^2}\right)^{1/2} = \left(\frac{u_I^2 / c_{sI}^2}{1 + \left(c_{sII}^2 / c_{sI}^2\right)} + \frac{u_{II}^2 / c_{sII}^2}{\left(c_{sI}^2 / c_{sII}^2\right) + 1}\right)^{1/2} = \left(\frac{y_{stI}^2}{1 + \bar{c}_s^{-2}} + \frac{y_{stII}^2}{1 + \left(1/\bar{c}_s^{-2}\right)}\right)^{1/2} \tag{3}$$

Calculation of the blade-jet speed ratio by formula (3) turned out to be more accurate results than by traditional method through reactivity of each stage instead of generalized, with some error, reaction ρt.

The gas turbine work is the sum of individual stages gas works

$$L_t = L_{stI} + L_{stII} \text{ or } \eta_t L_{t\,s} = \eta_{st1} L_{stI\,s} + \eta_{st2} L_{stII\,s} \tag{4}$$

Then, the coefficient of power efficiency

$$\eta_t = \frac{\eta_{st1} L_{stIs} + \eta_{st2} L_{stIIs}}{L_t} \tag{5}$$

where $L_{stI\,s} = \frac{k}{k-1} R T_0^* \left(1 - \pi_{stI}^{\frac{1-k}{k}}\right)$ and $L_{stII\,s} = \frac{k}{k-1} R T_{0\,II}^* \left(1 - \pi_{stII}^{\frac{1-k}{k}}\right)$ - isentropic gas flow work in the first and second stages respectively;

T_0^* and $T_{0\,II}^*$ - total temperature at the inlet of the first and second stage respectively;

$\pi_{stI} = p_0^*/p_{2\,I}$ and $\pi_{stII} = p_{0\,II}^*/p_2$ – pressure drop ratio in the first and second stages respectively; multiplication πst1 πst2 determines the pressure ratio πt of whole turbine;

η_{stI} and η_{stII} – coefficients of total-static efficiency for each stage that can be calculated by the formula [9]:

$$\eta_{st\,i} = 0.146 + 1.672 y_{st\,i} - 3.165\, y_{st\,i}^2 - 4.018(h_{ND}/D_{avr})_i + 65.827(h_{ND}/D_{avr})_i^2 + 0.0205\alpha_{1ef\,i} -$$
$$- 0.0011\alpha_{1ef\,i}^2 - 0.11(b/t)_{TW\,i}^2 - 0.002(s/h)_{ND\,i}^2 + 0.091 y_{st\,i}\,\pi_{st\,i} + 6.022\, y_{st\,i}\,(h_{ND}/D_{avr})_i +$$
$$+ 4.345(h_{ND}/D_{avr})_i\,(b/t)_{TW\,i} - 0.068\pi_{st\,i}\,\bar{\delta}_{edge\,ND\,i} + 0.12\,\bar{\delta}_{edge\,ND\,i}\,(b/t)_{TW\,i} -$$

$$- 0.006y_{st\,i}\,(b/t)_{TW\,i} + 0.195y_{st\,i}\bar{D}_{avr\,i} + 0.045y_{st\,i}(s/h)_{ND\,i} - 0.0019\pi_{st\,i}\,(s/h)_{ND\,i}$$
$$+ 1.106(h_{ND}/D_{avr})_i\,(s/h)_{ND\,i} + 0.033(b/t)_{TW\,i}\,(s/h)_{ND\,i} \quad (6)$$

The value of multiple coefficient of determination is equal to 0.967, which indicates a high correlation between effective and independent variables. But we should note that we obtained Eq. (6) on the basis of tests results of single-stage turbines. It is fully justified for the first stage because it does not take into account the effect of non-uniform parameters fields in axial gap on the efficiency of subsequent stages. However, it is quite acceptable to optimize the parameters using Eq. (5) and (6) at the stage of preliminary design. It was shown by the following results of verification of the surrogate model for an efficiency evaluation.

Surrogate models of the turbine size and weight were formed similarly by summing the length and mass of its stages.

Solution of the optimal design task of any technical system always presupposes the availability of different kinds of conditions and constraints. Implementation of these conditions indicates the compliance of output data to the requirements of technical task. The parametric constraints such as $X_{min} \leq X \leq X_{max}$ and functional limitations, which relate the parameters of vector X, are imposed in the object existence region.

The parametric conditions and constraints of the optimization problem were selected from the variation ranges of the independent variables x_j, which are given in [10].

There is a basic functional condition to provide a given mass flow rate G_{tu} through a turbine unit or to provide a given shaft power N_{tu}. Mass flow rate and power are defined through optimized parameters and some initial data:

$$G_{tu} = \frac{2.5 \cdot 10^{-3} \cdot m p^*_{in}\,\sigma_{in}\,\sigma_{throat} \cdot (h_{ND}/D_{avr})_I\,\bar{D}^2_{avr\,I}\,(\sin\alpha_{1ef\,I})\,q(\lambda_{throatND\,I})}{(1+\bar{\delta}_{edgeND\,I})\sqrt{T^*_{in}}}, \quad (7)$$

$$N_{tu} = G_{tu}\,L_{tu\,s\,p}\eta_{tu} \quad (8)$$

where $m = 0.0404 \left(\frac{kg \cdot K}{J}\right)^{-0.5}$ - a physical constant that depends on the properties of the working fluid;

σ_{in} - total pressure loss in the area from the entrance of the turbine to the turbine;

σ_{throat} - total pressure loss in the area from the nozzle diaphragm inlet to throat; approximately defined as $0.5(1+\sigma in)$ [5];

$p_{in}^* = p_0^*/\sigma_{in}$ and $T_{in}^* = T_{0\,I}^*$ - stagnation pressure and temperature at the inlet of turbine;

$q(\lambda_{throatND\,I})$ – relative flux density in the critical section (throat) of the nozzle unit of the first turbine stage;

$\eta_{tu} = \eta_{in}\eta_t\eta_{out}$ - coefficient of power efficiency, a multiplication of coefficients of the input merit ηin, the output merit ηout and the coefficient of turbine power efficiency η_t [5].

Calculated values G_{tu} and N_{tu} in the multimode turbine unit are not known in advance, but their possible values must be within the range that defined by the load diagram. Thus, one of the constraints given by inequality

$$G_{tu\,i\,min} \leq G_{tu\,design} \leq G_{tu\,i\,max} \text{ or} \quad (9)$$

$$N_{tu\,i\,min} \leq N_{tu\,design} \leq N_{tu\,i\,max} \quad (10)$$

In the case of single-mode turbine unit these conditions should meet the equations

$$G_{tu\,design} - G_{set} = 0 \text{ and} \quad (11)$$

$$N_{tu\,design} - N_{set} = 0, \quad (12)$$

where G_{set} and N_{set} - set values of the mass flow rate and of a turbine unit power that are specified in the technical task.

In addition, it is necessary to set equality $G_{st\,I} = G_{st\,II}$ for matching the flow paths of turbine stages at the mass balance condition.

Cyclogram of the designed turbine drive limits the operating ranges of rotation speed n. In this case, assumptions of surrogate mathematical models for evaluation criteria require a consideration of restrictions on the magnitude $y_{st\,\mathrm{i}} \leq 0.4$ in all operating modes.

Great importance in the design of ultra-low power turbine drive has a design and technological constraints. The main constructive restriction is the tolerance to nozzle diaphragm blades height:

$$0.5\ \mathrm{mm} \leq h_{\mathrm{ND}} \leq 5\ \mathrm{mm}, \quad (13)$$

where $h_{\mathrm{ND\,I}} = 50(h_{\mathrm{ND}}/D_{avr})_{\mathrm{I}}\,\bar{D}_M$ - the blades height of the first stage of ND;

$h_{\mathrm{ND\,II}} = 1.8\ h_{\mathrm{TW\,I}}$ - the blades height of the second stage ND;

The heights of turbine wheel blades of the first and second stages are defined by formula hTW i = 1.8 $h_{\mathrm{ND\,I}}$.

The blade height increasing of each subsequent by 1.8 times is caused by the fact that the optimum ratios of hub and shroud overhang $\Delta_{\mathrm{hub}\ i}$ and $\Delta_{\mathrm{shroud}\ i}$ (Figure. 1) to hND i are about 0.3 and 0.5 respectively.

Considering that

$$h_{\mathrm{TW}\,i} = h_{\mathrm{ND}\,i}\,[1+(\Delta_{\mathrm{hub}\,i}/h_{\mathrm{ND}\,i})+(\Delta_{\mathrm{hub}\,i}/h_{\mathrm{ND}\,i})] = 1.8\ h_{\mathrm{ND}\,i} \quad (14)$$

In addition to gas-dynamic, structural-technological and dimensional constraints, the strength limitations are usually taken into account while turbines optimizing. However, at the preliminary stage of design these factors could not be defined by the following reasons:

- strength factors do not play a determining role in the parameters selection for ultra-low power turbines that operate at low temperatures of the working fluid T^*_{in} during the limited operating time [11];
- parametric restrictions are imposed so as to minimize the probability of selecting the geometry of turbine that is not providing its static strength in wide ranges of mode parameters.

Strength limitations in the mathematical optimization model of ULPT do not included for reasons noted above.

3. Approbation of surrogate models while selecting rational parameters values of two-stage axial pressure-compounded turbine of an air starter

The largest application was received by turbine type air starters (turbine starters), which are interchangeable with electric starters. Typically, such turbine starters have two velocity stages and less often - pressure stages [12].

3.1 Initial data for designing.

The article deals with the design concept of a turbine driven air starter for diesel engine with the initial data corresponding to the standard atmospheric conditions: corrected rated power of turbine drive $N_{td\,corr} = 7.5$ kW; corrected rotation speed on the rated mode $n_{corr} \leq 20{,}000$ rpm. Inlet pressure p_{in}^* was kept constant and determined by the pressure drop ratio π_t, which ranged from 2 to 6 while optimizing process, and by the outlete pressure p_{out}^*. In the calculations we took the values $p_{\mathrm{out}}^* = 99$ kPa and $T_{\mathrm{out}}^* = 233$ K.

Cyclogram of turbine drive operating was not set because starter was designed for a conditional diesel engine whose parameters are not known. Turbine optimization was performed only under the nominal mode since a starter was calculated for a conventional diesel engine, the cyclogram of which is not known. Starter is supposed to run for 20 seconds at nominal mode [13].

3.2. Selection of criteria for efficiency evaluation.

According to the systematic approach for the turbine unit designing, the turbine parameters should be optimized on the evaluation criteria of efficiency for the entire turbine unit. Energy efficiency is characterized by specific mass flow rate through the turbine unit $G_{tu\,sp} = G/N_{tu}$ whereas mass efficiency is characterized by turbine unit (air starter in our case) specific mass $M_{tu\,sp} = M_{tu}/N_{tu}$. However, for the convenience of comparing the values of criteria based on the optimization results, it is desirable that they would have the same dimensionality. Therefore, instead of the specific air consumption, a criterion of the required specific mass of compressed air $M_{air\ sp} = M_{air}/N_{tu} = G\tau/N_{tu}$ was used for an air starter operation during operating time $\tau = 10\ s$. Then both criteria are measured in kg/W.

3.3 Selection of varying ranges of optimization parameters.
The ranges of varying operational and geometric parameters (which are included in Eq. (6)) were chosen on the recommendations [14]. The range corresponding the values of D_{avr}= 80…100 mm was chosen due to the analysis of the modern industrial air starters dimensions.

3.4 Features of optimization techniques.
Optimization was performed in MATLAB Optimization Toolbox package in the order that was described in detail in the paper [6].

We used following methods to the solutions searching: the goal attainment method as solver, two objectives ($M_{air\ sp}$ and $M_{tu\ sp}$), two vectors of goals values ($M_{air\ sp}^*$ and $M_{tu\ sp}^*$) and weights ($\Delta M_{air\ sp}$, $\Delta M_{tu\ sp}$), arrays with 16 inputs (divided into 8 parameters for each stage) lower bounds, upper bounds and start point and a special subroutine of nonlinear constraints that has been briefly described above.

3.5 The results of optimization.
At the first step of optimization, the optimal parameters have been found on criteria *Mair sp* and *Mtu sp separately*, i.e. single-criterion optimization was conducted. Next, the task of two-criterion optimization was set. Two-criterion optimization results are given in Table 1.

Table 1. Results of two-criterion optimization

No.of stage	Optimized parameter	Parameter value	Criterion value $M_{air\ sp}$, kg/kW	Criterion value $M_{tu\ sp}$, kg/kW
I	y_{st}	0.321	0.407	2.567
	π_{st}	2.039		
	h_{ND}/D_{avr}	0.014		
	α_{1ef}	14.69		
	$\bar{\delta}_{edgeND}$	0.178		
	$(b/t)_{TW}$	1.116		
	$(s/h)_{ND}$	7.59		
	$\bar{D}_{avr}$	2		
II	y_{st}	0.277		
	π_{st}	2.986		
	h_{ND}/D_{avr}	0.046		
	α_{1ef}	7.633		
	$\bar{\delta}_{edgeND}$	0.209		
	$(b/t)_{TW}$	1.471		
	$(s/h)_{ND}$	3.783		
	$\bar{D}_{avr}$	2		

Criteria deviation are calculated by the formula

$$\Delta y_{rp} = \frac{y_{rp} - y_{rp}^{min}}{y_{rp}^{min}} \cdot 100\% , \quad (15)$$

when y_{rp}^{min} - criterion value determined by results of single-criterion optimization.

Criteria deviation amounted to $\Delta M_{air\ sp}$ = 1.96% and $\Delta M_{tu\ sp}$ = 2.39%. The values of the deviations are small and fit within the range of error in determining the values of these criteria, which makes it advisable to use two-criterion optimization in this case.

According to the Equations (3-5), the design blade-jet speed ratio y_t = 0.298 and pressure drop ratio $\pi_t = \pi_{t\ st\ I}\ \pi_{t\ st\ II} = 6$. Design value of corrected rotation speed n_{corr} = 19,900 rpm.

4. Results of surrogate models validation

Validation of surrogate models has been conducted by comparing the optimization results and computational experiment data. Comparison was implemented by integral gas-dynamic parameters: coefficient of power efficiency η_t and corrected torque $M_{torq\ corr}$.

Computational experiments were performed using the commercial CFD software package Ansys CFX. The computational fluid flow domain corresponds to optimized turbine geometry. The turbine computational domain is shown in Figure. 3. We used the turbulence model *SST*, which is more suitable for the flow computations in small-sized turbines [15]. The flow parameters at the interface of the stationary and rotating elements transmitted through a cylindrical interface surface «frozen rotor», i.e. without averaging parameters in the circumferential direction. The flow models in ULPT were previously verified by field experiments [15].

Validation was conducted in two steps. At the first step we performed a comparison of CFD simulation data with the results of parameters selection for the design mode according to the developed method (Table 2).

Table 2. Comparison of calculation results and computational experiment

	$\eta_{st\ I}$	$\eta_{st\ II}$	η_t	$M_{torq\ corr}$, N m	$N_{t\ corr}$, W
Optimization results	0.568	0.647	0.607	3.6	7500
Computational experiments data	0.603	0.579	0.59	3.54	7396
Relative error, %	-5.8	11.74	2.88	1.69	1.41

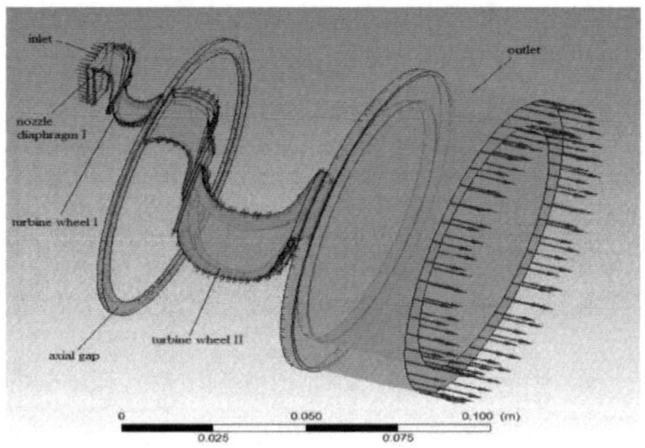

Figure 3. Fluid flow domain for CFD simulation of the optimized turbine geometry

As can be seen from Table 2, the computational experimental data practically coincide with the results of calculation, since the relative error fit within the experimental errors of the efficiency measurement [15]. However, the errors in determining the efficiency of individual stages turned out to be quite large. The reasons for this could be explained as follows. The higher experimental values η_{st} are caused by the influence of positive degree of reaction $\rho_{st\ I}$, which was not considered in the equation (6). At the same time, a decrease of $\eta_{st\ II}$ in comparison with calculation data is probably due to the additional axial gap losses by the uneven flow field. These losses are not taken into account in the surrogate optimization model. It is interesting that eventually these discrepancies cancel each other out so that the relative error of power efficiency determining by Eq. (6) has a very low value.

At the second step, we compared turbine characteristics that had been obtained on the optimization results by developed surrogate model with turbine characteristics by CFD simulation. Validation was carried out in the following sequence:

- computational experiments were carried out on the experimental design *in which the combinations of parameters* πt = 2; 3; 4; 5; 6 and n_{corr} = 13,000; 19,500; 26,000 rpm were varied;
- the characteristics $\eta_t = f(\pi t)$ and $M_{torq\ corr} = f(\pi_t)$ were described according to the results of these experiments;
- the torque Mtorq corr and power efficiency ηt values were calculated in the Matlab subroutine with the use of surrogate models by the developed method (based on the calculated parameters yst i , πst i and on the already known geometric ratios of turbine stages);

- the results were compared with each other (Figure. 4a and Figure. 4b).

Efficiency values that are calculated in MATLAB subroutine by the surrogate model have a very good agreement with the computational experiment data throughout the considered rotation speed range at the design value of turbine pressure drop ratio $\pi_t = 6$ (Figure. 4a).

However, within a deviation of π_t values from the design value, the divergences in most cases are significant and the characteristics curves are not congruent.

5. Comparison with results of an axial single-stage ULPT optimization

The method for selecting rational parameters of single-stage ULPT was first described in paper [6]. In this work we supplemented this method with surrogate models for two-stage axial pressure-compounded ULPT optimization. Now, let us compare the optimization results that were obtained in [6] for single-stage axial turbine and results we have gotten at present work for two-stage axial turbine, provided at the nominal mode with operating period $\tau = 20$ sec (Table 3).

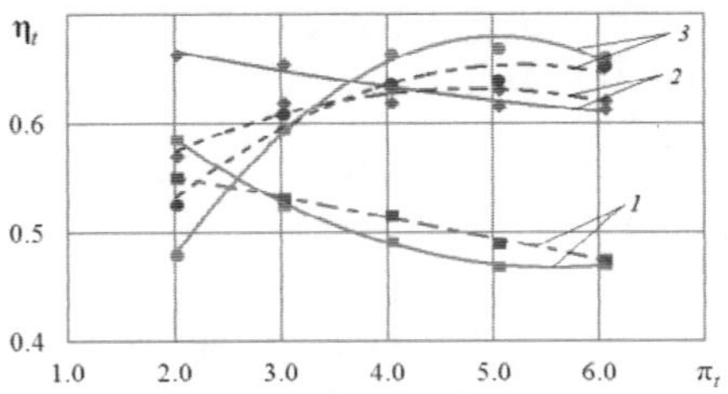

a

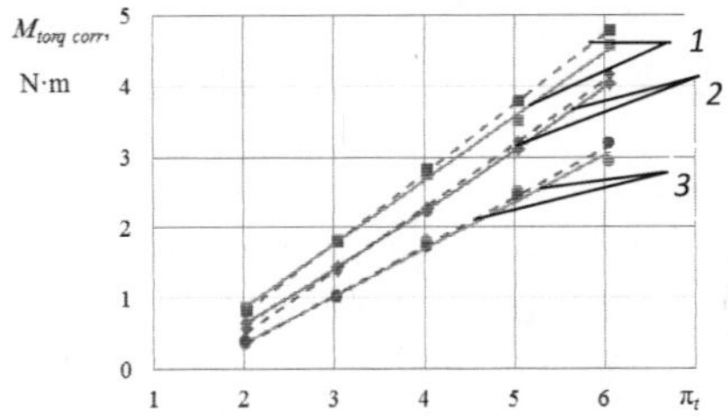

b

Figure 4. Calculated (———) and CFD-Simulated (– – –) efficiency (a) and torque (b) characteristics of the air starter ULPT (1 - n_{corr} = 13,000 rpm; 2 - 19,500 rpm;3 - 26,000 rpm)

Table 3. Single-stage axial ULPT vs. two-stage axial ULPT

Initial data	**Single-stage ULPT**	**Two-stage ULPT**
Inlet pressure p_0^*, kPa	594	592
Shaft power $N_{t\ corr}$, W	7500	7500
Torque $M_{torq\ corr}$, N m	3.6	3.6
Rotating speed n, rpm	19,900	19,900
Output data	**Single-stage ULPT**	**Two-stage ULPT**
Mass flow air consumption G, kg/s	0.147	0.119
Power efficiency η_t	0.455	0.607
Mass of air starter M_{air}, kg	4.91	16.28
Specific mass of a fluid (air) $M_{air\ sp}$, kg/W	0.754	0.407
Specific mass of an air starter $M_{tu\ sp}$, kg/W	0.312	2.567
Overall specific mass $M_{\Sigma\ sp}$, kg/W	1.066	2.974

As it can be seen from Table 3, we obtained different output results with identical initial data values. Energy efficiency was increased with the use of two pressure stages instead single one. But in this way the mass efficiency of turbine unit is several times worse. Thus, the overall efficiency of the two-stage ULPT will be the higher the longer the operating time on design mode. For short-term operating turbine units like air starters we would recommend to use the single-stage turbines.

6. Conclusions

Analysis of the obtained characteristics allows making some conclusions. The developed surrogate models for evaluating the energy efficiency criteria provide a fairly good agreement with the results of CFD simulation at the proviso of the optimization at a given operational mode when total inlet pressure is kept constant ($p_0^* = const$). This corresponds to the fact that the average degree of reaction of entire turbine ρ_t is equal to or close to zero.

If pressure drop ratio values π_t are differ from the calculated by more than 1.5 times (based on the estimates of errors from Figure. 4), we are not recommend this surrogate models for estimating the energy efficiency of a two-stage ULPT to avoid unreliable results. This means that the correction of surrogate model of turbine efficiency in view of degree of reaction influence is mandatory for an adequate selection the parameters of multimode turbine drive with $p_0^* = var$ during the operation. This will require the formulation and implementation of additional natural and computational experiments for further research.

So, the application of these surrogate models in the proposed method of parameter selection is most advisable for turbine units with maintaining a constant pressure at the pressurized air receiver like air starters turbines.

Acknowledgement
This study was supported by the Ministry of education and science of the Russian Federation in the framework of implementation of the Program of increasing the competitiveness of Samara National Research University among the world's leading scientific and educational centers for 2013-2020 years.

Nomenclature

a_{throat}	blade passage throat, mm;
b	blade chord, mm;
c	absolute flow velocity, m/s;
d, D	diameter, mm;
G	mass flow rate of an air , kg/s;
h	blade height, mm;
k	adiabatic index (k = 1.4 for air);
l	length, mm;
L	specific isentropic heat drop, J/kg;
M	weight, kg;
M	isentropic Mach number;
n	rotation speed, rpm;
N	shaft power, W;
p	flow pressure, Pa;
$q(\lambda)$	relative flux density
R	universal gas constant (for air it equals 287 J / kg K);
s	blade width, mm;
t	blade pitch, mm;
T	flow temperature, K;
u	blade linear velocity, m/s;
y	blade-jet speed ratio;
z	number of stages;
Greek	
α	angle of the absolute flow (nozzle angle), °C
β	angle of the relative flow (angle of a rotor blade), °C
δ	gap, mm;
δ_{edge}	blade edge thickness, mm;
Δ	blade overhang, m; difference or change of parameter, m;

Subscripts

a	axial direction
avr	average parameter
c	velocity
$corr$	corrected to the standard atmospheric conditions
ef	effective
i	No. of operational mode; No. of turbine stage
j	No. of optimization parameter
p	pressure; No. of a set initial uncertain data
q	No. of an optimized solution variant
r	radial direction; No. of efficiency criterion
s	isentropic parameter
sp	specified parameter
st	turbine stage
t	turbine
td	turbine drive
$torq$	torque
tu	turbine unit
design	design parameter value;
in	inlet
out	outlet
set	predefined (set) value of a parameter
ND	nozzle diaphragm
TW	turbine wheel
0	before a nozzle diaphragm
1	before a turbine wheel

ρ	degree of reaction;
σ	total pressure loss;
τ	operational time, s;
(b/t)	chord-pitch ratio;
(s/h)	relative extension of a blade (width-height ratio);
λ	reduced velocity;
η	coefficient of total-to-stati[c power efficiency
π_t	turbine pressure drop ratio
2	behind a turbine wheel
I	first turbine stage
II	second turbine stage

Superscripts

*	stagnation gas-dynamic parameter (total parameter); desired value (in optimization task)

References

[1] Grigoriev, V.A., Rad'ko, V.M., Kalabukhov, D.S. and Musatkin, N.F., 2010, "Analysis of the Problem of Efficiency Increase for Turbo Drive Midget Power and the Ways of Its Decision", Aerospace technique and technology, **7**(33), pp. 168-172 (*in Russian*).

[2] Zwyssig, C., Duerr, M., Hassler, D. and Kolar, J.W., 2007, "An Ultra-High-Speed, 500000 rpm, 1 kW Electrical Drive System", Power Conversion Conference, 7, IEEE, Nagoya, Japan, pp. 1577-1583.

[3] Kalabuhov, D.S., 2016, "The Determination of Rational Application of Axial, Centripetal and Centrifugal Ultralow Power Turbines by their Generalized Characteristics", Proceedings of Higher Education. Engineering, 9(678), pp. 53-61 (in Russian).

[4] Abiants, V. Kh., 1979, "Aircraft Gas Turbines Theory", Engineering, USSR, 246 p. (in Russian).

[5] Matveev, V.N., 1999, "A Method of Design Calculation of Multi-mode Turbine Drive that Formed from Centripetal Turbine", Samara State Aerospace Herald. Series: Problems and Prospects of Engines Development. Pt.2, 2, pp. 14-19 (in Russian).

[6] Kalabuhov, D.S. and Martynov A.A., 2017, "Method of Selection of Rational Variant of Very Low Yield Turbine in Turbine Aggregate System at the Stage of Concept Design", Herald of Mechanical Engineering, Russia, 1, pp. 44-56 (in Russian).

[7] Horlock, J.H., 1973, "Axial Flow Turbines: Fluid Mechanics and Thermodynamics", R. E. Krieger Pub. Co., Malabar, UK.

[8] Korpela, S.A., 2012, "Principles of Turbomachinery", John Wiley & Sons Inc., Hoboken, NJ.

[9] Kalabuhov, D.S., 2013, "About Formation of Mathematical Models of the Ultra Low Power Turbine Drive for Energy Efficiency Criteria Evaluating on the Basis of Computing Factorial Experiments", Proceedings of VII All-Russian Youth Scientific Conference "Mavlyutovskie chteniya", Ufa, Russia, pp. 99-100 (in Russian).

[10] Grigoriev, V.A., Rad'ko, V.M, and Kalabuhov, D.S., 2012, "Choosing Parameters of Small Turbo Drive of Major Circuit at the Experiments Designing", Herald of Ufa State Aviation Technical University, 2(47), pp. 35-44 (in Russian).

[11] Bykov, N.N. and Emin O.N., 1972, "Selection of parameters and calculation the low power turbines for the accessories drive", Engineering, Moscow, USSR (in Russian).

[12] "For Mine Haul Trucks, Anything Less Than a TURBOTWIN™ Starter is a Compromise", 2011, Mining Catalog, http://www.tdi-airstarter.com/products.

[13] Kvayt S.M., Mendelevich, J.A. and Chizhkov J.P., 1990, "Start-Quality and Starting Systems of Autotractor Engines", Engineering, Moscow, USSR (in Russian).

[14] Grigoriev, V.A., Rad'ko, V.M. and Kalabuhov, D.S., 2012, "Development and Study of Ultra-Low Power Turbine Volume and Mass Mathematical Models", Newsletters of Propulsion Engineering, Ukraine, 2, pp. 138-144 (in Russian).

[15] Kalabuhov, D.S., 2014, "Methods and Means To Improve The Efficiency of Ultra-Low Power Turbine Drives at the Initial Design", Thesis of Candidate of Technical Science, Samara State Aerospace University, Russia (in Russian).

IOP Conf. Series: Materials Science and Engineering **302** (2018) 012036 doi:10.1088/1757-899X/302/1/012036

Learning Data Set Influence on Identification Accuracy of Gas Turbine Neural Network Model

A V Kuznetsov[1]and G M Makaryants[1]

[1]Power Plant Automatic System Department, Samara National Research University, Moskovskoeshossest. 34, 443086 Samara, Russia

Email:a.v.kuznetsov91@mail.ru

Abstract. There are many gas turbine engine identification researches via dynamic neural network models. It should minimize errors between model and real object during identification process. Questions about training data set processing of neural networks are usually missed. This article presents a study about influence of data set type on gas turbine neural network model accuracy. The identification object is thermodynamic model of micro gas turbine engine. The thermodynamic model input signal is the fuel consumption and output signal is the engine rotor rotation frequency. Four types input signals was used for creating training and testing data sets of dynamic neural network models – step, fast, slow and mixed. Four dynamic neural networks were created based on these types of training data sets. Each neural network was tested via four types test data sets. In the result 16 transition processes from four neural networks and four test data sets from analogous solving results of thermodynamic model were compared. The errors comparison was made between all neural network errors in each test data set. In the comparison result it was shown error value ranges of each test data set. It is shown that error values ranges is small therefore the influence of data set types on identification accuracy is low.

1. Introduction

Increased attention to the accuracy and reliability of gas turbine engine (GTE) control systemshas heightened the need for investigations into the use of neural networks to identify the dynamic model of the engine [1-5]. Therefore, of particular interest and complexity is the problem of development the techniquefor tuning a dynamic neural network (DNN) GTE model.

Artificial neural network (NN) model is a mathematic abstraction of biological neuron. NN consists of weights matrix and activation functions. The aim of NN learning is calculating such weights coefficients to get neural network output equal target output.DNN is a special class of neural network to be depending not only from input vector but from time too. This characteristic allows simulating any nonlinear dynamic processes such as GTE process. There are different types of DNN but for regression task the greatest interest is the next networks: recurrent neural network with feedback connection from output layer (Jordan network) or from hidden layer (Elman network) to input layer.

There is a vast literature on GTE modeling using neural networks [6-12]. Generally, these paperspay particular attention only to the problem of the practical application of GTE neural networks models. For example, Tayarani-Bathaie et al.[6] developed DNN identification of GTE specifically for detecting engine defects.Fast et al. [7] predicted industrial turbine operation at steady state modes using feedforward neural network. Wolff et al. [8] applied artificial neural networks for misfiring detection in an annular pulsed detonation combustor.Sisworahardjo et al. [9] drew a comparison between conventional PI-regulator and a regulator based on neural network which controlled a micro-gas turbine power plant. Although considerable research has been devoted to apply the neural network models to identify the operation modes of GTE, rather less attention has been paid to the problems of DNN GTE model creating.The neural network parameter selection such as a neural network type, a number of hidden neurons, a number of neurons in the input and output layers and their activation function is carried out without any justification.

A training sample is one of the crucial parameters used for developing neural networks.The reference [7] the training sample is represented by time histories of power plant main parameters obtained from field experimentwhich was conducted under different external conditions and loads. The same approach for neural network training is demonstrated byNikpey et al. [10]. The DNN employed by Asgari et al. [11]identifies the starting process of GTE. The data used for DNN training was also collected from the real engine operating process under the change of ambient conditions. The literature shows thatthe data used for GTE neural network identification is obtained from full-scale study.However, these data can be obtained only from the engine which has already been constructed. Moreover, for knowing all of engine parameters (temperature, pressure) in different places you should have a lot of sensors in the engine and you should use special test bench for contact each of them during engine fire test.These factors lead to low financial efficiency of the GTE identification process. Even more, the other disadvantage of GTE identification based on experimental data related to the impossibility of such procedure at the stage of a new engine designing. Therefore, more promising is the development of the GTE identification technique based on the neural network with the use of engine thermodynamic mathematical model. This technique was demonstrated by Asgari et al. [12].The neural network model was created using 13 different training functions. However, the problem of training functions substantiation was not considered.

In this paper, thenew investigation of the training functions type influence into the GTE neural network model accuracy is presented.

2. Methods

The identification accuracyof the gas turbine engine dynamic characteristics depending on the training sample was estimated using the following sequence. Firstly, several types of functions were selected. Each types of the function used as training and as tested function. Then, a neural network model based on each training function was designed. Then, each of models was tested by means of all types of test functions.

The fuel consumption was used as the input signal of the model, and the GTE rotor speed was used as an output signal.Four types of training and test functions were used, as follows: a step function (type 1), a ramp function with "fast" and "slow" signal change (type 2 and 3 respectively) and a combination of previous three function (type 4).

The terms of "fast" and "slow" signal change is related to the ratio of signal acceleration and acceleration of the GTE rotor.The acceleration of fast signal is twice as much as the average acceleration of the rotor during the process of switching from idle mode to full throttle and back. The acceleration of slow signal is twice smaller than the average acceleration of the rotor during the injectivityprocess.Thus, the average value of the acceleration of the rotor is used as some reference value for selecting the acceleration of the sampling signal.A step signal was selected as thelimiting case to simulate a sudden change in the signal, which usually occurs when identifying the signal from the speed sensor in the presence of a noise component of the measurement.Combined or random signal includes all of these three signal types.Thus, four variants of the neural dynamic GTE model were developed, which were tested on four sets of test signals.

The simulation used a network with the following structure: one input neuron (fuel consumption), one hidden layer neuron and one output neuron (engine rotor speed).

3. Results and discussion

IOP Conf. Series: Materials Science and Engineering **302** (2018) 012036 doi:10.1088/1757-899X/302/1/012036

3.1. GTE rotor mean acceleration estimation

To determine the mean value of the acceleration of the rotor, a step fuel consumption signal is set from the value of the small gas to the maximum mode (Figure 1).

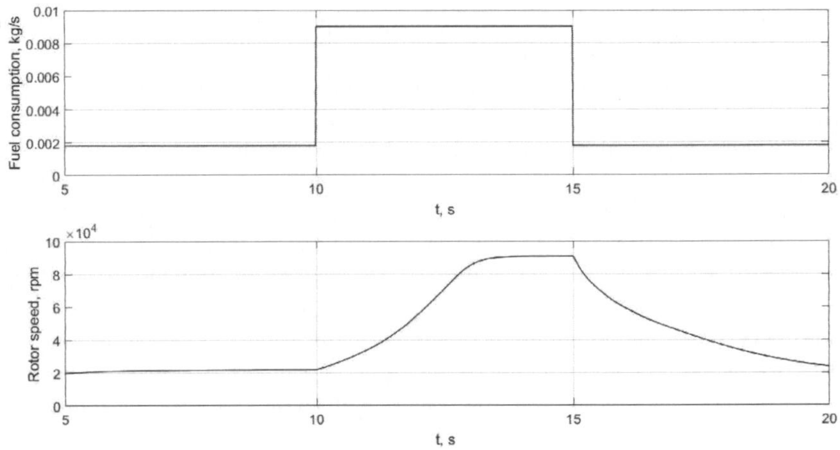

Figure 1. Step signal of fuel consumption (top) and rotor speed transition process (bottom)

The module of engine rotor acceleration is different for acceleration (transition stage from the minimum frequency to the maximum) and deceleration (transition stage from the maximum frequency to the minimum). Therefore different values of acceleration should be used. The average acceleration of the rotor during the transition stage from the minimum to the maximum mode is 18,000 rpm per s.The average deceleration during engine braking under similar conditions is 6,960 rpm per s.To simplify the work with the neural network, the mean acceleration values were normalized to the GTE speed at the maximum mode of 90,600 rpm (Figure 1).Then, the average reduced acceleration was 0.20 s^{-1}, and 0.08 s^{-1} during engine acceleration and braking respectively. Thus, the reduced signal acceleration at the input was obtained. The assumption that the acceleration of the signal at the output has the same value was made. Then, for the case of a rapid change in the signal during the engine speedup, the acceleration was 0.40 s^{-1}, and during the braking the acceleration was 0.16 s^{-1}.In the case of a slow signal change, the value of the reduced acceleration was0.10 s^{-1} and 0.04 s^{-1} during the engine speedup and braking, respectively.Figures 2-5 show four cases of input signals, including both the training sample and the test sample.

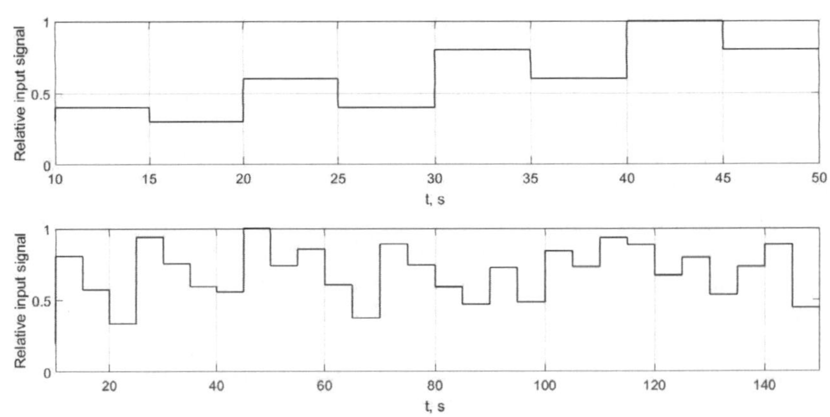

Figure 2. Step relative input signal for learning process (top) and test process (bottom)

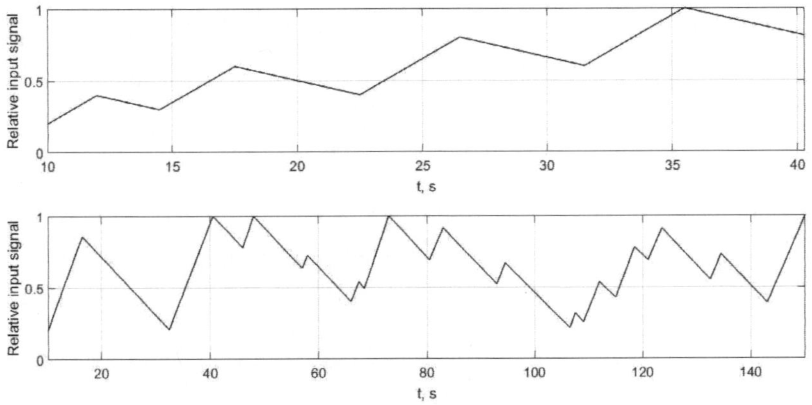

Figure 3. Slow relative input signal for learning process (top) and test process (bottom)

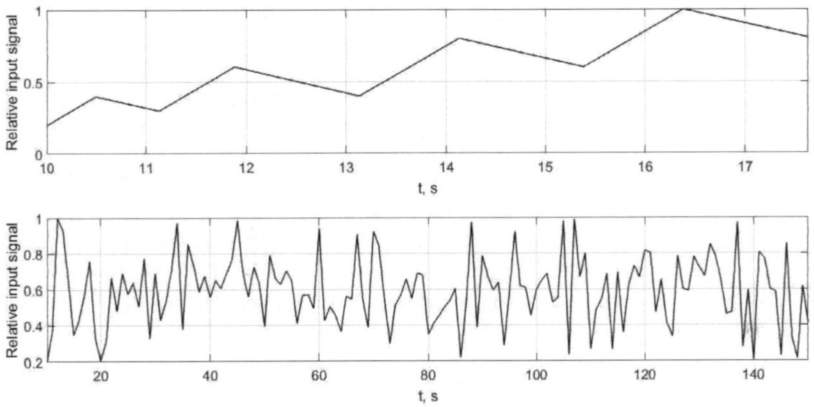

Figure 4. Fast relative input signal for learning process (top) and test process (bottom)

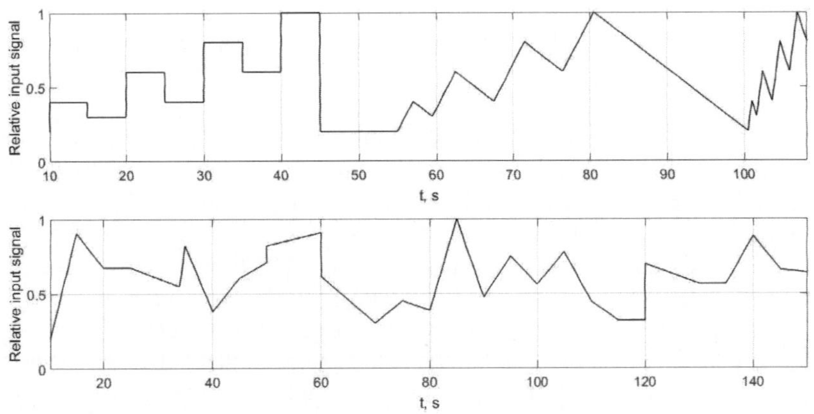

Figure 5. Random relative input signal for learning process (top) and test process (bottom)

3.2. Model testing and comparison of the results

As an example, Figure 6 depicts the comparison of GTE rotor speeds obtained fromthe neural network model which was tested on random relative input signal (type 4).

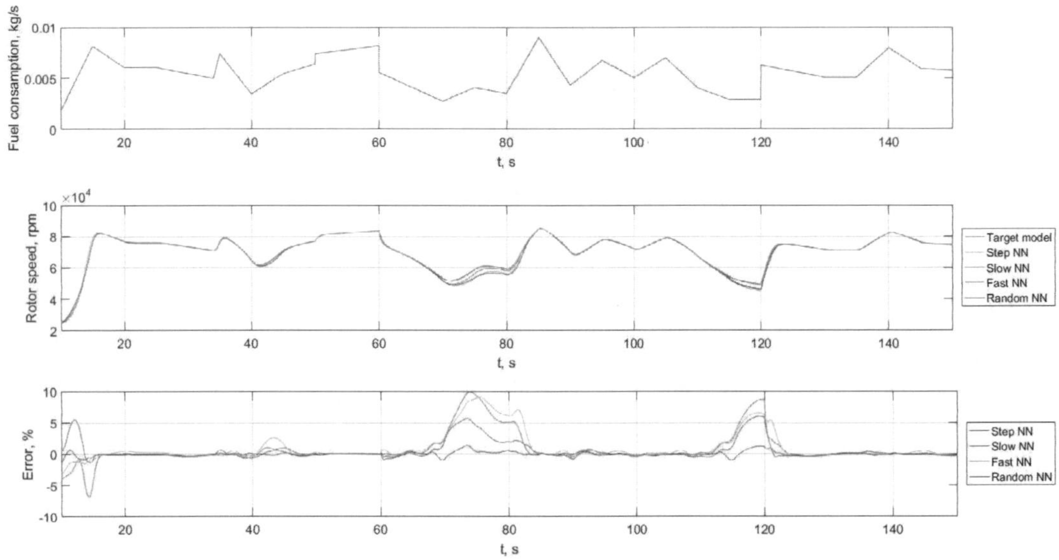

Figure 6. Results of testing different NN models on random signal.
Transition process (top) and identification errors (bottom)

A strong stratification of characteristics around the 80th and 120th seconds was due to the insufficient number of training data at low rotational speed.This stratification is true for all model types and has a local character regardless of NN model type (with the exception of the Random NN because it has more learning examples). In addition, this has little effect on the models average error magnitude over the entire operation time. The results of deviations for all neural network models are given in Table 1.

Table 1. Errors of NN models on different test data sets in %.

Test \Learn	Step NN model	Slow NN model	Fast NN model	Random NN model
Step test data set	0.7421	1.8770	0.8845	0.6414
Slow test data set	0.9451	0.4758	0.7944	0.6414
Fast test data set	1.7780	1.3810	0.7193	0.3035
Random test data set	0.8146	1.1980	1.2280	0.2047

3.3. Analysis

According to Table 1, for each of the test samples and the neural network models corresponding to them, it can be seen that the minimum error value corresponds to cases when the neural network model was tested on a sample of the same character as used for its training.

For the model trained on a type 1 sample, the error range was 1.036% for the type 2 it was 1.401% for the type 3 it was 0.5087 and for the type 4 it was 0.4367. The lowest value of the error was shown by a model trained on data of a type 4 (random type).The model, trained at slowly changing fuel consumption, showed the greatest value of the error. However, the difference in accuracy is negligible.

As can be seen, there is no fundamental difference between models trained on different types of input functions.Hence, when creating a neural network model, the researcher can choose the test method and those samples that will be most convenient for the modeling process.

The research was limited to the gas turbine scope, however, considering the universality of the structure and learning of the neural network models, it can be concluded that this study will show similar results on the other objects.

4. Conclusion

The results of the training function type influence to the accuracy of the dynamic neural network GTE model are obtained. It was found that the accuracy of the developed models is virtually independent of the sample on which the neural network was trained. The results are approximately the same for both a stepped input signal and a smoothly varying one.The average model error did not exceed 1.9%.In general, the increase in accuracy can be achieved by using the full spectrum of changes in all the parameters of interest in training samples, but often this is unnecessary and, as shown in this study, is not required to obtain sufficiently accurate models.

References

[1] Krupenich I N, Filinov E P, Ostapyuk Y A and Tkachenko A Y 2017 Investigation of theefficiency limits of the traditional gas turbine engines *International Journal of MechanicalEngineering and Robotics Research* **6(1)** 36-41
[2] Tkachenko A Y, Kuz'michev V S, Krupenich I N and Rybakov V N 2016. Gas turbine engineoptimization at conceptual designing MATEC Web of Conferences77
[3] Gafurov S, Rodionov L, Kryuchkov AandHandroos H 2016 HIL test bench for engine's fuelcontrol systems investigation 30th Congress of the International Council of the AeronauticalSciences
[4] Matveev V N, Popov G M, Baturin O V, Goryachkin E S and Kolmakova D A 2015 Workflowoptimization of multistage axial turbine 51st AIAA/SAE/ASEE Joint Propulsion Conference
[5] Rybakov V N, Kuz'Michev V S, TkachenkoA Y and Krupenich I N 2016 Thermodynamicmulti-criteria optimization of the unified engine core for the line of turbofan enginesProceedings of the ASME Turbo Expo
[6] Tayarani-Bathaie S Sina, Vanini Z N Sadough and Khorasani K 2014 Dynamic neural network-based fault diagnosis of gas turbine engines Neurocomputing125 153–165
[7] Fast M, Assadi M and De S 2009 Development and multi-utility of an ANN model for anindustrial gas turbine Applied Energy86 9–17
[8] Wolff S, Schäpel J S and King R 2016 Application of artificial neural networks for misfiringdetection in an annular pulsed detonation combustor mockupJournal of Engineering for GasTurbines and Power139
[9] Sisworahardjo N S, El-Sharkh M Y and Alam M S Neural network controller for microturbinepower plants Electric Power Systems Research78 1378–1384
[10] Nikpey H, Assadi M and Breuhaus P 2013 Development of an optimized artificial neuralnetwork model for combined heat and power micro gas turbines Applied Energy108 137–148
[11] Asgari H, Chen X Q, Morini M, Pinelli M, Sainudiin R, Spina P R and Venturini M 2016NARX models for simulation of the start-up operation of a single-shaft gas turbine AppliedThermal Engineering93 368–376
[12] Asgari H, Chen X Q,Menhaj M B and Sainudiin R 2013 Artificial neural network based systemidentification for a single-shaft gas turbine Journal of Engineering for Gas Turbines andPower135

IOP Conf. Series: Materials Science and Engineering **302** (2018) 012037 doi:10.1088/1757-899X/302/1/012037

The Optimization of Four-Stage Low Pressure Turbine with Outlet Guide Vane

V Matveev[1], O Baturin[2] and G Popov[3]

[1] Professor, Samara National Research University, Samara, Russia
[2] Associate professor, Samara National Research University, Samara, Russia
[3] Researcher, Samara National Research University, Samara, Russia

E-mail: grishatty@gmail.com

Abstract. The goal of the research was to improve efficiency of four-stage low-pressure turbine with outlet guide vane (LPT) at the design point by optimization the shape of all turbine stator vanes and the stagger angles of all rotor blades. The LPT mathematical model was created by using NUMECA FineTurbo software. Several constraints were imposed the mass flow rate through the LPT and the total pressure ratio at the design point may vary within ±0.5% from the original. Parameters of the stator blade shape and rotor blades stagger angles were obtained. This new geometry of the LPT blades produce a 0.8% increase in efficiency at the design point.

1. Introduction

The improvement of turbine efficiency is always an urgent task of the gas turbine industry [1]. Over recent years, the process of designing and development of efficient turbines made a big qualitative leap associated with the development of computational fluid dynamics (CFD) methods. They can reliably predict the characteristics of the turbine with the features of three-dimensional geometry and to evaluate the impact of geometrical or operational variation on the turbine characteristics.

Despite its apparent simplicity, using CFD methods is difficult [2]. The main reason is that the geometry of axial turbine stage is described by many variables. For a complete description of one stage, the number of independent variables reaches several dozen contradictory affecting the turbine characteristics.

Human is not capable to analyze the problem of such dimension. Also, the search time for optimal combination of variables would be prohibitively large. The problem becomes greater for the multistage turbine. In this case, the methods of mathematical optimization can help the engineer. Their application allows to automate the search of optimal combination of independent variables.

The search algorithm is the following. The optimizer, according to the algorithm for finding the extremum, generates a set of magnitudes of varying variables. Then, computational workflow model is formed in a special subroutine or CFD program. Then, the calculation of the flow in the generated variant is carried out in the CFD program. Its results are transmitted to the optimizer, which forms a new combination of independent variables. Described cycle is repeated until the desired extremum.

2. Subject of research and its numerical model.

With this algorithm, 4.5-stage (4 full stages and straightening blade row installed for the latter) shrouded axial turbine model designed and teste in NASA (Figure 1) was optimized [3].

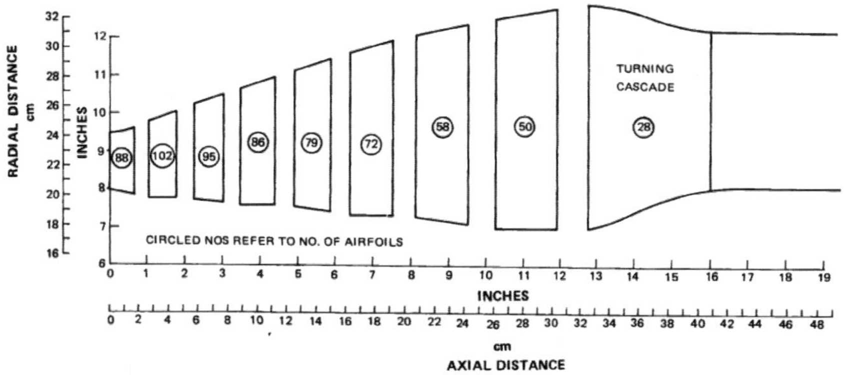

Figure 1. The appearance of investigated turbine [3].

Based on the original geometry of the turbine, a set of computational models was created. When creating computational models, the following assumptions were made (Figure 2).

1. The flow in each blade row (BR) and turbine cavities have the cyclic symmetry property. Therefore, all models contained only one blade passage with periodic boundary conditions on the sides.
2. The calculation was carried out in a stationary statement.
3. Deformation of turbine blades under the gas load was not considered in the calculation.
4. An ideal gas with properties of the kerosene combustion products was used as working fluid.
5. Specific heat and viscosity of the working fluid depend on its temperature.
6. Turbulence was considered as isotropic in all directions. Spalart-Allmaras model was used.
7. Heat transfer between the walls and the flow was not considered due to the process rapidness.

NGV was calculated in the stationary coordinate system. RW was calculated in the rotating coordinate system, the speed of which coincides with the rotor speed.

Radial distribution of total pressure p*, total temperature T*, flow angle α and turbulent viscosity were used as the boundary conditions at the turbine inlet (Figure 2). These parameters were taken as in turbine tests [3, 4]. The static pressure on the hub radius was set at the turbine outlet. The pressure on the rest radii was calculated automatically by the software according to radial-equilibrium equation. The value of the static pressure at the outlet of the turbine is appointed in accordance with the desired degree of expansion. To transfer data between NGV and RW, *Full Non Matching Mixing Plane* was used.

Three computational models with varying mesh density were created for each considered configuration of the turbine. The basic parameters of the mesh models are shown in Table 1.

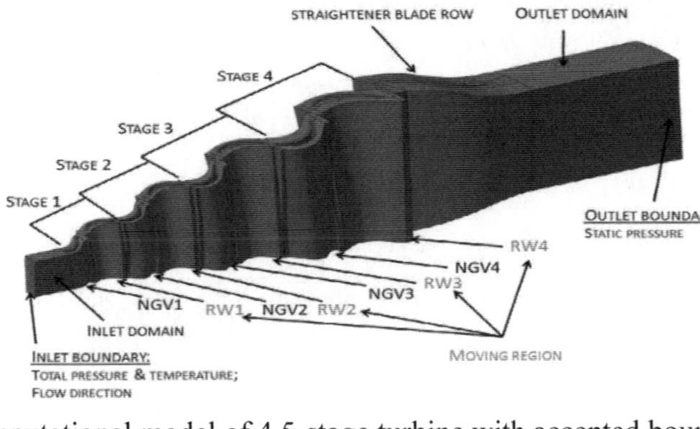

Figure 2. Computational model of 4.5-stage turbine with accepted boundary conditions.

Table 1. Basic parameters of created finite element models.

	number of elements in the 2D layer	number of layers along radial direction	total number of elements	y+ value
4.5St_Mesh 1	7755	49	3400000	7
4.5St_Mesh 2	14835	61	8100000	3
4.5St_Mesh 3	22015	81	16040000	1

3. Verification of the LPT numerical model.

Using created numerical models, the turbine characteristics (Figure 4) as efficiency and mass flow parameter dependences of the total pressure ratio in the turbine at a constant rotor speed were calculated [5,6]. Hereinafter, the following is meant by mass flow parameter:

$$A = \frac{G\sqrt{T^*}}{p^*} \quad (1)$$

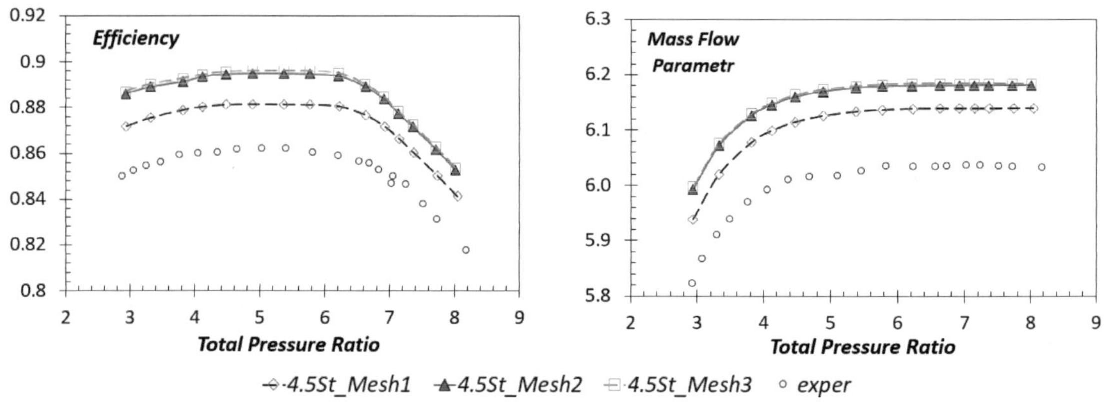

Figure 3. Comparison of calculated and experimental characteristics of the turbine.

It is clear from Figure 3 that the results obtained by medium and fine mesh slightly differ both qualitatively and quantitatively. This suggests that the mesh size ceases to affect the results when the number of finite volume per one blade row of more than 900 thousand. Thus, the grid convergence is achieved. Further mesh refinement will not affect calculation results.

The Figure 3 shows that turbine efficiency is overestimated by 2...3% (abs.) and the mass flow parameter by 2...3% (rel.). However, all created numerical models are qualitatively in a good agreement with test results. This is illustrated by the graph in Figure 5. It shows the relative mass flow parameter and relative efficiency dependences of the total pressure ratio in the turbine. The relative parameters are the ratio of the current value of the variable to its maximum value in considered range:

$$\bar{A} = \frac{A_i}{A_{max}} \quad (2)$$

$$\bar{\eta} = \frac{\eta_i}{\eta_{max}} \quad (3)$$

All four graphs in Figure 4 coincide with each other. This suggests that computational models evaluate with high accuracy the change of turbine parameters with the pressure ratio variation. For example, if the calculation shows the change in efficiency of 2% with some change in pressure ratio, in fact it is likely that the efficiency change will be the same even though calculation value of the efficiency is overestimated. This shows good quantitative agreement of calculation and experimental data.

Two numerical models were selected for optimization of turbine: 4.5St_Mesh1 provides a quick calculation of turbine operating parameters; 4.5St_Mesh2 provides grid convergence – increase in the number of elements does not lead to a significant change in the calculated parameters of the turbine.

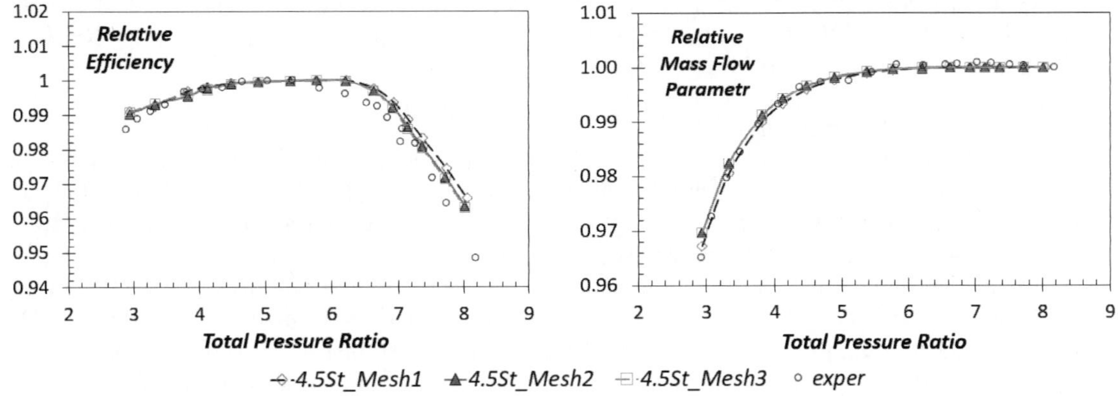

Figure 4. Comparison of relative calculated and experimental characteristics of the turbine.

4. Statement of the LPT optimization problem.

The algorithm for finding the optimal solution was built based on optimizer IOSO [7] and software NUMECA. The geometry of computational is constructed domain based on the array of independent variables generated by the optimizer using program «Profiler» [8].

The algorithm described in [9] was used for turbine optimization. Its main idea is that the thickness of the blade in the control section, and the nature of change along the chord remain unchanged during optimization. This approach can significantly reduce the number of variables, and keep the stress-strain state of the blades in the first approximation. This is because the ratio of the cross-sectional area of the hub and peripheral sections of the blade does not change in a large degree.

The shape of the camber line of the blade in the control section is not the only factor determining the turbine efficiency. Mutual position of sections, form of the leading edge, the shape of the meridian section of flow passage and other factors can affect it. These factors have been deliberately excluded from consideration, as they lead to the critical increase in the number of variables. Optimization process with more than 100 independent variables is not allowed with available license for the program IOSO.

Slightly modified approach from [7] was used to describe the shape of the camber line as a spline passing through 4 points. The end points corresponded to the centers of leading and trailing edges. The remaining points are evenly spaced along the spline. Changing the camber line of the rotor blades and relative position of the sections relative to each other was carried out by moving the spline middle points in the circumferential direction in the global coordinate system, as well by varying stagger angle γ.

NGV shapes were changed moving the first 3 points of the spline in the tangential direction. Also, shifts of the section in the axial and circumferential directions and the stagger angle of profile γ were varied. In total, the shape of the three sections (hub, mean, and shroud) was varied in this way in NGV.

Changing RW blade shape was performed only by the stagger angle for entire blade in a range of $\pm 2^0$ to save the stress-strain state of the rotor blade, as well as to reduce the number of variables.

Search for the optimal configuration of the 4.5-stage turbine was carried out by the following algorithm. The IOSO program generates block of input data based on which the Profiler program generates new blade geometry and transmits it to a Numeca as a text file. In the Numeca, computational model is created based on the received information, and calculation is produced. As a result, the efficiency and other parameters of the compressor are determined. IOSO forms a new combination of input data based on the calculation, as well as previous references to the numerical model, and the process is repeated until the desired extremum.

During optimization, the task of increasing the maximum turbine efficiency at the design mode was set. During the optimization process, the geometric parameters of RW and NG blades described above were varied. The total number of independent variables was 96.

The constraints associated with the peculiarities of the turbine functioning as part of GTE were set: mass flow of the working fluid through the turbine and the gas expansion ratio must not differ from the original values by more than 0.5%.

5. Results of the LPT optimization.

The optimization problem was solved using two numerical models 4.5St_Mesh1 and 4.5St_Mesh2 with an equivalent statement. Configurations of blades which provide maximum turbine efficiency under the above restrictions were found using these models. To obtain a solution, more than 1,000 references to computational model were required by program IOSO.

The results of solving optimization problem using 4.5St_Mesh1 and 4.5St_Mesh2 are qualitatively similar. Characteristics of optimal turbine variants were calculated using model 4.5St_Mesh 2. The results were compared with the characteristics of the original version (Figure 5). Comparison of the original and the optimized blade shape obtained using 4.5St_Mesh 2 are shown in Figure 6.

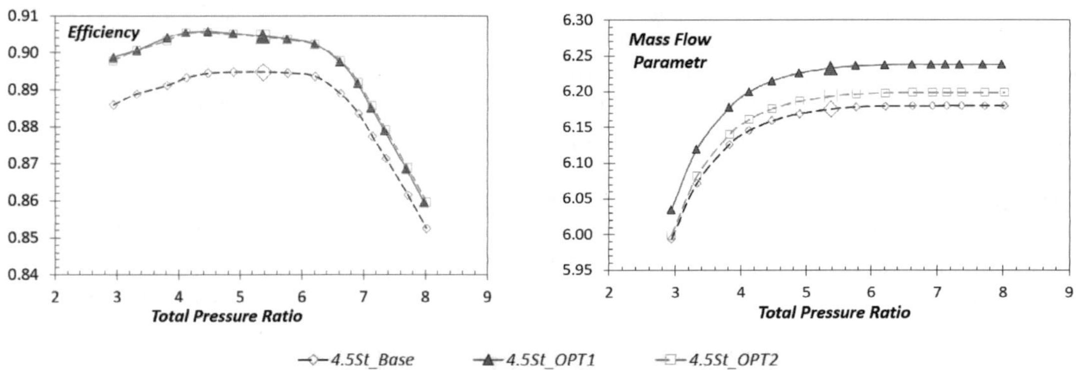

Figure 5. Comparison of characteristics of 4.5-stage turbine before and after optimization.

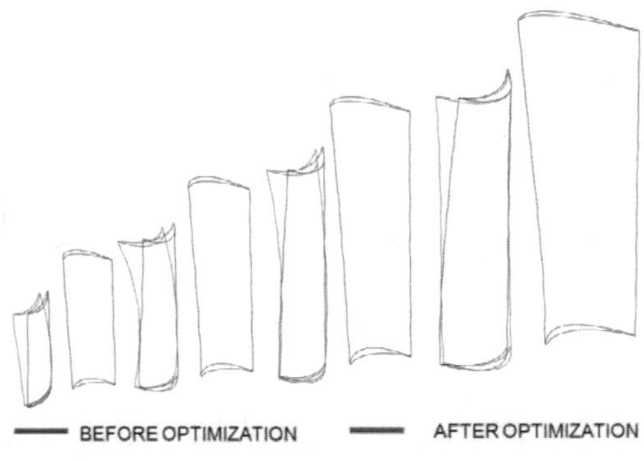

Figure 6. Comparison of the blade shapes of 4.5-stage turbine before and after optimization.

The found turbine configurations of provides the increase in efficiency by 0.8% at design mode (Figure 5). Figure 7 shows a comparison of the efficiency and work of turbine stages for initial and optimized variants at design mode. The optimization resulted in some redistribution of work between the stages: the first two stages were unloaded; the last stages were additionally loaded. Moreover, the specific work of the third stage was increased by 3%. Increase in the efficiency of caused the increase in the specific turbine work by 0.8% (rel.). The increase in stage efficiency is following: in the first stage - by 0.1%; in the second - by 1.5%; in the third - by 1.2%; in the fourth – by 0.5%.

The shape of the nozzle vanes was largely changed in optimization (Figure 6). They gained a complex spatial form. The shape of rotor blades was changed slightly. The main reason for the efficiency improvement was a decrease in secondary losses in NGV, as well as a matching in leakage angles.

6. Conclusions

Configuration of 4.5-stage axial turbine was found allowing the increase of its efficiency by 0.8% at the design mode. Increase of efficiency was achieved by a radical change in the shape of the nozzle guide vanes, which was carried out by using optimization methods and numerical simulation. The main reason for reduction in losses and increase in efficiency is the secondary losses decrease and negotiating the leakage angles of the flow in different blade rows.

To achieve this goal the computational model was developed and verified by comparing the calculated and experimental data for different number of stages. Also, the investigation of mesh convergence was conducted. It was shown that although the model provides overestimate efficiency of about 2.5%, it allows the obtaining of high-quality results which do not differ from the experimental and is suitable for finding the optimal solution.

Found reserve for the efficiency increasing of the turbine is not a limit. Because of limitations in the available license of optimizer IOSO, attention was focused only on NGV. There were not enough variables to describe the shape of the rotor blades. Perhaps, the change in shape of the rotor blades under the similar algorithm will improve the efficiency of the stages even more. In addition, the change in shape of the rotor blade will require verification of its stress-strain state at each step, which considerably complicates the algorithm for solving the problem. However, verification of strength state in the optimal configuration is a clear direction of the research development [10].

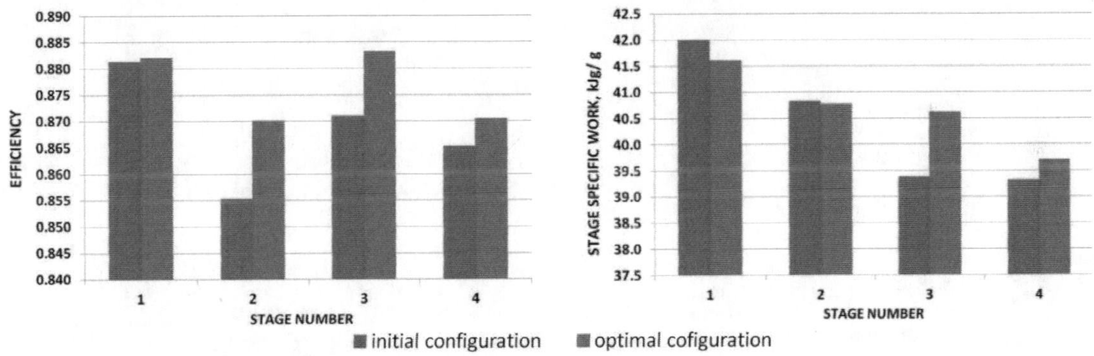

Figure 7. Comparison of the specific works and efficiency of the initial and optimized turbines.

Acknowledgments

The work was financially supported by the Ministry of education and science of Russia in the framework of the implementation of the Program of increasing the competitiveness of SSAU among the world's leading scientific and educational centers for 2013-2020 years.

References

[1] Kulagin V V 2002 *Teoria, raschet i proektirovanie aviacionnyh dvigateley i energeticheskih ustanovok. (Theory, calculation and design of aircraft engines and power plants)* (Mashinostroenie, Moskow, Russia) p 616

[2] Lomax H, Pulliam Th H and Zingg D W 2001 Fundamentals of Computational Fluid Dynamics (Springer, Berlin, Germany) p 265

[3] Webster P F 1976 Design of a 4 ½ stage turbine with a stage loading factor of 4.66 and high specific work output NASA CR-2659

[4] WhitneyW J, Behning F P, Moffitt T P and Hotz G M 1977 Cold-air investigation of 4 ½ stage turbine with a stage loading factor of 4.66 and high specific work output. I – Overall Performance NASA TM X-3498

[5] Lewis R I 1996 Turbomachinery performance analysis (Elsevier Science&Technology Books Pub) p 328

[6] Belousov A N, Musatkin N F, Rad'ko V M and Kuz'michev B V 2006 Proektny termogazodinamicheskiy raschet osnovnyh parametrov aviacionnyh lopatochnyh machin (The design of the thermodynamic calculation of the main parameters aviation impeller machines) (Samarskii gosudarstvennyi aerokosmicheskii universitet, Samara, Russia) p 316

[7] Kuzmenko M L, Egorov I N, Shmotin Yu N, Chupin P V and Fedechkin K S 2006 Proc. Of the 6th ASMO UK/ISSMO conference on Engineering Design Optimization, Oxford, UK

[8] Shabliy L S and Dmitrieva I B 2014 ARPN J. of Engineering and Applied Sciences 9 (10) 1849-53

[9] Goryahkin E, Popov G, Baturin O and Kolmakova D 2015 Proc. of the ASME Turbo Expo GT2015-43384

[10] Ermakov A I, Urlapkin A V and Fedorchenko D G 2014 The features of resonance stress scatter in turbine wheels with a weak connectivity of blade vibrations Research J. of Applied Sciences 9(11) pp 795-9

IOP Conf. Series: Materials Science and Engineering **302** (2018) 012038 doi:10.1088/1757-899X/302/1/012038

Mean-line Modeling of an Axial Turbine

A Yu Tkachenko[1], Ya A Ostapyuk[2] and E P Filinov[3]

Department of Aircraft Engine Theory, Institute of Engine and Power Plant Engineering, Samara National Research University, 34, Moskovskoye shosse, Samara, 443086, Russia

[1]tau@ssau.ru, [2]oya92@mail.ru, [3]filinov.evg@gmail.com

Abstract. The article describes the approach for axial turbine modeling along the mean line. It bases on the developed model of an axial turbine blade row. This model is suitable for both nozzle vanes and rotor blades simulations. Consequently, it allows the simulation of the single axial turbine stage as well as a multistage turbine. The turbine stage model can take into account the cooling air flow before and after a throat of each blade row, outlet straightener vanes existence and stagger angle controlling of nozzle vanes. The axial turbine estimation method includes the loss estimation and thermogasdynamic analysis. The single stage axial turbine was calculated with the developed model. The obtained results deviation was within 3% when comparing with the results of CFD modeling.

1. Introduction

The turbomachine development is very complex process that includes thermogasdynamic, structural, technological and economic investigations. Thermogasdynamic designing is iterative process, in which the calculations results performed at the conceptual design stage are subsequently improved with application the higher fidelity models [1, 2]. During the conceptual design stage the simple models such as the model of the one-dimensional estimation along the flow path mean line is used for performance prediction and preliminary optimization of the turbomachine design. These models are very important, because they allow fast estimation of velocity triangles, losses and parameters of the blades at mean line with acceptable accuracy.

In most cases stations of the turbine flow path are normal to the axis of the engine [3, 4]. This paper presents the method in which these stations are normal to the mean line (figure 1). Thus the radial velocity component can be excluded from consideration. The main advantage is capability for both axial and radial turbines modeling.

2. One-dimensional estimation of an axial turbine

2.1. Model description

The one-dimensional (1D) mathematical model is reasonable for the turbomachines performance calculation along the flow path mean line. This model (figure 2) describes the relation between thermodynamic and kinematic parameters and geometric parameters of blade row at various operation conditions.

A model of the axial turbine stage (figure 3) can be draw up using the turbine blade row model. The model of the axial turbine stage also includes the calculation elements of nozzle vanes, blades, stage overall performance as well as inlet and outlet data interface elements for an external model. The turbine stage model can take into account the cooling air flow before and after a throat of each blade row, outlet straightener vanes existance and stagger angle controlling of nozzle vanes, etc.

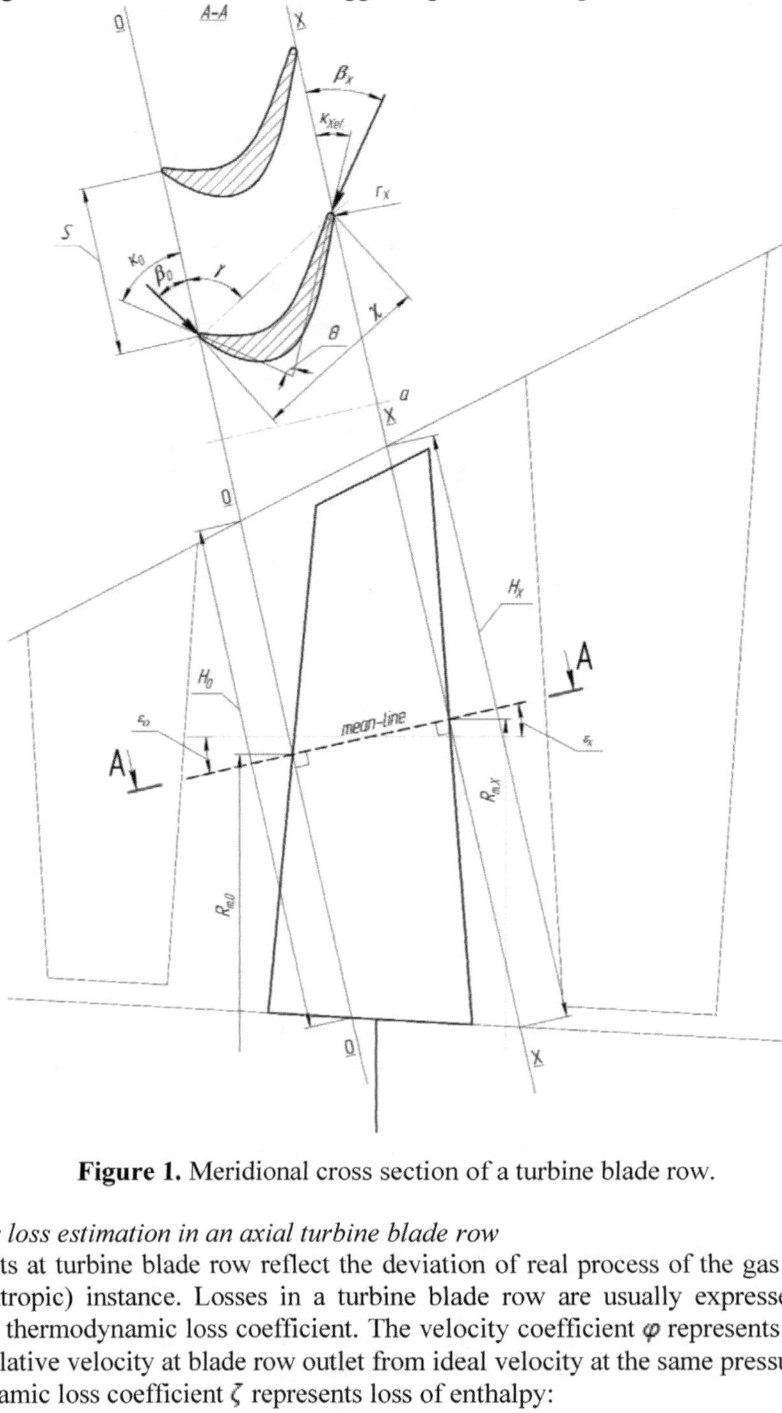

Figure 1. Meridional cross section of a turbine blade row.

2.2. Method for loss estimation in an axial turbine blade row

Loss coefficients at turbine blade row reflect the deviation of real process of the gas expansion from its ideal (isoentropic) instance. Losses in a turbine blade row are usually expressed in a velocity coefficient or a thermodynamic loss coefficient. The velocity coefficient φ represents deviation of the gas flow real relative velocity at blade row outlet from ideal velocity at the same pressure ratio.

The thermodynamic loss coefficient ζ represents loss of enthalpy:

$$\zeta = \frac{i_X - i^*_{Xs}}{i^*_{wX} - i_X} \quad (1)$$

where i_X – static enthalpy at outlet from blade row;
i_{Xs} – total isoentropic process enthalpy at outlet from blade row;
i^*_{wX} – total relative flow enthalpy at outlet from blade row.

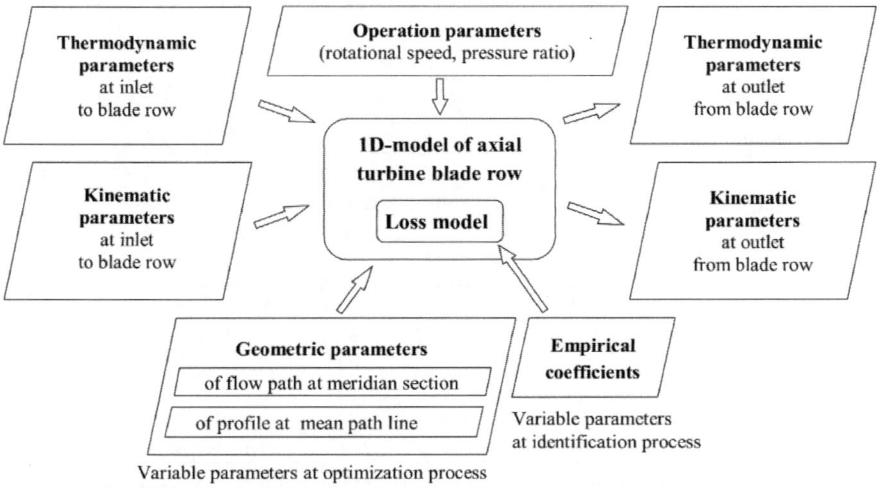

Figure 2. Data streams of the 1D-model of an axial turbine blade row.

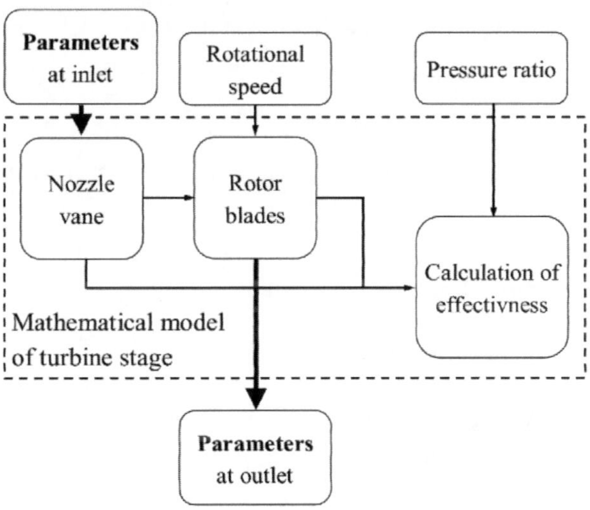

Figure 3. Turbine stage model.

The velocity coefficient is related with the thermodynamic loss coefficient:

$$\varphi = \sqrt{1-\zeta} \tag{2}$$

Total loss has complex structure and includes profile, secondary and cooling losses:

$$\zeta = \zeta_{pr.i.W} + \zeta_s + \zeta_c \tag{3}$$

The structure of the total loss is represented in figure 4, where $\overline{G_a}$ – relative cooling air mass flow rate, K_{KW}, K_V, K_ι – auxiliary coefficients. The cooling losses and other kind of losses are described in detail in [5, 6, 7].

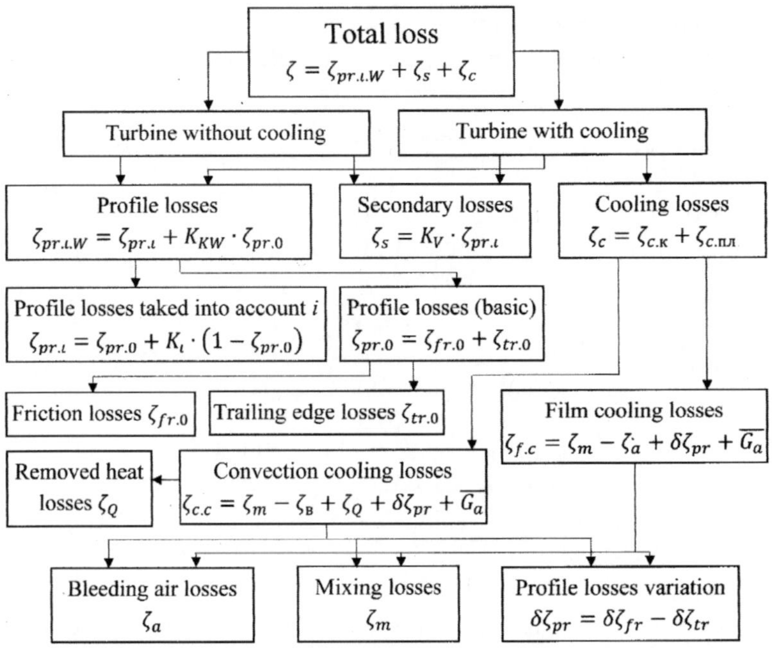

Figure 4. The structure of total loss taked into account in the 1D-model of an axial turbine blade row.

The function for determination of the loss coefficient ζ can be presented in the following general form:

$$\zeta = f(\{G\}, p_O, T_O, T_X, V_X, \lambda_{WX}, \iota, \kappa_O, \kappa_{Xef}, H_X, \chi, s, r_X, \zeta_c), \quad (4)$$

where G – working fluid parameter array;
p_O, T_O – static pressure and static temperature at blade row inlet;
T_X – static temperature at blade row discharge;
V_X – absolute velocity at discharge;
λ_{WX} – corrected relative velocity at discharge;
ι – incidence angle;
κ_O, κ_{Xef} – blade inlet angle and efficient blade outlet angle;
H_X – blade height at discharge;
χ – chord profile length;
s – blade spacing;
r_X – trailing edge radius;
ζ_c – cooling losses coefficient.

2.3. Method for the thermogasdynamic analysis of an axial turbine blade row

The method for the thermogasdynamic analysis of an axial turbine blade row (nozzle vane as well as rotor blade) along the flow path mean line allows to calculate the gas flow parameters at inlet and discharge stations and the cycle parameters depending on the gas flow parameters at inlet, rotational speed and M_{CX} parameter. The M_{CX} parameter describes the flow state at outlet.

Since the methods of the loss coefficients estimation and thermogasdynamic analysis needs the detail description, they will be published in future papers.

3. Model validation

The software modules for computer-aided environment for thermogasdynamic calculations and analysis ASTRA [8] were implemented to perform validation of proposed models. The testing was performed on high pressure turbine of the small-scale engine. The results of calculation are shown in the table 1.

Table 1. Results of the test calculation

G_O	kg/s	2,527	Gas flow rate
p^*_O	kPa	332,526	Total pressure at inlet to nozzle vane
T^*_O	K	1175	Total temperature at inlet to nozzle vane
n	rpm	40000	Rotational speed
$\varepsilon_{O.nv}$	°	0	Taper angle at inlet to nozzle vane
$R_{m.O.nv}$	m	0,094	Mean line radius at inlet to nozzle vane
$\alpha_{O.nv}$	°	0	Angle of entrance at inlet to nozzle vane
$\varepsilon_{X.nv}$	°	0	Taper angle at outlet from nozzle vane
$R_{m.X.nv}$	m	0,094	Mean line radius at outlet from nozzle vane
$r_{X.nx}$	m	0,001	Trailing edge radius of nozzle vane
$\kappa_{O.nv}$	°	0	Blade inlet angle of nozzle vane
$\kappa_{X.ef.nv}$	°	-66	Efficient blade outlet angle of nozzle vane
$\varepsilon_{X.i}$	°	0	Taper angle at outlet from rotor blade
$R_{m.X.i}$	м	0,09375	Mean line radius at outlet from rotor blade
$r_{X.i}$	м	0,00035	Trailing edge radius of rotor blade
$\kappa_{O.i}$	°	23	Blade inlet angle of rotor blade
η	—	0,8828	—
N	kW	431,1	—
Nozzle vane			
ζ_c	—	0	Cooling losses coefficient
ζ_{fr}	—	0,028	Friction losses coefficient
$\zeta_{pr.0}$	—	0,063	Profile losses coefficient (basic)
$\zeta_{pr.i}$	—	0,063	Profile losses coefficient taked into account the angle of incidence
$\zeta_{pr.i.W}$	—	0,0634	Profile losses coefficient taked into account the angle of incidence and flow velocity
ζ_{tr}	—	0,035	Trailing edge losses coefficient
ζ_s	—	0,048	Secondary losses coefficient
$\zeta_{pr.i.th}$	—	0,028	Coefficient of the profile losses before the throat area taked into account the angle of incidence
$\zeta_{pr.i.W.th}$	—	0,0282	Coefficient of the profile losses before the throat area taked into account the angle of incidence and flow velocity
$\zeta_{s.th}$	—	0,0214	Coefficient of the secondary losses before the throat area
ζ_{th}	—	0,0496	Losses coefficient at the throat area
ζ	—	0,1114	Total losses coefficient
φ	—	0,9426	The velocity coefficient
π	—	1,5343	Pressure ratio p^*_O/p_X
π^*	—	1,046	Pressure ratio p^*_O/p^*_X
Rotor blade			
ζ_c	—	0	Cooling losses coefficient
ζ_{fr}	—	0,042	Friction losses coefficient
$\zeta_{pr.0}$	—	0,059	Profile losses coefficient (basic)
$\zeta_{pr.i}$	—	0,0603	Profile losses coefficient taked into account the angle of incidence
$\zeta_{pr.i.W}$	—	0,063	Profile losses coefficient taked into account the angle of incidence and flow velocity
ζ_{tr}	—	0,017	Trailing edge losses coefficient
ζ_s	—	0,0315	Secondary losses coefficient

$\zeta_{pr.i.th}$	—	0,0433	Coefficient of the profile losses before the throat area taked into account the angle of incidence
$\zeta_{pr.i.W.th}$	—	0,0452	Coefficient of the profile losses before the throat area taked into account the angle of incidence and flow velocity
$\zeta_{s.th}$	—	0,0226	Coefficient of the secondary losses before the throat area
ζ_{th}	—	0,0678	Losses coefficient at the throat area
ζ	—	0,0945	Total loss coefficient

Basing on presented results it can be concluded that model provides the adequate solution and stable convergence in the numerical solution process. The obtained results deviation was within 3% when comparing with the results of CFD modeling.

4. Conclusions

The blade row model of the one-dimensional estimation along the mean line was developed for more detailed analysis and preliminary optimization of an axial turbine. The blade row model allows to make up a single axial turbine stage as well as a multistage turbine. The turbine stage model can take into account the cooling air flow before and after a throat of each blade row, outlet straightener vanes existence and stagger angle controlling of nozzle vanes.

In the presented method stations are normal to the mean line. It simplifies the computation algorithm because the radial velocity component may be excluded from consideration. The main advantage is capability for both axial and radial turbines modeling.

It can be concluded that proposed model provides the adequate solution and stable convergence in the numerical solution process. The detailed comparing of the 1D and 3D CFD modelling results will be possible after models integration and creation of an identification unit.

Acknowledgments

This work was supported by the Ministry of education and science of the Russian Federation in the framework of the implementation of the Program of increasing the competitiveness of Samara University among the world's leading scientific and educational centers for 2013-2020 years.

References

[1] Popov G M, Baturin O V, Kolmakova D A, Krivcov A V Improvement results of TK-32 turbocompressor turbine with gas-dynamics and strength CAE-systems 2014 *International Journal of Engineering and Technology* **6**(5) 2297-2303

[2] Matveev V N, Popov G M, Baturin O V, Goryachkin E S, Kolmakova D A Workflow optimization of multistage axial turbine 2015 *51st AIAA/SAE/ASEE Joint Propulsion Conference* 132015

[3] Ainley D G, Mathieson G C R 1951 *A method of performance estimation for axial-flow turbines* (London: Her Majesty's Stationery Office) p 31

[4] Schobeiri T One-dimensional methods for accurate prediction of off-design performance behavior of axial turbines 1992 *ASME* **1**

[5] Venediktov V D 1990 *Gas dynamics of a cooled turbines* (Moscow: Machine Industry Press) p 240

[6] Wei N 2000 *Significance of loss models in aerothermodynamic simulation for axial turbines* (Doctoral thesis) (Stockholm: Royal Institute of Technology) p 164

[7] Dahlquist A D 2008 *Investigation of losses prediction methods in 1D for axial gas turbines* (Master thesis) (Sweden: Lund University) p 192

[8] Kuz'michev V S, Ostapyuk Y A, Tkachenko A Y, Krupenich I N, Filinov E P Comparative analysis of the computer-aided systems of gas turbine engine designing 2017 *IJMERR* **6** (1) 28-35

Modeling Adsorption–Desorption Processes at the Intermolecular Interactions Level

Vera V Varfolomeeva and Alexey V Terentev

Department of Environmental Health and Safety, Institute of Engine and Power Plant Engineering, Samara National Research University, 34, Moskovskoye shosse, Samara, 443086, Russia

E-mail: varf2@ssau.ru

Abstract. Modeling of the surface adsorption and desorption processes, as well as the diffusion, are of considerable interest for the physical phenomenon under study in ground tests conditions. When imitating physical processes and phenomena, it is important to choose the correct parameters to describe the adsorption of gases and the formation of films on the structural materials surface. In the present research the adsorption-desorption processes on the gas-solid interface are modeled with allowance for diffusion. Approaches are proposed to describe the adsorbate distribution on the solid body surface at the intermolecular interactions level. The potentials of the intermolecular interaction of water–water, water–methane and methane–methane were used to adequately modeling the real physical and chemical processes. The energies calculated by the B3LYP/aug-cc-pVDZ method. Computational algorithms for determining the average molecule area in a dense monolayer, are considered here. Differences in modeling approaches are also given: that of the proposed in this work and the previously approved probabilistic cellular automaton (PCA) method. It has been shown that the main difference is due to certain limitations of the PCA method. The importance of accounting the intermolecular interactions via hydrogen bonding has been indicated. Further development of the adsorption-desorption processes modeling will allow to find the conditions for of surface processes regulation by means of quantity adsorbed molecules control. The proposed approach to representing the molecular system significantly shortens the calculation time in comparison with the use of atom-atom potentials. In the future, this will allow to modeling the multilayer adsorption at a reasonable computational cost.

1. Introduction

The most important role in ensuring the long-term and trouble-free functioning of space vehicles belongs to the resistance of construction material resistance to the influence of external factors. In conditions of space flight, the apparatus is affected by the environment, including the cosmic vacuum [1]. Space stations have a special equipment for collect and analyze pollutants. Contamination plume, which are the products of incomplete combustion, are collected by plate for a further analysis. The natural process that prevents the slick contamination formation on the surface of spacecraft is the particles desorption from the surface [2]. Therefore, it is important to study the contaminants composition and to select materials with the minimal sorption properties as regards to the products of incomplete combustion.

It is also known that under terranean conditions and depending on the type of physical phenomenon under study, it is possible to model various processes. In this respect, for the laboratory modeling of the mass losses, it is necessary to consider such processes as adsorption of gases and chemical interaction of the residual gas with surface materials. Modeling position of the residual gas molecules (molecules, atoms, ions, excited particles) is necessary in view its significant influence on the speed of the physicochemical reactions on the surface [3]. Laboratory modeling allows the engineer (researcher) to experiment in the cases when it is practically impossible or impractical to do this with a real object.

It should be noted that the solution of each practical problem depends on numerous input variables and parameters. When selecting the appropriate parameters, it becomes possible to adequate modeling of the actual physicochemical processes. Structural and energetic parameters of the molecules in the gas phase and adsorbed state, the physicochemical characteristics of solid, as well as the area per a molecule in a dense monolayer, and others can serve as such.

2. Theory and calculations

When modeling the adsorption-desorption processes, it is necessary to determine the number of molecules adsorbed on the surface per a time unit. According to the molecular-kinetic theory of gases [4], the number of collisions of gas particles with the surface per a time unit ν is described by the formula:

$$\nu = N_A P/\sqrt{2\pi MRT}, \tag{1}$$

where N_A – Avogadro's constant, P – gas partial pressure; M – molecular mass; R – universal gas constant; T – gas temperature.

The probability of the desorption process realization for each molecule W_{des} is described by the formula [5]:

$$W_{des} = \frac{t}{t_0}\exp\left(-\frac{E_{des}}{RT_{surf}}\right), \tag{2}$$

where t – time step size (adsorption lifetime); t_0 – the order of a lattice vibration period; E_{des} – desorption activation energy; T_{surf} – temperature at the surface.

The time step size and the order of a lattice vibration period are defined by the equation:

$$t = t_0 \exp\left(\frac{E_{des}^{min}}{RT_{surf}}\right), \tag{3}$$

$$t_0 = \frac{h}{kT_{surf}}, \tag{4}$$

where E_{des}^{min} – minimal desorption activation energy, h – Planck's constant, k – Boltzmann's constant.

Two series of experiments were carried out to determe the area size per one molecule of methane. They differed in the substance amount (0.1 mol and 1.0 mol, respectively), the adsorbent section size (17×17 Å^2 and 25×25 Å^2), the temperature of the gas and the surface (100 K and 300 K). In the each series, the calculations differed in the number and magnitude of optimization steps.

The period of oscillations and the adsorption-desorption periods was calculated automatically by the formulas (3) and (4), respectively: $4.8\cdot10^{-13}$ and $5.32\cdot10^{-12}$ sec at T = 100 K; $1.6\cdot10^{-13}$ sec, $3.57\cdot10^{-13}$ sec at T = 300 K. Modeling of the physical adsorption–desorption processes was terminated after the reached equilibrium. Graphite was chosen as the adsorbent, having the uniform surface. The heat of adsorption of methane on graphite was taken equal to 11.8 kJ/mol [6].

Modeling the adsorption-desorption process of water and methane two-component gas mixture (with the same molar ratio of components in the gas phase) was carried out using the intermolecular interaction potentials (figure 1). The interaction energies of water–water, water–methane and methane–methane molecules was calculated by the B3LYP/aug-cc-pVDZ method. The B3LYP method is well suits for hydrogen bonded systems [7–11].

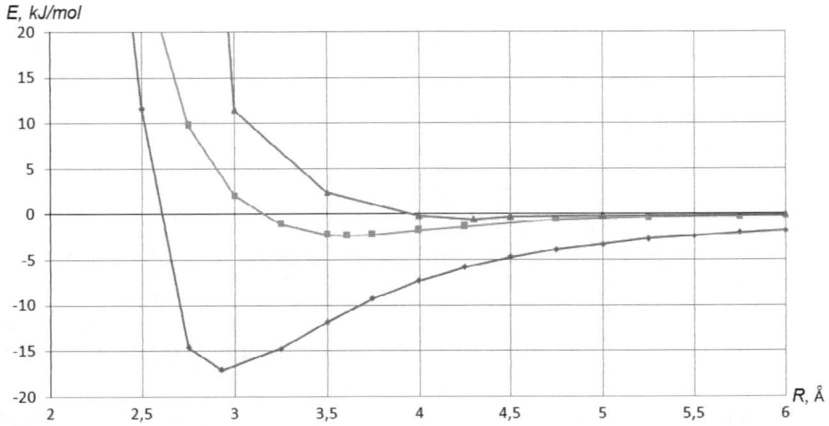

Figure 1. Energy dependence of the intermolecular interaction of water–water (blue), water–methane (green) and methane–methane (red) on distance.

The distance change between the O...O, O...C and C...C atoms was carried out in 0.25 Å step size for pairs of water–water molecules, water–methane and methane–methane molecules, respectively. At each step the system was optimized. Modeling the adsorption-desorption process was carried out using the energy values obtained. In this case, the potential energy function was created automatically by the system based on the linear interpolation between the values obtained by the B3LYP/aug-cc-pVDZ method. In our opinion, such approach is more effective than the atom-atomic potentials use, since all types of intermolecular interactions are accounted in one function. This makes it possible to significantly reduce the computation time at simultaneous maintenance of computational accuracy for small molecules. Such approach is not to be used to calculate the systems with extended and flexible molecules since possible intramolecular transitions are not taken into account here. However, the errors connected with this assumption should not be significant for small molecules, such as the components and products of the rocket propellants decomposition.

3. Results and discussion

3.1 Area per one molecule calculation

To verify the adequacy of physical adsorption used on the "gas-solid" surface computational model, the average area per molecule of methane in a dense monolayer was calculated. The calculation was carried out after reaching the equilibrium state. The results of two series of experiments are presented in table 1.

Table 1. The calculated values of the area per one molecule of methane.

N	Optimization step size, Å	Number of steps	Average area per molecule, $Å^2$	
			Series 1	Series 2
1	0.20	10	16.1	17.9
2	0.20	10	15.2	18.9
3	0.20	10	16.1	17.9
4	0.10	20	15.2	18.1
5	0.10	20	17.0	18.4
6	0.10	20	16.1	17.6
7	0.05	40	16.1	18.2
8	0.05	40	18.1	17.3
9	0.05	40	17.0	18.2

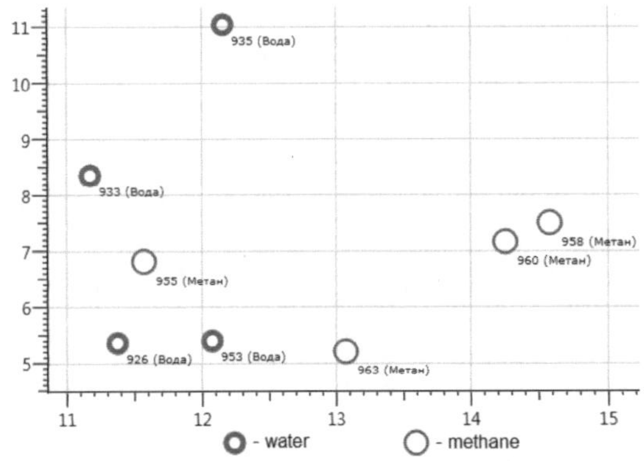

Figure 2. Graphite surface fragment at the initial moment of methane and water adsorption (axes are distances, Å).

According to the results of the computational series 1 and 2, the average area size per one molecule of methane is 17.2 Å^2. The results obtained well agree to the experimental data which fall into the range 16–18 Å^2 [12, 13].Good accordance with the experimental results (table 1) proves the adequacy of the model used. It is worth mentioning that the key factor for adequate modeling of the physical adsorption process is the use of correct structural and energetic characteristics describing the adsorbate-adsorbate and adsorbate-adsorbent interaction.

Earlier, we determined the area per molecule, using the probabilistic cellular automaton (PCA) method [14]. The use of this tool also made it possible to achieve a satisfactory agreement of the calculated values with the experimental data. In this case, only the adsorbate molecules structural characteristics and the interaction energy of the adsorbate with the adsorbent were taken into account. However, the PCA method has some limitations. So, if the sizes of molecules in a mixture differ widely, for example, benzene (S ≈ 50 Å^2 [6]) and oxygen (S ≈ 16 Å^2 [15]), then the used physical model can not be justified. The reason is the impossibility to account the differences in the molecules areas when specifying the dimensions of the PCA cell. As well as, today there is no possibility to take into account the changes of the area per one molecule as a result of the hydrogen bonds formation. Therefore, the PCA method usage at of modeling the adsorption-desorption processes for multicomponent mixtures is obstructed [14].

In our opinion, the proposed in this paper variant of the method to modeling adsorption-desorption processes using the potentials describing the intermolecular energy as a function of distance is quite promising.

3.2 Process modeling of physical adsorption of water and methane molecules on the graphite surface

To adequately describe the process of physical adsorption of a two-component gas on the graphite surface the intermolecular interaction energy potentials were used.

At the initial moment of gas adsorption on the graphite surface, the methane and water concentrations are approximately the same (figure 2, 3). This is explained by the fact that the initial

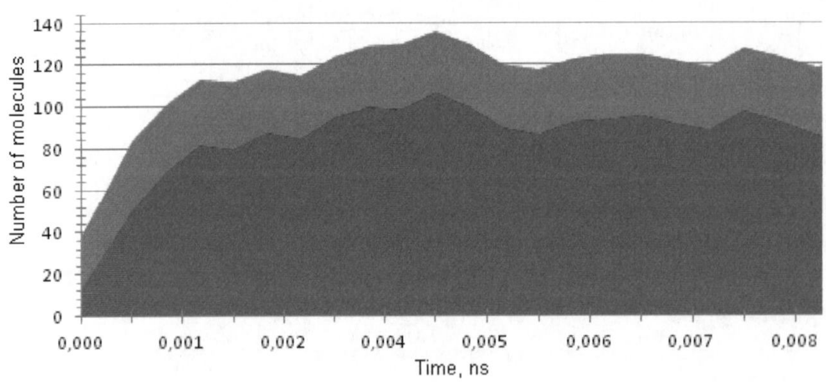

Figure 3. Number of adsorbed molecules of water (blue) and methane (red) dependence on time.

concentrations of the water and methane molecules in the gas phase are equal, and the molecular masses of the water (18 u) and methane (16 u) practically do not differ. Therefore, according to formula (1), the probabilities of adsorption of these molecules at the same temperature are close to each other.

Figure 2 shows that some molecules of water and methane find themselves at a distance of less than 1 Å from each other, that is, in the area of exchange repulsive short-range. Similar particle collisions can occur in the real adsorption process. After reaching equilibrium, they should not significantly affect the values of the system averaged macroscopic characteristics. However, in the initial phase, they can affect the probability of cluster formation. It should be noted such effects consideration cannot be fully taken into account when modeling by the cellular automaton.

As can be seen from figure 3, the proportion of water molecules in the adsorption layer increases with time. On the one hand, this may be due to the fact that water molecules can form hydrogen bonds with the graphite surface [16, 17]. However, the distances between most water molecules are about 2 Å (figure 4). This value corresponds to the hydrogen bond distance between the water molecules 1.9 Å [18]. Thus, due to a greater adsorbate-adsorbate interaction energy as compared to the adsorbate-adsorbent energy ($E_{HB}(OH\cdots\pi)$ = 7.5 kJ/mol [19] vs. $E_{HB}(OH\cdots O)$ = 13–30 kJ/mol [18, 20–22]), a hydrogen-bonded network of water molecules is formed and, as a consequence, there takes place a formation of a film on the surface. This well agrees with the already known experimental data [23]: the adsorption energy values of isolated water molecules are much less than the experimental values; in the monolayer on the graphite surface there exists a flat pack of associated molecules at the experiment temperature 293 K.

The proportion of the methane molecules does not change significantly over time. As seen from figure 4, methane molecules are in the chaotical position between water molecules. This is due to the fact that the methane molecules is much less inclined to the displays of lateral interactions.

It should be noted that in the model under consideration, the potentials describing the intermolecular interaction are used. They do not take into account the geometry of the adsorbate molecules. This can cause significant errors in modeling various systems, for example biological ones, where it is extremely important to consider the intramolecular hydrogen bond [24]. But most of the rocket propellants components and products of incomplete combustion have quite a compact molecular structure. Therefore, neglecting the geometry of the adsorbate molecules should not lead to a loss of physicality. The proposed variant of the molecular system representation significantly shortens the computation time in comparison with the use of atom-atomic potentials. In future, this will allow to modeling the multilayer adsorption at a reasonable computational cost.

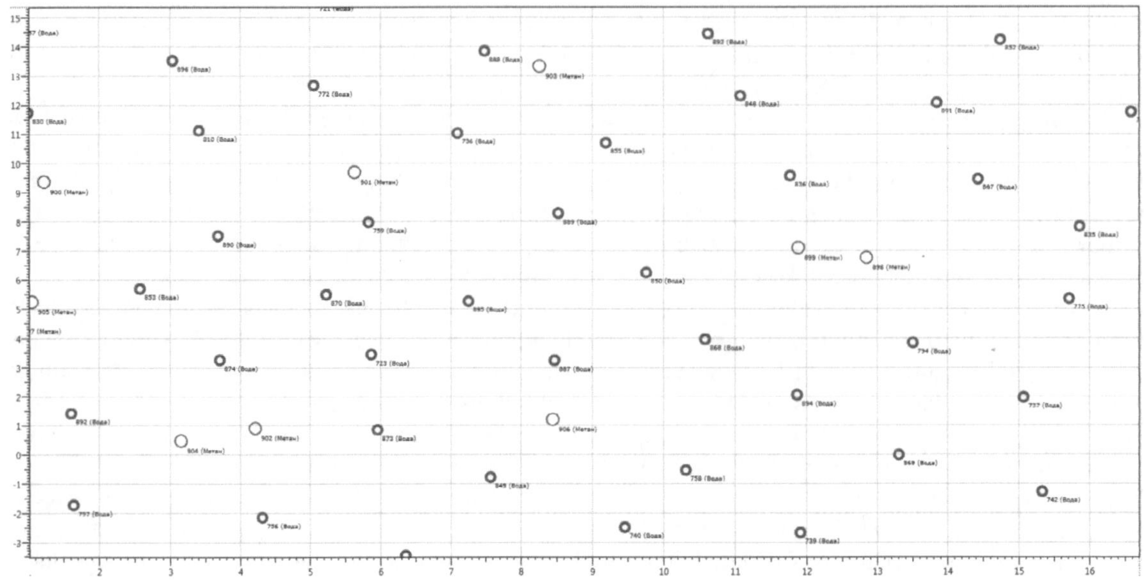

Figure 4. Fragment of the graphite surface after reaching equilibrium (axes are distances, Å).

4. Consolutions

We carried out the modeling of adsorption-desorption processes on the graphite surface. To account the diffusion the water–water, water–methane and methane–methane intermolecular interaction potentials, obtained by the B3LYP/aug-cc-pVDZ method, were used. The gained results of the area size per molecule and the distribution of water and methane molecules on the surface of a solid body well match with the known data. In the future, basing on the proposed variant of the modeling of adsorption-desorption processes on the intermolecular interactions level taking into account diffusion, there can be created the desorption process models of the multicomponent mixture, complicated by companion processes (for example, gas-surface chemisorption).

References

[1] Matveev S G, Orlov M Yu, Chechet I V and Semenov A V 2009 *Vestnik SSAU* **3** 210–4
[2] Akishin A I and Novikov L S 1983 *The impact of the environment on spacecraft materials*. – (Moscow: Znanie) p 64
[3] Semkin N D, Telegin A M and Kalaev M P 2013 *Space and its impact on the elements of spacecraft designs* (Samara: Samara State Aerospace University) p 46
[4] Oura K, Lifshits V G, Saranin A A, Zotov A V and Katayama M 2003 *Surface Science: An Introduction* (Springer) p 443
[5] Agafonov A N, Varfolomeeva V V, Eremina I N and Sanoyan A G 2007 *Vestnik SamGU* **9** 55
[6] Avgul' N N, Kiselev A V and Poshkus D P 1975 *Gas and Vapor Adsorptin on Homogeneous Surfaces* (Moscow: Khimiya) p 384
[7] Oliveira B G and Vasconcellos M L A A 2009 *Struct. Chem.* **20** 897-902
[8] Ünal A and Okur M 2017 *AIP Conf. Proc.* **1815** 100006-1-4
[9] Srivastava A, Mishra R, Kumar S, Dev K, Tandon P and Maurya R 2015 *J. Mol. Struct.* **1084** 55–73
[10] Nazarparvar E, Zahedi M and Klein E 2012 *J. Org. Chem.* **77** 10093
[11] Varfolomeeva V V and Terentev A V 2014 *Prot. Met. Phys. Chem. Surf.* **50** 715-22
[12] Bezus A G, Dreving V P and Kiselev A V 1964 *Russ. J. Phys. Chem.* **38** 59–67
[13] Hoory S E and Prausnitz J M 1967 *Trans. Faraday Soc* **63** 455–460

[14] Varfolomeeva V V and Terentev A V 2013 *Sorbtsionnye i Khromatograficheskie Protsessy* **13** 238–44
[15] Gregg S J and Sing K S W 1982 *Adsorption, surface area and porosity. 2nd ed.* (London: Academic press) p 303
[16] Varfolomeeva V V and Terentev A V 2011 *Carbon Black: Production, Properties and Uses* (Nova Science Publishers) pp 93–114
[17] Varfolomeeva V V and Terentev A V 2015 *Phys. Chem. Chem. Phys.* **17** 24282–93
[18] Pihko P M 2009 *Hydrogen Bonding in Organic Synthesis* (Wiley-VCH) p 397
[19] Suzuki S, Green P G, Bumgarner R E, Dasgupta S, Goddard W A and Blake G A 1992 *Science* **257** 942–5
[20] Scheiner S 1997 *Hydrogen Bonding: A Theoretical Perspective* (Oxford University Press) p 396
[21] Marechal Y 2007 *The Hydrogen Bond and the Water Molecule: The Physics and Chemistry of Water, Aqueous and Bio Media* (Elsevier) p 332
[22] Jeffrey G A and Saenger W 1991 *Hydrogen Bonding in Biological Structures* (Springer-Verlag) p 569
[23] Avgul' N N, Kiselev A V and Lygina I A 1961 *Izvestiya Akademii Nauk, Otdelenie Khimicheskikh* **8** 1404–11
[24] Varfolomeeva V V and Terentev A V 2017 *J. Struct. Chem.* **3** in press

IOP Conf. Series: Materials Science and Engineering **302** (2018) 012040 doi:10.1088/1757-899X/302/1/012040

Fiber-optic Sensor of Tactile Force for Anthropomorphic Robot Grips

SA Matyunin* and O G Babaev

Samara National Research University, Moskovskoe shosse, 34, Samara, 443086, Russia

E-mail: *s.a.matyunin@yandex.ru

Abstract. This paper examines an operating principle of a Fiber-Optic Tactile Force Sensor (FOTFS), describes a linearization method and temperature compensation on transfer characteristics (TC) of the FOTFS. The high linearity of the FOTFS TC (the nonlinearity of the TC does not exceed 0.01% within the range of tactile forces from 0 to 1.4 N) and the high temperature stability of the FOTFS (the temperature coefficient does not exceed 0.0025%/°C within the temperature range from 0 to + 50 °C) are experimentally-confirmed.

1. Introduction

At the present time, intensive research is being carried out to create intelligent anthropomorphic robots for various applications: for use in everyday life, at work, in space, for example, for repairing space vehicles. To perform a variety of works anthropomorphic robots require a wide variety of "sense organs", including "force control organs" of robot grips [1].

When developing robotic systems, contactless Fiber-Optic Tactile Force Sensors (FOTFS) are of special interest, as they are basically "non-electrical" with a closed optical channel (nonfouling in the area of optical channel). The FOTFS can be implemented on the basis of various effects: on the magneto-optical effect, on the effect of bending losses modulation in an optical fiber, on microbendings of a Bragg grating, on the basis of an external Fabry-Perrot interferometer, etc. [2-10].

In [2] stretch sensors for soft robots with elastomeric actuators are described. These robots designed for the sphere of human-robot interaction (HRI) such as direct contact with the heart for ventricular assist devices and wearable robots for exocolettes with increased strength. However, sensors of this type can not be used in space, where the temperature drops to −150 degrees. In [3] FOTFS for an industrial robot with a two-fingered arm are described. These sensors are capable of operating over a wide temperature range. The main problem of such FOTFS is the dependence of the operating characteristics of their elements (mainly semiconductor lasers) on external influences, primarily on temperature, and also the nonlinearity of the transfer characteristic (TC). An interesting solution is proposed in [4] - the use of stretchable flexible fibers to determine the deformation in the prosthetic hand. In [5, 6] fiber-optic sensors of tactile force on Bragg gratings, which can be used for anthropomorphic robot grips, are described. Advantages of this solution: high accuracy of measurement, small overall dimensions. Disadvantages: a complex scheme of optical signal processing, low speed of a multichannel optical signal processing system, expensive equipment (high-resolution spectrometer). As well as stretchable sensors, they can not be applied in space. To prospective developments can be attributed sensors on the effect of bend losses modulation in an optical fiber, described in [7, 8]. Advantages of such tactile sensors: a closed optical channel, small mass-dimensional parameters; low cost; wide operating temperature range. Disadvantages: the ambiguity of the measurement results - the sensors do not allow you to determine the location of the tactile load, but only the integral load along the entire length of the optical fiber. In [9, 10] many other options for implementing FOTFS are considered.

The main problem of such FOTFS is the characteristics dependence of their elements (mostly semiconductor lasers) on external effects, primarily on the temperature, as well as the nonlinearity of the transfer characteristics (TC) of the FOTFS caused by mode content variation of the emission for the sensors based on effect of the bending losses modulation, etc. [9, 10].

As noted above, according to minimization of dimensional-and-weight characteristics of the FOTFS, promising FOTFS are those, which are based on the effect of bending losses modulation in the optical fiber - the sensitive fiber diameter of such FOTFS is in the range of 0.125 mm without a protective coating and 0.8-0.9 mm with a protective coating.

However, the nonlinearity of TC and its dependence on temperature are typical for many FOTFS. Thus, there is a need to develop a linearization method and compensation of the temperature effect for the FOTFS TC.

2. Design and function of the FOTFS

This paper considers the linearization method of TC and compensation of the temperature effect on the FOTFS TC using the example of the FOTFS based on the modulation effect of bending losses in the optical fiber (this method can be applied practically unchanged also for fiber-optic sensors based on other principles of operation).

The schematic structure of the FOTFS based on the effect of bending losses modulation in the optical fiber is shown in Figure 1. The FOTFS consists of an intertangled optical fibers (OF) matrix forming a sensing element (SE), an optical emission source (ES) combined in a fiber-optic link (FOL), a reversal mirror (RM), a CCD matrix (CCDM) receiving emission and a microcontroller (MC). The temperature in the area of force exertion to the SE is measured by a fiber-optic temperature sensor TS, which can be used as one of the SE fibers, unaffected by tactile force.

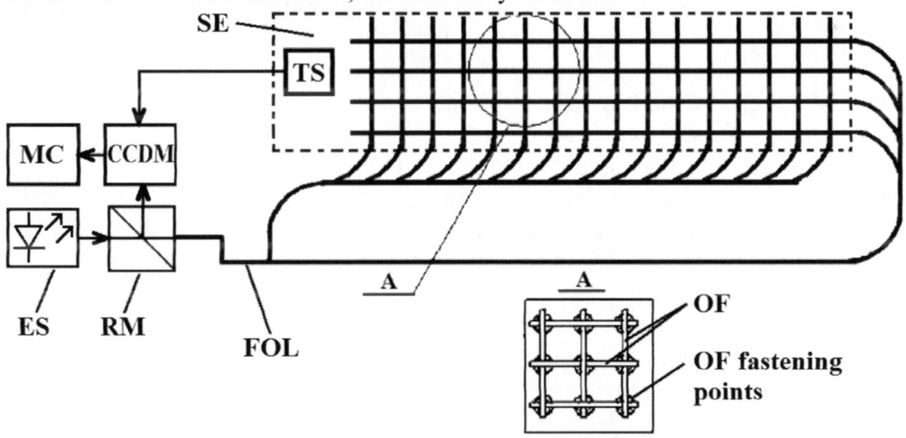

Figure 1. Schematic structure of the FOTFS.

The operation principle of the FOTFS is based on the use of bending losses modulation in the optical fiber when it deforms within the elastic limit [9, 10]. The ES emission (semiconductor laser) passed through the RM gets into the FOL (in the form of a tow) and then into the SE. The SE is a pieces of optical fiber settled as a rectangular grid. The optical signals reflected from the ends of optical fibers get back into the FOL and being reflected from the RM, go to the CCD matrix and then to the MC for further information processing.

Figure 2 shows the experimental TC of one SE channel of the FOTFS (curve 1 - experimental values, curve 2 - polynomial approximation). In figure 2, a sharp nonlinearity of the TC can be observed in the range of tactile forces q from 0 to 1.2 N (displacement x of optical fiber SE from 0 to 0.6 mm). It should be noted that the temperature coefficient of the output optical power change $P(q)$ (without stabilizing feedback) reaches a value of minus 1.2%/°C.

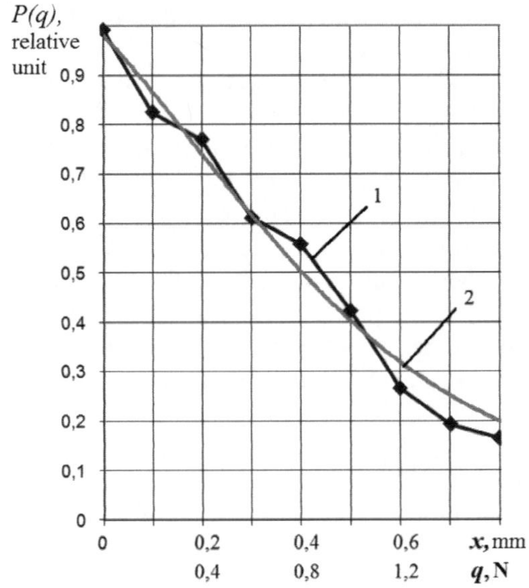

Figure 2. Experimental TC of one SE channel of the FOTFS.

3. Method of linearization and compensation of the temperature effect on the TC

The method consists of two stages:

Stage 1 – identifying of the FOTFS calibration function;

Stage 2 – temperature compensation and linearization of the TC.

The identifying of the calibration function involves the determining of the correspondence between the arguments (tactile force and temperature of the FOTFS SE) and the FOTFS output signal, processing the measurement results and obtained dependence recording to the memory of the microcontroller (MC). The method under consideration allows to reduce significantly the number of the FOTFS calibration function points, which is especially important in the temperature dependence calibration, as it allows to use points deviating in a random way from the nodal points with uniform spacing. Subsequent points interpolation to the nearest nodal ones and piecewise-surface approximation of the calibration function allows to reduce used memory size of the MC for calibration function storing and the time required to process the measurement results.

Stage 1. Identifying of the calibration function

The output signal value of one CCDM channel denoted via $U_K(q, t)$, where q is the value of the tactile pressure; T is the temperature, K is the number of the CCDM channel.

When the sensor is calibrated, the TC of each SE channel of the sensor is recorded into the MC memory for various values of the tactile force q and temperatures t. It should be noted that the calibration is performed for a relatively small number of points in temperature and tactile force (16 to 64 points for each variable q and t).

The grid nodes of the calibration matrix are practical to set with a constant spacing. It is obvious that the experimentally obtained values of the variables q and t cannot get exactly to the fixed grid nodes. Therefore, during the calibration procedure, a piecewise-surface (linear or nonlinear of a fixed order) interpolation (extrapolation) is performed to the nearest nodal points (Figure 3) of the calibration function, which are stored in the memory of the MC.

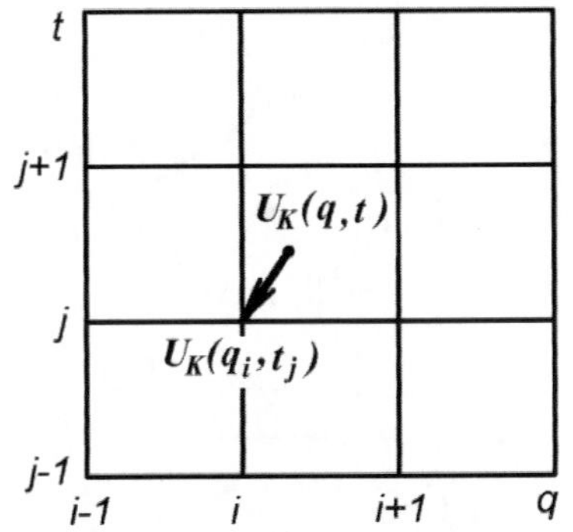

Figure 3. TC calibration field.

Stage 2. Temperature compensation and linearization of the TC

At stage 2, both the linearization of the FOTFS TC and the temperature correction of the measurement results are performed. For this purpose, it is necessary:

1. to measure the signal values of each CCDM channel - $U_K(q, t)$;
2. to measure the grip temperature sensor signal $U^T(t)$;
3. to determine the deviation of temperature from the nearest nodal point in temperature $U^T(t_j)$:

$$U^T(t) = U^T(t_j) + \frac{dU^T(t)}{dt}(t - t_j) \approx U^T(t_j) + \frac{U^T(t_{j+1}) - U^T(t_j)}{t_{j+1} - t_j}(t - t_j) \rightarrow$$

$$\rightarrow \Delta t = (t - t_j) = \frac{U^T(t) - U^T(t_j)}{U^T(t_{j+1}) - U^T(t_j)}(t_{j+1} - t_j). \quad (1)$$

4. Using, for example, linear interpolation, there are performed the temperature compensation and the linearization of the TC $U_K(q, t)$. For this purpose, the measurement signal for channel K is set as a deviation from the nearest node points (Figure 4):

$$U_K(q,t) = U_K(q_i + \Delta q, t_j + \Delta t) = U_K(q_i, t_j) + \frac{\partial U_K(q_i, t_j)}{\partial t_j}\Delta t + \frac{\partial U_K(q_i, t_j)}{\partial q_i}\Delta q \approx$$

$$\approx U_K(q_i, t_j) + \frac{U_K(q_i, t_{j+1}) - U_K(q_i, t_j)}{t_{j+1} - t_j}\Delta t + \frac{U_K(q_{i+1}, t_j) - U_K(q_i, t_j)}{q_{j+1} - q_j}\Delta q. \quad (2)$$

It follows that:

$$q = q_i + \Delta q =$$

$$= q_i + \frac{U_K(q,t) - U_K(q_i, t_j)}{U_K(q_{i+1}, t_j) - U_K(q_i, t_j)}(q_{i+1} - q_i) - \frac{U_K(q_i, t_{j+1}) - U_K(q_i, t_j)}{(U_K(q_{i+1}, t_j) - U_K(q_i, t_j))}\frac{\Delta t}{(t_{j+1} - t_j)}(q_{i+1} - q_i), \quad (3)$$

where Δt is determined from (1).

It is obvious that equation (3) is obtained from equation (2) by calculating the function of the inverse function $U_K(q,t)$, which ensures a linear dependence of the output parameter q on the measured value of the tactile force.

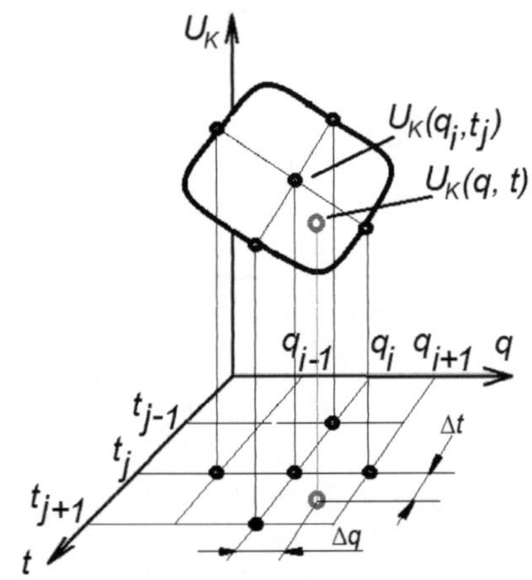

Figure 4. Measurement and linearization of the TC.

5. After that, the obtained values of the argument q are scaled to the appropriate range, (for example, to the range [0 ... 4095]). As a result, the microcontroller digital code q^* corresponding to the tactile force is calculated:

$$q^* = 4095 \frac{q - q_{\min}}{q_{\min} + q_{\max}}, \tag{4}$$

where $q_{\min}$ and q_{max} are the minimum and maximum values of the tactile force respectively.

4. Experimental research

Experimental research was carried out using the FOTFS samples, developed in the laboratory of NIL-53 of the Samara National Research University. The experimental research results are shown in figure 5. Figure 5 shows the FOTFS TC (curve 1) fed to the input of the electronic converter and the digital result code (curve 2).

Given approach (4) allows moving easily the operating range of the linearization of the TC to any part of the range.

The main characteristics of the experimental FOTFS samples are given in Table 1.

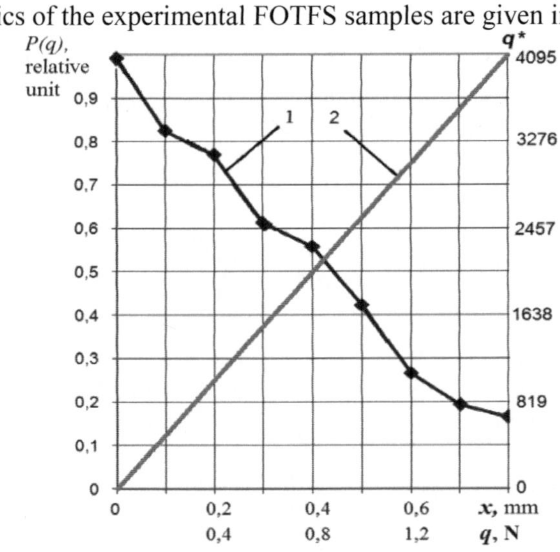

Figure 5. Experimental research: (1) experimental TC of the FOTFS; (2) TC of the FOTFS after scaled-up linearization.

Table 1. FOTFS main characteristics

Characteristics	Value
Dimensions of the sensing element, mm	10x15
Measured tactile force, N	from 0 to 1.4
Accuracy class	0.5
Complementary temperature error, %/°C	0.0025
Nonlinearity of TC, %	0.01
Operating temperature, °C	from 0 to +50
Length of fiber-optic cable, m	to 10
Fiber-optic cable	Ø0.9 SM
Optical connectors	FC/PC
Storage temperature, ° C	from − 40 to +85

5. Conclusion

1. Certain advantage of the developed method is the fact that scaling and linearization can be carried out within any measurement and temperature range. This allows to calibrate the FOTFS in a simple way using the control object and speeds up the calibration process, which is extremely important for mass production.
2. The high linearity of the TC is experimentally confirmed - the nonlinearity of the TC does not exceed 0.01% within the range of tactile forces from 0 to 1.4 N.
3. The high temperature stability of FOTFS is experimentally confirmed - the temperature coefficient does not exceed 0.0025%/°C within the temperature range from 0 to +50 °C.

Acknowledgements
The work is performed with financial support from the Ministry of Education and Science of the Russian Federation. Unique ID: Applied research and experimental work RFMEF157816X0209.

References

[1] Sareh P and Kovac M 2017 Mechanized creatures *Science* 355(6332) p 1379
[2] Li S, Zhao H and Shepherd R 2017 Flexible and stretchable sensors for fluidic elastomer actuated soft robots *MRS Bulletin* 42(2) pp 138-142
[3] Park Y-L, Ryu S C, Black R J, Chau K K, Moslehi B and Cutkosky M R Exoskeletal force-sensing end-effectors with embedded optical Fiber-Bragg-Grating sensors 2009 *IEEE Transactions on Robotics* 25(6) pp 1319-31
[4] Zhao H, O'Brien K, Li S and Shepherd R F 2016 Optoelectronically innervated soft prosthetic hand via stretchable optical waveguides *Science Robotics* eaai7529
[5] Shao L-Y and Albert J 2010 Compact fiber-optic vector inclinometer *OSA: Optics Letters* 35(7) pp 1034-6
[6] Fernandez A F, Berghmans F, Decréton M, Delchambre A 2001 Multicomponent force sensor based on multiplexed fibre Bragg grating strain sensors *Measurement Science and Technology* 12(7) pp 810-813
[7] Silva A S, Catarino A, Correia M V and Frazão O 2013 Design and characterization of a wearable macrobending fiber optic sensor for human joint angle determination *SPIE: Optical Engineering* 52(12) 126106
[8] Koyama Y, Nishiyama M and Watanabe K 2015 Multi-channel measurement for hetero-core optical fiber sensor by using CMOS camera *Proc. SPIE 9655 'Fifth Asia-Pacific Optical Sensors Conference'* (Jeju) 965525
[9] Grattan K T V and Sun Dr T 1999 Fiber optic sensor technology: an overview *Sensors and Actuators* 82 (2000) pp 40-61
[10] Matyunin S A 2016 Fiber-optical measuring system of angular movements of finger falangs of the anthropomorphous robots hand *Izvestiya SFedU. Engineering sciences* vol 1 pp 240-252

IOP Conf. Series: Materials Science and Engineering **302** (2018) 012041 doi:10.1088/1757-899X/302/1/012041

Burnishing Systems: a Short Survey of the State-of-the-art

I N Bobrovskij[1]

[1]Togliatti State University, Belorusskaya st. 14, Togliatti, Russian Federation

bobri@yandex.ru

Abstract. The modern technological solutions allowing to implement a new technology of surface plastic deformation are considered. The technological device allowing to implement the technology of hyper productive surface plastic deformation or wide burnishing (machining time is up to 2-3 revolutions of workpiece) is presented. The device provides the constant force of instruments regardless the beating, non-roundness and other surface shape defects; usable and easily controlled force adjustment; precise installation of instruments and holders toward the along the worpieces axis; automation of the supply and retraction of instruments. Also the device allowing to implement the technology of nanostructuring burnishing is presented. The design of the device allows to eliminate the effect of auto-oscillations.

1. Introduction

Surface plastic deformation (SPD) is a set of technologies for modifying the surface and near-surface layers of workpiece by plastic deformation. In the world literature, unlike the Russian Federation, this set of technologies is called "burnishing", by the name of one of the main technologies. The peak in the development of PPD occurred in the 1980s, and at present SPD technologies continue to develop. The main directions of development are concerned with the transition to advanced digital and intellectual production technologies for the transition to a new technological way.

In this paper, several technological solutions allowing to implement modern SPD technologies are considered. The choice of these technologies is specified by several factors: mandatory industrial approbation and use in batch production, as well as the presence of a key difference in the mechanics of the process. These characteristics make it possible to distinguish these technologies as the best available, which contributes to the priority of their choice in the production development. The implementation of the following technologies is considered: the technology of hyper productive SPD (wide burnishing) and the technology of nanostructuring burnishing.

2. Hyper productive surface plastic deformation processing or wide burnishing (WB)

The most prevalent technology in the industry, including aerospace industry, from among the SPD technologies is a ball burnishing [1], however in spite of high numerical values of microgeometry parameters it is difficult in application. The purpose of developing a wide burnishing was the creation of a technology with processing time up to 3-4 revolutions of workpiece in a mass production. This required increase of force during processing from 150-300N (standard burnishing) to 2500-3000 N.

The issues of thermal stress and compensation of the transverse direction of the burnishing force have a special importance for these technological parameters.

The problems of increased thermal stresses [2] and their modeling [3] were considered earlier. It was found that with an increase in speed and pressure on the treated surface, the average temperature rises up to 8-9 times, but even under extreme processing conditions of high-strength cast iron with globular graphite, the temperature did not exceed 400 °C. The revealed dependencies made it possible to establish technological limitations of the processing on two factors: to prevent negative structural-phase transformations, and to reduce the intensity of tool wear.

To eliminate the appearance of a force directed across the axis of the worpiece and causing its deformation, a new design of the device for burnishing was developed [4]. Providing a proportional distribution of the processing force makes it possible to create residual compressive stresses.

The constructions of devices allowing processing without deformation are known before and the instruments are coaxial. However, the presence of defects in the shape of the workpiece and the basing leads to an uneven compression of the instrument against the surface [5]. As a result, the quality of the treated surface deteriorates and the uneven wear of the processing instruments increases significantly during processing. Double-lever devices are known [6], as equipped with two instruments, which of them is loaded with a separate spring. This devices allow to compensate the beating and non-roundness of the workpiece, but also have a uneven loading of instruments and their uneven wear.

There are Double-lever devices with two instruments loaded by a single spring located between the opposite ends of the levers [7].

The devices do not allow quickly, conveniently and accurately adjust the force of tools pressing to the workpiece. The device is moved to the working position by moving the entire device across the workpieces axis, with a spring-loaded stop applied for the device centering, which contacts the workpiece during processing and, therefore, can degrade the surface roughness and creates additional friction forces. The accuracy of centering of such a device is small. The device does not allow processing the workpieces with a change in size along the axis, for example, acclivous cones and other shaped parts due to the pressing force of the instrument will not be constant.

None of the mentioned devices is possible to set the coincidence of the axes of the working part of the indenters (their location in one plane perpendicular to the axis of the worpiece). Meanwhile, it is necessary that the traces of the indenters represent a regular two-screw line with steady pitch, it will ensure an even distribution of roughness along the length of the product. The use of all these devices in automated machines is difficult. The known dependences for determining the state of a plastically deformed surface layer with taking into accounting the thermal effect of plastic deformation, present the need to provide this solution [8].

The developed device provides a constant pressure of the instruments, regardless of beating, non-roundness and other changes in the shape of the surface. It is also provides a convenient and easily controlled force adjustment, precise mutual installation of instruments and holders in the direction along the worpieces axis as well as automation of the supply and retraction of processes.

The goal is achieved by the fact that the device consist of a body and double-levers hinged on the body, as well as holders with instruments fixed to one end of the levers. The second ends of the levers are equipped with hinges, to which are respectively connected the rod and the body of the hydraulic cylinder installed between them and combining the functions of the feed and pressing of the instrument. The rotation axes of the levers are located at a distance equal to the diameter of the workpiece, symmetrically with respect to axis and the working surfaces of the instruments due to eliminate the component of the working force deforming the workpiece in a plane perpendicular to the line connecting the instruments.

For precise mutual installation of instruments and holders in the direction along the axis of the workpiece, the holders and the ends of the levers are provided with guides for rectilinear movement, the holders are movably mounted along the axis of the workpiece and each of them is provided with at least one adjusting screw and also a screw for securing it in an adjusted position. For precise adjusting of the holder the adjusting screws can be made micrometric. Figure 1 presents the side view of the proposed device.

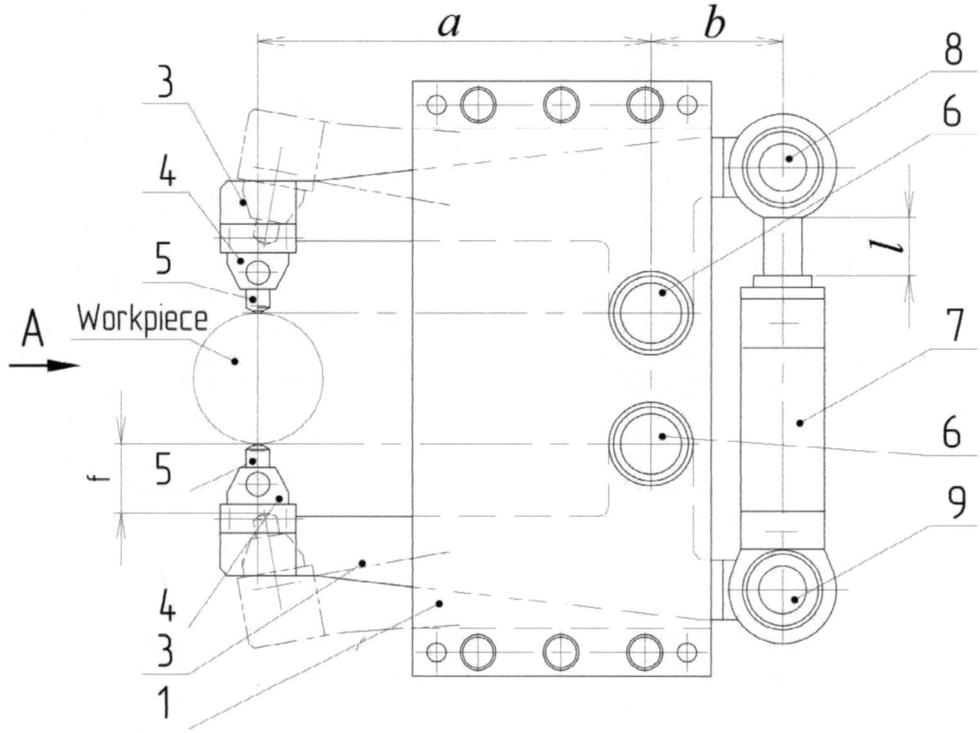

Figure 1. Design of the device.

The body of the device consists of two plates 1 and 2, connected together, for example, by screws (figures 1 and 2). Between the plates 1 and 2 two double-levers 3 (the upper and lower) are situated. The instrument holders 4 and instruments fixed in them, for example diamond burnishers - indenters 5 are situated at the one end of the levers. Each lever 6 is supported by a hinge in the body. The other ends of the levers are provided with hinges 8 and 9. The rod and the body of the hydraulic cylinder 7 are connected with these hinges respectively, which perform the functions of supply-retraction and pressing instruments to the workpiece. The hinge axes 6 are located at a distance equal to the diameter of the workpiece, symmetrically relative to its axis and instruments work surfaces.

To provide instrument retraction by a given amount f, the stroke of the hydraulic cylinder l is defined from the relation:

$$2\frac{b}{a}f < l \tag{1}$$

This feature is provided for two tasks:

Replacing the burnishing instrument;

free setting trajectory of workpiece.

At the end of the carrying the instrument lever 3, there is a rectilinear motion guide, for example, "dovetail" type or rectangular groove "B", as shown in Fig. 3. In this groove the shank of the instrument holder 4 is placed with the possibility of moving along the workpieces axis. The instrument (indenter) 5 is fixed in the holder with a screw 10. From the ends the groove B is closed by two slats 11 fixed to the lever 3 by screws 12. The adjustment screws 13 are screwed into the slats 11. It may be used only one adjusting screw, but then it is necessary to fix its head from the movement in the slat 11 and screw it into the hole made in the holder 4. For precise movement of the holder 4, the adjusting screw can be made micrometric. The instrument holder 4 is fixed to the lever 3 with screw 14.

The body of the device is mounted on a working unit designed to move along the axis of the workpiece. Then, the device is set as the common axis of the working surfaces of the instruments 5 in a retracted position crosses the axis of the workpiece. The body of the device is secured in this position. The screws 14 holding the holders 4 are loosen, and by manipulating the adjusting screws 13, the holders are moved along the axis of the workpiece, ensuring that the axes of the working surfaces of the instruments match. In the same way, it is possible to fine adjust the initial position of the processing instruments.

Then, the holders are again fixed using the screws 14. By adjusting the pressure in the part of the hydraulic system of the machine to which the hydraulic cylinder 7 is connected, set the required amount of pressing force of the instruments to the surface of the workpiece (the pressing forces determine the working burnishing force). After the starting the cycle, the product starts to rotate, hydraulic fluid is supplied to the hydraulic cylinder, and the levers are reduced by pressing the instruments on the work surface. Due to the placement of the hinge axes 6 at a distance equal to the diameter of the workpiece and symmetrical relatively its axis and the working surfaces of the instruments, the pressing forces are transferred to the workpiece strictly perpendicularly to the surface. Then the device is moved along the axis of the workpiece and processing is performed.

If the workpiece surface has beating (or this beating is due to inaccuracy in unit for fixing the workpiece), the instruments with levers move relative to the axis of rotation, while the piston also moves inside the hydraulic cylinder. This, however, does not lead to changes in the pressure of the instruments, since it are not determined by the position of the levers, but by pressure value in the working cavity of the hydraulic cylinder. Therefore, the proposed device allows processing not only the surfaces of straight circular cylinders, but also parts with significant non-roundness, and also cones, for example, simultaneously processing the neck of the shaft for sealing and facet before it. At the end of the processing, the machine control system (or the operator itself) gives the command to retract the levers and return the device to its initial position. The hydraulic cylinder 7 separates the levers 3, and the part can be removed.

Thus, the hydraulic cylinder 7 in the device performs two functions: the supply and retraction of the tool (idling), as well as the creation of a working force and keeping it constant, regardless of the beating and changings in the dimensions of the processed surface. It seems promising to use this type of device in titanium alloys processing.

The implementation of this device significantly improves the stability of the wide burnishing processing, increases the durability of instruments, improves the quality of the machined surface; It allows to process parts with significant beating and non-roundness of the surface, and also shaped surfaces and cones.

3. Nanostructuring burnishing

In this technology, an enlarged roller of the pushed metal is formed, due to indenters with increased coefficients of friction and heat resistance, which ensure a high level of shear deformations. This effect contributes to the process of dispersing the structure of the surface layer [9]. However, the growth of the roller can lead to destabilization of the dynamic system of burnishing and the appearance of self-oscillations. It should be noted that when implementing the technology of green (dry) burnishing performed at increased temperature loads, the author also encountered this effect, also called "Moire effect" [10]. In this case, finite element modeling allowed to identify patterns of changes in the stress-strain state and to reveal mechanisms of formation of a nanostructured layer [11, 12]. To eliminate the effect of self-oscillations, the following design of the burnishing tool was developed.

The indenter is designed to improve heat transfer and reduce the mass with a cone shaped blind hole with an end of a circular shape (to reduce the possibility of stress concentrators). The design of the indenter ensures that it performs the function of the plunger (pressing to the body). When the low-boiling liquid is poured into the case (for example, alcohol or acetone), the indenter is cooled down and it is possible the speed increase of burnishing (i.e. performance). The presence of the bush provides damping of self-oscillations arising when the burnishing speed is increased.

The burnisher consist of the following elements: body 1, indenter 2, set of disk springs 3, nut 4, rubber seal 5, low-boiling liquid 6, rubber bush 7. Body 1 is arranged vertically. The indenter 2 with a rubber bush 7 mounted on it is installed in the body 1. A set of disk springs 3 is installed in the body 1 and low-boiling liquid 6 is poured. A rubber seal 5 is fitted onto the nut 4.

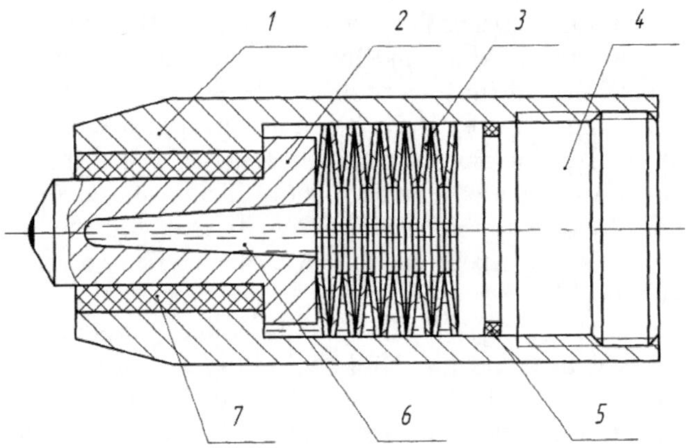

Figure 2. Burnisher design [13].

By screwing the nut 4 into the body 1, the set of disk springs 3 is pressed to the required burnishing force. The tare set of disk springs 3 is pressed by nut 4 to the required burnishing force. After the contact of the indenter 2 with workpiece the instrument is inserted into the workpiece until the indenter 2 abuts against the body 1. In this moment, the indenter 2 stops abutting the body 1 and begins to receive the force of the pre-pressed disk springs 3. When the burnisher moves, the indenter 2 is cooled down with low-boiling liquid 6, and the rubber bush 7 damps the self-oscillations of the indenter 2.

Thus, the considered device due to design features allows to implement the technology of nanostructuring burnishing in a wide range of operating modes. At the same time, increased friction provides the nanostructural burnishing cân be estimated in terms of contact phenomena [14, 15].

4.Conclusion

The development of burnishing technologies in direction of operational characteristics increasing of the processed parts must be accompanied by appropriate technological support. The development of new SPD technologies is associated with new distinctive features (for example, hyper productivity) and constraints. The presented technical solutions can be used not only in the implementation of the SPD technologies for which they are intended. For example, the authors' experience shows that the phenomena arising in nanostructuring burnishing also arise in green (dry) burnishing, and the general approach described in this publication can be applied to solving problems of minimizing the probability of manifestation of these phenomena. The presented device allows to implement constant pressure at wide burnishing regardless of beating and non-roundness. The device in the case of nanostructuring burnishing that allows to eliminate the effect of self-oscillations, which directly affects the parameters of the microgeometry of the processed parts is presented.

References

[1] Abul'khanov S R, Goryainov D S, Skuratov D L and Shvetsov A N 2015 Formation of the surface layer in diamond smoothing *Russian Engineering Research*. **35(2)** pp 147-9.

[2] Bobrovskij N M, Melnikov P A, Grigoriev S N and Bobrovskij I N 2015 Aspects of thermal field by wide burnishing *IOP Conference Series: Materials Science and Engineering*. **91(1)** 012035.
[3] Bobrovskij N M, Melnikov P A, Grigoriev S N and Bobrovskij I N 2015 Simulation of thermal fields using different types of wide burnishing *IOP Conference Series: Materials Science and Engineering*. **91(1)** 012034.
[4] Bobrovskij N M, Vilchik V A, Bokk V V, Maksimenko N N, Melnikov P A Device for surface plastic deformation processing. Patent, Russian Federation. MPC B24B 39/00 No. 2348504, asserted 21.01.2008; published 10.03.2009.
[5] Zakharov O V, Bobrovskij N M, Korolev A A, Bobrovskij I N, Kochetkov A V and Ivashchenko V A 2016 Optimal control method for the sphericity error using CMMs *2016 Dynamics of Systems, Mechanisms and Machines*, Dynamics art. 7819113.
[6] Golembievskij A I and Konovalov E G Device for rolling of outer spherical surfaces. Patent, USSR. MPC B 24 b 39/04. No. 324136, asserted 10.04.1970, published 23.12.1970.
[7] Azarevich G M, Gusyatskij I A and Savelieva L B Device for burnishing and hardening processing of outer cylindrical surfaces. Patent, USSR. MPC B 24 b 39/04. No. 445565, asserted 15.02.1972, published 28.11.1974.
[8] Khaimovich A I and Abulhanov S R 2014 Equations of state of plastically deformed polycrystalline medium considering plastic deformation thermal effect *ARPN Journal of Engineering and Applied Sciences*. **9(12)** pp 2867-75.
[9] Khaimovich A I and Balaykin A V 2014 Analysis of titanium alloys plastic properties under severe deformation conditions in machining *ARPN Journal of Engineering and Applied Sciences*. **9(10)** pp 1828-33.
[10] Grigoriev S N, Bobrovskij N M, Bobrovskij I N, Melnikov P A and Lukyanov A A 2017 Environmental Aspects Of The Green Surface Plastic Deformation Technology Of Car Parts *IOP Conference Series: Earth and Environmental Science*. **50** 012015.
[11] Kuznetsov V P, Smolin I Yu, Dmitriev A I, Konovalov D A, Makarov A V, Kiryakov A E and Yurovskikh A S 2013 Finite element simulation of nanostructuring burnishing *Phys Mesomech.* **16(1)** pp 62-72.
[12] Kuznetsov V P, Dmitriev A I, Anisimova G S and Semenova Yu V 2016 Optimization of nanostructuring burnishing technological parameters by Taguchi method *IOP Conference Series: Materials Science and Engineering*. **124(1)** 012022.
[13] Gorcog V G, Kuznecov V P and Gubanov V F Burnisher. Utility model, Russian Federation. MPC B 24 b 39/02. No. 62554, asserted 08.09.2006, published 27.04.2007.
[14] Grechnikov F, Khaimovich A and Alexandrov S 2016 Estimation of hot stamping lubricant efficiency under dynamic loading conditions *Journal of Materials Processing Technology*. **234** pp 300-8.
[15] Popova M A, Kuznetsov V P, Lesnikov V P, Popov N A and Konakova I P 2015 The structure and mechanical properties of single-crystal nickel alloys with Re and Ru after high-temperature holds *Materials Science and Engineering A*. **642** pp 304-8.

Optimization Parameters of Air-conditioning and Heat Insulation Systems of a Pressurized Cabins of Long-distance Airplanes

Sergey A. Gusev [1,2] and Vladimir N. Nikolaev[3]

[1] Institute of Computational Mathematics and Mathematical Geophysics Siberian Branch of Russian Academy of Sciences, pr. Academika Lavrenjeva, 6, Novosibirsk, 630090, Russia
[2] NSTU, pr. K. Marksa, 20, Novosibirsk, 630073, Russia
[3]S.A. Chaplygin Siberian Aeronautical Research Institute, Polzunov Street, 21, Novosibirsk, 630051, Russia

sag@osmf.sscc.ru; nikvla50@mail.ru

Abstract. The method for determination of an aircraft compartment thermal condition, based on a mathematical model of a compartment thermal condition was developed. Development of solution techniques for solving heat exchange direct and inverse problems and for determining confidence intervals of parametric identification estimations was carried out. The required performance of air-conditioning, ventilation systems and heat insulation depth of crew and passenger cabins were received.

1. Introduction

To find right solutions to various research problems in aircraft design and service, including the problem of effectiveness evaluation of air-conditioning, ventilation and heat insulation systems, it is required to determine thermal state of compartments.

A symbolic model of the system of a pressurized heat insulated compartment with air-conditioning systems and not pressurized and not insulated compartments can be represented by a system of one-dimensional equations of an insulating lining, windows and ordinary difference equations of convective heat transfer of the inner surface of the heat insulation of the lining and inner surface of the lining in not insulated compartments, seats, on-board equipment.

2. A symbolic model of the system of a pressurized

The equations of heat exchange of coverings we will present in the form of the one-dimensional equations of heat conductivity describing process of heat transfer in a multilayered structure:

$$C_{cv}(x)T_{cv,t} = (\lambda_{cv}(x,T_{cv})T_{cv,x})_x, \quad 0 < x < l; \tag{1}$$

$$\begin{aligned} \lambda_{cv}(x,T_{cv})\,F_{cv}\,T_{cv,x} = \alpha_{cv,out}(t)\,F_{cv}\,(T_e(t) - T_{cv}(t,x)) + Q_{cv,out} - \\ - c_0\,\varepsilon_{cv,out}\,F_{cv,out}\,T_{cv}^4(t), \qquad x = 0; \end{aligned} \tag{2}$$

$$\lambda_{cv}(x,T_{cv})F_{cv}T_{cv,x} = \alpha_{cv,in}(t)F_{cv}(T_{air}(t)-T_{cv}(t,x))+ \\ +\sum_j g_{j,cv}T_j^4/T_{ms}^4 - c_0\,\varepsilon_{cv,in}F_{cv,in}T_{cv}^4(t)+Q_{cv,in}; \qquad x=l; \tag{3}$$

$$T_{cv}(0,x)=T_0(x), \quad 0<x<l, \tag{4}$$

where $C_{cv}(x)=C_i$, $\lambda_{cv}(x,T_{cv})=\lambda_{i,0}+\lambda_{i,1}T_{cv}$ at $l_{i-1}\le x<l_i$, $(i=1,\ldots,k-1)$, $C_{cv}(x)=C_k$, $\lambda_{cv}(x,T_{cv})=\lambda_{k,0}+\lambda_{k,1}T_{cv}$ at $l_{k-1}\le x\le l_k$,i.e. factors C_{cv}, λ_{cv} depend on in what layer heat transfer is being considered.

Thus $0=l_0<l_1<\ldots<l_k=l$.

In the equations (1) - (4) the following designations are used:

$C_{cv}(x)$ - a volume thermal capacity of a multilayered structure (product of a specific thermal capacity on relative density); $\lambda_{cv}(l,T)$ - heat conductivity coefficient of a multilayered structure; $\alpha_{cv,out}$ – heat exchange coefficient in an outer surface of a covering; $\alpha_{cv,in}$ - heat exchange coefficient in an internal surface of a covering; F_{cv} - the covering area at outer and internal heat exchange; $Q_{cv,out}$ - thermal energy of external sources; c_0 - Stefan-Boltsman constant function; $\varepsilon_{cv,in}$ - radiation emissivity factor of an internal surface of a covering; m - quantity of blocks in a compartment; $g_{j,cv}$ - factor of a radiant exchange of system: j-element of a compartment - a covering; T_e - restoration temperature; t - time; T_{air} - temperature of the air environment in a compartment or regarding a compartment; $T_{cv}(x,t)$ - temperature of a multilayered structure; T_{ref} - reference temperature; T_j - temperature of j - element of a compartment; $T_{cv,x}$ - first derivative T_{cv} on x; $T_{cv,x,x}$ - second derivative T_{cv} on x; l - a thickness of a multilayered structure, $\varepsilon_{cov,\,out}$ - radiation emissivity factor of external surface of a covering.

Heat exchange coefficient $\alpha_{cv,out}$ an outer surface of a covering and heat exchange coefficient $\alpha_{cv,in}$ in an internal surface of a covering we will calculate by techniques, according to those described in [1] and [2].

The equations of heat exchange in pressurized heat-insulated partitions between pressurized heat-insulated compartments and not pressurized heat-insulated ones we will present in the form of one-dimensional equations of heat conductivity.

The equation of heat exchange of the onboard equipment we will present in the form of an ordinary differential equation describing its convective and radiant heat exchange.

Heat exchange coefficient of the onboard equipment at convective heat exchange we will calculate using the technique described in [3].

The equation of heat exchange of a person we will present in the form of an ordinary differential equation describing its convective and radiant heat exchange.

The equation of heat exchange of cargo we will present in the form of the ordinary differential equation describing its convective and radiant heat exchange.

The equation of heat exchange of designs we will present in the form of an ordinary differential equation describing its convective-radiant heat exchange

The equation of heat exchange of the air environment in a pressurized compartment we will present in the form of an ordinary differential equation describing convective heat exchange of an internal surface of a thermal protection of a covering of the onboard equipment, people, armchairs, luggage or cargo and enthalpy transfer from the air conditioning system:

The equation of heat exchange of the air environment in a nonheat-insulated unpressurized compartment we will present in the form of an ordinary differential equation describing convective heat exchange of an internal surface of a covering, of a surface of heat-insulated pressurized partition and cargo.

The factor of a radiant exchange in the equations of heat exchange is defined by Monte-Carlo method [4].

3. Forward modelling of a thermal condition of compartments

For forward and inverse modelling of a thermal condition of compartments, the equation for a covering is sampled on a spatial variable based on Galerkin's method using piecewise-linear basis. As a result of application of this method the solution of the equations (1) - (4) and the equations of heat exchange in pressurized heat-insulated partitions is reduced to the numerical solution of system of ordinary differential equations where unknown values are ones of temperature in knots of the set grid on a piece [0, l_z]. The ordinary differential equations received thus for multilayered structures (1) - (4) and the equations of heat exchange in pressurized heat-insulated partitions, the equations for the onboard equipment, passengers, cargo, structures and air system make one system of ordinary differential equations which generally can be written down in the following way:

$$Y_t = F(Y(t,\Theta)), t \in (0,t_t); \quad Y_t = Y_\Theta, F, Y \in R^S; \quad \Theta \in R^r, \ldots \tag{5}$$

where $\mathbf{Y} = [T_1, T_2, T_i, T_\Theta, \ldots]^T$ - a vector of parameters of a thermal condition of a compartment; $\mathbf{Y}_t$ - a vector of first derivatives Y on t; $\Theta = [\vartheta_1, \vartheta_2, \ldots, \vartheta_5]^T$ - a vector of factors of model; T - the top index designating operation of transposing.

Let the vector of measured parameters of system (5) $Z = [Z_1, Z_2, \ldots, Z_i]^T$ be the function from the solution of system (5) $Z = Z(Y)$.

4. Inverse modelling of a thermal condition of compartments

The problem to evaluate factors Θ models is reduced to minimisation of the weighed sum of squares nonviscous between measured in a course of experiment by values Z^* to the corresponding values $Z(Y(t, \Theta))$, received in calculating on the model equations:

$$\Phi(\Theta) = \sum_{k=1}^{N}\sum_{i=1}^{S} \Gamma_{k,i}(Z^*_{k,i} - Z_i(Y(t_k, \Theta)))^2, \tag{6}$$

where $\Gamma_{k,i}$ - weight factors; t_k - time moments at k = 1,..., N.

As it has been noted in [4], for minimisation of function (17) it is expedient to use Brojdena-Fletchera-Goldfarba-Shenno's quasi-Newton method in a combination to Newton's method which is realised according to the formula:

$$\Theta_{j+1} = \Theta_j + b_j S(\Theta_j), \tag{7}$$

where b_j - the factor characterising length of a step on j - iterations; S - the parameter specifying a direction of search of a vector Θ of the valid values of factors Θ.

Next direction S of search of j vector Θ in this algorithm is defined from system of the equations:

$$\nabla^2 \Phi(\Theta_j) = -\nabla\Phi(\Theta_j), \tag{8}$$

where $\nabla^2 \Phi - (r \times r)$ - the Gess matrix representing a square matrix of the second private derivatives of function Φ on a vector Θ_0:

Initial matrix $\nabla^2 \Phi(\Theta_k)$ in the equation (8) has been accepted by the individual.

For the decision of system of the equations (8) matrix $\nabla^2 \Phi(\Theta_j)$ is represented in the factorized to the form:

$$\nabla^2\Phi(\Theta_j) = L(\Theta_j)D(\Theta_j)L^T(\Theta_j), \tag{9}$$

where $L(\Theta_j)$ - low triangle matrix with an individual diagonal; $D(\Theta_j)$ - a diagonal matrix.

Matrixes $L(\Theta_j)$, $D(\Theta_j)$ receive decomposition Holessky matrixes $\nabla^2 \Phi(\Theta_j)$ on the algorithm described in [5].

For the decision of the equation (8) it is offered to use the following numerical scheme of Rosenbrock type of the second order of approximation for on-line systems [6]

$$\vec{Y}_{n+1} = \vec{Y}_n + aK_1 + (1-a)K_2; \tag{10}$$

$$K_1 = h(I - ah\vec{\Psi}_Y(\vec{Y}_n, t_n, \Theta))^{-1}\vec{\Psi}(\vec{Y}_n, t_n + ah, \Theta); \tag{11}$$

$$K_2 = h(I - ah\vec{\Psi}_Y(\vec{Y}_n, t_n, \vec{\Theta}))^{-1}\vec{\Psi}(\vec{Y}_n, t_n + aK_1, t_n + 2ah, \vec{\Theta}); \quad a = 1 - 1/\sqrt{2}, \tag{12}$$

where $\vec{Y}_n, \vec{Y}_{n+1}$ - the decision of the system received on n - and (n + 1) iterations, accordingly; $\vec{\Psi}$ - the right part of system; $\vec{\Psi}_Y$ – Yakobi matrix; I - an individual matrix; h - an integration step.

Confidential intervals of estimations of factors Θ nonlinear mathematical model of a thermal condition of a compartment of a kind (5) can be defined with the help dispersion matrix $P(\Theta)$ errors of estimations Θ required factors of model (the last characterise deviations of the calculated factors of model from the valid values). The method of displaying of joint confidential area of estimations on co-ordinate axes of space of factors [7] is thus used.

5. Object of research

As object of research the prototype of Brazilian main plane Embraer 190 has been accepted. Researches were spent according to Norms of flight validity Federal Aviation Regulations part 25, USA.

The basic criterion is achievement of temperature 290.15 – 298.15 K in a cabin of crew and salons of passengers no more, than through 1200 seconds with the ambassador of launch under condition of land preparation.

For cold type of a climate air reference temperature should correspond 283.15 K, for extreme warm dry – 315.15 K. Temperature of all elements of nonpressurized compartments is 228.15 K and 323.15 K. Temperature air on an exit from air central air in flight should not be more low 276.15 K and above 333.15 K.

There are 21 and 106 passengers accordingly.

Mass velocity of the air flow in pressurized heat insulated compartment was accepted as equal to 0.4 kg / (м^2 s), for nonpressurized compartments was calculated using the technique described in [8].

Parametrical identification or evaluation of a vector of factors Θ models of a thermal condition was spent on algorithm (6), (7).

Vector of factors of model

$$\Theta = [l_1, l_2, \ldots, l_5, G_1, G_2]^T \tag{13}$$

includes required central air characteristics (the expense on an exit of a central air G_1 and G_2) and values of a thickness of a thermal protection of a cabin of crew and salon of passengers $l_1, l_2, \ldots, l_5$ (Figure 1).

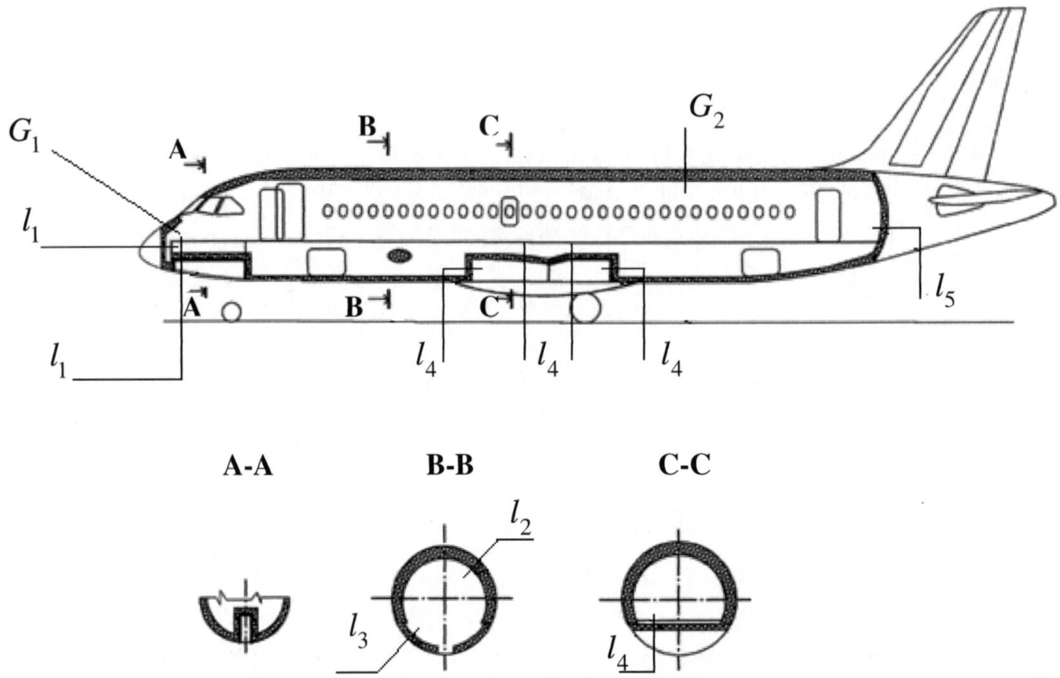

Figure 1 – Vector of factors of model: G_1 and G_2 – the expense on an exit of a central air; values of a thickness of a thermal protection of a cabin of crew and salon of passengers $l_1, l_2, \ldots, l_5$

Estimations of factors of model for cold and extreme warm dry climate types are accordingly equal to:

$$\Theta = [0.062\ 0.058\ 0.043\ 0.039\ 0.050\ 0.090\ 0.761]^T;$$

$$\Theta = [0.072\ 0.062\ 0.046\ 0.035\ 0.042\ 0.083\ 0.862]^T.$$

Confidential intervals of estimations of factors Θ make at confidential probability p = 0.99 accordingly

$$\Delta\Theta = [0.0001\ 0.0008\ 0.0006\ 0.0007\ 0.0009\ 0.0014\ 0.0122]^T;$$

$$\Delta\Theta = [0.0009\ 0.0009\ 0.0007\ 0.0008\ 0.0007\ 0.0016\ 0.0137]^T.$$

Thus, the mathematical model of a thermal condition of compartments of the main plane is developed; working out of methods of the decision of direct and return problems of heat exchange and definition of confidential intervals of estimations of parametrical identification is passed; are received required characteristics of a central air conditioning and ventilation system, and also thickness of a thermal protection in a crew cabin and passenger salon of a prototype of Brazilian plane Embraer 190.

REFERENCES

[1] G. I. Voronin. *Aircraft Air Conditioning Equipment*. - M: Mechanical engineering, 1973. – 443p.

[2] G. N. Dulnev, N.N.Tartakovsky. *Thermal Modes of Electronic Equipment*. - L: Energy, 1971. – 248 p.

[3] G. N. Dulnev, B.V.Polshchikov, A.Y.Potyagaylo. *Algorithm of Hierarchical Modelling of Heat Exchange Processes in Radio-Electronic Complex* //Radio Electronics. 1979. - № 11. - P. 49-54.

[4] N.Nikolaev, S.A..Gusev, O.A.Mahotkin. *Mathematical Modelling of convective-radiant Heat Exchange of Aircraft Nonpressurized Compartments* // Calculation on Durability of Airframe Elements. Issue: Aircraft Durability: Collection of Scientific and Technical Articles/ Novosibirsk: SibNIA.-1996 - Issue 1. - 98-108pp.

[5] P.Gill, E.Murray. *Quasi-Newton Methods for Unconstrained Optimization* // Journal of the institute of mathematics and its applications. - 1971. - v.9, № 1. - p. 91-108.

[6] S.S.Artemev, G. V. Demidov, E.A.Novikov. *Minimisation of Ravine Functions by Numerical Methods to Solve Rigid Equations*. - Novosibirsk, 1980. – 13p. - (CC Academy of Science USSR, № 74).

[7] V.N.Nikolaev, D.F.Simbirsky. *Confidence Intervals of Results of Parameter Identification of Heat Exchange Processes of the Airplane Onboard Equipment* // Research Methods and Tools of external influencing factors for the Aircraft Onboard Equipment. / Novosibirsk: SibNIA.-1991 – Issue 2. - 11-15pp.

[8] V.N.Nikolaev. *Performance and Theoretical Methods to Define Parameters of Heat Exchange of the Onboard Equipment of a Fighter Aircraft* //Algorithms and Software Support of Research of External Influences upon the Onboard Equipment of Airplanes and Helicopters. Issue: Aircraft Equipment: Research Digest / Novosibirsk: SibNIA .-1989.-Iss. 4.- 17-26 pp.

The Architecture Design of Detection and Calibration System for High-voltage Electrical Equipment

Y Ma[1], Y Lin[1], Y Yang[1], Ch Gu[1], F Yang[2] and L D Zou[3*]

[1] State Grid Shandong Electric Power Research Institute, Jinan, China 250002
[2] Shandong Zhongshi Yitong Group Co. Ltd., Jinan, China 250002
[3] Shandong University of Finance and Economics, Jinan, China 250014

zoulida@163.com

Abstract. With the construction of Material Quality Inspection Center of Shandong electric power company, Electric Power Research Institute takes on more jobs on quality analysis and laboratory calibration for high–voltage electrical equipment, and informationization construction becomes urgent. In the paper we design a consolidated system, which implements the electronic management and online automation process for material sampling, test apparatus detection and field test. In the three jobs we use QR code scanning, online Word editing and electronic signature. These techniques simplify the complex process of warehouse management and testing report transferring, and largely reduce the manual procedure. The construction of the standardized detection information platform realizes the integrated management of high-voltage electrical equipment from their networking, running to periodic detection. According to system operation evaluation, the speed of transferring report is doubled, and querying data is also easier and faster.

1. Introduction

As the increasing improvement on "intensified management of three components and integrated systems of five processes" of state grid [1], Electric Power Research Institute takes on the job of material sampling [2], test apparatus detecting [3] and field test [4]. These detection and calibration work provides protection for the secure and stable operation of electrical equipment from the aspects of equipment entrance, the source of testing data and equipment operating process.

At present, material sampling, test apparatus detecting and field test remain in the manual stage and paper archive level, which cannot satisfy the demands of field production. First, receiving and distribution of equipment rely on manual records. The attributes such as factory number, model and manufacturer need hand-copying. Most equipment is inspected once a year. Since there is no database, the information is registered more than once and workload increases. Second, the experimental data in testing report are from the table or card in handwriting. The error and uncertainty rate are computed and input into computer manually, which is un-efficient and error-prone. Moreover, an instrument corresponds to a testing report. For different instruments, their testing items are diverse. We need to add or delete relative testing table manually, which is heavy and complicated. Last, along with the development of state evaluation and production management system, it is necessary to manage all the

* Corresponding author

testing instruments from a high level. In the case, the computing error and error variation can be queried any time and state evaluation also can be more efficient.

Based on the above background, the construction of information platform is more important [5]. For material sampling and test apparatus detecting, their complicated process and manual work is simplified. For testing people, their testing reliability and accuracy are maintained. For condition-based maintenance of whole province, the platform can make sure the accuracy of state evaluation.

2. System framework

Detection and calibration system of high-voltage electrical equipment include material sampling, test apparatus detecting and field test. Material sampling is to carry on the quality detection for access network equipment through various experimental items, and give a test report. Test apparatus detecting is to check the high-voltage test instruments such as direct-current high voltage generator, voltage divider, DC resistance tester from power supply companies and power generation enterprises. Field test is to help district power distribution companies identify the possible causes of device defect or failure and give the further suggestions through field testing of technical personnel from electric power research institute. In the section we give the concrete flowchart of the three parts.

2.1. Material sampling

The module consists of six parts and its flowchart is shown in figure 1. The first is entering the handover form. The second is choosing the testing report, testing item and generate the final report. The third is generating the weekly report, monthly report and yearly report. The fourth is supporting the searching and counting function. The fifth is the circulation of testing report. The sixth is scanning and uploading the paper records.

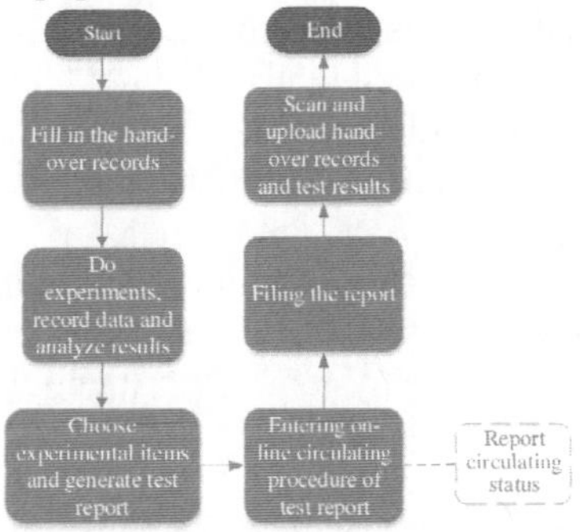

Figure 1. The flowchart of material sampling

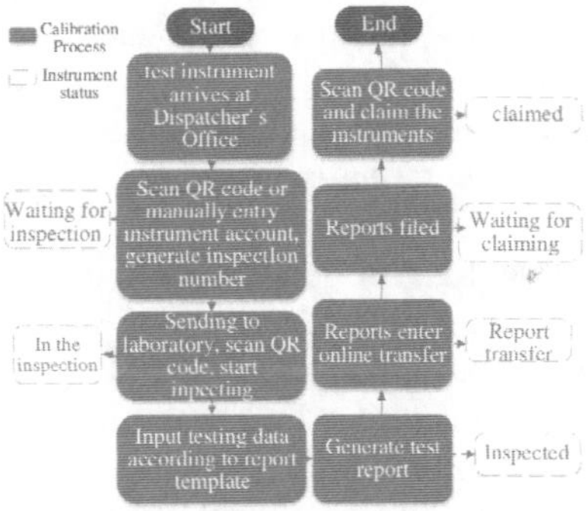

Figure 2. The flowchart of test apparatus detecting

1. Handover form include the attributes such as sample number, sample name, sample type, sample unit, client affiliation, handover time, testing type, notes. Sample name, sample type, sample unit, client affiliation and testing type are chosen from several options.

2. The template of testing report is loaded in advance. We integrate 11 kinds of templates, include dry type transformer, oil-immersed transformer, electric cable, pole-mounted circuit breaker, composite insulator, high-tension switch board, ring main unit, porcelain insulator, etc. The concrete testing report is generated dynamically according to the selected testing items [6]. If a testing report chooses three testing items, the report will show the tables of the three testing items. In addition, the testing instruments and detection criterion are brought into the report automatically. The table of contents is generated automatically, as well.

3. The weekly report, monthly report and yearly report can be generated and exported periodically. They include the attributes such as sample name, sample type, sample length (number), testing item, qualified or not, unqualified items, completion time, testing people. The above attributes are acquired from testing report or handover form.

4. The searching and counting function is also supported. Queries based on client affiliation, start date, end date and testing result are realized and Querying results can be exported in form of EXCEL.

5. The testing report needs three signatures: writer, reviewer and approver. Previously the signature is done offline [7]. Now we develop the system to circulate the report electronically. The flow is divided into five steps: generation, writing, reviewing, approval, filing. Different persons are granted different permissions. The person could see all the reports and only edit the granted reports. The rollback operation is supported. After the previous step is finished, a sending button triggers the next status of the report.

6. All the paper records need uploading and storing into database. We provide the uploading and downloading functions.

2.2. Test apparatus detecting

The flowchart of test apparatus detecting is shown in figure 2. First, after test apparatus is sent to dispatcher's office, acceptance forms are generated according to the types of equipment and each acceptance form is attached by several apparatuses. Each test apparatus has an inspection number.

Here we note that test apparatus is divided into four kinds: high-voltage metering type, electrical testing type, gas relay and safety equipment. High-voltage metering type has account management. Its attributes include product type, standard product name, product name, product factory, product model, product number, manufacturing date, product status, test date last time, test period. If the test apparatus is already in the database, the data is directly extracted. If it is not in the database, its account is added.

Also, each apparatus has a QR code [8], which is associated with base information of apparatus. When the apparatus is tested next time, we can just scan its QR code to get the record in database. If the apparatus is sent to test for the first time, we automatically generate a QR code, print the QR code and stick it on the surface. After the inspection number is given, the status of the apparatus is waiting for inspection.

Second, after finishing the inspection, each apparatus corresponds to a test report. The report is an instance of report template for this kind of apparatus. We input all kinds of report templates in advance. The report instance is generated according to its account and apparatus type. The report is modified online and all the computation is integrated with equation. After the report is saved, the apparatus is changed to the inspected status.

Third, another highlight of our system is to realize the electric signature and online transfer of the report. The nodes of a whole report include generating, writing, reviewing, approving, filing. Each identity has its own username and password of logging in the system. Each identity also has its own password of electric signature, which could maintain the safety of system. We also build a function of backlog items, which could remind them of the reports to be treated.

Last, after the report finishes transferring and is filed. The status of the apparatus is waiting for claiming. When the clients come to get their apparatus back, we scan the QR code and the apparatus is claimed.

The module also supports to query workloads of different persons, besides the general query and statistical computation, such as testing rate, qualification rate, the apparatus of exceeding the testing period, the apparatus close to testing period. The status of tested apparatus can also be queried by clients.

2.3. Field test

Figure 3 shows the flowchart of field test. It mainly includes three functions. The first is inputting the client's application form. The second is uploading the testing report, filed test record, and realizing the electronic signature. The third is querying and counting functions.

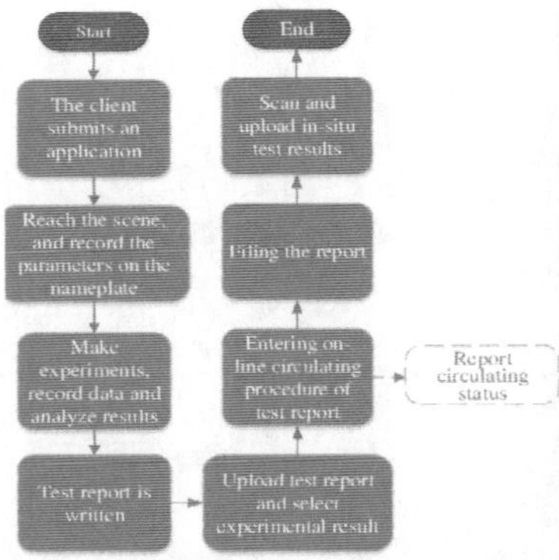

Figure 3. The flowchart of field test

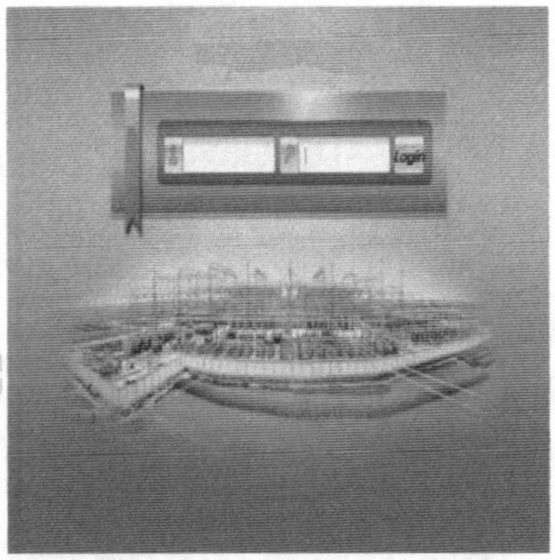

Figure 4. The login page

Figure 5. The first page

1. The client's application form include testing date, equipment type, testing item, client affiliation, substation, applicant and applicant phone. Testing date, equipment type, testing item and client affiliation are chosen from fixed items.

2. The testing report has various formats, which support uploading WORD form. After uploading is complete, electronic signature follows. Our electronic signature procedure needs to record the name of operator, status of testing report. It supports the rollback function. After the review and approval, the report is generated. It could connect the printer and directly print paper report. It also could upload the scanned report. When filing the report, the conclusion of the report is extracted as an attribute.

3. The model support querying and counting functions. The querying attributes include client affiliation, experimental date, testing conclusion. Querying results can be exported in EXCEL format.

3. System Implementation

In the section, we give some system screenshots to further introduce our system.

The login page is shown as figure 4. The system consists of four modules: backlog items, material sampling, test apparatus detecting and field test. After logging into system, it is like figure 5.

First we see the backlog items, which are divided into three parts: material sampling, test apparatus detecting and field test. Each part has its own backlogs. On the page, it shows the base information of reports to be treated by the login user. When the user clicks the report, it can directly go into the processing flow of the report, as shown in figure 6.

Figure 6. The backlog items

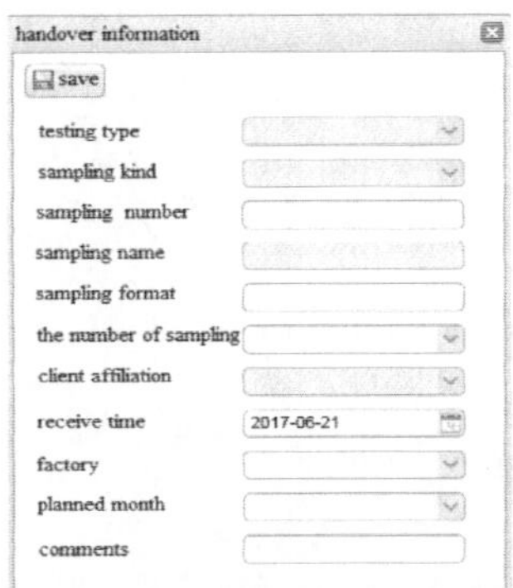

Figure 7. The handover record

Second we introduce the material sampling. Figure 7 shows how to record the handover information, including testing type, sampling kind, sampling number, sampling name, client affiliation, factory, etc. Figure 8 gives how to opening a report, how to sign a name and how to send a report to the next person. In the module the comprehensive query page is also designed, which could query the handover and report information according to different attributes.

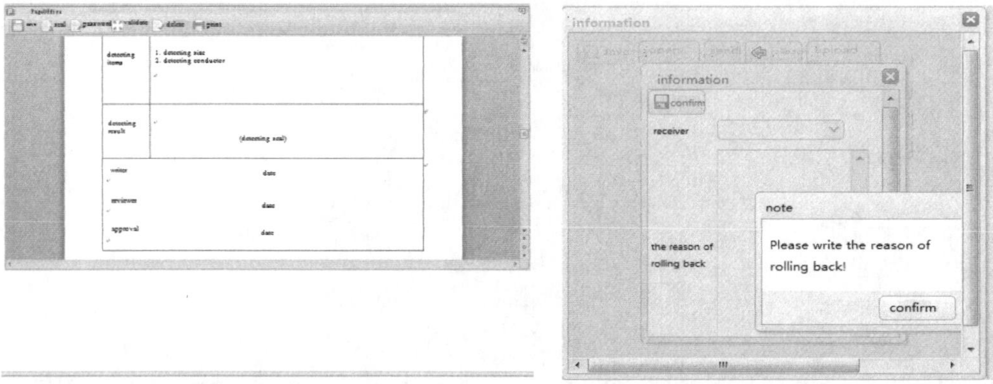

Figure 8. The report instance and report transferring

Third we next introduce the module of test apparatus detecting. Figure 9 gives the account management. We standardize all the factory name, apparatus type, apparatus model, and get them into database. Figure 10 shows how to build an acceptance form, which consists of four steps. First a new form is built, second the client affiliation is chosen, third the apparatus is added and last select the detecting time. For the testing report management function, we add the reminding button. When there is a red button before the report, the report processing deadline is close. Our testing faculty should accelerate work schedule. Figure 11 shows the workload statistics. It includes the times of acceptance, testing, writing, reviewing, approval, filing, claiming for each person. Figure 12 gives the apparatus claiming page. We just scan the QR code of the apparatus, and it is change to the claimed status.

Figure 9. Test apparatus account

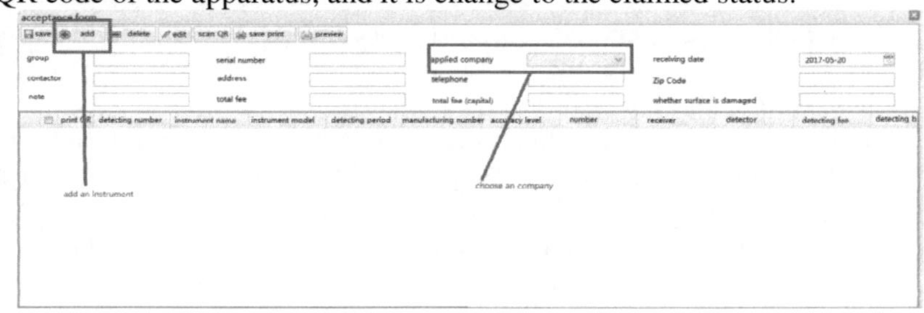

Figure 10. An acceptance form

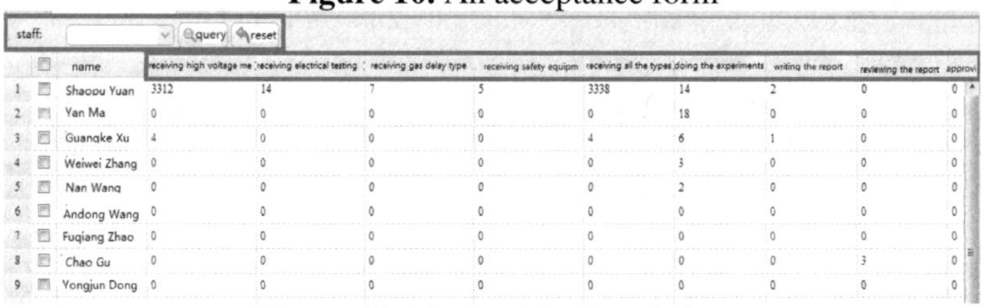

Figure 11. The workload statistics

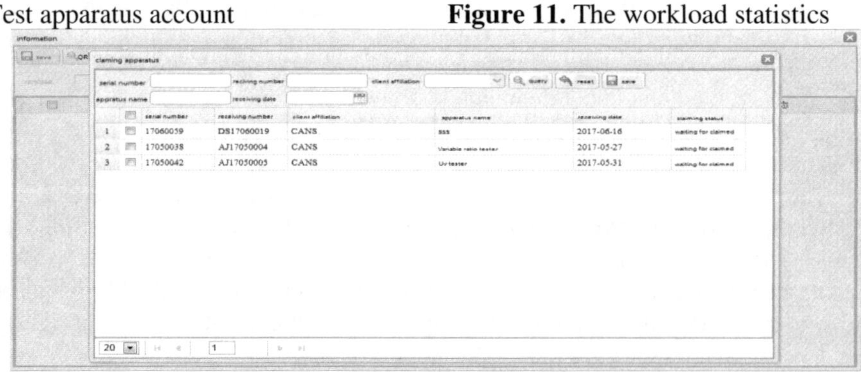

Figure 12. The apparatus claiming

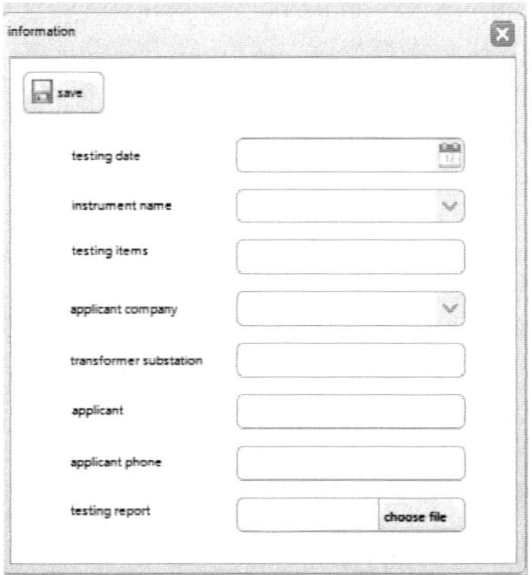

Figure 13. The client application form

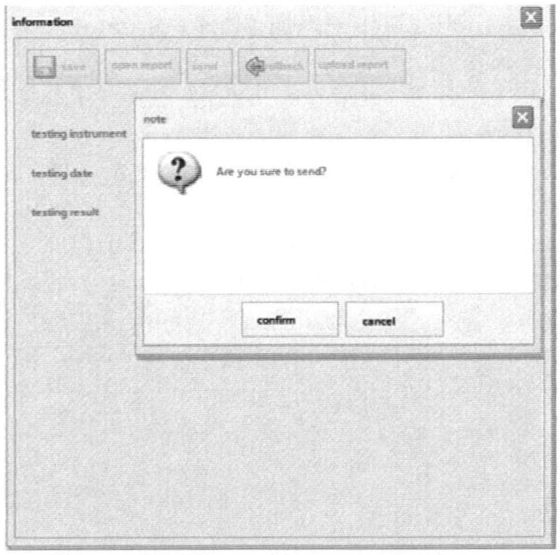

Figure 14. The online transferring of report

Last we introduce the field test module. Figure 13 shows how to generate the client application form, which is similar to that of material sampling. Figure 14 gives how to transfer the field testing report online. Since the report has no fixed template, we do not integrate report template into system. We instead directly upload the report, but the transferring of the report supports online format. All the signatures are signed electrically. Figure 15 shows the workload statistics function, which is similar to test apparatus detecting.

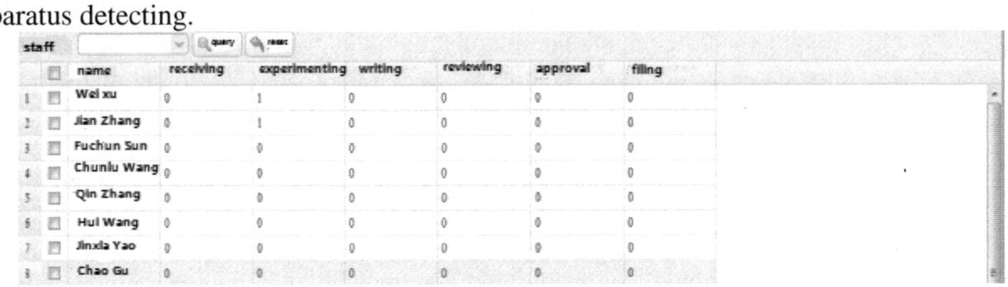

	name	receiving	experimenting	writing	reviewing	approval	filing
1	Wei xu	0	1	0	0	0	0
2	Jian Zhang	0	1	0	0	0	0
3	Fuchun Sun	0	0	0	0	0	0
4	Chunlu Wang	0	0	0	0	0	0
5	Qin Zhang	0	0	0	0	0	0
6	Hui Wang	0	0	0	0	0	0
7	Jinxia Yao	0	0	0	0	0	0
8	Chao Gu	0	0	0	0	0	0

Figure 15. The workload statistics of field test

4. Conclusions

In the paper we design and implement a detection and calibration system for high-voltage electrical equipment. Through system operation lasting close to two years, the test quality and job efficiency are improved largely, where report transferring is reduced from 3 days to 1.5 days. Our system not only guarantee the accuracy and reliability of the test data, also introduce bar code, electric signature, automatic report generation to perfect the information level of instrument detection. In the future, we will further add the input module of real-time testing data to record the whole testing process.

Acknowledgments

This work was supported by the institute-level project in State Grid Shandong Electric Power Research Institute (2017-12). The authors are grateful for the anonymous reviewers who made constructive comments.

References

[1] YU L J 2011 Analysis on intensive management in power grid enterprises based on theory of company value chain *Energy Tech. Eco.* **23** pp 1-5
[2] Li L D 2016 Construction of informatization based whole process management and control mode of material quality *J. Anhui Elec. Eng. Pro. Tech. Col.* **21** pp 63-66
[3] Shui H D 2010 Analysis and design of management system for high-voltage test instruments, *Ann. Info. Conf. Elec. Pow. Indus.* pp 155-158

[4] Huang Y 2008 The development of the pda-system in the local operation management *Zhejiang University*
[5] Qin Q 2010 Based on web technology laboratory information management system *Zhengzhou University*
[6] Jin L F and Zhou W X 2016 Auto generation of experiment report in WORD form *China Measurement Technology* **33** pp 112-115
[7] Cui W 2006 The research and implementation of the digital seal system based on the document transmission *Shandong University*
[8] Liu X P 2010 Research on provincial agencies equipment information receiving and dispatching system based on two dimensional bar code *South China University of Technology*

IOP Conf. Series: Materials Science and Engineering **302** (2018) 012044 doi:10.1088/1757-899X/302/1/012044

Modelling the EDLC-based Power Supply Module for a Maneuvering System of a Nanosatellite

A A Kumarin[1] **and I A Kudryavtsev**[1]

[1] Space research department, Samara University, Samara, Russia

E-mail: alky_samara@mail.ru

Abstract. The development of the model of the power supply module of a maneuvering system of a nanosatellite is described. The module is based on an EDLC battery as an energy buffer. The EDLC choice is described. Experiments are conducted to provide data for model. Simulation of the power supply module is made for charging and discharging of the battery processes. The difference between simulation and experiment does not exceed 0.5% for charging and 10% for discharging. The developed model can be used in early design and to adjust charger and load parameters. The model can be expanded to represent the entire power system.

1. Introduction

The development of modern electronic components allows to implement more and more complex missions via light classes of space vehicles. Nanosatellites (NS) have become widespread, mostly because of the introduction of CubeSat platform which unifies the requirements to launching containers. One of the crucial systems of a satellite is a power system (PS). It normally consist of an energy generator (usually a solar battery), electrical storage device (usually Li or Ni battery) and a control system.

Modeling is an essential part of design which allows to avoid some mistakes in early design. In this phase some parameters of the system are uncertain. Therefore, simulations allow to adjust parameters of a designing system. In this paper a model of a power supply module for a maneuvering system (MS) of a nanosatellite is described.

The MS is a propulsion engine for using in a nanosatellite. It is used for orbit correction. It create thrust by expelling reaction mass through the nozzles. Working mass is a mix of water and ethanol and is stored at low initial pressure of about 3 atm, which is not enough to create a suitable thrust. That is why the liquid flows through a heater where it partly vaporizes. It leads to the increasing pressure which creates greater thrust. The main problem of such concept is the high value of peak electrical power needed in the heater. It cannot be provided by regular PS unit.

Most PS of NS nowadays are based on Li batteries. Most of them cannot sustain current greater than 2C, where C denotes the capacitance in Ah, without permanent damage, which would result a reduction of the capacitance. Thermal calculations of the MS shows that the PS have to provide at least 200W peak power and 4kJ total energy. The maximum voltage of Li-Ion battery (LIB) is 4.2V which means, that the minimum capacitance of the battery would be around 24 Ah. For a battery of LIB with maximum voltage 8.4V the capacitance would be around 12 Ah, for 16.8V – around 6 Ah. Each case is unacceptable. That is why a buffer storage device is used in this paper.

IOP Conf. Series: Materials Science and Engineering **302** (2018) 012044 doi:10.1088/1757-899X/302/1/012044

Electrical double layer capacitors (EDLC) are used in this work [1, 2]. Their main features are higher energy density than of ordinary electrolytic capacitors and higher discharge current than of chemical batteries. However, they cannot be used as primary battery because their energy density is lower than of Ni-Mh batteries.

2. Methods

2.1. EDLC selection

An analysis of components of the shelf (COTS) EDLC was made for the developing system. The choice was based on such criteria as energy density, maximum current and the amount of cells in battery to fit energy requirements. Only EDLC for power applications were taken into account. The data from manufacturer's datasheets was combined in table 1.

Table 1. COTS EDLC characteristics

Partnumber	Capacitance, F	Maximum current. A	Length, mm	Diameter, mm	Stored energy, J	Cells in battery	Mass, g	Battery volume, cm^3	Energy density, J/g
BCAP350	350	34	61	33	1276	4	252	209	17.4
BCAP650	650	88	52	61	2369	2	320	320	12.8
BCAP1200	1200	110	74	61	4374	2	520	520	14.5
BCAP1500	1500	140	85	61	5468	1	280	280	16.8
BCAP2000	2000	170	102	61	7290	1	360	360	17.5
BCAP150	150	40	50	25	547	9	288	288	14.7
SCCV60B107MRB	100	53	60	18	365	13	260	260	15.7
SCCX50B207VSB	200	96	50	22	729	7	280	280	15.7
SCCY62B307VSB	300	104	62	35	1094	5	385	385	14.3
SCCY68B407VSB	400	174	68	35	1458	4	348	348	14.5

As can be seen from the table, different models have similar energy densities. The EDLCs with lower values of capacitance are impractical because of large amounts of cells in batteries, which are difficult to install. Cells with high capacitance are also impractical because of problems with installation. Thus, BCAP350 where chosen. The battery consists of 4-5 cells, because usable energy is lower with non 100% depth of discharge.

2.2. Math model

The math model is based on experimental data. A measurement equipment was designed to conduct the experiment. It is based on pulse current source. It charges the EDLC with constant current and monitor the voltage (U) on its terminals over the time.
Differential capacitance is determined by equation (1).

$$C = I\left(\frac{dU}{dt}\right)^{-1} \tag{1}$$

Since the current was maintained constant, it was only necessary to measure the value of the time derivative of the voltage. It is averaged during small time periods, where it can be considered constant. The value is obtained using least square linear regression method for N experimental data points as shown in equation (2).

$$\frac{dU}{dt} \approx \frac{N\sum_i U_i t_i - \sum_i U_i \sum_i t_i}{N\sum_i U_i^2 - \left(\sum_i U_i\right)^2} \tag{2}$$

Averaging results of multiple experiments shows that the capacitance depend on voltage linearly. A measurement, based on single charging process can give bias at about 3% from average as shown on figure 1. This is happening most likely because of difficulty for ions to penetrate pores of an activated carbon pores [3]. These processes cannot be predicted, therefore, the dependence is modeled as linear function – equation (3). Numerical values may significantly differ from cell to cell.

$$C = A_C U + B_C \approx 62.0U + 268.6\,. \tag{3}$$

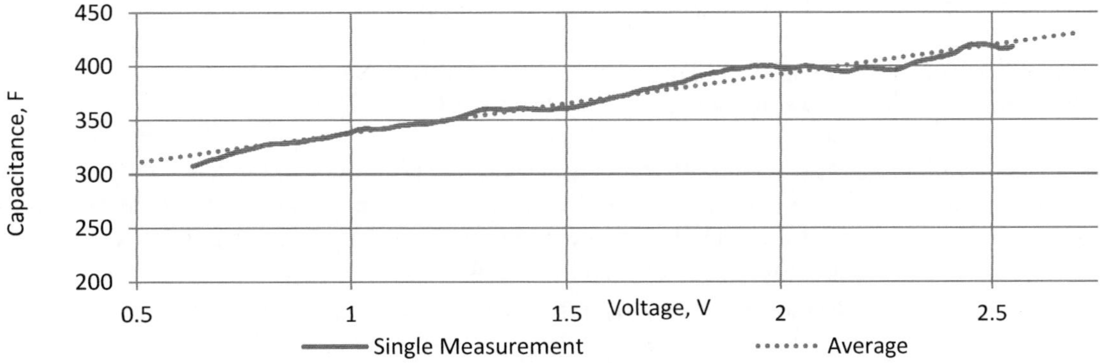

Figure 1. Capacitance over voltage of BCAP350

The ESR of the EDLC is considered to be constant in this paper since its value is lower than any other resistances in the circuit (about 3mOhm) and its dependence on voltage can be neglected [4].

Another factor in every process with EDLC is leaking. It is modelled in this paper via leak power which depends on the voltage of the cell. Experimental data shows that it could be approximated with exponential function. The data was acquired by measuring open circuit voltage (U) of the cell. The measurements were taken each 4 seconds during 35 hours. The cell leaked from 2.64V to 2.30V which corresponds to approximately 25.5% of total energy. Calculated values of leak power are represented on figure 2. Approximation function used for simulation is shown in (4).

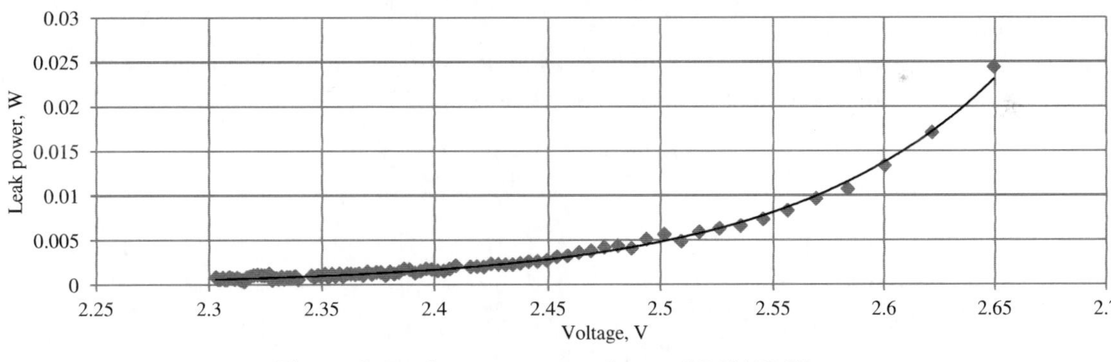

Figure 2. Leak power over voltage of BCAP350

$$P_l = A_l e^{B_l U} \approx 2.33 \cdot 10^{-14} e^{10.42U} \tag{4}$$

The finite element method is used to solve the problem. The time step is defined to fit the following condition: relative difference in voltage of the EDLC at the beginning and in the end of the time step should not exceed a predefined value ε. In this paper 10^{-4} was used as the value and equation (5) – as the time step estimation. Thus, the voltage of the EDLC can be considered constant with error estimation. Time step is variable, therefore processes with low current can be calculated faster with same accuracy.

$$\Delta t = \varepsilon \frac{U \cdot C(U)}{I}, \tag{5}$$

The state of EDLC is considered to be a function of the cell open circuit voltage. That is suitable since current changes slowly in all processes. Transition processes can be neglected because they last for less than a second while other processes last for minutes or even hours. Therefore, the inductance does not affect the results significantly [5]. Thus, the open circuit voltage of the EDLC on the next time layer is described by equation (6).

$$U_{OC}(t+\Delta t)=U_{OC}(t)+\left[\frac{I(t)}{C(U)}-\frac{P_l}{U\cdot C(U)}\right]\Delta t\,. \quad (6)$$

The voltage on the terminals of EDLC depends on current, because there is a voltage drop in internal series resistance r. The dependence is described by equation (7).

$$U(t)=U_{OC}(t)+rI(t)\,. \quad (7)$$

Equations before described a single EDLC cell. However, MS is supposed to have a series battery of EDLC. Therefore, for estimation simulations voltage can just be multiplied by the number of cells in the battery. The current I is the deference between current from charger and current of load defined by Ohm law. The value of current I_{in} from satellite's power grid with voltage U_{in} can be obtained using charger current I_C, EDLC voltage and the efficiency η of charger via equation .

$$I_{in}=\frac{I_C U}{\eta U_{in}}\,. \quad (8)$$

The electrical load of the system is a heater. Its resistance depends on temperature. The dependence is considered linear and the temperature resistance coefficient is used. The temperature of the heater is defined by equation (9). It was acquired by simple reworking of Fourier's law and the law of conservation of energy.

$$T(t+\Delta t)=T(t)+\left(\frac{U^2}{R(T(t))C_h}-\lambda\left(T(t)-T_{amb}\right)\right)\Delta t\,, \quad (9)$$

where U – heater voltage, R – heater resistance, C_h – heater thermal, λ – fixture thermal conductivity, T_{amb} –temperature of the fixture.

2.3. Experiment

The propriety of simulation results has to be proven by verification. Therefore, an experiment was conducted. The electrical scheme of experimental setup is represented in figure 3.

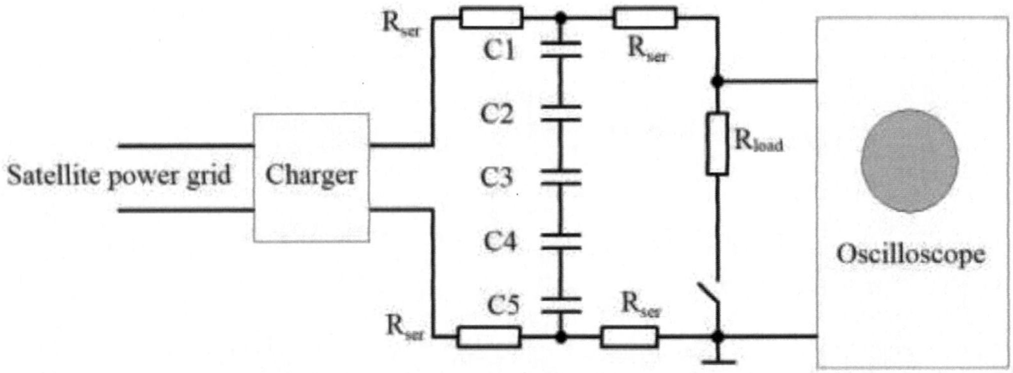

Figure 3. Electrical scheme of the experimental setup

The first mode of the setup is charging: the switch is closed, regulated current flows from charger to the battery. The second mode is discharge: the switch is open and current flows from battery to load. In both cases charging/discharging current does not flow through measurement circuit. An oscilloscope GWInstek GDS-72304 is used to measure voltage. The measurement error with respect to averaging sample does not exceed 0.2%.

2.4. Modeling results and verification

Simulation and experiment data for charging process are represented in figure 4. Relative error is shown in figure 5. The data for discharging process is shown in figure 6, relative error – figure 7.

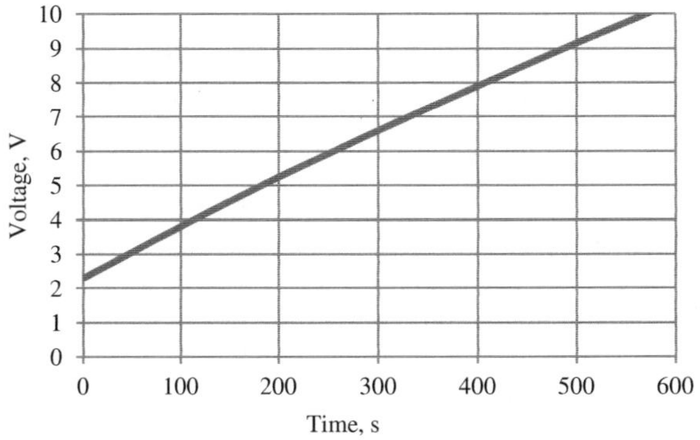

Figure 4. EDLC voltage in charging process

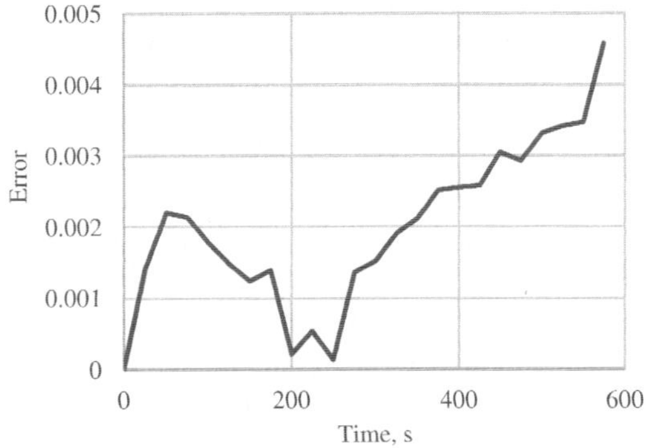

Figure 5. Relative error of simulation of charging process

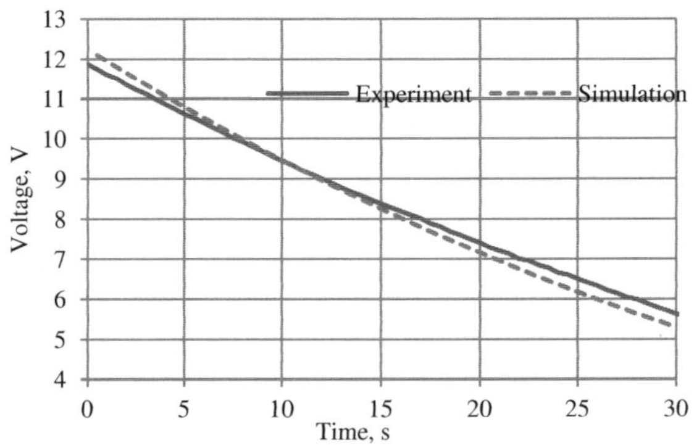

Figure 6. EDLC voltage in discharging process

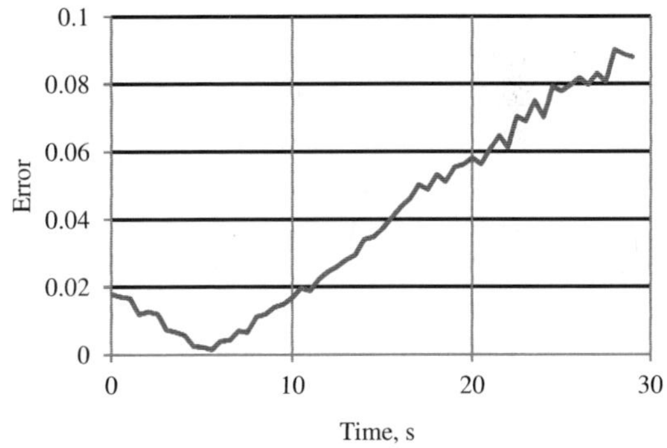

Figure 7. Relative error of simulation of discharging process

3. Discussion

The paper discuss power supply module for maneuvering system for a nanosatellite. The aim of creating a model was an opportunity of simulation of the processes in a module with set parameters. This could help to reduce the amount of mistakes in early design and in addition, to adjust the parameters of heater and charger to fit desired characteristics.

A comparison of simulation and experiment data has shown, that relative error does not exceed 0.5% for charging process, which allows accurate charger parameters adjust.

Relative error of discharging process does not exceed 10%. Such error is probably due to the difference between table values of thermal resistance coefficient, heat capacity and most importantly heat conductivity and their real values. These values have to be obtained from experiment or by detailed thermal modelling with respect to design features of the heater. However, 10% is an acceptable error for estimation simulations during early design, which was the purpose of the study.

The described model can be integrated in a model of the whole power supply system, which would allow to adjust all important parameters.

References

[1] Alkali M, Edries M Y, Khan A R, Masui H and Cho M 2015 Design considerations and ground testing of EDLC as energy storage components for nanosatellites *Journal of small satellites* vol 4 no 2 p 387 – 405

[2] Alkali M, Edries M Y, Khan A R, Masui H and Cho M 2014 Preliminary study of electric double layer capacitor as an energy storage of simple nanosatellite power system *65th international astronautical congress* p 1 – 7
[3] Huang J, Sumpter B G and Meunier V 2008 A universal model for nanoporus carbon supercapacitors applicable to diverse pore regimes, carbon materials and electrolytes *Chemistry a European journal* vol 41 p 6614 – 26
[4] Funaki T and Hikihara T 2008 Characterization and modelling of the voltage dependency of capacitance and impedance frequency characteristics of packed EDLC *IEEE transactions on power* electronics vol. 23 no 3 p 1518 – 25
[5] Ionescu C, Vasile A and Negroiu R 2015 Accurate modeling of supercapacitors for DC operation regime *IEEE 21st International symposium for design and technology in electronic packaging* p 303 – 306

IOP Conf. Series: Materials Science and Engineering **302** (2018) 012045 doi:10.1088/1757-899X/302/1/012045

Estimation of the Thermal Process in the Honeycomb Panel by a Monte Carlo Method

S A Gusev [1,2] and V N Nikolaev[3]

[1] ICM&MG Sib. Br. RAS, pr. Academika Lavrenjeva, 6, Novosibirsk, 630090, Russia
[2] NSTU, pr. K. Marksa, 20, Novosibirsk, 630073, Russia
[3]SibNIA named after S. A. Chaplygin, Polzunov Street, 21, Novosibirsk, 630051, Russia

sag@osmf.sscc.ru; nikvla50@mail.ru

Abstract. A new Monte Carlo method for estimating the thermal state of the heat insulation containing honeycomb panels is proposed in the paper. The heat transfer in the honeycomb panel is described by a boundary value problem for a parabolic equation with discontinuous diffusion coefficient and boundary conditions of the third kind. To obtain an approximate solution, it is proposed to use the smoothing of the diffusion coefficient. After that, the obtained problem is solved on the basis of the probability representation. The probability representation is the expectation of the functional of the diffusion process corresponding to the boundary value problem. The process of solving the problem is reduced to numerical statistical modelling of a large number of trajectories of the diffusion process corresponding to the parabolic problem. It was used earlier the Euler method for this object, but that requires a large computational effort. In this paper the method is modified by using combination of the Euler and the random walk on moving spheres methods. The new approach allows us to significantly reduce the computation costs.

1. Statement of the problem

One of the perspective directions in design of heat-shielding coatings is the use of honeycomb type panels. A honeycomb thermal protection panel (figure 1) is a construction consisting of two parallel plates of carbon fibre reinforced plastic, between which is a thin frame made of the same material in the form of bee honeycombs filled with a substance with low thermal conductivity, and this can be, for example, air.

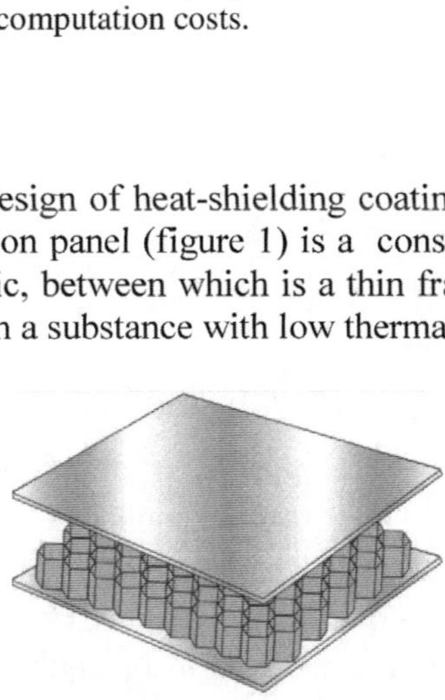

Figure 1. A honeycomb panel.

In the paper [1] it was suggested to use the method of stochastic differential equations (SDE) for solving heat exchange problems in honeycomb panels. This method is based on the probability representation of the solution of the boundary value problem for the parabolic equation. Such representations are well known and they can be found, for example, in [2-4]. Solving of the problem in this case is reduced to estimation the expectation of the diffusion process, which is the solution of the corresponding SDE system.

The heat transfer in the honeycomb panel is described by a boundary value problem for a heat equation with a discontinuous coefficient of the thermal diffusivity. In this regard some problems arise associated with the justification of existence and uniqueness of the solution of the corresponding SDE system. In this connection, it was proposed in [1] to use smoothing discontinuous coefficients on the basis of the integral averaging.

The numerical solution of the SDE system in [1] was carried out by the Euler method that is very computationally laborious. For this reason, we propose in this paper a new method which consists in the following. We apply the walk on moving spheres method [5] for modelling the diffusion paths in cells of the panel except for a given neighborhood of the framework and the boundary plates where we use the Euler method.

The area of space variables in $\mathbb{R}^3$ is a rectangular parallelepiped: $G=(-l_1,l_1)\times(-l_2,l_2)\times(0,l_3)$. The parallelepiped G is a union of two disjoint sets $G=G_1\cup G_2$, where G_1 is a subarea corresponding to the frame and the plates, bounding the panel, and G_2 is a union of subareas, corresponding to cells containing heat insulating substance. Every cell is an interior of a regular hexagonal prism whose lateral surface is parallel to the axis x_3. The process of heat transfer in the panel during the time interval $[0,T]$ is considered.

We describe the heat transfer in the honeycomb panel by a boundary value problem for the heat equation. In this problem, convection heat exchange occurs on the two sides of the panel, and the other four sides are thermally insulated.

$$\frac{\partial u}{\partial t}=\sum_{i=1}^{3}\frac{\partial}{\partial x_i}\left(b(x)\frac{\partial u}{\partial x_i}\right) \quad \text{where } b(x)=\begin{cases} b_1, & x\in G_1 \\ b_2, & x\in G_2 \end{cases} \tag{1}$$

$$u(x,0)=\varphi(x), \tag{2}$$

$$\left.\frac{\partial u}{\partial x_1}\right|_{x_1=-l_1}=0, \quad \left.\frac{\partial u}{\partial x_1}\right|_{x_1=l_1}=0, \tag{3}$$

$$\left.\frac{\partial u}{\partial x_2}\right|_{x_2=-l_2}=0, \quad \left.\frac{\partial u}{\partial x_2}\right|_{x_2=l_2}=0, \tag{4}$$

$$\lambda\left.\frac{\partial u}{\partial x_3}\right|_{x_3=0}=\alpha_1(t)\left(u-\psi_1(t)\right), \tag{5}$$

$$-\lambda\left.\frac{\partial u}{\partial x_3}\right|_{x_3=l}=\alpha_2(t)\left(u-\psi_2(t)\right). \tag{6}$$

In the formulas (1) – (6) the following notations are used: φ is the initial distribution of temperature in the panel; λ is the coefficient of thermal conductivity of carbon; b_1 , b_2 are the coefficients of thermal diffusivity of carbon and air, respectively; α_1 , α_2 are the heat transfer coefficients; ψ_1, ψ_2 are ambient temperatures at the top and bottom sides of the panel, respectively.

2. Obtaining the problem with smoothed coefficients

Since the thermal protection panel consists of two materials with different thermal properties, we have in the problem (1) – (6) the heat equation with the discontinuous coefficient of thermal diffusivity. A sufficiently complete investigation of boundary value problems for parabolic equations with discontinuous coefficients is given in the book [6]. For such boundary value problems the existence and uniqueness conditions of a generalized solution are given in [6] in the space of functions $V_2^{0,1}(Q_T)$ $\left(Q_T = G\times(0,T)\right)$ with the norm $|u|_{Q_T} = \max\limits_{0\le t\le T}\|u(x,t)\|_{L_2(G)} + \left(\int\limits_{Q_T} u_x^2 dxdt\right)^{\frac{1}{2}}$, possessing the property $\int\limits_0^{T-h}\int\limits_G h^{-1}(u(x,t+h)-u(x,t))^2 dxdt \to 0$ as $h\to 0$. As it is proved in [6] (Theorem 5.1, Chapter III), the generalized solution of the problem with discontinuous coefficients is stable in the space $V_2^{0,1}(Q_T)$ with respect to the coefficient variations. Applying this theorem to the problem (1) – (6), we can conclude that at $m\to\infty$ an uniformly bounded sequence $b^{(m)}$ convergences almost everywhere to b, then the generalized solutions with coefficients $b^{(m)}$ strongly converge in the norm of $V_2^{0,1}(Q_T)$ to the generalized solution of the problem (1) – (6).

So, we can obtain an approximate solution $u^{(m)}$ of the problem (1) – (6) with discontinuous coefficient of thermal diffusivity via changing this coefficient by its smooth approximation $b^{(m)}$. Smoothing the coefficient b is carried out by the integral averaging with infinitely differentiable finite kernel [7]

$$b^{(m)}(x) = \int\limits_{|x-y|<\rho_m} \omega_{\rho_m}(|x-y|)b(y)dy = \frac{1}{\rho_m^3}\int\limits_{|x-y|<\rho_m} \omega_1\left(\frac{|x-y|}{\rho_m}\right)b(y)dy \tag{7}$$

where $\omega(|\xi|)=0$ when $|\xi|>1$; $\int\limits_{|\xi|\le 1}\omega_1(|\xi|)=1$. It is assumed in (7) that $\rho_m\to 0$ as $m\to\infty$. Functions $b^{(m)}$ have derivatives of any order [7]. Because $b\in L_q(G)$ ($q>0$), then for any subarea $G'\subset G$ spaced from the boundary at distance no less than ρ_m the averaging $b^{(m)}$ convergence to b in the space $L_q(G')$, i.e. $\|b^{(m)}-b\|_{q,G'} \le \sup\limits_{|v|\le\rho_m}\left(\int\limits_{G'}|\mathrm{b(x-v)-b(x)}|^q dx\right)^{\frac{1}{q}} \to 0$ as $m\to\infty$. So, approximations $b^{(m)}$ converge to b in the norm of the space $L_q(G)$. It is known that the convergence in $L_q(G)$ implies the convergence in measure. And from a sequence that converges in measure, one can extract the subsequence convergences almost everywhere [8]. Also in [8] it is given a constructive method of finding such a subsequence. In what follows, we assume that this subsequence is $\{b^{(m)}\}$. We replace in (1) - (6) the coefficient of thermal diffusivity by $b^{(m)}$. After that we obtain the problem

$$\frac{\partial u^{(m)}}{\partial t} = b^{(m)}(x)\sum_{i=1}^{3}\frac{\partial^2 u^{(m)}}{\partial x_i^2} + \sum_{i=1}^{3}\frac{\partial b^{(m)}}{\partial x_i}\frac{\partial u^{(m)}}{\partial x_i}, \tag{8}$$

$$u^{(m)}(x,0) = \varphi(x), \tag{9}$$

$$\left.\frac{\partial u^{(m)}}{\partial x_1}\right|_{x_1=-l_1} = 0, \quad \left.\frac{\partial u^{(m)}}{\partial x_1}\right|_{x_1=l_1} = 0, \tag{10}$$

$$\left.\frac{\partial u^{(m)}}{\partial x_2}\right|_{x_2=-l_2} = 0, \quad \left.\frac{\partial u^{(m)}}{\partial x_2}\right|_{x_2=l_2} = 0, \tag{11}$$

$$\lambda \frac{\partial u^{(m)}}{\partial x_3}\bigg|_{x_3=0} = \alpha_1(t)\big(u - \psi_1(t)\big), \tag{12}$$

$$-\lambda \frac{\partial u^{(m)}}{\partial x_3}\bigg|_{x_3=l} = \alpha_2(t)\big(u - \psi_2(t)\big). \tag{13}$$

Further, we use the following notations: ∂G is the boundary of G; χ_A is the indicator function of a set A. We assume that the functions α_1, α_2, ψ_1, ψ_2 are such that for some $0<\alpha<1$ there exists a solution of the problem (8) - (13) in the Holder space $H^{3+\alpha,\frac{3+\alpha}{2}}(Q_T)$ [6] (Theorem 5.3, Chapter IV). It is proved in [9] that on the existence and uniqueness conditions of the solutions of the problems (1) - (6) and (8) - (13), for any $\varepsilon > 0$ there exists natural number m_ε such that for any $m > m_\varepsilon$ the following inequality holds $\operatorname{ess\,sup}_{Q_T} |u^{(m)} - u| < \varepsilon$. Consequently, choosing the averaging radius sufficiently small, it is possible approach $u^{(m)}$ to u at any accuracy.

3. Probability representation of the solution

Probability representations of solutions of parabolic boundary value problems with boundary conditions of the second and third kinds are very known [2-4]. To obtain the probability representation of the solution of the problem (8) - (13) we introduce the following SDE system

$$X_r = x + \int_{T-t}^{r} \sigma_v \chi_G(X_v) dW_v^{(3)} + \int_{T-t}^{r} a_v \chi_G(X_v) dv + \int_{T-t}^{r} n(X_v)\chi_{\partial G}(X_v) dK_v\,, \tag{14}$$

$$Y_r = 1 + \int_{T-t}^{r} \alpha_1(v) Y_v \chi_{X_3(v)=0} dK_v + \int_{T-t}^{r} \alpha_2(v) Y_v \chi_{X_3(v)=l_3} dK_v\,, \tag{15}$$

$$Z_r = \int_{T-t}^{r} \alpha_1(v)\psi_1(v) Y_v \chi_{X_3(v)=0} dK_v + \int_{T-t}^{r} \alpha_2(v)\psi_2(v) Y_v \chi_{X_3(v)=l_3} dK_v, \tag{16}$$

$$K_r = \int_{T-t}^{r} \chi_{\partial G}(X_v) dK_v \tag{17}$$

where $x \in G$ is an initial point for $X_.$; $W_.^{(3)}$ is three dimensional Wiener process; $\sigma_v = \left(2b^{(m)}(X_v)\right)^{\frac{1}{2}}$; $n(P)$ is the unit inner normal at the point $P \in \partial G$; K_v is a local time of the process $X_.$ on ∂G, i.e. it is a scalar increasing process which increases only when $K_v \in \partial G$.

The probability representation of the solution of the problem (8) - (13) is the following expectation of the functional of the solution of the SDE system (14) - (17)

$$u(x,t) = \mathrm{E}_{x,T-t}(\varphi(X_T)Y_T + Z_T) \tag{18}$$

where $\mathrm{E}_{x,T-t}$ is denoted the expectation with respect to the probability measure $\mathrm{P}_{x,T-t}$, corresponding to the random process starting at time point $T-t$ from the point x, i.e. $X_{T-t} = x$.

Approximations to $u^{(m)}$ can be obtained by the Monte Carlo method as result of modelling paths of the SDE system (14) - (17). For this purpose in [1], [9] the Euler method was used, and the modelling paths of the SDE's (14) - (17) by this method is performed by the scheme

$$X_{i+1}^{\Delta} = x_i + h^{\frac{1}{2}}\sigma_i \xi_i + h a_i, \tag{19}$$

$$x_{i+1} = X_{i+1}^{\Delta} + (\Delta_{i+1}K) n_i \chi_{R^3 \setminus \bar{G}}(X_{i+1}^{\Delta}), \tag{20}$$

$$y_{i+1} = y_i(1 + (\alpha_1)_i \chi_{\partial G}(x_i)\Delta_{i+1}K + (\alpha_2)_i \chi_{\partial G}(x_i)\Delta_{i+1}K), \tag{21}$$

$$z_{i+1} = z_i + y_i(\alpha_1)_i(\psi_1)_i \chi_{\partial G}(x_i)\Delta_{i+1}K + y_i(\alpha_2)_i(\psi_2)_i \chi_{\partial G}(x_i)\Delta_{i+1}K, \tag{22}$$

$$k_{i+1} = k_i + \Delta_{i+1}K \tag{23}$$

where i, $i+1$ are numbers of nodes of the time grid; ξ_i is the vector of tree independent $N(0,1)$ random numbers; $\Delta_{i+1}K = d(X^{\Delta}_{i+1})$ is the distance from X^{Δ}_{i+1} to ∂G when X^{Δ}_{i+1} is out of G.

4. Combination of the Euler and the walk on moving spheres methods

Because it is assumed it this work that the thermophysical characteristics of CFRP and air are constant values, then the motion of the random process $X_{\cdot}$ in uniform media corresponding to the frame or cell up to a constant factor coincides with the motion of the Wiener process. It is known that if the Werner process starts from the centre of the ball, then the point corresponding to the first exit time from the ball is distributed uniformly on its surface. And the exit point does not depend on the first exit time. In connection with this, to accelerate calculations, it is proposed to use the random walk on spheres method in the cells of the panel, and use the Euler method for moving around the frame and the boundary plates. When one solve non-stationary problem by walk on spheres, it is necessary besides modelling exit points on spheres also define corresponding first exit times. To model the first exit time from the ball of the radius R, it is proposed in the paper to determine the first hitting time of the level R by the Bessel process which is defined by the SDE [10]

$$S_t = S_0 + \int_0^t S_r^{-1}dr + W_t \tag{24}$$

where $W_{\cdot}$ is one-dimensional Wiener process. One-dimensional random process (24) is obtained as a result of applying the Ito's formula to the Euclidean norm of the tree-dimensional Wiener process starting from the point $x \in G$, i.e. $S_0 = \|x\|$. The first exit time of tree-dimensional process from the ball of radius R and the first hitting time of the level R by the process (24) have the same distribution. It is known explicit formulas for densities of the first hitting times by Bessel processes starting from points $S_0 = 0$ and $S_0 > 0$ [5]. But they are series that contain in their members the Bessel functions of the first kind and their positive zeros. Therefore, these formulas are extremely inconvenient for numerical statistical modelling. On the other hand, one can choose a curve depending on t such that the first hitting time by Bessel process of this curve has simple and convenient formula of the density which can be used for modelling. Let μ be sigma-finite positive measure. Let us define a function

$$r(t,x) \equiv q(t,0,x) - \frac{1}{a}\int_0^{\infty} q(t,y,x)\mu(dy) \tag{25}$$

where $q(t,y,x)$ is the transition probability density of the Bessel process; a is a numeric parameter. It follows from the feature of density q that the functions q and r satisfy to the parabolic equation

$$\frac{\partial v}{\partial t} = \frac{1}{2}\frac{\partial^2 v}{\partial x^2} - \frac{\partial}{\partial x}\left(\frac{1}{x}v\right) \tag{26}$$

Let $g(t)$ be a function such that on the curve $x = g(t)$ the equality $r(t,x) = 0$ holds. Then this curve can be considered as a boundary condition to the equation (25). The solution of the obtained boundary value problem is unique because the starting point of the Bessel process has been defined.

Let us denote τ_g the hitting time of this moving boundary, i.e. $\tau_g = \inf\{t > 0, S_t = g(t)\}$. It can be proved [5] that the density of the hitting time of the boundary $x\text{-}g(t) = 0$ satisfy to the equation

$$p_g(t) = -\frac{d}{dt}\left(\int_0^{g(t)} r(t,x)dx\right) \tag{27}$$

If one takes $\mu(dy) = y^2 dy$, then $g(t)$ and $p_g(t)$ are defined by the formulas [5]:

$$g(t) = \left(2t\ln\left(\frac{a}{\Gamma\left(\left(3/2\right)\sqrt{2}\right)}\right)\right)^{\frac{1}{2}}, \quad p_g(t) = \frac{g^3(t)}{2at} \tag{28}$$

where Γ is a notation of the Gamma Function. In this case a simple formula for modelling τ_g is obtained $\tau_g = \left(\frac{a}{\Gamma\left(\left(3/2\right)\sqrt{2}\right)}\right)^{\frac{2}{3}} e^{-H}$ where H is a gamma distributed random value with parameters $5/2$, $2/3$. In the process of calculations, after obtaining τ_g, the value of the radius of the sphere is defined by the formula $g(\tau_g)$ in (28). The ability to ensure that the simulated balls are inside the cells can be achieved by selecting the parameter a. When the distance to the cell border is less than the set value ε, the calculations are performed by the Euler method. It is evident that proposed combined method is more effective than the Euler method. The efficiency indicator depends on the physical and geometrical characteristics of the honeycomb panel, as well as on the step size in the Euler method. In some cases, results of calculations using the proposed method showed about a hundredfold increase in the computational speed in comparison with calculations only by the Euler method. When solving problems based on real physical data with a step size that provides an accuracy of the order of 1 K, the acceleration was approximately 5-6 times.

5. Conclusion

The paper proposes a numerically statistical method for estimating the thermal state of heat-protective structures of the honeycomb type. The proposed method is based on the probabilistic representation of the solution of the parabolic boundary value problem. The computations are performed using a combination of the Euler and the random walk on moving spheres methods. A significant acceleration of the counting time is obtained in comparison with the computations by only the Euler method.

Acknowledgment

This work was supported in part by Russian Foundation for Basic Research grant 17-01-00698 A.

Reference

[1] Gusev S A and Nikolaev V N. 2014 *Adv. Mater. Res.* **1016** 758 - 63
[2] Dynkin E B 1965 *Markov Processes* (Springer Berlin)
[3] Gikhman I I and Skorohod A V 1982 *Stochastic Differental Equtions and their applications* (Naukova dumka Kiev)
[4] Kushner H J 1977 *Probability Methods for Approximations in Stochastic Control and for Elliptic Equations* (Academic Press. New York)
[5] Deaconu M and Herrmann S 2013 Ann. Appl. Prob. **23**(6) 2259 - 89
[6] Ladyženskaja O A, Solonnikov V A and Ural'ceva N N 1968 *Linear and quasi-linear equations of the parabolic type* Translations of Mathematical Monographs, **23** (Providence, RI: American Mathematical Society)
[7] Sobolev S L 1988 *Applications of Functional Analysis in Mathematical Physics* (Nauka Moskow)
[8] Kolmogorov A N and Fomin SV 1968 *Elements of the theory of functions and functional analysis* (Nauka. Moskow)
[9] Gusev S A 2015 Numerical Analysis and Applications Springer **8**(2) 122 - 34
[10] Revuz D and Yor M 1999 *Continuous Martingales and Brownian Motion* (Springer Berlin)

IOP Conf. Series: Materials Science and Engineering **302** (2018) 012046 doi:10.1088/1757-899X/302/1/012046

Hybrid Hydro Renewable Energy Storage Model

Asit Kr Dey

Department: Process Energy & Environmental Technology Station (University Of Johannesburg)

kdey@uj.ac.za

Abstract. This paper aims at presenting wind & tidal turbine pumped-storage solutions for improving the energy efficiency and economic sustainability of renewable energy systems. Indicated a viable option to solve problems of energy production, as well as in the integration of intermittent renewable energies, providing system flexibility due to energy load's fluctuation, as long as the storage of energy from intermittent sources. Sea water storage energy is one of the best and most efficient options in terms of renewable resources as an integrated solution allowing the improvement of the energy system elasticity and the global system efficiency.

1. Introduction

Sea water storage hydropower systems utilize elevation changes to store off-peak electricity for later use. Water is pumped from a lower reservoir through wind turbine & tidal turbine to a reservoir at a higher elevation during off-peak periods. Subsequently, water is allowed to flow back down to the lower reservoir, generating electricity in a method similar to a conventional hydropower plant. A seawater storage energy power station is identical to a regular pumped-storage system. The main difference consists of the lower reservoir, which is the sea presenting an unlimited volume of water and storage (Dam). A storage facility is needed, in order to store the excess of energy produced. Under this project, sea water storage presented itself as a good option to overcome the storage problem. Hence the integration of renewable energies in the system needs to be compensated by a storage solution that will help to solve the problem of demand and supply. The supply always needs to be guaranteed, and with the intermittent characteristic of renewable sources.

2. Methodologies

The concept used in sea water storage arrangements is moderately not complex. Water in a reservoir (Dam) of a given height has potential energy. When the water flows down either through a stream, it has a velocity due to gravitational force and this flowing water has kinetic energy. This kinetic energy can then be harvested and converted into electricity by the use of a turbine generator.

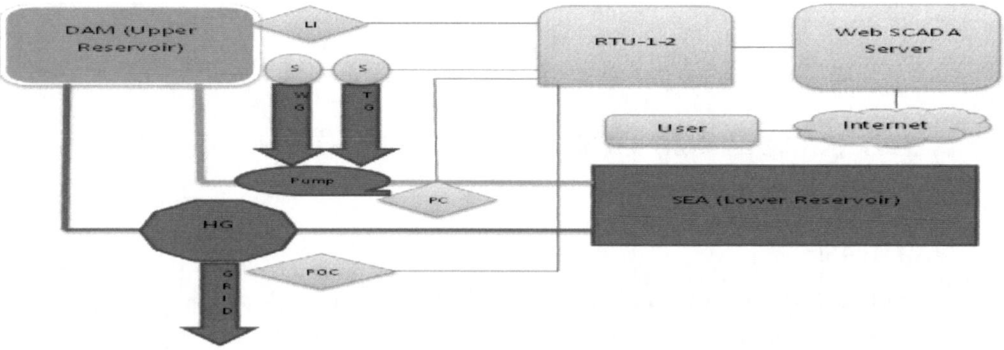

Figure1. Overall diagram for sea water storage energy modeling. HG: Hydro Generator, WG: Wine Generator, TG: Tidal Generator, S: Sensor, PC: Pump control, LI: Level indicator, PO: Power output control.

The height between where the water starts to flow and the turbine is known as the head height. At times of the high speed of wind or sea water current speed is high when the cost of electricity is low, pumps or pump turbines can be used to pump water into the Dam (upper reservoir) for later use. Besides just relying on off-peak electricity to fill the upper reservoir, excess energy generated by irregular renewable energy sources (wind & tidal) turbine can be used, thus making the facility act as storage for unused electricity [1]. Overall system shown in Figure 1.

A flow of sea water or wind speed with a given velocity passes through the turbine blades, which then turn the turbine that generates electricity [2]. That electricity feed to the pump for transfer sea water from low reservoir to high reservoir. Let's definite at the turbines output efficiency base on water level of the upper reservoir with integrated systems.

2.1. Turbines productivity

Tidal turbine: Various turbine designs have varying efficiencies and therefore varying power output. If the efficiency of the turbine Ɛ is known the equation below can be used to determine the power output of a turbine [3] [4]. The energy available from these kinetic systems can be expressed as: Pt = $\frac{\varepsilon \rho A V^3}{2}$

Where: P = the power generated (in watts). Ɛ = the turbine efficiency.ρ = the density of the water (seawater is 1025 kg/m³), A = the sweep area of the turbine (in m²), V = the velocity of the flow Relative to an open turbine in free stream, depending on the geometry of the shroud shrouded turbines are capable of as much as 3 to 4 times the power of the same turbine rotor in open flow.

Wind turbine: Let us consider total available wind power at wind speed v, passing through the cross-sectional area ρ, the power generated by a wind turbine is known by following expression: Pw = $\pi \eta \rho r^2 v^3 / 2$

Where ŋ is the overall efficiency of the system and r is the radius of the turbine [5].

2.2. Procedure of the system

To determine the number of wind turbine (N_{wind}), the number of tidal turbine (N_{tidal}), and the required dam water volume V_{water} is expressed as follows. The time limit t =t_0;t_1; ......t_T.Average powers P_{wind}(t) and P_{tidal}(t) respectively generated by WT and TT, and average power demanded by the load P_L(t), at every moment t in the time period [t_0;t_T]. The total power generate from outlet dam: $P_L(t) \leq N_{wind}P_{wind}(t) + N_{Tidal}P_{tidal}(t)$. The optimum water volume requires producing electricity from the generator: $P_{w-min}(t) = N_{wind}P_{wind}(t) + N_{Tidal}P_{tidal}(t) - P_L(t)$

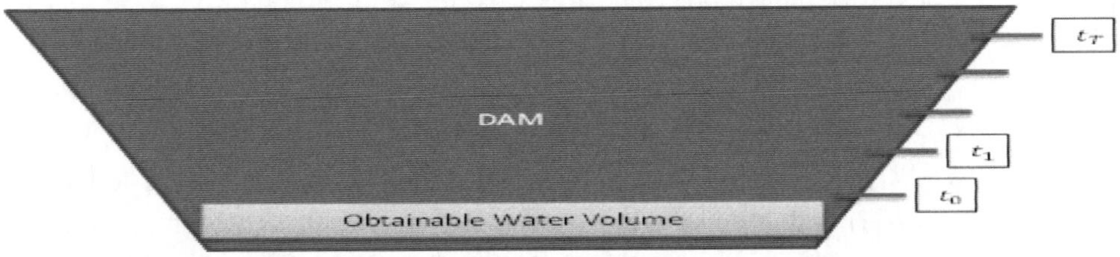

Figure 2. Optimal water volume indicator.

The optimal water volume of the dam determined by Obtainable Water Volume (OWV), shown in figure: 2. The available energy storage in the dam at the moment t: OWV(t) = OWV(t_0) + $\sum_{t=t_1}^{t} P_{w-min}(t)$ Where t =t_1;t_2; ......t_T. So total power generates from tidal turbine and wind turbine: $P_{L(t)} = N_{wind}\frac{\pi\eta r^2 v^3(t)}{2} + N_{Tidal}\frac{\varepsilon\rho A V^3}{2}$

To calculate the theoretical power available from falling water, equation was used: P= η × ρ × q × g × *h* Where P is power (W), η is turbine efficiency (%), ρ is density (kg/m3), q is water flow rate (m3/s), g is acceleration due to gravity (m/s2) and h is head height (m). As for the pump turbines to be used in the sites, the turbine theoretical output tool can be used to consider the control structure of pump turbines that is appropriate for the given sites [6] [7]. The number of turbines can be manipulated by having multiple penstocks connected to the mouth of the reservoir. In order to analyze the sea water storage energy system scenarios, daily based time setup transient control models for water flow were developed. Figure 3 shows the control diagram of the sea water transient model. The model works on a twofold level: water volume and energy parameter. At the water side, the water storage Dam will be account the water volume (W_{vol}) and water pressure in order to define the actual water level (W_{level}).

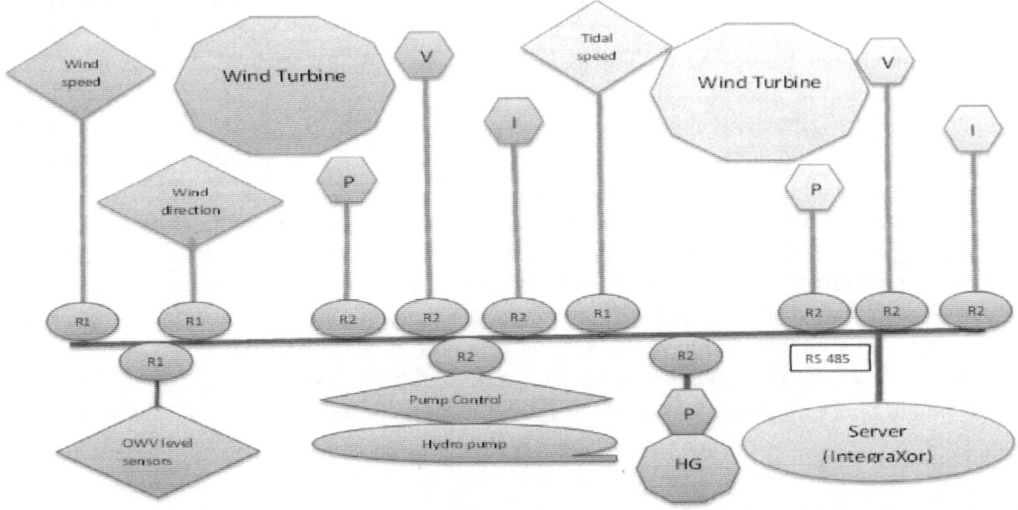

Figure 3. over all control diagram. (R1: RTU1, R2: RTU2, V: voltage, I: Current, P: Power, HG=Hydro Generator).

The control logic of the energy production is aimed at maintaining a daily autonomy water level in the Dam in case of power plant default, by indicating the power need to activate the power feedback unit PFU by (P_{feb}), shown in figure 4. To this end the storage capacity request will vary in depends on outsourcing water from outside. The energy control level defines the sources needed to supply the requested power to the PFU units (P_{feb}) by giving priority to the renewable source (P_{res}) in situation when wind and tidal turbine at on running mode. The initial water storage condition is settled to Obtainable Water Volume state as, whatever value is variable according to condition of the wind-tidal turbine [8].

2.3. Control configuration

The comprehensive hardware connection of the deliberate hybrid system containing two induction generators one is wind turbine (WG) and another is tidal turbine (TG). Power synchronize unit can combine two different forms of renewable energy together. The both power source will apply to the pump for store water (Energy) to the upper reservoir. The variable excitation capacitors C1 and C2 are respectively the self-excited capacitors of WG and TG and they can be effectively switched through the use of the controller of the PLC under various wind-tide speed and different loading conditions of the feeding pump. The fuses may serve as over-current and cut-off protective devices in case of faults while the contactors may execute parallel operation and isolation under dissimilar commands of the PLC. The digital power meters are installed for accessing the measured electrical data of the renewable energy power generating systems. The projected PLC also connects to a remote PC through site network. Hereafter, the proposed real-time monitoring and control system can use the load supervision arrangement to perform different functions such as wind-flow or tide-flow, data measurement and collection, pump status, real-time load monitoring and load-shedding control. The human-machine interface developed by software can collect, manage, and visualize both wind-flow or tide-flow data and information needed of the hybrid wind-tide system.

A Programmable logic controller is employed to control the Dam water level through wind-tidal turbine electricity powered pump system and two remote terminal units are employed to collect the data from sensors. A Delta made PLC is chosen in the implementation. Due the limitation of the number of Analog inputs and the capability to read the sensors, the data acquisition system used to control the wind-tidal-hydro turbine system. The RTUs are developed based on two microcontroller system by developing such microcontroller system, various sensor modules could be acquired easily. One microcontroller is used to obtain the environment data from solar ultrasonic sensor, integrated water level, wind speed sensor, wind direction sensor, and water current sensor. These sensors, except the wind speed sensor, are connected to the microcontroller via the Analog input. While the wind and sea water current, voltage and power sensor is connected via the digital input. All systems have their protection, control systems and power analyzer. Communication path starts at the data source device (generators, sensor, and power analyzer), it follows to a specific transducer (RS-232, RS-485 or others to TCP/IP), it continues to the network switch and finally to a PC. That configuration creates a network where only circulates relevant data circulates, but it also allows remote supervision. Each renewable energy source has different treatment for condition monitoring and fault diagnosis [9]. The system works with two kinds of models. First, mathematical models are used when the equations that represent equipment performance are known. That includes the speed of the wind module or the speed of the tide current; second, heuristic models are considered when only input-output signals are available and there is not enough information available about inside turbines behavior to build a mathematical model. After the implementation of each model, the following activities will carry out: simulation and calculation of standard deviations, model fitting, definition and refinement of fault analysis.

IOP Conf. Series: Materials Science and Engineering **302** (2018) 012046 doi:10.1088/1757-899X/302/1/012046

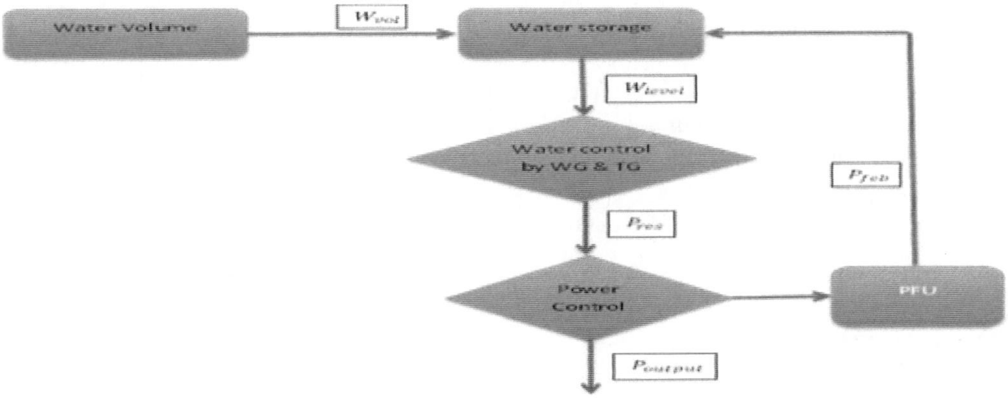

Figure 4. Flowchart for water-power control.

The control stagey decides on which situation last (nth) particular hydro pump will be switched ON or OFF and when a particular renewable energy source is to be connected or disconnected depending upon the water level of the Dam. Figure 5 show a flow chart, which reflect the logic information of all renewable energy source and water level condition of the dam. When wind-tidal turbine total power generation becomes less then set point power (P_{sp}), automatically switch off the last (nth) hydro pump. If 'no' then the total produce power by renewable energy sources is greater than set point power, for that condition last (nth) hydro pump will be enable (on state) and meet the demand for water level. On the other side when water level is high-high than all hydro pump must be off state.

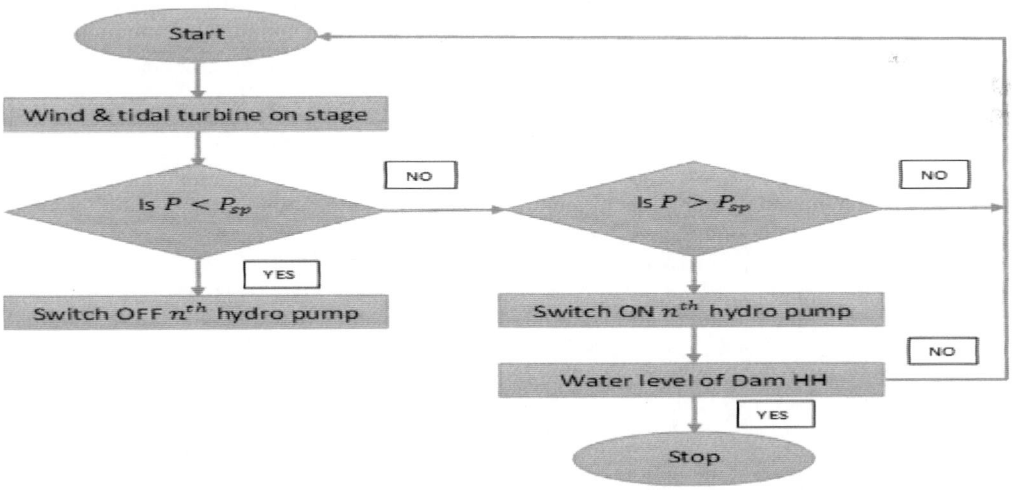

Figure 5. flowchart for Wind-tidal turbine running condition.

3. Conclusions

The present work presented a special storage system—the hybrid hydro energy storage system—which has proven to be a possible and a good energy option as described in the two case studies (i.e. based on renewable energy efficiency and sea-water control solutions). Renewable hydro-storage represents a technology of storing energy during periods of low demand. This technology is viable, since it uses electricity in low demand hours to pump water from sea and use it in hours of high demand. One of the goals was to verify and compare characteristic parameters that will lead to a successful implementation of such system. It was concluded that these systems present several positive characteristics, namely to store energy, to generate profit, to provide flexibility regarding start-ups and shut-downs and to compensate the energy load fluctuations. One of the major outcomes is the possibility to provide the storage of energy from intermittent sources (e.g. wind and tidal) making the whole system more flexible and reliable, with a better performance, and improving the elasticity and water-energy nexus most valorized. This project demonstrates how the profit can be maximized, being a strong alternative to fossil fuels. Studies regarding renewable energy penetration, investments, employment, decrease in imports and fuel costs and the reduction of CO^2 emission helped to prove this system viability.

Reference

[1] R. Thresher, P. Denholme, G. Hagerman, G. Heath, S. O'Neil, J. Paquette, D. Sandor, and S. Tegen, "Chapter 9. National Renewable Energy Laboratory. Renewable Electricity Futures Study," *Ocean Energy Technologies,* vol. 2, pp. 1 - 36, 2012.
[2] A. Williams, M. Nthontho, S. Chowdhury, and S. P. Chowdhury, "Modelling South African Agulhas marine current profile data for electricity generation," in *Power System Technology (POWERCON), 2012 IEEE International Conference on*, 2012, pp. 1-7.
[3] A. Goude and O. Ågren, "Simulations of a vertical axis turbine in a channel," *Renewable Energy,* vol. 63, pp. 477-485, 3 2014.
[4] F. O. Rourke, F. Boyle, and A. Reynolds, "Tidal Energy Update," *Applied Energy,* vol. 87, pp. 398 - 409, September 2016.
[5] A. M. Biadgo, A. Simonovic, D. Komarov, and S. Stupar, "Numerical and analytical investigation of vertical axis wind turbine," *FME Transactions,* vol. 41, pp. 49-58, 2013
[6] K. Rajagopalan and G. C. Nihous, "Estimates of global Ocean Thermal Energy Conversion (OTEC) resources using an ocean general circulation model," *Renewable Energy,* pp. 532-544, August 2015.
[7] J. N. Goundar and M. R. Ahmed, "Marine current energy resource assessment and design of a marine current turbine for Fiji," *Renewable Energy,* vol. 65, pp. 14-22, 5// 2014.
[8] S. H. Wright, S. P. Chowdhury, and S. Chowdhury, "A feasibility study for marine energy extraction from the Agulhas current," in *Power and Energy Society General Meeting, 2011 IEEE*, 2016, pp. 1-9.
[9] Basak, P.; Saha, A.K.; Chowdhury, S.; Chowdhury, S.P. "Microgrid: Control Techniques and Modeling." In Proceedings of the 2009 44th International Universities Power Engineering Conference (UPEC), Glasgow, UK, 1–4 September 2009; pp. 1–5.

IOP Conf. Series: Materials Science and Engineering **302** (2018) 012047 doi:10.1088/1757-899X/302/1/012047

Energy Efficiency Model for Induction Furnace

Asit Kr Dey

Department: Process Energy & Environmental Technology Station (University Of Johannesburg)

kdey@uj.ac.za

Abstract. In this paper, a system of a solar induction furnace unit was design to find out a new solution for the existing AC power consuming heating process through Supervisory control and data acquisition system. This unit can be connected directly to the DC system without any internal conversion inside the device. The performance of the new system solution is compared with the existing one in terms of power consumption and losses. This work also investigated energy save, system improvement, process control model in a foundry induction furnace heating framework corresponding to PV solar power supply. The results are analysed for long run in terms of saving energy and integrated process system. The data acquisition system base solar foundry plant is an extremely multifaceted system that can be run over an almost innumerable range of operating conditions, each characterized by specific energy consumption. Determining ideal operating conditions is a key challenge that requires the involvement of the latest automation technologies, each one contributing to allow not only the acquisition, processing, storage, retrieval and visualization of data, but also the implementation of automatic control strategies that can expand the achievement envelope in terms of melting process, safety and energy efficiency.

1. Introduction

Energy source and energy efficiency are the two main aspects for a sustainable growth of foundry sector. Renewable energy module of the foundry industry needs to develop up and implement sustainable practices based on Environmental adaptability. Energy resource constraint is an area which also needs to be focus, foundry industry depends on natural resources directly and indirectly PV solar is one of such best resources. To meet energy efficiency target, we need to utilize data acquisition technology to deliver castings at the desired quality level:

- Optimum heat losses from furnace.
- Reduce time delay during the process operation.
- Become Effective (productive), functional and fully perfect Castings process control.
- Linking sensor data from monitoring physical processes with analysis data.
- Alarm and emergency acknowledgements via the internet.
- At the lowest cost process.

To success in highly competitive global markets, foundry sector need to employ modern PV solar power source, heat preservation and data acquisition system that included energy enhance, supervision integrate process and concurrent engineering.

2. Methodology

Data acquisition, Solar energy source and heat preservation system can be a fundamental element in the modern foundry sector towards a sustainable energy future. Data acquisition system can contribute to improving the energy efficiency of foundry industry furnace heating system plants in different ways. Implementing better supervision, control and optimization strategies improves energy performance directly and indirectly, through better process control system practices that help to prevent and increase in energy use due to operation downtime and start up-shutdown process. The block diagram (Figure 02) shows to physique an induction heating furnace that produces efficient heat than the equivalent amount of electricity with modified source. Practically induction heating process required an impedance matching system which essential between the high frequency source and the output coil in order to ensure decent power flow. A control block diagram of monitoring system algorithm employed to control heating action and time heating cycle to ensure consistent output [1].

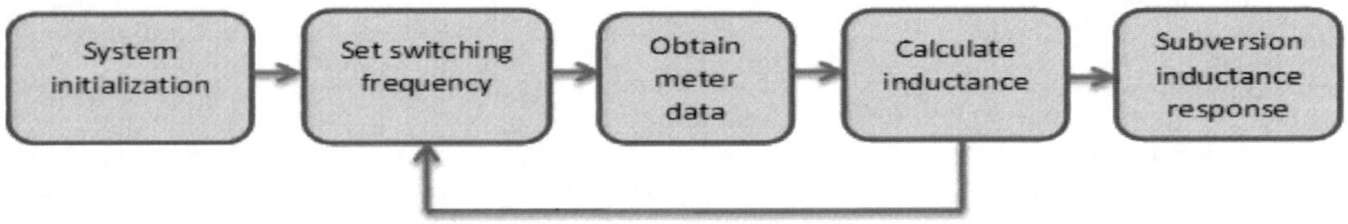

Figure 1. Block diagram for inductor characterization process.

A frequency control system is developed that automatically search for and operate at the natural load resonant frequency and continuously track this resonant frequency during the heating cycle as shown in figure1. This is accomplished by data acquisition system monitoring the load phase angle and altering the inverter operating frequency accordingly. Frequency is very important parameter in the design of induction heating power supplies because power components must be rated to operate at the critical frequency. In order to heat a solid piece of metal via induction heating require an incredible current to flow in the surface of the metal. However this can be contrasted with the inverter that generates the high frequency power. The inverter generally works better if it operates at fairly high voltage but a low current. Increasing the voltage and decreasing the current allows common switch mode MOSFETs to be used. During power MOSFET turn-off, voltage rise time is controlled by the ability of the gate drive circuit to sink the current being delivered to the gate by the gate-drain multi-capacitor as the drain voltage increases and thus charges the gate-drain capacitor. If the driver current exceeds the power MOSFET gate-drain feedback current during voltage switching, power MOSFET transistor will be off during the capacitive charging of the power MOSFET output capacitance. This charging is performed by an inductive element and charging is thus lossless. During power switched turn-on, the store capacitive charge in the power MOSFET output capacitor is however dissipated in the power MOSFET channel. At higher voltage levels or high switching frequencies such as in off-line applications, voltage switching losses can become a dominant loss factor.

During power MOSFET current turn-off, the negative current slop created by the decaying drain-source current, induces a negative voltage across the common source inductance. Since the common source inductance is also in series with the gate-source loop, the negative voltage induced across the common source inductance appears as a positive gate-source voltage on the power MOSFET. When power MOSFET gate-source voltage reaches the voltage level required to conduct the drain-source current, the system will be operating in its active linear mode having high drain-source voltage. Due to the negative feedback, the power MOSFET will effectively limit rate-of-current decay i.e. negative di/dt (DS) to a fixed upper limit. Any tendencies to exceed this upper limit will increase power MOSFET gate-source voltage further, causing power MOSFET to reduce the slop of the negative di/dt (DS), thereby effectively controlling maximum current switch off speed. In this mode, switch voltage levels never reach clamp voltage levels and converter current commutation is therefore controlled by power MOSFET maximum current turn-off di/dt. That maximum current turn-off speed in a power MOSFET is controlled by the common source inductance. Thus the lowest losses is obtained when converter current commutation is limited only by the power MOSFET maximum turn-off di/dt. The comparatively low currents make the inverter less sensitive to layout issues and stray inductance. This is the main work for matching inductor and working coil itself to transform the high-voltage/low current from inverter to the high current/low voltage required to the heat the workpiece efficiently.

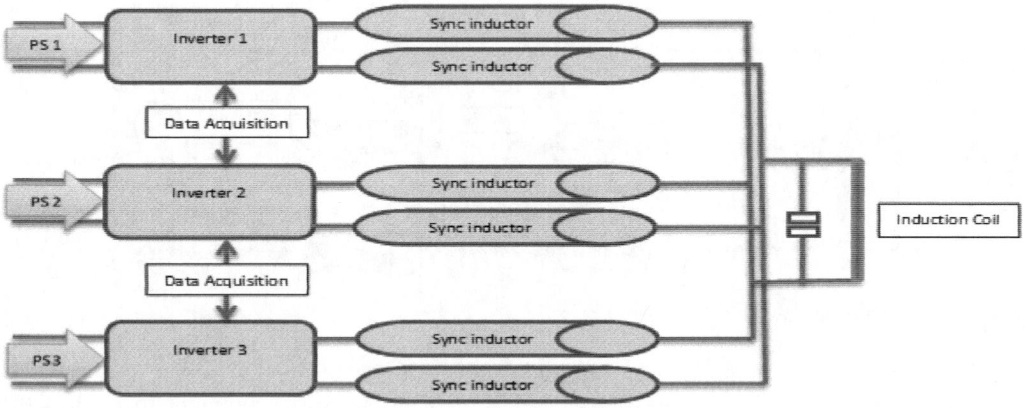

Figure 2. Induction furnace multiply power source (PS: power source)

Software outline: The main deal of the project is design and implementation of a Data Acquisition system for furnace efficient heating system of foundry, as shown in figure 3.

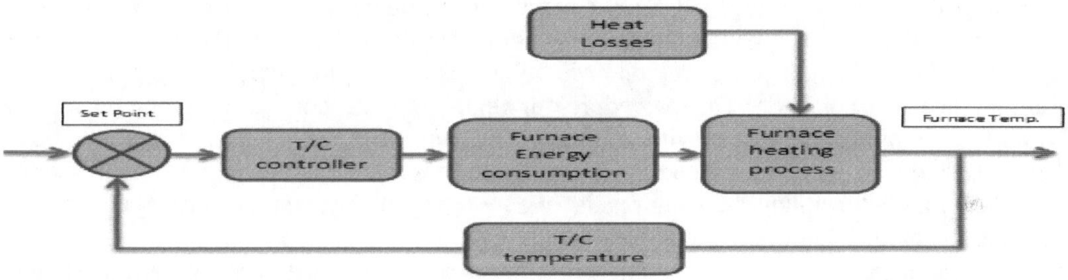

Figure 3. Process diagram for energy losses-heat control.

The system engaged with Schneider IGSS software and SICK sensors which is ideal for heat & frequency control monitoring system. It is use for reading data from over 1000 Analog and digital sensors, display synoptic images in a hierarchical structure, display events and alarms(over heating acknowledgement), command field devices, system diagnostics under user permission concept. Also allows web-based monitoring and command through Object linking embedding process control and web server by using the existing communication infrastructure. When the Manager of the acquisition system runs with different application; it should select the furnace number from overall system to read status of individual furnace, as shown in figure 4. Then manager sets the tolerance value and to calculate energy optimization according to heating conventional value. After that the data acquisition application provides the manager in the network result to make a decision based on outcome [2].

The total Energy consumption, P (t), of a solar induction furnace can be defined as: P (t) = V (t) I (t).

Where V (t) and I (t) are the solar source DC voltage and current respectively. With the voltage defined as: $V = RI(t) + K_v\,\omega(t)$

The power P (t) can be described as a function of the electrical current (I), and the angular velocity ω such as

$$P(t) = [RI(t) + K_v\,\omega(t)]I(t) = R^{I^2}(t) + K_v\,\omega(t)I(t). \qquad (1)$$

R is the resistance, Kv the electrical (back emf) constant, where this power implements by the voltage multiplier device [3].These analytical expressions are therefore very useful in designing and optimizing high input current converters for high conversion efficiency.

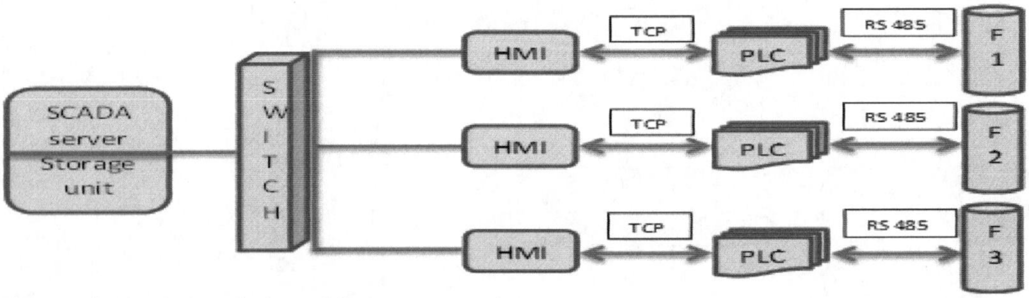

Figure 4. Hardware design of furnace control.

Voltage Multiplier Method: Here we are presenting an inductor-less solar power management system for induction furnace. A device is used some form of switching component to control the connection of voltages to the capacitor. Under low light intensity, the PV voltage is low and the voltage multiplier device steps up the voltage either for charging the battery or powering the induction furnace. At the same time, the optimal power tracking unit monitors the voltage multiplier device output power and determines the adjustment of the system operating parameter. Based on the adjustment decision, the PLC unit tunes the operating frequency of the device in order to maximize the system power output, shown in figure 5. Under a certain light intensity, the output of the PV cell behaves like a current source with a voltage limiter [4]. For different loadings, the PV cell will operate at different points, either in current source region or in voltage source region, where the output current or voltage almost keeps a constant. The procedure for regulating the converter output power via sustaining resonance is more convenient with annulling the phase difference the output voltage and current through the inductor than the procedure for correcting the phase between the output voltage and inductor voltage. Inductor voltage sensing results in a more significant phase error and also higher harmonic components are present in the inductor voltage.

The converter output power control with correcting the phase angles between the output voltage and current through the inductor. The presented phase control algorithm for the resonant converter shows that by supervision the output voltage or output current phase difference, for a particular resonant frequency, the converter output power can be controlled. Any deviation from the resonant frequency is structured by changing the phase angle between voltage and current of the converter. The voltage multiplier method proves that the iteration procedure is possible to perform adaptive regulating on the phase difference between output voltage and current of the converter and thus indirectly to achieve a new resonant frequency which corresponds to the changes for the resonant circuit values. Due to the effect of regulation, output current can affect the voltage stresses on multiplier's diodes and capacitors. Since regulation is directly proportional to output current, and as input voltage is usually increased to compensate for regulation, the diodes and capacitors near the input side of the multiplier will be subjected to higher voltage stress at the higher output current [5]. The programmable logic controller maintain constant output current. The controller essentially consists of an operational amplifier based proportional integral controller. The pulse width modulator generates the switch duty circle in response to the output voltage received from the controller. The current feedback circuitry consists of the necessary sensing and signal conditioning circuits required for closed loop control. The power controller also functions as a protection system for the voltage multiplier output stage, allowing the maximum current level and multiplier peak switching voltage to be set independently [6].

Load resonant converters allow for higher voltage multiplier switching frequencies to be achieved efficiently and also enable lower power rated switches to be employed due to the reduction in switching power losses. Due to the dynamic nature of induction heating loads, a considerable change in natural resonant frequency of the load circuit can occur during the heating cycle.

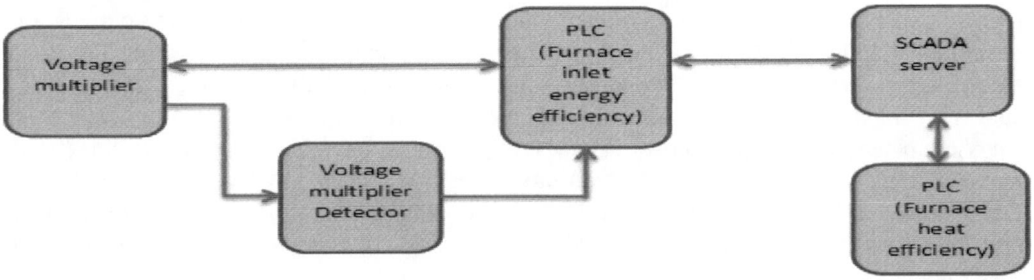

Figure5. Diagram for PLC control Voltage multiplier.

The PV output power also varies with loadings and the maximum output power point exists at the crossing point of the two regions. The voltage multiply device is a step-up voltage converter which consists of only capacitors and switches, as shown in figure 6. In order to prevent the reverse current through the switches of voltage multiplier when they are turned off, the switches size should be made small, and the driving capability of the gate control inverters should be strengthened [7]. At the same time, the size of the switches should guarantee the charge on the capacitors can be fully shared even when the device is operating at high frequencies. To supply maximum power to charge the battery or to directly drive the computation circuit, the data acquisition unit is used to monitor the amount of power flowing out of the device.

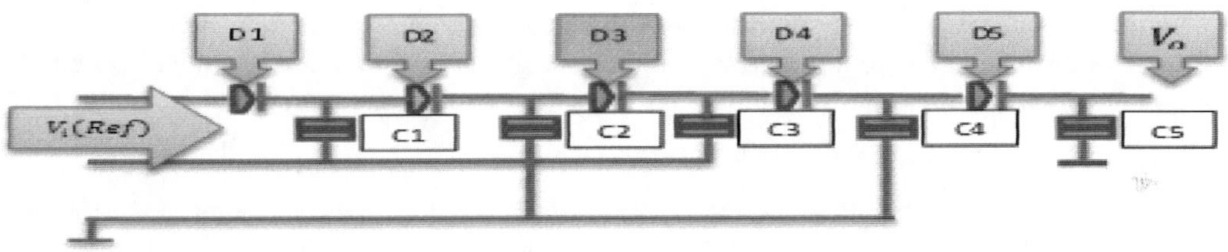

Figure 6. voltage multiplier diagram

The control unit is used to adjust the system operating parameters based on the decision from the programmable logic control unit in order to maximize the output power from the voltage multiplier device. Usually the period capacitors C_p are much smaller than the output capacitor C_o. Then, the output voltage changes very slowly and the circuit can be regarded as operating in the steady state. Here, steady state means that during each switching cycle, the charge being transferred into each period capacitor as well as the output is the same as the charge flowing out of them during the discharge period, and we denoted it as Q. In each switching cycle, let the voltage at the ith period capacitor be V_i after it is charged by the previous stage. At the second clock phase, the switch between the ith and $i+1^{th}$ stage capacitors is on and charge sharing occurs. After that, the voltages on both sides of the switches equals to V_{i+1}. At the same time, the voltage at the other plate of the ith stage capacitor, which is driven by the clock buffer equals to the voltage control oscillator power supply voltage V_{ph}. The voltage across the ith stage capacitor thus equals to (V_{i+1} - V_{ph}). For the ith capacitor:

$$C_p\,(V_{i+1} - V_{ph}) + Q = C_p V_i,\ i = 1,2,..\ n.$$
$$Q = C_p V_i + V_{ph} C_p - C_p V_{i+1} \qquad (2)$$
$$I_o = f_{clk} Q$$

There are Q amount of charge being transferred to the output, the input to the voltage multiplier device has to provide Q charge to each stage capacitors either directly or through the clock (f_{clk}) buffers. So total power feeding to the induction furnace is:

$P(t) = R\{f_{clk} Q\}^2 + Kv\ \omega\ (t)\ f_{clk}Q$. Where $Q = C_p V_i + V_{ph} C_p - C_p V_{i+1}$. (from equation no. 2).

The actual V-I characteristic of the PV cells is different from the ideal voltage and current source. The output current will increase with the decrease of the PV voltage, even at the current source region [8]. We can see in figure no 5 that in order to increase the voltage multiplier device output current, we should increase the PV cells output current i_{ph}, and reduce the frequency dependent current loss. The system output voltage is regulated by the battery and the voltage of the battery changes very slowly; we can assume that maximizing the system output power is equivalent to maximizing the system output current. Here, we use the current sensor to measure the voltage multiplier output current. Based on the measured current value, the data acquisition units check whether the system is at the optimal point. An inductor-less voltage multiplier device was used to step up the PV voltage for the battery charging. In order to maximize the power outputted from the system, the system operating behavior was theoretically analyzed.

3. Conclusions

In this project design was divided into two different section. Voltage multiplier use for operating frequency of the device in order to maximize the system power output. And additional circuits were used to link the generated control signals to the MOSFET transistors of the inverter circuit. The paper refereed to control of the manufacturing processes in foundry with aid of systems belonging to the data monitoring system. Application of a system in the foundry requires a special approach. Three main areas present the future evolution of this design: Heating optimization, energy efficiency and data acquisition system. A Circuit model for the optimal power control algorithm was discussed where low power techniques are employed to avoid the power deficiency and utilizes a converter topology from other applications to create a unique contribution to the induction furnace of foundry. Data acquisition system is very useful for foundry sector when it is necessary to control and supervise heating system, being able not only to work items in the system, but also to warn about various warnings. The use of this system enables constant power converter operation, energy saving and technological processes effectuating.

References

[1] Gandhewar, V.R., Bansod, S.V., Borade, A.B.(2011).Induction Furnace-A Review,International Journal of Engineering and Technology Vol.3 (4), 277-284.

[2] Sika, R., Ignaszak, Z., Rogalewicz, M. (2014). In Polska Metalurgia 2011-2014, Świątkowski, K. (Eds.), Specificity of application of the spc methods in complex manufacturing process on example of foundry (357-371). Krynica Zdrój: Akapit.

[3] Bala, K.C. (2005). Design Analysis of an Electric Induction Furnace for Melting Aluminum Scrap, AU J.T. 9(2): 83-88.

[4] Ignaszak, Z. & Sika, R. (2012). Specificity of SPC procedures application in foundry in aspect of Data Acquisition and Data Mining. Archives of Foundry Engineering. 12(4), 65-70. DOI: 10.2478/v10266-012- 0108-8.

[5] Yin Y., Regan Z.: Digital Phase Control for Resonant Inverters, IEEE Power Electronics Letters, Vol. 2, June 2004.

[6] K. Agbossou, J. L. Dion, S. Carignan, M. Abdelkrim and A. Cheriti, "Class D Amplifier for a Power Piezoelectric Load", IEEE Trans. On Ultrasonics, Ferroelectrics and Frequency Control, July 2000, vol. 47, no. 4, pp. 1036- 1041.

[7] Anaidhuno, P. U. (2014). Construction of an Electric Heat Treatment Furnace, M. Eng Thesis; Mechanical Engineering Department; Anambra State University, Nigeria

[8] A.I. Martinez and H. Calleja, "A simple, high-quality output PV system aimed at peak demand reduction," in Conference Publication of the VII IEEE international Power Electronics Congress, CIEP 2000, 15-19 Oct. 2000, pp.301-306.

Material Selection for Cable Gland to Improved Reliability of the High-hazard Industries

S P Vashchuk[1,a], **S M Slobodyan**[2], **V S Deeva**[3] **and D S Vashchuk**

[1] Samara University, ul. Moskovskoe tr. 34, Samara, 443086, Russia
[2] Omsk State Technical University, av. Mira 11, Omsk, 644050, Russia
[3] Tomsk Polytechnic University, av. Lenin, Tomsk, 634050, Russia

[a] Corresponding author's E-mail address: amurgermovvod@mail.ru

Abstract. The sealed cable glands (SCG) are available to ensure safest connection sheathed single wire for the hazard production facility (nuclear power plant and others) the same as pilot cable, control cables, radio-frequency cables et al. In this paper, we investigate the specifics of the material selection of SCG with the express aim of hazardous man-made facility. We discuss the safe working conditions for cable glands. The research indicates the sintering powdered metals cables provide the reliability growth due to their properties. A number of studies have demonstrated the verification of material selection. On the face of it, we make findings indicating that double glazed sealed units could enhance reliability. We had evaluated sample reliability under fire conditions, seismic load, and pressure containment failure. We used the samples mineral insulated thermocouple cable.

1. Introduction

Annual growth of atomic power stations that are being build and power supply units additionally brought into service in nuclear power stations (China, Iran, India, a new atomic station in Turkey) that are potentially hazard production facilities, results in increasing requirements of their functional safety. Different types of cable lines are used in nuclear and space industries to provide communication, power supply and control of power and nuclear plants, complex components, machines and vehicles, diagnostics of radiation-thermal condition of equipment components of hazard production facilities [1–4]. Solution of these tasks stipulates necessity to introduce pilot cables, radio-frequency cables, and control cables into the envelope of a hazard production facility. Cable glands are intended to provide tight cable introduction into the hazard production facility. They provide communication and information interaction of parts of the system or parts of different systems widely or nearly spaced.

Interaction of object parts is carried out via a great number of sealed cable glands of radio-frequency signals. Moreover, sealed cable glands are necessary for feeding circuits, for controlling and measuring inside the nuclear plants. Their operation provides key functions inside the reactor, and their insulation is to withstand pressure inside the nuclear containment. Organic polymers (epoxy resin) meet [1] general standards of insulation for sealed cable glands at nuclear plants of the first and the second generations. Experts in the sphere of nuclear power engineering are concerned that in case of severe accidents integrity of epoxy sealed cable glands can be damaged. Quality of technical condition of sealed cable glands, their efficiency characterized by the signal transfer ratio is critical when an emergency arises. It is exceptionally important to provide high stability of electrical, radiation, thermal, and thermodynamic elements – sealed cable glands (SCG) of pilot and power lines of power supply units of hazard production facilities [1,5]. This defines research objectives and analysis of tasks stated below.

2. Objectives and tasks of analysis

The problem of providing ideal transfer or close to ideal properties of sealed cable glands can be solved by means of searching a rational material composition and developing an optimum SCG unit design. Preceding long years of research followed stated results of searching ways and methods of solutions, procedures of variant choice that meet requirements and performance standards of reliability and SCG stability in critical and emergency situations at radiation and hazard facilities. In this paper, requirements for SCGs, research results and experience accumulated during development of reliable SCGs that are resistant to critical and emergency situations have been systematized.

3. Research fundamentals

Cable glands based on polymer compounds [5–6] are widely used nowadays. This tendency predominating in the world practice has some disadvantages. In particular, ageing processes take place under natural and artificial radiation. In operation conditions ageing action results in increase of mechanical stresses in polymer materials that cause material cracking, and as a result, loss of SCG tightness. In fire condition (extreme emergency), polymer components are subjected to inflaming when the temperature is above 300° C. It results in loss of sealing in SCG, loss of its durability and escaping of ionizing radiation fluxes into the environment that surrounds the nuclear plant. The search strategy of materials choice for elements of new SCG designs is determined by the engineering requirements based on their actual operating conditions in the structure of nuclear power plants. Increase of safety requirements of such objects (nuclear plants) results in establishing of particular specifications and their combinations in normative documents that regulate safety parameters for the containment systems and their components in case of non-project accidents. Imposing these requirements is reasonable, because SCGs are the components of containment systems that are critical for safety. Moreover, SCGs perform localizing and providing functions of technogenic-hazard object.
Main requirements imposed on materials in designing new radiation-hardened SCGs are high resistant to complex effect of ionizing radiation fluxes in a wide range of changes of environment effect in operational conditions of SCGs, to physical-mechanical actions of thermodynamic pattern, and also, to solution optimality of mass-dimensional indexes task.

Negative experience confirms this. Nuclear plant "Fukushima-1" in Tohoku region [6] was damaged by the 9-earthquake intensity on the Richter scale on 11 April 2011. This earthquake resulted in tsunami wave of up to 40.5m height. Tsunami caused catastrophes of three nuclear reactors of nuclear complex "Fukushima-1" where hydrogen explosion was recorded. This hydrogen accumulated beyond the outer reactor shielding after the cooling system failure.

Experts of the operator-company of nuclear plant "Fukushima-1" analyzed the accident and came to conclusion that in the accident, the temperature inside the protective shielding increased more than four times in comparison with the operating one. The design pressure exceeded more than two times. Exceptionally high level of temperature and pressure resulted in deformation of the whole set of epoxy sealed cable glands. This, in turn, caused leakage of highly explosive hydrogen. On completion of the investigation, it can be affirmed, that it was hydrogen leakage through the SCG that resulted to the explosion of "Fukushima-1". This proves once more the fact that SCGs play a key role in providing object safety. Taking into account experience with "Fukushima-1" accident [6] that happened because of seismic cataclysm it can be stated that SCG material in some cases is to be resistant to a wide range of thermal shock loads arising in critical and emergency situations. In other words, characteristics of SCG insulating basis greatly effect on plant safety and reliability. During operation of a cable line with SCG ageing of polymer insulation accelerated by electromagnetic impact of current through the cable causes electro physical performance degradation [5, 6].

High power current in the cable line generates thermo- and electrodynamic of factors effecting on alteration of volume of SCG elements. Appearing small mutual shifts of contact surfaces of SCG elements caused by thermomechanical factors and electromagnetic impact of high power Ampere current in the line can be effectively diagnosed by means of differential methods [7–8] of optical and laser engineering [9–11] with high or sufficient accuracy of measurement comparable with the wave length fraction of probe radiation. Propagation of damaging action of thermodynamic factors and mechanical deformation of SCG elements results in material properties destruction and rise of acoustic emission effect that can be recorded by means of developed methods of acoustic diagnostics [12].

Briefly, the essence of main requirements is the following [1, 4, 6]. SCG is: to have higher radiation resistance under action of gamma-radiation (accumulated radiation dose average on great number of accidents $\approx 5 \cdot 10^8$ rad); to consist of incombustible materials having chemical resistance under high temperatures (standard fire mode – up to 990°C); to withstand action of decontamination fluids and to meet natural conditions of corresponding State Standards.

To solve the problem of material radiation resistance, analysis of current condition of material properties research suited for SCG has been carried out. The results of large quantity of different authors' papers have been taken into account [1–10]. Moreover, research carried by the authors in cooperation with the group of scientists of Institute of Nuclear Physics of Uzbek Academy of Sciences. Results of generalized analysis of practical operation conditions of SCG and in the best interest of ionizing radiation factors effecting on material properties in critical situations displaying data for determination of extreme parameters of SCG operation modes are shown in the table 1.

Table 1. Operation modes

Parameters	Normal operation mode	Heat removal disturbance mode of the pressurized area	Minor leak mode	Maximum leak mode
1. Temperature, °C	15-60	< 75	< 90	< 150
2. Pressure (discharging) MPa (atm)	0,85-0,103 (8,5-1,03)	0,05-0,12 (0,5-1,2)	0,17 (1,7)	< 0,5
3. Humidity,%	90 at T=60 °C	100	Steam-air mixture	Steam-air mixture
4. Specific activity Ci/L (Bq/L)	$< 2 \cdot 10^{-6}$ $(7{,}4 \cdot 10^4)$	$< 2 \cdot 10^{-6}$ $(7{,}4 \cdot 10^4)$	$< 1{,}5 \cdot 10^{-4}$ $(5{,}56 \cdot 10^6)$	< 2,5 $(9{,}25 \cdot 10^{10})$
5. Absorbed dose rate, rad/h, Gy/h	< 100 (1)	< 100 (1)	< 100 (1)	$< 10^5$ (10^3)
6. Mode action time, h	–	< 15	< 5	< 10
7. Rate of mode occurrence	–	Once a year	Every other year	Once a life time
8. Post-accident temperature, °C	–	–	< 60	
9.Post-accident pressure, MPa	–	–	0,05-0,12	0,05-0,12
10. Action time of post-accident modes, day	–	–	30	30

4. Aspects of rational material choice

While analyzing properties of different materials, it is possible to come to conclusion that owing to specific character of chemical bonds some materials are not virtually affected gamma-radiation factors. These materials are copper, steel, nickel, molybdenum, solders based on copper and silver. These materials are radiation-resistant. Property of radiation resistance means that it is possible to use these materials for a long period under any gamma-radiation dose. Therefore, radiation-resistant solders based on copper and silver were chosen for designing of SCGs for nuclear power plants.

In SCGs designed by our group, cables with copper current-carrying conductor and electrical insulation based on magnesium oxide of MICSC (mineral insulated copper sheathered cable) type have been used. SCGs use thermocouple cables with mineral insulation and current-carrying conductors of alumel and chromel. Besides, cables with mineral insulation of MICSC type have copper or steel shell, and some current-carrying conductors sealed with magnesia insulation against the shell are inside this shell. Such radiation-resistant and mechanically-resistant cables are unaffected by burning (they resist temperature up to 1000 °C). When choosing insulating materials, three groups of materials were researched: ceramic materials, materials having glass components, and polymer-based materials. Among the mentioned materials, the following ceramic insulating materials have rather high radiation resistance: magnesia and high-aluminous ceramics, steatites and porcelains. Electro-technical glass has similar properties.

When analyzing opportunity of developing practical SCGs, the fact was taken into account that nowadays polymers are widely used in SCGs. Polymers are used for sealing ends of cable modules and for tight connection of cable modules with SCG flanges. Structures using polymers have some disadvantages. SCGs are subjected to gamma-radiation, oxygen, moisture, air, mechanical and electrical stresses, and thermal action. Gamma-radiation impact on polymer materials in combination with listed factors accelerates negative processes happening in polymers. With time, sealing units lose their tightness under gamma-radiation because of wide range of processes taking place in polymer components. These processes include: destruction, crosslinking, intramolecular bond formation, decomposition and formation of conjugate double bands, radiation oxidation, and crystallinity change. Free radicals having increased activity are also formed. Polymer materials radiation happens in the medium of very active chemical element – oxygen. Combined action of these factors results in decreasing mechanical strength and accelerates destruction, and as a result, break of polymer molecule chains and breaks of internal and intermolecular chemical bonds in cross-linked polymers [1]. Crosslinking, radiolysis and destruction processes are followed by element volume alteration, making it brittle, gas products emission. All this taken together results in loss of tightness. Besides, using cables with polymer insulation inside the sealed volume of sealed cable glands, long-term gamma-radiation action to polymer insulation results in pressure increase due to radiolysis gas products. In the paper [1] it is mentioned, that the lifetime of polymer materials under simultaneous action of gamma-radiation and high temperature is far less than the lifetime under the same, but successive actions. Gamma-radiation of polymers accelerates post-radiation processes caused by reactions of trapped free radicals. As an absorbed dose increases, crosslinking rate also increases. At the same time accumulation of fragments of disrupt molecules takes place, this worsens material properties and results in its destruction. Gamma-radiation do not practically influence on glass, ceramics and metals in contrast to polymers. Glass sealing in SCGs has some disadvantages – it is poorly protected from mechanical action that causes loss of tightness of a unit. Figure 1 shows overall view of the SCG based on thermocouple cables with mineral insulation.

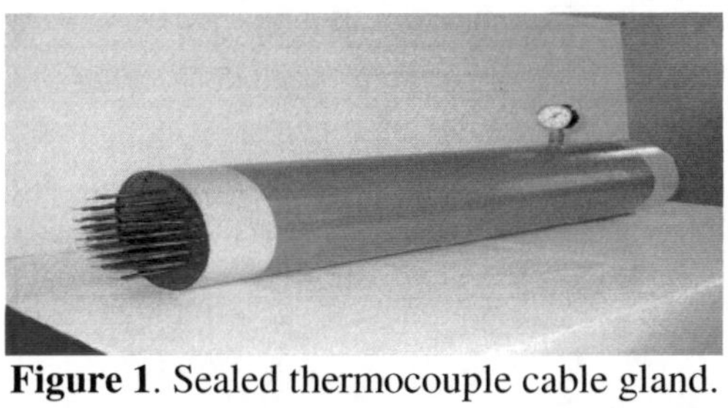

Figure 1. Sealed thermocouple cable gland.

Having analyzed practical and theoretical research of above mentioned three groups of materials and having compared given γ-radiation doses after exceeding of which substantial decreasing (by 25%) electro-physical material properties starts (for ceramics, glass and metals), it is possible to come to conclusion: radiation resistance of ceramics and metals is incomparably higher than of polymer materials.

In the process of reasonable choice of terminal cable ceramic insulators for test samples of sealed cable glands research of some kinds of metal-ceramic units (MCU) based on high-aluminous ceramics microlite were performed. MCUs have been researched before and after gamma-radiation action by $5 \cdot 10^8$ rad dose in a steam-air medium adding decontamination mixture at temperature 150 °C and pressure 0,5 MPa. As a result of research, it has been found that MCUs based on ceramics of Microlite type provides higher vacuum tightness (up to 85%) in comparison with MCUs based on ceramics, where vacuum tightness is lower (by 7–8%). Electrical resistance of MCU insulation under normal conditions is rather high $5 \cdot 10^{12}$–$5 \cdot 10^{13}$Om. After gamma-radiation action by $5 \cdot 10^8$ rad dose magnitude of electrical resistance of insulation medium changes slightly. If MCUs are placed into a steam-air medium and in case of creating modes of large and small break accident, MCU resistance falls dramatically up to 10^5–10^6 Om, at the same time MCU completely keeps its tightness. Following the research, it has been found that metal-ceramic insulators based on Microlite have higher characteristics. For high reliability of components and, in general, for SCG construction, components based on MCU [1, 13–21] have been chosen.

Metal-ceramic module is rather simple. It consists of a ceramic tube hermetically soldered with front and rear cuffs. The front cuff is hermetically joined with an input cable conductor, and the rear cuff is joined with an output cable shell. In the developed SCG thermocouple cables KTMC(XK)-2×0,9 or KTMC(XA)-2×0,9 are used. Cable ends are sealed with two-channel metal-ceramic insulators or glass tablets made of glass-ceramics. The equivalent circuit model to measure electrical characteristics of the SCG is demonstrated in Figure 2.

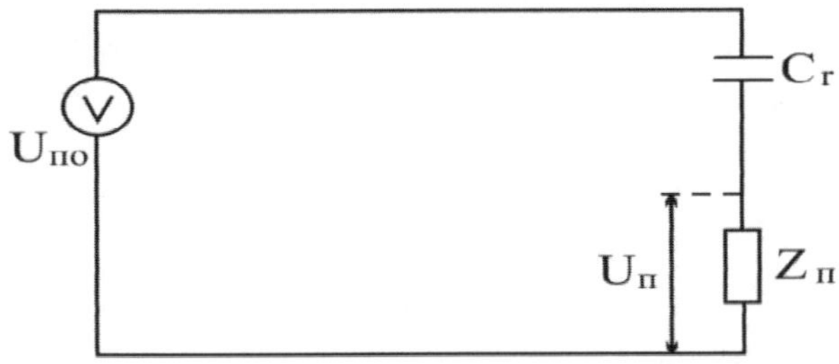

Figure 2. The schematic diagram of connections to qualify rating SCG and to determine the level of interfering signal in single conductor screened cable: V – voltmeter to measure voltage drop (interfering signal); U_{no} – test signal voltage; U_n – drop voltage due to the interfering signal through circuit resistance; C_r – cable capacitance per length; Z_n – impedance of the measuring circuit for interfering signal.

Some experimental results obtained during the research of electrical parameters of SCG are shown in the Table 2.

Table 2. Results of experimental measurements of electrical parameters of SCGs.

Cable type	C_p, pF/m	tg δ	Z, Om	α, dB/m	$(\varepsilon_r)^{1/2}$	ε_r
MICSC-1×4	475	$8\cdot10^{-4}$	15	$10,4\cdot10^{-3}$	2,15	4,6
MICSC-7×2,5	228	$16\cdot10^{-4}$	32	$4,5\cdot10^{-3}$	2,20	4,8

Calculated values of electrical parameters of MICSC-1×4 cable and SCG are obtained by means of methods stated in [1, 22–23].

For SCGs of SPCG type (sealed pilot cable gland) different steels have been used (for example, Steel-3), solders based on copper and silver, cables with copper or steel shell and mineral insulation (type MICSC), ceramic insulators. These materials have high radiation resistance. Moreover, they have high fire resistance, in contrast to polymers decomposing at high temperatures and releasing toxic substances. In addition, the materials are protected with special coverings with high resistance to action of damp tropic climate and decontamination solutions.

5. Conclusion

SCG based on metal-ceramic materials and cables with mineral insulation have enhanced heat stability under radiation and other emergency factors. The method to evaluate an effect of an electromagnetic field to parameters of circuits with SCG pilot cables of MICSC type has been developed. It has been proved that electrical parameters of MICSC type cables permit using them in SCG pilot cables at technical high-hazard objects.

Acknowledgments

This work was supported by the Ministry of Education and Science of the Russian Federation in the framework of the implementation of the Program «Research and development on priority directions of scientific-technological complex of Russia for 2014–2020»

References

[1] Kostyukov N S, Vashchuk S P, Pakulov G I and Kholodnyi S D 1993 *Atomic Energy* **75** 983
[2] Romanishina S A, Katyuk D Y, Deeva V S and Slobodyan S M 2015 *IEEE 35th Int. Conf. on Electronics and Nanotechnology* (*ELNANO2015*) – Conf. Proc. 7146848 p 116
[3] Vislov I S, Pishchulin V P, Slobodyan S M and Klad S 2016 *Thermal Eng.* **63** 581
[4] Rahn F J, Adamantiades A G, Kenton J E and Braun C 1984 *A Guide to Nuclear Power Technology* (New York: Wiley–Interscience)
[5] Deeva V S, Slobodyan S M and Teterin V S 2016 *Materials Science Forum* **870** 677
[6] Bol'shov P A 2016 *Atomic Energy* **121** 3
[7] Bol'shanin A A, Slobodyan S M, Yakovlev A R and Vasil'eva L A 1987 *Measur. Techn.* **30** 954
[8] Slobodyan S M 2006 *Measur. Techn.* **49** 1
[9] Volkov V F, Peshel' A, Slobodyan S M and Tyryshkin I S 1981 *Instrum. Exper.Techn.* **24** 1522
[10] Slobodyan S M 1992 *Proc. SPIE Acquisition, Tracking, and Pointing VI* **1697** 213
[11] Slobodyan S M 2003 *Measur. Techn.* **46** 28
[12] Slobodyan M S, Shishigin S A and Slobodyan S M 2008 *Measur. Techn.* **51** 798
[13] Mayer M 1926 *Sicherheit der Bauwerte und ihre Berechnung nach Granzkraften Statt nach zulassigen Spannungen* (Berlin: Springer Verlag)
[14] Arutyunov V A , Mel'nikov V G, Slobodyan S M, Chaporov D P and Popov O N 1983 *Measur. Techn.* **26** 641
[15] Slobodyan M S, Slobodyan S M and Tsupin A A 2008 *J. Opt. Techn.* **75** 301
[16] Deeva V S and Slobodyan S M 2017 *Tribology Int.* **105** 299
[17] Lukin V, Sazanovich V and Slobodyan S M 1980 *Radiophys. Quant. Electron.* **23** 484
[18] Deeva V S, Ivoylov E V, Slobodyan S M and Sekachev A F 2016 *Dynamics of Systems, Mechanisms and Machines*, Dynamics 2016 p 7818999
[19] Deeva V S, Slobodyan S M, Vashchuk S P and Voronina N V 2017 *Proc. of Mech. Eng. Res. Day* (Kampus Teknologi: UTeM) p 1
[20] Deeva V, Slobodyan S and Martikyan M 2016 *Materials Today*: *Proc.* **3** 3114
[21] GOST R 27.002–2009 (Moscow: Standartinform)
[22] Anders G J 2005 Rating of electric power cables in unfavorable thermal environment (New York: Wiley–IEEE press)
[23] Kholodnyi S D, Serebryannikov S V and Boev M A 2009 Test and diagnostics methods in electrical insulating and cable engineering. (Moscow: MEI)

Creating an Electronic Reference and Information Database for Computer-aided ECM Design

M.V. Nekhoroshev, N.D. Pronichev and G.V. Smirnov

Samara University, 34, Moscow Highway, Samara, 443086, Russia

E-mail: maxnogood@ssau.ru

Abstract. The paper presents a review on electrochemical shaping. An algorithm has been developed to implement a computer shaping model applicable to pulse electrochemical machining. For that purpose, the characteristics of pulse current occurring in electrochemical machining of aviation materials have been studied. Based on integrating the experimental results and comprehensive electrochemical machining process data modeling, a subsystem for computer-aided design of electrochemical machining for gas turbine engine blades has been developed; the subsystem was implemented in the Teamcenter PLM system.

1. Introduction

Currently, aviation engine manufacturers are being transformed to create innovative production processes. Such transformations are based on technical rearmament, organizational enhancements, and implementation of business management principles. Comprehensive informatization is the foundation of innovative production. This is a complex problem most enterprises are trying to solve step-by-step, which is fairly slow. The first stage there is usually about creating computer-aided pre-production design systems, which are built of modules. Some enterprises implement CAD/CAM systems that enable the creation of solid-state 3D models which are then used for automated design of processes and CNC programs. CAE systems are used for engineering analysis of processes occurring in technological systems.

State-of-the-art aviation engine manufacturing is a multifunctional process that contains multiple design objects; computer-aided preproduction design management requires use of PLM systems based on the storing and using product data collected at certain stages of its service life.

This research is to automate the preproduction processes relating to the implementation of electrochemical machining (ECM) of gas turbine engine (GTE) parts. Figure 1 shows the sequence and description of problems to be solved. These problems have been solved by various specialists working in this area of research.

2. Materials and methods

For instance, paper [1] gives detailed examples of physical and mathematical models that became the basis for a computer-aided ECM design module. The module enabled the simulation of processes occurring in an electrochemical electrochemical channel.

Paper [2] is about using a CAD or a CAE system that enabled 3D modeling of ECM processes. The models employed finite elements to analyze the distribution of electric field in the interelectrode gap while factoring in whatever could affect such distribution (e.g. the flow and the passivation effect of electrolyte). The results were made into a database.

Paper [3] is about using a processing simulation method based on finite elements with a spherical cathode to create an adequate numerical ECM control model. To do that, they had used ANSYS to 2D-model the distribution of the electric field and the current density. Results were verified by physical experiments. It was thereby what that the modeling method was on par with engineering calcuations in terms of accuracy.

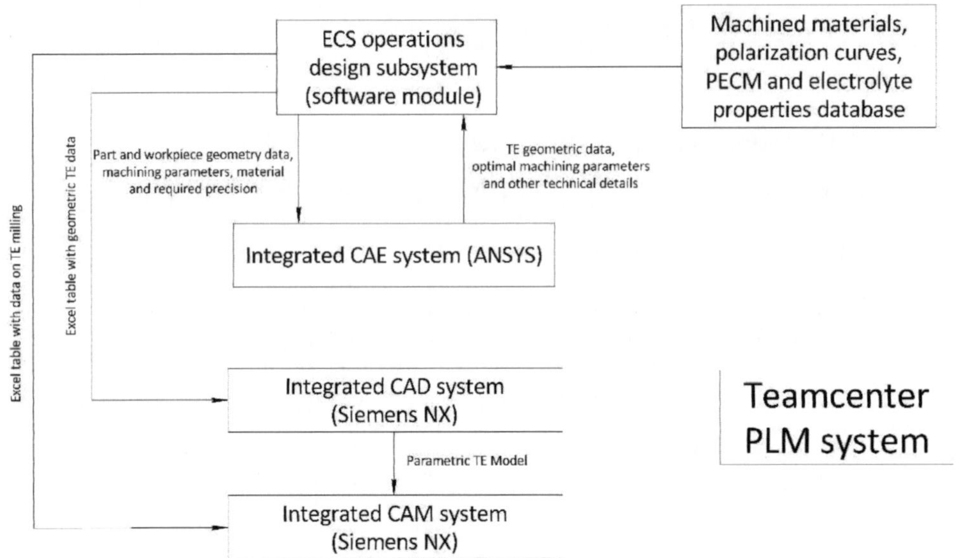

Figure 1. General computer-aided ECM design diagram

In paper [4], the authors designed an ECM process model for subsequent tool electrode profiling. This model was created was based on a COMSOL module and had a number of considerable assumptions to simplify calculations; however, it performed well.

Paper [5] covers the issue of optimizing the electrochemical machining of hard-alloy materials. Its authors used a mathematical algorithm of their own making and obtained mode parameters regression equations. They believed that this algorithm was superior to others in terms of how quickly it could return an optimal values, which is crucial for ECM design cost reduction.

Paper [6] describes an original ECM visualization method that employed a high-speed video camera synchronized with the characteristic dissolution points. The data were then confirmed by COMSOL simulation of the dissolution process. If used properly, visualization is a powerful tool that allows one to visualize the process at various stages to adjust them when designing an ECM operation.

In papers on similar topics [7,8,9,10], the authors solved the problems of modeling the ECM process in geometrically complex the interelectrode gap (IEG), an inter-blade gap of a mono-wheel, to describe the complex electrode trajectory. Simulation-obtained data were used in CAD/CAM systems for CNC machining. In the process of modeling, some parameters were assumed to be constant and would be determined by preliminary experiments. The solution optimized the electrolyte flow rate as well as the shape of the tool electrode.

In papers [11,12], the researchers determined the optimal blade airfoil machining parameters to minimize the labor intensity and maximize the cost-effectiveness of the process. This principle can be used to optimize the ECM process.

In particular, results of researching titan alloys [13] were used to define the parameters of machined materials.

To sum up, researchers in this focus area are actively trying to develop various components of computer-aided ECM design systems as well as to invent methodology to verify the process. This is still far from complete, and thus a relevant and demanded work.

3. Developing an electronic reference and information database for computer-aided ECM design

This paper dwells upon the implementation of computer-aided ECM design in a PLM system, which is important for integration of these data in the single information space of the enterprise.

Based on the peculiarities of ECM design, a classification system that includes the following information classes has developed:

- Technological Processes are standard and specific processes that include transition sets;
- Documentation is the set of route and operation maps, tooling documents, control programs, etc.
- ECM Electrolytes is the data on electrolyte compositions, machining parameters plus the array of polarization curve data.
- Tooling is the description of equipment and data on standard and changed-over tooling.
- Electrochemical description contains experimental data on dissolution rates for various electrolytes, materials, and parameters.

Data in the databases created in the system have been recorded by the developers themselves. Information obtained by CAD/CAM/CAE modeling should be added to databases automatically.

Electronic technological environment being created while developing a computer-aided ECM design technology for GTE blades can solve multiple important practical problems:

- it considerably reduces the time to launch the production of new products;
- all technological solutions are tested and refined more thoroughly thanks to process modeling and analyzing a large number of options for decision making;
- testing and refining technological processes takes less time and work;
- knowledge is accumulated and stored while also becoming accessible to young professionals.

The standard electrochemical shaping (ECS) design and operation diagram includes the following steps:

- choose the machining pattern;
- design the tooling;
- choose the electrolyte composition;
- determine the machining allowance values;
- determine the machining parameters;
- choose the equipment;
- design tool electrodes (TE);
- test and adjust the process.

These can be changed depending on the part geometry and the machining pattern chosen.

Machining parameters should be optimized for design to ensure high performance, specified precision, and quality of surfaces machined.

A number of tasks, such as the choosing the machining pattern, the electrolyte composition, the machining parameters, tooling and equipment can be rather easy solved by using standard solutions.

One of the problems the solution of which affects the precision of electrode profiling is about determining the minimum ECM allowance. In mechanical machining, minimum allowance is a sum of three components: shape error, surface roughness, and the depth of the defective layer left from earlier machining. In case of ECM, one has to take into account one more component that is due to the relatively low localization of stock removal (compared to mechanical machining). In mechanical machining, stock is removed exactly in the cutting zone, while in ECM, material can be dissolved in a fairly great range of interelectrode gaps. This depends both on the machining parameters and the localizing capacity of the electrolyte, i.e. its ability to limit the removal to a minimum gap range; This property of electrolytes is determined based on the experimental dependencies of stock removal on the interelectrode gaps while using various electrolytes; these data are presented in the ECM Electrolytes class mentioned above.

To solve the problems relating to determining the optimum allowance for pulse electrochemical machining (PECM) operations, comprehensive experiments on the electrolyte-dependent localization of the process has been carried out. Figure 2 presents the results of those studies. Analysis of results shows that the localizing capacity of electrolytes should be determined with such gaps wherewith the rate of anode dissolution is zero. This property can be recorded in the e-database to determine the machining allowance.

In our case, the $NaClO_4$ electrolyte showed maximum localizing capacity with other machining conditions being equal. For the EI-961 alloy, it had minimum active dissolution gap range.

The biggest problems of ECS design are related to the profiling of TE working surfaces as well as to choosing the machining parameters. How well these problems are solved affects the amount and labor intensity of finishing, the performance, and the precision of ECS.

The proposed methodology for the computer-aided 3D ECS design is based on mathematical models and analysing the spatial evolution of arbitrary surfaces as well as the physico-chemical processes in IEG and near-electrode layers. Figure 1 presents a general computer-aided ECM design diagram.

In the framework of this study, the following steps were done:

1. Modeled the electric fields in the electrolyte [14,15].
2. Experimentally studied the kinetics of electrode processes to determine the boundary conditions on electrode surfaces [14,15].
3. Developed a method for computer profiling of tool electrodes [14,15].
4. Created computer-aided ECM design applications.
5. Provided a parametric translation of ECM design output to the Siemens NX system.

An important part of this work is creating a Siemens NX information and parametric model for CNC milling of tool electrodes profiled based on the computer modeling of GTE blade ECM. This model is implemented in the environment of the calculation module developed earlier that is used to profile tool electrodes for pulse electrochemical machining of GTE parts, whereby the mathematical model developed is applied.

Based on integrating the results of comprehensive experimental studies in [14,15], a database (DB) of the physical properties of aviation material ECM has been desighed.

The software for determining the TE profile contains a Source Data Input module for the input of data on part geometry and physical properties of the process simulated.

The computational models takes those data from the relational database that contains experimental results. Storing the experimental data in a database that is not within the calculation system enables us to develop a versatile calculation module as well as a convenient system for the input and analysis fo experimental results.

The main method were used to design the relational database consisted in building a semantic ER diagram, see Figure 3.

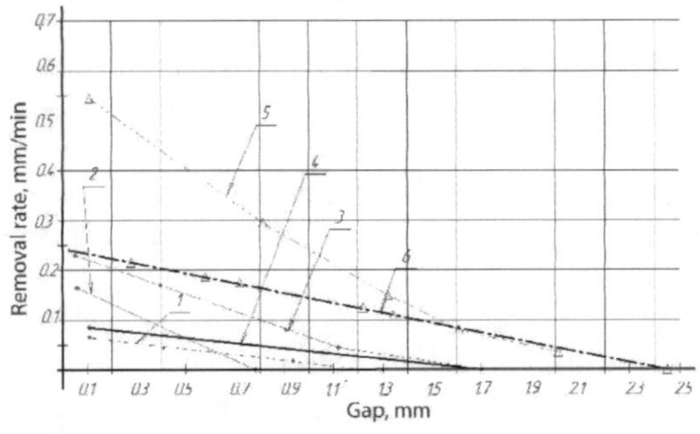

Figure 2. Experimental dependencies of the removal rate on the interelectrode gap for various electrolytes EI-961 alloy: 1 - *4%*$NaNO_3$; 2 - $NaClO_4$; 3 - *4%NaCl+15%*KNO_3; 4 - *1%NaCl+4%*$NaNO_3$; 5 - *10%NaCl*; 6 - *4%NaCl*

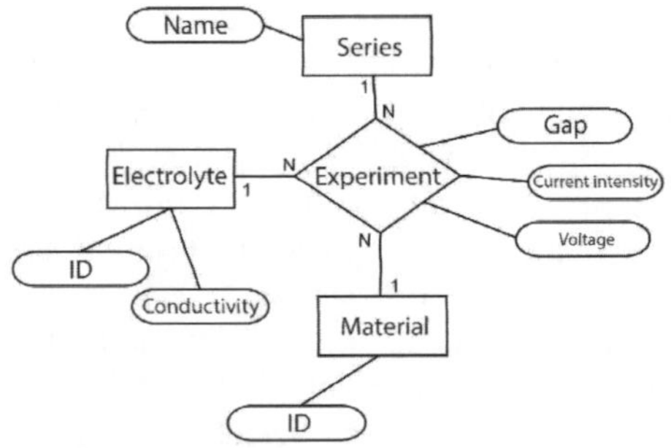

Figure 3. ER diagram of experiment parameters

IOP Conf. Series: Materials Science and Engineering **302** (2018) 012049 doi:10.1088/1757-899X/302/1/012049

The ER diagram contains the following entities: Electrolyte, Material (of the workpiece), Series (of the experiment). The Electrolyte and Material attributes determine the results of the Experiment. Indicators recorded in each experiment are the attributes of the Experiment link. Adding the Series entity to the diagram enables analysis of different experiments.

By applying a known ER diagram transformation algorithm, a chart of relational experiment database tables has been obtained, see Figure 4.

As the algorithm requires to reduce data redundancy, the diagram has been refined to specify the Experiment table which uniquely stores sample (material and electrolyte) IDs.

When determining the type of data for the Experimental Conditions fields which are related to Electrolytes, Machined Materials, as well as for the Experiment ID field, a value propagation tool that ensures the integrity of data in the related tables was used.

For the first step of the wizard, the Experimental Conditions table was used as the data source, see Figure 5.

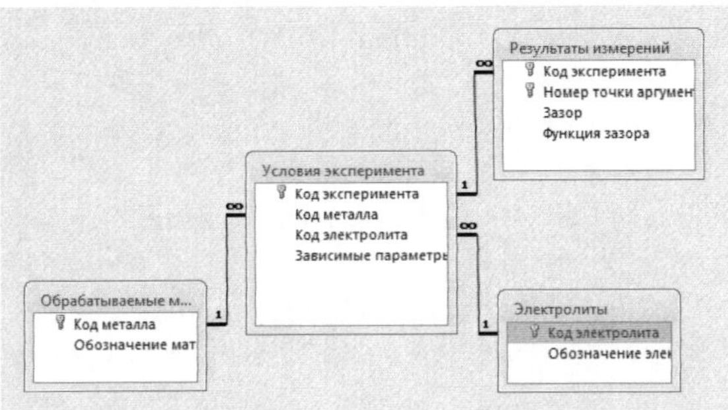

Figure 4. Diagram of the relational database

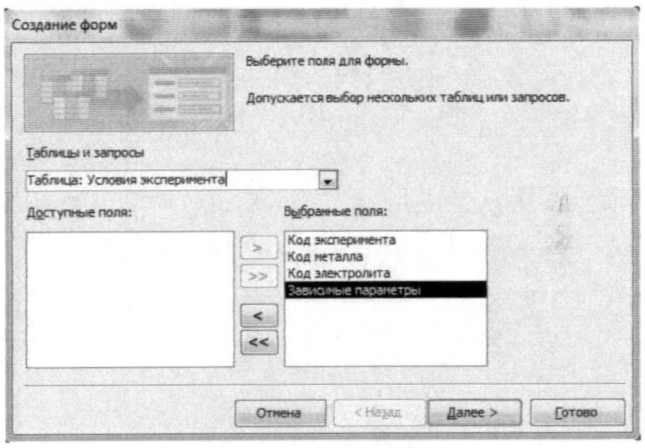

Figure 5. Choosing the Experimental Conditions table

When recording the results of experiments with materials and electrolytes that already have an ID in the database, the corresponding ID is chosen from the drop-down menu.

When adding new data to the database, new values can be put in the Machined Materials and Electrolytes tables automatically. The main data management form is supplemented with macros; double-click to run them.

Activating the Machined Materials form triggers Macro1 which opens an additional form that displays the parameters of materials recorded in the database.

The form can be used to add new material IDs. Double-click any field in the row of the material you need in the Machined Materials table to select the material and return to the main form to record the results of experiments with such new material.

In the Electrolytes table, values are added and selected in the same fashion.

The database is to be continuously updated as more materials and electrolytes are experimented with. It is to be included in the electrochemical shaping calculation software. That software is based on iterative ANSYS calculations and will be of use for technologists to optimize the machining parameters, to profile tool electrodes, and to reduce the pre-production time for new products.

4. Conclusion

This paper gives an analytical review on the issues of simulating and visualizing the ECS process; the reviews proves that this focus area is relevant. The paper lays the foundation for the model obtained as applicable to PECM. There were described the principles of research and relational database updates; the database was used to implement the ECS model developed. For that purpose, the characteristics of pulse current occurring in electrochemical machining of aviation materials were studied.

Based on the synthesis of research results, the experimental results, and comprehensive ECM process data modeling, a subsystem for computer-aided design of electrochemical machining for GTE blades has developed; the subsystem was implemented in the Teamcenter PLM system.

References

[1] Kozak J 2013 *Lecture Notes in Electrical Engineering* vol 170 p 95
[2] Lei W 2010 *Advanced Materials Research* vol 97-101 p 4061
[3] Yang Y, Kang M and Fu X Q 2011 *Key Engineering Materials*, vol 458 p 93
[4] Klocke F, Zeis M, Harst S, Klink A, Veselovac D and Baumgärtner M 2013 *Procedia CIRP* vol 8 p 265
[5] Mukherjee R and Chakraborty S 2013 *International Journal of Advanced Manufacturing Technology* vol 64(5-8) pp 781-791
[6] Rebschläger A and Kollmannsperger R 2014 *Procedia CIRP* vol 14 pp 418-423
[7] Zhao J, Wang F, Xu J and Liu Y 2013 *Hangkong Xuebao/Acta Aeronautica et Astronautica Sinica* vol 34(12) pp 2841-2848
[8] Wang F, Xu J and Zhao J 2010 *Hangkong Xuebao/Acta Aeronautica et Astronautica Sinica* vol 31(12) pp 2450-2456
[9] Zhu D, Zhu D and Xu Z 2012 *International Journal of Advanced Manufacturing Technology* vol 62(1-4) pp 147-156
[10] R Wu, Zhang D and Sun J 2011 *Research Journal of Applied Sciences, Engineering and Technology* vol 3(9) pp 1007-1013
[11] Korovin E M, Lunev A N and Tsareva V V 2012 *Russian Aeronautics* 55(1) pp 76-82
[12] Lunev A N, Moiseeva L T and Solomina M V 2007 *Russian Aeronautics* vol 50(2) pp 193-198
[13] Agapovichev A V, Balaykin A V, Smelov V G and Agapovichev A V 2015 *Modern Applied Science* vol 9(4) pp 151-159
[14] Nekhoroshev M V, Pronichev N D and Smirnov G V 2014 *The Open Mechanical Engineering Journal* vol 8 pp 436-440
[15] Nekhoroshev M V, Pronichev N D and Smirnov G V 2013 Proceedings of the Samara Scientific Center (Russian Academy of Sciences) vol 15 №6(4) pp 897-900

Mathematical Model of the One-stage Magneto-optical Sensor Based on Faraday Effect

O G Babaev*, V D Paranin and L I Sinitsin

Samara National Research University, Moskovskoe shosse, 34, Samara, 443086, Russia

E-mail: *orbaev@yandex.ru

Abstract. The aim of this work is to refine a model of magneto-optical sensors based on Faraday's longitudinal magneto-optical effect. The tasks of the study include computer modeling and analysis of the transfer characteristic of a single-stage magneto-optical sensor for various polarization of the input beam and non-ideal optical components. The proposed mathematical model and software make it possible to take into account the non-ideal characteristics of film polaroids observed in operation in the near infrared region and at increased temperatures. On the basis of the results of the model analysis it was found that the dependence of normalized transmission $T(\gamma_2)$ has periodic nature. Choosing the angle $(\gamma_3 - \gamma_1)$ makes it possible to shift the initial operation point and change the sensitivity $dT / d\gamma_2$. The influence of the input beam polarization increases with the increase of polaroid parameter deviation from ideal and shows itself as reduction of modulation depth and angular shift of the sensor conversion response.

1. Introduction

Magneto-optical sensors are often used for measuring electric and magnetic fields. The action of the sensors is based on Faraday longitudinal magneto-optical effect [1-5]. The magnetic field effect in sensors of this kind causes magnetic rotary polarization of transmitted light. This rotation is transformed into intensity modulation at the sensor output with the help of an analyzer.

Fiber-optic sensors of magnetic and electric fields are developed mostly for remote control in direct-current and alternating-current networks [3]. The need for these sensors is caused by the necessity of electric separation of high and low voltage of measuring tools. Stringent requirements to the dynamic range width, temperature interval (–20 ...+80 °C), and low measurement error (no more than 1%) are imposed on such sensors.

Three types of fiber-optic sensors are distinguished [4]: 1) interferometric single-mode fiber sensors with a magnetostrictive cladding; 2) sensors based on Faraday longitudinal magneto-optical effect in single-mode fiber; 3) sensors based on Faraday longitudinal magneto-optical effect in a bulk element – a magneto-optical rotator coupled with a multi-mode fiber waveguide. Sensors of the third type are discussed in this paper.

The Jones matrix method [6-7] applicable for polarized radiation analysis is one of the most efficient methods of calculating polarization optical circuits. The authors of [2] showed the use of Jones matrix method to describe various states of beam polarization and distribution of its resultant intensity in passing the optical system "polarizer – magneto-optical rotator – analyzer". However, the proposed model does not take into account the non-ideal analyzers, nor is it investigated.

The aim of this work is to refine a model of magneto-optical sensors based on Faraday's longitudinal magneto-optical effect. The tasks of the study include computer modeling and analysis of the transfer characteristic of a single-stage magneto-optical sensor for various polarization of the input beam and non-ideal optical components.

2. Mathematical modeling

The scheme of a magneto-optical sensor with a single-pass sensing element in the form of a magneto-optical film is shown in Figure 1. Passing through the polarizer P the light from the light source becomes linearly polarized with an azimuth angle γ_1. An yttrium-iron garnet (YIG) magneto-optical film turns the plane of polarization about an angle γ_2. An analyzer with an azimuth angle γ_3 is installed at the output of the circuit. The analyzer transforms the rotation of the polarization plane into intensity modulation.

The difference between the azimuth angles of the polarizer and analyser $(\gamma_3 - \gamma_1)$ determines the initial operation point of the transfer characteristic of the circuit. Thus, it is possible to optimize the sensor's sensitivity to variations of low or medium values of the magnetic field.

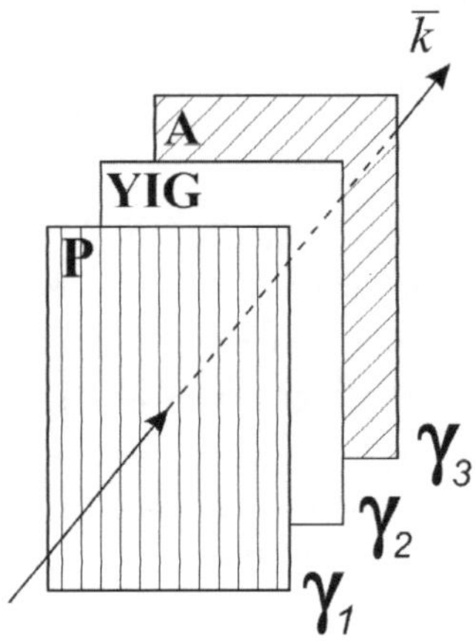

Figure 1. Figure with short caption (caption centred).

The Jones matrix method applicable for completely polarized input light is used to calculate such a system. Unlike the calculation technique based on Malus law the Jones matrix method takes into account the state of polarization of the input beam, elliptical in the general case, and makes it possible to calculate the transmission of an optical scheme in the case of using non-ideal polarizers. The Maxwell column is operated on within the framework of the Jones method. The elements of the column determine the amplitudes and phases of transverse components of the electric vector of the electromagnetic field:

$$\bar{E} = \begin{bmatrix} E_x \\ E_y \end{bmatrix} = \begin{bmatrix} A_x \exp(i\varphi_x) \\ A_y \exp(i\varphi_y) \end{bmatrix}, \quad (1)$$

where A_x, A_y are amplitudes of x - and y - components of the oscillating electric field, respectively, [V/m]; φ_x, φ_y are phases of x - and y - components of the electric field [rad].

To calculate the beam intensity the Maxwell vector is multiplied from left by its complex conjugated transpose [6]. In so doing we get:

$$I = \overline{E}^{*T}\overline{E} = \left[A_x e^{i\varphi_x} A_y e^{i\varphi_y} \right] \begin{bmatrix} A_x e^{i\varphi_x} \\ A_y e^{i\varphi_y} \end{bmatrix} = A_x^{\,2} + A_y^{\,2}. \tag{2}$$

If the response of the medium to the electromagnetic field effect is linear it can be expected that if the vector:

$$\overline{E^{in}} = \begin{bmatrix} A_x^{\,in}\exp\left(i\varphi_x^{\,in}\right) \\ A_y^{\,in}\exp\left(i\varphi_y^{\,in}\right) \end{bmatrix} \tag{3}$$

is a Maxwell column of incident radiation the output beam Maxwell column can be written as:

$$\overline{E^{out}} = \begin{bmatrix} A_x^{\,out}\exp\left(i\varphi_x^{\,out}\right) \\ A_y^{\,out}\exp\left(i\varphi_y^{\,out}\right) \end{bmatrix}, \tag{4}$$

with

$$\begin{bmatrix} A_x^{\,out}\exp\left(i\varphi_x^{\,out}\right) \\ A_y^{\,out}\exp\left(i\varphi_y^{\,out}\right) \end{bmatrix} = \begin{bmatrix} J11 & J12 \\ J21 & J22 \end{bmatrix} \begin{bmatrix} A_x^{\,in}\exp\left(i\varphi_x^{\,in}\right) \\ A_y^{\,in}\exp\left(i\varphi_y^{\,in}\right) \end{bmatrix}, \tag{5}$$

where the matrix J consisting of four elements J_{ij}, is called the Jones matrix of the optical device (magneto-optical sensor).

Thus, the matrix equation governing the polarization of an optical system according to the Jones method in concise form is as follows:

$$\overline{E^{out}} = J\,\overline{E^{in}}. \tag{6}$$

An non-ideal polarizer with an azimuth angle γ_1 to the axis x, with transmission amplitudes t_x and t_y along the axes x and y, respectively, can be represented as the product of matrices [6,7]:

$$M_1 = \begin{bmatrix} \cos\gamma_1 & -\sin\gamma_1 \\ \sin\gamma_1 & \cos\gamma_1 \end{bmatrix} \begin{bmatrix} t_x\exp(i\varphi_{x1}) & 0 \\ 0 & t_y\exp\left(i\varphi_{y1}\right) \end{bmatrix} \begin{bmatrix} \cos\gamma_1 & \sin\gamma_1 \\ -\sin\gamma_1 & \cos\gamma_1 \end{bmatrix}. \tag{7}$$

An analyzer at the circuit output with the azimuth angle γ_3, is described similarly:

$$M_3 = \begin{bmatrix} \cos\gamma_3 & -\sin\gamma_3 \\ \sin\gamma_3 & \cos\gamma_3 \end{bmatrix} \begin{bmatrix} t_x\exp(i\varphi_{x3}) & 0 \\ 0 & t_y\exp\left(i\varphi_{y3}\right) \end{bmatrix} \begin{bmatrix} \cos\gamma_3 & \sin\gamma_3 \\ -\sin\gamma_3 & \cos\gamma_3 \end{bmatrix}. \tag{8}$$

A magneto-optical Faraday rotator with the rotation angle γ_2 and the transmission coefficient t_2 is described with the following Jones matrix [8]:

$$M_2 = t_2 \begin{bmatrix} \cos\gamma_2 & \sin\gamma_2 \\ -\sin\gamma_2 & \cos\gamma_2 \end{bmatrix}. \tag{9}$$

The Maxwell column of the output beam for a single-stage circuit is determined:

$$\overline{E^{out}} = M_3 M_2 M_1 \overline{E^{in}} \tag{10}$$

The intensities of input and output radiation, according to (2) are written as follows:

$$I^{in} = \overline{E^{in}}^{*T}\,\overline{E^{in}} \tag{11}$$

$$I^{out} = \overline{E^{out}}^{*T}\,\overline{E^{out}} \tag{12}$$

and the expression for the intensity transmission coefficient (the transfer characteristic):

$$T = \frac{I^{out}}{I^{in}}. \tag{13}$$

Figures 2, 3 present the results of calculated transmission T for circular ($\varphi_y^{in} - \varphi_x^{in} = \pi/2$) and elliptical ($\varphi_y^{in} - \varphi_x^{in} = \pi/6$) polarization of incident beam on condition that $\varphi_{x1} = \varphi_{y1}$, $\varphi_{x3} = \varphi_{y3}$, $t_2 = \sqrt{0,15}$.

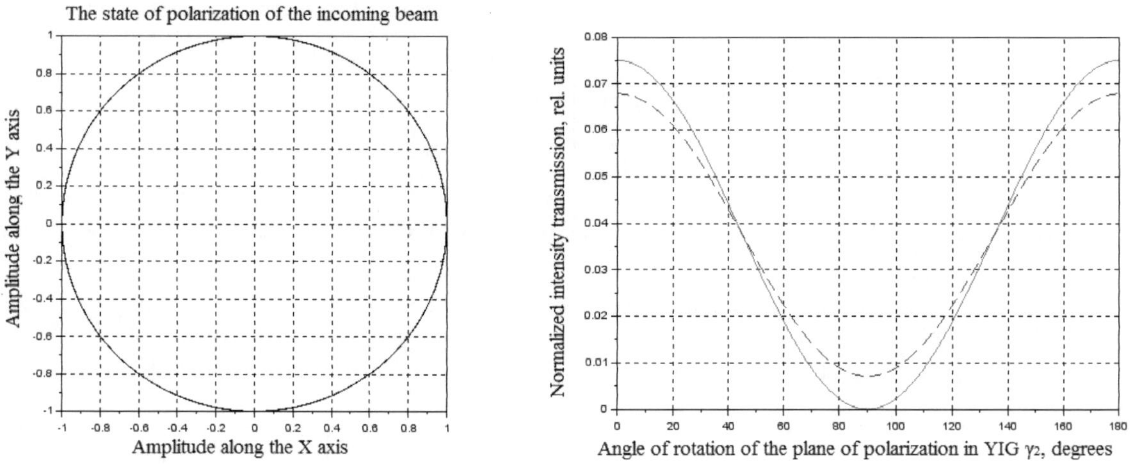

(a) (b)

Figure 2. Results of modeling: (a) circular polarization of the input beam; (b) dependence of normalized intensity transmission on the Faraday rotation angle γ_2 for ideal ($t_x = 1$, $t_y = 0$, solid line) and non-ideal ($t_x = \sqrt{0,95}$, $t_y = \sqrt{0,05}$, dotted line) polarizer and analyzer, $\gamma_1 = \gamma_2 = 0°$.

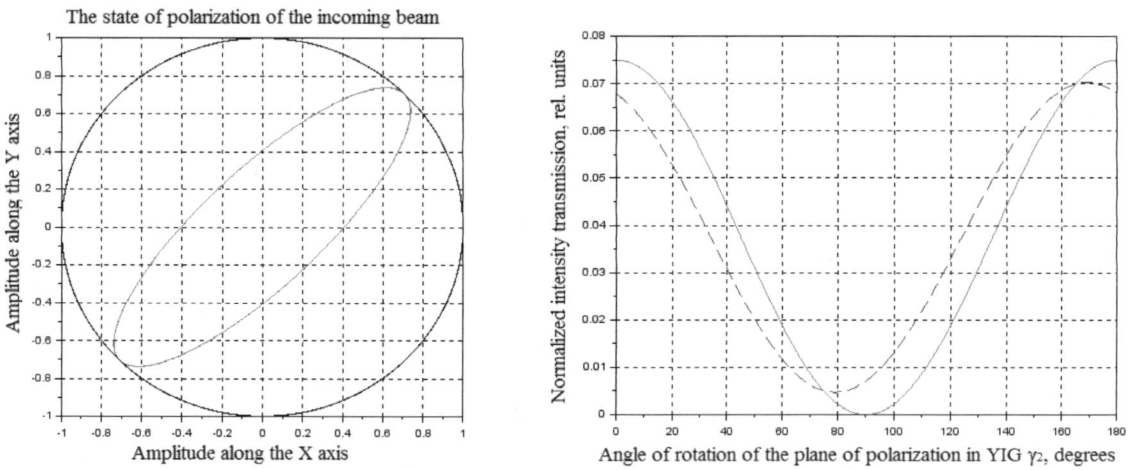

(a) (b)

Figure 3. Results of modeling: (a) elliptical polarization of the input beam; (b) dependence of normalized intensity transmission on the Faraday rotation angle γ_2 for ideal ($t_x = 1$, $t_y = 0$, solid line) and non-ideal ($t_x = \sqrt{0,95}$, $t_y = \sqrt{0,05}$, dotted line) polarizer and analyzer, $\gamma_1 = \gamma_2 = 0°$.

Figures 4, 5 present the results of calculated transmission T for elliptical ($\varphi_y^{in} - \varphi_x^{in} = \pi / 6$) polarization of incident beam on condition that $\varphi_{x1} = \varphi_{y1}$, $\varphi_{x3} = \varphi_{y3}$, $t_2 = \sqrt{0,15}$. These figures allow to estimate the effect of the alignment accuracy on the modulation depth of the transmission intensity.

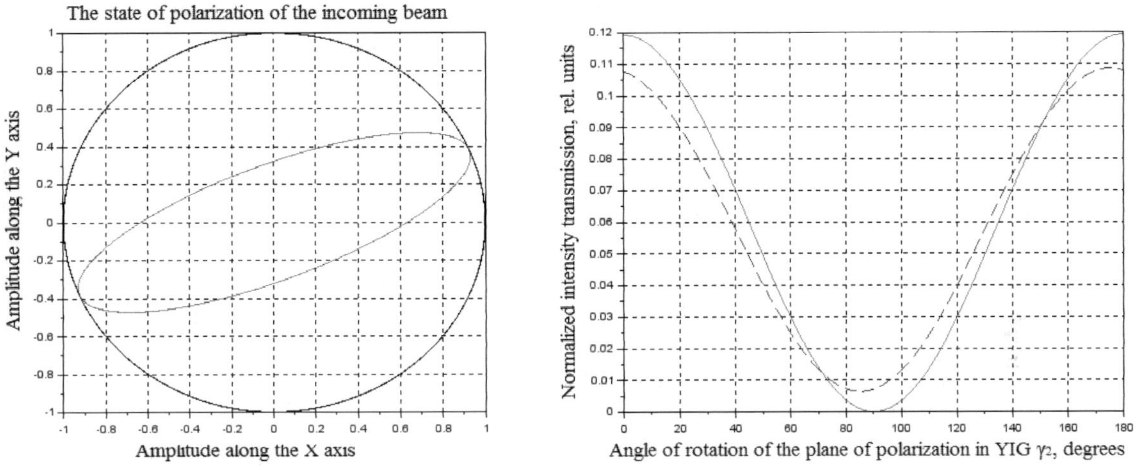

(a) (b)

Figure 4. Results of modeling: (a) elliptical polarization of the input beam; (b) dependence of normalized intensity transmission on the Faraday rotation angle γ_2 for ideal ($t_x = 1$, $t_y = 0$, solid line) and non-ideal ($t_x = \sqrt{0,95}$, $t_y = \sqrt{0,05}$, dotted line) polarizer and analyzer, $\gamma_1 = \gamma_2 = 0°$.

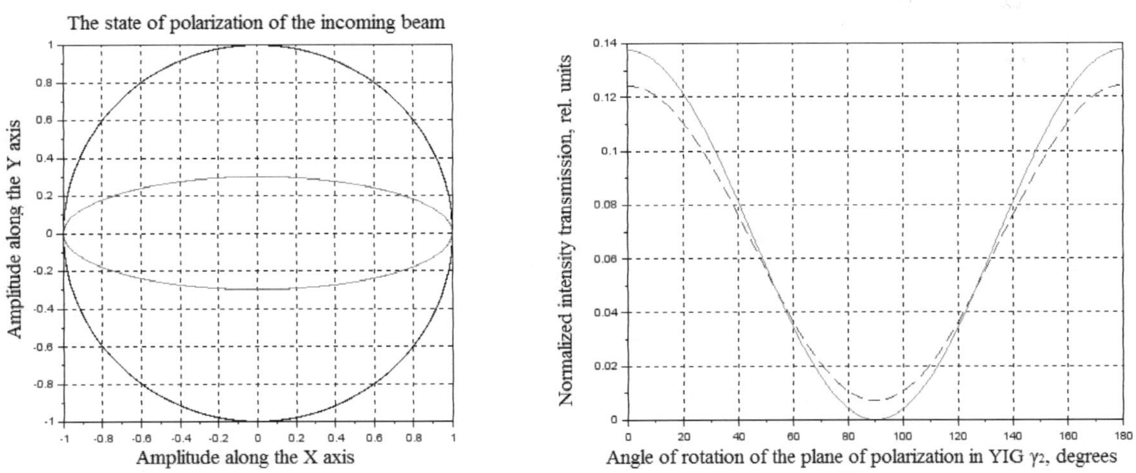

(a) (b)

Figure 5. Results of modeling: (a) elliptical polarization of the input beam; (b) dependence of normalized intensity transmission on the Faraday rotation angle γ_2 for ideal ($t_x = 1$, $t_y = 0$, solid line) and non-ideal ($t_x = \sqrt{0,95}$, $t_y = \sqrt{0,05}$, dotted line) polarizer and analyzer, $\gamma_1 = \gamma_2 = 0°$.

As can be seen from the figures, at a small angle of polarization rotation, a change in the modulation depth of the transmission intensity is observed. It can be used to select the optimal incidence angle of polarized beam. By varying the incidence angle of the polarized beam it is possible to achieve the required sensitivity. But it imposes certain limitations on the alignment accuracy of the entire chain.

It follows from the calculations that the sensor transmission for the ideal polarizer and analyzer does not exceed $0,5t_2^2$ of the input beam intensity. In the case of non-ideal polarizer and analyzer the output modulation depth intensity is reduced. The instability of the state of polarization of the input beam leads to a shift of the transmission characteristic on the scale of angles γ_2. Thus, it is necessary to stabilize both the intensity of incident radiation and the state of its polarization for magneto-optical sensors.

3. Conclusion

The proposed mathematical model and software make it possible to take into account the non-ideal characteristics of film polaroids observed in operation in the near infrared region and at increased temperatures. On the basis of the results of the model analysis it was found that the dependence of normalized transmission $T(\gamma_2)$ has periodic nature. Choosing the angle $(\gamma_3 - \gamma_1)$ makes it possible to shift the initial operation point and change the sensitivity $dT / d\gamma_2$. The influence of the input beam polarization increases with the increase of polaroid parameter deviation from ideal and shows itself as reduction of modulation depth and angular shift of the sensor conversion response.

Acknowledgements
The work was supported by the Ministry of Education and Science of the Russian Federation. Unique ID: Applied research and experimental work RFMEF157816X0209.

References

[1] Randoshkin V V and Chervonenkis A Ya 1990, *Applied magnetooptics* (Moscow: Energoatomizdat), 320 p
[2] Yevstafiev A I and Urakseev M A 2014 Application of Jones matrices to describe the state of light polarization in a magneto-optic current transducer, *Electrical and data processing facilities and systems* №3, v 10, pp 97-102
[3] Berikashvili V Sh *et al.* 2004 Fiber-optic sensor based on the domain structure YIG+Bi for measuring current strength, *Radio Engineering* № 12 - pp 78-81
[4] Paranin V D 2016 Investigation of optical properties of epitaxial yttrium iron garnet films *AIP Conference Proceedings* 1724, 020024
[5] Matyunin S A *et al.* 2015 Fiber-optical sensors based on mono-crystal films of garnet ferrites for mechatronic systems *Procedia Engineering 106* - pp 202-209
[6] Gerrard A and Berch D M 1978 *Introduction to matrix optics* (Moscow: Mir), 341 p
[7] Shercliff W 1965 *Polarized light* (Moscow: Mir), 264 p
[8] Tarasov L V 1981 *Physics of processes in coherent optical radiation generators* (Moscow: Radio and communication), 440 p.

Linearization of Positional Response Curve of a Fiber-optic Displacement Sensor

O.G. Babaev*, S.A. Matyunin and V. D. Paranin

Samara National Research University, Moskovskoe shosse, 34, Samara, 443086, Russia

E-mail: *orbaev@yandex.ru

Abstract. Currently, the creation of optical measuring instruments and sensors for measuring linear displacement is one of the most relevant problems in the area of instrumentation. Fiber-optic contactless sensors based on the magneto-optical effect are of special interest. They are essentially contactless, non-electrical and have a closed optical channel not subject to contamination. The main problem of this type of sensors is the non-linearity of their positional response curve due to the hyperbolic nature of the magnetic field intensity variation induced by moving the magnetic source mounted on the controlled object relative to the sensing element. This paper discusses an algorithmic method of linearizing the positional response curve of fiber-optic displacement sensors in any selected range of the displacements to be measured. The method is divided into two stages: 1 - definition of the calibration function, 2 – measurement and linearization of the positional response curve (including its temperature stabilization). The algorithm under consideration significantly reduces the number of points of the calibration function, which is essential for the calibration of temperature dependence, due to the use of the points that randomly deviate from the grid points with uniform spacing. Subsequent interpolation of the deviating points and piecewise linear-plane approximation of the calibration function reduces the microcontroller storage capacity for storing the calibration function and the time required to process the measurement results. The paper also presents experimental results of testing real samples of fiber-optic displacement sensors.

1. Introduction

Currently, the creation of optical measuring instruments and sensors for measuring linear displacement is one of the most essential problems in instrumentation. Of special interest are fiber-optic sensors (FOS) based on the magneto-optical effect [1-9]. They are essentially contactless (no mechanical contact with the controlled object), non-electrical (do not contain electrical components in the sensing element and do not require electrical power) and have a closed optical channel not subject to contamination [6-9].

The main problem of this type of sensors is the non-linearity of their positional response curve (PRC) due to the hyperbolic nature of the magnetic field intensity variation induced by moving the magnetic source mounted on the controlled object relative to the sensing element [5, 8]. Figure 1 shows the dependence of the output signal on the displacement of a typical FOS sensing element based on the Faraday magneto-optical effect in epitaxial films of yttrium iron garnet (YIG) [6-8].

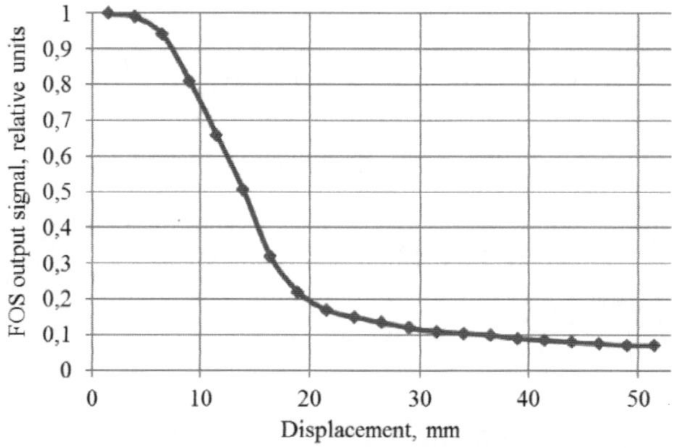

Figure 1.FOS output signal versus displacement.

2. The structure and operating principle of displacement FOS

Figure 2 shows a block schematic diagram of a differential FOS, based on the magneto-optical effect in epitaxial films of yttrium iron garnet (YIG). The fiber-optic sensor consists of a laser diode (LD), a Wollaston prism (WP), a YIG on a transparent sapphire substrate and two film polarizers oriented at angles of ± 45° relative to each other (FP1 and FP2). The LD has a local circuit of temperature stabilization and radiation power stabilization. A secondary converter (electronic converter - EC) of the fiber-optic sensor contains two fiber-optic photodiodes (PD1, PD2) that convert the sensor's optical signals into electrical ones, and amplifiers of their output signals. The converter is placed at a safe distance from the control object to avoid fire or explosion.

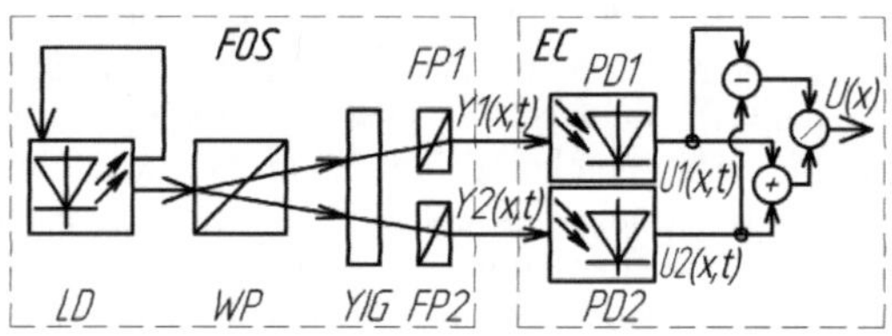

Figure 2. The block schematic diagram of a differential FOS.

Consider the operating principle of a fiber-optic sensor (Figure 2). The light beam from the light source LD with circular polarization travels to the WP via a fiber optical cable. The WP splits the beam into two beams with orthogonal polarization of 0° and 90°. Both beams pass through the YIG plate that rotates the polarization plane of each beam to an angle proportional to the intensity of the magnetic field induced on the FOS. Then the YIG beams pass through the polarizers FP1 and FP2, initially configured to +45° and −45° for each branch of the sensor. The resulting optical signals of the FOS are applied to the respective photodiodes of the EC wherein amplification and mathematical processing are performed.

The output signal of the FOS is typically determined according to the following expression [6-8]:

$$U(x) = \frac{U1(x,t) - U2(x,t)}{U1(x,t) + U2(x,t)}, \tag{1}$$

where $U(x)$ is the dependence of the FOS output signal on the displacement;

$U1(x,t) = Y1(x,t)\cdot k$, $Y1(x,t)$ are the output electrical and input optical signals of PD1;

$U2(x,t) = Y2(x,t)\cdot k$, $Y2(x,t)$ are the output electrical and input optical signals of PD2;

k is the coefficient of conversion of the optical signal to an electrical one by photodiodes PD1, PD2.

However, as shown by the simulation results (Figure 3), formula (1) is applicable only if the condition (2) (curves 1 and 2 in Figure 3) is fulfilled:

$$U1(x,t)+U2(x,t)=Const \tag{2}$$

If condition (2) is violated there is a sharp distortion of the FOS PRC (curves 3 and 4 in figure 3) and deterioration of the multiplicative error compensation using the expression (1). The algorithm described in this paper is free from these disadvantages.

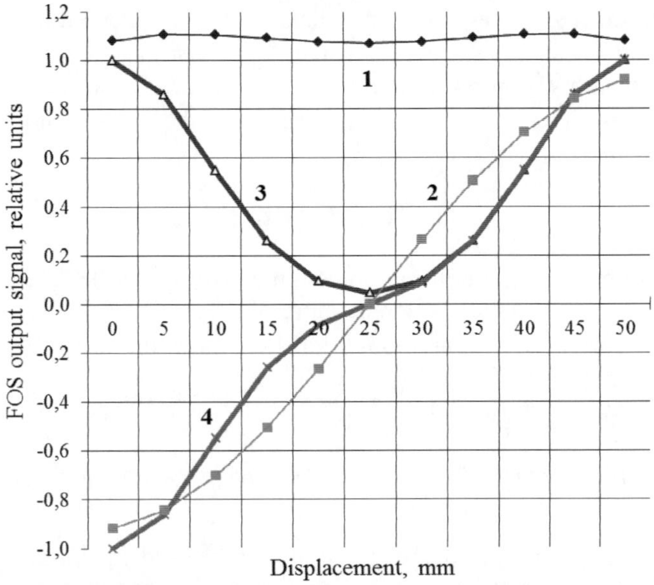

Figure 3. The PRC of the differential FOS: curves 1, 3 – function $U1(x,t)+U2(x,t)$ for cases $U1(x,t)+U2(x,t)=0{,}05...1{,}0$ and for $U1(x,t)+U2(x,t)=1{,}08...1{,}1$ respectively; curves 2, 4 – function (1) for cases $U1(x,t)+U2(x,t)=0{,}05...1{,}0$ and for $U1(x,t)+U2(x,t)=1{,}08...1{,}1$ respectively.

3. Method of calibration and linearization of FOS PRC

The calibration function determines the correspondence of the arguments (FOS displacement and temperature) to the FOS output signal, processing of measurement results and storing the obtained dependence in the microcontroller (MC) memory.

In the measurement mode FOC PRC linearization and temperature correction of the measurement results takes place. In addition, the algorithm under consideration significantly reduces the number of points of the calibration function due to the use of the points that randomly deviate from the grid points with uniform spacing. Subsequent interpolation of the deviating points and piecewise linear-plane approximation reduces the microcontroller storage capacity to store the calibration function and the time required for the processing of measurement results.

The method is divided into two stages: 1 – definition of the FOS calibration function, 2 – measurement and linearization of the positional response curve (including its temperature stabilization).

3.1. Definition of the calibration function of FOS

To calibrate the sensor the FOS positional response curve is recorded in the microprocessor memory. Calibration is carried out for a relatively small number of points (16 to 64 points for each variable). Consider FOS calibration for some portion of the PRC (Figure 4).

Let us assume the following points of the PRC were obtained experimentally: $U^R_{i,j}(x^R_{i,j};t^R_{i,j})$, $U^R_{i+1,j}(x^R_{i+1,j};t^R_{i+1,j})$, $U^R_{i,j+1}(x^R_{i,j+1};t^R_{i,j+1})$, $U^R_{i+1,j+1}(x^R_{i+1,j+1};t^R_{i+1,j+1})$, where $U^R_{i,j}(x^R_{i,j};t^R_{i,j})$ is the output voltage of the amplifiers connected to the PD1, PD2 photodiodes located in the FOS electronic converter, [mV]; $x^R_{i,j}$ is the predetermined value of linear displacement, [mm]; $t^R_{i,j}$ is the predetermined value of temperature, [°C]; i, j is the number of points on the displacement and temperature axis, respectively; R is the index indicating the experimental data.

Obviously, the experimental values may not coincide with the specified nodes of the measurement grid (grid nodes are usually assigned at a constant step $\Delta x = const$, $\Delta t = const$).

Therefore, during the calibration procedure piecewise surface (linear or non-linear) interpolation (or extrapolation) of a preset order is carried out and values of the calibration function $U_{i,j}[x_i, F_{i,j}(t_j)]$ in the adjacent grid points are calculated (Figure 4):

$$U^R_{i,j}(x^R_i, t_j) \rightarrow U_{i,j}(x_i, t_j) \tag{3}$$

Since the FOS does not contain a built-in temperature sensor, and there is no way of determining the temperature explicitly, the variation of temperature of the FOS sensing element can be taken into account implicitly (2) according to the variation of function value $F_{i,j}(x_i, t_j)$ (for i, j - grid points):

$$F_{i,j}(x_i, t_j) = U1(x_i, t_j) + U2(x_i, t_j) \Rightarrow t_j = F^{-1}_{i,j}[U1(x_i, t_j) + U2(x_i, t_j)], \tag{4}$$

where $F^{-1}_{i,j}$ is the inverse function of $F_{i,j}$. Hereinafter the arguments are replaced with numbers of grid points i, j for simplification.

The operating temperature range (0...50) °C is normally used in the development of industrial sensor applications. In the case of determining the calibration function of PRC with a step of 10 °C we obtain 6 reference points on the temperature scale $j = 0...5$. For the displacement range (0...31) mm and a 1 mm quantization step we obtain 32 reference points on the displacement scale $i = 0...31$. In this case, to store the PRC calibration function in the space of two variables x_i, t_j 192 points of the function values $U2_{i,j}$ are required.

3.2. PRC measurement and linearization algorithm

Both the linearization of the FOS PRC and temperature correction of the measurement results are carried out in the measurement mode:

1. Voltage values $U1^R(x)$, $U2^R(x)$ of the electronic converter amplifiers connected to the FOS are measured.
2. The nearest grid points of i, j for the measured voltage values are defined:

$$U^R_{i,j}(x^R_i, t_j) \in [U_{i,j}(x_i, t_j), U_{i+1,j+1}(x_{i+1}, t_{j+1})]. \tag{5}$$

3. The temperature of the nearest grid point is calculated from equation (4).
4. Temperature deviation from the nearest grid point is determined:

$$\Delta t_{i,j} = (t^R_{i,j} - t_{i,j}). \tag{6}$$

5. Using linear interpolation temperature compensation of the voltage values is implemented:

$$U1^R(x,t) \rightarrow U1^c(x),\ U2^R(x,t) \rightarrow U2^c(x), \tag{7}$$

where $U1^c(x)$, $U2^c(x)$ are the measured voltage values after temperature correction.

6. The PRC is calculated from the calibration curve using the inverse function of PRC:

$$x = x_i + U^{-1}[\Delta U_{1,2}(x)], \tag{8}$$

where $\Delta U_{1,2}(x) = U1^c(x) - U2^c(x)$. Thus, linearization and temperature correction of the PRC are carried out.

7. Then the range of the obtained argument values is scaled to the full measurement range $x \in [x_{\min}, x_{\max}]$ (for example, to the range [0 ... 4095]). The result is the ADC digital code $ADCcode(x)$ of the converter microcontroller corresponding to the displacement:

$$ADCcode(x) = 4095 \frac{x - x_{\min}}{x_{\max} - x_{\min}}. \tag{9}$$

4. Experimental study

Experimental studies were carried out on displacement FOS samples SAM.FOS.YIG-2/50 developed at the NIL-53 laboratory of Samara University (Figure 5). The experimental results are shown in Figures 1, 3, 6. Figure 6a shows the FOS PRC (curve 1) applied to the input of the electronic converter, and the digital code of the result (curve 2). There linearization was carried out in the area of displacement [20 ... 32 mm] (the area of displacement can be anything in the range [0 ... 32 mm]). Figure 6b shows the same FOS PRC after linearization on an enlarged scale. The algorithm makes it easy to move the operating range of PRC linearization to any part of the displacement range including the entire range. Designed FOS (SAM.FOS.YIG-2/50) has the following operating characteristics (table 1).

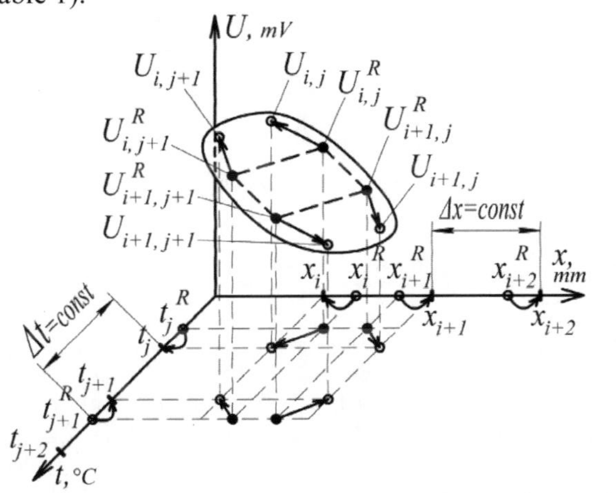

Figure 4. Determination of calibration values of FOS PRC.

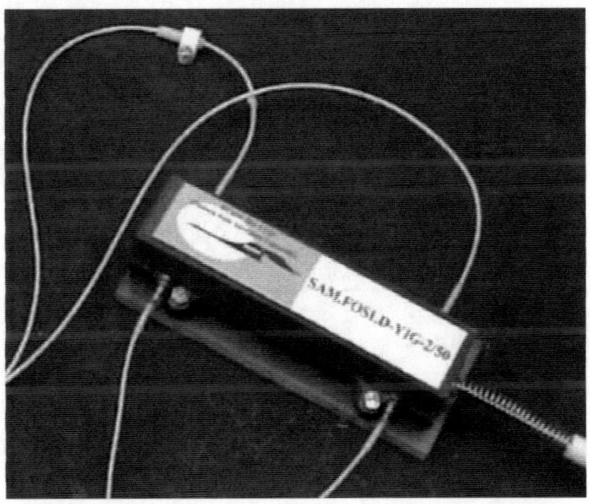

Figure 5. External view of SAM.FOS.YIG-2/50 sensor.

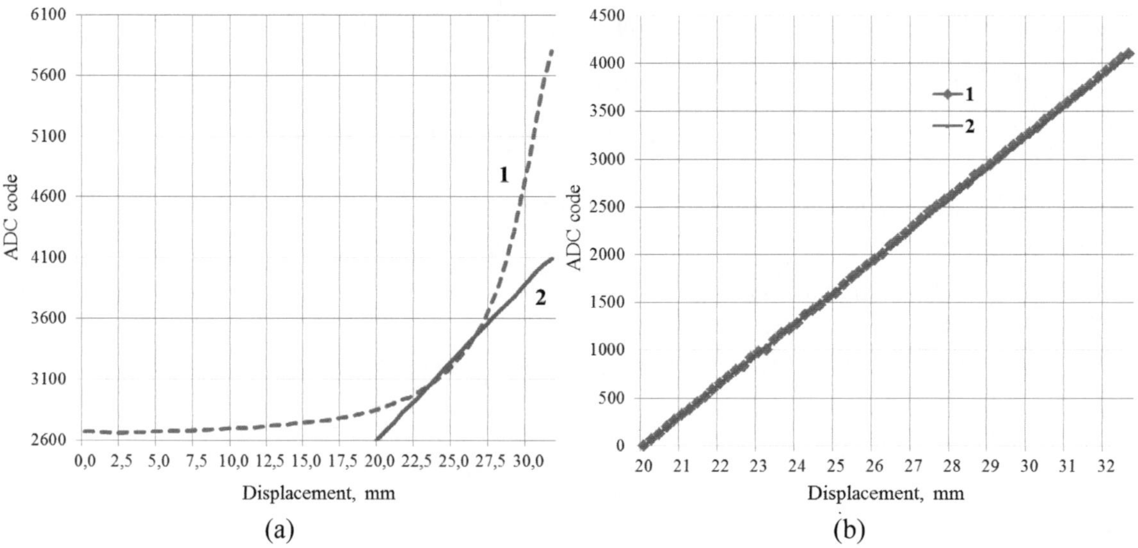

Figure 6. (a) Experimental FOS PRC; (b) FOS PRC after linearization on an enlarged scale.

Table 1. Operating characteristics of SAM.FOS.YIG-2/50.

Distance (m)	Velocity (ms^{-1})
Dimensions of the sensing element, mm	Ø3.5 x 30
FOS dimensions, mm	79 x 18(30) x 13
Stroke of moving rod, mm	from 0 to 50
Accuracy class	0.5
Additional temperature error, %/°C	0.0025
PRC nonlinearity, %	0.01
Operating temperature, °C	from 0 to +50
Length of the fiber- optic cable, m	to 10
Type of the fiber -optic cable	Ø0,9 MM (62.5/125)
Type of the optical connectors	FC/PC
Storage temperature, °C	from –40 to +85

5. Conclusions

1. The apparent advantage of the developed technique and the algorithm is that linearization and scaling can be performed in any range of the measured displacements and temperatures. This makes FOS calibration on the controlled object easy and speeds up the calibration process, which is essential for mass production of FOS.
2. High linearity of PRC is experimentally confirmed – PRC nonlinearity does not exceed 0.01% in the range of displacement from 0 to 50 mm.
3. High temperature stability of FOS is experimentally confirmed – the temperature coefficient does not exceed 0.0025%/°C in the temperature range from 0 to +50 °C.

Acknowledgements
The work is conducted with financial support of the Ministry of Education and Science of the Russian Federation. Unique ID: Applied research and experimental work RFMEF157816X0209.

References

[1] Li C, Yoshino T and Cui X 2007 Magneto-optic sensor by use of time-division-multiplexed orthogonal linearly polarized light *Appl. Opt.* 46(5) pp 685-8

[2] Müller G M, Yang L, Frank A and Bohnert K 2014 Simple fiber-optic current sensor with integrated-optics polarization splitter for interrogation *Proc. Imaging and Applied Optics 2014 (Seattle)* paper AM4A.3

[3] Sohlström H and Svantesson K 1992 The performance of a fibre optic magnetic field sensor utilizing a magneto-optical garnet *Proc. International Conference on Optical Fiber Sensors 1992 (Monterey)* paper P25

[4] Deeter M N 1996 Fiber-optic Faraday-effect magnetic-field sensor based on flux concentrators *Appl. Opt.* 35(1) pp 154-7

[5] Garzarella A, Shinn M A and Wu D H 2015 Responsivity optimization in magneto-optic sensors based on ferromagnetic materials *Appl. Opt.* 54(26) pp 7904-11

[6] Matyunin S A, Fedotov Y A, Babaev O G, Wirchenko M K, Gusev M Y and Neustroev N S 2015 Fiber-optical sensors based on mono-crystal films of garnet ferrites for mechatronic systems *Procedia Engineering* 106, pp 202-9

[7] Matyunin S A, Babaev O G 2015 Contactless fiber-optic vibration sensors for explosive manufacturings *Proc. 22nd International Congress on Sound and Vibration (Florence)*, p 150

[8] Matyunin S A, Stepanov M V, Babaev O G 2016 Simulation of the characteristics of a magneto-optical displacement transducer *Measurement Techniques* 59(8), pp 832-7

[9] Babaev O G, Matyunin S A, Stepanov M V 2017 Simulation of Contactless Fiber-optic System for Valve Status Monitoring *Procedia Engineering* 176, pp 2-11

IOP Conf. Series: Materials Science and Engineering **302** (2018) 012052 doi:10.1088/1757-899X/302/1/012052

Overall Equipment Effectiveness Implementation Criteria

I G Abramova and D A Abramov

Samara University, 34, Moscow Highway, Samara, 443086, Russia

E-mail: abramova-ig@ssau.ru, abi_ssau@inbox.ru

Abstract. This article documents the methods applied in production control technics specifically focused on commonly used parameter OEE (Overall Equipment Effectiveness). The indicators of extensive and intensive use of equipment are considered. Their appointment this is comparison in the same type of production within the industry and comparison of single-type and / or different types of equipment in terms of capacity. However, it is shown that there is no possibility of revealing the reasons for the machine's operation: productive / unproductive, with disturbances. Therefore, in the article reveals the approaches to calculating the indicator characterizing the direct operation of the equipment. The Machine Load coefficient is approaching closely to the indicator of the efficiency of the use of equipment. Methods analysis is proceeded through the historically applied techniques such as "Stopwatch" and "Motion" studies. Was performed the analysis of the efficiency index of OEE equipment using the comparable indexes performance of equipment in the Russian practice. An important indicator of OEE contains three components. The meaning of each of them reflects historically applicable indicators. The value of the availability of equipment indicator is close to the value of the equipment extensibility index. The value of the indicator of the efficiency of work can be compared with the characteristic of the capacity of the equipment and the indicator of the quality level can meet the requirements for compliance with the manufacturing technology. Shown that the sum of the values of the coefficient of "Availability" of the equipment and the value of the "Factor of compaction of working hours" are one. As well as the total value of the indicator "level of quality" and the coefficient of marriage given in the result unit. The measurability of the indicators makes it possible to make a prediction about efficiency of the equipment.

1. Introduction

Production process effectiveness is the most common criteria for the last 50-60 years. Lead time of the production process verification is not a new idea as well as loss criteria accounting for confirmation of effectiveness. Time loss criteria is the most commonly used, the most recent methods define production process through the following parameters:

1. plant and machinery, means of labor (equipment),
2. manufacturing process, the process of labor,
3. product with its accessories, the object of labor, which turns into finished products.

Every element has individual parameters, including effectiveness verification process.

Production facility as individual element accounted as "fixed manufacturing assets". Effectiveness of "fixed asset" accounted by the multiple criteria divided in to the following categories:

I. General assesses: fixed-asset turnover, capitalization ratio, capital-labor ratio, net profit margin as well as capital-labor ratio effectiveness parameters verses net profit margin at the same period.
II. Main assets transaction coefficients: input, renewal, outrange, process increments and termination, replacement and extension of production lines.
III. Assists amortization and replacement criteria
IV. Manufacturing machines deployment coefficients: intensive and extensive, integral and detrimental, shifts or singular use.

Production process includes such parameters as shift norms, maintenance, labor intensity and production capacity.

Above mentioned Production facility and Production process parameters are well known from the 50-60 years ago.

In this regard an interesting result gives a comparison of a planned manufacturing process in USSR compare to USA customer delivery oriented activities. Manufacturing process were commonly designed on a mass production level for 600-800 thousand norm hours/year. Production design was mainly focused on one man operation per one machine or type of operation. Therefore in the past a full load parameter was a main criterion of production effectiveness.

The article presents the indicators characterizing the main means of production. Among the many indicators highlighted one that uses as features of the equipment loos time, load, and quality of work. Calculation of the elements of this indicator is shown in comparison with the indicators used in the Russian practice. Thus, it is shown two different approaches to the definition of the coefficient of efficiency of the equipment.

2. Comparability of the element base, concepts and terms

Production equipment operation parameters mainly defined by the following coefficients:

- Extensive production equipment load coefficient:

$$C_{Ext} = \frac{F_{act}}{F_{eff-schedule}} \tag{1}$$

where F_{act} – actual time the equipment over the planning period. $F_{eff-schedule}$ – Effective time of equipment working according to the schedule.

The actual time the equipment over the planning period reflects that the equipment "worked" for a certain time and does not answer the question about a full load of equipment, quality products without defects, derogate from the standards, the organization of the maintenance and movement of transport streams. C_{Ext} is used for determination of standard loss level and applied for comparison of the similar production processes within same industry/category. It is widely applied due to its simplicity, however does not give information about cause effect of the detected loss level. In practice it is hard to detect production loss items due to several factors: 1) no motivation of employees of loss recording. 2) no tools for loss recording, 3) no organizational structure for recording. 4) no resources (financial / operational/ infrastructural) for recording. For metal processing equipment C_{Ext} falls within 0.92-0.95 parameters. For equipment with more than 25 years of operations C_{Ext} = 0.85~0.88.

- Next parameters is a coefficient of intensive load of manufacturing equipment.

$$C_{\ Intensive} = \frac{P_{real}}{P_{declared}} \tag{2}$$

where *Preal* — production machines productivity real; *P declared* — declared productivity according technical specifications.

This parameter is applied for production process evaluation within industry/ manufacturers with similar production category, however it does not determine routes of the production capacity reduction or change.

Coefficient that includes both C_{Ext} and $C_{\ Intensive}$ defined as $C_{Integral}$.

$$C_{Integral} = C_{Ext} \cdot C_{Intensive} \tag{3}$$

Value of $C_{Integral}$ is always lower than it's components since it includes negative and positive aspects of both coefficients.

Next useful parameter - the shift coefficient of equipment operation - is defined by the following formula:

$$C_{Sh} = \frac{MH_1 \cdot MH_2 \cdot MH_3}{N_{wb}} \tag{4}$$

where MH_1, MH_2, MH_3 - number of Machine hours in shift 1, 2 and 3 respectively; N_{wb} – number of work benches / machines.

In general, the number of worked machine-hours is a positive indicator; it shows the effective use of the general composition of the workshop equipment. However, in this indicator it is difficult to understand of the reasons for the poor performance, the reasons of the downtime. From this general indicator it is possible to go to a private - for one machine and as an indicator only one particular machine is considered. In this case, the "shift coefficient of equipment operation", as a factor for the machine, is already close to the indicator of "Machine Load coefficient".

Machine Load coefficient:

$$C_{Load} = \frac{L_{\Sigma}}{F_{eff-schedule}} = \frac{N_{wb\,(calc)}}{N_{wb\,(real)}} \tag{5}$$

где L_{Σ} is a total laboriousness all parts, $Nwb\ calc$ is a calculated number of work benches $N_{wb\,real}$ - is a real number of work benches implemented in the production process - it is accepted, rounded parameter of $Nwb\ calc$.

$$N_{wb\,(calc)} = \frac{\sum_{i=1}^{m}(t_{item-calc\,i} \cdot N_{ai} \cdot C_{incomplete})}{F_{eff-schedule} \cdot C_{plan\,compl} \cdot Ne/wb \cdot 60} \tag{6}$$

where $t_{item-calc\,i}$ - item-calculated time, is a time from production table for development of item/part of unit the product with count number i, with preparation and ending time. This is base time labor intensity.

N_{ai} - annual program per production plan, counted in units,

C_{incomp} – coefficient of in complete production,

$C_{plan\,completion}$ - coefficient of the production plan completion,

$N_{e/wb}$ - number of employees per 1 work bench,

m – number of operations.

Labor coefficient is widely used for calculations. This coefficient calculated by two methods:

1. "Stopwatch study" method,
2. "Motion study" method (Production daily log).

Both methods are used for more than 50 years [1], however still applied in TQMS or other management systems.

First method is focused on determination of $t_{op\ i}$ the operational time of production equipment. Elements of the method are shown in Figure 1 as below.

"Motion study" method based on determination of type of operations. Maintenance condition of production machines were identified in two categories: time of operations and time of standby. Based on that time of standby were subjected for cause analysis that lead for correction.

Time compression rate coefficient drawn from the motion study methods as below [2, 3]:

$$Ccomp.r = \frac{Tshift-(t_{p.e}+ t_{op} + t_{maintenance} + t_{op.m}+ t_{wp})}{Tshift} \tag{7}$$

where $C_{comp.r}$ time compression rate coefficient. $Tshift$ - observation time or time of one shift.

This coefficient helps to determine proportional workforce productivity.

$$C_{\Uparrow WP} = \frac{C_{comp.r}}{1-C_{comp.r}} \tag{8}$$

or

$$C_{\Uparrow WP} = \frac{t_{loss}}{T_{shift} - t_{loss}} \tag{9}$$

where t_{loss} - down time loss, $t_{loss} = Tshift - (t_{p.e} + t_{op} + t_{maintenance} + t_{op.m} + t_{wp})$

In the mordent techniques OEE parameter (Overall Equipment Effectiveness) is the most commonly used [4, 5]. OEE has been proposed in 1960s by Seiichi Nakajima [6], however become widely applied in 80s and was populated in 90s by Robert Hansen. [7].

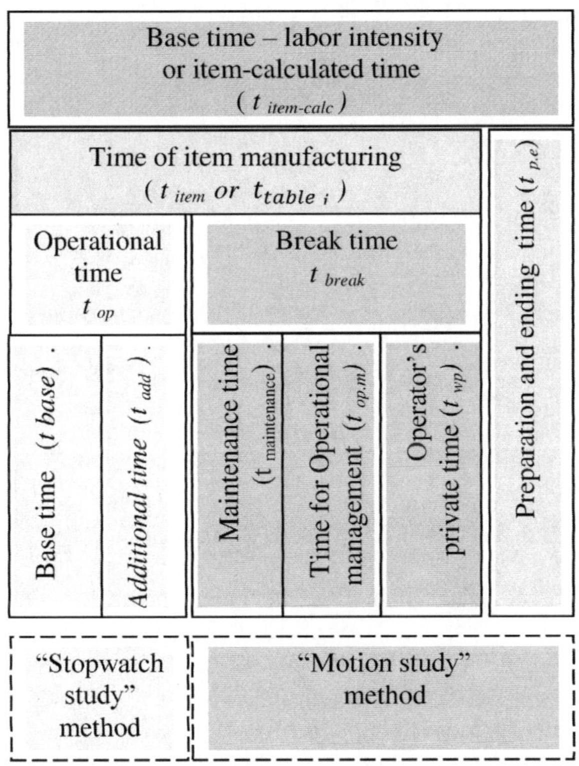

Figure 1. The scheme of composition of elements of Base time – labor intensity or item-calculated time (t item-calculated) and methods that help to identify them.

General parameter of effectiveness (OEE) determined by 3 following items:

- A-Availability: (percentage of scheduled time that the operation is available to operate.
- P-Performance: (speed at which the Work Center runs as a percentage of its designed speed).
- Q-Quality: (Good Units produced as a percentage of the Total Units Started).

$$OEE = A \cdot P \cdot Q \tag{10}$$

Each parameter can be represented by coefficients: For instance if A = 0,87; P = 0,50; Q = 0,95, than OEE = 0,87 · 0,50 ·0,95 = 0,42

Availability is calculated by [6, 7, 8, 9]:

$$A = \frac{OT}{PPT} = \frac{PPT - DTL}{PPT} \quad (11)$$

Where *OT - Operating Time* (reducing stand by time), *DTL* - (*Down Time Loss*) all stand by or non-operational time by all reasons. *PPT* - (*Planned Production Time*).

Formula (12) can be drawn in accordance with the parameters commonly used in Russian manufacturing practice:

$$A = \frac{T_{shift} - t_{loss}}{T_{shift}} \quad (12)$$

It is clear to show a correlation of “Availability”.

$$A = \frac{T_{shift} - t_{loss}}{T_{shift}} = 1 - \frac{t_{loss}}{T_{shift}} \quad (13)$$

$$Ccomp.r.. = \frac{Tshift - (tp.e + t\ opr + tmaintenance + twp)}{Tshift} = \frac{T_{shift} - t_{effectiv}}{T_{shift}} = \frac{t_{loss}}{T_{shift}} \quad (14)$$

$$A = 1 - Ccomp.r. \quad (15)$$

$$A + Ccomp.r. = 1 \quad (16)$$

From formula (16) and Figure 2 as below it is clearly shown that both A and Ccomp.r. are completely opposite by meaning and at the same time compose one integration unity.

If we analyze OEE’s parameter of *Performance* (P) [3], that is determined as

$$Performance = P = ICT / (OT/TP) \quad (18)$$

or

$$P = (TP/OT) / IRR \quad (19)$$

where: *ICT* (*Ideal Cycle Time* for production of 1 item and *IRR* – *Ideal Run Rate*, theoretical maximum of production, reverse to *ICT*; *OT* – (*operating time*), *TP* – (*Total Pieces*), produced by the time of *OT*.

Stand by time should be removed, including low speed operations. If we try to draw above parameters with commonly used theoretical base applied in Russia it can be shown as below:

$$P1 = ICT / (OT/TP) = \frac{t_{item-calc.}}{t_{base(i)} / 1\ part} = \frac{t_{item-calc}}{t_{base.i}} \quad (20)$$

In total, while for production (N_{reali}) of items (i) of work benches :

$$P_N = ICT / (OT/TP) = \frac{\Sigma t_{item-calc\ i}}{\Sigma t_{basei} / \Sigma N_{real\ (i)}} = \frac{\Sigma N_{real\ (i)} \cdot t_{item-calc(i)}}{\Sigma t_{base\ (i)}}$$

$$P_N = (TP/OT) / IRR = \frac{N_{real(i)} / \Sigma t_{base(i)}}{\Sigma t_{item-calc(i)}} = \frac{\Sigma N_{real(i)} \cdot t_{item-calc(i)}}{\Sigma t_{base(i)}}$$

or

$$P = ICT/(OT/TP) = (TP/OT)/IRR = \frac{\Sigma N_{real(i)} \cdot t_{item-calc(i)}}{\Sigma t_{base(i)}} \quad (21)$$

Third parameter of OEE defined a quality level or (Q), by the following formula [1, 2]:

$$Q = \frac{N_{produced\ (i)} - N_{non-confomirng\ (i)}}{N_{produced(i)}} \quad (22)$$

where $N_{produced\ (i)}$ is a total production by defined time cycle (i), $N_{non-confomirng\ (i)}$ – is a number of defective or non-conforming units by the same period.

In Russia a similar parameter "output level" is applied together with "level of non-conformities" (q):

$$q = \frac{N_{non-confomirng\ (i)}}{N_{produced\ (i)}} \quad (23)$$

$$Q = \frac{N_{produced\ (i)} - N_{non-confomirng\ (i)}}{N_{produced(i)}} = 1 - q \quad (24)$$

Or

$$Q + q = 1 \quad (25)$$

On Figure 3 correlations of those parameters drawn on pie chart:

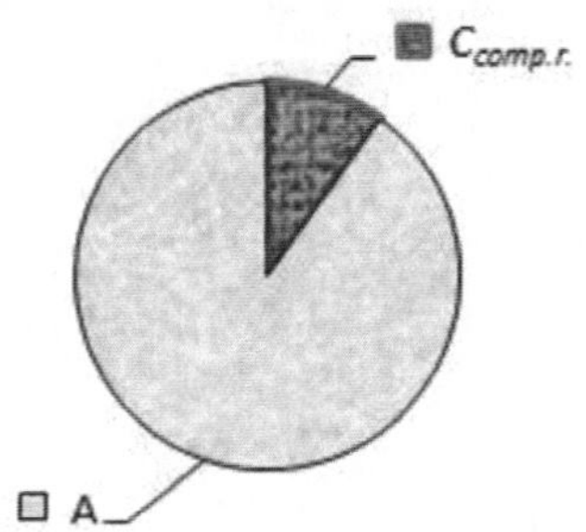

Figure 2. Schematic determination of OEE's parameter of Availability and Ccomp.r. - time compression rate coefficient

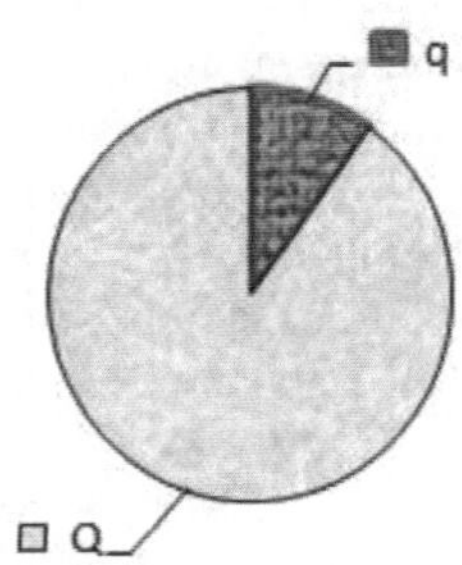

Figure 3. Correlation of q (applied in Russia) and Q according to OEE.

Given the formulas (15), (21), (24) the expression (10), which determines the parameter OEE will be of the form

$$OEE = (1 - Ccomp.r..) \cdot \left(\frac{\sum N_{real\ (i)} \cdot t_{item.calc(i)}}{\sum t_{base\ (i)}}\right) \cdot (1 - q) \quad (26)$$

3. Conceptual approaches to the prediction of equipment effectiveness criteria:

Parametrical analysis shows that applied in the past "Motion study" method is based on data collected by the direct verification of production activities. For instance there is no additional explanation or training required for operators to count non-conforming products or standby time Application of OEE parameter is based on determination of the loss and cost deduction methods.

Determination of loss and cost deduction lead us to a next question, how to calculate a time loss. Calculation of "Availability" parameter does not give us methods of loss time accountability.

Stand by time determination is also missing. We should understand that OEE methods is commonly applied with support of IT and digitalization of production process. By the integrated production statistical technics those parameters are automatically calculated. However not all manufacturers could afford those It/automatic calculation software, therefore an additional loss parameters might be added in the future. According to "Motion study" methods structure is clear defined. Same time OEE does not give us an output for labor intensity/manpower calculation.

It was shown that OEE and some commonly used parameters applied in Russian manufacturing practice are assimilated each other, are uniting same core parameters and can be applied together for determination of the full picture of manufacturing process management.

It was shown that by applying OEE, Q, q, Ccomp.r parameters all together production activities can be assessed from different angle by focusing same target of continues effectiveness improvement of production process.

When the value of the performance index of equipment is projected, it is necessary to have a clear classification of the elements of productive time and down time loss.

4. Summary:

Analysis and comparison of the elemental base of indicators of the efficiency of equipment in the works of American and Russian researchers showed the presence of indicators reflecting the state of efficient and inefficient operation of equipment.

Methods “Stopwatch study” and “Motion study”, which allowed determining the elements of the effective operation of equipment, had the goal of rationing the time of work on machines, in order to subsequently pay the main production workers.

"Normalizing" the time of inefficient operation of the equipment requires its classification and purpose.

Strict separation of the effective time of equipment operation time and loss of time associated with the implementation of quality products and the fulfillment of marriage, are shown in figures 2, 3, involves the creation of a clear classification of the elements of effective time loss of time in accordance with the requirements of technology and time, does not comply with these requirements.

Classification of elements of time must reflect the relationship of two directions: the machine work outside the machine, to show the connection with the means of labor and labor process. This will enable you to analyze and control the manufacturing process.

References

[1] Organization of mechanical engineering production, edited by B.Ya. Katzenbogen. Approved VKDHS under SNK of the USSR as a textbook for machine-building universities. M. L., SOSTI, 1941. 648p. (Before the title the authors Abramovich, Dumler, Katzenbogen, Kornitsky and Kheifets)

[2] Abramova, I.G. Basics of the organization of production of machine-building enterprises [Electronic resource]: (Lecture course and workshop): tutorial / I. G. Abramova, Samara State Aerospace University named after S.P. Korolev (National Research University) (SSAU). - Electronic text data. – Samara, 2011. (CD-ROM).

[3] Organization and planning of production at the machine-building enterprise. [Textbook for high schools on specialty "Economics and organization of machine-building industry"]. Under the editorship of Professor V. A. Letenko. M., “High school”, 1972.

[4] OEE Primer, Calculating OEE, http://www.oee.com/calculating-oee.html

[5] Thomas R. Cutler. Improving overall equipment effectiveness with lean and value-stream mapping. 10/26/2010. http://www.qualitydigest.com/inside/twitter-ed/improving-overall-equipment-effectiveness-lean-and-value-stream-mapping.html#

[6] Shopin, A.G., OEE and downtime management: from theory to implementation in SIMATIC IT / A.G. Shopin, I.V.Zanin, “Automation in industry”, 2006, 09, p. 24 – 29. http://mescenter.ru/images/abook_file/avp_shopin1.pdf

[7] Hansen, Robert C. Overall Equipment Effectiveness (Print-On-Demand Edition) Powerful Production/Maintenance tool for Increased Profits. 290 Pages, Softcover Published: January, 2001. ISBN: 9780831102180. http://new.industrialpress.com/overall-equipment-effectiveness-1.html

[8] Antonenko I., Kryukov I., Shestopalov P. The efficiency of the production equipment. . http://www.cfin.ru/management/manufact/oee.shtml

[9] Optimal scheduling planned preventative maintenance — a way to improved using of equipment / G. Medvedeva, A., Museridze, Tikhonova, A. Kryukov, A. Zaidullin // CAD and graphics, 2013 №1, c. 84 – 89., http://www.sapr.ru/article.aspx?id=23579&iid=1085

IOP Conf. Series: Materials Science and Engineering **302** (2018) 012053 doi:10.1088/1757-899X/302/1/012053

The Development and Calculation of an Energy-saving Plant for Obtaining Water from Atmospheric Air

D A Uglanov[1], K E Zheleznyak[2] and P A Chertykovsev[3]

[1]Samara National Research University, Russia
[2]Samara National Research University, Russia
[3]Samara National Research University, Russia

[1]dmitry.uglanov@mail.ru, [2]zheleznyak999@yandex.ru, [3]p.a.chert@mail.ru

Abstract. The article shows the calculation of characteristics of energy-efficient water generator from atmospheric air. This installation or the atmospheric water generator is the unique mechanism which produces safe drinking water by extraction it from air. The existing atmospheric generators allow to receive safe drinking water by means of process of condensation at air humidity at least equal to 35% and are capable to give to 25 liters of water in per day, and work from electricity. Authors offer to use instead of the condenser in the scheme of installation for increase volume of produced water by generator in per day, the following refrigerating machines: the vapor compression refrigerating machines (VCRM), the thermoelectric refrigerating machines (TRM) and the Stirling-cycle refrigerating machines (SRM). The paper describes calculation methods for each of refrigerating systems. Calculation of technical-and-economic indexes for the atmospheric water generator was carried out and the optimum system with the maximum volume of received water in per day was picked up. The atmospheric water generator which is considered in article will work from autonomous solar power station.

1. Introduction

At present, there are problems of energy saving in various areas of human life and in power complexes of industrial plants [1, 2, 3].

According to the World U.N. report about development of water resources, many countries already achieved limit opportunities of water use: consumption of fresh water for the last half a century increased triply. In extensive regions of developing world unequal access to safe drinking water, water purification for production of provisions and processing of drainage water remains. Nearly five billion people will be left without drinking water to 2030 if people will not undertake nothing, and it is about 67% of the population of the planet. Problems with drinking water and water resources have led to creation of atmospheric water generators. The atmospheric water generator is the unique mechanism which produces safe drinking water by extraction it from air. For the first time the atmospheric water generators were developed in the 1990th. They were similar to system which is used for dehydration air in refrigerators.

In this article the analysis is made of the energy-efficient installation working from the photo-electric converter (installation power W = 200V) with use of refrigeration installations, such as the VCRM, the TRM and the SRM. The analysis allowed to estimate fully overall operational efficiency of the generator and quality of water, and also promoted adoption of the design solutions increasing the volume of received water by installation in per day.

2. The atmospheric water generator working at VCRM basis

Achievable temperatures in installations, refrigerating capacity and expenses of mechanical work significantly depend of kind and properties of refrigerant coolants. Refrigerant coolants need to have ability to absorb a large amount of warmth at evaporation, to have small specific volumes of steam, low critical temperatures, viscosity and density, high coefficient of heat dissipation and heat transfer, to be harmless, fireproof, available and inexpensive. Therefore, for work two refrigerant coolants were chosen: R134A and R502.

In this work for calculation and analysis of a cycle of the VCRM the CoolPack [4] program was used. It is the program for design, calculation, analysis and optimization of refrigerating systems.

We will consider the return Rankine cycle for each refrigerant coolant in autumn, spring, summer and winter seasons. Example is shown on the figure 1. Values of average temperatures and humidity in the Samara region for every season, and also the consumed power by installation from the photo-electric converter are presented in table 1.

Table 1. Initial data for calculation of the VCRM

	winter	spring	*summer*	autumn
Temperature(° C)	-15	+8	+24	+10
Moisture content(%)	85	65	50	75
Power consumption plant N (W)			200	

Such levels of temperature of evaporation at calculation are caused by need of choice of the optimum mode which corresponds to the maximum quantity of the received water. All values received in the program are tabulated for each refrigerant coolant and season.

For definition of moisture content at each mode we will use I-d diagram of humid air [5].

Schedules of dependences of moisture content d, mass and volume consumption (m and V) of refrigerant coolant, refrigerating coefficient COP, heat input Qe and heat output Qc from temperature T for each refrigerant coolant were constructed. We will consider dependences for refrigerant coolant R134A on the figures 1-4.

IOP Conf. Series: Materials Science and Engineering **302** (2018) 012053 doi:10.1088/1757-899X/302/1/012053

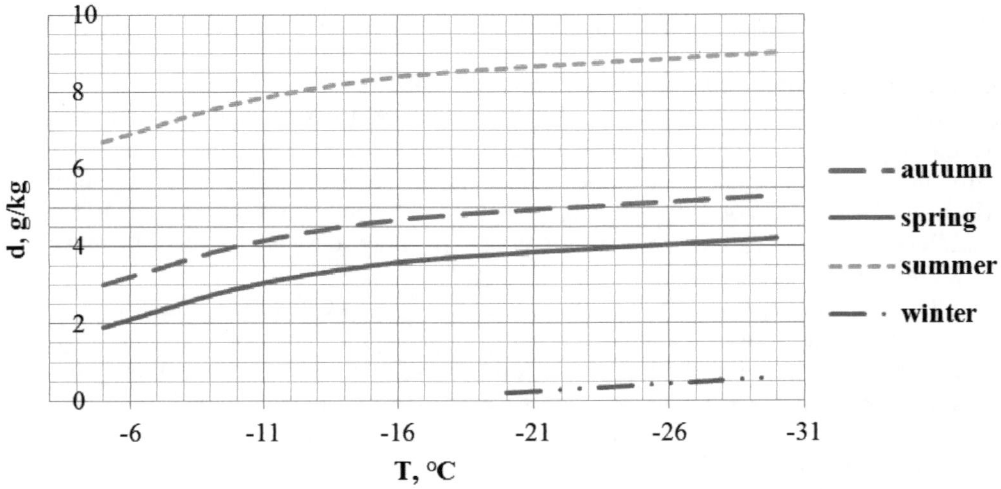

Figure 1. Addiction of moisture content d from temperature T for refrigerant coolant R134a

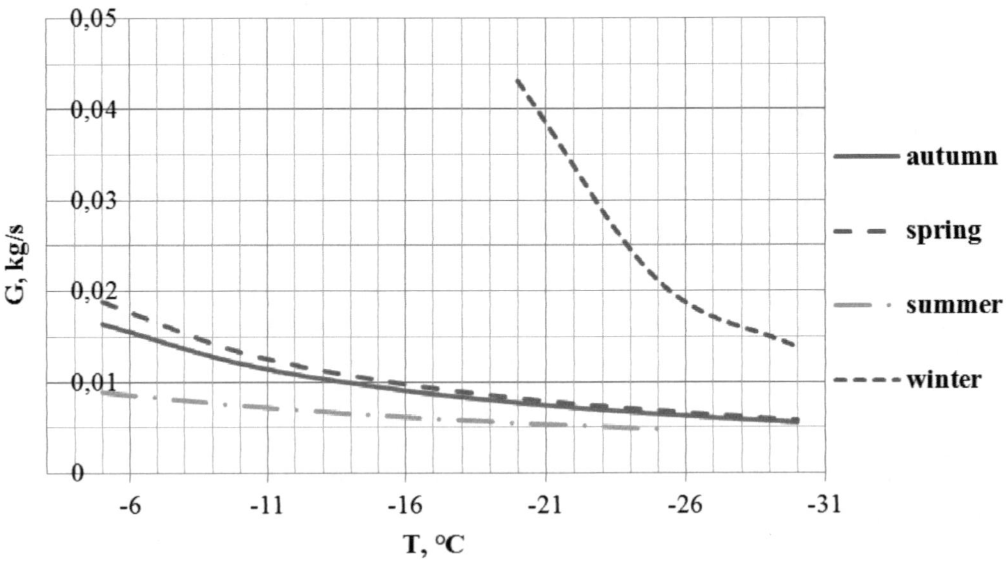

Figure 2. Addiction of mass consumption from temperature T for refrigerant coolant R134a

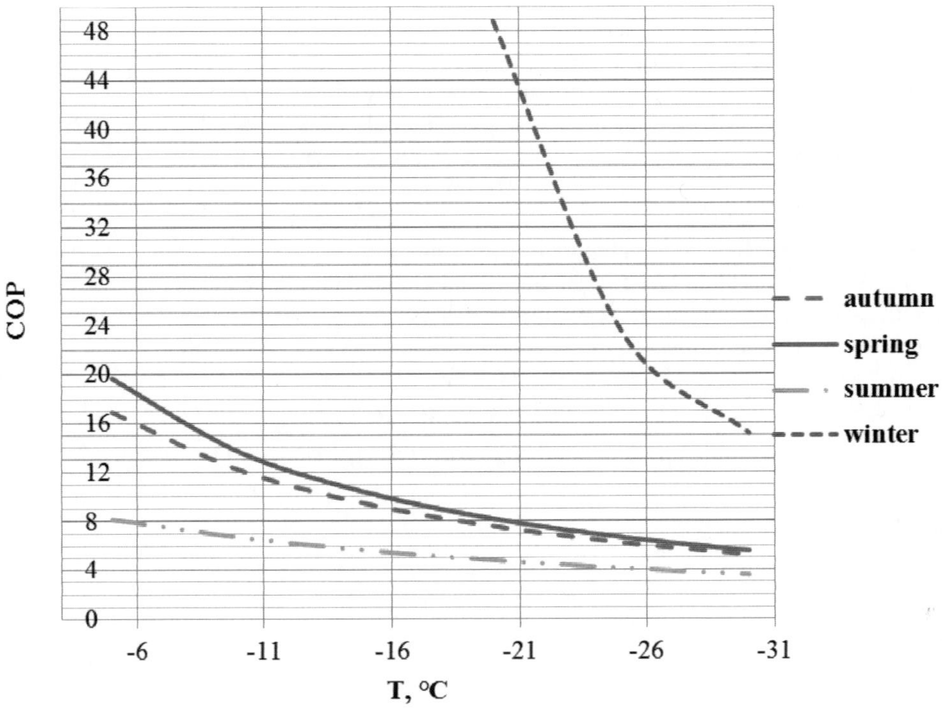

Figure 3. Addiction of refrigerating coefficient COP from temperature T for refrigerant coolant R134a

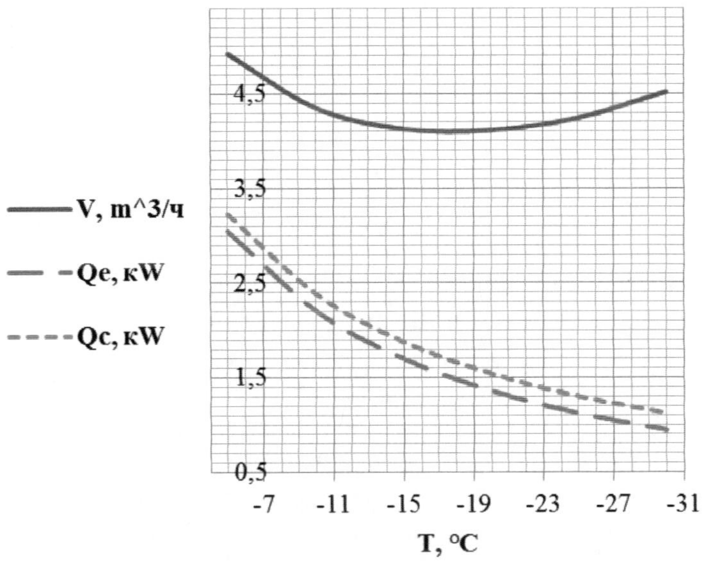

Figure 4. Addiction of volume consumptionV, heat input Qe and heat output Qc from temperature T for refrigerant coolant R134a

The analysis of received values showed that during the summer period at temperature of evaporation -30 °C in the VCRM on refrigerant coolant R134a the maximum quantity of moisture content is developed. But for further calculation we choose the autumn period which corresponds to average value of received moisture content.

Further by known methods [6], main elements of the VCRM were calculated: compressor, condenser and evaporator. Values of coefficients of heat transfer are defined k_{ev}=39,1 W/m^2*K and k_{cond}=23 W/m^2*K, heat dissipation α=41,31 W/m^2*K and α=23 W/m^2*K, effective refrigerating coefficients ε=2,24, Qe=0,942 kW, Qc=1,122 kW and geometrical parameters of evaporator F=0,27 m^2, condenser F=0,18 m^2 and compressor Ne=0,309 kW.

3. Calculation of the Stirling-cycle refrigerating machine

Calculation was carried out by Schmidt's method [7].

For calculation the following main equations were used:

- Equations of motion

$$V_h = 0{,}5 \cdot V_{OX} \cdot (1+\cos\varphi) \tag{1}$$

$$V_c = 0{,}5 \cdot K \cdot V_{OX} \cdot (1+\sin\varphi) \tag{2}$$

$$0{,}5 \cdot K \cdot V_{OX} \cdot (1+\sin\varphi) + 0{,}5 \cdot V_{OX} \cdot (1+\cos\varphi) + X \cdot V_{OX} \tag{3}$$

$$V_P = X \cdot V_{OX} \tag{4}$$

- Energy balance equations

$$L_c = Q_1 - Q_2 \tag{5}$$

$$Q_1 = Q_2 \cdot \tau \tag{6}$$

$$\varepsilon = \frac{Q_2}{L_c} \tag{7}$$

- Material balance equations

$$m_z = m_h + m_c + m_M \tag{8}$$

$$\tau = \frac{T_h}{T_c} \tag{9}$$

$$K = \frac{V_h}{V_c} \tag{10}$$

$$S = \frac{2 \cdot X \cdot \tau}{\tau + 1} \tag{11}$$

$$\delta = \frac{\sqrt{\tau^2 + k^2 + 2 \cdot \tau \cdot k \cdot \cos\alpha}}{\tau + k + 2 \cdot S} \tag{12}$$

- Gas equation

$$m_h = \frac{P_h \cdot V_h}{R \cdot T_h} \tag{13}$$

$$m_c = \frac{P_c \cdot V_c}{R \cdot T_c} \tag{14}$$

$$m_M = \frac{P_M \cdot V_M}{R \cdot T_M} \tag{15}$$

Calculation of the regenerative heat exchanger was carried out. Parameter found

$$\eta_P = \frac{N_1}{N_1 + 1}\left[1 - \frac{1}{9\left(\frac{C_H m_H}{C_p m_3}\right)^2}\right] \tag{16}$$

and losses for approach, the hydraulic resistance and total losses from ratio l/d, hot and cold heat exchangers, external loading.

4. Calculation of the thermoelectric refrigerating machines

Calculation of parameters of the thermoelectric refrigerating machines was carried out by method [8]. Calculation of the refrigerating thermobattery in mode of the maximum refrigerating coefficient.

Further the software product «KRYOTHERM» [9] was used for calculation and choice of standard modules. Initial data for calculation in program are Tr = 283 K, Tx= 243 K, Qo = 25 W. The module TB–127–2,0–1,65 was picked up for initial data and characteristics of chosen module were constructed (figure 5, 6).

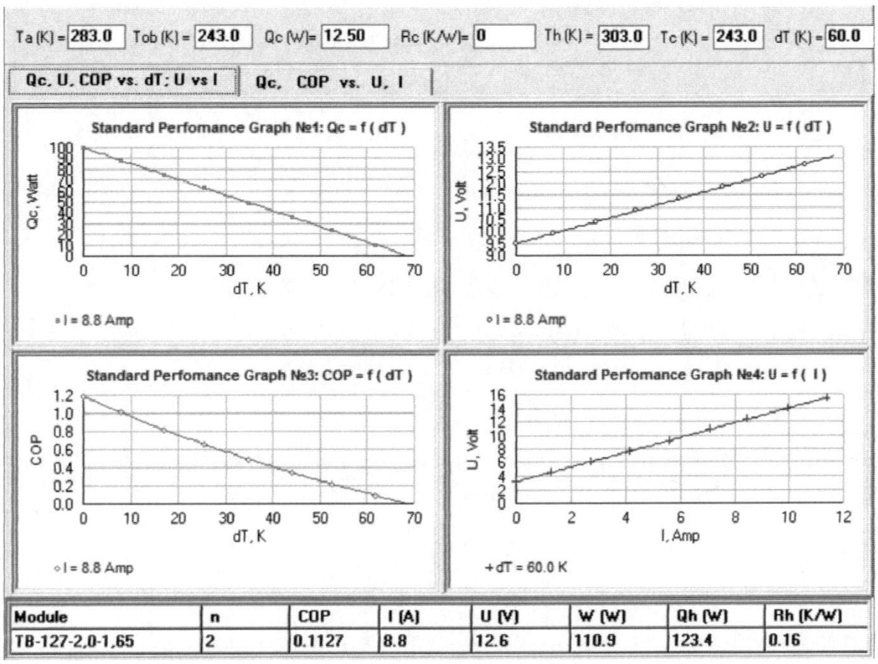

Figure 5. Standard performance Qc = f(dT),U = f(dT), COP = f(dT), U = f(I) module TB - 127–2,0–1,65

IOP Conf. Series: Materials Science and Engineering **302** (2018) 012053 doi:10.1088/1757-899X/302/1/012053

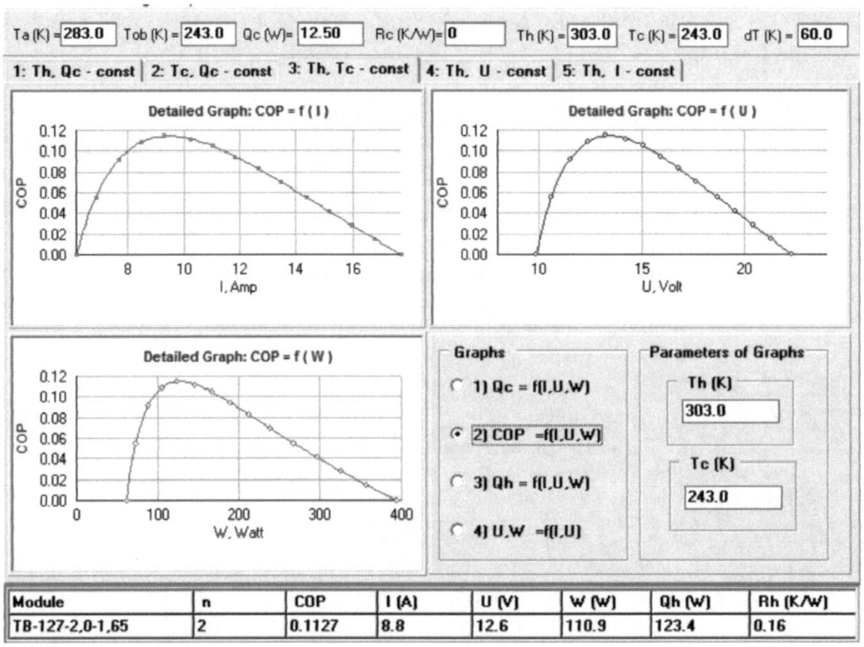

Figure 6. Detailed performance Qc = f(dT),U = f(dT),COP = f(dT), U = f(I) module TB – 127–2,0–1,65

5. Determination of efficiency of installation

The produced quantity of water in per day by the atmospheric generator with various refrigerating machines was determined by the following method:

$$m_w = \frac{G_a \cdot d \cdot 24}{1000} \tag{17}$$

where G_a - air-mass flow through automatic ventilating machine; d - humidity.

Air-mass flow through automatic ventilating machine is equal:

$$G_a = V_{fan} \cdot \rho_a \cdot n \tag{18}$$

where V_{fan} - air-quantity flow through automatic ventilating machine; ρ_a - density of air; n - quantity of automatic ventilating machines.

We will define the necessary quantity of automatic ventilating machines: $n = \frac{F_{out}}{F_{fan}}$ and air-quantity flow through automatic ventilating machine V_{fan}, knowing F_{out} and F_{fan}.

6. Results

The produced quantity of water in per day by the atmospheric generator with refrigerating systems table 2.

Table 2. The produced quantity of water in per day by the atmospheric generator with refrigerating systems

	N, W	*n, pcs.*	*Nn, W*	*d, g/kg*	*mw, l/d*
VCRM	180	12	10,08	5,3	84
SRM	180	7	5,88	5,3	48
TRM	180	1	0,84	4,8	2

Number of technical-and-economic indexes was entered for determination of efficiency, total power consumption, energy and monetary cost 1 l/days of water table 3.

Table 3. Technical-and-economic comparison of installation with various refrigerating systems

	N, *W*	*N Σ*, *W*	*Nn*, *W*	*m*, *kg*	*mw*, *l/d*	*Q c*, *W*	*Kq0*, *l/d*kW*	*Km*, *l/d*kg*	*P*, *rub*	*Kc*, *l/d*rub*	*W Σ*, *kW*h*	*C*, *rub/l*
VCRM	180	190,1	10,08	13,5	84	942	0,09	6,22	47 000	0,002	4,6	0,13
SRM	180	185,9	5,88	11	48	1134	0,04	4,36	82 000	0,001	4,46	0,23
TRM	180	180,8	0,84	1,3	2	25	0,08	1,54	14 000	0,0001	4,34	5,58

For technical-and-economic comparison we offer to use the following coefficients:

- coefficient of efficiency depending on its refrigerating capacity, mass, price, and also energy and monetary cost 1 l/days of water.

$$K_{q0} = \frac{m_w}{Q_c}, \frac{l}{d \cdot kW} \tag{19}$$

- coefficient of efficiency depending on mass and price of refrigerating system is determined by the following formulas

$$K_m = \frac{m_w}{m_\Sigma}, \frac{l}{d \cdot kg} \tag{20}$$

$$K_c = \frac{m_c}{P}, \frac{l}{d \cdot rub} \tag{21}$$

- the total power consumed by installation

$$W_\Sigma = \frac{N_\Sigma}{1000} \cdot 24, kW \cdot h \tag{22}$$

- We will use the following formulas for determination of energy and monetary cost 1 l/days

$$C_{en} = \frac{W_\Sigma}{m_w}, \frac{kW \cdot h}{l / d} \tag{23}$$

$$C = C_{en} \cdot i, \frac{rub}{l / d} \tag{24}$$

Electric rate is $i = 2,57 \frac{rub}{kW \cdot h}$.

7. Conclusion

This article describes methods of calculation and selection of main elements of refrigerating systems used in the atmospheric generator for increase of volume of received water. As refrigerating system the vapor compression refrigerating machines, the thermoelectric refrigerating machines and the Stirling-cycle refrigerating machines are considered.

The return Rankine cycles were constructed in the program CoolPack for the vapor compression refrigerating machine. Calculation of cycle parameters is carried out for autumn, spring, summer and winter seasons for refrigerant coolants R134a and R502. On the basis of received results the optimum mode - the autumn season was chosen ($t_u = -30°C$, $t_к = 10°C$, $Q_e = 0,942 kW$, $Q_c = 1,122 kW$, $d = 5,3 g / kg$) which is used in further calculations. Calculation was carried out for evaporator
($F = 0,27 m^2$), condenser ($F = 0,18 m^2$) and compressor ($N_e = 0,309 kW$).

In the program Kryotherm the thermoelectric module TB-127-2,0-1,65 (n = 2 pcs) was picked up and was calculated for the thermoelectric refrigerating system. Calculation was carried out for regenerator ($\eta_p = 0,98$, $\Delta Q_\Sigma = 115,63W$), heat exchangers of external loading of compressor and expander cavities, volumes of hot and cold cavities ($Q_1 = 78,3J$, $Q_2 = 67,5J$) for the Stirling-cycle refrigerating machine.

The automatic ventilating machine DEEPCOOL XFAN 120 was picked up ($V = 44m^3/h$, $N = 0,84W$). The necessary quantity of automatic ventilating machines for the VCRM is n=12 pcs, for the SRM is n=7 pcs, for the TRM is n=1 pcs. The quantity of received water in per day was calculated. So, the atmospheric generator with the VCRM can generate $m_w = 84\ l/d$, with the SRM - $m_w = 48\ l/d$, with the TRM - $m_w = 6\ l/d$

Coefficients of efficiency were calculated for each system depending on its refrigerating capacity K_{q0}, mass K_m, price K_c, and also energy C_{en} and monetary C cost 1 l/days of water. Monetary cost 1 l/days of water received by the atmospheric generator with the VCRM is $C = 0,13\frac{rub}{l/d.}$, by the atmospheric generator with the SRM is $C = 0,23\frac{rub}{l/d.}$, by the atmospheric generator with the TRM is $C = 5,58\frac{rub}{l/d.}$.

The optimum refrigerating machine with the maximum efficiency of water in per day is the vapor compression refrigerating machine for work in fixed conditions.

References

[1] Kerdphol, T., Qudaih, Y., Mitani, Y.Optimum battery energy storage system using PSO considering dynamic demand response for microgrids.(2016) International Journal of Electrical Power and Energy Systems, 83, pp. 58-66.

[2] Kakumoto, Y., Koyamatsu, Y., Shiota, A., Qudaih, Y., Mitani, Y. Application of Geographic Information System to Power Distribution System Analysis (2016) Energy Procedia, 100, pp. 360-365.

[3] Burtsev, S.A., Karpenko, A.P., Leontiev, A.I. A method for distributed production of liquefied natural gas at gas-distribution stations. (2016) High Temperature, 54 (4), pp. 573-576.

[4] http://www.en.ipu.dk

[5] Sokolov E Y, Brodyansky V M Energy basis of heat transformation and cooling processes. -M .: 1967. 336 p.

[6] Mysin V M, Tatarenko Y V Selection and calculation of the main elements of a single-stage vapor compression refrigerating machine. - SPb .: University of ITMO, 2015. 64 p.

[7] Walker G Stirling Engines / Clarendon Press, Oxford, 1980. 145 p.

[8] Shostakovskii P P Thermoelectric sources of alternative power supply. Components & Technologies, 2010, no. 12, pp. 131-138.

[9] http://kryothermtec.com/ru/

IOP Conf. Series: Materials Science and Engineering **302** (2018) 012054 doi:10.1088/1757-899X/302/1/012054

Modelling the High-speed Injector for Diesel ICE

V V Buryuk[1], S S Kayukov[1], A A Gorshkalev[1], A V Belousov[1], R E Gallyamov[1] and V A Zvyagintsev[2]

[1] Heat engineering department, Samara National Research University, 443086 Samara, Russian Federation
[2] Department of Aircraft Production and Quality Control in Mechanical Engineering, Samara National Research University, 443086 Samara, Russian Federation

E-mail: Agorsh@bk.ru

Abstract. The article describes the results of research on the option of improving the operation speed of the electro-hydraulically driven injectors (Common Rail) for diesel ICE. The injector investigated in this article is a modified serial injector Common Rail-type with solenoid. The model and the injector parameters are represented in the package LMS Imagine.Lab AMESim with the detailed description of the substantiation and background for the research. Following the research results, the advantages of the proposed approach to analysing the operation speed were detected with outlining the direction of future studies.

1. Introduction

In the fuel injection systems of diesel internal combustion engines (ICE), there is a problem of overcoming the contradiction between the fuel injection laws and ensuring the fuel atomization quality. The characteristics of injection and spraying of fuel depends on the design of the nozzle. The characteristics such as: the geometric characteristics of flow of sprayed fuel, the structure of the fuel flare, the fineness of fuel spraying, a number of other parameters of the fuel supply process. To ensure high-efficiency combustion of fuel and, consequently, to obtain high technical and economic indicators of the diesel, it is necessary to organize a rational mixture formation. The quality of fuel spraying is characterized by its fineness and uniformity. It is necessary to organize the process of jet disintegration so as to obtain a high quality of spraying at all stages of injection, including the initial and final, when there is a low level of fuel pressure in front of the nozzle. A quality atomization requires continuous flow at a high pressure upstream the holes in the atomizer. Compliance with the fuel injection law, i.e. distribution of the crankshaft angle cycle charge in different ICE modes requires a wide flow variation range. The assumed priority of the requirements of the fuel injection law in systems employs variable pressure, including in the Common Rail systems [1,2]. Electrohydraulic injectors (EHI) in these systems are intended for implementing a multicycle injection with the number of cycles and duty ratio being limited by the speed of the injector. Partial limitation of the operation speed is due to the injector flexibility in terms of its operating pressures. The consequence of insufficient operating speed is deviation of the real injection conditions from the requirement [3-5].

2. Constant pressure high-speed injection nozzle

The electrohydraulic injectors include:

- an actuator in the form of a needle with a multiplier representing a hydraulic cylinder with the working chamber diameter exceeding the needle diameter and constituting a control chamber, and the needle spring representing a sealing device in the absence of the operating pressure;
- a pressure control device in the control chamber of the multiplier in the form of a slider or a valve with a solenoid drive.

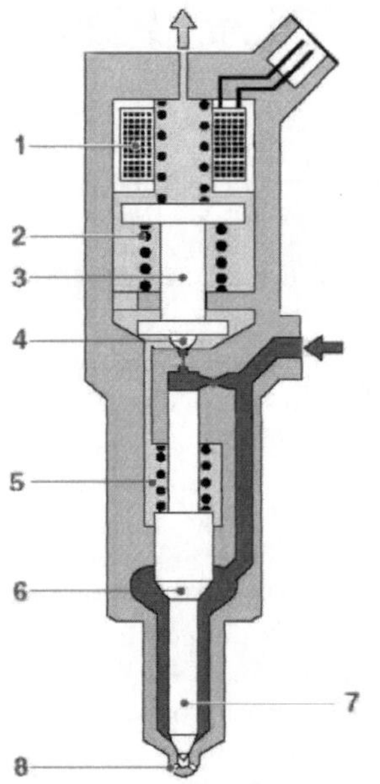

Figure 1. The EHI diagram: 1 - solenoid, 2 - auxiliary spring, 3 - solenoid armature, 4 - valve, 5 - atomizer spring, 6 - atomizer needle cone, 7 - needle, 8 - atomizer orifices.

The valve 4 is small-sized, and therefore it is a high-speed valve since the main flow of the injected fuel does not pass through it. On the other side, the movement of the needle 7 is driven by the hydraulic force of the valve. The valve is located on the discharge side, which facilitates its layout, preserves the possibility of cooling, reduces the requirements for its hydraulic tightness and reduces the power necessary for control. The end solenoid 1 with the disk armature 3 opens the ball valve 4 by overcoming the spring 2. The pressure from the top of the multiplier decreases, and the needle opens the passage to the injector orifices 8. After the solenoid 1 is deenergized and the valve resetting, the pressure to the right of the multiplier is restored through the atomizing jet (Figure. 1).

The following components may be referred to the structural features of the injector: Spring 5, multiplier, needle 7. The needle spring for the injector operation is mandatory and shall be installed around the multiplier to prevent the valve from reopening due to pressure waves when resetting. The closing multiplier increases the closing force acting on the needle. Unlike pressure-driven closing when the pressure is applied to the full area of the needle, this design allows for a large range of closing forces with the same pressure change. The inertia of the multiplier slightly impairs the system response. Also, the multiplier forms a precision pair with the case and increases the dimensions. Its main purpose is that increasing the ratio of the multiplier diameter to the diameter of the needle makes the process more controllable, accurate, and shorter.

When fuel is injected, the multiplier, with its upper end, tends to close the axial orifice leading to the ball valve 4. It may not be fully realized, since high pressure is restored through the jet and the multiplier is force up to the orifice. However, there is a real benefit from this equilibrium stabilization process: fuel flow for control becomes limited (from the battery through the jet, valve to drain).

The multiplier is a needle extension with a spring. From the point of view of the operating speed, it is an increase of the moving mass. The need to use the multiplier is stipulated by its suitability for a wide range of operating pressures.

The needle closing multiplier results increasing the closing force with the same pressure acting in the control chamber and, thus, accelerating the the needle closing. Even considering the weighting of the mass moving with the needle, the operating speed increases. The EHI is unserviceable for a high-speed diesel if the closing multiplier is omitted. Even if it is possible to ensure its operability, the delivery stability is low with respect to the wave processes, process deviations, etc.

The increase in the speed of the EHI ensures a shorter delivery time and, consequently, mean injection pressure and control flow, a reduction of the minimum stable delivery. At first glance, closing of the needle without the multiplier may be accelerated by tightening the spring, but the fundamental difference of this solution is the constant closing force, i.e. with the accelerated closing, opening will be decelerated as well. Therefore, an equally efficient structure may not be obtained by increasing the mechanical closing force of the needle as compared to the multiplier; and in case of low effort, the EHI becomes unstable.

According to the known results of the study [6] of the EHI the speed limitation is conditioned by a wide range of injection pressures, as well as by the deviation of the real supply conditions from the requirement. However, the above statement is invalid regarding the EHI operating at a constantly high pressure. Transition to a constant pressure allows considering several options for increasing the injector operating speed:

1) increasing the rigidity and tension of the return spring, which gives it an additional compensating function as applied to the growing needle opening force acting on the tip of the needle;

2) decreasing the diameter of the multiplier piston with the corresponding decrease in the volume of the control chamber;

3) reducing the cross-sections of the bypass throttle and the sealing section of the control valve, with a corresponding additional reduction in traction force and an increase in the operating speed of the solenoid valve.

To analyse the above possibilities for increasing the operating speed, the one-dimentional LMS Imagine.Lab AMESim [7,8] package was selected. Bosch 0445110190 electrohydraulic injector with a solenoid control valve and the dimensions of the seat, contact part of the needle and the outlets (Table 1), corresponding to the atomizer of the YAZDA 33.1112010-03 serial injector for the KamAZ-7409 engine was selected as the initial design of the model (Fig. 2). According to the assumed method, simulation was performed using models with an operating inlet pressure of 100 MPa.

Table 1. Injector model parameters

Model component in LMS AMESim		Bosch	A modified model
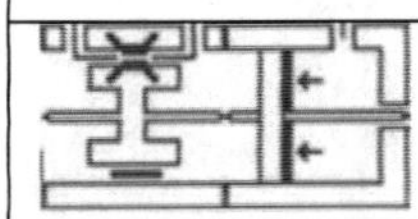 BAP12-2 + BAF01-1	Multiplier diameter, mm	4.3	2.5
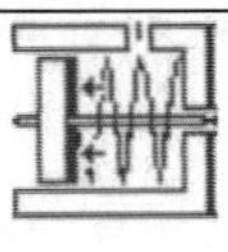 BAP015-1	spring: Pretensioning, N Force, N/mm	 30 13	 60 15000

Detailed parameters of the model are given in table 2,

Table 2. Detailed injector model parameters

Model component in LMS AMESim	Component description	A modified model	Bosch

MAS005-1	mass	3 g	
	lower displacement limit	0 mm	
	lower displacement limit	0.052 mm	
BAP24-1	seat diameter (hole)	0.75 mm	
	seat semi-angle	57.5 degree	
	ball diameter	1.4 mm	
BAP12-2	piston diameter	2.5 mm	4.3 mm
	rod diameter	0 mm	0 mm
BAF01-1	external piston diameter	2.5 mm	4.3 mm
	clearance on diameter	0.003 mm	0.003 mm
MAS005-1	mass	6 g	
	lower displacement limit	0 mm	
	lower displacement limit	0.05 mm	
BAP015-1	piston diameter	4 mm	4 mm
	rod diameter	0 mm	0 mm
	spring force	60 N	30 N
	spring stiffness	15000 N/mm	13 N/mm
BAF02-1	external piston diameter	4 mm	
	clearance on diameter	0.01 mm	
BAP12-1	piston diameter	4 mm	
	rod diameter	3 mm	
BAN004-1	diameter of poppet	3 mm	
	diameter of sac chamber	2.99 mm	
	number of orifices	6	
	orifice diameter	0.12 mm	

IOP Conf. Series: Materials Science and Engineering **302** (2018) 012054 doi:10.1088/1757-899X/302/1/012054

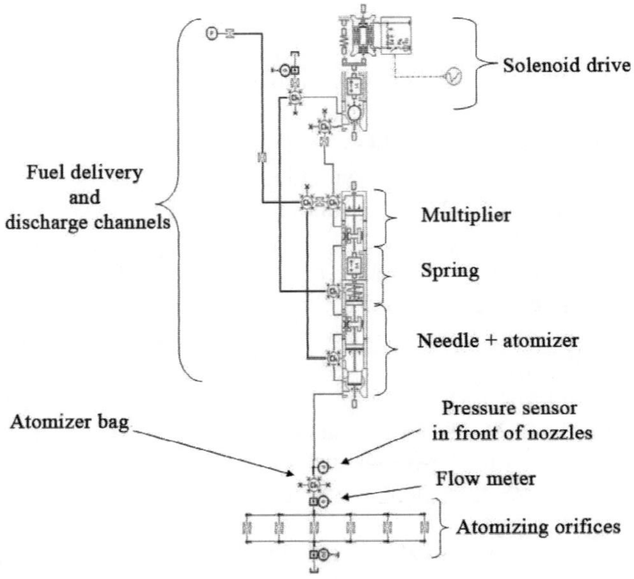

Figure 2. Structural view of the EHI model

Following the calculation results, the diagrams were obtained as shown in fig. 3 and fig. 4. The data are compared against the cycle dose value for the same control signal, moreover, it should be considered that the higher the dose, the faster the response.

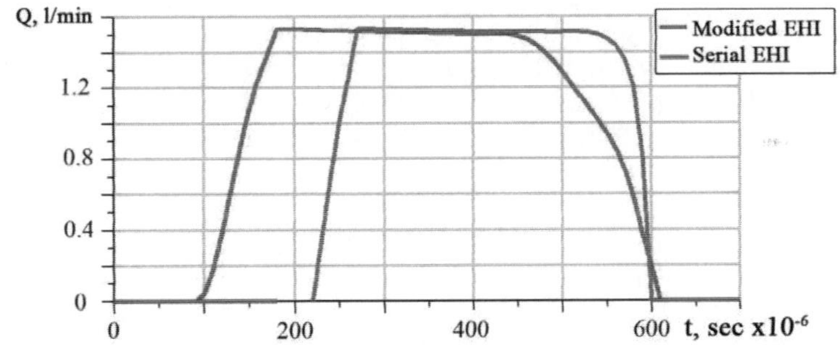

Figure 3. Comparison of EHI characteristics at flow diagram

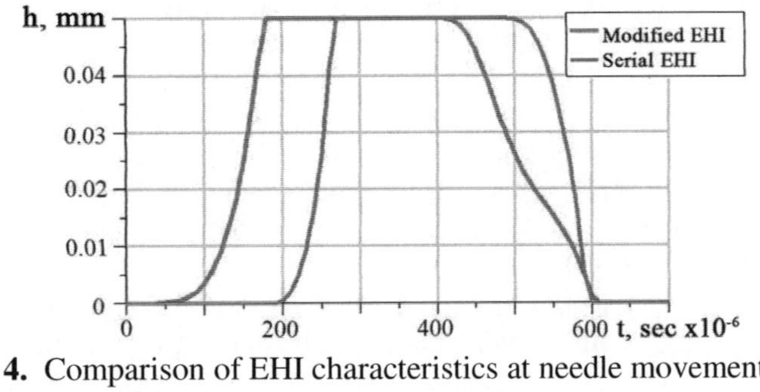

Figure 4. Comparison of EHI characteristics at needle movement diagram

Following the results of measurements given in the diagrams at fig. 3 and fig. 4, the injector operating speed has observably increased. In this regard, further research will consist in analysing the influence of other factors and optimizing their values in order to obtain the maximum operating speed.

3. Conclusions

Further studies to improve the speed of the injector will be made using the described method. The process efficiency will be assessed by the full short-time opening mode of the Bosch injector and the modernized injector. The modernized injector should provide a smaller cycle dose in the modes of short-time full opening, while it will be possible to reduce the duty ratio between the injections.

Based on the simulation results, it can be concluded that the cycle dose should be reduced in the short-time full opening mode, while it will be possible to reduce the duty cycle between injections. In this connection, the research shall be continued with respect to the rest of the factors and optimization of their values to obtain the maximum operating speed.

The most promising for increasing the speed of injectors are the following activities:

1) Coordination of forces acting on the needle in order to reduce the required effort of the electromagnet. This matching allows the use of well-designed low-power high-speed electromagnets and their power devices in high-performance injectors. The peculiarity of this coordination process is its essential dependence on the tightness requirements, since the force of sealing the needle along the saddle is, in fact, the calculated force of the electromagnetic drive and can be significantly reduced. For example, the force of the nozzle solenoid valve control valve can be reduced by an order of magnitude.

2) Elastic coupling of the drive to the needle. This connection allows you to achieve the effect of multiplication, when large needle strokes without the use of mechanical or hydraulic transmission devices are provided by small moves of the electromagnet drive. At the same time, a reduction in the working stroke of a drive of any type ensures an increase in its speed.

3) Application of springs of rod type. Such springs provide the maximum level of material utilization efficiency for the accumulation of deformation energy. Rod springs provide minimum values of moving masses and extend the possibilities of arrangement of injectors.

Acknowledgements

This work was supported by the Ministry of Education and Science of the Russian Federation in the framework of the implementation of the Program «Research and development on priority directions of scientific-technological complex of Russia for 2014– 2020».

References

[1] Reif K. (Hrsg.) 1998, 2002, 2004, 2012 *Dieselmotor-Management: Systeme, Komponenten, Steuerung und Regelung 5 Auflage. Vieweg+Teubner* (Verlag Springer Fachmedien Wiesbaden) p.544

[2] Reif K. (Hrsg.) 2010 *Dieselmotor-Management im Überblick einschließlich Abgastechnik Vieweg+Teubner* (Verlag Springer Fachmedien Wiesbaden) p.172

[3] Gorshkalev A A, Kayukov S S, Korneev S S, Urlapkin V V 2017 Modeling a VR-type piston engine as the power plant IOP Conference Series: Materials Science and Engineering 177(1), 012066

[4] Biryuk V V, Kayukov S S, Zvyagintsev V A, Lysenko U D Ways of speed increase for internal combustion engine fuel injectors Research Journal of Applied Sciences 9 (11), p. 721-724

[5] Kurchatkin I V, Gorshkalev AA, Blagin E V Methods of the working processes modelling of an internal combustion engine by an ANSYS IC Engine module Journal of Physics: Conference Series 803(1), 012085

[6] Grekhov L V, Ivashchenko N A and Markov V A 2013 *Fuel equipment and diesel engine control system: Textbook for Universities* (Publishing House of the Legion-Avtodata) p.344

[7] LMS Imagine.Lab / Web-page: http://www.plm.automation.siemens.com/ru_ru/ products / lms / imagine-lab / amesim /

[8] LMS Imagine.Lab Amesim 2015. *LMS Amesim 14 Tutorial guide*. p.181

IOP Conf. Series: Materials Science and Engineering **302** (2018) 012055 doi:10.1088/1757-899X/302/1/012055

Developing the Model of Fuel Injection Process Efficiency Analysis for Injector for Diesel Engines

M Yu Anisimov[1], S S Kayukov[1], A A Gorshkalev[1], A V Belousov[1], R E Gallyamov[1] and Yu D Lysenko[2]

[1] Heat engineering department, Samara National Research University, 443086 Samara, Russian Federation
[2] Department of Aircraft Production and Quality Control in Mechanical Engineering, Samara National Research University, 443086 Samara, Russian Federation

E-mail: Agorsh@bk.ru

Abstract. The article proposes an assessment option for analysing the quality of fuel injection by the injector constituting the development of calculation blocks in a common injector model within LMS Imagine.Lab AMESim. The parameters of the injector model in the article correspond to the serial injector Common Rail-type with solenoid. The possibilities of this approach are demonstrated with providing the results using the example of modelling the modified injector. Following the research results, the advantages of the proposed approach to analysing assessing the fuel injection quality were detected.

1. Introduction

The injection performance and fuel atomization depend on the injector design: geometric characteristics of atomized fuel jets, fuel spray structure, fuel atomization fineness, and a number of other parameters of the fuel delivery process [1, 2]. To ensure high-efficiency combustion of fuel and, consequently, obtain high technical and economic indicators of the diesel engine, it is necessary to organize a rational injection. The quality of fuel atomization is characterized by its fineness and uniformity. The process of jet disintegration shall be organized so as to obtain high quality of spraying at all stages of injection, including the initial and the final stage, when there is a low fuel pressure upstream the injector.

Presently, three-dimensional modelling of fuel delivery units and systems of diesel engines is being actively employed [3]. However, three-dimensional calculaiton is a complex and labor-consuming task. Therefore, its application in the calculation of injectors is limited to and concentrated on the analysis of fuel atomization and spray geometry. In some researches, a rapid evaluation of fuel delivery characteristics is required, as well as a study of the dynamics and qualitative characteristics of fuel atomization, for example, to verify or compare different types of injectors.

In solving some tasks of estimating the compared options by their quality of fuel atomization, flow diagrams are not always possible to be used. Also, to improve the injection process, it is necessary to increase the injection pressure, which is accompanied by an increase in the flow rate of the fuel through the atomizing orifices (resulting in the flow turbulence) and the spreading velocity of the fuel jets in the CC. The disintegration of the fuel jet begins immediately at the atomizing orifice. These factors lead to the improved quality of fuel atomization, which indicates the need to study the dynamics of atomizers. Thus, there is a challenge in qualitative and quantitative estimation of fuel dispersion, which is presently a labor-intense and an energy-consuming task. It is not possible to obtain accurate fuel atomization characteristics without perofrming three-dimensional modeling, which requires a high and an advanced level of knowledge of the calculation packages.

2. Determining the cycle doze along the pressure spectrum

In solving some tasks of estimating the compared options by their quality of fuel atomization, flow diagrams or pressure diagrams upstream the atomizing orifices may be sufficient. For example, the assessment of the influence of the electrohydraulic injector (EHI) parameters may be performed using the value of cycle supply and the moments of the injection beginning and end of with the same control signal. The option of the injector with a higher cycle delivery rate, provided the equal delay of the moments of the injection beginning and end relative to the moment of the control signal beginning and end, is faster, and, as a consequence, more preferable, since it is known in advance that the fuel delivery quality improves with increasing EHI operating speed. However, in most cases, there is no information on the correlation between the modelling output parameters and the fuel atomization quality, which requires changing the approach to forecasting it.

As one of the tools for assessing the fuel atomization quality based on modeling the dynamics of atomizers, the distribution of the cycle dose over the pressure spectrum upstream the atomization orifices is used. Due to such a tool we are capable of qualitatively assessing the process of fuel injection into the CC. By knowing the pressure at which the fuel dose is injected, we can assess the atomization quality. Also, another advantage of this tool is the possibility of performing a detailed analysis of the fuel delivery process and its features, deriving the effect caused on it by particular parameters of the fuel delivery system elements. The tool model in the LMS Imagine.Lab AMESim [4, 5] software package is represented in fig. 1.

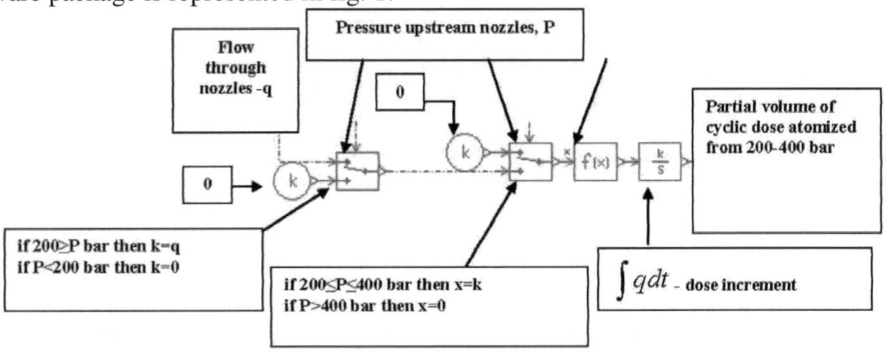

Figure 1. The model of the tool for cycle dose distribution according to the pressure spectrum (from 200 to 400 bar) upstream the outlet nozzles in LMS Imagine.Lab AMESim

If it is necessary to assess the process of injection according to the criteria for the emergence of the fuel ignition conditions in the combustion chamber, it is possible to use the incremental time-volume diagrams for the cycle doses atomized at the pressures of the assigned spectral ranges. When using the same modelling block, this tool is able to demonstrate the pressure, time, and portion of fuel, which is injected to the CC. The obtained dependence is also able to show the nature of ignition, since the pressure at which a portion of fuel enters the cylinder at the time of the injection beginning, is very important.

2.1 Changes in the mean pressure values upstream the orifices in the injector atomizer

To study the various fuel atomization processes, it is necessary to obtain the numbers or integral components serving as the basis for an assessment, as well as an analysis of the regularities of various parameters. Therefore a tool was developed within the LMS Imagine.Lab AMESim package allowing to study change of the mean pressure value upstream the orifices in the injector atomizer.

A particular feature in determining the mean pressure upstream the nozzles is the necessity to consider the features of different injection modes, in which the needle can not only oscillate, but also set in the seat, and therefore the time of non-zero flow through the atomizer orifices may not coincide with the cycle time. Therefore, the mean pressure is proposed to be calculated by dividing the integral from the pressure upstream of the orifices of the atomizer in time by the time integral corresponding to a non-zero flow through the orifices of the atomizer (1).

$$Pcp = \frac{\int P dt}{\int dt} \quad (1)$$

The package LMS Imagine.Lab AMESim is capable of representingn the model results in the form of the curve of caluclated values versus time. Therefore, it is possible to select several parameters to be used for comparison of the systems' options and building regression models based on the analysis of the mean pressure variation upstream the orifices as full injection cycle characteristics, for example:

- a finite or a maximum mean pressure value upstream the nozzles;
- the time for reaching the maximum mean pressure value upstream the nozzles

Building these curves depending on the parameters varying through the modelling process extends the result analysis possibilities to the design stage. The tool model in the LMS Imagine.Lab AMESim software package is represented in fig. 2.

General view of the injector with the above blocks is given in fig. 3

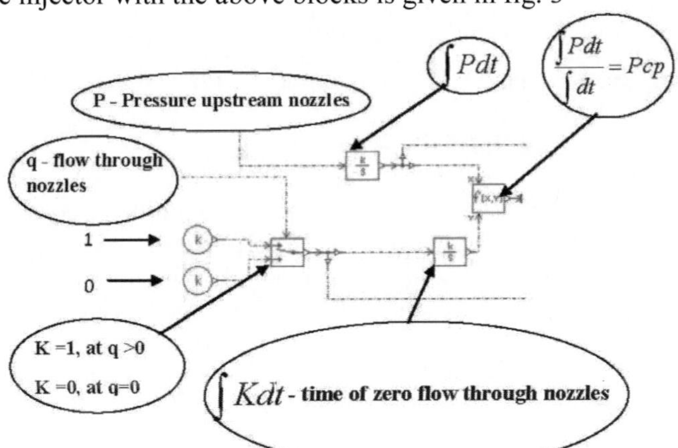

Figure 2. The calculation block for delivery ratio and mean pressure of the injection cycle

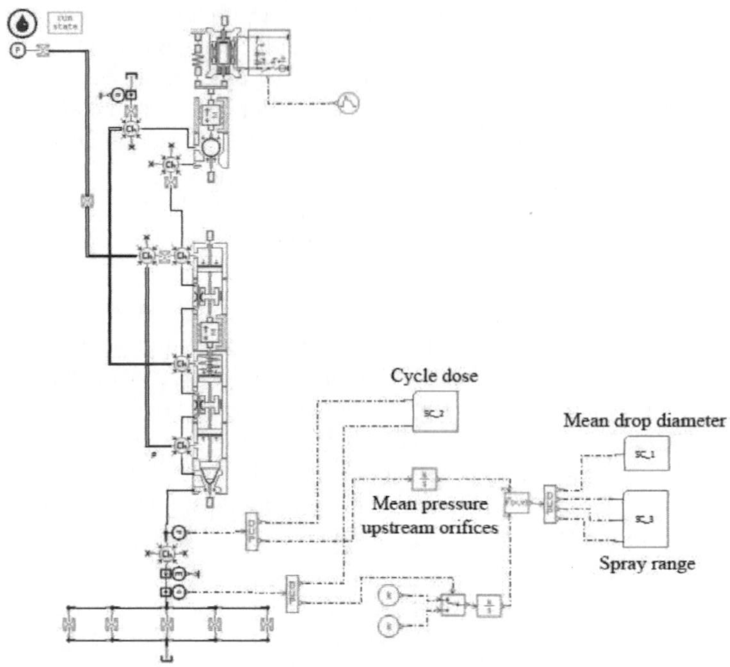

Figure 3. General view of the model with analysis blocks

2.2 Determining Sauter mean drop diameter and jet range

The fuel atomizing quality is directly related to the geometry of the fuel spray, the nature of the jet disintegration and the diameter of the drops. Because of disintegration, the fuel jet is transformed into a fuel spray with a certain drop size distribution. Furthermore, the drop diameter is a random continuous quantity, i.e. it can take any value in the interval of its definition. In this regard, the fuel spray should be considered not only as an accumulation or a set of drops, but as a static aggregate of drops of various sizes.

According to the distribution laws considered, at least two parameters are required to study the drop size spectrum of the atomized fuel: the average drop diameter and the jet distribution characteristics (dispersion, range).

A Sauter mean drop diameter method was selected. It represents a ratio of the total volume of all the drops to their total surface.

Fuel atomization depends on the Weber number, which is a dimensionless quantity establishing a relationship between three parameters that influence the fineness of liquid atomization and found from the expression (2):

$$We = \Delta P d_c / \sigma ; \tag{2}$$

ΔP -Pressure drop, H/m^2 between the nozzle orifice and medium to which the injection is performed d_c -nozzle orifice diameter, m ; σ – fuel surface tension, N/m

One of the basic laws of capillary phenomena influencing the fineness of atomization is the Laplace law. A dimensionless criterion links four parameters that influence the fineness of fuel atomization (3):

$$L_P = (\rho_T d_c \sigma)/\mu_T ; \tag{3}$$

ρ_T - fuel density kg/m^3 between the nozzle orifice and medium to which the injection is performed; d_c - nozzle orifice diameter, m ; σ – fuel surface tension, N/m, μ_T – fuel dynamic viscosity, Ns/m^2

IOP Conf. Series: Materials Science and Engineering **302** (2018) 012055 doi:10.1088/1757-899X/302/1/012055

A significant effect on the fuel atomization is caused by the rate at which the liquid flows out of the nozzle orifice. The rate of flowing out fuel at which the disintegration begins immediately at the nozzle orifice depends on a number of factors (4):

$$\vartheta_T = f(\rho_T;\rho_b;\mu_T;\mu_b;\sigma;d_c); \quad (4)$$

where ρ_T and ρ_b – are density of fuel and air respectively μ_T and μ_b – fuel and air dynamic viscosity; σ –fuel surface tension; d_c - nozzle orifice diameter.

Therefore, the formula takes the form (5):

$$\vartheta_T = \sqrt{2 \cdot \Delta P / \rho_T}\ ; \quad (5)$$

ΔP - mean pressure upstream the orifices, Pa, ρ_T - fuel density, kg/m^3

To determine Sauter mean drop diameter the formula is used (6):

$$d_{32} = d_c \cdot 1.68 \cdot (\rho_k \cdot We)^{-0.266} \cdot L_P^{-0.073}; \quad (6)$$

Based on the selected method, a model was simulated in the LMS Imagine.Lab AMESim software package with the diagram represented in figure 4.

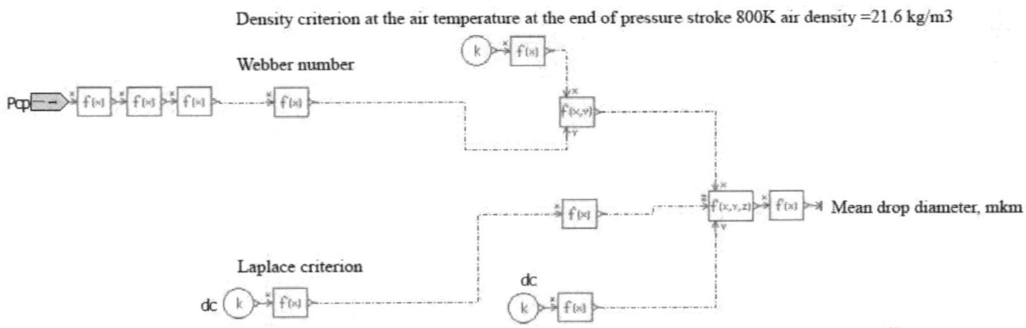

Figure 4. Calculation block of the Sauter mean drop diameter in LMS Imagine.Lab AMESim.

Also during the injection process, it is important to aim at the formation of small fuel drops with simultaneously ensuring the ability of drops to penetrate deeply in all directions in the compressed air medium in the combustion chamber. Penetration of drops into the compressed air medium depends on the range of the fuel jet. With the reduced range of the jet, the fuel drops may not penetrate into the remotest parts of the combustion chamber, incomplete combustion occurs, the specific fuel consumption increases, the engine power decreases.

The range of the spray from the nozzle orifie may be determined from expression (7):

$$L_F = \frac{d_C}{1.2}\left(\frac{\vartheta_D \cdot \tau}{d_c}\right)^{0.5} \cdot \frac{We^{0.105} \cdot M_x^{0.08}}{1.7 \cdot \rho_k^{0.5}}; \quad (7)$$

d_C - Nozzle orifice diameter, m; ϑ_D – actual fuel discharge rate m/s; τ – time of spray movement from the injector s; We – Weber criterion; M_x - Mach criterion (ratio of the fluid flow rate to the sound velocity; ρ_k - density criterion (ration of the air density to the fuel density).

This method for determining the spray range was also modeled in the LMS Imagine.Lab AMESim software package, whose flowchart is given in fig. 5

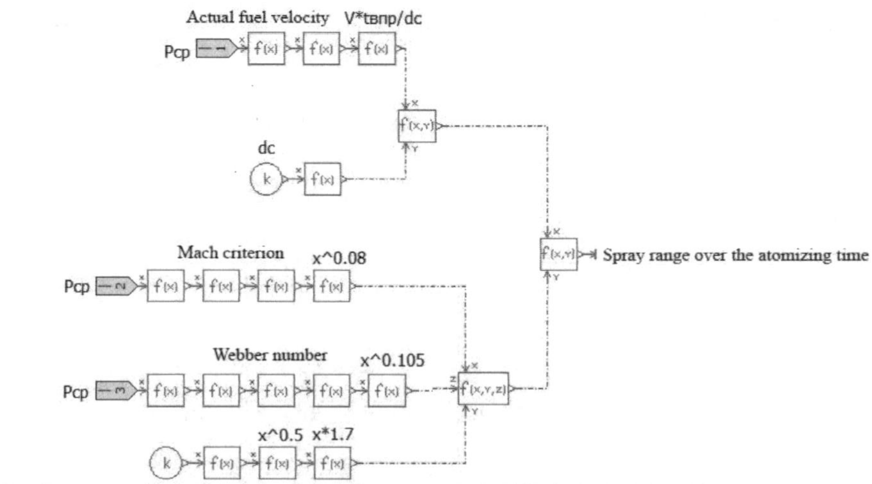

Figure 5. The block of spay-range calculation using LMS Imagine.Lab AMESim.

The modelling results are given in fig. 6 and in table 1.

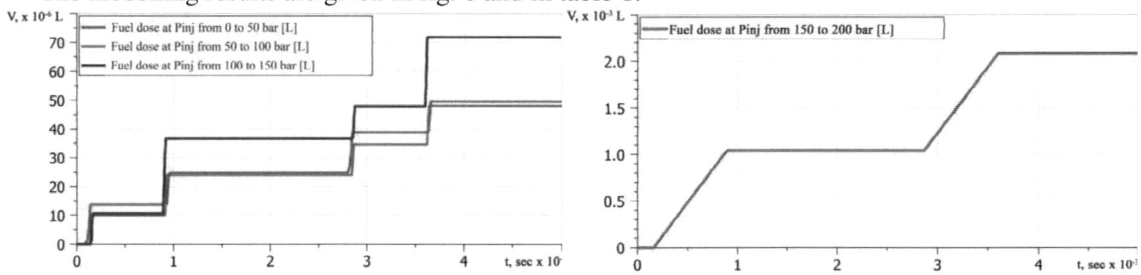

The injected cycle fuel dose

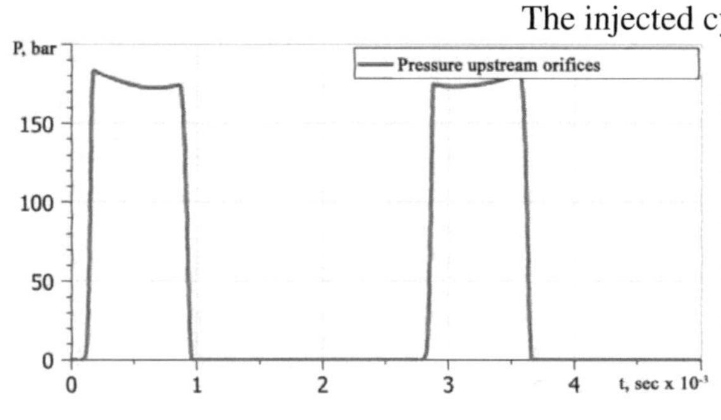

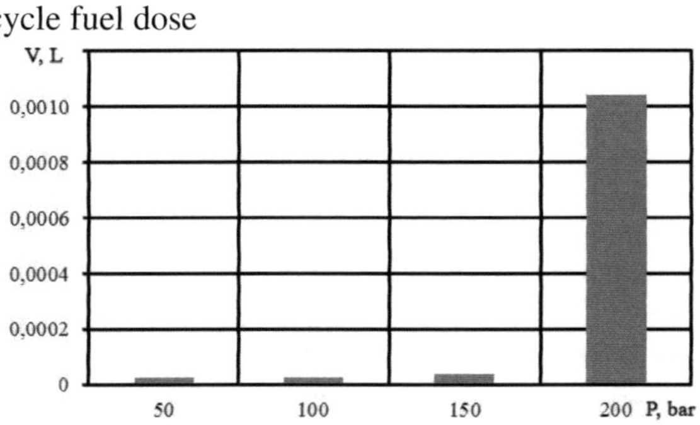

Pressure upstream orifices

Distribution of cycle dose volumes over the range upstream the outlet nozzles

Figure 6. Modeling results

Table 1. Numeric values of the parameters obtained	
Mean pressure upstream the orifices, bar	158.7
Mean drop diameter, micron	97.29
Spray range, m	0.03616
Injector inlet pressure, bar	1000

3. Conclusions

1. It has been demonstrated that when modeling diesel injectors in the ICE in LMS Imagine.Lab AMESim it is possible to preliminarilarily assess the quality of fuel atomization following the results of the analysis of processes in these systems without three-dimensional modeling of fuel spray formation processes.

2. The following design characteristics are proposed as the tools for analyzing the effect of processes in fuel systems caused on the fuel atomizing quality of the injector:

- distribution of the shares of cycle delivery over the pressure spectrum upstream the injector nozzles;
- change of the mean pressure values upstream the injector nozzles.

3. The blocks for calculating the mean diameter of drops and the range of the spray were modeled as the tools for preliminary assessment of fuel atomization quality.

4. The represented tools allow avoiding the use of labor-intensive three-dimensional calculation and analysing the fuel atomization quality based on modelling the dynamics of atomizers.

Based on the simulation results, it can be concluded that it is necessary to increase the injection pressure to reduce the drop mean diameter and increase the spray range. In this connection, the research shall be continued with respect to the rest of the factors and optimization of their values to obtain the maximum high fuel injection quality.

Acknowledgements
This work was supported by the Ministry of Education and Science of the Russian Federation in the framework of the implementation of the Program «Research and development on priority directions of scientific-technological complex of Russia for 2014– 2020».

References

[1] Reif K. (Hrsg.) 1998, 2002, 2004, 2012 *Dieselmotor-Management: Systeme, Komponenten, Steuerung und Regelung 5 Auflage. Vieweg+Teubner* (Verlag Springer Fachmedien Wiesbaden) p.544

[2] Grekhov L V, Ivashchenko N A and Markov V A 2013 *Fuel equipment and diesel engine control system: Textbook for Universities* (Publishing House of the Legion-Avtodata) p.344

[3] Harun M I, Hoon K Ng and Suyin G 2012 Evaluation of non-premixed combustion and fuel spray models for in-cylinder diesel engine simulation *Applied Energy* vol 90 pp 271-279

[4] LMS Imagine.Lab / Web-page: http://www.plm.automation.siemens.com/ru_ru/ products / lms / imagine-lab / amesim /

[5] LMS Imagine.Lab Amesim 2015. *LMS Amesim 14 Tutorial guide*. p.181

IOP Conf. Series: Materials Science and Engineering **302** (2018) 012056 doi:10.1088/1757-899X/302/1/012056

A Refined Model for Calculation of the Vortex Tube Thermal Characteristics

V.V. Biryuk[1], A.A. Gorshkalev[1], D.A. Uglanov[1], V.V. Urlapkin[1] and S.S. Korneev[1]

[1]Samara National Research University, 34, Moskovskoe shosse, Samara, 443086, Russian Federation

E-mail: Agorsh@bk.ru

Abstract. The article deals with the main types of vortex tubes, provides a brief description of the fundamental principles of the vortex interaction hypothesis. A physical process is represented reflecting the physical essence of the gas flow energetic separation process in the vortex tube due to the intensive turbulent heat exchange from the forced vortex to the free one. A method for refinement of the design characteristics for the cold and hot gas temperatures in a vortex tube through the employment of the gas-dynamic and thermodynamic corrections is proposed. A refined calculation method allows reaching close agreement between the cold gas temperature and the experimental values.

1. Introduction

The vortex effect of energetic gas separation discovered by J. Ranque in 1931, and then studied by R. Hilsch to a considerable detail, has been drawing close attention of scientists for 86 years. It should be noted that since before the Ranque's discovery, the works of Maxwell, Schmidt, K. Strakhovich had provided theoretical background for gas energetic separation in a swirling flow. Due to the studies carried out by the Russian scientists, optimal designs of the main types of vortex tubes were obtained. [1,2,3,4,5,6,7,8] They include: a separating tube, a cooled tube, a self-vacuuming tube, and a double-circuit vortex tube.

2. The essence of the hypothesis

A separating vortex tube, where the compressed gas is separated into two flows: a cold flow and a hot flow.

A cooled vortex tube cooling the gas flow through the walls using the external heat carrier to the lower temperature as compared to the temperature of the external heat carrier.

A self-vacuuming vortex tube allows transforming the vortex kinetic energy to the pressure energy using a rotating diffuser and thus sharply reducing the pressure at the energy separation chamber axis with substantially increasing a cooling effect in it.

A two-circuit vortex tube with the additional flow performs efficiently at $\mu>1$ thus ensuring high heat exchange intensity between the main and the additional flow.

In spite of the fact that the separating vortex tube (VT) has an extremely simple design, the processes occurring in it are so complex and paradoxical that it is only now that it is recognized that the vortex interaction hypothesis developed by professor A.P. Merkulov describes the physical essence of processes occurring in the energetic separation chamber to the fullest extent possible. [1, 2, 5]

The essence of this hypothesis is the following: after outflowing from the tangential nozzles, the gas flow forms a free vortex spreading to a certain radius and displacing along the tube axis to the throttle. A free vortex is resistant to the internal friction forces and is not destroyed by them. The vortex may become breaking down only at its radial boundaries due to the wall friction and the interaction with the near-axial elements. Furthermore, its swirling intensity decreases due to the decrease of the circumferential velocity when the vortex moves along the tube to the throttle, the static pressure radial gradient in the vortex flow decreases and the vortex moves closer to the axis. [5, 6, 7]

The decrease of radial gradient, in turn, results in the static pressure axial gradient, which forces the gas entering the near-axial region changing its axial motion initial direction to the opposite one, and moving to the nozzle section. While moving to the near-axial region, gas elements are becoming intensively turbulized. High turbulent viscosity forces the near-axial flow to swirl according to the solid body law. The reverse near-axial flow, with the advancement to the nozzle section, is swirled to form an increasingly intensive free vortex.

Apart from transferring the kinetic rotation energy from the free vortex to the forced vortex, an intensive turbulent heat exchange occurs between them with a high static pressure gradient being normal to the average flow velocity.

A radial displacement of the turbulent gas element is due to the radial turbulent pulsating velocity. When entering a higher or a lower pressure area, the element will shrink or expand adiabatically.

If after the displacement, the gas element temperature is different from the temperature of the surrounding elements, then heating or cooling of the elements will occur, i. e. turbulent elements pass refrigeration cycles by giving heat to the peripheral layers; the turbulence is the source of mechanical energy.

Turbulent heat transfer will cease (q=0) upon reaching the isentropic radial temperature distribution $\frac{dT}{dr} = \frac{k-1}{k} \cdot \frac{T}{p} \cdot \frac{dp}{dr}$.

This hypothesis makes it possible estimating the limiting possibilities of the vortex effect under the assumption that the energy exchange between the vortices has fully completed in the nozzle section.

By using the radial velocity distribution laws for free and forced vortices, the conditions for gas mechanical equilibrium, the isentropic radial distribution of the static temperature of the forced vortex, the mechanical and thermal conjugacy of the vortices, the expressions for distributing the parameters of the nozzle section radius may be obtained, and in case of the defined geometry the mean-integral temperature of the cold and hot flows of the vortex tube as a function of the main parameters of the vortex tube may be determined. For the relative temperature of the cold and hot flows of the vortex tube, these dependencies are as follows: [3, 4 ,8]

$$\theta_x = \theta\left(\pi, \mu, \overline{F}_c, \overline{d}_Д\right); \quad \theta_Г = \theta'\left(\pi, \mu, \overline{F}_c, \overline{d}_Д\right). \tag{1}$$

They provide a qualitative correspondence with the experimental characteristics of the vortex tube as well as describe the reversing phenomenon in the vortex tube (θ_x>1 at $\mu \to 0$) and are indicative of the possibility of cooling at μ=1 in the cooled VT or at μ>1 in the double-circuit VT or at μ=0 in self-vacuuming VT. Furthermore, the design and experimental values of temperature are considerably different (Figure. 1).

To refine the design characteristics of the VT, the losses of the total gas pressure when outflowings through the nozzle inlet (gas dynamic calculation), as well as the heat recovery in counter-flow motion of the cooled near-axial and heated peripheral gas flows in the energy separation chamber (thermodynamic analysis) shall be considered. The vortex interaction hypothesis about the existence of two vortices - a free vortex with the potential flow $V_\tau \cdot r = const$ and a forced vortex $\omega = \frac{V_\tau}{r} = const$ - is assumed as the basis of the gas dynamic calculation of the VT.

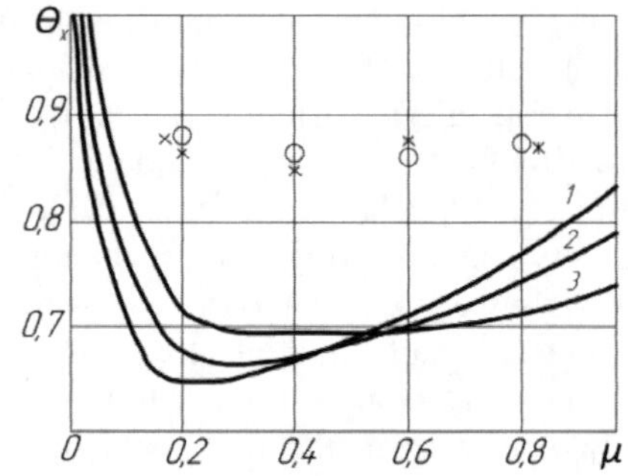

Figure 1. The limiting design characteristics of the VT:
The experimental data
K=1.4; π=5; $\bar{F}_c$=0.1; x – Merkulova; * - Gulyayeva;
O – Martynova; 1-$\bar{r}_д$=0, 4; 2-$\bar{r}_д$=0.45; 3-$\bar{r}_д$=0.5.

The momentum equation for a non-viscous gas may be integrated

$$\frac{dp}{dr} = \rho \cdot \frac{V_r^2}{r} \tag{2}$$

with obtaining the radial gas parameter variation laws, p, T, ρ.

The limiting condition for solving this equation with respect to p is determined based on the solution of the flow-rate equation for the diaphragm section, which lies in the VT nozzle section.

$$\mu G_1 = \int_0^{\bar{r}_д} \rho \cdot V_z \cdot 6,28 \cdot r \cdot dr. \tag{3}$$

Here μ is the fraction of the cooled gas flow,

$\bar{r}_д$ - the diaphragm opening radius,

V_z is the axial speed in the diaphragm opening,

G_1 is the gas mass flow through the VT.

The axial speed, in turn, is determined based on the expression,

$$V_z = \pm \sqrt{\frac{2k}{k-1} R T_1 \left(\frac{p}{p_1}\right)^{\frac{k-1}{k}} / \left[1 - \left(\frac{p_x}{p}\right)^{\frac{k-1}{k}}\right]}, \tag{4}$$

where p_1 is the pressure at the energy separation chamber periphery.

The gas flow direction and the sign of the expression V_z are determined depending on the ratio of the gas pressure p and the pressure p_x of the medium into which the cooled gas flows out.

The value p_1 is obtained under the condition of radial adiabatic distribution of the static gas parameters within the expression as follows:

$$p_1 = \frac{\mu G_1 R T_1}{6.28 \cdot \left[-\int_0^{r_a} \left(\frac{p}{p_1}\right)^{\frac{1}{k}} V_z r dr + \int_{r_a}^{r_д} \left(\frac{p_x}{p_1}\right)^{\frac{1}{k}} V_z r dr \right]}. \tag{5}$$

Here r_a is the radius at which the axial component of the velocity V_z changes its sign. The limiting condition in the equation for the gas temperature at the energy separation chamber radius, $T=T_1$ is obtained from the energy equation for the gas flow in a free vortex near the chamber wall:

$$T_1 = \frac{T_1^*}{1 + \frac{k-1}{2}(M_1^2 - M_2^2)} . \quad (6)$$

The tangential velocity M_1 at the periphery of the energy separation chamber is expressed through the ratio of pressures

$$M_1 = \sqrt{\frac{2}{k-1}\left[\left(\frac{p_k^*}{p}\right)^{\frac{k-1}{k}} - 1\right] - M_z^2} . \quad (7)$$

The relative axial velocity M_z is found from the equation of flow conservation through the free vortex region assuming the constancy of the radial axial velocity

$$M_z = \frac{G_1}{\sqrt{kRT_1} \cdot 6.28 \cdot \int_{r_2}^{r_1} \rho r dr} . \quad (8)$$

Full gas pressure on the walls of the energy separation chamber pk* is found considering the pressure losses in a tangential nozzle

$$p_k^* = p_1^* \left(\frac{\alpha_c}{\varphi_c}\right)^{\frac{k-1}{k}} , \quad (9)$$

Where the speed factor φc may be found by the flow factor αc based on the expression

$$\frac{\alpha_c}{\varphi_c} = \frac{1 - \frac{k-1}{k+1}\lambda_c^2}{1 - \frac{k-1}{k+1}\varphi_c^2 \lambda_c^2} . \quad (10)$$

The latter expression comprises the relative speed of the gas in the outgoing section of the tangential nozzle inlet. It is connected with the tangential component of the gas velocity component at the periphery of the energy separation chamber. The form of the expression λ_c on M_1 is obtained based on the momentum-of-momentum equation

$$\lambda_c = \frac{M_1}{\sqrt{\frac{2}{k+1} \cdot \frac{k-1}{k+1}(M_1^2 + M_z^2)}} \cdot \frac{r_1}{r_c} . \quad (11)$$

The gas flow G1 is determined considering the flow factor αc using the relative velocity λc:

$$G_1 = \alpha_c \frac{p_1^* F_c \sqrt{k} \left(\frac{2}{k+1}\right)^{\frac{k+1}{2(k-1)}} \bar{q}(\lambda_c)}{\sqrt{RT_1^*}} . \quad (12)$$

where the gas dynamic function of the flow of the medium velocity in the nozzle

$$\bar{q}(\lambda_c) = \frac{\int_0^h q(\lambda_c) b dh}{F_c} . \quad (13)$$

The last undetermined value necessary for closing the set of equations is the vortex separation radius r2, which is assumed to be determined based on the variation principle of the flow maximum entropy:

$$G_1 R \cdot \ln \frac{p_1^*}{p_k^*} = \max . \quad (14)$$

The system includes an empirical value, i.e. the flow factor of the tangential nozzle input, for whose calculation the following experimental dependence was obtained:

$$\alpha_c = 1 - \frac{1.63}{\mathrm{Re}^{0.25}}(1.03 - \lambda_c^4). \tag{15}$$

The represented set of equations is solved by the iteration method and allows determining the gas flow through the VT, the radial distribution of gas parameters in the nozzle section, and the averaged total gas temperature in the diaphragm section (the so-called theoretical limiting temperature).

The design temperature of the cooled gas is determined by the expression

$$\tilde{T}_{XT} = \frac{\int_0^{r_A} \rho V_z \left(T + \frac{V_r^2}{2C_p} \right) r dr}{\int_0^{r_A} \rho V_z r dr}. \tag{16}$$

To determine the pressure of the heated gas, the semiempirical formula is proposed

$$p_\Gamma = \frac{1}{2}\left[(p_{oc} + p_2) + \mu (p_1 - p_{oc}) \right], \tag{17}$$

assuming the linear behaviour of the dependence of the fraction of the cold gas flow on the pressure of the heated gas flow and connecting the pressure values at the characteristic radii (r=0, r=r2, r=r1) of the nozzle section.

The thermodynamic analysis of the VT performance consists in determining the temperature of the cooled and heated gas by finding and considering the conditions resulting in the adiabatic distribution of the static gas parameters the along the nozzle section radius. With these conditions, a relation between the temperature $\tilde{T}_{XT}$ and the cooled gas temperature T_X in the idealized process approximating the heat transfer process in the VT, may be found.

The idealization of the heat transfer process in VT is based on the assumption that this process occurs according to the regenerative pattern and proceeds in the following way.

The gas, entering the VT, moves along the energy separation chamber periphery to the VT throttle and receives some quantity of heat q from the near-axial gas layers, thus heating to the temperature T_Γ. The gas is separated in two parts near the throttle: one its fraction (1-μ) goes out through the throttle as a heated flow with the temperature T_Γ, and its second fraction μ, while moving to the diaphragm, gives the heat q along its way to the periphery gas layers with cooling to the temperature T_X. Furthermore, the pressure of near-axial gas layers is decreased from p_Γ near the throttle to p_X downstream the diaphragm. This is the first thermodynamic process in the VT. Without considering the heat transfer mechanism, the temperatures T_Γ and T_X may be calculated by using the thermodynamic analysis method.

Let's consider that the adiabatic distribution of the parameters in the nozzle section is achieved when such a quantity of heat is diverted from the peripheral layers, which is supplied to them in the actual process in the BT from the near-axis layers.

Let's consider the other thermodynamic process. The gas with flow rate G_1, while moving to the throttle, acquires heat q from the near-axial layers and heats up to the temperature of the heated gas T_Γ in the real energy separation process. A quantity of heat q is isobarically removed outside from this gas flow; furthermore the flow temperature is reduced from T_Γ to some temperature T_z. A part of the gas with a flow of μG_1 forms a near-axial zone and, while moving counter-flow to the peripheral flow, gives it heat q due to the vortex effect.

The temperature of this portion of gas is decreased from T_z to the theoretic temperature T_{XT}. Furthermore, the gas pressure is decreased from the heated gas pressure p_Γ to the cooled gas pressure p_X. This is the second thermodynamic process considered.

Heat removed per unit time to the outside of the peripheral flow considering the isobaricity of the process is determined

$$q = -G_1 C_p (T_Г - T_z) . \tag{18}$$

On the other hand, the heat transferred from the near-axial gas layers to the peripheral layer is equal to

$$q' = G_x \int_{S_z}^{S_{XT}} T dS \tag{19}$$

or is approximated by the expression

$$q' = G_x \frac{T_z + T_{XT}}{2} R \cdot \ln\left[\frac{p_Г}{p_x}\left(\frac{T_{XT}}{T_z}\right)^{\frac{k}{k-1}}\right] . \tag{20}$$

The temperature Tz may be found by the iteration method from the balance of enthalpies for the second thermodynamic process:

$$G_1 C_p T_1 - G_Г C_p T_z - G_x C_p T_x + q' = 0 . \tag{21}$$

The equation q and q' is used to determine the temperature of the hot flow TГ in the second thermodynamic process, which is equal to the temperature of the hot gas flow in the first thermodynamic process

$$T_Г = T_z + \frac{k-1}{2k}\mu (T_z + T_{XT}) \ln\left[\frac{p_Г}{p_x}\left(\frac{T_{XT}}{T_z}\right)^{\frac{k}{k-1}}\right] . \tag{22}$$

The cooled gas temperature TX of the first thermodynamic process will be determined from the balance equation of its enthalpies.

$$G_1 C_p T_1 - G_Г C_p T_z - G_x C_p T_x = 0 . \tag{23}$$

The proposed calculation method allows reaching a close agreement between the value of the cooled gas temperature and the experimental values (Figure. 2).

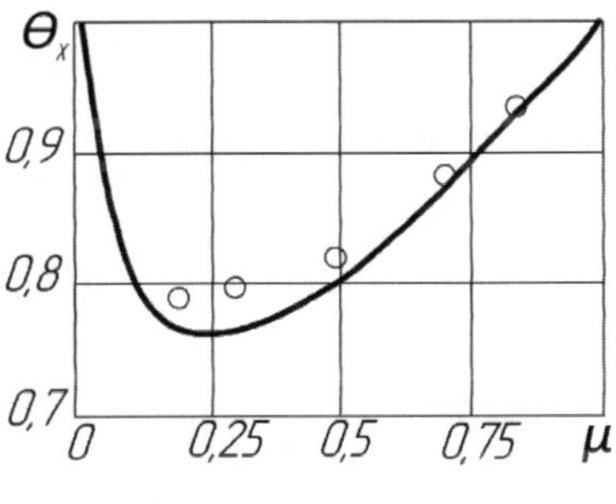

Figure 2. A refined temperature characteristics of the SVT:
- design; o – experiment at π=5

The above thermodynamic analysis may be applied to more complex cases of realizing the vortex effect, for example, a cooled VT, a self-vacuuming VT, and a two-circuit VT.

3. Conclusion

As a result of studying the VT performance, a method for refining the calculation of characteristics for the SVT and other types of the VT with the use of gas-dynamics, thermodynamics and thermal corrections was developed.

Consideration of empirical corrections allows performing mathematical modelling of the VT processes in aircraft vortex cooling systems. According to the proposed methods, the techniques and

IOP Conf. Series: Materials Science and Engineering **302** (2018) 012056 doi:10.1088/1757-899X/302/1/012056

programs for calculating the actual characteristics of vortex tubes for aircraft cooling systems were developed. [4, 7, 8-10]

References

[1] A.P. Merkulov Vortex Effect And Its Application in Engineering. M.: Machine building, 1969. - 182 p.
[2] A.V. Martynov V.M. Brodyansky. What is the vortex tube? - M.: Energiya,1972.- 152 c
[3] A.D. Suslov S.V. Ivanov, A.V. Murashkin, Yu.V. Chizhikov Vortex apparatus. - M.: Machine building, 1985. - 256 p.
[4] V..V. Biryuk The Vortex Effect of Energy Separation of Gases in Aviation Technology And Engineering //News of Higher Educational Institutions. Aviation technology / KAI - Kazan: 1993. No. 2 - p. 20-23.
[5] Sh..A. Piralishvili. The Vortex Effect, T.1 (Physical phenomenon, experiment, theoretical modeling): - M .: LLC Nauchtekhlitizdat, 2012. - 342p.
[6] Sh..A. Piralishvili. S.V. Veretennikov A.A. Shaykina. Thermophysics of Energy Separation Processes in the Flow Processes of Vortex Tubes and the Leontief tube (article) .: News of the Academy of Sciences. Power Engineering. Moscow.-2017, No. 1.- p.139-148.
[7] Burtsev S.A., Leontiev A.I. Study of the influence of dissipation effect on the temperature stratification in gas flow.: High Temperature, 2014, vol. 52, is.2, pp. 297-307
[8] V.V. Biryuk S.V. Veretennikov, A.I. Guryanov, Sh.A. Piralishvili. The Vortex Effect. Technical Annexes. T.2 (part 2) - M.: LLC Nauchtekhlitizdat, 2014.- part 1 - 288p.
[9] Biryuk V V, Tsapkova A B and Shimanov A A 2016 Key Engin. Materials 685 153-157
[10] Biryuk V V, Gorshkalev A A, Tsapkova A B, Shimanov A A and Blagin E V 2017 Journal of Physics: Conference Series 803 1 012022

IOP Conf. Series: Materials Science and Engineering **302** (2018) 012057 doi:10.1088/1757-899X/302/1/012057

Efficiency Assessment of the Power Supply System of an Industrial Enterprise Through the Assessment of the Modes of Electric Power Consumption

Gennadii V Miatishkin and Anastasiia S Filinova

Department of Thermal Engineering and Heat Engines, Institute of Engine and Power Plant Engineering, Samara National Research University, 34, Moskovskoye shosse, Samara, 443086, Russia

E-mail: Anast.prokhorova@gmail.com

Abstract. The paper reviews the terms of the rational consumption and distribution of the energy resources at an enterprise. The authors analyze the significance and the effect of the change of the profile of the energy consumption per hour by the enterprise. The text explores the factors influencing the discipline of the planned electricity and power consumption. The authors present a calculation of the weighted average deviations for consumers and the terms of their assessment. The authors make conclusions concerning the rationality of the means of defining the energy efficiency system through the assessment of the modes of electric energy consumption of an industrial enterprise.

1. Introduction

In 2009-2010, the energy consumption of the world economy increased because of the effect of the economic crisis of 2008. The countries that suffered from the crisis most were Russia, Japan and the EU states where the crisis lead to the increase of the energy consumption in the economies. The energy efficiency of India and African states suffered less. Starting from 2011 the impact of the factors of the crisis of 2008 has been going down and the rates of the decrease of the energy consumption of the world economy have been demonstrating positive dynamics: in 2011 the decrease of the energy consumption of world gross product was 1.3%, in 2012 – 1.0%, in 2013 – 0.6%. In 2012-2013, it was only in the USA that the GDP energy consumption increased largely due to the increase in 2013 of the heating season degree-day by 23% as well as due to the structural changes of the US economy, which resulted in the increase of the power consuming industries.

In case of absence of the due planning of the production based on minimizing energy consumption, there might occur in the activity of the enterprise a possibility of unfavorable situations and consequences, financial losses and losses of potential benefit, i.e. the possibility of risks. Energy risks are a mixed type of risks including industrial, economical, commercial, resource and organizational types of risks. Energy risks should be managed through the continuous monitoring and the use of statistical methods in the energy input management system (EIMS). As a result, the possibility of risk is decreased (or eliminated) in three cases: if the event (factor of risk) takes place, if the result of this event is made predictable and if the risk factor is eliminated. Efficient risk management should solve a range of problems from defining and monitoring the risk to its cost estimate.

2. Assessment of the modes of electric power consumption

The rational consumption and distribution of the power resources at an enterprise depends on the external and internal conditions of resource supply. The external conditions of power supply include the requirements defined in the regulatory documents that concern resource circulation. The internal conditions include the reliability of the enterprise power supply comprising the factors of power limitations / power outages due to external causes, emergency cases and the breakdown of resource-receiving and resource-transmitting equipment, as well as the management of power consumption.

Power consumption uniformity of an enterprise defines the profile of consumption per hour. The consumption per hour profile is defined by the characteristics of the electric energy and power price formation on the wholesale power market (WPM) and on the retail power market (RPM) in the Russian Federation [1-5]. On these markets, the kWh weighted average price is defined by the consumption per hour and the prices at this time, according to the price bracket calculation method [2]. Thus, power consumption per hour uniformity helps to minimize overcosting in the conditions of the volatility of the electric energy prices on WPM.

The consumption per hour profile has economical, technical and technological significance. The profile of consumption per hour reflects the quality of planning as well as the following energy (power) consumption at the enterprise. An increasingly even graph of consumption per hour (with minimized consumption peaks) results in a more efficient energy use regardless the situation on WPM and RPM (there occurs an effect of consumption peaks smoothing). At the same time, the current Russian legislation in the field of electric energy (power) circulation increases the significance of this index through the process of the overall price-formation, as well as the cost of the electric energy (power) paid for by the consumer.

It is worth mentioning that the decrease of the power consumption per hour does not contradict but is a co-directional vector in the direction of the increase of the efficient power use. Both indexes serve as variables in the function of energy consumption of an enterprise. That is why it is worthwhile using them separately in the course of energy efficiency assessment.

To assess energy consumption per hour, a calculation of contracted power utilization time (CPUT) is made on the basis of the consumption per hour data for the last calendar year given by the consumer. For the 3rd-6th price category (PC) consumers [2], WPM consumers or consumers who have contracts with independent electric distribution companies (EDCs) the contracted power utilization time is calculated using specified methods (1) [6-7].

$$CPUT = V_{act} / P_{max} \quad (1)$$

where V_{act} stands for the actual contracted electric energy consumption at the enterprise in the preceding regulatory period in kilowatt-hours (kWh);

P_{max} – the actual (contracted) maximum power consumption in kilowatts (kW).

The maximum power consumption (P_{max}) is calculated on the basis of the power consumption per hour data according to the current rules of the retail power market [2]. In the absence of the consumption per hour data or in case of impossibility to directly calculate maximum power consumption, the following methods are used for the calculation (given in descending order of usage priority):

- hourly measurement of power in the winter / summer in operating day (24 hours);
- hourly measurement of power during 24 hours in 2 or more regulatory periods (months);
- definition of maximum power consumption according to the Certificate of Delineation of Balance Sheet Attribution (CDBSA);
- according to the typical daily load curve for the consumer on the working days and days-off.

For the 1st-2nd PC consumers, the CPUT used is equal to the calculation value of the default provider guaranteeing energy supply in the region of the consumer.

The discipline of the planned energy and power consumption of an enterprise is characterized by the deviation of the actual energy (power) consumption from the planned consumption. The deviation of the actual energy (power) consumption per hour from the planned consumption per hour is the second most significant factor of the increase of the energy (power) resulting costs in the conditions of the current WPM and RPM regulations [1-5]. Consumption deviations result in the risks of overspending on energy and power on the WPM and RPM, the creation of transition modes of the local power generation (the increase of specific consumption of the fluid), as well as the wearing of the basic and auxiliary facility.

In order to assess the index characterizing the discipline of the planned energy and power consumption, the deviation of the actual consumption per hour from the planned consumption per hour is calculated on the basis of the data of the interval metering system (IMS) of the enterprise. For the enterprises that are supplied by a default provider, this calculation is done as for the 3rd-6th PC consumers. For the 1st-2nd PC consumers, the deviation value is 5% because of the relaying of costs of the default provider from the WPM to the RPM [2-3]. In the absence of the interval metering devices and (or) in the absence of the planned hourly energy (power) consumption value, the deviation value is considered to be 5%. The calculation of the weighted average deviation value (Δ) for consumers with the interval measuring and planned consumption is performed according to the formula (2):

$$\Delta = \frac{\sum_{t=T} \frac{abs(N^t_{fact} - N^t_{plan})}{N^t_{plan}}}{t} \quad (2)$$

where N^t_{fact} is the actual power consumption during an hour t (kW);

N^t_{plan} – the planned power consumption during an hour t (kW);

T – the sum of hours in the annual amount of working days.

The assessment of the average deviation value is based on the data on the planned and actual consumption during the year preceding the assessment or on the data from the records of the enterprise for the previous 2 years. For the consumers of the WPM, the calculation of the balance value is done according to the WPM regulations.

For the enterprises with the local power generation, the deviation assessment value is based on the analysis of the fluctuations of the specific fuel consumption for the generation of 1 kWh. To carry out the comparative assessment, the identical periods are selected for comparison, the most significant factors being the season of the year, the production plan of the enterprise, the quality of the fluid as well as the unscheduled shutdown of the generation equipment. No less than three identical periods are selected for assessment.

3. The assessment of deviations (profile) in energy (power) consumption

For the assessment of deviations (profile) in energy (power) consumption it is necessary to define the quality index that shows the specific deviations of the actual from the planned energy (power) consumption per hour at the enterprise - $\eta(P_2)1\!-\!2$.

The index characterizes the discipline of the planned energy and power consumption. The deviation of the actual energy (power) consumption per hour from the planned consumption per hour is the second most significant factor of the increase of the energy (power) resulting costs in the conditions of the current WPM and RPM regulations. Consumption deviations result in the risks of overspending on energy and power on the WPM and RPM, the creation of transition modes of the local power generation (the increase of specific consumption of the fluid), as well as the wearing of the basic and auxiliary facility.

The deviation of the actual consumption per hour from the planned consumption per hour is calculated on the basis of the data of the interval metering system (IMS) of the enterprise. The calculation of the weighted average deviation value (Δ) for consumers with the interval measuring and planned consumption is performed according to the formula (2).

For the enterprises that are supplied by a default provider, this calculation is done as for the 3rd-6th PC consumers. For the 1st-2nd PC consumers, the deviation value is 5% because of the relaying of costs of the default provider from the WPM to the RPM [2 - 3]. In the absence of the interval metering devices and (or) in the absence of the planned hourly energy (power) consumption value, the deviation value is considered to be 5%.

For the enterprises with the local power generation, the deviation assessment value is based on the analysis of the fluctuations of the specific fuel consumption for the generation of 1 kWh. To carry out the comparative assessment, the identical periods are selected for comparison, the most significant factors being the season of the year, the production plan of the enterprise, the quality of the fluid as well as the unscheduled shutdown of the generation equipment. No less than three identical periods are selected for assessment.

The resulting deviation value is a sum of deviation indices (Δg) divided by the volume of the actual consumption (generation) of electric energy (Vg) in the given identical periods (g):

$$\Delta = \frac{\sum_g \Delta g \times Vg}{\sum_g Vg} \quad (3)$$

4. The manageability of energy resources consumption

The manageability of energy resources consumption is an index characterizing the manageability of the natural and cost indices of energy resources consumption. A complex of interrelated documents regulating the accounting, planning and energy (power) consumption discipline united into a system allows to speak about the management of energy consumption as well as the guaranteeing of its efficient use.

The level of the manageability and the efficient use of energy resources depends on: the existence of the energy consumption management system (ECMS), the introduction and the updating of the ECMS so that it corresponds to the Russian National Standard ISO 50001-2012 "Systems of energy management. Requirements and application guidelines", the certification of the system as well as its meeting the requirements and the special features of the energy circulation according to the Russian Federation regulations.

Systemic control and the integral management of energy resources, including the cases of the low rating of the index of specific consumption for each produced item, nonetheless, is a positive trend and serves as a basis for the sustainable development and growth of energy efficiency.

In order to assess the index of manageability of energy resources consumption the enterprise (organization) is to provide the data on the current state of things in the field of energy management in the form of a questionnaire. The state of the energy management system is defined by the existence and the level of development of the following elements:

Energy survey (energy budgeting) or the updating of data on the consumption of each of the energy resources (in a declaration or by means of internal audit);

Registration, accounting and analysis of data on the consumption of each of the energy resources;

Energy consumption planning, analysis and monitoring of the consumption discipline of each of the energy resources;

Competence and number of specialists involved in energy management (the management of each of the energy resources);

Efficiency indices of each of the energy resources use at the enterprise (organization) and the existence of the measurable target results in what concerns the achievement of these indices;

Paperwork management, information (data) exchange, standardizing and choosing those in charge from the staff;

The existence of a policy in the field of energy efficiency meeting the requirements of the energy accounting and circulation regulations of the Russian Federation and the requirements of the Russian National Standard ISO 50001-2012;

A certificate of the Russian National Standard ISO 50001-2012, an energy performance certificate (declaration), a rank in the rating of energy efficiency.

Each element is given a mark in the range 1÷11 for each type of the energy resources. The ECMS is then assessed for each type of the energy resources by means of expert review.

5. Resource supply reliability

Reliable energy supply is the factor of efficient and continuous use of energy resources. Its absence can lead to production (operation activity) stoppage and result in the increase of production costs and economic losses. Breaks in energy supply, whether caused by accidents, off-schedule repairs, power limitations or power outages as a result of overdue payment to the resource supplier, lead to transition modes of consumption or emergency power supply. Al this can have a negative impact on the rational (efficient) use of energy resources.

In order to assess this index, the major aspects of reliability (events / incidents) for each type of the energy resources are analyzed:

Electric energy (power) - $\eta(p_2)_4$;

Energy supply stoppage as a result of contract breach by the consumer according to the requirements of the energy (power) supply regulations [1 - 7];

Emergency stoppage of energy supply because of external / internal factors;

Off-schedule repairs of electric mains and / or of the consuming equipment of the enterprise (organization);

The use of back-up power sources as a result of the maximum power consumption exceedance.

The events are characterized by the frequency of occurrence and the graveness of the economic and technical consequences for the consumer. The economic consequences result from a large number of factors of the operation activity of the enterprise. In fact, defining the possibility of the event and its consequences means defining a risk.

In order to assess the index, a maximum-permissible possibility of an event is set, which can lead to the economic consequences, the growth of expenses for the energy resource or its back-ups as well as the operation activity and production costs. In the assessment of the index it is hypothetically assumed that the events in question are not taking place simultaneously and are independent. On the basis of the questionnaire of the enterprise (organization), an analysis is carried out of events that took place during the last 2-3 years that lead to the risk of unreliable energy supply in a month during the analyzed period. The resulting value is averaged either over the events or arithmetically (in the absence of data on the losses) or is calculated with respect to the other factors in the total sum of the economic losses if there is a value of losses resulting from these events.

6. Conclusions

Having defined electric energy (power) consumption uniformity and having calculated the weighted average deviation value, we can assess the modes of electric power consumption of an industrial enterprise and compare them to the consumption modes of the competitors. This will make it possible to assess the energy efficiency of the enterprise and to define the domains of power consumption that require special attention during the energy audit.

Acknowledgments

This work was supported by the Ministry of education and science of the Russian Federation in the framework of the implementation of the Program of increasing the competitiveness of Samara University among the world's leading scientific and educational centers for 2013-2020 years.

References

[1] Federal Law No. 35-FZ of March 26, 2003 (as amended on March 26, 2003) "On Electric Power Industry"

[2] Decree of the Government of the Russian Federation No. 442 dated 04.12.2012 "On the functioning of retail electricity markets, full and (or) partial restriction of the regime of electricity consumption"
[3] Decree of the Government of the Russian Federation No. 1179 of December 29, 2011 (as amended on December 23, 2016) "On the definition and application of unregulated prices for electricity (capacity) by guaranteeing suppliers"
[4] Decree of the Government of the Russian Federation No. 1172 of December 27, 2010 (as amended on December 23, 2016) "On Approval of the Rules for the Wholesale Electricity and Capacity Market and on Amending Certain Acts of the Government of the Russian Federation on the Organization of the Wholesale Electricity and Capacity Market"
[5] Decree of the Government of the Russian Federation No. 1178 of December 29, 2011 (Edited on January 20, 2017) "On pricing in the field of regulated prices (tariffs) in the electric power industry"
[6] Decree of the Government of the Russian Federation No. 861 of December 27, 2004 (Edited on December 24, 2016) "On approval of the Rules for non-discriminatory access to electricity transmission services and the provision of these services, the Rules for non-discriminatory access to dispatching services in the electric power industry and rendering Of these services"
[7] Order of the Federal Tariff Service of Russia No. 20-e / 2 of August 6, 2004 (as amended on 14.04.2014, as amended on September 16, 2014) "On approval of the Methodological recommendations for calculating regulated tariffs and prices for electrical (thermal) energy at retail (Consumer) market"

IOP Conf. Series: Materials Science and Engineering **302** (2018) 012058 doi:10.1088/1757-899X/302/1/012058

Correction of Dynamic Characteristics of SAR Cryogenic GTE on Consumption of Gasified Fuel

V A Bukin[1], A G Gimadiev[2] and G Gangisetty[2]

[1] PJSC «Kuznetsov», Samara
[2]Institute of Engine and Power Plant Engineering, Samara National Research University, Moskovskoye shosse, 34, 443086, Russia

gimadiev_ag@mail.ru;gopga321@student.liu.se

Abstract: When the gas turbine engines (GTE) NK-88 were developed for liquid hydrogen and NK-89 for liquefied natural gas, performance of the systems with a turbo-pump unitary was improved and its proved without direct regulation of the flow of a cryogenic fuel, which was supplied by a centrifugal pump of the turbo-pump unit (TPU) Command from the "kerosene" system. Such type of the automatic control system (SAR) has the property of partial "neutralization" of the delay caused by gasification of the fuel. This does not require any measurements in the cryogenic medium, and the failure of the centrifugal cryogenic pump does not lead to engine failure. On the other hand, the system without direct regulation of the flow of cryogenic fuel has complex internal dynamic connections, their properties are determined by the characteristics of the incoming units and assemblies, and it is difficult to maintain accurate the maximum boundary level and minimum fuel consumption due to the influence of a booster pressure change. Direct regulation of the consumption of cryogenic fuel (prior to its gasification) is the preferred solution, since for using traditional liquid and gaseous fuels this is the main and proven method. The scheme of correction of dynamic characteristics of a single-loop SAR GTE for the consumption of a liquefied cryogenic fuel with a flow rate correction in its gasified state, which ensures the dynamic properties of the system is not worse than for NK-88 and NK-89 engines.

1. Introduction

Aviation gas turbine engines operating on cryogenic fuel have a number of significant advantages. Therefore, this research area became more popular in recent years [1-3]. However, when creating such kind of engines, there are problems associated with the selection of a schematic diagram of the fuel supply and the control system, ensuring the required dynamics of the SAR. The usage of cryogenic fuel in GTE necessarily assumes to be in gasified state, which leads to delay in the functioning of SAR [4]. If the heat exchange-gasifier (HEG) enters into the motor control loop, the delay has a negative effect on the quality of the regulation [5-7]. The use of cold resources of cryogenic fuel for oil cooling of the lubrication system of GTE supports in the fuel, oil cooler further increases the delay in the system, which can leads inability to provide the required indicators for stability and speed. Single-loop systems with direct control over the flow of cryogenic fuel are quite acceptable either with a small delay in maintenance, or with permissible long transient processes. When using this type of a circuit, the required acceleration time is less than 10s on the above engines.

The use of gasified fuel consumption for the regulation of gas turbine engines is possible in two ways. The first method is based on regulating the consumption of a gas dispenser, which can be carried out if there is any sufficient gasified fuel involves the installation of the receiver of a large volume. Such scheme has been successfully implemented at the domestic main gas turbine engine GTE [8]. The second method involves correction in the flow of cryogenic liquefied fuel at the expense of its gasified state. The second method can allow us without making significant changes in the system to improve the stability and performance of SAR engine.

In the practice of operating complex SAR, including with delay links, cascaded control is often used when an internal loop with a static regulator is introduced, and its controlled by an external primary circuit [9,10]. The introducing of an internal circuit with an additional adjustable parameter can significantly improve the quality of transient processes [6]. Corrective local feedbacks are used to compensate for the inertia involved in the internal circuit of the elements, which makes their use often preferred [11, 12]. The additional introduction of the relaxation circuit in the rigid feedback loop of the internal circuit, corresponding to the introduction of a speedup element, also contributes to improving the quality of regulation [10]. This article explains the detailed mathematical model of SAR GTE with regulation of the consumption of cryogenic fuel, and also the possibility of improving the dynamic quality of the system by the flow rate correction in its gasified state and its proved successfully.

2. Frequency characteristics of the high-pressure rotor and gas turbine heat exchanger as control objects

Theoretical and experimental studies of the NK-88, NK-89 engine systems are shown that the high-pressure rotor (HPR) and HEG can be described by the fourth-order degree transfer functions [7]. To simulate the dynamic processes in the SAR GTE, the transfer functions of the HPR and HEG, consists of series aperiodic links:

$$W_{HEG}(s)=\frac{\Delta\bar{m}_g(s)}{\Delta\bar{m}_l(s)}=\frac{1}{(T_{1HEG}s+1)(T_{2HEG}s+1)^3}; \quad (1)$$

$$W_{HPR}(s)=\frac{\Delta\bar{n}_{HPR}(s)}{\Delta\bar{m}_g(s)}=\frac{1}{(T_{1HPR}s+1)(T_{2HPR}s+1)^3}, \quad (2)$$

where $\Delta\bar{m}_g$, $\Delta\bar{m}_l$ - are the relative changes in the flow rate of gasified and liquefied cryogenic fuel; $\Delta\bar{n}_{HPR}$ - relative change in the rotational speed of the gas turbine engine GTE; s - an operator in the Laplace transform; T_{1HEG}, T_{1HPR} - is the time constants of the HEG and HPR of GTE respectively and characterizes their inertial properties; similarly T_{2HEG}, T_{2HPR} - is the time constants of HEG and HPR of GTE respectively and characterizes their delay. Moreover, the component of the delay HEG is Several times greater than that of the HPR GTE and this relationship is determined in the frequency tests of engines [6].

The time constant values of the GTE HEG (1) and HPR (2) at the maximum regime are: $T_{1HEG}=0.4\,s$; $T_{2HEG}=0.044\,s$; $T_{1HPR}=0.44\,s$; $T_{2HPR}=0.01\,s$. Figure 1 shows the characteristics of logarithmic amplitude-frequency $20\cdot\lg|W_{HEG}(\omega)|$, $20\cdot\lg|W_{HPR}(\omega)|$ and phase-frequency $\arg W_{HEG}(\omega)$, $\arg W_{HPR}(\omega)$ respectively for HEG and GTE high pressure HPR, here ω represents angular frequency of oscillation, $j=\sqrt{-1}$.

Frequency characteristics analysis is shown in Figure 1 and it is clear that the influence of the component of the delay of the heat exchanger will be much more higher. Moreover, as the oscillation frequency increases the delay.

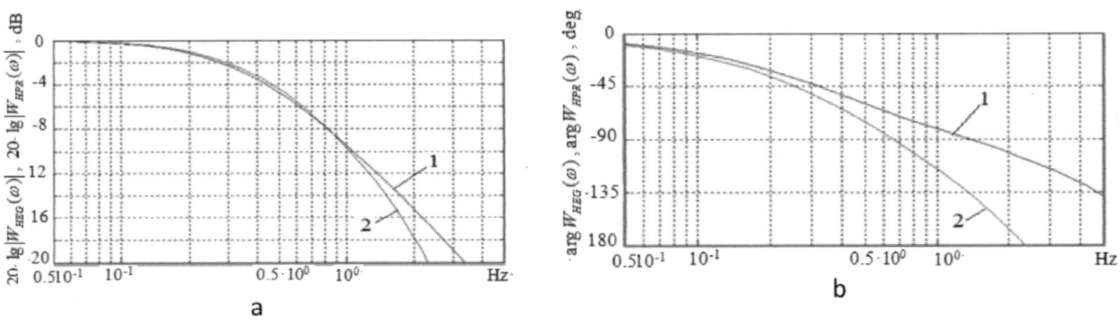

Figure 1. The logarithmic amplitude-frequency (a) and phase-frequency characteristics (b) of the objects of regulation: 1 - the high pressure rotor GTE; 2 – heat exchange-gasifier GTE.

So at 0.8 Hz the negative phase shifts are for the HPR - 74^0, for the HEG -101^0, that is less for 27^0, and for the frequency 2 Hz - accordingly less for 64^0.

3. Model SAR GTE with direct regulation of the consumption of cryogenic fuel and correction of its consumption in the gasified state

The presence of HEG in a closed SAR GTE, naturally, worsens its stability and speed with the direct regulation of cryogenic fuel (prior to its gasification). Improving the quality of system regulation can be achieved by correcting the regulator (dispenser) of the liquid flow of cryogenic fuel $\dot{m}_l$ by its consumption in the gasified state $\dot{m}_g$, that is, by introducing a correcting circuit (Figure 2

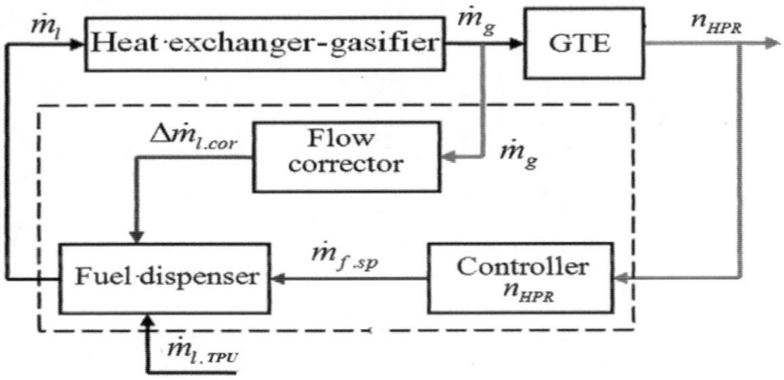

Figure 2. Functional diagram of the two-loop SAR GTE with correction of the cryogenic liquid fuel consumption $\dot{m}_l$ over its gasified state $\dot{m}_g$.

Modeling of dynamic processes in SAR GTE was conducted in three variants. The first option (a) is a basic SAR with the regulation of the flow rate of kerosene PI-controller GTE. The second option (b) is SAR with single-loop flow control of a cryogenic fuel PID-regulation. The third option (c) dual-circuit system with flow control, cryogenic fuel PID-regulation and correction of the flow rate in the gasified condition. This corrector is represented by a periodic link with a gain of 1.7 and the time constant 0.3 s.

Figure 3 shows the indicated different variants of the structural diagrams of the SAR GTE control blocks are made by using the simulation software MATLAB/SIMulink [13].

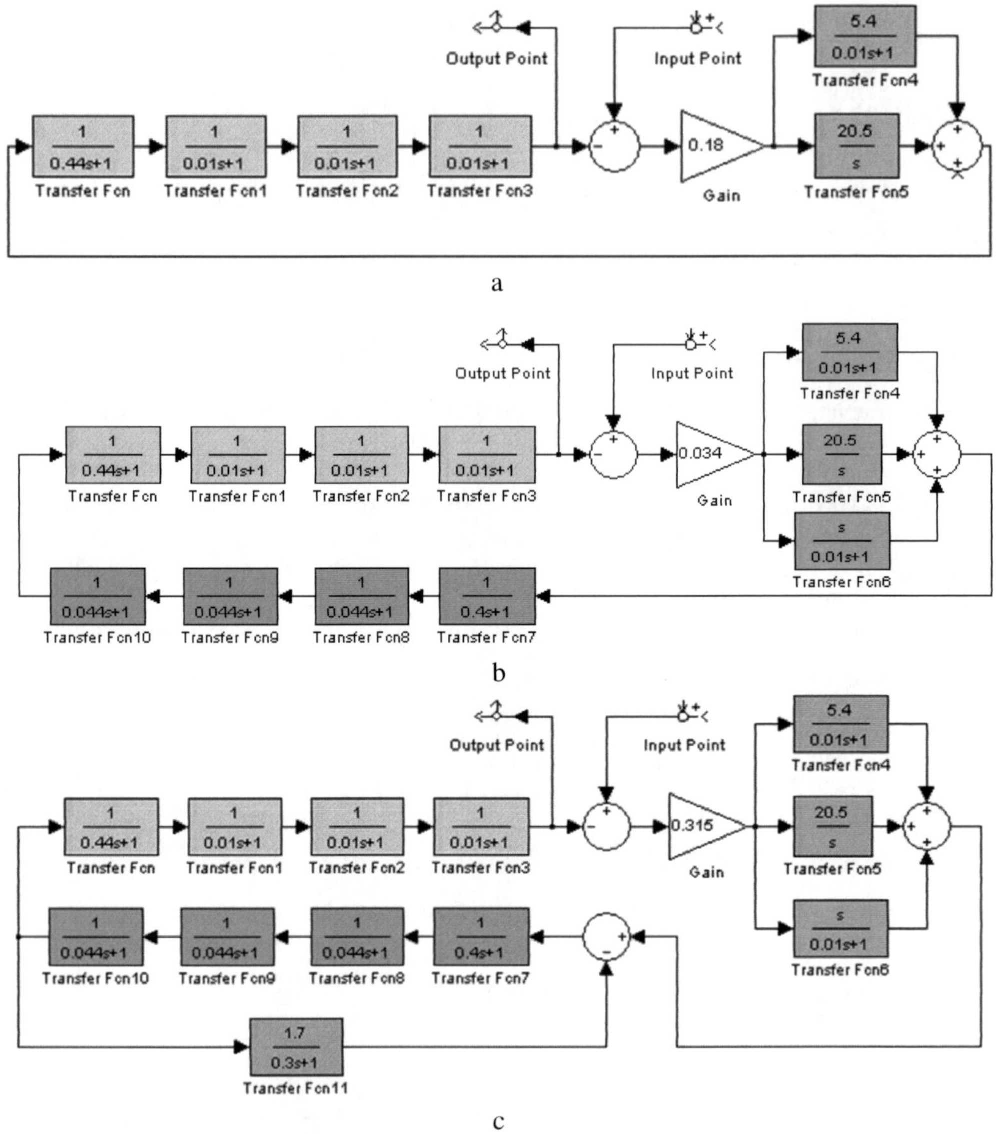

Figure 3. Structural diagrams of the GTE SAR for modeling in the software package MATLAB / SIMulink: a - basic SAR with the regulation of kerosene flow by the PI-controller; b - with an adjustment of the flow rate of the liquefied cryogenic fuel by the PID-controller; c - with the regulation of the flow rate of the liquefied cryogenic fuel by a PID-controller with correction for its consumption in the gasified state, ● - HPR; ● - HEG; ● - controller; ● - fuel consumption corrector.

4. Analysis of the efficiency of a correcting characteristics of a cryogenic gas turbine engine with respect to the gas consumption of a fuel

The parameters of the gas turbine controllers for all variants and the flow corrector in the third variant are selected in such a way that the overshoot of the HPR speed is 5 %. Figure 4 shows the parameters of the transient process with a single step-like action. As it can be seen in Figure 4, reaching 95 % of the regulated parameter in a single-loop cryogenic SAR with direct regulation of the cryogenic fuel occurs 4 times slower with respect to the basic "kerosene" system (2.76 s and 0.68 s).

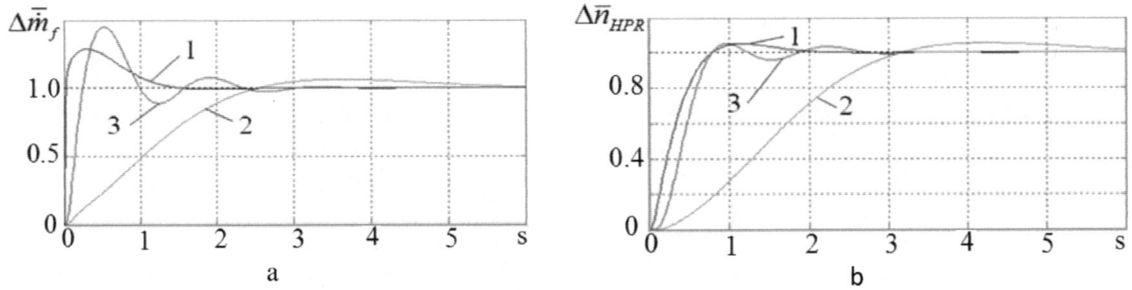

Figure 4. Transient processes in the SAR GTE on fuel consumption (a) and on the rotation frequency of the HPR (b): 1 - basic system with the regulation of kerosene flow PI-controller; 2 - single-loop system with regulation of the flow of liquefied cryogenic fuel by a PID-controller; 3 - a system in the regulation of the flow rate of a liquefied cryogenic fuel by a PID-regulator and correction for the consumption of cryogenic fuel in the gasified state.

In a system with direct control of the flow rate of the cryogenic component by a PID-controller and correction by the consumption of gasified fuel, the transient process is close to the process in the base system. The time to reach the specified control parameter with a deviation not more than 1 % has the same tendency of 1.93 s in the base system, 6.0 s in the single-circuit system and 2.6 s in the system with the PID-controller and correction for the gasified fuel consumption.

5. Conclusions

A scheme for correcting the dynamic characteristics of an aircraft GTE with the regulation of the flow rate of a liquefied cryogenic fuel by a PID-controller and correcting for its consumption in a gasified state is proposed. As a result of simulation of dynamic processes in SAR, it has been established that when the internal correcting circuit is introduced for the consumption of gasified fuel and the use of the PID-controller, the system performance is significantly increased. The duration of the transient in SAR with correction is more than three times less than the duration of a single-loop system with regulation of the consumption of liquefied fuel.

Nomenclature:

$\dot{m}$	mass flow
n	shaft speed

Indices:

g	gas
l	liquid
f	fuel
cor	correction

Abbreviations:

GTE	the gas turbine engines
HEG	the heat exchange-gasifier
HPR	the high pressure rotor
SAR	the system of automatic regulation
TPU	the turbo-pump unit

References:

[1] Koroneos C, Dompros A, Roumbas G, Moussiopoulos N. Advantages of the use of hydrogen fuel as compared to kerosene. *Resource Conservation Recycle* 2005; **44**:99-113

[2] Bhupendra Khandelwal, Adam Karakurt, Paulas R. Sekaran, Vishal Sethi, Riti Singh. Hydrogen powered aircraft: The future of air transport. *Progress in Aerospace Sciences* 60 (2013). pp.45–59.

[3] Srinivasana C B, Subramanian Dr R Hydrogen As a Spark Ignition Engine Fuel Technical Review // *International Journal of Mechanical & Mechatronics Engineering IJMME-IJENS* Vol:14 No:05 111 144205-7575-IJMME-IJENS (2014)

[4] Osobennosti sistem toplivopitaniya i regulirovaniya aviatsionnykh gazoturbinnykh dvigateley na kriogennom toplive // V P Shorin, S M Ignachkov, Ye V Shakhmatov i dr. Samara: *Izd- vo SGAU,*1998.-148 s. (in Russian).

[5] Bukin V A Metody inzhenernogo sinteza SAR GTD, rabotayushchikh na kriogennom toplive // *Vestnik SGAU,* №3 (27), chast' 2, 2011. S. 68-77. (in Russian).

[6] Bukin V A Eksperimental'nyye dinamicheskiye kharakteristiki teploobmennikov-gazifikatorov vodorodnogo GTD NK-88 i NK-89 // Mezhdunarodnyy nauchno-tekhnicheskiy forum, posvyashchennyy 100-letiyu OAO «Kuznetsov» i 70-letiyu SGAU, Samara, 5-7 sentyabrya 2012 goda: Sbornik trudov v 3 tomakh. Tom 2. *Samara: Izd. SGAU,* 2012. S. 191-193. (in Russian).

[7] Bukin VA, Gimadiyev A G Dinamika kontura regulirovaniya chastoty vrashcheniya rotora turbonasosnogo agregata vodorodnogo GTD NK-88 / *Izvestiya Samarskogo nauchnogo tsentra Rossiyskoy akademii nauk,* t. 18, № 4(6), 2016. S. 1234-1241. (in Russian).

[8] Bukin V A Modernizatsiya sistemy avtomaticheskogo upravleniya magistral'nogo gazoturbovoza na szhizhennom prirodnom gaze / *Journal of Dynamics and Vibroacoustics,*tom 1, *№2* (2014). S. 1-7. (in Russian).

[9] Guretskiy K H Analiz i sintez sistem upravleniya s zapazdyvaniyem. Per. s pol'skogo. M.: *Mashinostroyeniye,* 1974. 328 s. (in Russian).

[10] Besekerskiy V A, Popov Ye P Teoriya sistem avtomaticheskogo regulirovaniya. Izdaniye tret'ye, ispravlennoye. M.: *Izdatel'stvo «Nauka», Glavnaya redaktsiya fiziko-matematicheskoy literatury,*1975. 768 s. (in Russian).

[11] Shirokiy D.K., Kurilenko O.D. Raschot parametrov promyshlennykh sistem regulirovaniya. Spravochnoye posobiye. «*Tekhnika*», 1972. 232 s. (in Russian).

[12] Kulakov G T, Kulakov A T, Korzun M L, Basalay D V Strukturno-parametricheskaya optimizatsiyasistem avtomaticheskogo regulirovaniyas differentsirovaniyem romezhutochnogosignala // Strukturno-parametricheskaya-optimizatsiya-sistem-avtomaticheskogo regulirovaniya-s-differentsirovaniem-promezhutochnog- signala(1).pdf Adobe Reader. (in Russian).

[13] D'yakonov V P Simulink 5/6/7. M.: *DMK-Press*, 2008. 784 s. (in Russian).

IOP Conf. Series: Materials Science and Engineering **302** (2018) 012059 doi:10.1088/1757-899X/302/1/012059

Optimizing Virtual Network Functions Placement in Virtual Data Center Infrastructure Using Machine Learning

I P Bolodurina[1] and D I Parfenov[1]

[1]Department of Applied Mathematics, Orenburg State University, Pobedy avenue, 13, Orenburg, Russia

parfenovdi@mail.ru

Abstract. We have elaborated a neural network model of virtual network flow identification based on the statistical properties of flows circulating in the network of the data center and characteristics that describe the content of packets transmitted through network objects. This enabled us to establish the optimal set of attributes to identify virtual network functions. We have established an algorithm for optimizing the placement of virtual data functions using the data obtained in our research. Our approach uses a hybrid method of visualization using virtual machines and containers, which enables to reduce the infrastructure load and the response time in the network of the virtual data center. The algorithmic solution is based on neural networks, which enables to scale it at any number of the network function copies.

1. Introduction

Today, considerable volumes of data circulate in modern telecommunication networks. The modern paradigm is need new solutions effective control of traffic in networks and, at the same time, do not demand cardinal changes of the existing infrastructure of a data center. Traditional data centers use the concept of resources virtualization for various infrastructure facilities (network, computing nodes, systems of storage etc.) to achieve the goal. The application of a complex approach to virtualization is reflected in the architecture of virtual data centers (VDC). This architecture was used for the placement of multi cloud platforms. Multi cloud platforms use hybrid methods of virtualization based on software-defined components. It enables to increase the efficiency of computing resources use and, thus, to reduce the economic cost of maintaining the infrastructure of traditional data centers. However, the conception of resources virtualization is not quite effective. It allows abstracting the processed and transmitted data flows from physical devices [1]. But, nowadays, the problem of the effective placement of key components of the virtual network environment in a multi cloud platform is not solved [2].

One of the approaches applied in virtual data centers apart from the virtualization of traditional objects of network infrastructure, is the use of software realization instead of traditional hardware solutions, such as firewall, load balancer, NAT, routers and others. In practice, such solutions are based on the technology of network function virtualization (NFV) [3]. The NFV technology provides more flexible deployment and enables to control the virtual objects of a multi cloud platform, which perform the roles of hardware network devices, more effectively [4]. As a rule, the NFV technology is applied together with the software-defined network (SDN) and enables to exercise adaptive traffic control. However, the technology of network function virtualization has a number of disadvantages. The main problem is the lack of effective methods of planning for placing virtual objects in physical computing nodes. The review of research shows that existing solutions for placing the NFV in the infrastructure of data center use the approaches based on virtual machines or containers. The existing solutions do not deal with resource intensity of each virtual network function and its functional purpose for multi cloud infrastructure of the VDC. We have developed the approach that allows us to cluster the existing virtual and physical objects of infrastructure and, then, to place virtual network functions. The main idea of our solution is to estimate the consumption of resources by each element of the network. Besides, we will use the hybrid method of virtualization based on the simultaneous use of virtual machines and containers to create a flexible solution. It will enable to optimize the placement of the technology of network function virtualization in the infrastructure of a VDC.

Our approach is relevant, since it represents the combination of two modern innovative technologies in the field of the organization of network functioning and virtualization of its components for resource and data flow control in the SDNs based on the technology of network function virtualization. The goal of our investigation is to improve the quality of service (QoS) for applications and services of the multi cloud platforms placed in a VDC. Besides, we use the methods of intellectual data analysis to process information about the state and load of key objects as well as the flows between network devices received from the systems of computing nodes monitoring in the software-defined infrastructure of the multi cloud platform. It enables to receive the consolidated assessment of the QoS and to predict uninterrupted operation and operability of the software-defined infrastructure of a multi cloud platform and the entire virtual data center.

Further, we will describe our approaches to the optimization of the placement of virtual network functions in the multi cloud environment of VDC. Section 2 gives information about modern research and the existing approaches to working with network functions. In Section 3, we describe the methods and approaches applied within a framework of our solution. Besides, we describe the main stages of its implementation. The neural network work model, which is a basis for the formation of cards of placement of network functions in the multi cloud environment of virtual data center, is presented in section 4. Section 5 and 6 we gives the algorithmic solution and experimental investigation in network environment of VDC.

2. Related work

The variety of physical network devices of various vendors increases both capital expenditure and operational costs for the maintenance of data centers. The technology of network function virtualization allows solving this problem by the realization of network functions as software. The application of the NFV implies the use of the technology of network objects virtualization, which functions as software and particular computing processes, or as complex infrastructures of cloud computing instead of hardware solutions.

The group of scientists headed by Min Chen [5] analyzes the architecture and mechanisms of the interaction between the technology of NFV and the SDNs. As noted in the research, if the number of the users who launch applications in the multi cloud network environment increases, there is a competition for resources. Each user request is described by the requirements to network environment from the viewpoint of productivity, safety, and the effective control of objects in virtual infrastructure.

Scientists from Arizona State University [6] have studied a multi cloud system. They offer an approach to the creation of network architecture based on the NFV technology as alternatives for traditional hardware network devices. However, this research does not solve the main problem concerning the methods of NFV placement on computing nodes.

The NFV technology also has other a number of disadvantages. One of them is associated with the organization of the coordinated control of the entire network infrastructure of the VDC. To solve this problem, scientists from University of Wisconsin-Madison have proposed an approach, which is a framework of the OpenNF [7]. This approach provides the effective coordinated control of both the internal state of network functions and the state of data transmission network. But, this decision does not solve another important issue associated with the overall performance and load on the controller of network functions.

The scientists from Nation Chung Cheng University have investigated the reduction of load on the controller to ensure the work of the NFV technology [8]. As a rule, in case of the SDN use, the controller classifies the traffic received from the ports on network nodes to define a path to network objects, which play a role of the virtual network functions. This process generates large volumes of traffic in the plane of control. The authors offer the expansion for the architecture based on the SDN to reduce the load created by network traffic on the controller due to the use of the NFV technology. The solution represents two-layer classification of traffic based on the OpenFlow protocol. Network events are analyzed in the plane of data instead of the plane of control.

The group of scientists headed by Hassan Jameel Asghar has developed a scalable system to work with the technology of network function virtualization in the multi cloud environment [9]. The authors have offered an abstract model of the standard network function distributed between several cloud platforms. They system enables to increase the speed of package processing considerably in comparison with similar hardware solutions. An essential disadvantage of the concept is the resource intensity of this system. The overhead costs of the deployment and work of the Splitbox network functions demand the same quantity of computing resources as in case of hardware network devices.

Thus, it is established that the technology of NFV and the existing cloud solutions on based on the software-defined infrastructure of the VDC has a number of advantages, which enable to improve the quality of service in data transmission networks. However, today, there are adequate and effective solutions, that would enable to control the placement of the NFV on physical and virtual computing nodes in the data center.

3. Methods and approaches

Nowadays, neural networks are the most effective and high-speed method for forecasting, parameter identification, clustering and classifications in various fields of knowledge. Today, we see many successful examples of the application of a neural network approach for the creation of intellectual information systems [10]. Besides, the advantage of the neural networks use is the possibility of adaptive self-training with the use of additional methods of approaches. We have used an iterative approach based on a group of methods associated with the optimization of placement of VFN on the objects of the software-defined infrastructure of the VDC.

First of all we will present all network objects of the multi cloud platform placed in the VDC as a communication graph. The graph is based on the topology the physical network switching. Each network object is the graph vertex. It can be described by a basic set of parameters, which characterize each element of the network and influence productivity. We have chosen the following characteristics as parameters: volume of memory, volume of disk space, the frequency and quantity of kernels, etc. Further, we will use these characteristics as the input parameters acting as the training set for a neural network during the study of data.

A multi cloud platform supports the placement of various applications and services. Therefore, to identify the flows of traffic passing through infrastructure facilities of the VDC is an important task for the placement of virtual network functions. In this research, we have used the method based on the analysis of the known network ports for popular applications to obtain this information. This method enables to make an integrated classification of traffic flows; however, since there are non-standard network solutions applied in service-oriented applications, there obtained data will not be enough for the effective control of traffic flows. For a deeper analysis of traffic flows as a data source, we offer to use the method of decoding the protocols of communication based on the analysis of contents of the transferred packages. However, since this approach has rather high resource intensity, it will be used only at a low level of the analysis, for more exact identification of traffic flows of similar applications. The third method uses the sample approach based on the specific signatures located in protocol heading for the identification of the application. The fourth method is based on machine training. This method uses the accumulated data obtained by the above-mentioned methods and applies to them the algorithms of machine training to identify the applications based on characteristic packages and the saved-up statistics of data flows. The advantage of this approach is that algorithms can be trained in real time that will allow reconfiguring software-defined infrastructure of the VDC on the fly.

The proposed solution is based on an integrated approach to the collection of data on the traffic flows circulating in a multi cloud platform. It will allow optimizing the placement of network functions on computing nodes of the VDC. To achieve the goals of the research, we have created a neural network system to predict the placement of network function virtualization in the multi cloud environment of the virtual data center.

4. Model of the clustering of infrastructure facilities of the virtual data centers for the placement of network functions

We have chosen Kokhonen's network as a neural structure for modeling, since it is the most efficient in the clustering and classification of objects. Another important factor is the visualization of results; it enables to improve the understanding of the structure and character of data at early stages and to specify a neural network model further. Due to the peculiarities of network function virtualization, the support of classification in Kokhonen's network can be used to identify the uniform elements in network for further optimization of their placement. Kokhonen's network is trained by a method of consecutive approximations. Starting from the initial placement of objects selected randomly, the algorithm gradually improves it to supply the data clustering. Another advantage of Kokhonen's network is the opportunity to identify new clusters. The trained network detects clusters in the training data and refers all the data to certain clusters. If the network meets a set of data, which differ from any known samples, it will independently reveal a new cluster of elements then. This feature is very relevant, since it allows entering new network functions into the architecture of virtual data center without the actual change of algorithms of their distribution on physical and virtual computing nodes.

The principle of creating a neural network system to optimize the placement of network function virtualization in the multi cloud environment of the virtual data center is as follows. We have selected a number of criteria using the data obtained from the systems of virtual data center monitoring. This enables to both identify the virtual network function and assess its load on computing nodes. Criteria are formulated so that the answer could be always represented in the binary form. i.e. 1 is "Yes" or 0 is "No". The obtained data enable to form the vector of signals $E = \{e_1, e_2, \ldots, e_n\}$, which is placed at the entrance of a neural network. The vector of output values is similar, it has binary components.

The neural network is a two-dimensional matrix of neurons of dimension n (the number of inputs of each neuron) per m (the number of neurons). The number of inputs of each neuron is determined with respect to the number of criteria established earlier. The amount of neurons m coincides with the required number of classes and corresponds to the number of the unique network functions used in work of a multi cloud platform. The importance of each of the entrances to neuron is characterized by the numerical size called by weight. It is set in the form of matrix:

$$X = \begin{pmatrix} x^{11} & x^{12} & \ldots & x^{1j} \\ x^{21} & x^{22} & \ldots & x^{1j} \\ \ldots & \ldots & \ldots & \ldots \\ x^{i1} & x^{i1} & \ldots & x^{ij} \end{pmatrix} \tag{1}$$

With the vectors of weight coefficients of connections $x^{ij} = \{x_1^{ij}, x_2^{ij}, \ldots, x_n^{ij}\}$ as its elements.

Kokhonen's network consists of three layers of neurons. The basis of the network is a covert Kokhonen's layer. However, in this research, we have offered a changed scheme of output neurons of Kokhonen's network to obtain the results to identify destination and, simultaneously, to find critical loading on a calculating node (Fig. 1).

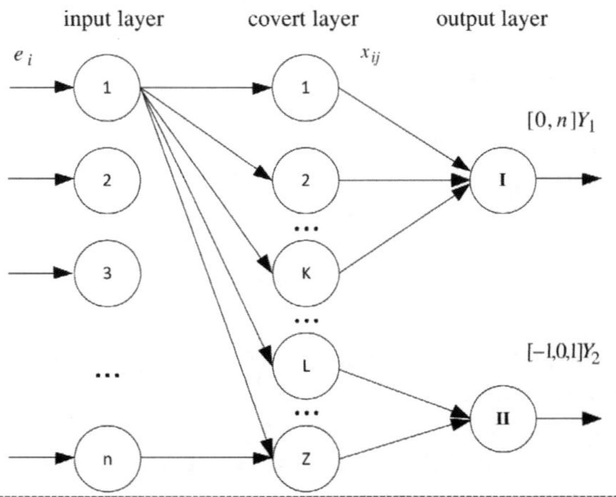

Figure 1. A neural network model of VFN clustering.

We offered to divide the covert layer of Kokhonen's neural network into two sets. The first set of neurons [1…K] is responsible for the identification of a network function placed in the VDC. The work of neural network changes input scales at the exit layer activates the linear function Y_1, which takes the value [0…n], where 0 means that the network object under study has no signs of the VFN, and values from 1 to n correspond to a particular network function identified by a neural network model. The second set of neurons [L…Z] analyzes the loading of the network object under study and initiates the function Y_2 at the exit, which take the values [-1,0,1], where 0 is a normal state, -1 means that the network function is idle or does not perform its functions, and 1 means that the network function is overloaded.

The basic criteria used as input data to detect virtual network functions are network records and events, data of the time of packages going through a network object, time of packet input and output, memory loading, the use of CPU, the intensity of dataflow, TTL and others. The data collected in the network are placed at the entrance of a neural network and create a full neural network. That being the case, we should simplify and verbalize a neural network by excluding some elements without the significant reduction of the detection quality.

To train a neural network, we have used the data obtained from the system of monitoring of virtual data center of the Orenburg State University. We have chosen the most popular virtual network functions for the experimental study such as vRouter, vNAT, vFirewall, vProxy, vSwitch, vDPI.

The application of the developed neural network system enables to identify virtual network functions correctly in 93-98% of cases, while the resource intensity is insignificant. This increases the efficiency of optimization for placing them in the infrastructure of a multi cloud platform.

5. The algorithm for optimizing the placement of virtual network functions in the data center infrastructure

Our model of identification of virtual network functions allows optimizing their placement in the virtual data center. We will optimize the placement of the network functions found in the virtual data center by using Kokhonen's network by the following criteria: the current load created on computing nodes; resource intensity of a network function; the number of flows going through computing nodes. The main objective of virtual network functions placement is to choose the optimum number of the nodes to implement required functionality as a software solution. Thus, there is a problem of resource planning. Planning is of particular relevance in the organization of dynamic topology in the virtual data center, since the load on computing nodes can change over a wide range at rather short intervals of time and depends on the chosen type of specific network functions placement. To solve the task of optimization, we have developed the algorithm for monitoring the infrastructure of the virtual data center as well as for the placement and launch of network functions. In comparison with the available analogs, the algorithm uses the heuristic analysis of traffic flows and their classification depending on the type of a network function.

The general algorithm has the following sequence of steps.

Step 1. To identify the arrangement of virtual network functions taking into account the topology of the network infrastructure of the virtual data center.

Step 2. To estimate the number of the launched copies of each virtual network function and to range them according to the popularity of the network infrastructure. The popularity is estimated towards traffic flows, which run through the launched copies of a virtual network function.

Step 3. To define the load on physical and virtual computing nodes created by each copy of function.

Step 4. To compare data and to define virtual network functions, which demand scaling or folding, using the data obtained in steps 1 and 2.

Step 5. To reconfigure the topology on the controller of the software-defined network, stop and release the occupied resources of virtual data centers for network functions, which require folding.

Step 6. To evaluate a method of placement for virtual network functions, which require scaling and creation of the maximum load on infrastructure. To distribute the most loaded network function using a hybrid way of placement (the containers developed in the virtual machine). To transfer network functions, which are less loaded but require scaling, to the operating mode of the virtual machine.

Step 7. To provide the migration of the virtual machines with network functions on the least loaded computing nodes.

The approach in this algorithm for controlling the placement of virtual network functions enables to take into account a method of placement and to organize the work of the virtual data center with account for circulating traffic flows and regulate the number of the launched copies of each function.

The purpose of experimental investigation is to define the efficiency of using the developed algorithm for placing the virtual network functions in infrastructure of the virtual data center.

We have created the templates of container placement and the images of virtual machine for the deployment in the software-defined infrastructure of data centers for each virtual network functions.

We have created the flows based on statistical information using the generator of requests to imitate the work of network environment in the data centers. To estimate the productivity of virtual network functions, we have used the flows with varying intensity: in the first case, the flows created minimum permissible load. This enabled to estimate the time of a response and delays brought by the infrastructure of the virtual data center (exp. 1). In the second case, we have created the working load on the virtual network functions in the infrastructure of the virtual data centers. This enabled to estimate the time of response for applications in network environment (exp. 2). In the third case, an experimental study evaluated the operation of a virtual network environment of a multi cloud platform, using the developed algorithm for optimizing the location of virtual network functions (exp. 3). In this case, we have established the consumption of resources for each of the launched copies of virtual network function. This enabled us to predict the required resources in network environment. The results of the experiment are provided in Fig. 2.

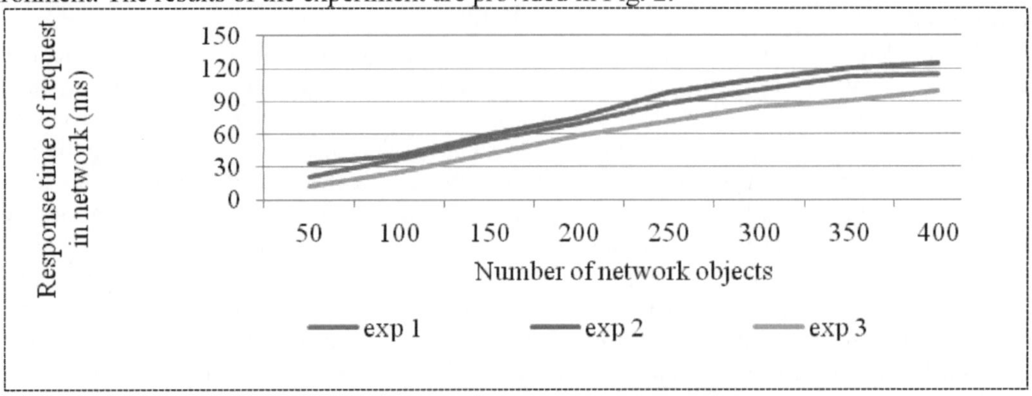

Figure 2. Diagram of dependence of response time in the multi cloud platform

The research has shown that the static placement of containers on physical nods is inefficient, since it does not enable to redistribute loading quickly. Besides, the transfer of a container to another computing nod leads to the loss of the current connections. The placement of virtual network functions on the basis of virtual machines was more efficient due to flexible load balancing. However, the load on computing nods has considerably increased due to additional overheads for the use of virtual machines. In our research, the most efficient placement was the placement with the use of containers in virtual machines. It has enabled to increase the density of the placement of applications and services in virtual data centers. Besides, it has enabled to place the containers and applications of services and data in direct network proximity from each other and to reduce the time of response to user requests from applications and, thus, to increase the overall performance of a system.

6. Conclusion

Our research enabled to create a neural network model of virtual network function identification. The proposed model of classification based on the statistical properties of the flow defined a systematic approach to the selection of the optimal set of attributes of the traffic flow. The developed algorithmic solutions, which are based on hybrid methods of virtualization and use the data of a neural network model, enabled to optimize the placement of network functions in software-defined segments of the data center network. The results show that the optimization of the number of NVF enables to improve the quality of service by 20-25% by reducing the response time and load on physical network devices. It became possible due to the identification of applications at the initial stage by using the data of the NVF placement in the network of the virtual data center. Further, we are going to assess our approach with a larger number of NVF, since it will allow us to assess the accuracy of our solution.

Acknowledgments

The research work was funded by Russian Foundation for Basic Research, according to the research projects No. 16-37-60086 mol_a _dk and 16-07-01004, and the President of the Russian Federation within the grant for state support of young Russian scientists (MK- 1624.2017.9).

References

[1] Bolodurina I, Parfenov D et al 2017 A model of cloud application assignments in software-defined storages // Journal of Physics: Conference Series — Vol. 803.

[2] Bolodurina I, Parfenov D 2017 Development and Research of Models of Organization Distributed Cloud Computing Based on the Software-defined Infrastructure // Procedia Computer Science — Vol. 103 —P. 569–576.

[3] Juliver de Jesus Gil Herrera, Juan Felipe Botero Vega 2015 Network Functions Virtualization: A Survey. — IEEE.

[4] Parfenov D, Bolodurina I 2016 Development and research of models of organization storages based on the software-defined infrastructure 39th International Conference on Telecommunication and signal processing: materials of conference 27-29 June 2016, Vienna, Austria. - P. 1-6. . — IEEE.

[5] Yong Li, Min Chen 2015 Software-Defined Network Function Virtualization: A Survey. — IEEE.

[6] Ruozhou Yu, Guoliang Xue, Vishnu Teja Kilari, Xiang Zhang 2015 Network function virtualization in the multi-tenant cloud. — IEEE.

[7] Aaron Gember-Jacobson, Raajay Viswanathan, Chaithan Prakash, Robert Grandl, Junaid Khalid, Sourav Das, Aditya Akella 2014 OpenNF: Enabling Innovation in Network Function Control. — IEEE.

[8] Ying-Dar Lin, Po-Ching Lin, Chih-Hung Yeh, Yao-Chun Wang, Yuan- Cheng LaiAn 2015 Extended SDN architecture for network function virtualization with a case study on intrusion prevention. — IEEE.

[9] Hassan Jameel Asghar, Luca Melis, Cyril Soldani, Emiliano De Cristofaro, Mohamed Ali Kaafar, Laurent 2016 MathySplitBox: Toward Efficient Private Network Function Virtualization. — IEEE.
[10] Bolodurina I, Parfenov D 2015 Approaches to the effective use of limited computing resources in multimedia applications in the educational institutions // WCSE 2015-IPCE. — IEEE.

IOP Conf. Series: Materials Science and Engineering **302** (2018) 012060 doi:10.1088/1757-899X/302/1/012060

Possibility of Thermomechanical Compressor Application in Desalination Plants

E.V. Blagin[1,*], A.A. Shimanov[1], D.A. Uglanov[1] and S.S. Korneev[1]

[1]Heat Engineering and Heat Engines Department, Samara National Research University, Samara, Russia

*Corresponding author: evgenyblagin@gmail.com

Abstract. This article deals with estimation of thermocompressor operating possibility in desalination plant with mechanical vapour compressor. In this plant thermocompressor is used instead of commonly used centrifugal compressor. Preliminary analysis shows that such plant is able to operate, however, power consumption is 3.5-6.5 higher in comparison with traditional MVC plant. In turn, utilization of thermocompressor allows avoiding usual high-frequency drive of centrifugal compressor. Drives with frequency of 50 Hz are enough for thermocompressor when centrifugal compressor requires drives with frequency up to 500 Hz and higher. Approximate thermocompressor dimensions are estimated.

1. Introduction

Lack of the fresh water became a problem in recent years. According to researches, in 2025 about half of the states will suffer fresh water deficit [1]. Thus, generation of the fresh water is one of the main tasks which must be solved to ensure ecological safety of the Earth population. There are several ways of fresh water generation. One of the most promising is seawater desalination or waste water treatment. There are several ways to desalinate seawater. They include reverse osmosis (RO) [2], electrodialysis (ED) [3], freezing [4] and distillation [5-7]. Distillation plants in turn can be divided on Multi-effect distillation plants (MED) [8], multi-stage flash distillation plants (MSF) [9], multi-effect distillation plants with thermal vapour compression (MED-TVC) [10] and plants with mechanical vapour compression (MVC) [11]. Every type of has its own advantages and disadvantages. For example, MED plants can utilize waste heat of thermal power station, however, their efficiency is lower in comparison with RO plants. RO plants have a higher efficiency, however, they require often change of the membranes. ED plants are applicable on the water with relatively small salinity. Freezing plants are currently under development but in future they will allow using cold energy of liquid natural gas for desalination purposes. MVC plants have satisfactory energy consumption but they require careful design of the steam compressor which operation is crucial for plant.

Thermomechanical compressor is a device suited for gas compression by the heat adding in the closed volume [12]. From design point of view thermocompressor can be made with internal [13] or external regenerator [14] which can be represented in the form of annular gaps [15] and porous heat exchanger.

Thermomechanical compressor can be used in distillation plants as a substitution of mechanical compressor or as efficiency increase of other distillation plants.

2. Thermocompressor working principle

Technically thermocompressor consists of such elements as: piston with regenerator inside; piston rod; heat exchangers to compensate non-perfection of regenerator; input and output valves, cold and hot chambers. Principal scheme of the mechanical thermocompressor is presented on figure 1.

Working cycle of thermocompressor starts from high dead centre when both valves are closed and volume of the cold chamber is maximal (point 1). Piston starts to move downside pushing working fluid through regenerator to a hot chamber. Due to Amontons's Law of Pressure-Temperature, pressure is increasing because mean temperature of the working fluid inside the thermocompressor is increasing. When the pressure is high enough, output valve opens and working fluid with given pressure starts to be charged to a customer. When piston reaches low dead centre, output valve closes (point 3) and working fluid stops to be charged to a customer. Piston starts to move upside pushing working fluid back to a cold chamber. Pressure is decreasing due to decrease of the mean temperature inside the thermocompressor. When pressure drops low enough, input valve opens, charging new portion of working fluid inside the thermocompressor. Working cycle of thermocompressor is presented on figure 2.

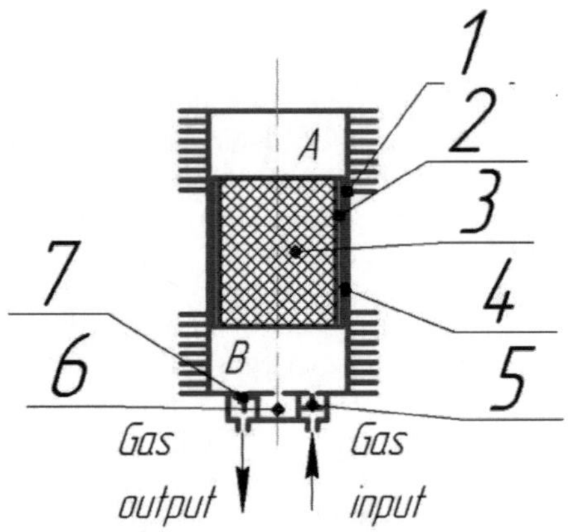

Figure 1. Principal scheme of the mechanical thermocompressor:
1 – thermocompressor shell; 2 – displacer-regenerator shell;
3 – regenerator; 4 – gap; 5 – input valve; 6 – rod;
7 – output valve; A – hot chamber; B – cold chamber.

Thus, for one working cycle thermocompressor rise the pressure of certain amount of gas Δm, which enters the TC through the input valve and leaves the TC from the output valve during one cycle.

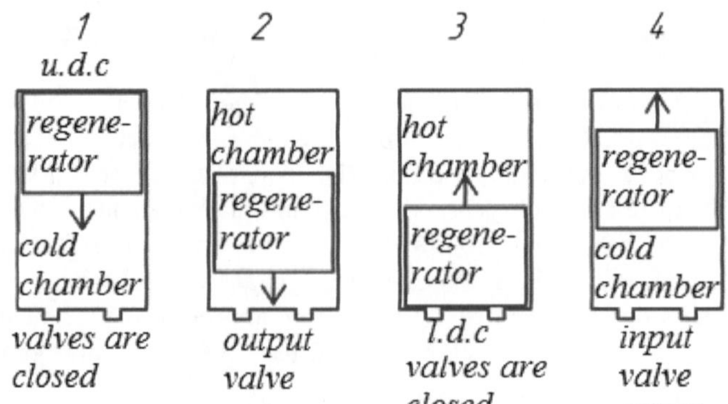

Figure 2. Working cycle of the thermocompressor.

3. Thermocompressor coupled work with distillation plants

The main feature of the MVC plant is requirement of only one energy source which is electricity. Electric energy is transformed in thermal energy and potential energy of pressure which are further used to execute consequential phase changes seawater → steam → distillate. Design scheme of the plant with thermocompressor is presented on figure 3.

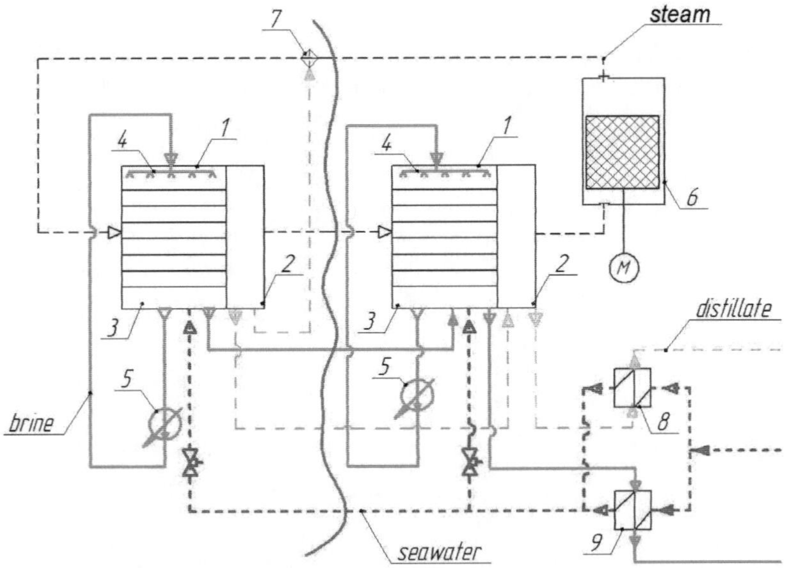

Figure 3. Desalination plant with thermomechanical compressor 1 – evaporator stage, 2 – distillate chamber, 3 – brine chamber, 4 – water sprayer, 5 – circulation pump 6 – thermocompressor, 7 – mixer heat exchanger, 8 – water-distillate heat exchanger, 9 – water-brine heat exchanger

In this case steam enters thermocompressor instead of mechanical compressor. During displacement through regenerator both its temperature and pressure are raised. Let's compare amount of energy necessary to raise the pressure in the same degree. Steam parameters of the last stage will be taken from the plant currently developed by Samara University together with "Metallist-Samara" OJSC. These parameters include: steam temperature 60 °C, steam pressure 20 kPa, required compression ratio is 1.6, steam flow rate is 0.111 kg/s. Other compression ratios are also added to compare the behavior of consumed work on bigger pressure range.

For the case of adiabatic expansion, amount of energy consumed by compressor will be determined as:

$$N_{comp} = G\frac{k}{k-1}RT_{evap}\left(p^{I}\right)\left[\left(\frac{p^{II}}{p^{I}}\right)^{\frac{k-1}{k}} - 1\right], \quad (1)$$

where Ncomp – power consumption, G – steam flow rate, k – specific heat ratio, R – specific gas constant, Tevap – evaporation temperature, pI – gas pressure in the input of the compressor, pII – gas pressure at the output of the compressor.

For more real expansion, enthalpy difference is used:

$$N_{comp} = G(i^{II} - i^{I}), \quad (2)$$

where iII and iI – are enthalpies of steam before and after the compressor.

Finally, for the case of thermocompressor, it is necessary at first determine required temperature of gas to ensure required pressure:

$$T^{II} = T_{evap}\left(\frac{p^{II}}{p^{I}}\right), \quad (3)$$

where TII is a temperature after thermocompressor.

In this case required power can be calculated as:

$$N_{comp} = G \cdot c_v \cdot (T^{II} - T_{evap}), \quad (4)$$

where cv – isochoric heat capacity of the steam.

Results of the power consumption comparison are presented on figure 4.

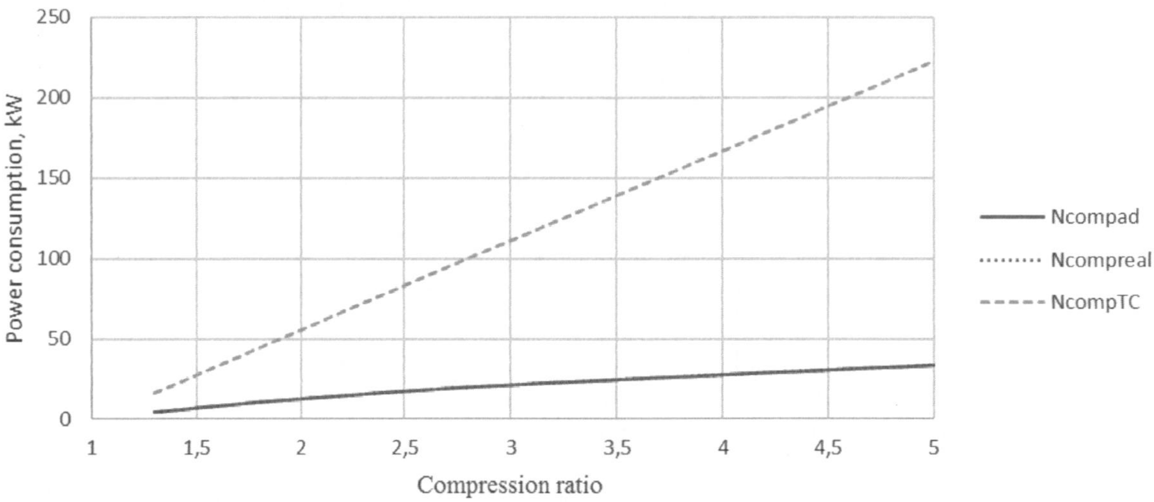

Figure 4. Comparison of the power consumption depending on compression ratio

It can be seen that power consumption in thermocompressor is 3.5-6.5 higher and this relation is increasing with increase of compression ratio. However, despite this fact, thermocompressor has some advantages over centrifugal compressors which are commonly used is such desalination plants. These advantages include independence of compressor working parameters (frequency, dimensions) on compression ratio. In the meantime, operation of centrifugal compressor is determined by both dimensions and frequency and in case of aforementioned conditions its frequency reaches the values of 30000 r/min which negatively affects bearings.

Approximate dimensions of the proposed thermocompressor can be estimated. According to [12] relation of the flow rate to maximal mass contained in thermomechanical compressor depends on several parameters: temperature relation, regeneration efficiency, specific dead volumes. Working frequency also must be included in thermocompressor dimension analysis. Initial data is presented in table 1.

According to calculation non-dimensional mass output for given condition is about 0.18. That means that only about 18% of mass contained in thermocompressor will be charged per cycle.

Required mass flow rate can be determined as:

$$m_c = \frac{G}{f}. \quad (5)$$

Thus, maximal mass of gas inside the thermocompressor will be determined as:

$$m_{\max} = \frac{m_c}{(dm / m)} \quad (6)$$

Maximal mass in thermocompressor will correspond to position when cold chamber volume is maximal. This position also corresponds to an input pressure.

$$p_{in} = \frac{m_{max} R}{\frac{\varepsilon_h V_d}{T_h} + \frac{2\varepsilon_{reg} V_d}{T_h + T_c} + \frac{(1+\varepsilon_h) V_d}{T_c}}. \quad (7)$$

Table 1. Initial data for estimation

Parameter	valu e
Input temperature T_c, K	333
Hot chamber temperature T_h, °C	900
Initial pressure p_{in}, kPa	20
Required pressure p_{out}, kPa	32
Regeneration efficiency η_{reg}	0.95
Specific dead volume of hot chamber, ε_h	0.05
Specific dead volume of cold chamber, ε_c	0.05
Specific dead volume of regenerator chamber ε_{reg}	0.1
Frequency f, Hz	50

Solving of this equation along displaceable volume Vd gives the value of 0.085 m3, which is acceptable.

4. Conclusions

Application of the thermocompressor in desalination plants allows avoiding high-frequency centrifugal compressor which allows easing bearing loads. However, specific power consumption of thermocompressor driven MVC plant is higher in comparison with traditional MVC plant. Approximate dimensions of the thermocompressor for developed MVC plant are estimated and displaceable volume is equal to 0.085 m^3. More accurate investigation and optimization of the desalination plant working parameters are required for more accurate estimation of thermocompressor application potential in desalination plants.

Another direction towards which investigation can be carried out is numerical modeling of working process in software suited for fluid flow modeling such as ANSYS Fluent. Recent successes in both modeling of compressor operation [16-17] and piston machines modeling [18-19] allow assuming a possibility of successful modeling of working process in thermocompressor because it combines elements of both compressor and piston machines design.

Acknowledgments

This work was carried out by lead performer of research and development effort with financial support of Russian Federation Ministry of Science and Education with the realization of Goverment decree №218 from 09.04.2010 according to contract about providing and using subsidy № 02.G25.31.0150 from 01.12.2015.

References

[1] Deficiency of fresh water: problems and methods of solving http://thewallmagazine.ru/lack-of-fresh-water/
[2] Sarai Atab M, Smallbone A J and Roskilly A P 2016 Desalination 397 174-184
[3] Largier T D, Wang D, Mueller J, & Cornelius C J 2017 Journal of Membrane Science 531 103-110
[4] Cao W, Beggs C and Mujtaba I M 2014 Desalination 355 22-32
[5] Maghsoudi K, Aliasghari M and Mehrpanahi A 2016 Desalination and Water Treatment 57(38), 17707-17721

[6] Alasfour F N, Darwish M A and Bin Amer A O 2005 Desalination 174(1) 39-61
[7] Zimerman Z 1994 Desalination 96 51-58
[8] Fathalah K and Aly S E 1991 Energy Conversion and Management 31(6) 529-544
[9] Darwish M A 1987 Desalination 63(C) 143-161
[10] Maghsoudi K, Aliasghari M and Mehrpanahi A 2016 Desalination and Water Treatment 57(38) 17707-17721
[11] Darwish M A, Jawad M A and Aly G S 1990 Desalination 78(3) 313-326
[12] Bush V US Patent 2157229
[13] Edwards M J and Peterson R B 2007 Proceedings of the Institution of Mechanical Engineers 221(4) 459-472.
[14] Kornhauser A A 1996 Paper presented at the Proceedings of the Intersociety Energy Conversion Engineering Conference 2 1331-1336
[15] Lin W, Wu X, Yang J and Yang L 2013 International Journal of Refrigeration 36(4), 1376-1387
[16] Popov G, Goriachkin E, Kolmakova D and Novikova Y 2016 Paper presented at the Proceedings of the ASME Turbo Expo 2C-2016
[17] Baturin O V, Popov G M, Goriachkin E S and Novikova Yu D 2017 Journal of Physics Conference Series 803(1)
[18] Saigakov E A, Gorshkalev A A, Kayukov S S and Blagin E V 2014 Research Journal of Applied Sciences, 9(10) 669-673
[19] Biryuk V V, Gorshkalev A A, Kayukov S S and Uglanov D A 2014 Open Mechanical Engineering Journal 8 441-444.

IOP Conf. Series: Materials Science and Engineering **302** (2018) 012061 doi:10.1088/1757-899X/302/1/012061

Investigation of the Hydraulic Characteristics of Capillary Elements of the Injector Head of Jet Engines under Conditions of Isothermal Flow of A Liquid

V E Nigodjuk[1] and A V Sulinov[1]

[1]Institute of engines and power plants, Samara University, Samara, Russia

Email: abcsamara@yandex.ru

Abstract. The article presents the results of an experimental study of the hydraulic characteristics of capillary elements of the injector head of jet engines in isothermal fluid flow and the proposed method of their calculation. The main geometric dimensions of the capillaries in the experiment were changed in the following range: Inner diameter from 0.16 to 0.36 mm, length from 4.3 to 158 mm and relative length from 25 to 614 and the inlet edge of the capillaries: sharp or smooth the leading edge. As the working fluid during the tests were distilled water, acetone and ethyl alcohol. Based on the results of a study of the dependences for calculation of ultimate losses in laminar and turbulent flow regimes in capillary tubes with smooth and sharp edges input. The influence of surface tension forces on loss of input on a sharp cutting edge. Experimentally confirmed the possibility of calculating the linear coefficient of hydraulic resistance of capillary tubes with a diameter of 0.16-0.36 mm in isothermal stable during the known dependencies that are valid for hydrodynamically smooth round tube.

1. Introduction

Jet thrusters as fluidic devices for supplying fuel into the combustion chamber using a round capillary tube tube of small diameter large relative lengths: inner diameter (0, 15-0, 35) mm and the relative length of more than 20. Round capillary tube compared to conventional round tubes have a number of features as in an isothermal fluid flow and flow with heat supply [1]. Therefore to calculate the hydraulic characteristics and the choice of the geometric dimensions of the capillary nozzle elements of the thrusters was an experimental study of the hydrodynamics of flow in capillary tubes and the method of calculation of their hydraulic characteristics in an isothermal liquid flow. The proposed calculation method of the capillary nozzle elements based on the known materials of calculation of hydraulic characteristics with the features identified by the authors in the experimental study of capillary tubes.

2. Facilities and experimental conditions

A hydrodynamic study was performed in capillaries made of stainless steel. A pilot study was conducted on the installation from cold flow. Install the provided pressure flow of the working fluid in the capillary with the aid of compressed air, whose pressure was controlled by pressure sensor, the flow rate of the working fluid was measured either by weighing – or volumetric method using a level gauge, the filling time is fixed by an electric stopwatch. As working fluids in prolivka distilled water, acetone and ethyl alcohol. The main geometric dimensions of the capillaries during the experiment were changed in the following range: inner diameter from 0.16 to 0.36 mm, length from 4.3 to 158 mm and the relative length from 25 to 614. The capillaries were carried out with sharp or smooth edge of the input.

3. Hydraulic characteristic of the capillary

As the main hydraulic characteristics of the capillary was used, the flow characteristic in common to the hydrodynamics of the fluid flow [2]:

$$\Delta p = \left(\lambda \frac{l}{d} + K \right) \frac{\rho u_0^2}{2} , \tag{1}$$

where Δp - the pressure drop in the capillary; ρ - the density of the liquid; u_0 - the fluid velocity at the entrance to the capillary; λ - linear coefficient of hydraulic resistance; l and d – the length and inner diameter of the capillary; K - end loss.

The expression (1) can be converted with correlation:

$$\xi = \lambda \frac{l}{d} + K, \tag{2}$$

$$\xi = \frac{1}{\mu^2}, \tag{3}$$

$$\frac{\rho u_0^2}{2} = \frac{8 \dot{m}^2}{\pi^2 \rho d^4} \tag{4}$$

to the next view:

$$\Delta p = 8 \frac{\xi \dot{m}^2}{\pi^2 \rho d^4} = 8 \frac{\dot{m}^2}{\pi^2 \mu^2 \rho d^4} , \tag{5}$$

where $\dot{m}$ - mass flow rate of the liquid; ξ – hydraulic resistance coefficient capillary; μ - flow coefficient capillary.

Linear coefficient of hydraulic resistance λ, generally, depends on the regisma flow - Reynolds number Re, roughness, wall temperature, etc. In the research process, as shown hereinafter, for the description of the hydrodynamics of fluid flow in the capillaries can be used known formulas to calculate the coefficient of hydraulic resistance in laminar flow - Poiseuille's formula [2], in turbulent flow regime (up to $\mathrm{Re} = 10^4$) for hydrodynamically smooth pipes (the roughness is not more than ~ 1 μm)- the formula of Blasius [2].

4. Input loss

End for the pressure loss K at the initial part of the capillary which the rebuilding of the velocity profile, and the acute front edge and the turbulent mode, in addition, the compression and expansion of the flow, and depend on the flow regime and type of inlet edge.

In laminar flow, due to the influence of the initial hydrodynamic phase, it is convenient to consider using functions $K_{\bar{z}}$ from dimensionless capillary length $\bar{z} = l/(d\,\mathrm{Re})$. In this regard, conducted the experimental determination $K_{\bar{z}}$ of the capillary with a diameter of 0.16 mm with a sharp entrance edge and a smooth entrance. The results of the experiment are presented in figure 1a.

Processing of the obtained experimental results allowed us to obtain the following approximating dependence [3]:

$$K_{\bar{z}} = 1 + 1{,}2[1 - 0{,}61 \exp(-94{,}8\,\bar{z})] \tag{6}$$

which can also be represented as a function of dimensionless flow $\overline{m} = \dot{m}/(\eta\, l)$ (where η - is the dynamic viscosity coefficient of fluid):

$$K_{\bar{z}} = 2{,}2 - 0{,}726 \cdot \exp(-74{,}5/\overline{m}). \tag{7}$$

According to (6), (7) just as in the case of a smooth inlet edge and for the case with a sharp entrance edge (figure 1 a) $\bar{z}$ > 0.003 and $\overline{m}$ < 260.

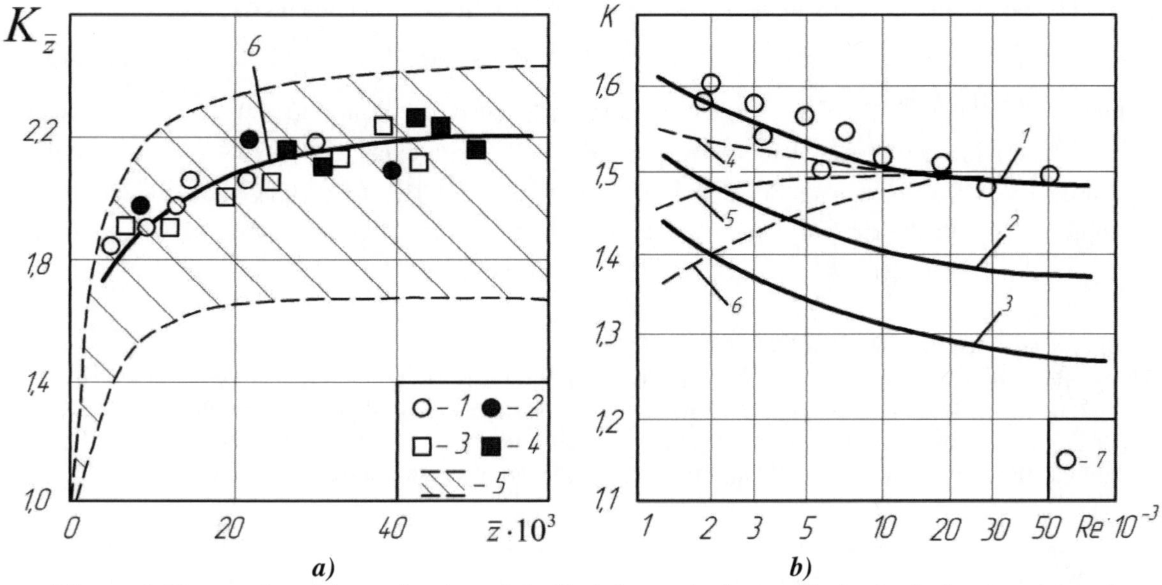

Figure 1. The results of determination of the limit losses in the capillaries in the laminar (*a*) and turbulent (b) flow regimes: *a*) 1, 2 – sharp entrance edge, 3, 4 – input smooth edge, 1, 3 – water, 2, 4 – alcohol; 5 – area of known results [3]; 6 – approximating the dependence (6); b) 1 - We=∞, 2 - We=100, 3 - We=20, 4 – Lp=$1\cdot10^4$; 5 – Lp=$5\cdot10^4$; 6 – Lp=$20\cdot10^4$; 7 – experimental data if We=∞ [4].

In turbulent flow regime limit the loss for a smooth entrance is small, due to only rebuild the velocity profile and can be taken into account using the proposed in [5] dependencies:

$$K = 1 + 2{,}65\lambda \tag{8}$$

At the sharp front edge into the capillary as the jet is compressed during expiration the diaphragm and then expands to form an isolated annular cavity, in which the pressure is considered equal to the pressure in the compressed section of the jet and lower than the pressure in the place of adhesion of liquid to the wall. Using the Bernoulli equation and the Board for the mechanism of flow was obtained the formula for calculating entrance losses:

$$K = 1 + \xi_0 = \frac{1}{\mu^2} - 2\left(\frac{1}{\varepsilon} - 1\right) \tag{9}$$

where ξ_0 - input loss; ε - the compression ratio of the jet efflux through the aperture.

As an isolated annular cavity has a phase interface, then the value of the input loss at small sizes of the capillaries should have an impact force of surface tension. Analysis of the results of experimental study of the influence of Weber numbers We for the coefficients of the diaphragm [2] shows that the relative change of the flow coefficient can be represented as:

$$\overline{\mu}_{\sigma} = \frac{\mu_{\sigma}}{\mu} = \overline{\mu}_{\sigma}^{\Delta p} \cdot \varepsilon_{\sigma} \ , \tag{10}$$

where μ - the expense ratio in the absence of the influence of surface tension forces; μ_{σ} - the ratio consumption in the presence of the influence of surface tension forces; $\overline{\mu}_{\sigma}^{\Delta p}$ - coefficient taking into account the change of flow rate of a diaphragm for reducing the effective differential pressure on it and determine by calculation; $\overline{\varepsilon}_{\sigma}$ - coefficient taking into account the increase in the area of the compressed section of the jet efflux through the aperture and determined experimentally.

Taking into account the experimental values $\mu(\mathrm{Re})$, $\varepsilon(\mathrm{Re})$ and $\overline{\varepsilon}_{\sigma}(We)$ [2, 4], where the Weber number $We = (u_0^2 d\rho)/\sigma$, σ - surface tension, calculated according to (9) and (10) change $K(\mathrm{Re}, We)$. The results (figure 1 b) show good coincidence with experiment at $We \rightarrow \infty$ and significant influence of surface tension on entry-nye losses We < 103. The dotted lines show the change of the limit for loss of the capillary when you change the flow rate (Reynolds number Re and the Laplace number $Lp = (\rho\sigma d)/\eta^2$).

5. Linear coefficient of hydraulic resistance

Due to the complexity of measuring the pressure drop on the section of stable flow, a linear coefficient of hydraulic resistance was determined by pressure drop over the capillary with the given limit losses according to the formulas (6)...(10). Thus the calculated diameter of the capillary is chosen so that λ in laminar flow regime was consistent with the dependence of Poiseuille [2]:

$$\lambda = \frac{64}{\mathrm{Re}} \tag{11}$$

The same method was used in [6], its validity is confirmed by good coincidence of experimental and calculated values of the coefficient of hydraulic resistance:

$$\xi = 16\pi \frac{\eta l}{\dot{m}} + 2{,}2 - 0{,}726 \left(-74{,}5 \frac{\eta l}{m} \right) \tag{12}$$

for laminar regime. The deviation of experimental points from the calculated curve is observed in case of turbulence, which allows to determine the transition flow regime and to find critical Reynolds number:

$$\mathrm{Re}_{cr} = \frac{4\,l}{\pi\, d\, \overline{m}_{cr}} \ . \tag{13}$$

The obtained experimental results confirmed (figure 2) applicability (1)...(5) for descriptions of the hydrodynamic characteristics of the capillaries, not only in laminar but also in turbulent (to Re = 104) flow regimes capillaries are hydrodynamically smooth pipes (the roughness less than ~ 1 μm) and amount is determined by the dependence of the Blasius [2]:

$$\lambda = \frac{0{,}3164}{(\mathrm{Re})^{1/4}} \tag{14}$$

In figure 2 shows the following notation of experimental points: 3 – capillary d=0.26 mm, l/d=613; 4-6 – capillary d=0.35 mm, l/d=238; 7, 8 – capillary d=0.33 mm, l/d=438; 9 – capillary d=0.17 mm, l/d=232; 10-12 – capillary d=0.36 mm, l/d=211; 13-15 – capillary d=0.33 mm, l/d=158; 16, 17 – capillary d=0.35 mm, l/d=150; 18-22 – capillary d=0.16 mm, l/d=60; 25, 26 – capillary d=16 mm, l/d=26; 3, 4, 7-11, 13, 16 -19, 25, 26 – sharp input edge; 5, 6, 12, 14, 15, 20-24 - smooth the input edge; 3-5, 7, 9, 10, 13, 14, 16, 20, 23, 25 - water; 6, 8, 11, 12, 15, 17, 18, 21 - acetone; 19, 22, 24, 26 - alcohol.

At Re > 104 (figure 2) the value of the linear coefficient of hydraulic resistance λ is practically unchanged and amounts to (0,031…0,029). In the case of a sharp inlet edge, the loss of stability of laminar flow, determined by the change λ, occurs at Re > 1,5·103, in the case of smooth input – Re > (2,5...5)·103, the transition to a developed turbulent flow – in Re=3·103 and Re = (4...6)·103, respectively, even in the case of a smooth entry at Re > 7·103 no plot with a laminar boundary layer. Similar to the transition region is obtained for changing the appearance of the flowing stream.

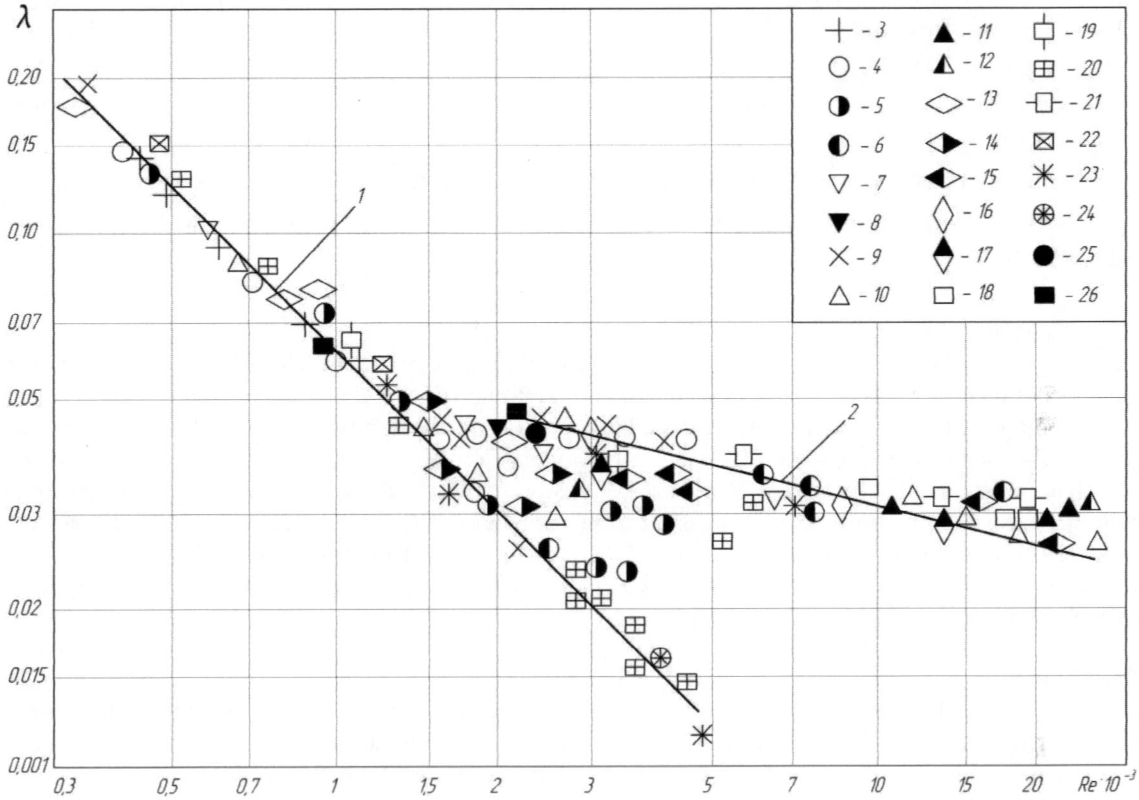

Figure 2. The dependence of the linear coefficient of hydraulic resistance of the capillaries under conditions of isothermal flow: 1 – the dependence of Poiseuille; 2 – the dependence of Blasius; 3...26 - experimental data obtained in prolivka capillaries

6. Flow of fluid from the capillary

The expiration of a short capillary with a sharp entrance edge has a number of features, due to the formation of an end of an isolated cavity. In the experiment determined the minimum relative length of capillary *l/d* in which it ceases to affect the impact of the end of an isolated cavity on the discharge and the dependences for estimating the boundaries of the transition to stall at $\mathrm{Re} < 10^4$ and cavitation $\mathrm{Re} > 10^4$ the conditions of the flow through the cavitation number:

$$\aleph_{cav} = \xi\left(\frac{p_0 - p_1^{\min}}{p_0 - p_2}\right) = \frac{1}{\mu^2}\,, \tag{15}$$

as well as the boundaries of the transition to shear mode expiry Re > 104:

$$\aleph_{ct} = \xi\left(\frac{p_0 - p_1^{\min}}{p_0 - p_2}\right) = \left[\frac{\mu^2\,(2+\xi_0)}{\xi}\right]^{-1}. \tag{16}$$

In expressions (15), (16) p_0 and p_2 - the pressure at the inlet and at the outlet of the capillary, respectively; p_1^{min} - the pressure in the narrow section of the capillary in the presence of an end of an isolated cavity. Comparison of calculated with experimental data [2, 4, 7] numbers of cavitation theoretically defined according to the relations (15) and (16), shows their satisfactory agreement.

Important when using jet schemes of mixing is to estimate the lower limits of applicability of these schemes - estimation of the minimum cost at which the jet is realized the regime of flow of the capillaries. The analysis shows that a jet is formed if the velocity head greater than the pressure created on the interface surface tension:

$$0{,}5\,\rho\,u_0^2 > 4\frac{\sigma}{d} \quad \text{or} \quad We = \rho\,u_0^2\,\frac{d}{\sigma} > 8\ . \tag{17}$$

Experimental test performed when the capillary was poured with water, confirmed the validity of (17), however, showed that the degeneration of the jet occurs at smaller numbers We, and education with large numbers We.

7. Conclusion

Thus, a pilot study was proposed the method of calculation of hydraulic characteristics of capillary elements of mixing heads jet thrusters in an isothermal liquid flow, and also investigated the boundary cavitation and jetting regime of flow of the capillaries. Based on the results of a study of the dependences to calculate the end losses in laminar and turbulent flow regimes in capillary tubes with smooth and sharp edges input. The influence of surface tension forces on the input losses on the sharp front edge. Experimentally confirmed the possibility of calculating the linear coefficient of hydraulic resistance of capillary tubes with a diameter of 0,16...0,36 mm at isothermal stable during the known dependence of Poiseuille and Blahsiusa, fair round for hydrodynamically smooth pipes (the roughness less than ~ 1 μm).

References

[1] Godlewski V E, Nigodjuk V E and Sulinov A V 2009 Features of calculation of hydraulic characteristics of capillary nozzle elements of liquid rocket engines of small thrust *Vestnik of the Samara State Aerospace University* (Samara) No 3 (19), part 2, pp. 241-247.

[2] Stoczek N P and Shapiro A C 1978 Hydraulics LRE (Moscow Mashinostroenie) p 118.

[3] Godlewski V E, Nigodjuk V E and Sulinov A V 1983 Features of calculation of hydraulic resistance and heat transfer for single-phase and two-phase flows in capillary tubes Engineering-physics journal of (Minsk) T. 45, No 2, pp 327. (Manuscript, 16 p, ill., bibliographer. 17 the name. DEP. 1.04.83 in VINITI, No 1670).

[4] Geller Z I and Skobeltsyn Y A 1963 The expiration of a real fluid from a long and an extremely short external cylindrical nasdaw Proceedings of the Universities: Oil and gas No 8, pp. 77-88.

[5] Altshul D A and Kiselev P G 1975 Hydraulics and aerodynamics (MoscowStroyizdat) p 323.

[6] Subbotin V I, Ibragimov M H, Taranov G S, Gusakov V I 1977 Turbulent flow (Moscow Nauka) pp 64-69.

[7] Siow B N 1968 Flow of fluid through the nozzles in the media backpressure. (Moscow Mashinostroenie) p 216.

Methods and Algorithms for Computer-aided Engineering of Die Tooling of Compressor Blades from Titanium Alloy

A I Khaimovich[1] **and I N Khaimovich**[1]

[1] Samara National Research University, 34, Moskovskoe shosse, Samara, 443086, Russia

E-mail: kovalek68@mail.ru

Abstract. The articles provides the calculation algorithms for blank design and die forming fitting to produce the compressor blades for aircraft engines. The design system proposed in the article allows generating drafts of trimming and reducing dies automatically, leading to significant reduction of work preparation time. The detailed analysis of the blade structural elements features was carried out, the taken limitations and technological solutions allowed to form generalized algorithms of forming parting stamp face over the entire circuit of the engraving for different configurations of die forgings. The author worked out the algorithms and programs to calculate three dimensional point locations describing the configuration of die cavity.

1. Introduction

It is known that the design of 3-D models of reducing and cutting dies for volumetric punching of blades is not a fully formalized process, which in many ways stands on the experience of the designer and technologist depending on the applied practice in production, and is not always amenable to automation [1-3]. In this connection, the purpose of this work is to generalize the experience of developing CAD die tooling, which would implement a balanced approach to the combination of a formalized automated process of die geometry determination with the possibility of its interactive correction based on production experience and the possibility of geometry optimization, for example, using CAD-systems (Deform 3D, Qantorform et al. [4-8]).

The worked out system consists of three consequential calculations of units: development of die forging, calculation of generic mathematical model (GMM) of final reducing die and unit of technological equipment engineering of the second kind (test pattern and others). The first unit provides for creation of standard drawing of die forging without 3-D model with the possibility to display it on hard copy. Such representation of calculation results allows, without violating the tradition of enterprises, to agree on the shape and dimensions of the forging with all interested services and industries. In order to increase the versatility of the program, the size of the allowances on all blade surfaces is not related to normative documents, which differ markedly in different enterprises. Therefore, allowances can vary within wide limits and are set by the user in the initial data. The necessary overall dimensions and point location of the blade section are entered from the design drawing of the blade in manual and automatic modes. To increase the visibility of input and reduce subjective errors, the system provides a series of slides. Assignment of allowances, stamping inclinations, alignment block (pad), and creation of forging drawing with all sizes are performed automatically.

IOP Conf. Series: Materials Science and Engineering **302** (2018) 012062 doi:10.1088/1757-899X/302/1/012062

The received dimensions of the forging design in automatic mode are transferred to the unit for , calculation of generic mathematical model (GMM) of final reducing die. CAD has an "input validation" unit, which allows you to apply manual input, excluding the forging design block, to change any work size after automatic input, optionally enter the missing parameters. The described principles of CAD design of volumetric stamping of GTE blades are implemented in software.

2. Methods and algorithms of research

According to the results of the practice of using the system and its technical approbation, it is possible to single out several key points concerning the processes of formalized determination of the die geometry and construction of its GMM.

1, Select the shape of the gutter geometry. The form of the gutter linker along the blade edges in the system under consideration is adopted in the form of a natural extension of the profile in each section. In this case, the linker is tapering and does not have sharp pips along the edge with the forgings, which increases the durability of the die. The width of the linker and its final thickness are variables and are set in manual mode.

2, Solve the task of selecting a line of connector dies around the blade shank and alignment block (pad), which is the most complex and multivariate. The analysis of the large nomenclature of forgings has shown that, depending on the blade design, three basic variants of blade crossing with the tractor's shank surface can be distinguished. Algorithms and programs have been developed to implement each of these options.

In practice, three types of connector dies are used: parallel, stepped, inclined. Inclined and stepped connectors allow you to adjust the depth of the die cavity in the upper and lower liners. The stepped connector simplifies the manufacture and control of the stamp. The gradient of the step levels is carried out at a small angle - 5-7 degrees. However, this connector complicates the stamp setting and worsens the quality of the burr cut. A small gap in the vertical section of the connector (0.1 ... 0.3 mm) often leads to a collision of dies with a slight displacement.

The proposed algorithm provides for the use of only two kinds of connectors: parallel or inclined.

The end shank surface from the side of the input edge of the blade in the lower liner (control base) is formed by turning the forgings in the stamp by an angle of TAY3. A similar surface in the top liner is at an angle (ALNEY + TAY3), where ALNEY is the selected stamping gradient. In the calculations, this angle is assumed to be constant. This design of inclines leads to the formation of a "cover" on the forging of a variable width, which is partially removed when burring.

Stamping incline in the lower liner on the shank end from the side of the blade output edge is always assumed to be constant (ALNEY). To control the displacement of dies in the transverse direction, the stamping gradient in the upper liner is assumed to be overestimated.

Here and further all profile surfaces and die connector surfaces are calculated taking into account the "hot" dimensions of forgings, which are determined by the coefficient of linear expansion in the range of forging temperatures.

The key problem in the system of die tooling is to choose die lines around the blade root and alignment block (pad), which is the most complicated and multivariate. The analysis of a great number of forging showed that depending on the blade construction it is possible to point out three main variants of blade intersection with path surface of shank end. There are algorithms and programs for realization each of these variants.

The design of parting face starts from the side of input edge of the blade (Fig.1).

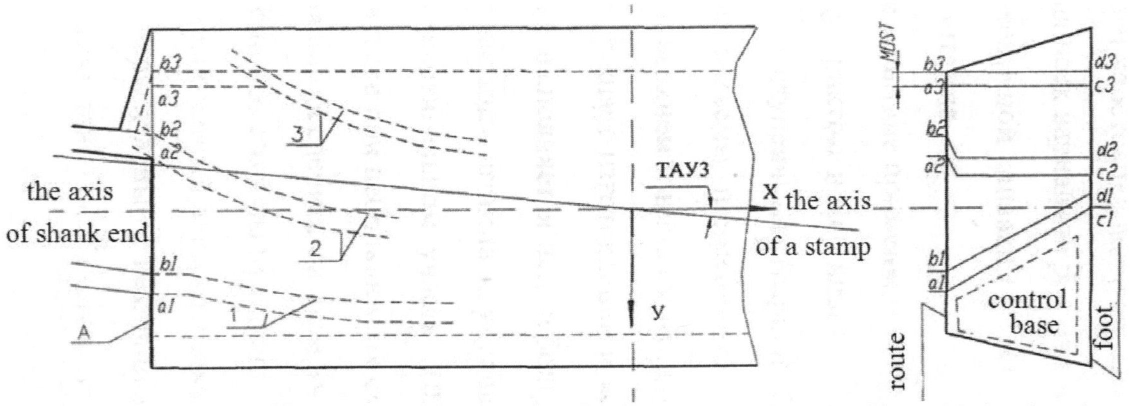

Figure1. Choosing parting of a die along route and shank end from the side of input blade edge: A – the surface of lock edge, 1,2,3 – the variants of blade crossing the route surface of lock.

1 variant. Limiting points of die impression blade stay within the surface of the shank end (surface A, (fig.1)). In this case the connector along route is set as parallel to middle plane of shank end and break the surface A in the point of al (lower insert) and b1 (upper insert). The further direction of a connector lies in axis of the stamp.

2 variant. Limiting points of die impression blade exceed the limits of the surface of the shank end. In this case the connector along route (points a2 и b2) lies at the intersection of A surface.

3 variant. Limiting points of die impression blade exceed the limits of route. This is the most complex variant of connector. In order to upload the configuration of the shank, it is provided for reduce the blade width, while ensuring the necessary allowance along the edge. The connector along the route runs parallel to the axis of the shank at the points of a3 and b3.

For all variants of connector there is the same thickness of flash gutter linker (MOST), accepted in the initial stamp design data.

In practice, three types of connector dies are used along the shank end between rout and foot: parallel (variant 3), stepped (variant 2), inclined (variant 1). Inclined and stepped connectors allow you to adjust the depth of the die cavity in the upper and lower liners. The stepped connector simplifies the manufacture and control of the stamp. The gradient of the step levels is carried out at a small angle of 5-7 degrees. However, this connector complicates the setting of the stamp and worsens the quality of the burring. A small gap in the vertical section of the connector (0.1 ... 0.3 mm) often leads to a collision of dies with a slight displacement. While burring it is practically impossible to provide optimized gap and the form of cutting edge, especially from the side of a top die.

The proposed algorithm provides for the use of only two kinds of connectors: parallel or inclined. Using an inclined connector allows solving two problems: create a sufficient surface (control base) on the end face of the lock, which the forging to be reliably set in the control device; ensure a greater depth of the die cavity in the lower insert compared with the top one.

The end surface of the shank in the lower insert (control base) forms a natural draft due to the turn of the forging in the stamp by an angle of TAY3. In this case side surface of a shank in the top insert is formed at an angle (ALNEY + TAY3) to axis of a shank, where ALNEY is the selected stamping gradient (fig. 3). In the calculations, this angle is taken as constant for route surface of a shank. This design of inclines leads to the formation of a "visor" on the forging of a variable width, which is partially removed when burring.

Then a connector of dies is formed along the shank end from the side of the leading edge of the blade.

The variants of intersection of a blade with the surface of shank end from the side of trailing edge are similar to those examined earlier for leading edge. Figure 2 schematically shows three possible options. Unlike the leading edge, parting of a die along route and further in flash gutter is formed parallel to the die axis, i.e. at the angle of TAУ3 to the shank axis. Thus, the output of the connector on the lock end is determined by the points of a4 - b4 (variant 1), a5 - b5 (variant 2), a6 - b6 (variant 3). The position of connector die along the shank end i.e. connector output to the foot is selected from the following conditions:

- if connector die along the shank end is located over the shank axis (for example, variants 2 and 3 at the figure 2), then connector at the end is parallel;
- if connector die is lower the shank axis, then connector is inclined opening onto the shank axis at the point of c4 (for example, variant 1 at the fig.2).

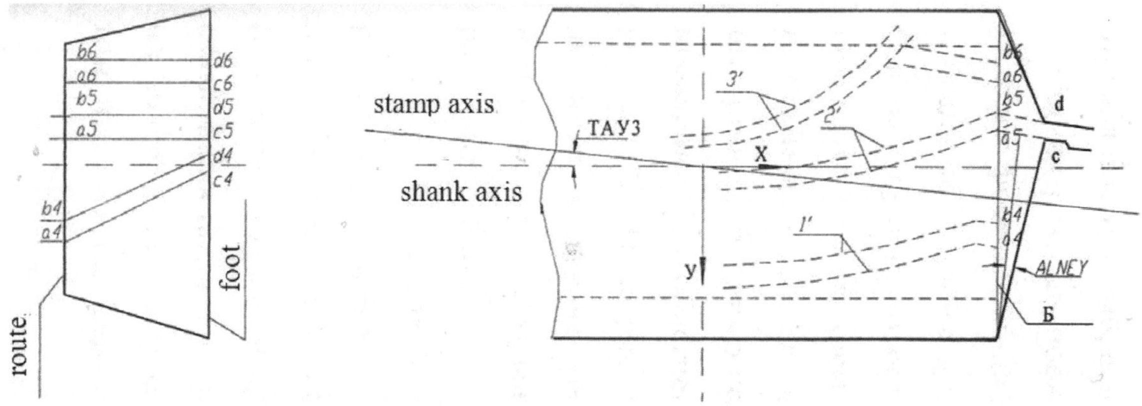

Figure 2. Choosing parting of a die on the shank end from the side of trailing edge of a blade airfoil: Б – the surface of shank lock, 1',2',3' – variants of intercrossing blade and route surface of a lock.

Stamping incline in the bottom insert in the shank end (fig. 3) is assumed to be constant (ALNEY), regardless the variant under consideration. To control the displacement of dies in the transverse direction during stamp setting, it is accepted that lower die cavity (point at the edge c in fig. 2) must lie in one vertical plane with upper die cavity (point at upper die d in fig. 2). In this case stamp draft in the upper die cavity will always be more than ALNEY. At the same time if connector of shank end is parallel, then the angle is constant, if it is inclined, then the angle is alternate.

The connector line runs along the foot of the shank. Taking into account the accepted conditions for the formation of the connector at the of the shank ends, we find out that the die connector at the foot in most cases is not parallel to the shank axis. However, the depth of the cavity in the lower die cavity will be greater than the corresponding depth in the upper die cavity. Considering this circumstance, the stamping gradient in the lower die cavity is assumed to be constant (ALNEY). When installing stamps, it is necessary to control the mutual displacement in the longitudinal direction. To make it easier to control, it is desirable that the cutting edges of the burr (B and H in Figure 3) are in a vertical plane. This is achieved by assigning a variable angle of the stamping bias in the top insert, which is larger than the preset one and smoothly changes along the die splitting line. The described algorithms for forming a die connector on the shank ends uniquely determine the connector on the foot: in all cases the depth of the cavity in the lower die will be greater than the corresponding depth in the upper one. Given this circumstance, the stamping gradient in the lower liner is assumed to be constant (ALNEY), and in the upper it is variable. This approach allows combining the stamp edges and using the foot of the forging shank to control the displacement of the liners in the longitudinal direction.

Let's consider the formation of parting plane around a technological pad. The technological pad is situated outside the working part of a blade. To simplify the design of stamp the connector around the pad is made flat. However in this case it is necessary to apply vertical (stepped) connector moving from a blade to a pad, which has major deficiencies. When developing CAD-system for stamp another decision concerning the formation of connector surface is suggested and carried out. The key point lies in the fact that aerofoil of top section of blade is spread into the area of pad.

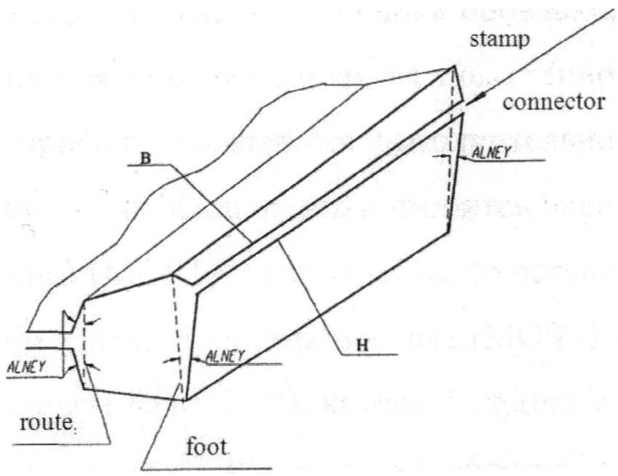

Figure 3. The formation of stamp inclines at foot and the route of shank end.

Using such method the stamp connector will be curvilineal, and its exit onto the side surface of pad will depend on the angle of top section turning. To obtain the estimated dimensions of the linker of the burring groove, it is necessary to make a reduction in the blade thickness around the pad due to suction or pressure side. The choice of the appropriate option depends on the size of 11 (Fig. 4), which determines the site of the forging control base. It is accepted that: if 11≤0.4K, where K is the thickness of the pad, then the thickness of the blade is reduced due to the suction side located in the lower liner. In this case, the profile of the pressure side without additional twist extends to the entire pad, and the profile of the suction side is built equidistantly, at a distance of the thickness of the linker (MOST). If 11> 0.4K, the profile of the suction side remains unchanged. The required linker thickness (MOST) is created by changing the profile of the t pressure side (Figure 5), which is constructed equidistantly to the profile of the suction side. To maximize the overlapping edges of the burr trim, it is possible to slightly reduce the width of the pad (C) by 1 ≤ 2 mm (Figure 4.5). To implement all the described conditions and limitations, appropriate algorithms and software tools have been developed.

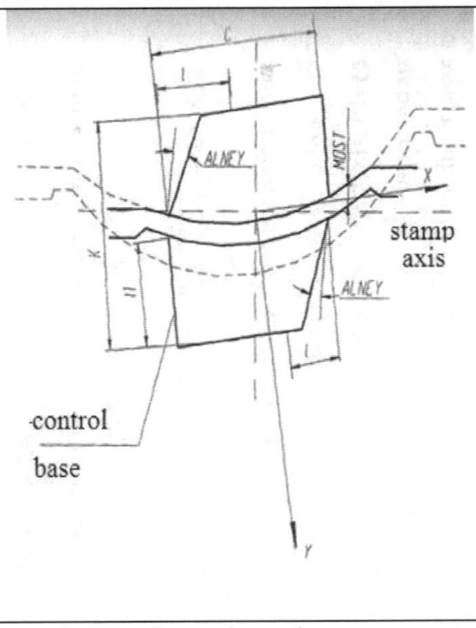

Figure 4. The formation of stamp connector on the pad with insufficient area of control base.

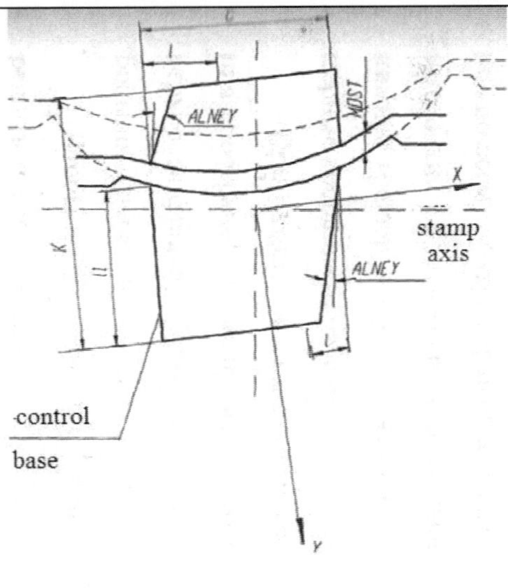

Figure 5. The formation of stamp connector on the pad with sufficient area of control base.

In the case of a curved connector of dies (forming the surface of a pad by prolonging the blade), its exit to the side surfaces of the pad depends on the angle of twist of the top section of the blade. To improve the quality of the burr cut, it is provided to reduce the thickness of the blade around this structural element by truncating the profile from suction or pressure side. If the area of the forging control base on the pad is not sufficient, then the thickness of the blade is reduced due to the suction side. In this case, the profile of the pressure side without additional twist extends to the entire pad, and the profile of the suction side is built equidistantly, at a distance of the thickness of the linker (MOST). Otherwise, the profile of pressure side remains unchanged. The required thickness of the linker is created by changing the profile of the trough.

3. Conclusion

Analysis of the blade from titanium alloys [9, 10] structural elements features, taken limitations and technological solutions allowed to form the surface of parting of a die throughout the engraving contour. Algorithms and programs were developed for calculating the three-dimensional coordinates of points describing the configuration of the die cavity, the base surfaces for the installation of control templates, the shape and overall dimensions of the top and lower inserts of the stamp. As a result, a generic mathematical model of the final stamp was obtained in the form of a three-dimensional array of base points joined by edges in the form of straight lines or splines. This model is the basis for the creation of design documentation for technological equipment and its monitoring tools.

References

[1] Khaimovich I N 2014 *Russ. Aeronaut.* **57** 169-174
[2] Kostyshev V A, Khaimovich A I 2015 *Russian Journal of Non-Ferrous Metals.* **56 (2)** 181-189
[3] Grechnikov F V, Khaimovich A I 2016 *Metal Science and Heat Treatment.* **57 (11)** 726-730
[4] E G Demyanenko *et al* 2017 *IOP Conf. Ser.: Mater. Sci. Eng.* **177** 012122
[5] Khaimovich A I, Surkov O S, Khaimovich I N 2016 *Russ. Aeronaut.* **59** 112-117
[6] Erisov Y A, Grechnikov F V and Surudin S V 2016 *Structural Eng. and Mech.* **58** 677-687
[7] Dmitriev A M and Vorontsov A L 2004 *Approximation of hardening curves of metals KSP OMD* ***1*** *23-26*
[8] Haimovich I N and Frolov M A 2016 *Key Eng. Mat.* **684** 487-507
[9] Demyanenko E G and Popov I P 2016 *Key Eng. Mat.* **684** 253-256
[10] Yakovlev S S, Remnev K S and Tregubov V I 2012 *Izvestiya TulGU. Technical Sci.* **4** 98-102

Method to Evaluate Measuring Uncertainty of Probe Radius Compensation with Account for Variation in Measuring Surfaces

Nikolay V. Ruzanov[1], Vadim A. Pechenin[1,*] and Michael A. Bolotov[1]

[1] Samara National Research University, Institute of Engine and Power Plant Engineering, Moskovskoye shosse, 34, Samara, Samara region, 443086, Russian Federation,

*E-mail: v.a.pechenin@ssau.ru

Abstract. This article describes a methodology of evaluating the uncertainty of the contact measurement of freeform surfaces with coordinate measuring machines, taking into account the variation in the form of measured surfaces. This variation is due to machining uncertainties, which appear during production using a particular machining method and technology. Probabilistic estimates of errors of determination of contact points for freeform surfaces were obtained. Use of this model makes it possible to obtain interval estimates of errors with the required confidence level of probability and mathematical expectation. The relations of the probabilistic error estimates at any of the measured points on surface are obtained.

1. Introduction

In industry, coordinate measuring machines (CMMs) are widely used for the measurement of surfaces; they can be equipped with contact and non-contact measuring systems. CMMs are flexible measurement tools which allow one to measure various parts, including a free-form surface (The accuracy of which can be difficult to obtain, and for the production of which are used in the technologies of injection molding [1] and additive technologies [2]) with high accuracy. In addition to production necessity, precise measurements of parts allow different calculations for their work [3, 4].If the accuracy of freeform surface measurement is the decisive factor for a product (e.g., blades of gas turbine engines (GTE)), the measurement is usually carried out by CMM implementing a contact measurement method. In the practice of controlling freeform profile parts, there are significant measurement uncertainties, which are difficult to confirm experimentally. In connection with this, it is actual to develop a method for estimating measurement uncertainties, which are necessary for the development of control programs for CMMs that provide the required accuracy.

The contact measuring method is performed by a contact of the probe tip with the surface of the measured part or product, which as a rule, has a spherical tip. After contact, the centre point of the measuring tip is fixed and then touch point is calculated. As a result of the set of measurements, a cloud of points is created. The cloud of points is used to define the surfaces, planes, lines and points of the parts, which are required to determine the geometric parameters of the product. When we measure surfaces with complex geometry, the accuracy of probing on CMMs is determined by many factors. One of the key factors affecting the accuracy of measurement is the radius of the probe tip (group of coordinate measuring system hardware) and curvature of part surface (a group of the workpiece features).

The greater the curvature of the measured surface and the larger the radius of the probe tip, the greater the measurement error of the determination of the actual part points is. The magnitude of the measurement error can be estimated by experiment (using an etalon surface) and by the method of mathematical modelling of the measuring process [5, 6].

Machined free form surfaces of work-piece are always different from their CAD-models by the value of form and location deviations [7]. Because of these variations, the coordinates of the probe tip, which are calculated to normal to CAD model from the centre point of the probe tip, are determined with an deviation. Figure 1 presents a measurement circuit for free-form surface with form and location deviations.

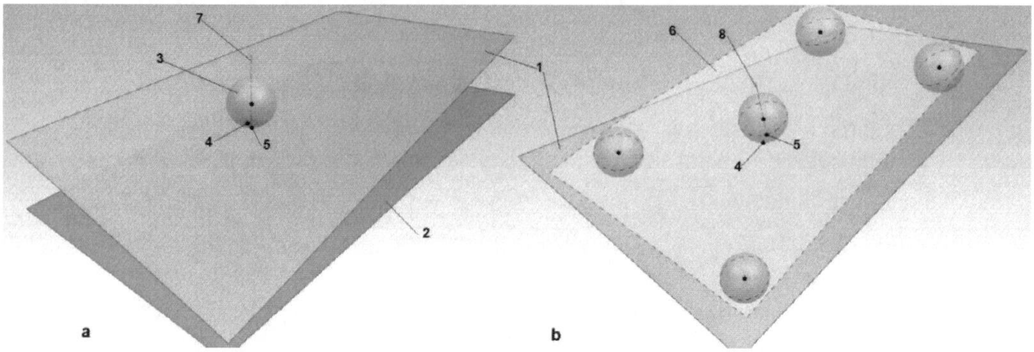

Figure 1. The error of surface 1 measurement by the probe tip 3 (4 -a touch point, 5- a measured (calculated) point) caused by: a) discrepancy between the nominal direction of the normal 7 to CAD surface 2 and real direction at the point on surface; b) mismatch between the normal vector 8 to offset surface 6 and the measured surface 1.

In the practice of coordinate measurements of geometric parameters of parts, there are situations when the standard of comparison (the CAD-model) for measured surfaces is absent (Figure 1.b). In the case, the probe tip radius compensation is carried out on the normal to the offset surface as defined by the centre points of the probe tip [8].

This paper describes the method of evaluating the uncertainty of contact measurement of freeform surfaces (blade feather) with coordinate measuring machines (CMM). This method takes into account variations in the form of measured surfaces. This variation is due to machining uncertainties, which appear in the process of manufacture by the method of machining and technology. Our developed method allows simulating the form and location deviation of the surface and calculating the coordinates of touch points, probe tip centres and measured points.

2. The method of evaluating measurement error with consideration to variations in the form of measured surfaces

We developed a method to assess the accuracy of measurement in accordance with stochastic principles. Figure 2 shows a flowchart of the method that takes into account the degree of dispersion of the parameter of form and location deviations for free form surfaces. The method uses the Monte Carlo method with a sequential repetition of steps 3, 4 and 5. Further sections describe the steps of this procedure.

2.1. Mathematical description of surfaces

To build the coordinate measurements model it is necessary to describe mathematically the details of the measured surface and the respective CAD-model. Nevertheless, the condition of smoothness at all points of the described surfaces is required to be fulfilled. In this paper we use the tool, which is the standard for description of free-form surfaces with the CAD programs, the so-called NURBS-surface [9] and its equation can be written as:

$$P(u,v)=\sum_{i=1}^{n}\sum_{j=1}^{m}h_{ij}\cdot P_{ij}\cdot N_{ik}(u)\cdot N_{jl}(v)\Bigg/\sum_{i=1}^{n}\sum_{j=1}^{m}h_{ij}\cdot N_{ik}(u)\cdot N_{jl}(v),\tag{1}$$

where u is an option for calculation of spline points coordinates varying in the range of $[t_k\text{-}1,\ t_n+1]$;

h_i are homogeneous coordinates of setting points; P_{ij} are the coordinates (x_i, y_i, z_i) of the i-th setting point in three-dimensional space; $N_{ik}(u)$, $N_{jl}(v)$ are basic spline functions (or fitting) in the parametric direction; t are nodal values of parameter whereat the functions $N_{ik}(u)$ are not zero; k is the degree of the piecewise spline in the direction of the parameter u; v is an option for calculation of the coordinates of the spline points varying in the range of $[s_{l-1};\ s_{m+1}]$; s are nodal parameter values where the functions $N_{jl}(v)$ are different from zero; l is the degree of the piecewise spline in the direction of the parameter v.

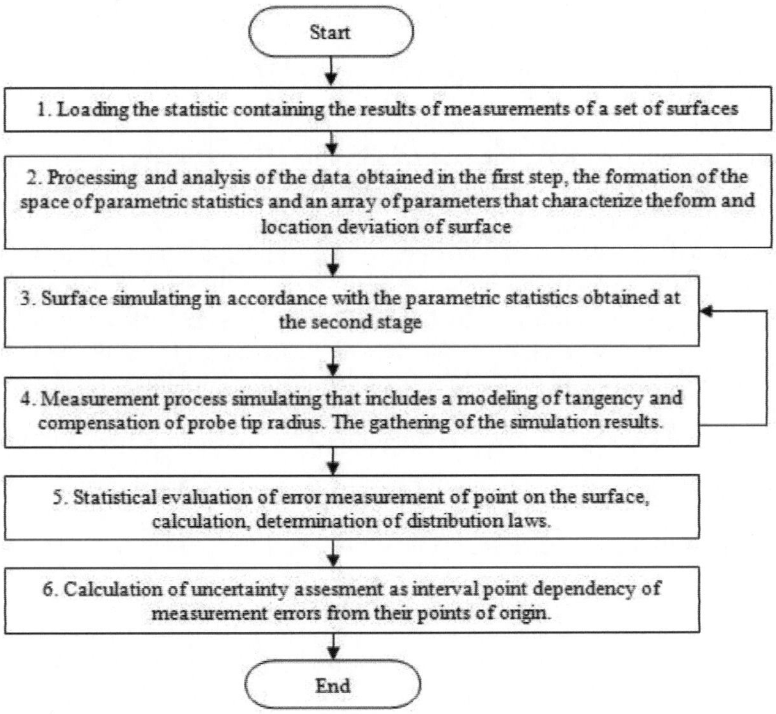

Figure 2. A flowchart of estimate method of measurement error with consideration of variation in measured surface form.

2.2. Simulated deviation of form and location

The points of the measured surfaces were simulated from the nominal by adding the form and location deviations to the coordinates of the nominal surfaces points. The error of the surface form at a high milling speed depends on the mechanical properties of the process material in the cutting area, which can be determined by the method described in [10]. The values of the form and location deviations were chosen on the basis of statistical observations for the series of blades.

Thus, the point coordinate of the measured surface can be expressed by the equation:

$$\vec{P}_{meas}=(\vec{P}_{CAD}+\vec{N}_{1\times3}\cdot\delta_F)\cdot R_{3\times3}+\vec{T}_{1\times3},\tag{2}$$

where $\vec{P}_{meas}$, $\vec{P}_{CAD}$ are coordinate vectors (x, y, z) of measured (simulated) and nominal surface, respectively; $\vec{N}_{1\times3}$ is the normal vector at the $\vec{P}_{CAD}$ point; δ_F is form deviation at $\vec{P}_{CAD}$ point; $R_{3\times3}$, $\vec{T}_{1\times3}$ are rotate matrix and the translate vector coordinates of a $\vec{P}_{CAD}$ point.

The matrix $R_{3\times3}$ contains three angles of rotation around the coordinate axes. The $\vec{T}_{1\times3}$ vector contains the values of displacement along the coordinate axes.
In parts production practice there are two components of the form deviation, such as, systematic and random errors. Consequently, these two types of deviations were included in the model of the measured surface. Systematic deviations consist of the sinusoidal form deviation and the machining form deviation [11].

Thus, the total form deviation at each point can be written as the sum of three components:

$$\delta_F = \delta_s + \delta_m + \delta_r . \quad (3)$$

Consider each of these components.

2.2.1. Sinusoidal form deviation. Harmonic form deviation can be approximated using the composition of the sine and cosine functions by the next equation:

$$\delta_s = A \cdot \sin(w_x \cdot x + w_y \cdot y + \varphi_{\sin}) + B \cdot \cos(w_x \cdot x + w_y \cdot y + \varphi_{\cos}), \quad (4)$$

where A and B are the amplitudes of the sine and cosine components, respectively,
w_x is given by $\omega_x \cdot 2\pi / L_x$, w_y is given by $\omega_y \cdot 2\pi / L_y$;
ω_x, ω_y are frequencies of harmonic components along the x and y direction, respectively,
L_x, L_y are bearing lengths along the x and y direction, respectively;
$\varphi_{\sin}$, $\varphi_{\cos}$ are angle phase of the sine and cosine.

*2.2.2. Machining form deviation.*The form deviation due to the machining process is the result of curvature measurement of the processed object when it is in contact with the surface of the cutting tool. The distribution of the error can be calculated basing on the values of the average curvature:

$$\delta_m = A_m \cdot (i_s - 0.5), \quad (5)$$

where A_m is the maximum error of machining process.

The index i_s is calculated by the equation:

$$i_s = (H - H_{min})/(H_{max} - H_{min}), \quad (6)$$

where H, $H_{\min}$ and $H_{\max}$ are values of the average curvature at the point on the surface, the minimum and maximum values of the average curvature [12] over the biparametric surface, respectively.

2.2.3. Random error. Random errors caused by numerous factors are small in their individual impact on the results and they cannot be taken into account during an experiment. Through multiple repeated measurements of the same non-random quantity the random measurement errors result in different instrumental data. Dispersion of the results of these instrumental data is usually subject to the Gaussian law. For the form deviation the value of the random component of the deviation δ_r is calculated irregularly with maximum amplitude of 0.005 mm.

2.3. Computation of touch point coordinates and probe tip center
We use a mathematical model permitting calculation of the point coordinates in measuring the gear surfaces in a contact coordinate-measurement machine [13]. Deflection of the coordinates of the point on the surface relative to the coordinates of the measured point characterizes the error of compensation of probe tip radius at the i-th measurement point:

$$\delta_{Rcomp.} = \|C_i - D_i\|, \tag{7}$$

where C_i are coordinates (*x*, *y*, *z*) of the real point of contact of the probe tip with the surface;

D_i are coordinates (*x*, *y*, *z*) of the measured point calculated by the equidistant surface or by CAD-model of reference surface.

3. Simulation Results

Simulation of the measurement process using the developed technique was carried out by the example of the suction and pressure sides of the blade of the GTE compressor.

A pen of GTE compressor blade is a blade airfoil portion placed in the stream of air (gas). The pen has two major surfaces: the suction and the pressure sides. The surfaces represent freeform surfaces.

Measurement surfaces and respective nominal surfaces of blade feather are simulated in MATLAB software package for a variety of coordinate points measured by CMMs DEA Global Performance 07.10.07 and the corresponding points taken from the CAD-model of the blade. The radius of the probe tip for measurement and experiments is 1 mm. One hundred surfaces of suction and pressure sides were modeled; the amplitudes of its form deviations were varying in the normal law. On the basis of statistical data of this type components measurement it was found that the maximum form deviation is not more than 0.2 mm. The range of *A* and *B* amplitudes (Equation 4) was 0.08 mm, the range of values of A_m (Equation 5) equals 0.1 mm. The phase and frequency (Equation 4) of harmonic component vary in the Equiprobable law. The phases range from 0° to 360°, the frequency is 1, 2, 3 and 4. Transposition matrix components changed under the normal law; displacement along the x-axis is ± 0.02 mm; the displacement along the y-axis is ± 0.2mm; the displacement along the z-axis is ± 0.15 mm. The angles of rotation surface vary in the normal law; the angle around the x-axis is ± 9 '; the angle around the y-axis is ± 3 '; the angle around the z-axis is ± 6 '.

Series of measurements and calculation of errors of the probe tip radius compensation at points were carried out with the use of the developed model for the 100 pressure sides and 100 suction sides. In measurement of surfaces we used the strategy of uniform distribution of control points on the measured surface, which is the simplest and widely used in practice. The distribution of errors at the points is subject to two basic laws: the normal and beta-distribution. Thereafter, expectation was calculated at each measured point, as well as, upper and lower estimate of the error of measurement of point coordinates at a confidence level of 99.73%. According to the obtained pointwise values of the middle, lower and upper boundaries of probe tip radius compensation error for the surfaces were defined using bilinear interpolation (Figure 3) in *u*-*v* parametric coordinates (Equation 1).

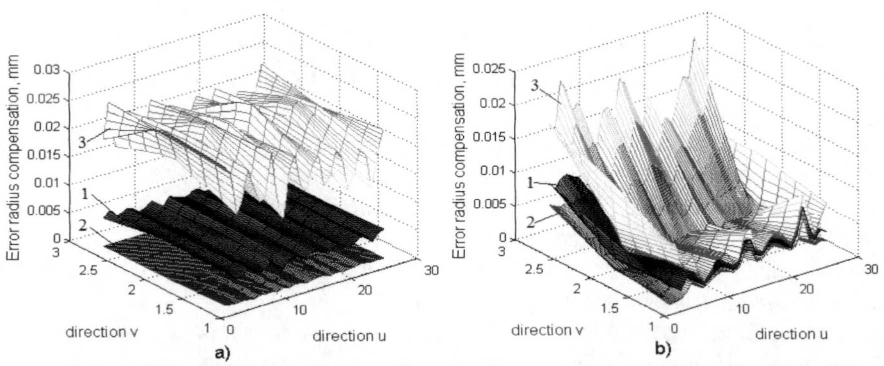

Figure 3. The probability evaluation of error values for the surface of the suction side a) with the use of CAD б) with the use of equidistance; 1, 2, 3 are the middle, lower, upper surfaces of random function propagation.

Thus, during the simulation we obtain dependencies, which allow determining the average, lower and upper propagation boundaries of the error of the probe tip radius compensation for any measured point on the surface. Table 1 shows the marginal probability values of errors of determination of the measured points coordinates obtained by the processing of statistical data (the expectation, the lower and upper limits of error variation) for the considered surfaces.

Table 1. Probability values of errors of the measured points determination.

The object of measurement	The number of measured points, n_u×n_v	Error with CAD, µm			Error without CAD, µm		
		Mean	The lower bound	The upper bound	Mean	The lower bound	The upper bound
The suction side of blade	28×3	7,0	0,5	26,7	10,7	6,8	27,2
	56×3	8,3	0,6	33,3	9,9	4,6	25,8
The pressure side of blade	28×3	6,0	0,5	28,2	39,5	35,3	48,0
	56×3	7,8	0,7	30,3	28,5	21,2	35,8

As the table shows, the maximum value of errors of the determination of the compensation vector by the method of probe tip radius compensation using a CAD model increases. It is due to the fact that the probability to get to the area with a strong deviation from the standard increases. In case of computation of the coordinates of the measured points on the equidistant surface the limit errors value is reduces with increasing points and the method becomes comparable in accuracy with the measurement by the method of CAD.

4. Conclusion

This article describes the developed method of obtaining probabilistic estimates of the errors of the probe tip radius compensation for the measurement of freeform surfaces with the coordinate measuring machine, which takes into account variations of the measured surfaces. The developed method can be used for propagation of measurement errors of geometrical parameters of engineering parts surfaces and also for the formation of the optimal sampling to control the geometry as a part of the optimizer.

Acknowledgments

This work was supported by the Ministry of Education and Science of the Russian Federation in the framework of the implementation of the Program ''Research and development on priority directions of scientific-technological complex of Russia for 2014– 2020.

References

[1] Stepanenko I S and Khaimovich, A I 2017 GTE blade injection moulding modeling and verification of models during process approbation *IOP Conference Series: Materials Science and Engineering* vol 177 no 1 012093

[2] Smelov V G, Sotov A V and Agapovichev A V 2016 Recovery technology features of aerospace parts by layering synthesis *Key Engineering Materials* vol 684 pp 316-322

[3] Zubanov V, Shabliy L and Krivcov A 2014 Centrifugal kerosene pump CFD-modeling *Research Journal of Applied Sciences* vol 9 no 10 pp 629-634

[4] Zubanov V, Shabliy L, Krivcov A 2014 Centrifugal kerosene pump CFD-modeling *Research Journal of Applied Sciences* vol 9 issue 10 pp 629-634

[5] Zakharov O V, Bobrovskij I N and Kochetkov A V 2016 Analysis of methods for estimation of machine workpiece roundness *Procedia Engineering* vol 150 pp 963 – 968

[6] Zakharov O V, Bobrovsky N M, Kochetkov A V, Grigoriev S N and Bobrovsky I N 2016 A sphericity measurement method based on the minimum measuring zone *AIP Conference Proceedings (Electronic Materials* vol 1785) 040094

[7] Savio E, De Chiffre L and Schmitt R 2007 Metrology of freeform shaped parts *CIRP Annals - Manufacturing Technology* vol 56 no 2 pp 810-835.

[8] Mayer J, Mir Y, Trochu F, Vafaeesefat A and Balazinski M 1997 Touch probe radius compensation for coordinate measurement using kriging interpolation *Proceedings of the Institution of Mechanical Engineers, Part B: Journal of Engineering Manufacture* vol 211 no 1 pp 11-18.
[9] Piegl L and Tiller W 1997 *The NURBS Book* (Berlin: Springer-Verlag)
[10] Rajamohan G, Shunmugam M S and Samuel G L 2011 Practical measurement strategies for verification of freeform surfaces using coordinate measuring machines, *Metrology and Measurement Systems* vol XVIII no 2 pp 209-222.
[11] Rogers D F and Adams J A 1990 *Mathematical Elements for Computer Graphics* (2nd ed., McGraw-Hill Publishing Company)
[12] Shunmugam M S and Radhakrishnan M S 1976 Comparison of difference methods for computing the two-dimensional envelope for surface finish measurements *Computed Aided Design* vol 8 no 2 pp 89-83
[13] Bolotov M A, Pechenin V A and Ruzanov N V 2016 Uncertainties in measuring the compressor-blade profile in a gas-turbine engine *Russian Engineering Research* vol 36 no 12 pp 1058-1065

IOP Conf. Series: Materials Science and Engineering **302** (2018) 012064 doi:10.1088/1757-899X/302/1/012064

Production of Space-rocket Technique: Psychological Factor

S P Vashchuk[1,a], O A Sviderskiy[1], O N Ezhova[2] and V V Rovenskaya[3]

[1] Samara University, st.Moskovskoe tr. 34, Samara, 443086, Russia
[2] Samara Law Institute of the Federal Service for the Execution of Punishments of Russia,st.Rylskaya 24 «V», Samara, 443022, Russia
[3] Donbass State Engineering Academy, st.Akademicheskaya 72, Kramatorsk, 84313, Ukraine

[a] Corresponding author's e-mail address: amurgermovvod@mail.ru

Abstract. The article is devoted to the issues of studying the mental and physical condition of the shop workers who assemble carrier rockets and ways of processing their internal conflicts. It is shown that the complexity of the ongoing labor processes, the intensity of production activities and the responsibility for its end result lead to a high level of neurotic workers. The tendency to a long experience of a stressful situation helps them to increase the level of various forms of aggression and create a negative assessment of their official capabilities and successes. It is established that the duration of experiencing a stressful situation depends on coping strategies in the behavioral and emotional spheres.

1. Introduction

The history of the cosmonautics development is inextricably linked with the city of Samara.

A special contribution to the production of rocket and space equipment was made by the Progress plant which incorporated the Central Specialized Design Bureau (CSDB) in 1996 and is now called the Joint Stock Company "Rocket Space Center Progress" (Progress RSC JSC).

During its existence, the employees of the Progress RSC JSC developed, produced and operated spacecrafts of both scientific and applied value, many of which are unique and have no other analogues, currently it specializes in the production of middle-class launch vehicles that are used to launch manned and transport spaceships.Launch of carrier rockets is carried out both from the Baikonur cosmodrome and from the new Russian spaceport Vostochny (the first in the Russian Federation that has the status of a civilian cosmodrome), so there is no doubt that the quality of products of Progress JSC is a task of the state importance since the implementation of international and commercial space programs depends on it.

The quality of products is largely determined by the use of innovative developments and the scientific potential of the space industry. However, in our opinion, with all the development and automation of production it is impossible to underestimate the importance of the human factor, namely, their resistance to stress and ability to withstand the great neuropsychic overload associated with stressful nature of the work. On the one hand, the use of information technologies significantly increases the work efficiency, and on the other hand, it increases the amount of information that the employees when assembling carrier rockets should process per day since the quality of the final product depends on the quality of the performed operations (at Progress RSC JSC flow production is used). This all leads to the emergence of psychological stress.

2. Objectives and tasks of analysis

The problem of psychological stress leads to the fact that a person does not cope with the task, does not have time to make the right decisions at the required pace, with a high degree of responsibility for their actions. The purpose of our study is to determine the mental state of employees and ways of processing their internal conflicts. The paper considers a psychodiagnostic complex of techniques that showed the enormous influence of emotions on the well-being, activity and mood of a person, which emphasizes the need to master the ways of regulation of the emotional sphere.

3. Research fundamentals

While getting a job, vocational training is highly assessed but insufficient attention is paid to their psychological training. The work intensity of the Progress RSC employees (often an irregular working day, uneven load) leads to the increase their psychological stress level resulting from a mismatch between the load and available resources, accompanied by negative emotions (fear, anger, frustration etc.). Psychological stress has a dual nature and is divided into information stress and emotional stress. Information stress occurs in situations of information overload, when a person can't cope with the task, does not have time to make the right decisions at the required pace with a high level of responsibility for his/her actions. Emotional stress appears in situations of threat, danger, resentment etc.

Professional stress is a diverse phenomenon, manifested in mental and physical reactions to stressful situations in the work activity of a person [1]. We conducted a study in which employees of the workshops took par, where assembly of carrier rockets (22 people in all) took place. We prepared a psychodiagnostic complex of methods, which include the following methods:

1. HAM methodology, this technique allows to determine the state of health, activity, mood at the moment. While carrying out this procedure, it was told in the instruction that the participants should note their most characteristic condition during the last 6 months.
2. Test of determining the duration of stressful situations.
3. L I Wasserman's method for diagnosing the neuroticism level (studying the features of the nervous system and its propensity for a neurosis).
4. N Pezeshkian's differential analytic questionnaire, from which we took only the block relating to ways of processing an internal conflict. For the convenience of results evaluating we introduced a "conflict conversion factor" (CCF), which allows us to determine the way in which the internal conflict is being processed. The value of the CCF shows whether the person is trying to resolve the emerging conflict (CCF = 5) or trying to compensate it by "running away" from it in one of the four activities (bodybuilding (exercise or illness), work, communication, fantasy) wherein the CCF values vary from 2 to 4 and from 6 to 8.
5. A Bass and A Darki's methodology for diagnosing the indicators and forms of aggression (A K Osnitsky's adaptation).
6. K Maslach and S Jackson's questionnaire MBI ("Professional burnout") adapted by N E Vodopyanova contains statements about the feelings and experiences associated with working activity on which basis three sub-districts (emotional exhaustion, depersonalization and personal achievements) are singled out.
7. Definition questionnaire of behaviour coping strategies respectively in the behavioural, cognitive, emotional spheres.

The HAM methodology showed that the variance in indicators of health, activity and mood in the group was small, on the average, the level of well-being was just above the mean (5.1), the mood slightly below the mean (4.9), activity – 4.3. We tested the relationship between mood and well-being in a group. The correlation coefficient was equal to 0.855 that confirms direct dependence of health status and mood. The test results for the duration of the stressful situation showed that 60% of the examinees had been experiencing a stressful situation for a long time, the rest 40% had recovered after stress quickly. The analysis of the results (Table 1) obtained by L I Wasserman's method made it possible to distinguish the neuroticism level.

Table 1. Diagnosis of the neuroticism level

Neuroticism level	Number of persons	Percentage
Low	7	32
High	15	68

The data in Table 1 show the neuroticism level among employees is quite high. It means that they have a pronounced emotional excitability which manifests itself in anxiety, tension, disquiet, confusion and irritability. This may also indicate difficulties in communication, social timidity, dependence and lack of initiative which are associated with dissatisfaction of desires are an alarming symptom for further professional activity.

Analysis of the CCF values allowed us to conclude that only 27% of respondents try to resolve the emerging conflict, for the others (73%) the desire to "escape" from the emerging internal contradictions (CCF from 5.75 to 6) is characterized. While a person is young and strong, the "flight" helps him to distract himself and cope with stress. However, the more often a person uses this way of internal conflict processing over a long period of time, the lower his effectiveness is as in each of these areas there are other stresses and conflicts which ultimately affect his health.

Analysis of the results of A Bass and A Darki's technique (A K Osnitsky's adaptation) made it possible to single out various indicators and forms of aggression.

Table 2. Results of the methodology for diagnosing indicators and forms of aggression

Indicators and forms of aggression	Low level	High level
Physical aggression	47%	53%
Verbal aggression	26%	74%
Indirect aggression	40%	60%
Negativism	46%	54%
Irritation	33%	67%
Suspicion	26%	74%
Resentment	40%	60%
Guilt	20%	80%

The results presented in Table 2 indicate that 74% of employees have a high level of verbal aggression. This means that employees allow sharp statements and insulting treatment in relation to other people. A high level of indirect aggression is noted in 60% of employees. The desire to see the bad in everything (high level of negativity) is noted in 54% of employees. 67% of employees are often very irritated. There is a high level of suspicion among 74% of employees. The employees have a high level of guilt – 80% (unfortunately, it is difficult to say whom they feel this feeling to: themselves, their family, their co-workers, convicts or other people). Thus, more than 50% of the staff that assembles carrier rockets has high rates of all forms of aggression.

The MBI questionnaire ("Professional Burnout") showed that 20% of respondents had psycho emotional exhaustion. This symptom is manifested in emotional and physical fatigue, indifference towards others (in some cases accompanied by depression or irritability). Deformation of relations with other people (depersonalization) is clearly expressed (30%). In some cases, this may mean increasing dependence on other people, in other cases it reflects the growth of negativity, the decrease in the number of contacts with others, irritability and intolerance in communication.

The highest values were obtained by the factor "Reduction of personal achievements" (32%). It can manifest either in a tendency to negative self-evaluation, their professional achievements and success, negativism about their official merits and opportunities, or in reducing their own dignity, limiting their abilities, obligations towards others which negatively affect the final result.

In foreign psychology the term coping is used to denote the specifics of these behaviour types. Stimulated or suppressed reactions that help restore the balance (coping) can be more or less appropriate, i.e. adaptive if they help achieve equilibrium and further reduce or eliminate harmful side effects.

Among the whole arsenal of behavioural individual responses, adaptive behaviour is highlighted on the basis of the concrete definition of the behavioural goal. Behavior can be called adaptive if its main goal is to establish productive interaction between the individual and the environment. In this context, the concept of "adaptive" does not mean that such behaviour necessarily ensures the optimal individual adaptation: the chosen strategy can be more or less successful. Nevertheless, adaptive behaviour is always launched in conditions of inconsistency of interacting systems and aims to achieve some coordination between them.

The problem of copying a person with difficult life situations arose in psychology in the second half of the 20th century. By copying R Lazarus [2] implies constantly changing cognitive and behavioural attempts to cope with specific external or internal requirements that are estimated as tension or exceed human resources in their overcoming. Following the approach of R Lazarus, N A Sirota and V M Yaltonsky consider the coping as an activity of the individual to maintain the balance between the environmental requirements and the resources that meet these requirements [3].

The coping ways mean constructive activity: to go through the situation, to survive the event, not to evade troubles. The subject of the coping psychology as a special field of study is the study of the mechanisms of emotional and rational behavioural regulation by a person aimed optimal interaction with life circumstances or their transformation in accordance with his intentions.

There is a sufficiently large number of different classifications of coping behaviour strategies.

At the stage of adaptation of a new employee to professional activity, E Yu Zarubko and N S Vakarina divide the coping strategies into constructive ones (seeking public support, positive interpretation and growth, active copying, planning, suppression of competing activities) and destructive (use of alcohol and drugs , Negation and mental detachment) [4].

We will consider coping strategies of behaviour in three planes (E Hale):

- behavioural (8 variants of behavioural strategies);
- cognitive (10 variants of cognitive strategies);
- emotional (8 variants of emotional strategies).

Adaptive variants of coping behaviour strategies include:

- in the behavioural sphere – cooperation, treatment, altruism (in all three cases, the behaviour of a person is aimed at constructive interaction with people, and sometimes a person has a desire to help people in overcoming their difficulties);
- in the cognitive sphere – problem analysis, setting the own value, self-control (these behaviour forms are aimed at analyzing the difficulties and possible ways to overcome them, increasing self-esteem and self-control, a deeper understanding of their own value as a person, belief in one's own resources by overcoming difficult situations);
- in the emotional sphere – optimism, protest, reflecting the emotional state of active outrage and protest in relation to difficulties and confidence in the possibility of getting out of any even the most difficult situation.

Relatively adaptive variants of coping behaviour strategies which constructiveness depends on the significance and severity of the overcoming situation include:

- in the behavioural sphere – compensation, abstraction, constructiveness – the behaviour forms characterized by the desire for temporary withdrawal from problems with the help of alcohol, drugs, immersion in a favorite business, travelling, fulfillment of their cherished desires;
- in the cognitive sphere – relativity, giving meaning, religiosity – the behaviour forms aimed at assessing difficulties in comparison with other problems, giving special meaning to their overcoming, believing in God and persevering in faith when facing complex problems;
- in the emotional sphere – emotional unloading, passive cooperation – the behaviour that aims to relieve tension associated with problems, emotional response, or to transfer responsibility for solving difficulties to others.

Non-adaptive variants of coping behaviour strategies include:

- in the behavioural sphere – active avoidance, retreat, which presuppose the behaviour that causes the avoidance of thoughts about troubles, passivity, solitude, isolation, the desire to get away from active contacts, refusal to solve the problem;
- in the cognitive sphere – humility, confusion, dissimilation, ignoring – the passive behaviour forms with the refusal to overcome difficulties due to unbelief in their own strengths and intellectual resources with deliberate underestimation of troubles;
- in the emotional sphere – suppression of emotions, obedience, self-blame, aggression – the behaviour characterized by a depressed state of hopelessness, submissiveness, anger and guilt on themselves and others.

If at least in one sphere a person chooses a non-adaptive coping behaviour strategy, then this leads to an internal conflict which means incompatibility, a clash of conflicting personal relationships. Consequences that are inherent in conflict become sources of mental and physical disability only when they are central to the personality relations system and when the conflict cannot be reworked so that pathogenic tension disappears and a rational, productive way out of the situation is emerged.

The number of adaptive, non-adaptive and relatively adaptive coping behaviour strategies in emotional, cognitive and behavioural spheres is presented in Table 3. The ratio of adaptive, non-adaptive and relatively adaptive coping behaviour strategies in the emotional sphere is 16-20-8 coping strategies, in the cognitive sphere 24-20-10, in the behavioural field 22-8-14 coping strategies respectively.

Table 3. Number of coping behaviour strategies

Type of coping strategy	Sphere	Amount
Adaptive		16
Non-adaptive	emotional	20
Relativelyadaptive		8
Adaptive		24
Non-adaptive	cognitive	20
Relativelyadaptive		10
Adaptive		22
Non-adaptive	behavioural	8
Relativelyadaptive		24

In the emotional sphere the number of non-adaptive coping behaviour strategies (20) is more than the adaptive behaviour strategies (16) that is reflected in complains about well-being, and is explained by the emotional work intensity of the Progress JSC employees in the production of space technology. A large number of relatively adaptive coping behaviour strategies in the behavioural sphere (24) indicates the presence of compensations but not active actions which is confirmed by a low activity index (4, 3).

If there are adaptive and non-adaptive coping strategies in the emotional sphere, this indicates a pronounced instability of the emotional sphere (there is the pendulum principle i.e. extremely positive and extremely negative emotions). This can be compensated by adaptive coping strategies in the behavioural sphere (in this case a person takes active actions aimed at reducing the stress level due to "flight" in the field of fantasy.

4. Aspects of rational material choice

The results of the research showed that the duration of the experience of a stressful situation depends on coping strategies in the behavioural and emotional spheres (i.e. on the emotions that a person experiences in a stressful situation and on the ways to overcome them). The duration of experiencing a stressful situation also depends on the way a person processes his internal conflict (the CCF value).The results of diagnostics showed once again the enormous influence of emotions on the well-being, activity and mood of a person, which emphasizes the need to master the ways of regulation of the emotional sphere.

In this regard, in order to improve the quality of products and the production efficiency of medium-sized launch vehicles special attention should be paid to the formation of emotional and stress resistance among Progress JSC employees. The stability of a person can be defined as a special, maximally effective organization of the functioning of a person as a system. Accordingly, stress resistance can be considered a special case of mental stability, which manifests itself under the action of stress factors [5]. Stress-resistance is the organization of the functioning of the personality as a system based on the psychological mechanisms of regulation, which allows him to act most effectively when exposed to stress factors.

According to B H Vardanyan [6], stress-resistance is provided due to either internal or external way of mental regulation. Internal methods include those methods of overcoming the emotional situation that help change attitudes towards it, give it a new meaning and cope with negative emotions. The methods of external overcoming are aimed at eliminating the external conditions of the emergence of an emotiogenic situation. This can be facilitated by training employees in self-regulation which mastery will allow an employee to cope with the negative stress consequences, prevent professional deformation that will have a beneficial effect on professional activity.

Summing up the above material, it can be said that the solution of the problem of the production and operation of space technology should be of a systemic nature, while paying attention not only to the technical and innovative improvement of production processes, but also to the psychological training of employees participating in this production since trainings of new staff are economically conditioned by the complexity of the technological processes.

5. Conclusion

The complexity of technological processes, the labour intensity and responsibility for its result lead to a high level of neurologization among the workers in the space industry that contribute to their increased aggressiveness of various forms and to the formation of a negative assessment of their job merits and opportunities because of a tendency to prolonged experience of stressful situations and lack of skills and abilities to effective conflict processing.

Acknowledgments

This work was supported by the Ministry of Education and Science of the Russian Federation in the framework of the implementation of the Program «Research and development on priority directions of scientific-technological complex of Russia for 2014–2020»

References

[1] Shcherbatykh Yu V 2006 *Psychology of stress and correction methods* (St. Petersburg: Peter) p 132
[2] Lazarus R 1970 The theory of stress and psychophysiological studies *Emotional stress* ed L Levi (Leningrad: Meditsina Publ.) pp 178–208
[3] Sirota N A and Yaltonskiy V M 1994 Coping behavior and psycho-prophylaxis of psychosocial disorders in adolescents *Review of psychiatry and medical psychology* **1** pp 63-74
[4] Zarubko E Yu and Vakarina N S 2014 Professional adaptation and coping strategies *Bulletin of the Tyumen State University* **9** pp 246-53
[5] Ezhova O N 2015 *Psychological training of penitentiary system staff* (Samara: Samara Law Institute of the Federal Penitentiary Service of Russia) p 86
[6] Vardanyan B H 1983 Mechanisms of adjusting of emotional stability *Categories, principles and methods of psychology. Psychical processes* (Moscow: Science) pp 542-3

Development Algorithm of the Technological Process of Manufacturing Gas Turbine Parts by Selective Laser Melting

A V Sotov [1,*], A V Agapovichev[1], V G Smelov [1] and R R Kyarimov [1]

[1]Samara State Aerospace University (National Research University), Moskovskoe sh. 34, Samara, 443086 Russia

[*]e-mail: SotovAnton@yandex.ru

Abstract. The technology of the selective laser melting (SLM) allows making products from powders of aluminum, titanium, heat-resistant alloys and stainless steels. Today the use of SLM technology develops at manufacture of the functional parts. This in turn requires development of a methodology projection of technological processes (TP) for manufacturing parts including databases of standard TP. Use of a technique will allow to exclude influence of technologist's qualification on made products quality, and also to reduce labor input and energy consumption by development of TP due to use of the databases of standard TP integrated into a methodology. As approbation of the developed methodology the research of influence of the modes of a laser emission on a roughness of a surface of synthesized material was presented. It is established that the best values of a roughness of exemplars in the longitudinal and transversal directions make 1.98 μm and 3.59 μm respectively. These values of a roughness were received at specific density of energy 6.25 J/mm^2 that corresponds to power and the speed of scanning of 200 W and 400 mm/s, respectively, and a hatch distance of 0.08 mm.

1. Introduction

Additive technologies (the three-dimensional printing) appeared in the late 80's of XX century. ASTM F2792-12a [1] defines the term additive technologies in the standard of the American Society for Testing and Materials, which like « a process of joining materials to make objects from 3D model data, usually layer upon layer, as opposed to subtractive manufacturing methodologies».

One of dynamically developing directions of the additive technologies (AT) is the process of selective laser melting (SLM) of metal powder. This method is based on layer-by-layer formation of a product by laser scanning of powder layer applied on a substrate.

SLM technology makes it possible to produce the parts from powders of aluminum [2, 3], titanium [4, 5], heat-resistant alloys [6, 7, 8] and stainless steels [9, 10, 11].

Interest in production of functional GTE details by AT method is synthesizing. In this regard the issue of development of the technological process methodology for manufacturing gas turbine parts from metal powders, in particular heat tubes of combustion chambers. This methodology includes optimization of technological modes in order to obtain products with the given production characteristics. Development of the technological process methodology will allow to exclude influence of qualification of the technologist on quality of the made products, and also to reduce labor input and energy consumption by development of TP, selection of the optimum processing conditions at the expense of the created databases of the SLM parameters integrated into the methodology.

2. Development algorithm of the technological SLM process

Development of TP of manufacture of details by SLM method is a complex, multivariate task that requires the consideration of a large number of various factors [12].

Proposed technique assumes work in two modules (CAD and CAE modules), as well as the use of developed TP database of typical details consisting of databases on allowances for machining, support material design and heat sinks cones, and the modes of melting. The algorithm for sequence designing of SLM technology is shown in figure 1.

As input parameters the design drawing of the part (1), which requires the development of TP appears. The assessment of a possibility of manufacture of a part with use of SLM technology can be detected only by the results of the technological analysis of the part drawing (2). At this stage appointment and operating conditions of a part is defined, the design and geometrical characteristics is described, the analysis of processability on material, the analysis of processability on accuracy of the sizes, forms and arrangements of surfaces, the analysis of processability on quality of the surface layer is carried out. These results will be input data for a task of the direction of printing of a part (3), and also for the feasibility study (4) uses of the SLM method.

Results of the carried-out technological analysis and the feasibility study should be formulated in a technical specification (TS) design of TP of a part (5) [13]. The requirements formulated in TS will allow carrying out the analysis of the standard parts which are available in the TP database (6) and by its results pass to formation of the working file of manufacture of it, or to model operation of preparation of the synthesizing part (7).

Process of workpiece modeling includes CAD and CAE modules, and also use of the developed databases, the appeal to which happens immediately when developing TP. The CAD module, in charge of preparation of a 3D model of a part for synthesizing process, includes such works as modeling of additional allowances (8), design of support material and heat sinks (9).

Final properties of selected material manufactured part will immediately depend on the choice of laser melting modes (10). For determination of parameters of the processing conditions it is necessary to solve the inverse problem of heat conductivity which will allow creating requirements to distribution of power influencing energy flows. This problem is planned to be solved in the CAE module with specification of part material, the processing conditions and boundary conditions in a preprocessor (11), with calculation of distribution problems of temperature profiles, residual stresses and an assessment probability of destruction in the processor (12), the analysis of the received results in the post-processor (13). The post-processor will allow estimating distribution of temperature profiles in a powder layer, and also will allow estimating the penetration of the powder particle and will give the chance to choose parameters of a laser melting. The problem of this module consists in achievement of the required qualities of a product, excepting neither optimum, non economic methods, in particular, a trial and error method which is often used at selection of laser parameters of SLM technology.

The developed algorithm for selective melting modes will allow selecting the optimal ratio of laser parameters for formation of required material structures formulated in the technical specification for part. The optimum ratio of parameters is characterized by the amount of transmitted energy by means of the specific density of energy (SDE). The SDE is equal to the ratio of emission power to the scan speed and laser spot diameter, expressed as, [J/mm^2].

After setting the range of SLM laser parameters and preparing the CAD model for synthesizing process, the final CAD model of the part (17), converted to the STL format, is formed. After formation of final CAD model of detail of SLM process (18) follows directly. After manufacture of workpiece part one of the important stages is additional machining (19) [14, 15]. This includes removing support material and finishing the surface.

The part (20) received during 3D printing has to undergo control (21) of the required surfaces according to the appointed TS. Only after that the stage of bench tests (22) in the conditions which are brought really closer to operating conditions of the made part follows. The analysis of test data defines a possibility of use of the synthesizing part in this or that external environment. Bench tests determine the functionality of the manufactured part, where, with a negative result, it is necessary to return to the stage of modeling the workpiece.

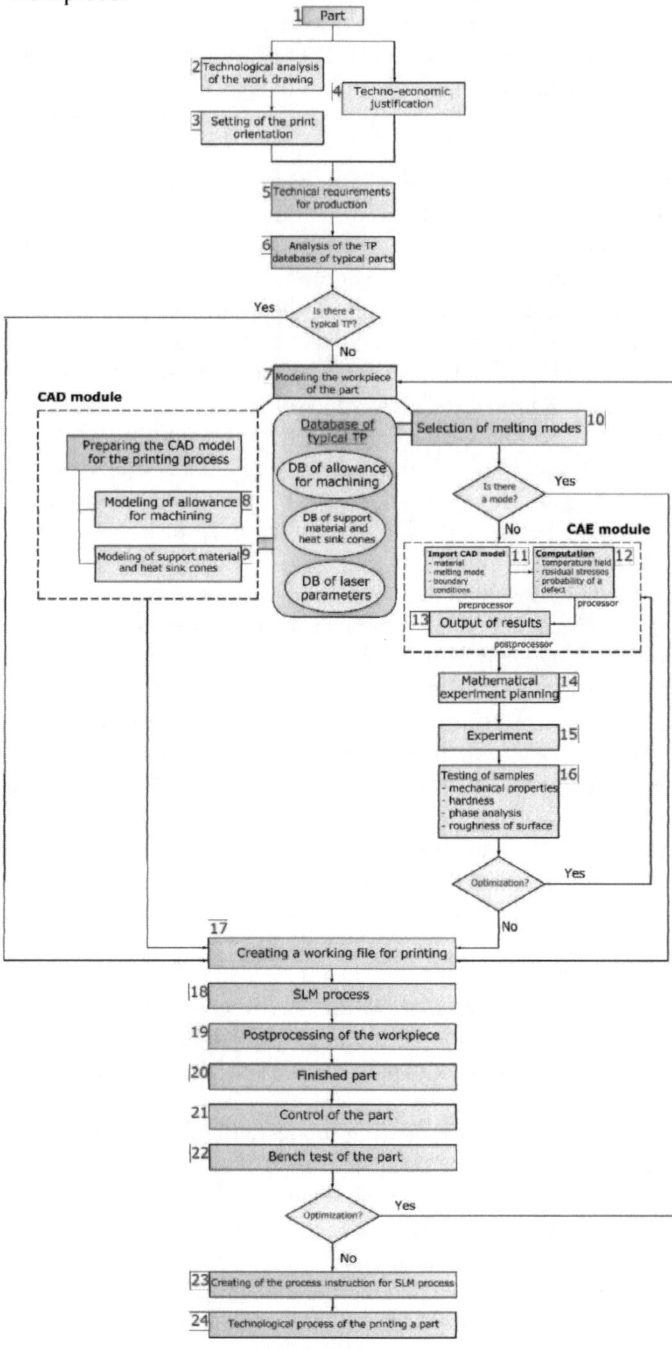

Figure 1. Algorithm of TP process development.

After completing all the above-described development stages of TP part manufacture the stage of execution of technological instructions (23) where specify the sequence of workpiece processing, the main SLM technological parameters, accuracy of the dimensions to be performed, the surface roughness value and the equipment used are indicated. The result of the developed technique is the designed TP for manufacturing the part by SLM method.

Approbation of the developed algorithm of SLM process was carried out for a heat tube of combustion chamber of a small-sized GTE. Process and result of manufacture is presented on figure 2. The separation of part from build platform was made with use of electro-erosive processing technology.

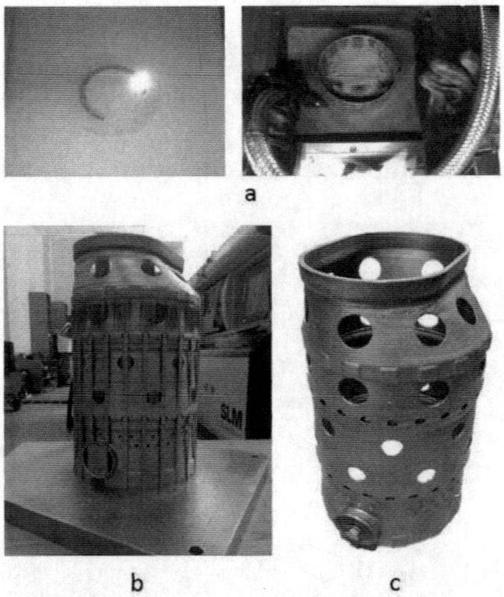

Figure 2. Process of heat tube production by SLM method (a), result of production after printing (b) and after machining (c)

Considering the high demands placed on roughness of a surface of heat pipes of GTE combustion chamber, as the experimental results the research of influence processing conditions of a laser emission on a roughness of the surface layer of synthesized material is presented. This research corresponds to the block 16 of the developed algorithm of projection of TP (figure 1).

3. Research methodology

The research of a roughness of a surface of synthesized material was conducted on exemplars in the transversal and longitudinal directions. Exemplars were made on 3D SLM 280HL machine. Print time at build layer of 50 μm made 3.5 hours. Production of parts was carried out on a build platform with size of 100x100 mm.

As the operated modes of laser scanning of a surface of metal powder parameters were taken: power of laser emission, P, W; scanning speed of powder surface, V, mm/s and hatch distance, H, mm.

For the studied exemplars values of specific density of energy in the range of 3.13…6.25 J/mm^2 were taken, the hatch distance for values of an overlapping coefficient equal to 1, 0.5 and 0 made 0.04 mm with laser beam spot diameter equal to 0.08 mm.

An investigation of the influence of above parameters was carried out for synthesized contour of a part. It is considered optimum to outline at first borders (contour) of the made product, and then to scan an interior. Values of the technological modes for the studied samples are reduced in table 1.

Table 1. Technological modes of SLM process.

№ sample	Power (W)	Scan speed (mm/s)	Hatch distance (mm)	SDE (J/mm^2)	Overlapping coefficient
1			0		1
2	100		0.04	3.13	0.5
3			0.08		0
4			0		1
5	150	400	0.04	4.69	0.5
6			0.08		0
7			0		1
8	200		0.04	6.25	0.5
9			0.08		0

The measurement of a roughness of surface was carried out in three directions: transverse and longitudinal directions and melted plane. For each plane about three measurements were carried out.

4. Experimental results and discussion

Outer examination of the made samples did not reveal apparent defects on a surface material. Figure 3 shows the appearance of a sample surface. On the surface of synthesized and on a lateral area of an exemplar unmelted particles of metal powder were found, and on a lateral area of similar particles it is more. Results of measurement surface roughness of synthesizable samples are shown in table 2.

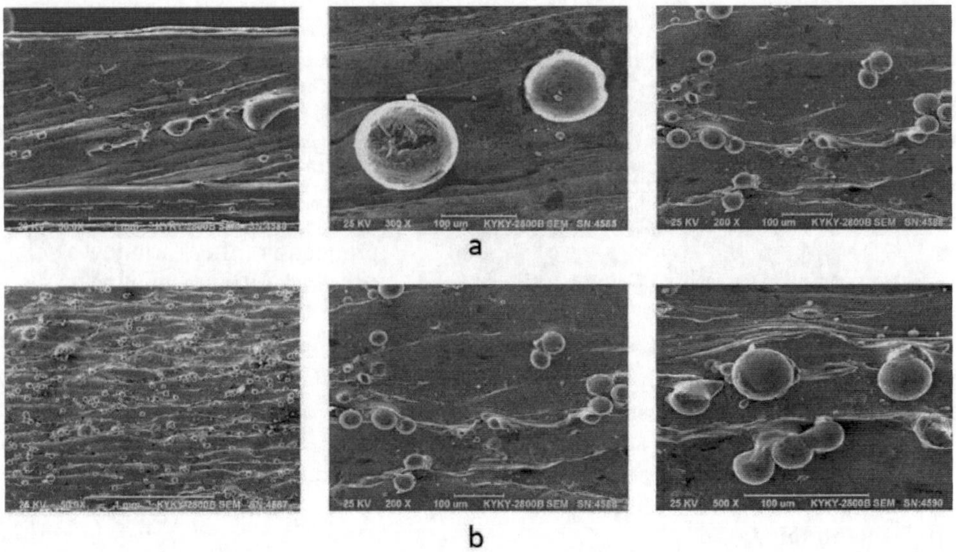

Figure 3. Appearance of a sample surface in melted plane (a) and on the lateral surface (b)

Table 2. Values of a surface roughness.

№ sample	Transverse direction (μm)	Longitudinal direction (μm)	Melted plane (μm)
1	13.08	13.72	16.58
2	11.55	9.38	23.91
3	12.48	8.26	17.36
4	4.90	2.67	9.03
5	5.79	3.07	9.37
6	4.78	3.81	8.13
7	4.56	3.19	8.95
8	3.98	2.33	9.32
9	3.59	1.98	7.90

For the obtained data graphs of dependence of a surface roughness in longitudinal and transverse directions for all planes from specific density of energy and a hatch distance were plotted. The received graphs are represented in figures 4.

Analysis of the results showed that at number 9 in the longitudinal and transversal directions the best values at specific density of energy of 6.25 J/mm^2 have an sample roughness that corresponds to power and scanning speed of 200 W and 400 mm/s with respectively, and hatch distance of 0.08 mm.

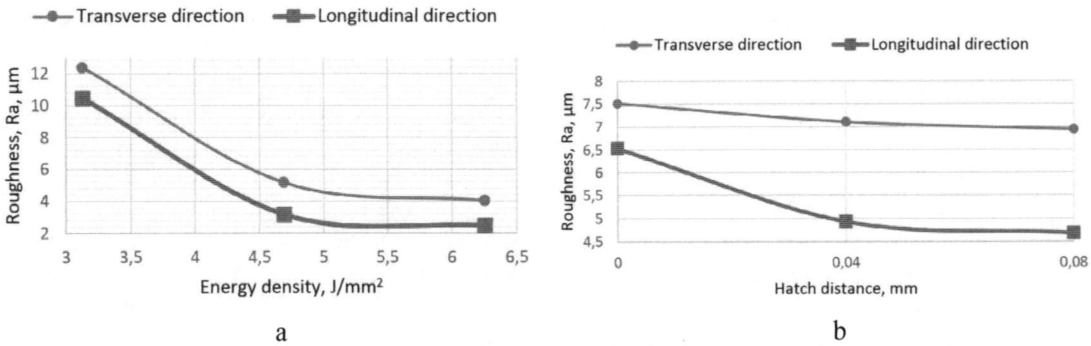

Figure 4. Dependence of the surface roughness (a) on the SDE and (b) on the hatch distance

5. Conclusion

The algorithm developed in this work reflects all main stages that need to be realized when testing of experienced TP SLM at manufacture of the functional parts. The created algorithm will allow reducing labor input and energy consumption by development of TP due to use of the databases of standard TP integrated into methodology.

As an approbation of the developed methodology the research of influence of the modes of a laser emission on a surface roughness of synthesizable material was presented. It is established that the best values of a roughness of samples in the longitudinal and transversal directions make 1.98 µm and 3.59 µm respectively. These values of a roughness were received at specific density of energy of 6.25 J/mm^2 that corresponds to power and scanning speed of 200 W and 400 mm/s with respectively, and hatch distance of 0.08 mm. The research of influence synthesizes modes allowed to determine the Pareto area of optimal values of a surface roughness.

Acknowledgements

The research was supported by the Ministry of Education and Science of the Russian Federation (Grant № 9.1299.2017/4.6).

References

[1] Standard Terminology for Additive Manufacturing Technologies 2013 – ASTM f2792-12a

[2] Kang N, Coddet P, Liao H, Baur T, Coddet C 2016 Wear behavior and microstructure of hypereutectic Al-Si alloys prepared by selective laser melting *Applied Surface Science*. – Vol. 378 pp 142-149

[3] Lin-zhi Wang, Sen Wang, Jiao-jiao Wu 2017 Experimental investigation on densification behavior and surface roughness of AlSi10Mg powders produced by selective laser melting *Optics & Laser Technology*. – Vol. 96 pp 88-96

[4] Agapovichev A V, Kokareva V V, Smelov V G and Sotov A V 2016 Selective laser melting of titanium alloy: investigation of mechanical properties and microstructure *IOP Conf. Series*: *Materials Science and Engineering*. Vol. 156 pp 1-6

[5] Yadroitsev I, Krakhmalev P, Yadroitsava I 2014 Selective laser melting of Ti6Al4V alloy for biomedical applications: Temperature monitoring and microstructural evolution *Journal of Alloys and Compounds*. Vol. 583 pp 404–409

[6] Kanagarajah P, Brenne F, Niendorf T, Maier H J 2013 Inconel 939 processed by selective laser melting: Effect of microstructure and temperature on the mechanical properties under static and cyclic loading *Materials Science & Engineering*. Vol. A 588 pp 188–195

[7] Sufiiarov V Sh, Popovich A A, Borisov E V, Polozov I A 2015 Selective laser melting of heat-resistant nickel alloy *Tsvetnye Metally*. Vol. 1 pp 79-84

[8] Jia Q, Gu D 2014 Selective laser melting additive manufactured Inconel 718 superalloy parts: High-temperature oxidation property and its mechanisms *Optics & Laser Technology*. Vol. 62 pp 161–171
[9] Riemer A, Leuders S, Thöne M, Richard H A, Tröster T, Niendorf T 2014 On the fatigue crack growth behavior in 316L stainless steel manufactured by selective laser melting *Engineering Fracture Mechanics* Vol. 120 pp 15–25
[10] Liverani E, Toschi S, Ceschini L, Fortunato A 2017 Effect of selective laser melting (SLM) process parameters on microstructure and mechanical properties of 316L austenitic stainless steel *Journal of Materials Processing Technology* Vol. 249 pp 255-263
[11] Smelov V G, Sotov A V, Agapovichev A V, Tomilina T M 2017 Implementation of the Additive Technology to the Design and Manufacturing of Vibroisolators with Required Filtering *Procedia Engineering* Vol. 176 pp 540-545
[12] Van Elsen, M. Complexity of selective laser melting: a new optimization approach: Phd dissertation / Maarten Van Elsen. – Heverlee, 2007
[13] Pechenin V A, Bolotov M A and Ruzanov N V 2014 Development of a method of ICP algorithm accuracy improvement during shaped profiles and surfaces control *International Journal of Engineering and Technology* Vol. 6, No. 5 pp 2229-35
[14] Nekhoroshev M V, Smirnov G V, Pronichev N D 2014 Computer simulation of high-speed anodic dissolution processes of geometrically complex surfaces of GTE details *The Open Mechanical Engineering Journal.* Vol. 8 pp 436-440
[15] Alexeev V, Balaykin A, Khaimovich A 2017 Influence of the direction of selective laser sintering on machinability of parts from 316L steel *IOP Conf. Series: Materials Science and Engineering* Vol. 177 doi: 10.1088/1757-899X/177/1/012120

IOP Conf. Series: Materials Science and Engineering **302** (2018) 012066 doi:10.1088/1757-899X/302/1/012066

How to Select the most Relevant Roughness Parameters of a Surface: Methodology Research Strategy

I N Bobrovskij

[1]Togliatti State University, Belorusskaya st. 14, Togliatti, Russian Federation

bobri@yandex.ru

Abstract. In this paper, the foundations for new methodology creation which provides solving problem of surfaces structure new standards parameters huge amount conflicted with necessary actual floors quantity of surfaces structure parameters which is related to measurement complexity decreasing are considered. At the moment, there is no single assessment of the importance of a parameters. The approval of presented methodology for aerospace cluster components surfaces allows to create necessary foundation, to develop scientific estimation of surfaces texture parameters, to obtain material for investigators of chosen technological procedure. The methods necessary for further work, the creation of a fundamental reserve and development as a scientific direction for assessing the significance of microgeometry parameters are selected.

1. Introduction

Increased number of parameters describing quantitatively surfaces structure is caused by requirements for stability and production quality being raised up. Also, increased amount of considered methodology parameters is caused by technological processes control capabilities rise which is required by processing surfaces durability increasing. With that there would be possible to acquire better technical and economic parameters of manufacture (incl. ICE engine sleeves surface and inner textures of aerospace technics long-length shafts).

The present article examines foundations for new methodology creation which provides solving problem of surfaces structure new standards parameters huge amount conflicted with necessary actual floors quantity of surfaces structure parameters which is related to measurement complexity decreasing.

By the concept of article [1] there is Rz variable used, while it would be possible to describe composing structure fully only with parameters complexity applying. That sort of method is being used by most of mechanical engineering investigators (incl. aerospace ones). Specialists on metrology use reverse method which is rise of applying parameters number.

Nevertheless, parameters number increasing is not well-defined factor at all. In 1982 in article [2] Whitehouse D.J. suggested some reckless point in parameters number applying. Over 30 years later, in 2011 he proposed surfaces structure parameters types which are supposed to be used in micromeasurement to nanomeasurement transfer in monography [3] (which is adapted partly to V.N. Krootikov tutorial handbook [4]).

Nowadays the surfaces structure parameters number is not decades but hundreds ones. Partially it could be explained by both need for different surfaces morphology considering and various applying spheres, also relational simplicity of new wide range parameters types mathematic conclusions. That fact is provided by huge variety of surfaces microprofile filtering methods.

It is quite rare event to contemplate different parameters importance; there are none of optional parameter unitary estimation methods at the moment. There are private efforts to determinate significant parameters of different technological processes known, f.e. article [5] presents to us estimation of rolling structures parameters. In present project "FoncRug3D" 29 investigators from different scientifical institutions take part; they analyzed 56 surface textures parameters in sphere of rolling and accepted Vmc as most significant parameter. Applying of Vmc could be useful in specialists practice in relation to aerospace cluster [6, 7]. Although, great variety of mechanical engineering technologies and ways to apply them requires new types of methodology.

The methodology examined in present article allows us to create practically necessary foundation, to research scientific direction of surface textures parameters estimating, also to obtain materials for technological processes investigators. Our task is devoted to problem solving of new standards excessive parameters amount conflicted with fact of surface structures parameters amount certain minimum.

2. Profiling of machines components surfaces

The authors point following research stages (figure 1):

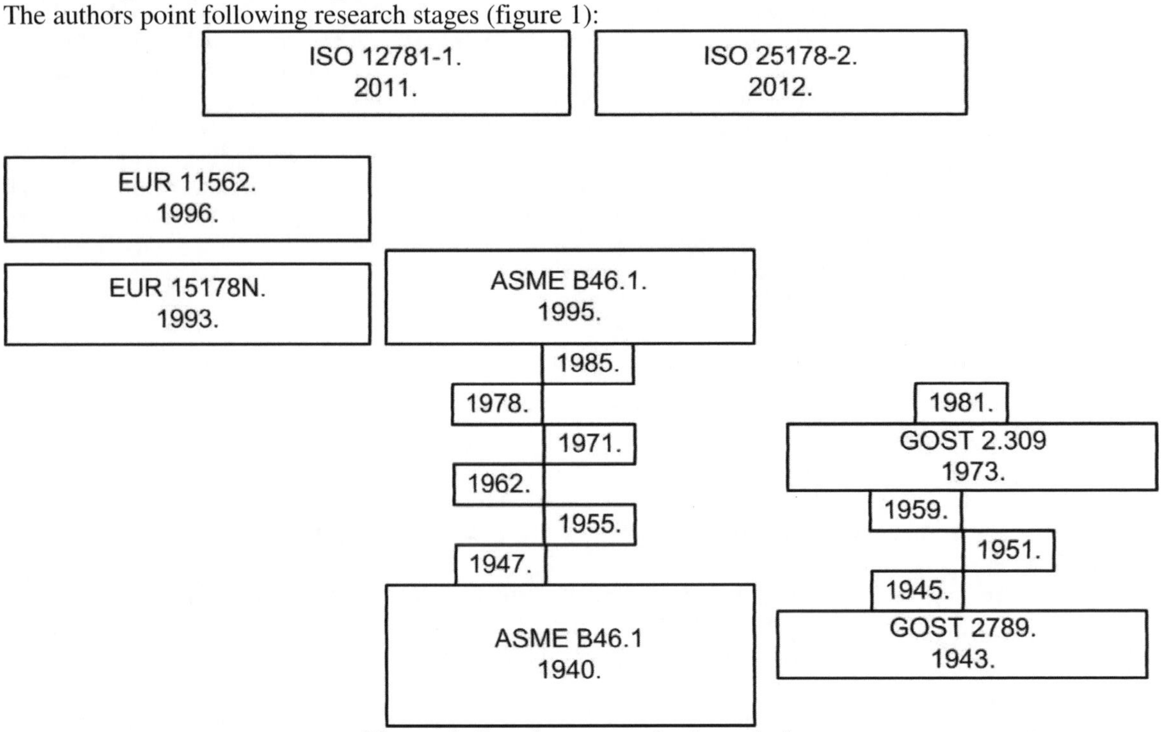

Figure 1. Roughness standards evolution.

I.

(a) ASME B46.1. 1940. Surface Roughness. Updated in 1947, 1955, 1962, 1971, 1978, 1985, 1995, 2002, 2009.

(b) GOST 2789 (since 1973 - 2.309). 1943. Surface Roughness. Description and parameters. Updated in 1945, 1951, 1959, 1973, 1981.

In presented standards there were well-known 2D parameters examined (such as Ra, Rz, etc.). Nowadays GOST 2.309 is not being actual anymore (in contrast to microhardness parameters GOST P 8.748-2011 based on ISO 14577-2002). The term "roughness" is relevant.

II.

EUR 15178N. 1993. The development of methods for the characterisation of roughness in three dimensions. Commission of the European Communities.

One of the first standards to estimate 3D roughness.

III.

(a) ASME B46.1. 1995. Surface Texture: Surface Roughness. New York: Waviness, and Lay, American Society of Mechanical Engineers.

(b) ISO 11562: 1996, Geometrical Product Specifications (GPS) – Surface Texture: Profile Method -- Metrological Characteristics of Phase Correct Filters (International Organization for Standardization, Geneva, 1996).

Entering of the terms "texture", "structure". Components surface structure is described by its 2D profile and 3D topography (which is cluster of connected profiles) and determines its texture. H.i.a. term "roughness" will be replaced by the term "surface texture".

IV.

(a) ISO 12781-1:2011 Geometrical Product Specifications (GPS) - Flatness - Part 1: Vocabulary and parameters of flatness.

(b) ISO 25178-2:2012 Geometrical product specifications (GPS) - Surface texture: Areal - Part 2: Terms, definitions and surface texture parameters.

New standards differentiation in two by the plane and texture parameters.

There are some efforts to extend new parameters knowledge in Russia proposed, not on standards position, but on position of methodological notices. There are terms "surface structure" and "surface texture" (already using by foreign investigators) were presented by The All-Russian Research Institute for Optical and Physical Measurements Federal State Unitary Enterprise; also methodological notices of surface topography standardization was formed [4].

Committee B46 ASME is forming standard at the moment [8].

It would be practical to use all stages parameters during new methods research cause of insolvent applying of GOST 2.309-73 parameters only. That can be shown by following example (figure 2-3).

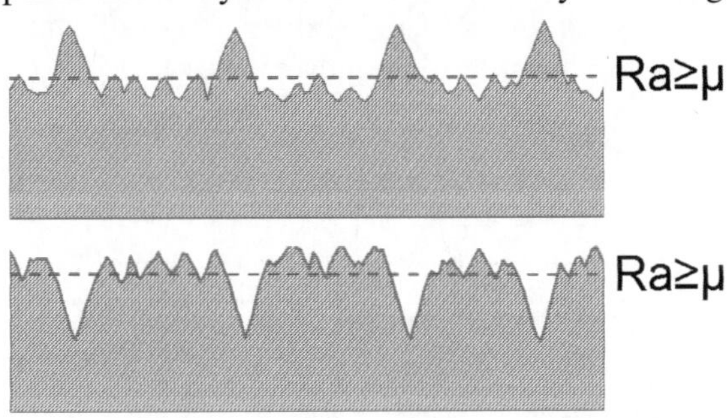

Figure 2. Same value of Ra parameter - different surfaces.

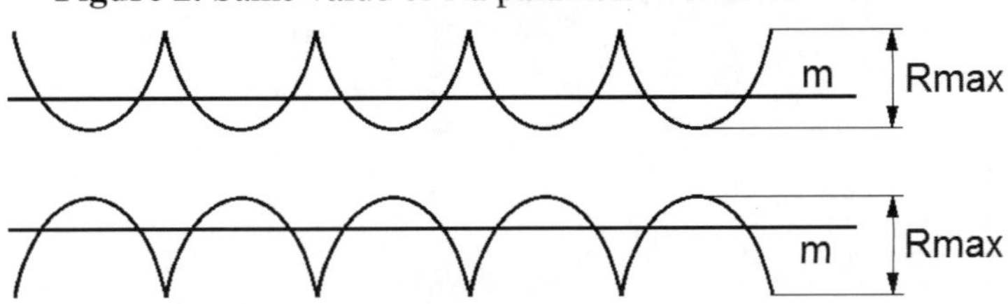

Figure 3. Same value of R max parameter - different surfaces.

Standards of stages II-III determine new parameters types; ISO 25178 determine 30 new parameters, EUR15178N determine 30 new parameters also (only several of them are identical to ISO 25178). Entering another time there are 16 parameters of sample rate.

Standards of stages IV determine 7 new parameters by each one; parameters of ISO 25178-2 are identical to parameters of ASME B.46.1 (which took 6 years to adapt american standart in Europe).

IOP Conf. Series: Materials Science and Engineering **302** (2018) 012066 doi:10.1088/1757-899X/302/1/012066

Here is methodology realization notion associating all types parameters presented; it allows to select most significant ones.

3. Methodology concept

The indication of methodology research ways providing the theoretical and experimental arguments for surface structure parameters minimal quantity cluster determination which is proved by confirmation with surface processed by any technology (such as summit honing). The determination of significant parameters would be executed in two stages by solving of directional and reverse tasks (figure 4-5).

Methodology research stages:

1. Directional task solving (quantitatively estimated 3D profile): acquiring of model mathematical description after model processing by chosen technology with following algorithmic and programmed realization for surface textures parameters values obtaining;

2. Reverse task solving (parameters values as 3D profile creation basis): comparing of synthetic surface characterization parameters values (as results of directional task) to consecutive variation of characterization parameters amount findings until asked validity completion (which is determined on base of functional properties accordance, f.e. on oil absorption of native and synthetic surfaces);

3. Comparison of obtained by solving both tasks parameters; selection of parameters with values variation most wide range (with appropriate rate of variation).

Nowadays there are analysis methods of surface response used for influence ascertainment of technological parameters to surface textures ones. Nevertheless, only parameter Ra is considered at the moment [9, 10].

It is supposed to be significant that processed surfaces texture new parameters values obtained on base of developing methodology and its mathematical description will be analyzed with computer-intensive methods incl. bootstrap (detalized while solving directional task). Estimation of forming synthetic processed surface based on surface textures new parameters value allows to obtain methodology accuracy conclusions (while solving reverse task).

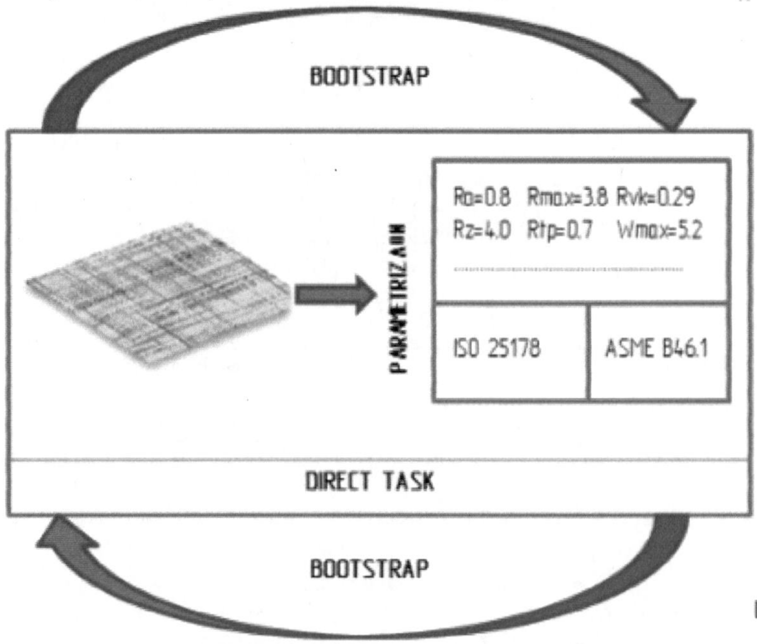

Figure 4. Methodology realization scheme – direct task.

IOP Conf. Series: Materials Science and Engineering **302** (2018) 012066 doi:10.1088/1757-899X/302/1/012066

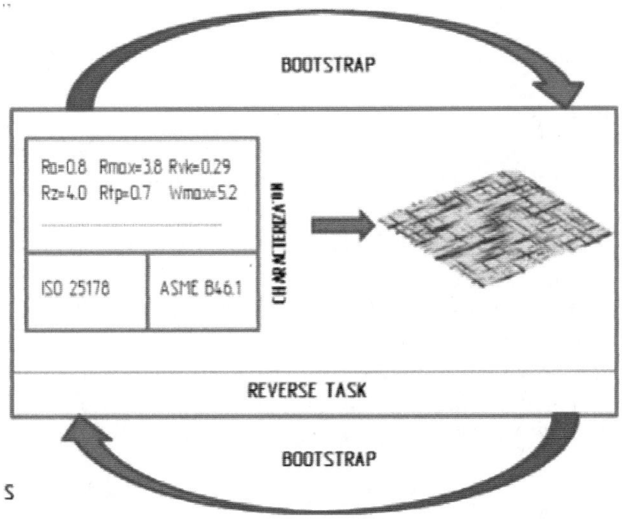

Figure 5. Methodology realization scheme – reverse task.

4. Notices and methods

There is applying of computer-intensive methods acquired which relate to Monte Carlo processes various types incl. methods of unparametric statistics (stcs. bootstrap). It caused by hard chance of experiment authentic reassignation obtaining (surface textures forming conditions) due to stochasticity of structures parameters creation (if processing type forming regular surface texture is not being used, f.e. vibration smoothing; also if quasiregular microsurface is created due to low accuracy of technological process, f.e. beating).

Unlike usual asymptotic method, the modeling of experimental spreading in that method will be acquired by generation of repetitive samples. The forming of quasisamples allows to obtain required texture wanted parameter properties such as mathematical expectation value, dispersion, confidence interval (CI). These methods include three particles: randomization, bootstrap and jackknife. It would be relevant to apply bootstrap itself.

The present method allows to create new cluster of values instead of original one; also the repeating of synthetic values statistics analysis procedure compared to original values analysis conclusions (method was presented in 1992-1993: [11, 12]; b. Russian edition in 2013: [13]); the present method was applying by french and british investigators during research of most significant parameter d. the sheet-metal rolling. One of the authors, Efron B., professor at the Stanford University, has proved progress of computer-intensive methods in article [14] in 1979. Speaking of executed analysis authors admit that resampling methods applying (not using of asymptotic methods type) is appropriate way to research present conception methodology.

5. Conclusion

The approval of presented methodology for aerospace cluster components surfaces allows to create necessary foundation, to develop scientific estimation of surfaces texture parameters, to obtain material for investigators of chosen technological procedure (f.e., force-controlled honing, surfaces plastic deformation).

By authors thought, processed surfaces textures new parameters values obtained on base of mathematical description are supposed to be analysed by computer-intensive methods incl. bootstrap. The estimation of forming synthetic processed surface based on surfaces textures new parameters values allows to obtain methods accuracy data.

High requirements for components surface layer applying in aerospace cluster need to be detected with new methods for processing modes assignment. Present being developed conception could be used in that sphere of mechanical engineering after approval.

References

[1] A process model for force-controlled honing simulations, Procedia CIRP. Italy, Naples: 3rd CIRP Global Web Conference on Production Engineering Research: Advancement Beyond State of the Art, CIRPe 2014.

[2] Whitehouse D J 1982 The parameter rash - is there a cure *Wear.* **83** 75-8.

[3] David J Whitehouse 2011 *Handbook of Surface and Nanometrology*, Second Edition. (CRC Press). ISBN: 978-1-4200-8201-2.

[4] Krootikov V N 2011 *Metrological support of nanotechnologies and nanoindustry products.* (Moscow: Logos) 592 p.

[5] Deltombe R, Kubiak K J and Bigerelle M 2013 How to select the most relevant 3D roughness parameters of a surface *Scanning.* **36(1)** pp 150-60.

[6] Abulkhanov S R, Skuratov D L, Khaimovich A I and Shvetcov A N 2014 Analytical approach to determining of parameters which characterize surface layer quality of the parts hardened by a traveling diamond sphere *International Journal of Engineering and Technology.* **6(5)** pp 2191-200.

[7] Khaimovich A I and Balaykin A V 2014 Analysis of plastic properties of titanium alloys under severe deformation conditions in machining *International Journal of Engineering and Technology.* **6(5)** pp 2184-90.

[8] Widder Ed ASME B46-PT32 Functional Standards Collection *ASME Webcast* May 8, 2012

[9] Tripathi B N, Singh N K and Vates U K 2015 Surface roughness influencing process parameters & modeling techniques for four stroke motor bike cylinder liners during honing: Review, *International Journal of Mechanical and Mechatronics Engineering.* **15(1)** pp 106-12.

[10] Jack Feng C -X, Yu Z –G S, Kingi U and Pervaiz Baig M 2005 Threefold vs. fivefold cross validation in one-hidden-layer and two-hidden-layer predictive neural network modeling of machining surface roughness data, *Journal of Manufacturing Systems.* United States, Peoria, IL: Dept. of Industrial and Manufacturing Engineering and Technology, Bradley University. **24.**

[11] Hall P 1992 *The Bootstrap and the Edgeworth expansion.* (New York:Springer. Verlag).

[12] Efron B and Tibshirani R J 1994 *An Introduction to the Bootstrap.* (New York: Chapman and Hall). 456 p. ISBN 9780412042317

[13] Vladimir K Shitikov and Gennady S Rozenberg 2014 *Randomization and Bootstrap: Statistical Analysis in Biology and Ecology with R.* (Togliatti: Cassandra). 314 p.

[14] Efron B 1979 Computers and the theory of statistics: thinking the unthinkable *SIAM Review.* **21(4)** pp 460-80.

IOP Conf. Series: Materials Science and Engineering **302** (2018) 012067 doi:10.1088/1757-899X/302/1/012067

Optimization of Selective Laser Melting by Evaluation Method of Multiple Quality Characteristics

A I Khaimovich[1], I S Stepanenko[1] and V G Smelov [1]

[1]Department of Aircrafts Engines' Construction Technologies, Samara National Research University, 34, Moskovskoye shosse, Samara, 443086, Russia

E-mail: stepanenko@ssau.ru

Abstract. Article describes the adoption of the Taguchi method in selective laser melting process of sector of combustion chamber by numerical and natural experiments for achieving minimum temperature deformation. The aim was to produce a quality part with minimum amount of numeric experiments. For the study, the following optimization parameters (independent factors) were chosen: the laser beam power and velocity; two factors for compensating the effect of the residual thermal stresses: the scale factor of the preliminary correction of the part geometry and the number of additional reinforcing elements. We used an orthogonal plan of 9 experiments with a factor variation at three levels (L9). As quality criterias, the values of distortions for 9 zones of the combustion chamber and the maximum strength of the material of the chamber were chosen. Since the quality parameters are multidirectional, a grey relational analysis was used to solve the optimization problem for multiple quality parameters. As a result, according to the parameters obtained, the combustion chamber segments of the gas turbine engine were manufactured.

1. Introduction

At the present time additive production methods are becoming increasingly widespread in the aerospace industry. Selective laser melting (SLM) makes it possible to manufacture parts of complex shape, but in the process of selective laser melting the temperature changes and distribution of temperature loads, which leads to the appearance of residual stresses [1, 2]. In addition, as a result of this thermal history, an anisotropic microstructure is formed. The final properties of the material (e.g. endurance strength and tensile strength) are directly related to the microstructural features of the part, and therefore the presence of residual stresses is unfavorable [1, 3]. The effect of the melting direction on the residual stresses was studied in [4, 5]. As a result of residual stresses from thermal loads, the laser induced synthesis of thin-walled parts generates significant deformations resulting in defective parts. In this respect, the actual task is to develop compensation methods for the residual stresses.

The compensation methods may be conditionally divided into 4 groups:

1. Optimization and correction of the melting process's operating parameters, including the laser power, the laser beam velocity, the beam width, scanning strategy;
2. Optimization of heat removal into the base platform through a substrate;
3. Topological optimization of the structure by adding removable technological elements such as supports to increase the rigidity and to prevent distortion from the thermal stresses effect;
4. Compensation of residual stress effect by introducing the pre-correction of part geometry.

The application of the first method is complicated by the fact that it is not always possible to select the sintering mode quickly, as it is necessary to take into account not only the properties of the basic material, but also part's design features. The second method is carried out by selecting the quantity, the cross-section geometry and the location of the heat sink. The latter two methods are particularly effective in growing thin-walled parts, where ensuring rigidity is most important. The preliminary correction can be performed by initial estimating the resultant deformations. In that case part configuration changes by inverting the distortion values multiplied by the configuration correction degree.

2. Materials, Methods and Equipment

The SLM build-up of the combustion chamber segments from the nickel-chromium metal-powder composition was performed on a selective laser melting machine SLM 280HL.

The measurements were carried out on a coordinate measuring machine DEA GLOBAL Performance. Numerical modeling of the SLM process was performed using the Simufact Additive© v1 software (trial ver. [6])

For SLM build-up, a metal powder with a fraction size of 40 μm of VV751P mark, the chemical composition of which is given in Table 1.

Table 1. Chemical composition of metal powder VV751P

Fe	*C*	Si	*Mn*	*Ni*	*S*	P	*Cr*	*Ce*	*Mo*	*W*	*V*	*Co*	*Nb*	*Ti*	*Al*	*B*	*Pb*	*Mg*	*La*
up to 1	0.04 – 0.08	up to 0.3	up to 0.3	50.09 – 50.86	up to 0.009	up to 0.015	10 -12	up to 0.01	4 - 5	2.5 - 3.5	0.4 – 0.8	14 - 16	3 – 3.5	2.5 – 3.1	3.7 – 4.2	up to 0.015	up to 0.07	up to 0.01	up to 0.02

The detailed experiment of a part's production by SLM method can be found in the works [5, 7].

For the study, the following optimization parameters (independent factors) were chosen: P is the laser power, v is the scanning velocity; two factors for compensation of residual thermal stresses, M is the scale factor for preliminary correction of the part configuration and t is the indicator for the introduction of additional stiffeners ($t = 0$ means stiffeners are absent, $t = 1$ – lengthwise stiffeners are introduced, $t = 2$ – lengthwise and transverse stiffeners are introduced). As dependent factors (quality parameters), the values of maximum and average deformations from the residual stresses effect were selected in 9 surface areas of the combustion chamber segment. The second quality parameter is a value of the ultimate strength for the melting conditions corresponding to the experiments. The values of distortion for various matched melting conditions were obtained by numerical methods using a verified experimental model.

To determine the tensile strength σ depending on the operating parameters of laser melting we used a second-order regression model (1).

$$\bar{\sigma} = 0,25 + 1,0627\bar{v} + 0,1532\bar{p} + 1,5324\frac{\bar{p}}{\bar{v}} - 1,3121\bar{v}^2 - 1,294\left(\frac{\bar{p}}{\bar{v}}\right)^2, \quad (1)$$

Where normalized parameters are written as: $\bar{\sigma} = \dfrac{\sigma - 891}{241}$, $\bar{v} = \dfrac{v - 710}{100}$, $\bar{p} = \dfrac{P - 225}{100}$

Since it was assumed that the main results of the experiments will be obtained by numerical methods, a full-scale experiment on combustion chamber segment growing was conducted to verify the numerical model of the SLM process and to estimate the calculation error. It was found that the calculated and measured distortion values have a good correlation. The calculated distortion values were further used as process quality parameters to estimate the distortion values on various melting modes.

3. Taguchi Quality Method

The key principle of Taguchi quality measurement methods is minimization of changeability (variability) in the technology of part manufacturing in response to noise factors and maximization of variability in response to control factors. The noise factors *N* are those that are not under the control of the technology, or are not pertain to the optimization problem. Signal factors *S* are those factors that are established or controlled by the technological process for a specific optimization problem. The goal of the quality improvement can be stated as an attempt to maximize the signal-to-noise ratio (S/N) for the corresponding product. When the task is nominal-is-better, it means to minimize the development of some undesirable characteristics of the product. In our case such characteristics are deformations caused by residual stresses. The S/N ratio should be calculated as follows [8]:

$$\xi = -10\log\frac{1}{n}\sum_{i=1}^{n}\frac{1}{y_i^2} \tag{2}$$

where y_i is the observed value in the experiment, n is the total number of experiments.

For our research we chose orthogonal array L9 (4×3) according to Taguchi method [8]), which represented 9 sets of experiments and contained 4 three-level factors (Table 2).

Table 2. Quality parameter values

№	P, W	v, mm/s	M	t	Maximum distortion value on areas of combustion chamber segments, mm									σ, MPa
					A1	A2	A3	A4	A5	A6	A7	A8	A9	
1	290	700	-0,80	0	3	2,80	1,80	2,80	1,80	1,80	4,00	1,8	1,8	1020
2	290	760	-1,00	1	0,317	0,81	1,07	0,51	0,39	0,39	0,34	0,57	0,65	1120
3	360	760	-0,80	2	1,2	1,2	1,2	1,4	1,4	1,4	1,2	1,2	1,4	1053
4	325	700	-1,00	2	0,4	0,60	0,80	0,30	0,30	0,30	0,60	0,6	0,6	966
5	360	800	-1,00	0	1	1,37	1,42	0,86	0,54	0,54	1,41	1,08	0,9	1025
6	360	700	-1,20	1	0,64	1,13	1,44	0,78	0,59	0,59	0,54	1,024	0,88	864
7	325	760	-1,20	0	1,2	1,39	1,44	0,97	0,67	0,67	1,67	0,32	0,99	1107
8	290	800	-1,20	2	0,68	0,87	1,17	0,58	0,44	0,48	0,46	0,704	0,605	1046
9	325	800	-0,80	1	1,6	1,00	1,60	1,60	1,20	1,60	1,00	1,6	1,2	1054

4. Main Effect Analysis

For main effect analysis we used the following dependence:

$$\Delta F = \max(\bar{F}_1, \bar{F}_2, \ldots, \bar{F}_n) - \min(\bar{F}_1, \bar{F}_2, \ldots, \bar{F}_n), \quad \bar{F}_i = \frac{1}{m}\sum_{j=1}^{m} y_{ij} \tag{3}$$

where $\bar{F}$ is the mean S/N ratio of the i-th level of factor F, m is the number of the i-th level of each factor, y_{ij} is the jth S/N ratio of the i-th level, ΔF is the value of the main effects of factor F, and n is the number of the level of each factor. The higher a factor's main effect value, the greater this factor's impact on the system will be.

In accordance with factor's main effect value where $\Delta F = 1{,}62375$ is for laser out power (P), $\Delta F = 0{,}43756$ is for scanning velocity (v), $\Delta F = 7{,}78723$ is for configuration correction scale factor

(M) and $\Delta F = 5,08169$ is for additional stiffeners (t) taking into account the grade of factor's impact on the quality, factors may be ranged in the following manner: $1 - M$, $2 - t$, $3 - p$, $4 - v$

5. Analysis of Variance (ANOVA)

ANOVA helps to identify the effects of the individual factors on distortion value as quality characteristics. ANOVA results are presented in Table 3. Scale factor of configuration correction is excluded due to its low integrity according to parameters value F and p based on preliminary analysis. The lower F value and the higher p value, the more adequate is the assessment of the factor's effect, estimated by the *SS* deviation value. *SS* factors are ranged by value in the descending order: $1 - t$, $2 - P$, $3 - v$.

Scale factor of configuration correction M due to its low integrity according to F criterion did not was not analysed.

Table 3. ANOVA results without the effect of corrective scale factor

	SS	df	MS	F	p
1 – Laser power	4.42766	2	2.21383	0.046462	0.955601
2 – Velocity	0.35226	2	0.17613	0.003697	0.996317
4 – Stiffness factor	40.09485	2	20.04742	0.420740	0.703858
Residual stresses	95.29594	2	47.64797		

The use of the F test compensates for the defects of the Taguchi method experiments, such as failure to identify the effects that the different experiments may have on quality characteristics or the level of experimental errors.

Figure 1 shows a graph of the signal-to-noise ratios of quality parameters in orthogonal experiment L9 according to Taguchi.

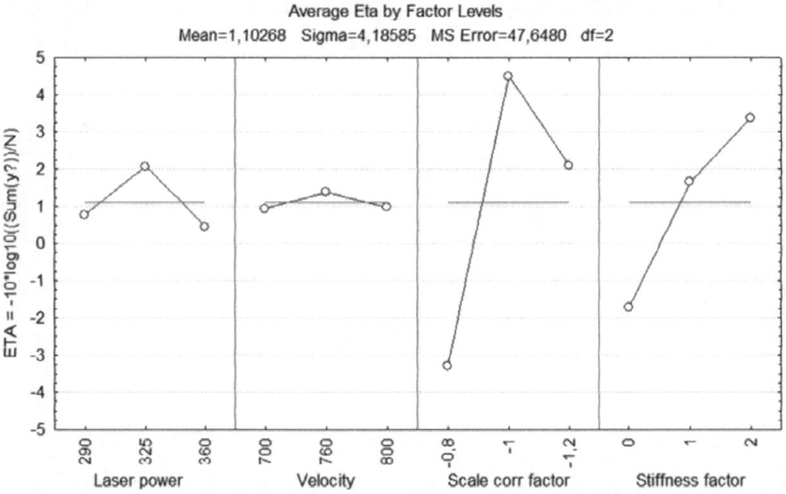

Figure 1. Average η depending on Control Factor Levels

Response graph (Figure 1) of average quality parameter η shows that the optimal operating SLM conditions of combustion chamber segment corresponds to the parameter value: laser power – 325 W, scanning velocity – 760mm/min, scale factor value of geometry correction -1 and introduction of additional stiffing elements relevant to maximum level.

6. Grey Relational Analysis

For a high quality part an average (and maximum) values of distortion must be minimized and the ultimate tensile strength values must be at maximum. As in this case Taguchi's lower-the-better analysis for quality characteristics is inapplicable due to their diversity, we used grey relational analysis. Grey relational analysis is a method of measurement in the grey system theory, which analyzes the degree and the level of the ratio of heterogeneous parameters for their discrete sequence.

The initial experimental data were first normalized in the range from 0 to 1 in accordance with the principle "the smaller-is-better" for distortion value using the following function:

$$x_{ij} = \frac{y_{ij} - \min_j y_{ij}}{\max_j y_{ij} - \min_j y_{ij}}, \quad (4)$$

$y_{ij} = \max_n y_{ij}^n$ is the most negative quality characteristic (distortion value) among $n = 1....9$ concerned areas of developed combustion chamber. The best normalized result corresponds the value 1.

For maximum strength which is characterized by the-higher-the-better normalization was carried out using the following dependence:

$$x_{ij} = \frac{\max_j y_{ij} - y_{ij}}{\max_j y_{ij} - \min_j y_{ij}}, \quad (5)$$

where y_{ij} is the value of maximum strength for j-th experiment, calculated using the dependence (1).

Grey relational coefficient ξ_{ij}, which is calculated to define ratio between ideal and factual experiment results may be as follows [8]:

$$\xi_{ij} = \frac{\min_i \min_j |x_i^0 - x_{ij}| + \zeta \max_i \max_j |x_i^0 - x_{ij}|}{|x_i^0 - x_{ij}| + \zeta \max_i \max_j |x_i^0 - x_{ij}|}, \quad (6)$$

where

x_i^0 is the ideal result (i.e. the best normalized result =1) for i - quality characteristics, $\zeta = [0,1]$ is the distinguishing coefficient, which target is to weaken the effect of $\max_i \max_j |x_i^0 - x_{ij}|$, when it becomes too high and therefore increases the value difference of the grey relational coefficient. In general, its value is taken to be 0.5 if all process parameters have equal weight.

Table 4 presents the grey relational coefficients for every experiment of orthogonal array L9. The weighting method is then used to integrate the obtained values of the grey relational coefficients ξ_{ij} for each experiment into an integrated relational evaluation. The overall assessment of multiple quality characteristics is based on an integrated relational evaluation, which is determined by the following dependence:

$$\gamma_j = \sum_{i=1}^{m} \frac{l_i}{\sum_{i=1}^{m} l_i} \xi_{ij}, \quad (7)$$

where l_i is the value level of i -h quality characteristic.

The resulting integrated relational estimates are shown in Table 4. The higher the integrated relational rating, the better the result of the experiment, the closer it is to the ideally normalized value.

IOP Conf. Series: Materials Science and Engineering **302** (2018) 012067 doi:10.1088/1757-899X/302/1/012067

The effect of each parameter of manufacturing process on the integrated relational evaluation at different parameter levels is taken into account separately, whereas the experiment plan is orthogonal.

In Table 4 y_1, y_2 are the maximum values for every experiment according to an average and maximum distortion, taken in 9 areas, y_3 is the value of maximum strength, x_1, x_2, x_3 are normalized by dependences (4) and (5) factors.

Table 4. Results of grey relational analysis

	y_1	y_2	y_3	x_1	x_2	x_3	ξ_1	ξ_2	ξ_3	γ
1	2,20	4	1020	0	0	0,6078	0,3333	0,3333	0,5604	0,44685
2	0,84	1,067	1120	0,7533	0,9166	1	0,6696	0,857	1	0,90039
...										
8	0,95	1,17	1046	0,6972	0,8844	0,7098	0,6228	0,8122	0,6328	0,69409
9	1,20	1,6	1054	0,5556	0,75	0,7412	0,5294	0,6667	0,6589	0,642205

The values of the relational estimates ξ_i were calculated from the dependence (6) on the vector of the desired normalized quality parameters $x_i^0 = (1,1,1)$, which corresponds to the maximum quality parameters obtained in all numerical experiments. The integral relational estimation γ was determined by the formula (7), while for the maximum value of the average distortion value in 9 areas (x_{1j}) it was accepted that $l_1 = 30\%$, for the greatest distortion value in 9 areas (x_{2j}) $l_2 = 70\%$, for the ultimate strength limit (x_{3j})- $l_3 = 100\%$.

Based on the calculated values γ, it is seen that the maximum value of the integral relational estimate (0.90039) has a second experiment, Table 4. These values of the factors best correspond to the vector of desirability of the quality parameters. According to the combination of the both analysis's results, and in accordance with the degree of factor values effect on the quality characteristics for growing of the combustion chamber segment, the following modes were adopted: laser power - 290 W, scanning velocity - 760 mm / min, additional stiffness elements -2 (maximum level). The segment of the grown combustion chamber is shown in Figure 2.

Figure 2. Manufactured combustion chamber segment.

7. Conclusion

This research is focused on the evaluating the process of selective laser melting of thin-walled not rigid parts prone to thermal warpage. This study combines the Taguchi method and orthogonal experiment plan consisting of 9 experiments to evaluate part quality. The Taguchi method was applied to optimize such manufacturing factors as radiation power, hatching speed, scale factor of geometry correction, adding of additional stiffness elements. Based on the experiment analysis results, we found the optimal values of the laser melting modes of the combustion chamber segment of a gas turbine engine from a heat-resistant nickel-chromium powder. These technological factors were ranged according to the degree of their impact on quality parameters. It is necessary to note that the Taguchi method is applicable mostly to optimize the single quality characteristic as distortion values from residual stresses during SLM process. It does not provide an agreed solution for the case when it is necessary to use multiple quality characteristics, such as distortion value and mechanical features. For this case grey relational analysis was adopted. The results of this method applying at the same experiment plan as the Taguchi method were different from the latter, as it takes into account more characteristics, in particular the maximum strength. The degree of consistency of multiple quality characteristics was taken into account based on the adopted desirability vector and the coefficients of impact and the level of accepted significance levels of the factors.

Acknowledgements

This work was supported by the Ministry of Education and Science of the Russian Federation in the framework of the implementation of the Program "Research and Development on Priority Directions of Scientific-Technological Complex of Russia for 2014– 2020".

References

[1] P Rangaswamy, M L Griffith, M B Prime, T M Holden, R B Rogge, J M Edwards and R J Sebring 2005 Residual stresses in LENS® components using neutron diffraction and contour method, *Mater. Sci. Eng.: A.* vol 399(1) pp 72–83

[2] F Liu, X Lin, G Yang, M Song, J Chen and W Huang 2011 Microstructure and residual stress of laser rapid formed Inconel 718 nickel-base superalloy Opt. Laser Technol. vol 43 pp 208–213

[3] K Dai and L Shaw 2002 Distortion minimization of laser-processed components through control of laser scanning patterns Rapid Prototype. J. vol 8 pp 270–276

[4] N Shamsaei, A Yadollahi, L Bian and S M Thompson 2015 An overview of Direct Laser Deposition for additive manufacturing; Part II: Mechanical behavior, process parameter optimization and control Additive Manufacturing vol 8 pp 12-35

[5] V P Alexeev, A V Balyakin and A I Khaimovich 2017 Influence of the direction of selective laser sintering on machinability of parts from 316L steel IOP Conference Series: Materials Science and Engineering vol 177(1) p 012120

[6] https://www.simufact.com/simufact-additive.html

[7] A V Agapovichev, V V Kokareva, V G Smelov and A V Sotov 2016 Selective laser melting of titanium alloy: investigation of mechanical properties and microstructure IOP Conf. Ser.: Mat. Sci. and Eng. vol 156(1) p 012031

[8] B Acherjee, A S Kuar, S Mitra and D Misra 2011 Application of grey-based Taguchi method for simultaneous optimization of multiple quality characteristics in laser transmission welding process of thermoplastics The Int. J. of Adv. Manuf. Tech. vol 56(9) pp 995-1006

IOP Conf. Series: Materials Science and Engineering **302** (2018) 012068 doi:10.1088/1757-899X/302/1/012068

Information Pattern in Imaging of a Rough Surface

S R Abul'khanov[1] **and N L Kazanskiy**[2]

[1] Associate Professor of Department of Engine Production Technology of Samara University,
[2] Director of Image Processing Systems Institute of the Russian Academy of Sciences, Professor at Technical Cybernetics Sub-Department of Samara University

Abulhanov58@mail.ru

Abstract. In this paper, we have proposed a method of parametrization of a rough surface image based on its information pattern. We have determined that the image information pattern makes it possible to keep track of any variations in the number of pixels in the image of the controlled rough surface of at least 0.192 per cent of the total number of image pixels. The offered method permits to compensate a non-linear perception of the controlled surface by a human eye. We have determined a ratio of the number of these pixels to the total number of image pixels. Such ratios, was treated as a certain square area. We packed this squares without intercrossings in the square of 2. This type of squares packing was designated as an information pattern. Using the information pattern, the parameter value was obtained. We have determined that the parameter value can keep track of any variations of the number of pixels in the image of the rough surface from at least 0.192 percent.

1. Introduction

In a framework of the descriptive approach in the technical literature, the influence of roughness character on some performance properties of component parts, operating under different conditions, was consistently reviewed [1, 2]. This approach allows us to draw certain conclusions about the interrelationship between the roughness parameters regulated by normative documents and the surface properties such as, for example, corrosion resistance, durability, fatigue endurance, etc.

Existing specified roughness parameters, being defined, for instance, by standards, can ambiguously characterize the controlled surface [3, 4]. For this reason, the numerical estimate of microroughnesses by a profilogram of the controlled surface is considered to be limiting [5, 6].

We know some works [7-9] where profilogram values of the rough surface are considered as numerical series. Such series are analyzed by means of spectral correlation methods. The disadvantage of these techniques is the difficulty in establishing relationships between the obtained results and the current experience of using standard roughness parameters.

In some papers, the authors determined a correlation value of profilogram fragments [10] or rough surface fragments [11] via their scaling. Disadvantages of fractal methods for engineering application are the following factors: fractal dimensions of a set being analyzed can hardly be compared with standard parameters of the rough surface; fractal dimensions for self-affine fractals cannot clearly be defined [12].

We think it is not practical to use any statistical methods to identify the most unique characteristics of profilogram fragments or controlled surface fragments and also to assess the stability of certain properties. Rough surface profilogram or images have finite sizes that restrict a sample length. A representative sample for reliable calculating, for example, of Hurst coefficient should contain several thousand measurements [13, 14].

The authors [15, 16] point out that new roughness parameters being developed must correspond to the standard roughness parameters. This means that parameters of three-dimensional surface microroughnesses must have profile prototypes which are regulated by normative documents.

We propose to characterize the rough surface by building its information pattern through imaging microroughnesses on the controlled surface and using the parameter value which is computed by the information pattern.

The purpose of this paper is to create visual and numerical tools to analyze the rough surface by its imaging.

2. Problem description

Human vision perceives the light brightness nonlinearly and has a limited response at very high or at very low intensities of the light flux being observed. Besides, a human eye can badly perceive small details of the image and the total image as a whole at the same time [17]. For these reasons, a person cannot keep track of any small changes in the image of the rough surface.

In order to eliminate the influence of physiological individualities of human perception of two-dimensional images, we propose to use the information pattern. When analyzing a raster image of the rough surface to create the information pattern, we used an image histogram. We considered a black-and-white image of the rough surface (Fig. 1). We referred the number of pixels, which conform to a certain color index in the image, to the total number of image pixels. We treated any obtained dependence as a square area. The number of such squares is always finite, since computer graphics palettes contain a finite number of colors. In accordance with paper [18], a finite sequence of squares (squares of the same area are assumed), the total area of which equals to one, may be arranged without overlapping within the square of 2. We propose this kind of arrangement of the squares, each area of which corresponds to the number of pixels with a certain colors index in the image, be considered as the information pattern of the rough surface image.

We consider that this information pattern does not depend on the number of colors in the image and on the selected indexed computer graphics color palette. We think that this information pattern may be used to study time changes of the rough surface (provided that the images of the same surface fragment obtained at different time intervals will have the same sizes).

In our opinion, among the profile parameters regulated by the National State Standard GOST 2789-73, a relative reference profile length (*tp*) may be considered as a prototype of the information pattern. In accordance with ISO 1302-2001 Indication of Surface Texture, the profile parameter *Rmr(c)* may be considered as a prototype of the proposed information pattern.

3. Research procedure

For our research, we used a titanium alloy sample BT9 (Ti – 6.8 Al – 3.2 Mo – 2.0 Zr – 0.3 Si). The sample surface was treated by polishing (Fig.1).

Figure 1. The surface of the titanium alloy sample after having been polished (×100).

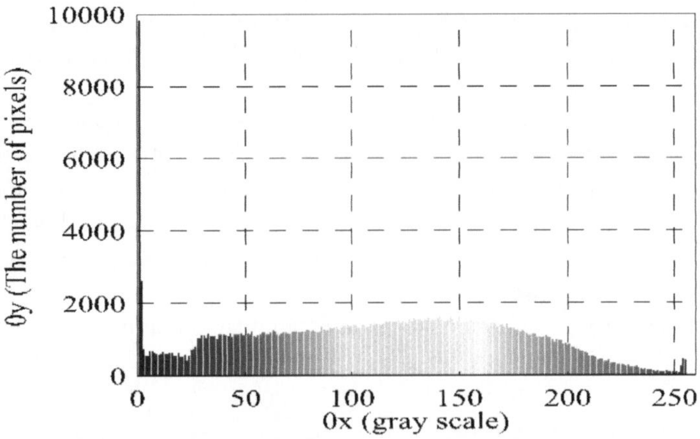

Figure 2. The histogram of the rough surface image presented in the RGB color palette.

The image of the rough surface 515×529 pel in size was obtained by means of an electron microscope Quanta 200.

A human eye can perceive different colors in the result of addition of three basic colors with different color intensities [19]. For this reason, we selected the RGB color palette in which different colors are formed in a similar way. We built up the image histogram shown in Fig. 1 using a Matlab based software environment. For this purpose, we put each index of the gray scale in correspondence with an appropriate RGB palette index (see Fig. 2).

The squares obtained from the histogram were arranged in the square of 2 in accordance with the following algorithm [20 p. 515] as follows:

1. The square S1, corresponding to the first element of the histogram, is set up at the bottom left corner of the square of 2 (hereafter referred to as the square S0). The rest of the square S0 is divided into two rectangles. A rectangle, whose height is greater than its length, is called a "vertical" rectangle. A rectangle, whose height is less than its length, is called a "horizontal" rectangle (Fig. 3).
2. In step k, the square Sk is set into one of the formed vertical and horizontal rectangles. To arrange the square Sk, we select the rectangle in which either the height (for the horizontal rectangle) or the length (for the vertical rectangle) differs slightly from the side of the square Sk.
3. The square Sk is set into the selected rectangle at the bottom left corner. If the selected rectangle is vertical, the top side of the square proceeds to its intercrossing with the left vertical side of the rectangle. If the selected rectangle is horizontal, the right side of the square Sk proceeds to the top horizontal side of the rectangle. In both cases, new horizontal and (or) vertical rectangles are formed.
4. The algorithm's execution returns to item 2 until the squares Si are settled.

Figure 4 shows the arrangement of the squares, which were calculated in accordance with section 2 of the histogram shown in Fig 2. This type of the arrangement will thereafter be called a packing.

4. Research results

To determine the sensitivity of the proposed method to changes in the image of the rough surface, we have built the information patterns for the image in Fig. 1 without first and last lines and without first and last columns (Fig. 5).

Figure 5 shows that the information pattern is well sensitive to any losses within one line or one column in the image in Fig.1. The loss within the sizes of one line amounts to 0.192 per cent (the line is 529 *pel* in length) and of one column – to 0.194 per cent (the column is 515 *pel*) of the whole image of the rough surface. We consider it necessary to denote that Figure 5 shows also the method's sensitivity to sizes of the image of the controlled rough surface.

The loss of a line or a column results not only in the loss of some number of pixels from a certain gray scale. Figure 6 shows the histograms of lost lines and columns and also a percentage value of lost pixels of the certain gray scale. It is apparent from Figure 6 that the proposed method is sensitive to a set (a variety) of small variations in the number of pixels of the certain gray scale.

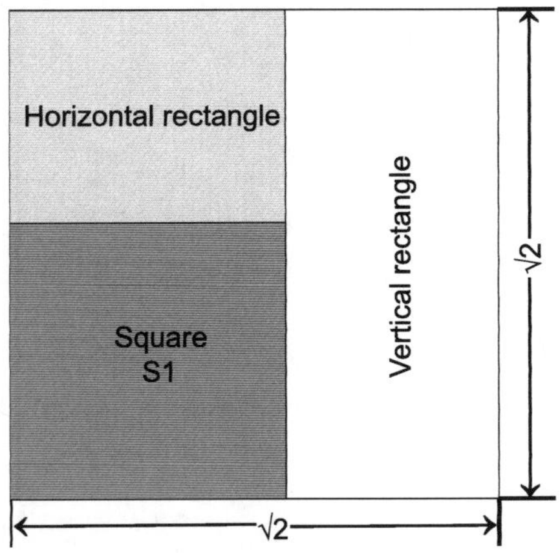

Figure 3. The arrangement of the first square, corresponding to the first element of the histogram, in the square of 2.

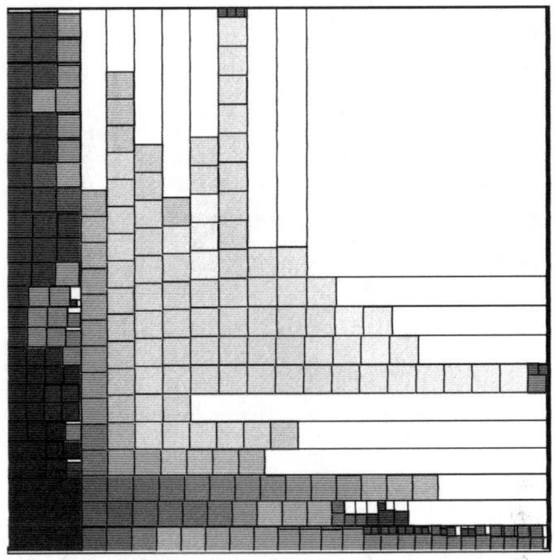

Figure 4. Packing of the finite sequence of the squares, corresponding to the histogram shown in Fig. 2, into the square of 2.

Figure 6 shows that for the image given in Fig. 1 the loss of one line or one column may cause the change in the number of pixels, corresponding to the certain gray scale, in the range from 1 to 13 pixels. Figure 5 also shows that lost pixels of the certain gray scale may be estimated at from 0.1 to 15.3 per cent from the total number of similar pixels in the image in Fig.1.

a

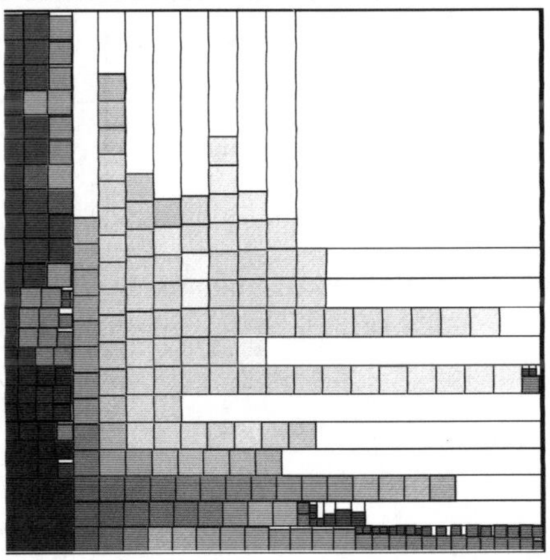

b

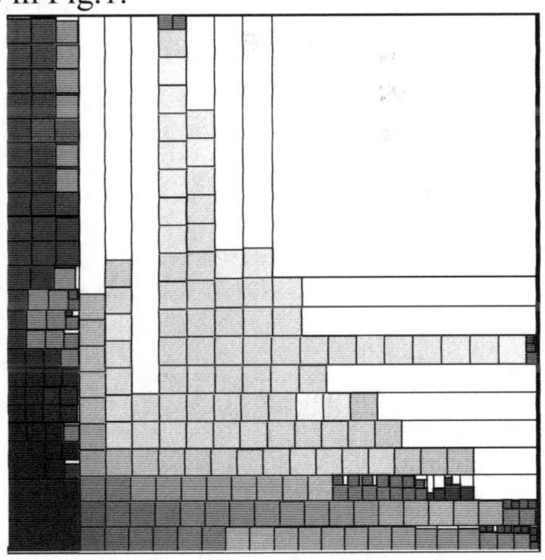

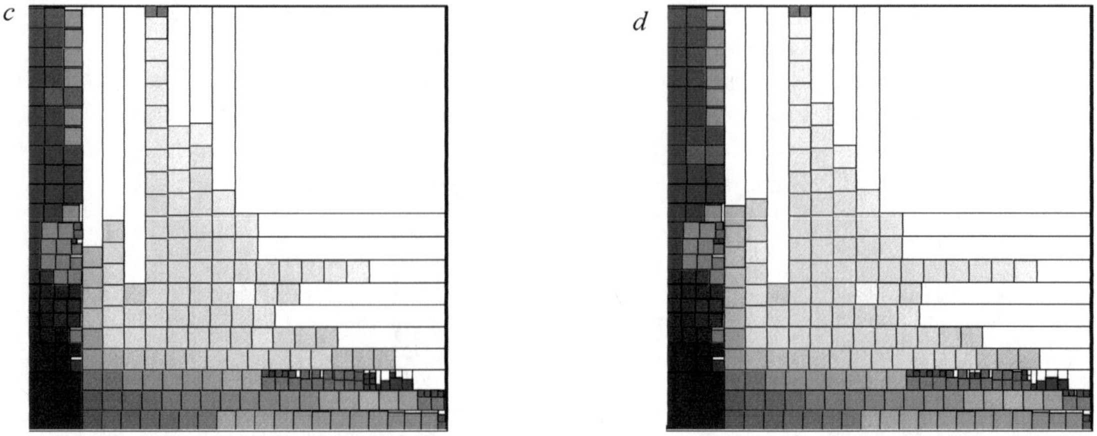

Figure 5. Information patterns of imaging of the rough surface with different losses: a – no first line and b – no last line; c – no left column and d – no right column.

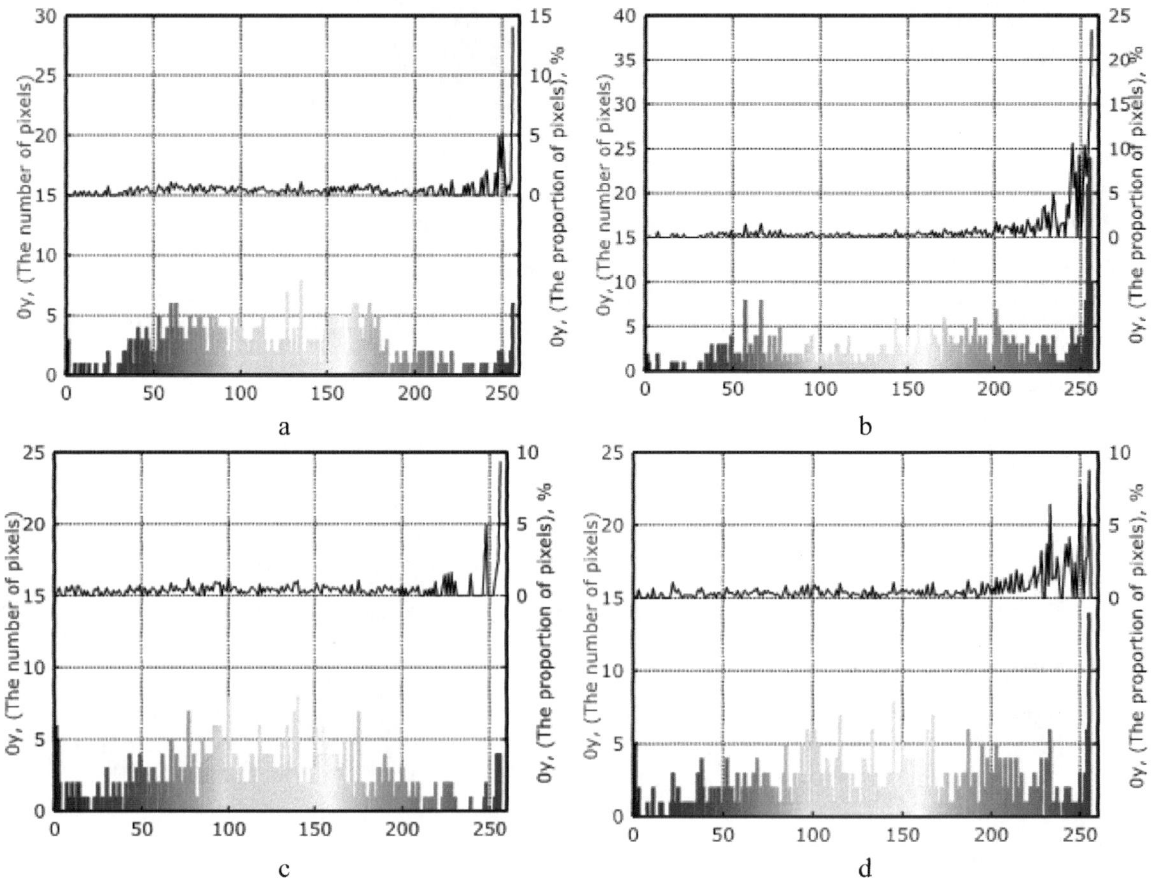

Figure 6. Histograms of lost lines and columns and the percentage value of deleted pixels with some gray scale steps: a – the first line and b – the last line; c – the right column and d – the left column.

5. Parameter value of packing of the image histogram

The applied method of packing the squares, whose area corresponds to the number of pixels of the certain gray scale, forms horizontal and vertical rectangulars in the range of the square of 2 (Fig. 7).

We consider that a ratio of the aggregate squares of vertical and horizontal rectangulars can serve as a numerical characteristic of the information pattern of the roughness image in Fig. 1. Table 1 shows the values of the proposed parameter for the packings in Fig. 3 and Fig. 5 *a, b, c, d*.

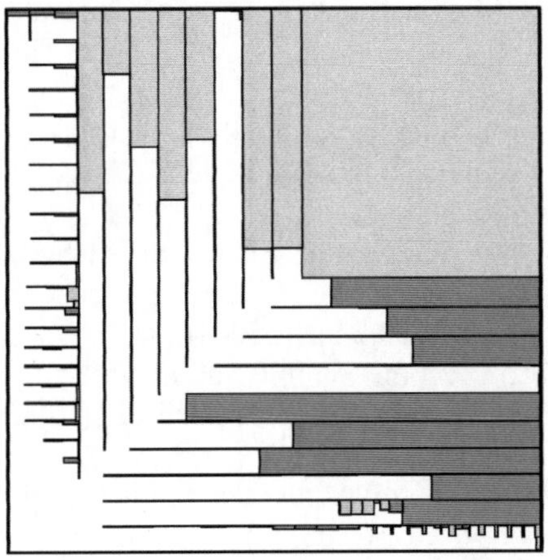

Figure 7. Horizontal and vertical rectangulars of packing of the roughness image histogram in Fig. 1.

Table 1. Parameter values for packings

	Packing				
	Fig. 3	Fig. 4 *a*	Fig. 4 *b*	Fig. 4 *c*	Fig. 4 *d*
Parameter value	$\frac{0.324}{0.685} = 0.4729$	$\frac{0.7641}{0.2648} = 2.8985$	$\frac{0.7271}{20.2914} = 2.495$	$\frac{0.7488}{0.2894} = 2.5874$	$\frac{0.7463}{0.2799} = 2.6663$

Table 1 shows that the parameter value is varied in the range of 429-513 per cent when the number of pixels in the image is changed within the range of 0.192-0.194 per cent.

6. Conclusions

We consider that the conducted research of the proposed method of building the information pattern of the rough surface has made it possible to determine the following: the information pattern visually displays changes of the roughness image in the range of at least 0.192 per cent of the total number of image pixels; the change of the number of pixels from 0.1 to 15.3 per cent in the groups of image pixels with some code of the gray scale results in the change of the parameter value within the range from 429 to 513 per cent (in this case, the number of lost pixels is not less than 0.192 per cent of the total number of image pixels); the information pattern can keep track of the dynamics of small variations (at least no less than 0.192 per cent) of the number of pixels in the image.

References

[1] Vázquez R and Torre D 2013 The Effect of Surface Roughness on Efficiency of Low Pressure Turbines *ASME. J. Turbomach.* **6 A** pp V06AT36A007-16

[2] Yashchuk V V, Samoylova L V and Kozhevnikov I V 2015 Specification of x-ray mirrors in terms of system performance: new twist to an old plot *Opt. Eng.* **54 2** p 025108

[3] Whitehouse D J 2011 *Handbook of Surface and Nanometrology* (Boca Raton, London, New York : CRC Press) p 955

[4] Whitehouse D J 2002 *Surfaces and their Measurement* (Boca Raton, London, New York : CRC Press) p 395

[5] Tabenkin A N, Tarasov S B and Stepanov S N 2007 *Sherohovatost, volnistost, profil Mezhdunarodnyiy opyit* [Roughness, waviness, profile. International experience] ed N A Tabachnikova (Sankt-Peterburg : Politehnicheskiy universitet Publ.) p 133

[6] DeGarmo E P, Black J T and Kohser R A 2003 *Materials and Process in Manufacturing* (New York, Chichester : John Wiley & Sons) p 355

[7] Gong Y, Scott T. Misture S T, Gao P. and Mellott N P 2016 Surface Roughness Measurements Using Power Spectrum Density Analysis with Enhanced Spatial Correlation Length *J. Phys. Chem. C.* **120 39** pp 22358 – 64
[8] Salazar F and Barrientos A 2013 Surface Roughness Measurement on a Wing Aircraft by Speckle
Correlation *Sensors* **13** pp 11772-81
[9] Yoshitomi K, Ishimaru A and Hwang J-N 1993 Surface roughness determination using spectral correlations of scattered intensities and an artificial neural network technique *IEEE Transactions on Antennas and Propagation* **41 4** pp 498-502
[10] Zhou Y, Li Y, Zhu H, Zuo X and Yang J 2015 The Thee-Point Sinuosity Method for Calculating
the Fractal Dimension of Machined Surface Profile *Fractals* **20 2** pp 1550016-26
[11] Ichida Y 2012 Fractal Analysis of Micro Self-Sharpening Phenomenon in Grinding with Cubic Boron Nitride (cBN) *Wheels Scanning Electron Microscopy* ed V Kazmiruk (Rijeka : INTECH) pp 393-410
[12] Outer A D, Kaashoek J F and Hack H R G K 1995 Difficulties with using continuous fractal thery for discontinuity surfaces *International Journal of Rock Mechanics and Mining ciences & Geomechanics Abstracts* **32 1** pp 3-9
[13] Cattaneo G, Ciucci D and Bianucci D 2008 Entropy and Co-entropy of Partitions and Coverings with Applications to Roughness Theory Granular *Computing Studies in Fuzziness and Soft Computing* **224** pp 55-77
[14] Pascoal R and Monteiro A M 2014 Market Efficiency, Roughness and Long Memory in PSI20 Index Returns: Wavelet and Entropy Analysis *Entropy* **16 5** pp 2768-278
[15] Leach R K and Giusca C L 2012 Determination of the metrological characteristics of optical surface topography measuring instruments *Proc. SPIE* **8430** pp 84300Q-7
[16] Leach R K 2014 *Fundamental Principles of Engineering Nanometrology* (Norwich, NY : William Andrew Publ./American Elsevir) p 384
[17] Neri P 2009 Nonlinear characterization of a simple process in human vision *Journal of Vision* **9 1** pp 1-29
[18] Jaglom I M 1968 *Kak razrezat kvadrat ?* [How to cut a square ?] (Moscow : Nauka Publ.) p 112
[19] Solomon S G and Lennie P 2007 The machinery of colour vision *Nature Reviews Neuroscience* **8** pp 276-286
[20] Korte B and Vygen J 2008 *Combinatorial Optimization* (Berlin, Heideberg : Springer Sciens) p 720

Impact of Grinding Wheel Position on Flute Profile of End mill and Cutting Process

Alexey N Zhidyaev*

Samara National Research University, Institute of Engine and Power Plant Engineering, Moskovskoye shosse, 34, Samara, Samara region, 443086, Russian Federation

E-mail:* a.n.zhidyaev@ssau.ru

Abstract. The article provides the research on flute profile of end mills. The flute profile of a specified end mill design is affected by grinding wheel position relative to end mill workpiece in grinding operation. Different profiles were obtained, their rake surfaces and cross section areas were compared. Temperatures and cutting forces were evaluated via finite element analysis for these profiles in case of titanium alloy and heat-resistant alloy machining. Finally, the impact of flute profile on cutting process was assessed.

1. Introduction

The manufacturing process of jet turbine engine (JTE) parts is constantly improving. The improvements include the increase of productivity and decrease of production costs. The productivity can be augmented by cutting conditions and special design of cutting tools but it is limited by the properties of so-called hard-to-machine materials [1, 2]. For example, properties of titanium alloys are also developing [3] which is likely to lead to their poorer machinability. The enhancement of manufacturing process [4] implies improvement of milling operations. The most complicated milling operations occur while 5-axis machining of free-form parts [5], i.e. blades, blisks.

The solid carbide end mills are extensively used in all metal cutting operations during the manufacturing process of JTE parts made of hard-to-machine materials: 3-axis and 5-axis; finishing, semi-finishing and roughing operations. Researchers have paid attention to end mill designing and constructional parameters choice [6, 7] as well as to special end mills design [8]. Some research is devoted to perspective designs of milling cutters [9, 10] for increased stability of the cutting process varying helix angle and pitch angle of the end mill. Also, there are articles on design and tool wear and theirs impact on temperature in tool-workpiece interface [11].

The design of the solid carbide end mill or the range of end mills for hard-to-machine materials begins with the selection of constructional parameters. For the cylindrical part of an end mill the parameters are outer diameter, number of teeth, rake angle, helix angle, core diameter and clearance angle. Manufacturers' standards or some national standards may serve as the source of constructional parameters. For a wide group of materials and applications the recommendations for the design and constructional parameters are usually of general character. Besides the set of these constructional parameters does not allow to evaluate a unique flute profile. A specified set of these parameters may result in a number flute profiles. The flute profile is affected by constructional parameters and also manufacturing technology. In case of using an NC tool grinding machine manufacturing technology determines grinding wheel type, dimensions and position relative to the workpiece. The workpiece is usually a cylindrical solid carbide rod ground to h6 or h5 tolerance.

Numerous research is devoted to the adjustment of grinding wheel position. Article [12] represents the method of flute grinding with the standard forms of grinding wheel enabling to obtain the desired flute width. The method of wheel position calculating obtained in the article [13] provides a more precise method of rake angle evaluation. Some research focuses on calculating the grinding wheel profile to obtain the desired flute, for instance, article [14] describes such method for special flutes. Also, a flute is a complex surface, thus the appropriate method of flute measurement should be considered providing high precision [15].

To summarise there is a gap in research on the impact of grinding wheel position on the flute profile in case of predefined constructional parameters. Moreover, the effect of various flute profiles on the cutting process has not been estimated. Specifically it concerns the effect of the rake surface of flute (surface in the environment of the cutting edge).

The purpose of this paper is to evaluate the impact of design parameters and grinding operation of end mill on the rake surface profile and evaluate the impact of this profile on the cutting process.

2. Method

The process of designing a range of end mills starts with the choice of the representative end mill with predefined outer diameter. Other end mills would be generated based on the representative mill. The outer diameter of the representative end mill is chosen equal to 12 mm. It is necessary to follow the certain algorithm of choosing and conforming parameters to design the end mill. Also, it is essential to know that the design is directly connected to manufacturing technology. Therefore the design of the end mill is produced via the "System for NC codes preparation". This system is provided with the "VIZAS" VZ-630F4 tool grinding machine.

At the first step, the system requires values of constructional parameters of the end mill to be specified. At this stage, end mill outer diameter, number of teeth, rake angle, helix angle, core diameter, clearance surface width and clearance angle should be determined in order to define flute profile. Other constructional parameters do not affect the flute. At the second step the types and dimensions of the grinding wheels' set are entered. In case of standard wheel's application for flute grinding 1A1 or 1V1 the wheel types can be applied. 1V1 type was applied for further research. The third step considers the cutting conditions to be determined.

Then the modelling of cross section and whole end mill is executed. This stage implies modelling of flute profile and flute's rake surface which is likely to have a crucial impact on the cutting process. The system provides adjusting the core diameter and the swivelling angle of grinding wheel relative to workpiece axis. Core diameter should not be changed as a constructional parameter. Swivelling angle is the B angle (angle of B axis) of the prescribed machine tool and it generally differs from helix angle of the end mill. Figure 1 represents the respective wheel and workpiece position in flute grinding. Finally, the NC code can be generated for NC tool grinding machine.

IOP Conf. Series: Materials Science and Engineering **302** (2018) 012069 doi:10.1088/1757-899X/302/1/012069

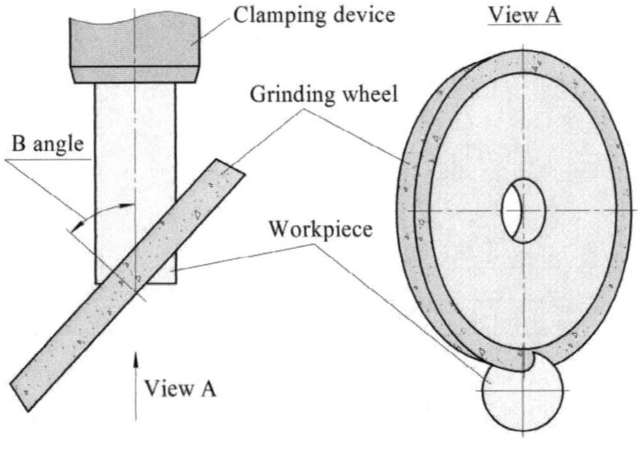

Figure 1. Wheel position in flute grinding operation.

Adjusting swivelling angle six different profiles (six sections) with the same rake angle and core diameter were generated. Rake angle was 9° whereas B angles were 42.4°, 45.0°, 47.4°, 50.0°, 52.5° and 55.5°. The area of each cross section was assessed. The seventh section was modelled as a cutter with constant rake angle. The profile for 45.0° B angle is shown in Figure 2a.

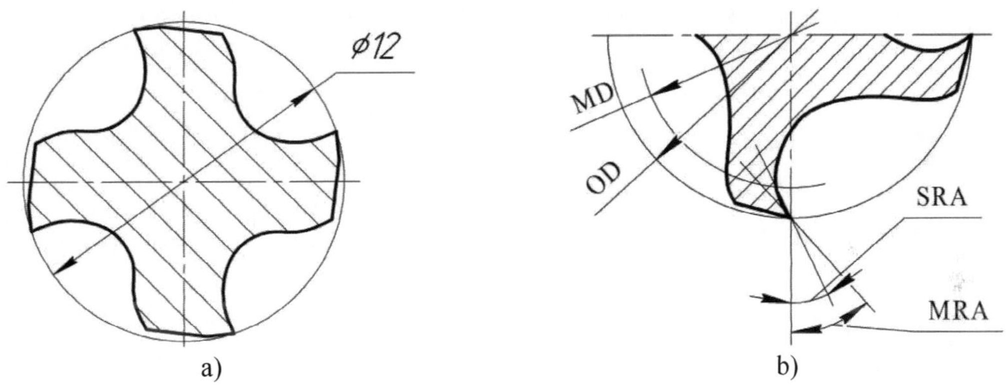

Figure 2. Generated profile and rake angle measurement: a) Profile generated for representative end mill; b) rake angle measurement (OD – outer diameter of the end mill, MD – diameter of measurement, SRA – specified rake angle, MRA – measured rake angle)

Rake angle measuring method should be taken into account. Figure 2b illustrates the difference between specified rake angle and measured rake angle which can derive from the difference in diameters. In "System for NC codes preparation" rake angle is specified at the outer diameter (OD) while Walter tool measuring machine provides measuring on a certain diameter. For instance, if OD equals 12 mm then the diameter of measurement is set as 11 mm. The additional eighth section with 12° rake angle on OD was modelled. For this section the rake angle on the diameter of measurement equals 9°.

After that, the process of milling titanium alloy Ti6Al4V and heat-resistant alloy Inconel 718 was modelled using "Deform" finite element analysis software. Peak (maximum) temperatures and peak tangential cutting forces were evaluated. The model of orthogonal cutting was modified for milling process, i.e. the depth of cut was linearly decreasing from the feed value to zero on the whole length of cut. Length of cut was calculated on the basis of mill diameter and depth of cut. Feed per tooth was 0.08 mm/tooth, milling depth was 2 mm for both materials while cutting speed was 60 m/min for

Ti6Al4V and 90 m/min for Inconel 718. It is also essential to use the verified cutting model for more reliable results [16, 17].

Two additional tests were performed with the feed 0.16 mm per tooth for sections 3 and 8.

3. Results

Rake angles on different measuring diameters and cross section areas were evaluated for obtained profiles. The angles are presented in Table 1.

Table 1. Measured rake angles.

		Diameter of measurement, mm					Cross section
Section	B angle	12.0	11.5	11.0	10.5	10.0	area, mm^2
1	42.4°	9.0	7.6	5.7	3.5	1.1	70.5
2	45.0°	9.0	6.8	5.1	3.6	2.4	68.7
3	47.4°	9.0	6.5	4.4	2.6	1.2	67.3
4	50.0°	9.0	7.5	6.7	4.1	1.9	65.1
5	52.4°	9.0	7.8	5.8	3.6	1.3	61.0
6	55.0°	9.0	5.9	4.1	2.2	0.6	61.1
7	–	9.0	9.0	9.0	9.0	9.0	–
8	47.4°	12.0	10.5	9.0	7.2	4.9	66.5

Section 7 is the cutter with constant rake angle therefore cross section area and B angle could not be evaluated. After that finite element modelling was conducted for these sections. Figure 3 and 4 demonstrate the cutting process of Ti6Al4V and Inconel 718 respectively for section 3 after 1.75 mm of cut with the feed 0.08 mm/tooth.

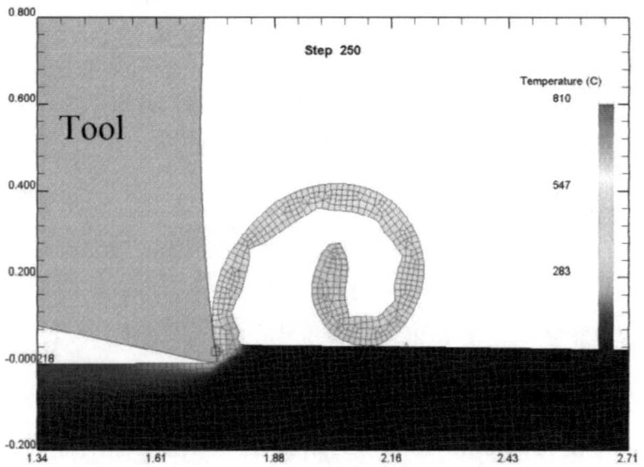

Figure 3. The modelling process of cutting Ti6Al4V with the tool derived from section 3.

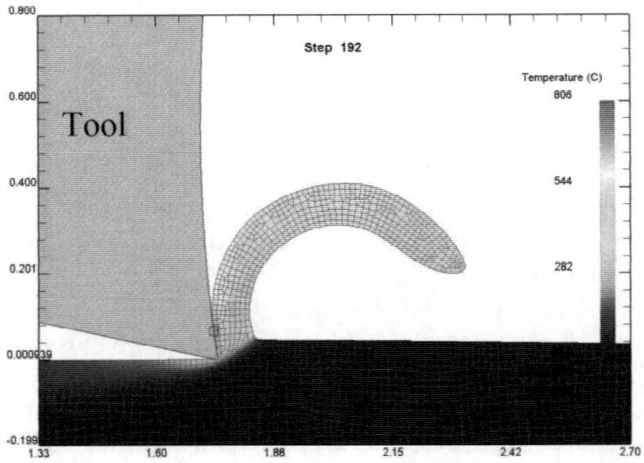

Figure 4. The modelling process of cutting Inconel 718 with the tool derived from section 3.

Peak temperatures were calculated in the tool-chip interface. Tangential cutting forces applied to the end mill's tooth were assessed. The modelled peak temperatures and tangential forces are represented in Table 2 for both materials.

Table 2. Peak temperatures and tangential forces obtained in modelling.

	Ti6Al4V		Inconel 718	
Section	Peak temperature, °C	Peak tangential force, N	Peak temperature, °C	Peak tangential force, N
1	786	106.5	764	154.6
2	795	108.0	774	162.3
3	777	106.7	786	162.2
4	785	106.2	768	157.2
5	783	102.4	783	156.3
6	773	109.4	767	159.2
7	788	106.8	787	154.3
8	780	104.4	757	148.3
3 (0.16 mm/tooth)	888	178.9	867	317.2
8 (0.16 mm/tooth)	888	177.1	857	286.9

Maximum contact length (CL) was measured for processes with 0.16 mm/tooth feed. In case of titanium alloy machining maximum CL was 0.09 mm, while in heat-resistant alloy machining CL equaled 0.12 mm.

4. Conclusion and discussion

As could be seen from Table 1 the measured rake angles decrease with descending of the measuring diameter. This evolution of rake angle results in different profiles and rake surfaces. The first derivation from the evolution is the diverse cross section areas. Section area tends to rise with the decrease of B angle. Thus the rigidity of cutting tool rises. The second derivation is the possible impact on the cutting process.

Peak temperatures given in Table 2 for Ti6Al4V alter from 773 °C to 795 °C for sections obtained by varying B angles, for the cutter with constant rake angle (section 7) and for section 8 in case of 0.08 mm/tooth feed. The difference in temperatures is only 3%. For the process with 0.16 mm/tooth feed the temperatures are equal. The research on temperatures reveals that there is no any crucial impact of rake surface in the chosen range of angles on the peak temperature in Ti6Al4V machining.

The peak temperature varies within the range of 4% during heat-resistant alloy machining. Temperature tends to be higher when rake angle decreases faster (for rake angles assessed at 11.5 OD).

The impact of rake surface in Inconel 718 machining is more noticeable in comparison to Ti6Al4V machining.

For Ti6Al4V maximum tangential cutting force for sections from 1 to 8 varies in the range of 6% with a weak correlation between it and rake angle change. This can be due to the small dimension of contact length compared to the length of rake surface where measurements occurred. In other words, on the measured contact length rake angle appears to be constant for different profiles. The considerable increase in feedrate might have an impact on forces obtained for different sections but increased federate is unlikely to be used in real machining practice for titanium alloys. Higher feedrates could be implemented in aluminum alloys machining.

Peak tangential cutting force in Inconel 718 machining demonstrates the fall while rake angle on 11.5 mm OD remains higher than 7° and conversely force increases with decrease of rake angle. For modelled sections 1-8 the force range is within 9%. This can be due to higher contact length in comparison to titanium alloy. Also cutting conditions affect the correlation between rake angle and peak temperatures or cutting forces. Preliminary tests run on Inconel 718 with 60 m/min cutting speed revealed no significant correlation.

In conclusion, it should be noted that simple specification of constructional parameters values does not enable to obtain the most effective end mill design. Temperatures and cutting forces can be affected by rake surface under certain cutting conditions. Higher rigidity can be achieved altering wheel position in flute grinding operation.

Using this method of the end mill design improvement the most sustainable design can be achieved and also metal cutting process can be enhanced. It is particularly important for stable hard-to-machine materials milling.

Acknowledgments

This work was supported by the Ministry of Education and Science of the Russian Federation in the framework of the implementation of the Program "Research and development on priority directions of scientific-technological complex of Russia for 2014-2020".

References

[1] Ezugwu EO, Bonney J, Yamane Y. An overview of the machinability of aeroengine alloys. J Mater Process Technol 2003;134(2):233-253.

[2] Arrazola P-J, Garay A, Iriarte L-M, Armendia M, Marya S, Le Maître F. Machinability of titanium alloys (Ti6Al4V and Ti555.3). J Mater Process Technol 2009;209(5):2223-2230.

[3] Grechnikov FV, Khaimovich AI. Raising the fatigue resistance of titanium alloys by high-speed deformation in the range of polymorphic transformations. Met Sci Heat Treat 2016;57(11-12):726-730.

[4] Vdovin RA, Smelov VG, Bolotov MA, Pronichev ND. Paths of Improving the Technological Process of Manufacture of GTE Turbine Blades. IOP Conference Series: Materials Science and Engineering [Internet]. IOP Publishing; 2016 Aug;142:012073. Available from: http://dx.doi.org/10.1088/1757-899x/142/1/012073

[5] Khaimovich IN, Surkov OS, Khaimovich AI. Practice in the application of specialized technologic process patterns of 5-axis machining operation of blade rings of blisks. Russ Aeronaut 2016;59(1):112-117.

[6] Jiang Z, Sun J, Xiong Q, Wang L, Li J. Structural design of groove and micro-blade of the end mill in aluminum alloys machining based on bionics. Int J Adv Manuf Technol 2017;88(9-12):3343-3356.

[7] Skuratov DL, Zhidyaev AN, Sazonov MB. Solid carbide end mills tool life increase in titanium alloys machining by design development and rational choice of geometrical parameters. Res J Appl Sci 2014;9(11):767-770.

[8] Liu X, Fan M, Ji W, Wang G, Chen T. Research on models of design and NC manufacturing for ellipsoid end mill. Int J Adv Manuf Technol 2016;85(9-12):2729-2744.

[9] Suzuki N, Ishiguro R, Kojima T. Design of irregular pitch end mills to attain robust suppression of regenerative chatter. CIRP Ann Manuf Technol 2016;65(1):129-132.
[10] Yusoff AR, Sims ND, Turner S. Experimental Validation of Chatter Stability for Variable Helix Milling Tools. IOP Conference Series: Materials Science and Engineering [Internet]. IOP Publishing; 2011 Dec 8;26:012010. Available from: http://dx.doi.org/10.1088/1757-899x/26/1/012010
[11] Evdokimov DV, Skuratov DL, Fedorov DG. Influence of tool wear on the heat-flux distribution and temperature at the contact surfaces in the end milling of OT4 titanium alloy. Russ Eng Res 2016;36(4):324-327.
[12] Ren L, Wang S, Yi L, Sun S. An accurate method for five-axis flute grinding in cylindrical end-mills using standard 1V1/1A1 grinding wheels. Precis Eng 2016;43:387-394.
[13] Nguyen V-H, Ko S-L. A New Method for Determination of Wheel Location in Machining Helical Flute of End Mill. J Manuf Sci Eng Trans ASME 2016;138(11).
[14] Li G. A new algorithm to solve the grinding wheel profile for end mill groove machining. Int J Adv Manuf Technol 2017;90(1-4):775-784.
[15] Pechenin VA, Bolotov MA, Ruzanov NV, Yanyukina MV. Optimization of measurements of the geometry of parts with complex surfaces. Meas Tech 2015;58(3):261-268.
[16] Malakizadi A, Hosseinkhani K, Mariano E, Ng E, Del Prete A, Nyborg L. Influence of friction models on FE simulation results of orthogonal cutting process. Int J Adv Manuf Technol 2017;88(9-12):3217-3232.
[17] Bil H, Kiliç SE, Tekkaya AE. A comparison of orthogonal cutting data from experiments with three different finite element models. Int J Mach Tools Manuf 2004;44(9):933-944.

IOP Conf. Series: Materials Science and Engineering **302** (2018) 012070 doi:10.1088/1757-899X/302/1/012070

Softening of Low-alloyed Titanium Billets with Laser Annealing

Serguei P. Murzin[1]* and Nikolay L. Kazanskiy[2,3]

[1]Department of Automatic Systems of Power Plants, Samara National Research University, Samara, Russia

[2]Image Processing Systems Institute of RAS – Branch of the FSRC "Crystallography and Photonics" RAS, Samara, Russia

[3]Department of Technical Cybernetics, Samara National Research University, Samara, Russia

*Corresponding author: murzin@ssau.ru

Abstract. An experimental study of work-hardened metal release conditions and recrystallized structure formation with laser heating, which showed the possibility of these processes in low-alloy titanium was performed. Studies of treated materials properties showed that using of laser annealing enhances ultimate tensile strain and reduces minimum bend angle for cold deformation of rolled sheet metal parts of these materials. Using of laser annealing for local softening before shaping billets of titanium alloys allows to improve the accuracy of parts manufacturing by reducing their springy action. This provides an opportunity to expand their shaping options without additional heating.

1. Introduction

Laser beam machining is a progressive process for improving of selected materials properties and service characteristics. The feasibility and advantages of the use of lasers in material processing are determined by the possibility of a non-contact, strictly dosed and intense supply of energy to the product surface. Laser technologies can be used to process remote as well as local areas of objects in the absence of vibrations and other negative effects on the material. These and other significant advantages lead to greater potential for the usage laser processing in present time as well as in the future. Due to its extremely localized, concentrated energy input of focused laser radiation, laser material processing delivers a higher energy density as any other heat source to processed components. Thus, laser material treatment can be used not only for laser welding or cutting, but for modification of physical and mechanical properties of materials. Current status of achievements in the area of physical characteristics of laser beam processing is presented in various papers and monographs [1-3]. Numerous reference books [4-6] describe in detail applications of laser devices in diverse production technologies. It is shown that such important characteristics of metallic materials, as tensile strength, fatigue strength and wear resistance are structurally-sensitive, i.e., can be controlled by proper changes of material structures due to laser processing. Only few works are devoted to investigations of softening materials by controlled changes in the structure of these materials [7-10]. Even the term «laser annealing», which in a broader sense means changing the structure of solids by laser radiation of varying duration in the literature usually refers to a pulse oriented crystallization of semiconductor structures by laser radiation of nanosecond duration.

One of the most effective means for increasing the strength of machine parts from sheet and round blanks is to use cold-worked materials which have high hardness and low plasticity. While improving the structural condition of the material by annealing, a significant improvement in the plastic flow of the metal and the localization of the deformation can be realized. [7-9] demonstrated the possibility of alloys recrystallization at high temperatures and short exposure times during laser heating. The width and location of the annealing zone on the blank are determined by the accuracy of forming operations and allowable deformation without fracture. Except the laser, no other known local heating method can be used to obtain the desired width of the processing zone because of their lower power density. In [10] the method of local laser annealing for metallic sheet materials was developed. Conditions for the removal of cold-hardening and formation of recrystallized structure were determined. The method was implemented using a CO_2-laser. In the optical transformation system, the focusing mirror was a diffractive optical element [11-13]. The power density was increased at the periphery of the rectangular-shaped heat source, which was moving at a constant speed. This resulted in uniform thermal conditions across the width of the heat affected zone. The local laser annealing of plastic deformation areas provides an increase in the ultimate elongation, and a decrease in the minimum bend radius.

Titanium alloys are used in forming production of details of aircraft and their engines for the purpose of increase in specific durability and rigidity of products [14, 15]. Medium and high-alloyed titanium alloys practically don't give in to a cold shaping because of the high resistance of metal of deformation, intensive hardening, tendency to cracking and tearing. The parts are manufactured with billets heating using traditional and special techniques at existing stamping production equipment. Billets heating is performed in resistance furnace, radiation and electrocontact heating units, heat conduction equipment [16, 17]. Low- alloyed titanium alloys also have lower ductility characteristics, and because of that the deformed metal should be heated. The most important features that determine processing characteristics of such structural materials, along with limited abilities of cold deformation are their low heat conduction and high interaction activity with the environment. The laser heating is preferred because it enables a significant reduction of metal materials residence time at intense oxidation and surface layers gas saturation temperatures. The purpose of this paper is to determine the possibility of laser annealing using for local softening of low-alloy titanium Ti-2Al-1.5Mn and Ti-4Al-1.5Mn billets before shaping.

2. Material under study and experimental equipment

Titanium alloys Ti-2Al-1.5Mn and Ti-4Al-1.5Mn are related to an alloy group with a prevalence of α -solid solution and a small amount of β -phase (pseudo- α -alloys) and have polymorphic transformation temperature of 1180–1220 K and 1190–1230 K respectively. These alloys are suitable for welded assemblies, parts (including thin-walled) and products continuously operating at temperature up to 570 K, and with operating time up to 2000 hours at temperature of 620 K. Shaping with heating is the main method of low-alloy titanium forming. These alloys have satisfactory technological plasticity at temperature range of 760–870 K for Ti-2Al-1.5Mn and 820–970 K for Ti-4Al-1.5Mn. However, heating in air at temperatures above 770 K forms defective oxide and gas-containing layers on billets surfaces, reducing the parts operational strength and deteriorating the material formability.

One of the main features of laser technologies for processing parts of aircraft engines with adjustable spatial distribution of beam power are increased requirements for the accuracy of maintaining the parameters of processing regimes. This makes it necessary to perform laser processing of parts by continuous radiation in an automatic mode. To provide the required set of characteristics of machined parts in the design of technological processes, it is advisable to apply mathematical modeling methods, it is necessary to modernize the optical system of the process equipment, use a computer control system for the processing of parts and non-destructive express control.

When monitoring temperature fields on the surface of objects, an element base was used from an infrared radiometer-a unit for visualizing the temperature field in the working area of a thermal imager with analog-to-digital converters, an image input / output unit and a personal computer with the appropriate image processing software. The correspondence of the emissivity of the surface of the investigated object to the values of the level and range of the investigated temperatures was established in the calibration mode of the signal. The temperature distribution over the surface of the object was fixed in the form of an image with different brightness of the sections or in the conventional colors of the RGB palette. Processing of information about thermal processes was carried out using the developed software, functioning in a Windows environment. The analysis of the thermal images was performed both in the automatic mode of operation and in the command interactive mode.

Laser treatment of the parts was performed with CO_2-laser processing equipment BYSTAR 2512 with BTL 1800 laser that has CW output power range 100–1800 W. The beams transport and formation systems including diffractive optical element were used for directive change of the high-intensity laser beam power density distribution [18-21]. The figure 1 shows CO_2-laser processing equipment BYSTAR 2512 with a technological optical device comprising a diffractive optical element and a system for contactless diagnostics of thermal processes.

Figure 1. CO_2-laser processing equipment BYSTAR 2512 with a technological optical device comprising a diffractive optical element and a system for contactless diagnostics of thermal processes.

The procedure of mechanical properties testing was conducted using tensile-testing machine Zwick/Roell Z050. The size of the samples was determined in accordance with state standards. A series of stretching tests of rolled sheet Ti-2Al-1.5Mn material specimens with thickness 2 mm were conducted to determine the ultimate tensile strain that characterizes alloys ductility. Rolled sheet titanium alloys samples were tested on bending. The estimation of minimum bend angle was carried out on test stand with force of 200 kN, by continuous increasing the load on the sample until first crack. The support bending radius was 15 mm and rounding mandrel diameter was 6 mm for plates of Ti-2Al-1.5Mn and Ti-4Al-1.5Mn with thickness of 2 mm.

3. Results and discussion

The experimental study of work-hardened metal removal and recrystallized structure formation conditions with laser heating showed the possibility of these processes in low-alloy titanium recrystallization temperature range (for Ti-2Al-1.5Mn: 990–1110 K, and for Ti-4Al-1.5Mn: 1030–1130 K). A nonlinear heat conduction problem was considered in order to determine treatment parameters during heating of the sample with surface heat source. The calculation was made for the sample that is moving at a constant linear velocity in the three dimensional Cartesian coordinate system placed at the heat source center. The assumption was made that the heat source with specified geometry and spatial power distribution is surface source and occupies an area on the sample.

Nonlinear conditions of heat transfer with the environment were specified on the rest of the surface. The treatment parameters were set as follows: laser power 450 ± 5 W; processing speed 0.6 ± 0.01 m/s.

Figure 2 shows the structure of low-alloy Ti-2Al-1.5Mn rolled sheet titanium with thickness 2 mm after laser annealing. The initial material has a fibrous structure. The annealing zone has a granular structure represented by α -phase and a small amount of β -phase. Recrystallization of deformed structure followed by stable recrystallized structure formation occur during laser annealing.

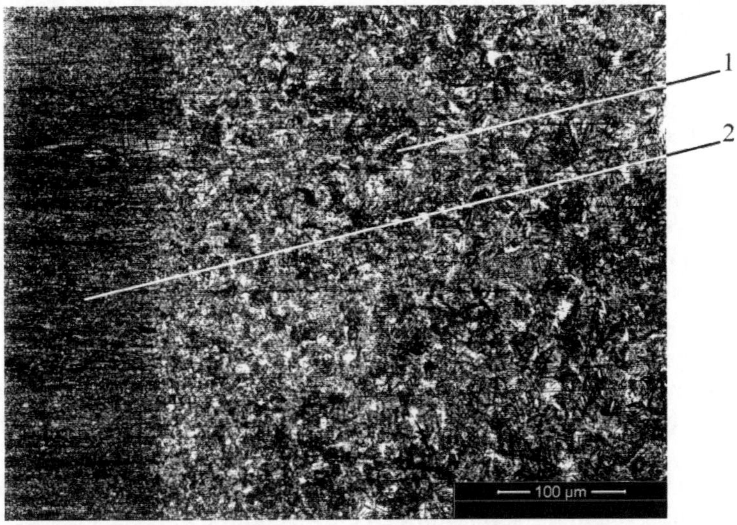

Figure 2. The rolled sheet titanium alloy Ti-2Al-1.5Mn structure after laser annealing: 1 - annealing zone, 2 - initial structure.

The results of stretching tests are shown in table 1. The results of bending tests are presented in table 2. Thus, laser annealing enhances ultimate tensile strain and reduces minimum bend angle for cold deformation of rolled sheet metal parts of these materials, that provides an opportunity to expand their shaping options without additional heating.

Table 1. Stretching tests results for Ti-2Al-1.5Mn rolled sheet material

Material condition	Elongation after fracture, %	Uncertainty of the result, %	Enhance of ultimate tensile strain, %
Before laser annealing	8.2 (7.4–8.9)	7.0	-
After laser annealing	21.5 (18.5–23.5)	7.3	10–15

Table 2. Bending tests results for Ti-2Al-1.5Mn and Ti-4Al-1.5Mn titanium alloys rolled sheet samples of 2mm thickness

Material condition		Minimum bend angle, rad	Uncertainty of the result, %	Reduction of the minimum bend angle, %
Before laser annealing	Ti-2Al-1.5Mn	0.94 (0.9 … 0.97)	7.6	-
	Ti-4Al-1.5Mn	0.77 (0.73 … 0.81)	6.8	
After laser annealing	Ti-2Al-1.5Mn	0.6 (0.56 … 0.64)	5.1	30–40
	Ti-4Al-1.5Mn	0.51 (0.47 … 0.53)	6.2	30–40

Stamped parts precision depends on a vast number of factors that cause error formation. These factors include: stamp type and billet fixing method, stamp manufacturing precision and its wear, the technological process structure, i.e. the number and sequence of operations, which quantity increase raises the accumulation of errors, etc. When bending accuracy is heavily dependent on the elastic and plastic material properties, determining the amount of elastic deformations (elastic springing expressed in angular coordinates) and causes manufactured parts forms and dimensions errors.

Comparative tests of Ti-2Al-1.5Mn parts with thickness 2 mm and bending angle 1.04 rad were conducted to estimate springing angle. The prototypes were manufactured with a local laser annealing of bending space in accordance with the developed approach. Control samples were made with same party material, their bending was carried out with the core technology without laser annealing. As a result, the springing angle without laser annealing of bending space was ~0.052 rad, with a local laser annealing was 0.028 (0.026 –0.03) rad. The use of a local laser annealing reduces the springing angle after bending low-alloy titanium parts by 40–50%. The comparison of laser annealed parts fatigue strength was conducted. The tests were carried out for the first oscillation mode of cantilever fitted parts using step loading at temperature 293 K with the number of test cycles $5\cdot10^6$. The comparison showed that the sample parts made using laser annealing are highly competitive with the serial ones.

Thus, using of laser annealing for local softening before shaping billets of titanium alloys allows to improve the accuracy of parts manufacturing by reducing their springy action. Interoperational annealing for work-hardened material removal of alloys having a low critical strain rate is also advisable to be carried out with laser heating. Annealing zone width of the billet and its placing is determined by shaping operations accuracy and conditions of failure strain course. The desired treatment zone width cannot be obtained by any other method except local laser heating as they have no high energy density for heat transfer.

Presumably, the annealing zone formation of a variable width will allow stamping parts with variable bending angle. Combining laser cutting and shaping operations with technological operation of local laser heating is a promising application for specialized equipment. Schemes of stress and strain conditions are different for various shaping operations, but the suggested approach to the development of the combined manufacturing method for laser annealing and stamping operations can be extended to drawing, swaging, flaring, shaping operations, etc. Titanium alloys are used for axial compressors rotor blades manufacture for gas turbine engines. It is known that the equiaxed globular structure of titanium alloy ensures maximum durability, while the needle-type structure ensures maximum heat resistance. Creating structures on local sites of the parts depending on their purpose and operational conditions may be an additional opportunity to increase the reliability of products.

4. Conclusions

The experimental study of work-hardened metal removal conditions and recrystallized structure formation with laser heating, which showed the possibility of these processes in recrystallization temperature range (for Ti-2Al-1.5Mn: 990–1110 K, and for Ti-4Al-1.5Mn: 1030–1130 K) was conducted. The treatment parameters were set as follows: laser power 450 ± 5 W; processing speed 0.6 ± 0.01 m/s. The annealing zone has a granular structure represented by α -phase and a small amount of β -phase. Recrystallization of deformed structure followed by stable recrystallized structure formation occurs during laser annealing.

Using of laser annealing enhances ultimate tensile strain by 10–15% and reduces minimum bend angle by 30–40 % for cold deformation of Ti-4Al-1.5Mn and Ti-2Al-1.5Mn titanium alloys rolled sheet parts. The use of a local laser annealing reduces the springing angle during bending parts of the low-alloy titanium by 40–50%. This provides an opportunity to expand their shaping options without additional heating.

Titanium alloys are used to manufacture working blades of axial compressors of gas turbine engines. It is known that the equiaxial globular structure of titanium alloys provides maximum endurance, and the structure of needle type - maximum heat resistance. The creation of structures depending on the purpose and working conditions of the parts can be an additional reserve to increase their reliability.

Acknowledgments
This work was supported by the Ministry of Education and Science of the Russian Federation as part of the Program "Research and development on priority directions of scientific-technological complex of Russia for 2014-2020" within the project RFMEFI57815X0131.

References

[1] Lawrence J, Pou J, Low D K Y and Toyserkani E 2010 *Advances in laser materials processing technology: technology, research and application* ed E Toyserkani (Cambridge, UK: Woodhead Publishing / CRC Press)

[2] Ion J C 2005 *Laser processing of engineering materials: principles, procedure and industrial application* (Amsterdam, NL/Oxford, UK. Elsevier/Butterworth-Heinemann)

[3] Schaaf P 2001 *Laser processing of materials: fundamentals, applications and developments.* ed P Schaaf (Berlin Heidelberg, DE: Springer-Verlag)

[4] Ready J F, Farson D F and Feeley T 2001 *LIA handbook of laser materials processing* ed D F Farson (Orlando, US: Laser Institute of America)

[5] Dahotre N B and Harimkar S P 2008 *Laser fabrication and machining of materials.* (New York, US: Springer Science + Business Media)

[6] Steen W M and Mazumder J 2010 *Laser material processing* (4th ed. London, UK: Springer)

[7] Bergmann H W 1987 Short Term Annealing by Laser Treatment *Proc. SPIE 0801 High Power Lasers: Sources, Laser-Material Interactions, High Excitations, and Fast Dynamics* p 296

[8] Minamida K, Kido M, Ishibashi A, Mogami S and Sasaki S 1990 Surface Annealing of Steel Wires for Automotive Tires by CO_2 Laser with Cone Shaped Focusing Mirror. *Proceedings of the Laser Materials Processing Conference (ICALEO '90)* (Orlando: Laser Institute of America / Bellingham) p 460

[9] Nolan S *R Method for Laser Annealing* US Patent 7,063,755 pub. june 20 2006.

[10] Murzin S P 2016 Local laser annealing for aluminium alloy parts *Laser. Eng.* **33(1-3)** 67

[11] Volkov A V, Kazanskiy N L, Moiseev O Ju and Soifer V A 1998 A method for the diffractive microrelief forming using the layered photoresist growth *Opt. Laser. Eng.* **4-5(29)** 281

[12] Bezus E A, Doskolovich L L and Kazanskiy N L 2011 Evanescent-wave interferometric nanoscale photolithography using guided-mode resonant gratings *Microelectron. Eng.* **2(88)** 170

[13] Pavelyev V S, Borodin S A, Kazanskiy N L, Kostyuk G F and Volkov A V 2007 Formation of diffractive microrelief on diamond film surface *Opt. Laser Technol.* **6(39)** 1234

[14] Peters M., Kumpfert J, Ward C H and Leyens C 2003 *Titanium alloys for aerospace applications in titanium and titanium alloys: Fundamentals and applications*, in: eds C Leyens, M Peters (Wiley-VCH Verlag GmbH & Co. KGaA, Weinheim) pp 333-350

[15] Sibum H 2003 Titanium and titanium alloys from raw material to semi-finished products *Adv. Eng. Mater.* **5(6)** 393

[16] Adamus J 2009 Stamping of titanium sheets *Key Eng. Mat.* **410-411** 279

[17] Ozturk F, Ece R E, Polat N, Koksal A, Evis Z and Polat A 2013 Mechanical and microstructural evaluations of hot formed titanium sheets by electrical resistance heating process *Mat. Sci. Eng. A-struct.* **578** 207

[18] Murzin S P 2016 Formation of structures in materials by laser treatment to enhance the performance characteristics of aircraft engine parts *Computer Optics* **40(3)** 353

[19] Murzin S P, Bielak R, Liedl G 2016 Algorithm for calculation of the power density distribution of the laser beam to create a desired thermal effect on technological objects *Computer Optics* **40(5)** 679

[20] Murzin S P, Shakhmatov E V, Igolkin A A, Musaakhunova L F 2015 A study of vibration characteristics and determination of the conditions of nanopores formation in metallic materials during laser action *Procedia Engineering* **106** 266

[21] Murzin S P, Kryuchkov A N 2015 Influence of Conditions of the Samples Fixation on the Intensity of the Nanoporous Structure Formation in the Metallic Material by Laser Action with Thermocycling *Procedia Engineering* **106** 272

IOP Conf. Series: Materials Science and Engineering **302** (2018) 012071 doi:10.1088/1757-899X/302/1/012071

New Metamaterials with Combined Subnano - and Mesoscale Topology for High-efficiency Catalytic Combustion Chambers of Innovative Gas Turbine Engines

Yu A Knysh[1] and G G Xanthopoulou[1,2]

[1]Department of Theory of Aircraft Engines, Samara National Research University, 34 Moskovskoe shosse, 443086 Samara, Russia
[2]Institute of Nanoscience and Nanotechnologies, NCSR "Demokritos", Agia Paraskevi 15310 Athens, Greece

knysh194@mail.ru

Abstract: The object of the study is a catalytic combustion chamber that provides a highly efficient combustion process through the use of effects: heat recovery from combustion, microvortex heat transfer, catalytic reaction and acoustic resonance. High efficiency is provided by a complex of related technologies: technologies for combustion products heat transfer (recuperation) to initial mixture, catalytic processes technology, technology for calculating effective combustion processes based on microvortex matrices, technology for designing metamaterials structures and technology for obtaining the required topology product by laser fusion of metal powder compositions. The mesoscale level structure provides combustion process with the use of a microvortex effect with a high intensity of heat and mass transfer. High surface area (extremely high area-to-volume ratio) created due to nanoscale periodic structure and ensures catalytic reactions efficiency. Produced metamaterial is the first multiscale product of new concept which due to combination of different scale level periodic topologies provides qualitatively new set of product properties. This research is aimed at solving simultaneously two global problems of the present: ensure environmental safety of transport systems and power industry, as well as the economy and rational use of energy resources, providing humanity with energy now and in the foreseeable future.

1. Introduction

The modern fleet of engines, including gas turbines, are various designs of thermal machines that use the kinetic and potential energy of the working body, which is a heated up to high temperature gas mixture of combustion products and the volume of air passing through the engine. For the next 30-50 years, there is no other alternative for heat engines, especially in the field of aviation equipment and ground-based power gas-turbine installation.

The combustion chamber, in which the fuel combustion process is carried out, is an indispensible part of these engines. The thermal effect of combustion depends from how efficient the combustion process will be, i.e. how effective the hydrocarbons oxidation in the fuel-air mixture stream will be. More complete combustion of fuel provides not only high thermal efficiency, but also reduces the release of harmful combustion products (CO and NOx) into environment.

Macroscale combustion process correctly organized in the combustion chamber ensures the uniformity of the gas flow temperature front at the exit of the combustion chamber and significantly increases the efficiency of the all engine.

Topology optimization has been recognized as one of the most effective design tools with the least mass and performance, especially in aeronautical and aerospace engineering [1].

Most of the existing research on topology optimization focuses on the design of mono-scale structures. In other words, the structures under consideration are made of homogeneous materials. In recent years, heterogeneous and metamaterials with increased specific mechanical and other performance characteristics, such as fibrous composite, structured porous metal, have been increasingly used. Despite the complex topology of such materials of micro and submicro levels, from the point of view of structural analysis these materials can be considered as homogeneous and traditional approaches to design are conventionally applied to them.

At the same time, the optimization of multiscale structural topology has a significant impact on their properties. Therefore, for calculation of effective mechanical designs, its microscopic heterogeneity is taken into account for an accurate estimation of structural indicators. In this connection, there is a need to design a topology of materials of both scale levels - a macroscopic structure and a microscopic structure simultaneously. Simultaneous optimization design of both the structure and the topology of materials at the macro level can be found in the early works [2,3]. Modern multi-scale analysis is outlined in work [4]. The traditional approaches to combustion processes optimizing in combustion chambers of known designs have practically exhausted their effectiveness. In particular, the use of modern software for finite element modeling of gas-dynamic flows in combustion chambers allows optimizing the temperature front and increasing the completeness of combustion, but they do not solve the fundamental problem of creating an environmentally safe combustion chamber. New approaches for chamber design needed for the combustion processes intensification.

As a material with a combined subnano - and mesoscale topology, in the context of this project we mean a metamaterial that has a periodic structure at each level. The structure of the mesoscale level provides the combustion process with the use of a microvortex effect with a high intensity of heat and mass transfer. High surface area (extremely high area-to-volume ratio) created due to nanoscale periodic structure and ensures catalytic reactions efficiency. Such metamaterial will be the first multiscale product of new concept which due to combination of different scale level periodic topologies provides qualitatively new set of product properties.

Thus, the object of research is related to the following critical areas and technologies:

- metals and alloys with special properties;
- technologies for creating membranes and catalytic systems;
- technologies for creating energy-efficient engines and propulsors for transport systems.

The subject of scientific research are:

- experimental research and modeling of combustion processes using microvortical matrices realized with the help of geometry of periodic structure;
- experimental studies of the surfaces catalytic activity produced by 3D printing and by thermal coating of catalysts deposited on developed surfaces;
- designing the topology of structures that realize the microvortical effect by computer simulation methods;
- development of technology for production of metamaterial by Selective Laser Sintering (SLS);
- development of a digital metamaterial model and experimental verification of the model;
- development of technical solutions prototypes for the design of the combustion chamber;
- development of technological solutions prototypes for growing a combustor by the SLS method from metal powder materials;
- development of technological solutions for the thermal spraying of catalytically active coatings.

In the context of the methods outlined above, it should be emphasized that the development of combustion chambers with intensification of the combustion process will permit to burn heavy hydrocarbons fuels (product of recycled biowaste conversion) optimally. Such combustion chambers will be possible to include in the structure of power installation operating on recycled wastes of biological origin.

Development of innovative microvortex combustion chambers designs oriented to the use of metamaterials with a multiscale topology will allow to carry out combustion processes of low-grade hydrocarbon fuels with low-emission pollutants (CO and NOx emissions will be less than 5 ppm). Low emission of pollutants is provided due to high completeness of combustion. The thermal efficiency of combustion will be more than 99%.

Development of technologies for creating metamaterials with a multiscale topology of submicro and meso levels using selective laser alloying of metal powder compositions will ensure creation of the material by a 3Dx2 topology level technique . Each scale level of metamaterial regular topological structure provides specific functional properties. In sum, such material provides a complex of properties, which has a cumulative character, superior to the simple addition of effects taken separately.

As a result of the project, the following local objectives will be achieved:

- providing scientific and research organizations with new and effective methods and tools for conducting research of the combustion processes analysis under conditions of its intensification with the help of a microvortex effect, catalytic reaction and acoustic resonance;
- obtaining significant scientific results allowing to proceed new types of catalytic vortex combustion chambers production;
- mastering of new methods, principles of metamaterials creation with 3Dx2 multiscale level topology, allowing to achieve significant results in the future at serial production by SLS method.

Thus, the global goal is to bring new scientific and technical products to the market, develop world-class technologies for the design and manufacture of a new generation of high-efficiency, catalytic vortex combustion chambers from metamaterials with a multiscale topology of the 3Dx2 level. Such products will reduce the environmental burden on nature through the optimization of combustion processes. The research aimed at organizing the optimal combustion cycle in the combustion chamber, the design and specially developed metamaterial from which it is manufactured, which together provide the multilevel intensification of the combustion process. This research is relevant, it is aimed at solving simultaneously two global problems of the present: ensure environmental safety of transport systems and power industry, as well as the economy and rational use of energy resources, providing humanity with fuel and energy now and in the foreseeable future.

2. Experimental

The microvortex catalytic combustion chamber produced by a 3D printing method by direct computer modelling. Further the model under the set program is divided into layers of the certain thickness. An operating computer of 3D printer received information about each layer and regulates sintering according program until complete construction of article. For 3D printing the metal powder of stainless steel 316L was used with the particles (spherical form) size up to 50 microns, with density of a powder 4, 74 g/cm^3.

Construction of a product realized on installation SLM 280 with the construction chamber 280x280x350 mm. As an energy source used the infra-red fiber laser (a source of electromagnetic radiation), by length of a wave of radiation 1075 nanometers, capacity 400 W with a cross-section direction of a laser beam distribution. This installation gives possibility of complex research for optimization of products selective laser synthesis parameters. Products were built in nitrogen atmosphere in the working chamber which is filled with protective gas before construction. The platform, on which the articles were synthesized, was warmed up to 150°C. [5].

3. Results and discussion

In the proposed project for the implementation of the combustion processes intensification, the microvortex catalytic combustion chamber is realized as a design of a two-level multi-scale metamaterial.

At the meso level, the combustion chamber is a metamaterial realized in the form of concentrically arranged annular structures shown in figure 1.

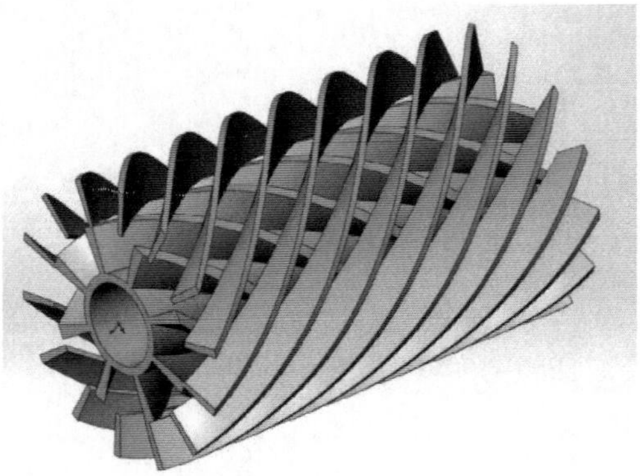

Figure 1. Model of the microvortical combustion chamber.

Each structure is formed by a family of helical ribs equally distributed around the circumference of the perimeter, the working surfaces are spirally along the generatrix of the cylinder. The direction of twisting of the spirals of two adjacent concentric structures is opposite, respectively, right and left.

Figure 2 shows a SEM microphoto of the finned working surface prepared for catalytic coating application.

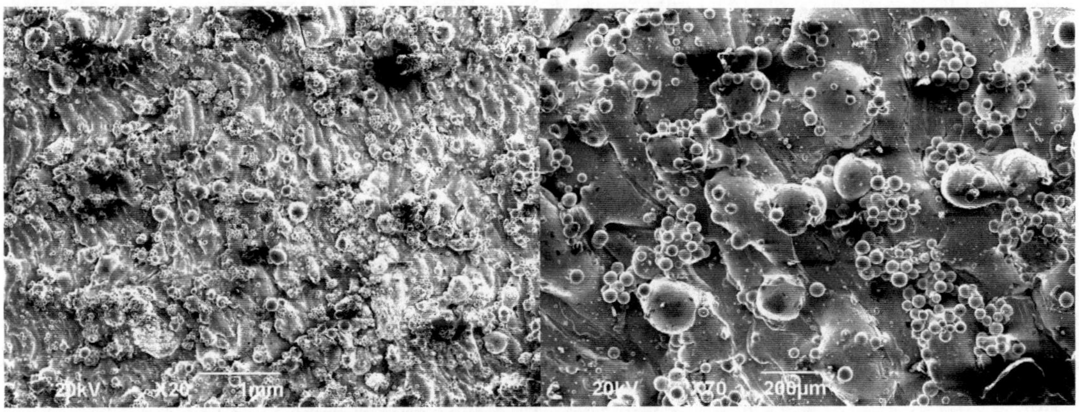

Figure 2. SEM microphoto of the finned surface for catalytic coating application.

Metamaterial is, first of all, a material whose properties are due not so much to the properties of the constituent elements as to the artificially created periodic structure [6]. The topology of the combustion chamber at the mesolevels perfectly suits this definition, it has a periodic structure which propagate in two directions - along the axis and radially, and its functional properties are determined mainly by its topology. The preparation of the ribs surface for the catalytic coating provides a multiple increase in the surface area.

During combustion, the gas mixture moves along the helical channels between the ribs, and a micro vortex matrix is formed at the point where two coplanar flows of the gas mixture which move along the two adjacent structures meet. Periodically located in the meeting places, microvortical matrices provide an intensification of the combustion processes due to the vortex effect.

Figure 3 presents the results of numerical simulation of the process of a linear vortex interaction intersecting transversely located cells through which conjugate flows occur.

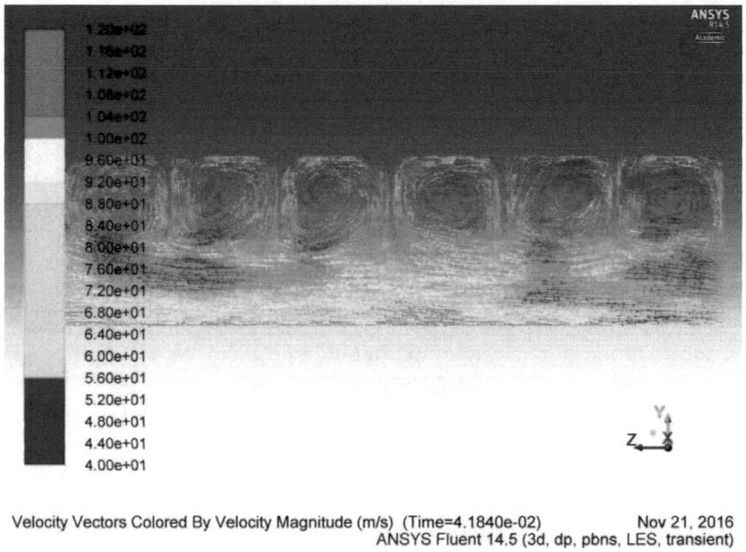

Figure 3. Results of numerical simulation of the process of a linear vortex interaction.

Interaction of intersecting streams generates a system of micro-vortex structures in which the level of mass forces generated by the twisting of flows is many hundreds and thousands of g, where g is the acceleration of gravity. As a result of centripetal accelerations, the mixture of combustion products on the catalytic surface having the highest temperature (and the lowest density), quickly float to the centers of the vortices and are carried downstream to the exit from the combustion chamber.

Fig. 4 shows the amplitude-frequency characteristic of pressure pulsations in the center of the microvortex cell.

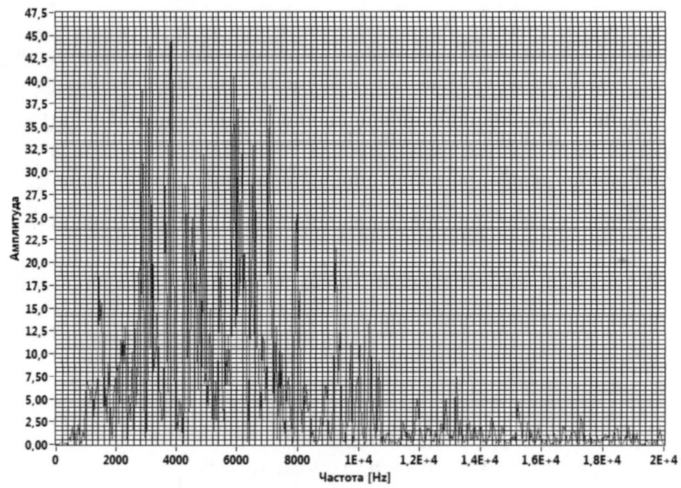

Figure 4. Amplitude-frequency characteristic of pressure pulsations in the center of the microvortex cell.

IOP Conf. Series: Materials Science and Engineering **302** (2018) 012071 doi:10.1088/1757-899X/302/1/012071

Local pressure pulsations form the parameters of acoustic emission as a whole in the micro vortex combustion chamber and contribute to the integration of heat exchange and combustion processes.

One of the main difficulties in implementing such a design is the technology. The screw structures of the metamaterial at the meso level form a spatial grid, which is an inseparable connection that is unsuitable for manufacturing by machining technologies. From the technological point of view a rational way of making such design is a new additive technology-selective laser sintering (SLS) of metal powder compositions.

Another way of intensifying combustion processes and ensuring high completeness of the oxidation reaction is provided in the work- a catalytic reaction on the surface of the catalyst fins. A certain experience of using the catalytic effect with respect to the combustion processes of hydrocarbon fuels is set forth in works [7,8]. For the first time in the world were designed and printed in 3D technique catalysts on the base of steel powder (figure 5a) and Ni-Al alloys.

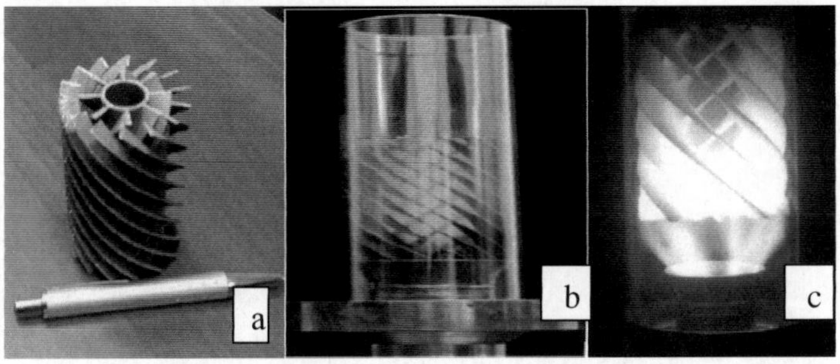

Figure 5 a) Catalyst 3D (from steel) block with intricate configuration and coplanar channels, b) propane-butane and (c) methane combustion in 3D microvortical combustion chamber.

Heat- transfer paths with intricate configuration and coplanar channels of the flow pass section for heat transfer enhancement and extending of heat transfer surface were produced (fig.5a). As a result of the gas-dynamic research proposed a fundamentally new way of organizing the catalytic oxidation in combination with combustion in the gas phase. The technology is put in heat-known scheme of coplanar channels, which provides high rates of heat transfer. 3D printed blocks were used as catalysts for methane oxidation, CO oxidation and methane combustion processes (fig.5). High flow rate of the working body - more than of 60 -100 kg / s, and respectively flow rate 30-70 m/s reduces the effectiveness of the known technical solutions to unsatisfactory hydraulic losses values. In the center of the gas-dynamic research proposed a fundamentally new way of organizing the catalytic oxidation in combination with combustion in the gas phase.

The activity of catalysts in heterogeneous catalysis strongly depends on the size and state of their surface, so the method of the catalyst manufacturing is important. With respect to the present design, the surfaces of the helical ribs of the catalytic combustion chamber along which the gas mixture moves is the catalyst surface or surface which can be coated by catalyst.

In the case of microvortex chambers, an increase in the area of the catalyst can be achieved by using the properties of a metamaterial with a topology of submicroscopic dimension. This is the second level of the metamaterial structure for the combustion chamber. To grow the combustion chamber by the SLS method, it is suggested to use powders with the size of the fraction up to 60 microns, which makes it possible to obtain periodic structures of microcells (surface roughness substructure with a regular profile of protrusions and valleys) on the working surface of the ribs due to the use of special technological methods. This will allow multiplying the active area of the catalytic reaction.

As a result, the developed metamaterial for the vortex catalytic combustion chamber has a two-level multiscale topology. Each of the two described levels of metamaterials carries its own functional load. A cumulative effect is achieved on the intensification of combustion processes, which by efficiency exceeds the sum of the positive effects of each scale level taken separately.

4. Conclusion

The new catalytic combustion chamber that provides a highly efficient combustion process through the use of effects: heat recovery of combustion products, microvortex heat transfer, catalytic reaction and acoustic resonance is developed.

Such material is the first product of multiscale metamaterials new concept which due to a combination of periodic topologies on different scale level provides qualitatively new set of product properties.

Acknowledgments

This work was supported by Ministry of Education and Science of the Russian Federation.

References

[1] *ZHU J, ZHANG W* and *XIA L* 2015 *Topology optimization*in *aircraft* and *aerospace structures design*", *Archives*of *Computational Methods* in *Engineering*, **23**, p 595

[2] Bendsøe, M.P. and Kikuchi N 1988 Generating optimal topologies in structural design using a homogenization method *Computer Methods in Applied Mechanics and Engineering*, **71** (2), pp 197-224

[3] Huang X, Zhou SW, Xie Y M and Li Q 2013 Topology optimization of microstructures of cellular materials and composites for macrostructures *Computational Materials Science*, **67**, pp 397-407

[4] Xia L, Breitkopf P 2015 Multiscale structural topology optimization with an approximate constitutive model for local material microstructure *Computer Methods in Applied Mechanics and Engineering*, **286**, pp 147-167

[5] V.G.Smelov, A.V.Sotov, S.A.Kosirev 2014 *ARPN Journal of Engineering and Applied Sciences*, **10** ,pp 1854-1858

[6] Engheta N and Ziolkowski R 2006 Metamaterials: Physics and Engineering Explorations , John Wiley & Sons & IEEEPress, p 440

[7] Xanthopoulou G G, Knysh Y A and Amosov A P 2015 SHS-produced catalysts for in-situ generation of syngas in combustion chambers of heat engines: A review , *International Journal of Self-Propagating High-Temperature Synthesis,* **24**(2), pp 56-62

[8] Xanthopoulou G G, Knysh Yu A, Amosov A A, and Dmitriev D N 2015 *Proc. Int. Symposium SHS 2015*(Antalia, Turkey) pp.36-37

IOP Conf. Series: Materials Science and Engineering **302** (2018) 012072 doi:10.1088/1757-899X/302/1/012072

Formation of A Non-detachable Welded Titanium-aluminium Compound by Laser Action

Serguei P. Murzin*

Department of Automatic Systems of Power Plants, Samara National Research University,
Samara, Russia

*Corresponding author: murzin@ssau.ru

Abstract. Progressive in the welding of dissimilar materials is the use of laser technology. With the use of the ROFIN StarWeld Manual Performance laser, an aluminium alloy AK4 and a titanium alloy VT5-1 were welded. Processing regimes have been determined, the realization of which during melting of materials in the zone of thermal influence makes it possible to obtain a homogeneous structure without voids and shells, which indicates a potential sufficiently high serviceability of the welded joint. To create the required power density distribution in the cross section of the laser beam, it is expedient to use diffractive optical elements.

1. Introduction

Materials based on aluminium are widely used in aircraft designs. The receipt of their high-strength and reliable welded joints with other metals and alloys is one of the topical tasks of the aerospace industry. Welding of aluminium and its alloys with other materials is accompanied by a number of difficulties: a large difference in thermal conductivity leads to intense heat removal towards aluminium, which prevents the formation of a quality welded joint. The use of so-called traditional methods for the production of integral welded joints of these materials presents significant difficulties and, often, does not provide the required quality of the connection. The intermetallides formed during the crystallization of such seams lead to embrittlement of the material [1]. For dissimilar metal materials, several welding methods are known: argon-arc welding with a non-consumable electrode [2], welding using an intermediate metal [3], welding with pressure [4], diffusion welding [5]. However, none of the above methods, as well as the use of a combination of different types of welding, do not exclude the appearance of intermetallic interlayers [6]. Welding by explosion allows to weld only sheet materials.

Laser welding is an attractive technique in comparison with the traditional methods of joining dissimilar metals, because the processing speed and accuracy are high, while the heat supply is very low, this method does not require the use of additional materials, special processing of the edges [7, 8]. This method provides the opportunity to obtain an ultrathin diffusion zone and avoid the appearance of defects. Redistribution of the radiation power density with displacement of its maximum value towards one of the welded metals makes it possible to obtain the required structure of the welded joint [9, 10]. Locality is provided for the depth and area of the physical processes flowing in the thermal effect zone while maintaining the original properties of the material in the remaining volume and the absence of significant deformations of the machined parts [11, 12]. To create the required power density distribution in the cross section of the laser beam, it is advisable to use diffraction optical elements [13-15]. At the same time, a high rate of melting and crystallization of the material is achieved, which is caused by intensive heat removal from a small-volume molten bath into the surrounding material of the article. The purpose of this study is to study the features of the formation of high-strength and reliable titanium-aluminium welded joints by laser action using diffractive optical elements.

2. Results and discussion

The main reason complicating the welding of titanium alloys with each other and with other materials is the active interaction of titanium with atmospheric gases at elevated temperatures, especially in the liquid state. In addition, during the welding thermal cycle structural transformations in the seam and the weld zone lead to the formation of brittle intermetallic and carbide phases. Using the known methods of fusion welding, it is difficult to obtain high-strength and reliable compounds of titanium alloys with each other and with aluminium. Sufficient strength and heat resistance of all-in-one joints does not allow soldering of titanium alloys. The physical and mechanical properties and heat resistance of solders are much lower than in titanium alloys, which leads to a lowered thermomechanical reliability of the compound. When soldering titanium and aluminium alloys, the choice of solder, an equally good wetting contacting surface, is complicated. In the welding of high-temperature corrosion-resistant titanium alloys with products made from aluminium-based materials, the use of laser technologies is progressive.

In the quality of the materials to be welded during the pilot studies, sheet materials were chosen: aluminium alloy AK4 and titanium alloy VT5-1 2 mm thick. The chemical composition of materials is presented in tables 1 and 2. This combination of materials is used in the manufacture of a number of aircraft units.

Table 1. Chemical composition of the alloy AK4,%

Al	Fe	Si	Mn	Ni
basis	0.8…1.3	0.5…1.2	unto 0.2	0.8…1.3
Ti	Cu	Mg	Zn	Impurities
unto 0.1	1.9…2.5	1.4…1.8	unto 0.3	Other, each 0.05; Only 0.1

Table 2. Chemical composition of alloy VT5-1,%

Ti	Fe	C	Si	V	N
basis	unto 0.3	unto 0.1	unto 0.12	unto 1.0	unto 0.05
Al	Zr	O	Sn	H	Impurities
4.,3…6.0	unto 0.3	unto 0.15	2.0…3.0	unto 0.015	Other 0.3

The deformable alloy AK4, belonging to the group of high-temperature aluminium alloys of special purpose, is used for manufacturing deformable semi-finished products. Alloy VT5-1 refers to deformable titanium alloys. Immediately before welding, the bonded surfaces of the parts were thoroughly washed in solvents in order to remove various kinds of contaminants (dust, grease films, etc.) from them.

Laser welding was performed on the ROFIN StarWeld Manual Performance, the applied system "Sweet Spot Resonator" provides the output of constant power beam to the material being processed. During welding to improve the parameters of the material processing process, the shape and duration of the radiation pulse can be changed using the "Pulse Shaping" program. Processing modes are determined, at which minimal thermal loads are achieved, the thermal cycle of heating and cooling of the processed material improves, which ensures a high quality of welding. The beams transport and formation system was used including diffractive optical element [16-19].

The structure of the seam of the butt joint without the bevel of the edges of the aluminium alloy AK4 and the titanium alloy VT5-1 was studied. For metallographic studies, a Phenom-ProX scanning electron microscope was used, with an integrated energy-dispersive analysis system and a CeB6 (cerium hexaboride) electron source. The range of magnifications of the electronic microscope is in the range of 80-45000x. The energy dispersive analysis system for determining the elemental composition at a point (Elemental Identification) is built into the image acquisition module. The device uses a silicon drift detector SDD with thermoelectric cooling without liquid nitrogen. The active area of the detector is 25 square millimeters. The energy resolution (Mn Ka no more than 140 eV) allows to determine the elements from C (carbon) to Am (americium) with high accuracy. Multichannel analyzer, 2,048 channels (energy 10 eV per channel). Reading speed 300000 pulses per second. In figure 1 shows the structure of the weld metal in the cross-section of the laser welding zone between the aluminium alloy AK4 and the titanium alloy VT5-1.

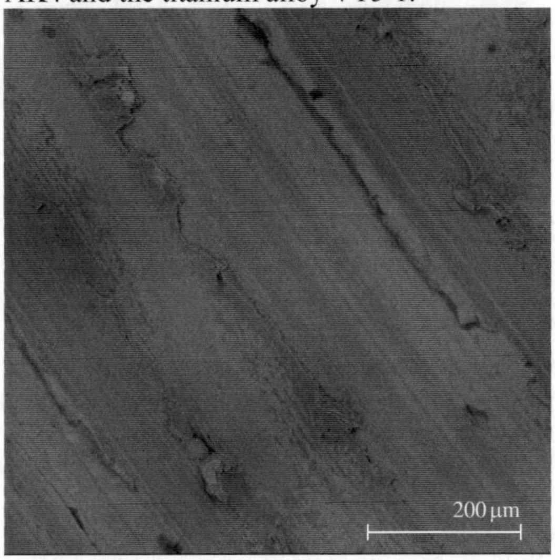

Figure 1. The structure of the weld metal in the cross section of the laser welding zone between the aluminium alloy AK4 and the titanium alloy VT5-1; Zoom. x380.

It is determined that the cast weld zone formed after melting has a homogeneous structure without voids and shells, which indicates a potential sufficiently high serviceability of the welded joint. The weld metal zone of dissimilar materials formed during welding is characterized by an individual structure, crystal orientation, and phase distribution.

The seam break was investigated. As a result of studying the structure of the kink, the following is established. Formed in the process of destruction, the fibers are stretched along the main crack. The microstructure of the fracture zone is characterized by the presence of large and small pits, and fibrousness is observed at the bottom of the wells. An image of the fracture surface of a welded joint is shown in figure 2. When studying the fracture surface in the field of view of an optical microscope at various magnifications, it is established that the structure of the fracture is homogeneous. The surface area of the fracture is characterized by fibrousness. The surface of the fracture does not have a metallic luster, there are no defects in the form of pores, nonmetallic inclusions. Fine-grained fibrous fracture without gloss characterizes the potentially good enough ductility and high impact strength of the metallic material. Metallographic studies of microsections make it possible to draw a conclusion about the formation of an actual contact between welded surfaces.

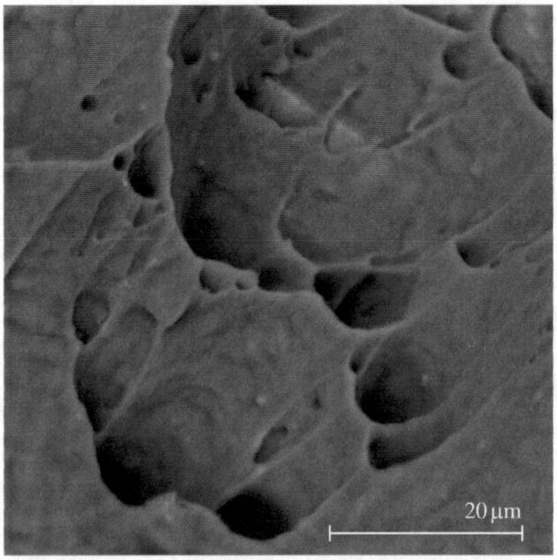

Figure 2. An image of the fracture of a welded joint obtained by means of an analytical scanning electron microscope Phenom-ProX; Zoom. x4700.

The chemical composition of the welded joint was studied on the material sections in the cross section of the laser welding zone. Figure 3 shows the designation of zones of investigation of the chemical composition on the section of material in the cross section of the laser welding region. In Fig. 4 presents the results of an analysis of the elemental composition of the material in the zone adjacent to the titanium alloy. The presence of oxygen in the diagram of the percentage of chemical elements is due to the oxidation of the surface of the thin sections when they are exposed to air for a long time after welding. The results of the analysis of the elemental composition in the zones on the section of the material in the cross section of the laser welding region of the aluminum alloy AK4 and the titanium alloy VT5-1 are shown in table 3.

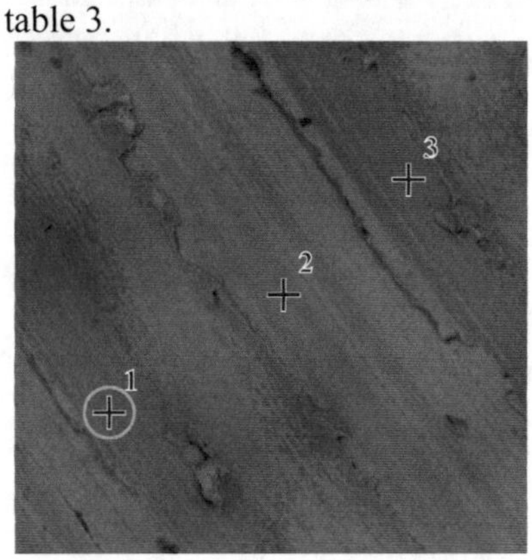

Figure 3. Designation of zones of investigation of the chemical composition on the section of material in the cross section of the laser welding region.

As a result of the conducted investigations, it has been established that during the laser welding, the weld metal is mixed. In this case, the concentration gradient of the chemical composition remains in the cross section of the laser welding zone. In the process of welding, a new material is formed in which chemical elements of both welded alloys, such as the aluminium alloy AK-4 and the titanium alloy VT5-1, are present.

IOP Conf. Series: Materials Science and Engineering **302** (2018) 012072 doi:10.1088/1757-899X/302/1/012072

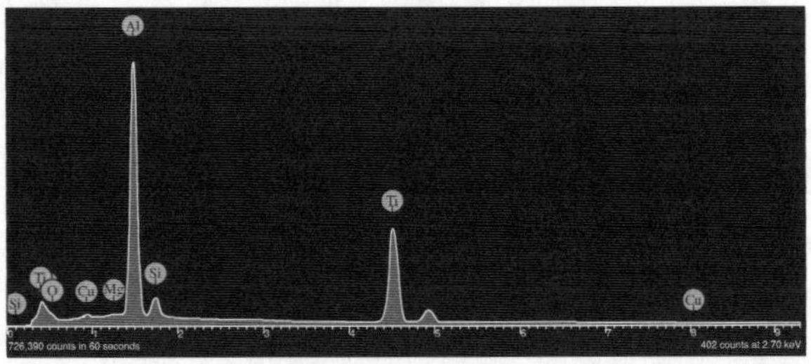

Figure 4. Results of analysis of the elemental composition of the material in the zone adjacent to the titanium alloy.

Table 3. Results of analysis of elemental composition in% in zones on the section of material in the cross section of the laser welding region of aluminium alloy AK4 and titanium alloy VT5-1

Study area number	Al	Ti	Si	Mg	Cu
1	60.8	36.7	1.4	0.5	0.5
2	63.6	32.4	1.3	1.0	1.5
3	75.2	21.6	1.0	0.5	1.6

As a result of metallographic studies it was determined that the diffusion zone formed during the laser action has a homogeneous structure without voids and shells, which indicates a potential sufficiently high performance of the welded joint. The weld metal zone of dissimilar materials formed during welding is characterized by an individual structure, crystal orientation, phase distribution. The laser action with its clearly defined and precise localized energy input makes it possible to largely control the growth of intermetallic compound layers.

3. Conclusions

The obtained results of experimental studies have shown the expediency of forming high-strength and reliable welded joints of materials based on aluminum with other metals and alloys by laser action. Due to the much more precise control of the parameters of the thermal source in the heat affected zone, laser welding of dissimilar materials can have significant advantages. The use of laser welding of aluminium-based materials avoids the formation of an Al_2O_3 oxide film, which is a fairly common defect in fusion welding. A feature of the application of laser welding technologies are high achievable speeds, which makes it possible to reduce the size of the intermetallic interlayer formation zones and minimize the thermal impact zone in materials with high thermal conductivity. However, the predictability of the wetting behaviour, diffusion processes and the thickness of the resulting intermetallic phase layer leaves much to be desired. Additional efforts are needed to better understand the physical mechanisms of laser welding of dissimilar materials. It is promising to carry out modelling of laser welding of dissimilar materials, which should allow increasing the reliability of the process with a view to its introduction into industrial production.

Integration of laser welding of dissimilar materials into industrial production lines remains a difficult task due to the need for an accurate local energy supply, as well as increasing requirements for sample preparation, systems for fixing and positioning the samples. Nevertheless, the use of compounds of dissimilar materials in a wide range of numerous industrial applications is constantly increasing because of the potential for reducing the weight of structures and saving material costs. To implement such technological processes, special optics should be used. More accurate metering of the laser beam energy supply is required locally in the thermal effect zone with the possibility of power density redistribution. Such a redistribution of the energy of the laser beam can be obtained with the help of diffractive optical elements, the use of which in the technological operations of welding heterogeneous materials opens the prospects not only for solving the specific problems described, but also for other applications.

Acknowledgments
This work was supported by the Ministry of Education and Science of the Russian Federation as part of the Program "Research and development on priority directions of scientific-technological complex of Russia for 2014-2020" within the project RFMEFI57815X0131.

References

[1] Lyushinskiy A V 2006 *Diffusion welding of dissimilar materials* [In Russian] (Moscow: Publishing Center «Academy»)
[2] Borrisutthekul R, Mitsomwang P, Rattanachan S and Mutoh Y 2010 Feasibility of using TIG welding in dissimilar metals between steel/aluminum alloy *Energy Research Journal* **1(2)** 82
[3] Saida K, Ohnishi H and Nishimoto K 2010 Fluxless laser brazing of aluminium alloy to galvanized steel using a tandem beam - dissimilar laser brazing of aluminium alloy and steels *Welding international* **24(3)** 161
[4] Shubhavardhan R N and Surendran S 2012 Friction welding to join dissimilar metals *International journal of emerging technology and advanced engineering* **2(7)** 200
[5] Wang J, Li Y, Wu H, Ren J. 2001 Micro-image analysis in the diffusion-bonded zone of Fe_3Al/Q235 carbon steel dissimilar materials *Bulletin of materials science* **25(5)** 367
[6] Ryabov B P 1983 *Welding of aluminum and its alloys with other metals* [In Russian] (Kiev: Naukova Dumka)
[7] Meco S, Pardal G, Ganguly S, Williams S and McPherson N. 2015 Application of laser in seam welding of dissimilar steel to aluminium joints for thick structural components Opt. Laser. Eng. 67 22
[8] Wang P, Chen X, Pan Q, Madigan B and Long J. 2016 Laser welding dissimilar materials of aluminum to steel: an overview *International Journal of Advanced Manufacturing Technology* **87(9-12)** 3081
[9] Hong K-M and Shin Y C 2017 Prospects of laser welding technology in the automotive industry: a review *Journal of Materials Processing Technology* **245** 46
[10] Fan J, Thomy C and Vollertsen F 2011 Effect of thermal cycle on the formation of intermetallic compounds in laser welding of aluminum-steel overlap joints *Physics Procedia* **12(PART 1)** 134
[11] Kazanskiy N L, Murzin S P and Tregub V I 2010 Optical system for realization selective laser sublimation of metal alloys components *Computer Optics* **34(4)** 481
[12] Murzin S P, Tregub V I, Melnikov A A and Tregub N V 2013 Application of radiation focusators for creation of nanoporous metal materials with high specific surface area by laser action *Computer Optics* **37(2)** 226
[13] Doskolovich L L, Kazanskiy N L, Soifer V A and Tzaregorodtzev A Ye. 1995 Analysis of quasiperiodic and geometric optical solutions of the problem of focusing into an axial segment *Optik - International Journal for Light and Electron Optics* **101(2)** 37
[14] Kazanskiy N L, Kotlyar V V and Soifer V A 1994 Computer-aided design of diffractive optical elements *Optical Engineering* **33(10)** 3156
[15] Pavelyev V S, Borodin S A, Kazanskiy N L, Kostyuk G F and Volkov A V 2007 Formation of diffractive microrelief on diamond film surface *Opt. Laser Technol.* **6(39)** 1234
[16] Murzin S P 2016 Formation of structures in materials by laser treatment to enhance the performance characteristics of aircraft engine parts *Computer Optics* **40(3)** 353
[17] Murzin S P, Bielak R, Liedl G 2016 Algorithm for calculation of the power density distribution of the laser beam to create a desired thermal effect on technological objects *Computer Optics* **40(5)** 679
[18] Murzin S P, Shakhmatov E V, Igolkin A A, Musaakhunova L F 2015 A study of vibration characteristics and determination of the conditions of nanopores formation in metallic materials during laser action *Procedia Engineering* **106** 266
[19] Murzin S P, Kryuchkov A N 2015 Influence of Conditions of the Samples Fixation on the Intensity of the Nanoporous Structure Formation in the Metallic Material by Laser Action with Thermocycling *Procedia Engineering* **106** 272

IOP Conf. Series: Materials Science and Engineering **302** (2018) 012073 doi:10.1088/1757-899X/302/1/012073

Elastic, Frictional, Strength and Dynamic Characteristics of the Bell Shape Shock Absorbers Made of MR Wire Material

Lazutkin G V, Davydov D P, Boyarov K V and Volkova T V

Samara State Aerospace University, 34, Moskovskoye shosse, Samara, 443086, Russia

lazutkin.gennadij@mail.ru;dpdavydov@yandex.ru;konst90@gmail.com; twigggy@yandex.ru

Abstract. The results of the mechanical characteristic experimental studies are presented for the shock absorbers of DKU type with the elastic elements of the bell shape made of MR material and obtained by the cold pressing of mutually crossing wire spirals with their inclusion in the array of reinforcing wire harnesses. The design analysis and the technology of MR production based on the methods of similarity theory and dimensional analysis revealed the dimensionless determined and determining parameters of elastic frictional, dynamic and strength characteristics under the static and dynamic loading of vibration isolators. The main similarity criteria of mechanical characteristics for vibration isolators and their graphical and analytical representation are determined, taking into account the coefficients of these (affine) transformations of the hysteresis loop family field..

1. Introduction

Mechanical characteristics of the MR shock absorbers [1] shall be based reasonably on the physical modeling of shock absorbers as the complex systems of structural damping [2, 3]. In this case, the interaction of all elastic frictional linkages in vibration isolators is considered and the problem of its deformation is solved in a closed form with the set of hypotheses and assumptions. However, this approach requires additional experimental confirmation of the proposed hypotheses and assumptions, the proof of which requires a lot of time and costs. At that the positive result may be not achieved.

Another approach is for the MR material presentation as a quasicontinuous environment. However, the determination of elasticity modulus non-linear characteristics and so on for MR material requires a large volume of experimental work, as well as for the case of geometric nonlinearity concerning the complex forms of MR products (e.g., bell-shaped ones) [4 -...8].

Therefore, one may consider an experimental method of the mechanical properties determination using the methods of similarity theory and dimensional analysis as the most relevant one.

The anisotropy of strength and elastic frictional characteristics (UFC) of MR material makes the provision of the aircraft engine isolation systems more difficult, especially under the influence of the spatial load. This disadvantage may be substantially eliminated by the inclusion of MR material in an array of high-resistant reinforcing element (HRRE), which is made from a special wire harness. The harness is a strand wrapped with a spiral tension coil to coil (type 1) (Fig. 1) or filament wire (type 2).

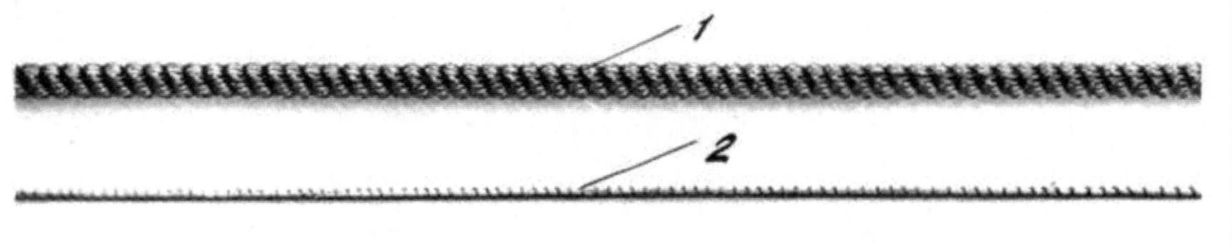

Figure 1. Special harness types: 1 - type 1; 2 - type 2

The increased damping capacity of elastic-damping elements (EDE) is ensured by the presence of additional energy dissipation at the AE boundaries of with the MR material array and within AE. The high strength of shock absorbers is achieved by the harness coverage of all mounting holes in the EDE.

On the basis of the described method MR UDE reinforcement method the modification of the "double bell reinforced" (DBR) shock absorbers is developed. This shock absorber is shown in Fig. 2.

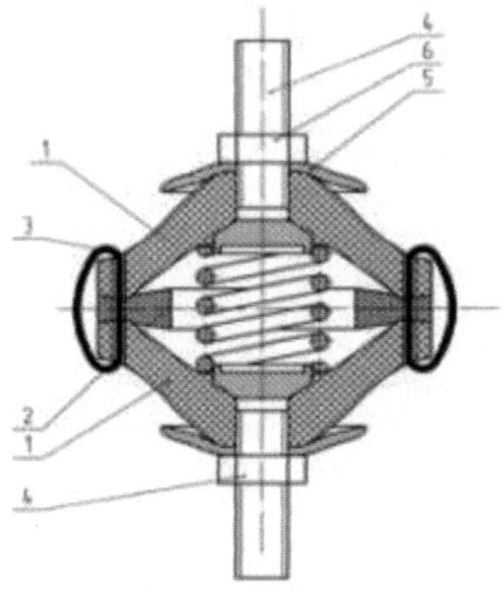

Figure 2. DBR shock absorber: 1- elastic-damping element; 2 - spacer; 3 - stitched wire; 4- mounting studs; 5 - limiting washer; 6- nut

The manufacturing process of the reinforced EDE consists of three main phases (Fig. 3 a, b, c). During the first step two bell-shaped EDE workpiece are made the spiral wire (see. Fig. 3a). During the second stage AE is produced from harness, which, if necessary, may be compressed according to the final product shape (see. Fig. 3b). During the third stage of AE is inserted in an array of spiral wire EDE blanks and wrapped over with a spiral wire. Then the final EDE shaping is performed by cold extrusion (see. Fig. 3c).

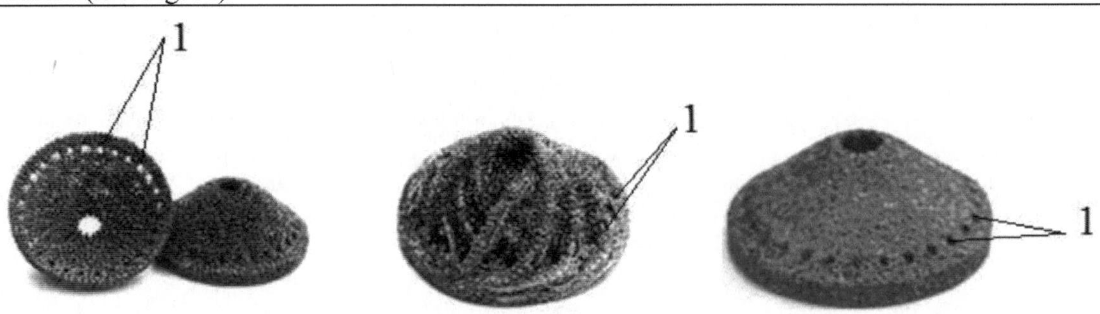

a)	b)	c)

Figure 3. Main stages of reinforced EDE production: a – EDE blanks; б - AE; в – finished EDE; 1 – fixing holes

2. Methods

2.1. Determination method of generalized UFH

The most important parameters determining the properties of simplest DBR shock absorbers without relief springs and a spacer (see. Fig. 2) are as follows: EDE diameter; the average density of EDE material ρ; the wire material density ρ_u and its diameter δ_u; spiral diameter d_u; fluidity limits σ_{mu}, σ_{m0} and elasticity modulus E_u of the wire in EDE blanks and within AE; brought to the sphere volume with the diameter D_K , AE density $\rho_Ж$ and all blanks $\rho_з$; δ_c diameter and the fluidity limit σ_{mc} of the stitched wire; diameter d_c brought along the stitch length and the number of stitching coils $n_в$.

The research of these parameters influence on the elastic frictional and strength characteristics of shock absorbers was carried out using the methods of similarity theory and dimensional analysis. The similarity according to UFH at the tension and compression of shock absorbers was determined by an affine similarity of hysteresis loop families with different A amplitudes and q tightness. The coefficients of these transformations were chosen as the sections cut off with congruent processes of loading and unloading from the coordinate axes (T_n - according to the load R, a_n - according to the deformation X).

In this case, the determined similarity criterion is the dimensionless load $\eta = R/T_n$. The determining criteria are the dimensionless strain $\xi = X/a_n$, the amplitude $\xi_A = A/a_n$ and the tension $\xi = q/a_n$. The experiments also showed that the similarity criteria are presented by the values composed of the shock absorber parameters:

Boundary condition criterion

$$\Pi_1 = \left(\frac{\sigma_{mu}}{E_u}\right)^{0,5} \left[\frac{\sigma_{mc}}{E_u}\left(\frac{\delta_c}{D_K}\right)^2 \frac{\delta_c}{d_c}\right]^{0,12} \left(\frac{d_{cn}}{d_c}\right)^{1,5}, \quad (1)$$

Where

$$d_{cн} = 0,7 D_K \left(\frac{\rho_з}{\rho} + 0,16\right); \quad (2)$$

damping criterion

$$\Pi_2 = \frac{d_u}{\delta_u}\left(1 - \frac{\rho_{ж}}{\rho_з}\right) + 24,5\left(\frac{\sigma_{mc}}{\sigma_{mu}}\right)^{0,5}\left(\frac{\rho_{ж}}{\rho_з}\right); \quad (3)$$

EDE buckling loss criterion

$$\Pi_3 = \frac{\rho_з}{\rho}. \quad (4)$$

The ranges of value change determining the similarity criteria were as follows: $\xi \in [-3,45;3,45]$; $\xi_A \in [0,14;3,45]$; $\xi_q \in [-2,07;2,07]$; $\Pi_1 \in [0,75 \cdot 10^{-2};1,3 \cdot 10^{-2}]$; $\Pi_2 \in [8,5;24,4]$; $\Pi_3 \in [0,05;1,32]$

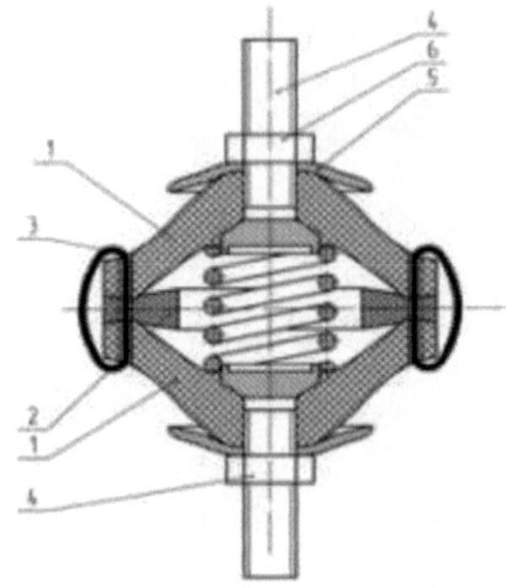

Figure 4. Generalized field of hysteresis loops

The research of the defining criteria effect for generalized UFH (Fig. 4) showed that the processes of DBR shock absorber deformation within the coordinates dimensionless load - dimensionless deformation are essentially nonlinear ones. The absorption coefficients ψ depending on the deformation amplitude ξ has its maximum (fig. 4). At that the most intense increase of the dimensionless average rigidity $\overline{C}_{cp}$ is observed at $\xi_A < 0,25$ (fig. 5). At the large values of ($\xi_A > 1,6$) $\overline{C}_{cp}$ is not changed almost. These circumstances are caused by the fact that during the amplitude decrease the shock absorber deformation most of the contact wires in the MR material as in the structural damping system (e.g., of a multi-layer cantilever beam) interact with each other elastically without slipping. This leads to a sharp decrease of the friction forces operation and increases the rigidity $\overline{C}_{cp}$. With the strain amplitude increase ($\xi_A \in [0,1;0,2]$)the most part of the contact wires interacts with slipping, causing a stiffness decrease $\overline{C}_{cp}$ and the friction force operation increase. However, at the further increase of $\xi_A > 0,25$ the increment of the friction forces is smaller than the increment of the elastic forces, making the reduction of ψ.

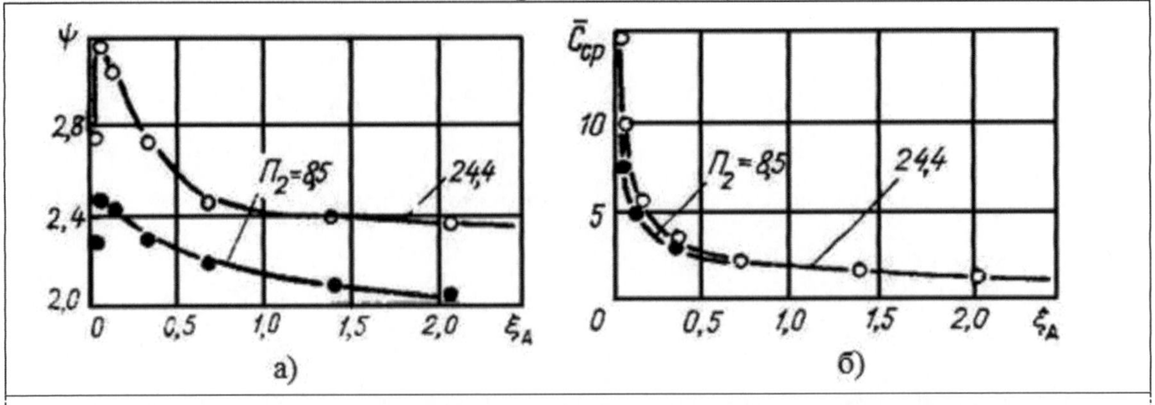

Figure 5. Dependencies: a – of absorption coefficient, б - average rigidity on the dimensionless amplitude and damping criterion

ξ_q criterion influences rather weakly on the nonlinear UFH nature. The strongest influence of ξ_q is performed on ψ and $\bar{C}_{cp}$. At the compaction of the shock absorber ψ is increased for fixed ξ_A, and is decreased at $\bar{C}_{cp}$. $\bar{C}_{cp}$ is increased during the extension, and ψ is decreased.

The influence of Π_1 criterion is especially evident at the shock absorber extension: while Π_1 is increased $\bar{C}_{cp}$ is increased. This is due to the compressive loads increase at EDE periphery, preventing the AE displacement relative wire spiral arrays. With the reduction of Π_2 criterion ψ is decreased. The UFH nonlinearity grade is decreased, estimated by the intensity of $\bar{C}_{cp}$ changes according to the amplitude ξ_A. Π_3 criterion physically characterizes the degree of MR material crimping and determines the EDE profile width. At a small profile width and large shock absorber deformations the load processes have flat areas and the period of EDE resistance loss may occur.

Along with the determination of the approximate similarity criteria Π_1, Π_2, Π_3 influence degree on the UFH, the functional connections were established experimentally of such transformation coefficients with the dimensionless complexes composed of the shock absorber parameters according to the first part of the P-theorem:

$$\frac{T_n}{E_u D_K^2} = 0{,}066\Pi_1^{2,2}\left(\frac{\rho_з}{\rho_u}\right)^{1,2}\left[\left(1-\frac{\rho_{ж}}{\rho_з}\right)+3{,}9\frac{\sigma_{m0}}{\sigma_{mu}}\frac{\rho_{ж}}{\rho_з}\left(0{,}8-\frac{\rho_{ж}}{\rho_з}\right)\right]; \tag{5}$$

$$\frac{a_n}{D_K} = 0{,}195\Pi_1\Pi_2. \tag{6}$$

The relations obtained in the aggregate with the data presented in Fig. 4, allow you to define any UFH of any DBR shock absorber at its cyclic loading, including the newly designed one, provided that the geometric similarity of its elements and tooling components is kept (profiles of punches and matrices, the scheme of harness stacking, etc.).

The requirements imposed on UFH, define the characteristics of the vibration-proof system with the dynamic action, for example, the resonance frequency of the system, the dynamic gain ratio, the allowable free-running and others.

2.2. Generalized dynamic characteristics determination method

Let's consider the harmonic kinematic excitation of a vibration-proof system (VPS) with one degree of freedom. The differential equation of the mass motion in the absolute coordinate system has the following formula:

$$m\frac{d^2x_1}{dt^2} + \Phi'\left[(r+x), x_k, sign\dot{x}, z_1, \ldots, z_m\right] = Q, \tag{7}$$

where x_1 is an absolute VPS transition; $x_1 = (x+r) + a_0 \sin\omega t$; m – system mass; r - the offset of dynamic equilibrium center; Q - constant force equal to the system weight; a_0 - exciting oscillation amplitude; ω - forced oscillations frequency; t - time.

Using the values of $\omega_q = \sqrt{\frac{T_q}{a_q m}}$ and a_q as such transformation coefficients of the variables t and x_1, x, r, x_k respectively, we obtain the generalized differential equation of motion

$$\frac{d^2\xi_1}{d\theta^2} + \Phi''\left[(\xi_r+\xi), \xi_k, \Pi_1, \Pi_2, \Pi_3\right] = \bar{Q}. \tag{8}$$

Here $\xi_1 = \xi_r + \xi + \xi_0 \sin\upsilon\theta$; $\xi_r = \frac{r}{a_q}$; $\xi_0 = \frac{a_0}{a_q}$; $\upsilon = \frac{\omega}{\omega_q}$; $\theta = \omega_q t$; $\bar{Q} = \frac{Q}{T_q}$.

The generalized amplitude-frequency characteristics (AFC) of the DBR shock absorbers were defined as the dependencies of the dynamic coefficient μ_q on the frequency υ of disturbing oscillations by varying the parameters ξ_0, Π_2 and $\bar{Q}$.

Fig. 6 shows the results of the AFC research, and Fig. 7 shows the results of the study. As can be seen from Fig. 7, the dimensionless natural frequency υ_p and the dynamic amplification factor at the resonance μ_p depend strongly on the excitation amplitude; at that the function ($\mu_p = f(\xi_0)$)has a pronounced minimum, located at the area of $\xi_0 = 0,15 \ldots 0,2$. One may also note, that while Π_2 criterion increases μ_p criterion decreases.

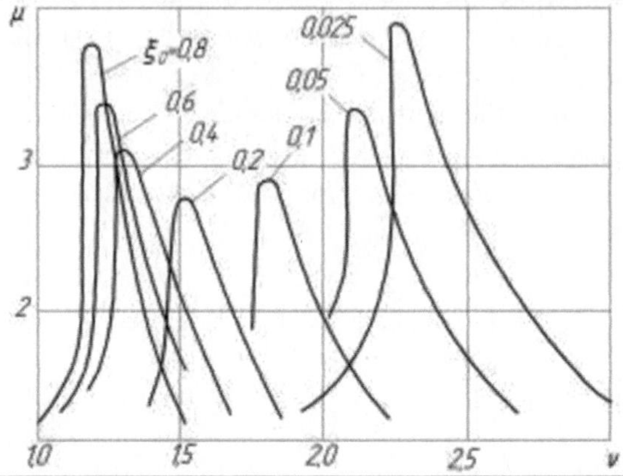

Figure 6. Generalized amplitude-frequency characteristics at Q=0,9

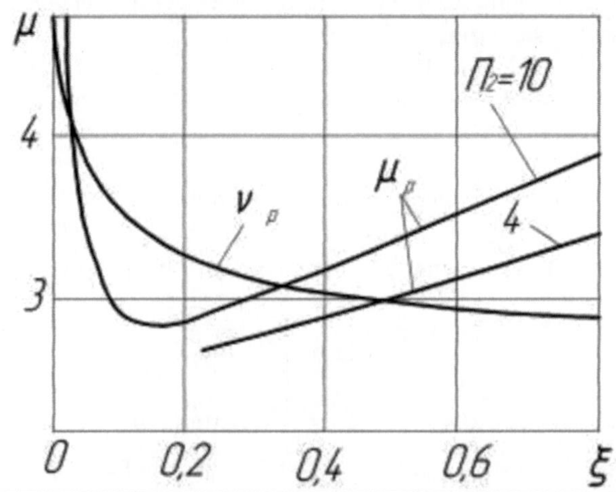

Figure 7. Resonance characteristics of shock absorbers

2.3. The method of strength characteristic research

The research of the DBR schock absorber strength characteristics was carried out under the static, vibration and shock loading.

The static strength of shock absorbers was determined by the P_p load value, causing EDE destruction. In this case, the EDE strength at break may be regarded as a total strength of EDE and AE workpieces. It is easy to show that the AE theoretical breaking strength is proportional to the complex $\frac{\rho_{ж}}{\rho_з}\frac{\sigma_{в0}}{E_и}$, where $\sigma_{в0}$ is the strength limit. The strength of MR material array made of coils may be determined experimentally on DK shock absorbers without AE. However, the actual working conditions of the schock absorber require the consideration of such factors as the uneven loading of the harness in AE, its interaction with an array of coils, the presence of stress raisers at the intersections of the strands. This may be accounted by using the experimental K coefficient:

$$\frac{P_P}{E_u D_K^2} = K\frac{\rho_{\text{ж}c}}{\rho_u}\frac{\sigma_{в0}}{E_u} + 0{,}76\cdot10^{-5}\frac{\rho_з}{\rho_u}\left(1-\frac{\rho_{\text{ж}c}}{\rho_u}\right)\left(\frac{D_K}{d_u}\right)^{0,5}, \tag{9}$$

where K=0,058 for type 1 strand and K=0,098 for type 2 strand.

The practiced expression analysis shows that the static strength of DBR shock absorbers may be changed 7-10 times without any noticeable changes in other characteristics.

Three types of destruction take place at the shock absorber vibration loading:

1) The emerging of cracks along the EDE periphery developing in the radial direction;
2) The breaking of stiching coils;
3) The emerging of radial cracks on EDE.

The last type of damage is typical for large relative deformation amplitudes ($\varepsilon_A = \frac{A}{D_K}; \xi_A > 0{,}12$) and it is the most dangerous one, as it causes a catastrophic reduction in the static strength of the vibration isolator. The long-term cyclic loading of the vibration isolator is accompanied by UDE vibration heating and the wire element wear in MR material. The intensity of these processes largely determines the vibration isolator resistance. The significant temperature of vibration heating (up to 673 K) is observed within AE areas located on the EDE periphery at $\varepsilon_A > 0{,}05$, $T_n > 2\cdot10^{-7}E_u D_K^2$ and the loading frequency of more than 20 Hz. The smaller the temperature (up to 473 K) is observed on the EDE surface. The DBR shock absorbers resistance is largely dependent on the relative amplitudes ε_A and Π_1 and Π_2 criteria. At the increase of Π_2 criterion from 8,5 to 15,0...18,0 at ceteris paribus the vibration isolator resistance increases 5-10 times and makes more than $N_{Ц} = 10^6$ operation cycles. With the increase of Π_1 criterion the vibration isolator resistance may be decreased in several times, at that the most characteristic is the first type of fracture that occurs due to the stress concentration increase around the stitched holes. With the small diameters of stitched wire and the low fluidity limit σ_{mc} the strength of the stitched wire is insufficient and leads to the second type of fracture. It should be noted that in comparison with the existing DK modification the DBR shock absorbers have a significantly higher vibration resistance (Fig. 8). Moreover, there is a reduction of the DBR damping capacity to 50% and the stiffness increase up to 60% during the operation time.

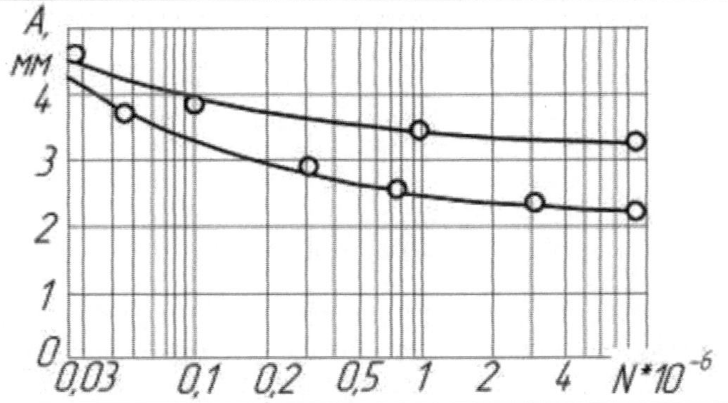

Figure 8. Dependence of vibration isolators endurance: 1 - DK-38-2/18; 2 - DBR -38-0,6/40

3. Conclusion

Thus, DBR isolators have the increased strength and damping properties. The performed studies confirm the possibility of DBR use within the vibration protection systems, operated under intense mechanic loadings.

The obtained results may serve as the basis for the design of various DBR vibration isolator sizes for a wide range of their mass loads and the intensity of vibration exposure. It should be noted that there may be a mathematical description of the loop field families, for example, by orthogonal polynomials and also by deformation processes at an arbitrary loading. This, together with the known approximate analytical methods of nonlinear differential equations solutions concerning the motion vibration isolation systems reveals broad prospects for the dynamic characteristics of the shock absorbers study at shock and random vibration loading, various laws of harmonic excitation from the frequency, the complex effect of static and dynamic loads, etc.

The solution of this type problems requires a more detailed study of the load and unload processes behavior at the arbitrary loading of the vibration isolator. Furthermore, due to the substantial non-linearity not only elastic, but also non-elastic forces in the MR material it is necessary to test the known approximate analytical methods concerning the possibility of their use during the solution of a wide class of dynamic problems.

This work was supported by the Ministry of Education and Science of the Russian Federation in the framework of the implementation of the Program ''Research and development on priority directions of scientific-technological complex of Russia for 2014– 2020.

References

[1] Ponomarev, Yu.K., Ermakov A.I., Simakov O.B., Mikhalkin I.K. Metallic counterpart of rubber: a material for vibration and shock protection. Metallovedenie I termicheskaya obrabotka metallov. # 1 (691). January, 2013. PP 8 – 13 / doi: 10.3103/S1068799809020111.

[2] Jiang, H.-Y., Hao, D.-G., Xia, Y.-H., Ulanov, A.M., Ponomarev, Y.K. Damping characteristics calculation method of metal dry friction isolators (2008) Journal of Beijing Institute of Technology (English Edition) 17 (2) PP. 173 – 177 / doi: 10.3103/S1068799809020111.

[3] Xia, Y., Yan, H., Jiang, H., Ao, H., Ulanov, A.M. Determination of elastic modulus of ring-like metal rubber isolator (2005) Run Hua Yu Mi Feng/Lubrication Engineering (3) PP. 34 - 36+39 Cited 6 times.

[4] Ao, H., Jiang, H., Ulanov, A.M. Dry friction damping characteristics of a metallic rubber isolator under two-dimensional loading processes (2005) Modelling and Simulation in Materials Science and Engineering 13 (4) PP. 609 - 620 Cited 11 times / doi: 10.1088/0965-0393/13/4/011.

[5] A.M.Ulanov, Yu.K.Ponomarev. THE CALCULATION OF VIBRATION ISOLATORS WITH COMPLEX SHAPE ELASTIC-DAMPING ELEMENTS MADE OF MR MATERIAL BY FINITE ELEMENT METHOD. Izvestiya of Samara Science Center of Russian Acadimia of Science, vol.11, No 5, 2009, pp. 222 – 226. [In Russian].

[6] Yu.K.Ponomarev, A.M.Ulanov. A COMPARISION OF RUSSIAN AND FOREIGN VIBRATION ISOLATORS MADE OF WIRE DAMPING MATERIALS. Izvestiya of Samara Science Center of Russian Acadimia of Science, vol. 11, No 3, 2009, pp. 214 – 218. [In Russian].

[7] A.M. Ulanov, Yu.K. Ponomarev. LIMIT OF WEARING OF MR MATERIAL FOR DIFFERENT LOADING. Izvestiya of Samara Science Center of Russian Acadimia of Science. Vol. 10, No 3 (25), 2008, pp. 849-852. [In Russian].

[8] A.M. Ulanov, Yu.K. Ponomarev. DESIGN OF VIBRATION PROTECTION SYSTEMS WITH ELASTIC ELEMENTS MADE OF MR MATERIAL. Izvestiya of Samara Science Center of Russian Acadimia of Science. Vol. 10, No 3 (25), 2008, pp. 857 – 864.[In Russian].

Influence of Porous Material MR Structure on its Flow Characteristics

Zhizhkin A M, Lazutkin G V, Davydov D P and Volkova T D

Samara State Aerospace University, 34, Moskovskoye shosse, Samara, 443086, Russia

azhizhkin@yandex.ru;lazutkin.gennadij@mail.ru;dpdavydov@yandex.ru; twigggy@yandex.ru

Abstract. Existing presentation of MR (metal rubber) material structure by its hydraulic (average) diameter of inner space for hydrodynamic equations is non-complete for understanding of influence of structure on hydrodynamic characteristics of MR material. Analysis of existing results of MR structure researches for porous structures shows that relative geometrical size of porous structure provides large influence on its average diameter. Therefore, it is proposed in a present research to determine a thick-wall and thin-wall porous details made of MR material. It is shown that placement of elements of solid phase is connected with value of parameter of density of pore distribution by size α, which is typical for gamma-distribution. Thus, it is proposed to use the dimensionless parameter α as characteristics of non-uniformity of MR material macrostructure. It is proved that for hydraulic equations the structure of porous detail made of material MR for direction perpendicular to liquid flow is determined univalent by effective hydraulic diameter proposed in the present paper. This diameter depends not only on average pores diameter but on parameter α too. Generalized dependency is obtained; it is possible to use it for calculation of hydraulic losses both in thick-wall and thin-wall details made of MR material by well-known and new technologies of manufacturing..

1. Introduction

Effective application of porous materials for hydro-gas-dynamic and heat-transport units is provided by choice of rational inner structure of materials which can improve its transport properties. It is evident that clear understanding of dependency of hydrodynamic characteristics of porous details of its structure features is necessary for it [1–6]. In the present time, a MR (metal rubber) material is used widely for manufacturing of porous details of engines and aircrafts. This material is made of metal wire spirals pressed as a shape of ready product by different ways proposed, for example, in [7–10]. The part made of MR material is shown in the Figure 1.

Figure 1. Porous part made of MR material

In [7, 8] details and units made of MR material for application for damping systems are presented. In [8, 9] ways of porous details manufacturing of MR material and research methods of these detail's characteristics are given. Details manufactured by these technologies are applied as hydrodynamic elements of engines and power plants pipelines [3].

However, existed methods of obtaining of hydrodynamic characteristics of porous details made of MR material often provide large calculation errors [11].

The aims of the present research are analysis of existed equations for calculation of hydraulic losses in details made of porous MR material, and obtaining of parameters of this material structure, which determine a process of filtration of working liquids.

It is well-known that hydraulic losses for porous details depend on physical properties of working media, kinematic characteristics of flow and inner features of porous structure. For common case a dependency between all these factors is presented as equation

$$\Delta p / L = f(V_x, D_x, \rho, \mu) \tag{1}$$

here Δp = p1 – p2 is changing of pressure; p1 is enter pressure; p2 is exit pressure; L is length of porous detail; Vx is characteristic velocity of working media flow; Dx is characteristic size of detail porous structure; ρ is density of working media, μ is coefficient of dynamic viscosity for liquid [3, 6, 11 etc].

There are many researches of influence of different parameters on value Δp/L. For example, it is proved in [3] that laminar, turbulent and transitional from laminar to turbulent conditions of flow may be presented in details made of MR material.

In [3, 6 and 11] two dimensionless complexes were obtained by methods of theory of similarity and analysis of dimensionality:

$$\pi_1 = \frac{\Delta p D_x}{L V_x^2 \rho}; \tag{2}$$

$$\pi_2 = \frac{\mu}{V_x D_x \rho}. \tag{3}$$

Analogously to pipeline hydraulic, it is possible to transform these complexes to equations

$$\xi = \eta = \frac{2\Delta p D_x}{L V_x^2 \rho}; \quad Re = \frac{V_x D_x \rho}{\mu} \tag{4}$$

here ξ or η are coefficients of friction resistance, and Re is Reynolds number [3].

Thus, it is possible to present the equation (1) as criterial: for laminar flow

$$\xi = \frac{A_1}{\mathrm{Re}} \tag{5}$$

for transitional and turbulent flow (V>Vk) for Re > 28

$$\xi = \frac{A_2}{\text{Re}} + B \tag{6}$$

Constant coefficients A1, A2, B for dependencies (5), (6) are obtained experimentally [3].

Authors of [3, 11] suggested to choose the hydraulic diameter of porous media as characteristic linear size of MR material structure.

Structure of MR material has no blind alleys, and all inner surface of material is moistened by liquid. Therefore, authors of [3, 11, 12 and 13] suggested choosing in absence of obliteration as characteristic size of porous media the hydraulic diameter of inner space

$$D_x = d_g = d_c = \frac{\Pi d_{\Pi}}{(1-\Pi)} \tag{7}$$

here d_g and d_c are hydraulic and average diameters of porous space of MR material; d_{Π} is wire diameter, Π is porousness of material.

An important peculiarity, as author of [3] think, is determination of hydraulic diameter in equation (7) mostly by structure characteristics of porous media: porousness Π and wire diameter d_{Π}.

Thus, by results of [3], it is possible to characterize the structure of MR material by any constant coefficient A and hydraulic diameter d_g (7), which determines the size of pores of porous detail.

However experimental results for MR material presented on Figure 2 contradicts to this statement and shows that coefficient A for equations (5) and (6) is non-constant value.

Thus, it is possible to suppose that coefficient A1 of dependency (5) is parameter of structure too. Therefore, it is necessary to solve two important problems.

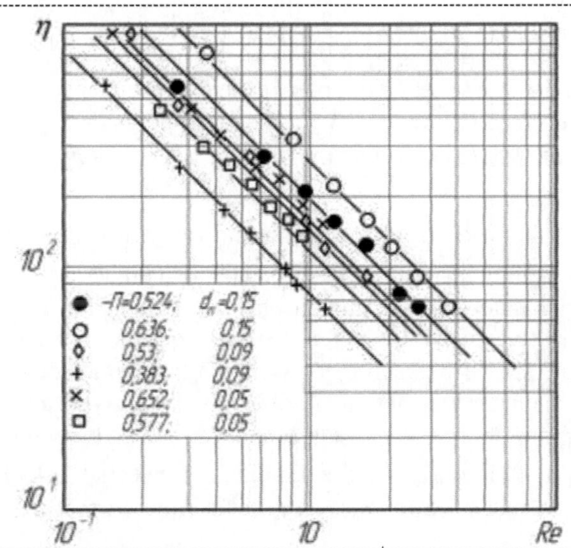

Figure 2. Dependency of η on Re [7]

Problem 1. On a basis of analysis and synthesis of results of existing works on calculation of hydraulic losses in porous details made of MR material to obtain the geometrical parameter for more correct describing of MR structure.

It is possible to divide methods of porous materials structure on direct ones and indirect ones (it is described, for example, in [2]). A development of software and computing allows numerical modelling of porous structures [14]. It allows obtaining of structures with preliminary required properties. MR material has stochastic structure. For correct description of MR structure, it is possible to obtain the most reliable data with direct methods of structure research. These methods provide immediate data which doesn't need additional interpretation and treatment. It does not lead to additional errors for data, and obtained results are more reliable.

Let we consider of MR material inner structure research. Cylindric and plane details made of MR material were used as the researched specimens. Porousness П (measured in %) of specimens was 55 – 85%. Statistic method of micro-section studying was used for research. It allowed obtaining of pores distribution by size for section of porous specimen with different structures. A distance between contours of elements of solid framework of porous space was taken as diameter of pore, as in [15]. The distance between little parts of framework contour was considered as random value equal to length of line segment d between two points on a surface of solid phase.

Obtained data allows calculation of statistic average distance dcs and statistic dispersion Dc by equations [15]:

$$d_{cs} = \sum_{i-1}^{k} d_{ic} f_i^* \Delta d_i \tag{8}$$

$$D_c = \sum_{i=1}^{k} (d_{ic})^2 f_i^* \Delta d_i - d_c^2 = \sigma_c^2 \tag{9}$$

Here dic is average value of distance for interval i; σc is mean square deviation for value d. The value of average distance for each interval was obtained as $d_{ic} = (d_{i-1} + d_i)/2$, and $\Delta d_i = d_i - d_{i-1}$. Relative frequency fi* of appearance of size d in interval [di-1; di] was calculated by equation

$$f_i^* = \frac{m_i}{N(d_i - d_{i-1})} = \frac{\Delta F_i^*}{\Delta d_i} \tag{10}$$

here mi is quantity of size value from interval i; di-1; di are boundaries of interval i; $N = \sum_{i=1}^{k} m_i$ is total number of measurement; ΔFi* is frequency of appearance of d в интервале [di-1; di].

Analysis of results of research shows that average size of pores dc (7) is equal to average distance dcs (8) (Figure 3) for detail with relative thickness $\delta_o = \delta_f / S_c > 1$, here δ_f is minimal thickness of detail; Sc is diameter of wire spiral.

It was found that for details made of MR material and for relative thickness of wall $\delta_f / S_c < 1$ the average distances dcs are not coincide with dependency (7) (Figure 4).

It is necessary to notice that for obtaining of average diameter by dependency (7) areas of moistening surfaces of wire spirals of MR material were taken into account, but areas of moistening surfaces of case in which the porous detail situated were not taken into account.

Therefore, the average diameter was obtained in [3, 7] for porous structure of MR material only. If to suppose that porous detail in the case is absent (condition П=1), by equation (7) the value $d_c \to \infty$. It is not right, because in this situation the hydraulic diameter is equal to hydraulic diameter of hole in the case, which contains the porous detail. For example, for ring channel $d_c = d_к = 2\delta_f$.

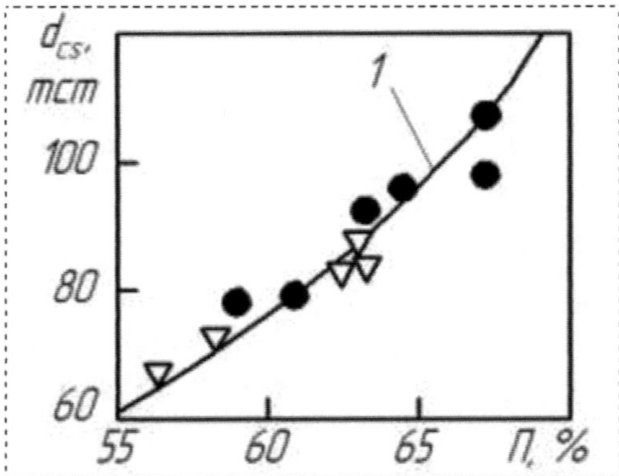

Figure 3. Average distance between little parts of framework of details made of MR material for $\delta_f / S_c > 1$: ∇ is average distance between little parts for radial section of specimen, • is average distance between little parts for axial section of specimen; 1 is calculated dependency (5)

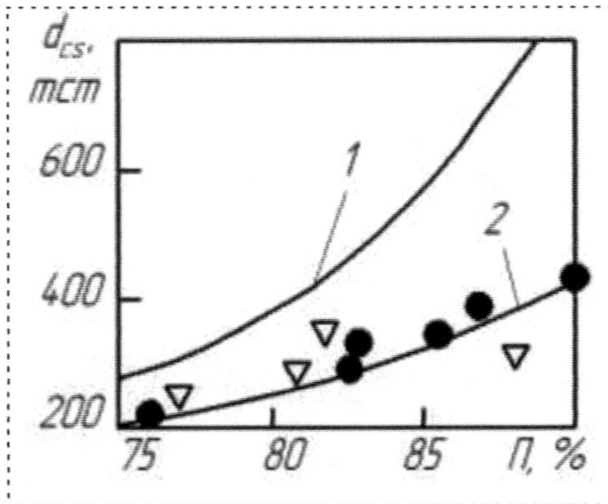

Figure 4. Average distance between little parts of framework of details made of MR material for $\delta_f / S_c < 1$: ∇ is average distance between little parts for radial section of specimen, • is average distance between little parts for axial section of specimen; 1 is dependency (5), 2 is dependency (9

If to take into account the area of moistening surface of case contacted with MR material, the equation for hydraulic dimeter of ring-type porous structure will be as

$$d_g = d_{cs} = \Pi d_\Pi / (1 + d_\Pi / 2\delta_f) - \Pi \qquad (11)$$

If porous element is absent (П= 1), value of average diameter will be equal to hydraulic diameter of hole in the case contain the porous structure. For example, for ring channel it will be equal $d_c = d_\kappa = 2\delta_f$, it coincides with reality and with a rule of boundary transition.

Results of calculation by dependency (11) and experimental data are presented on Figure 3. As it is seen from Figure 4, dependency (11), in a difference on dependency (7), presents the experimental results for condition $\delta_f / S_c < 1$ more precisely.

Thus, it is possible to distinguish two types of porous details made of MR:
- thick-wall, for condition $\delta_f / S_c > 1$;
- thin-wall, for condition $\delta_f / S_c < 1$.

It is necessary to notice that for condition $\delta_f / S_c > 1$ calculation fulfilled for value of average diameter dc by equations (7) and (11) for $\delta_f / D_c = 1$, have difference between their values not more than 9%. If ratio of parameters increases, difference of values obtained by equations (7) and (11) approaches to 0.

Thus, dependency (11) generalizes both of cases. It is more general equation for middle diameter of pores in MR material both for thick-wall and thin-wall details.

Subsequently in the present paper the value of average diameter of pore dc is calculated by equation (11).

By results of structure researches the histograms were obtained for different types of researched specimens and different relative frequencies fi*(d), calculated by equation (10).

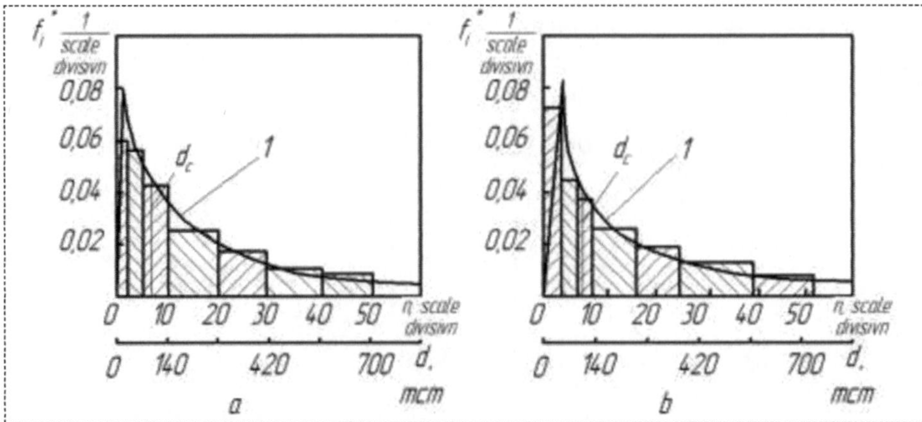

Figure 5. Density of distribution of distance d for plane specimens (П = 0,6; dП = 50 micrometers; dc = 75 micrometers): a)is radial section; b)is axial section; 1is density of γ-distribution, dc is average distance between contours of elements of solid framework

An example of these histogram is presented on Figure 5.

By the shape of histograms, it is possible to conclude that density of pores size distribution is non-symmetric.

If to compare the shape of histogram and a shape of curve 1 on Figure 4, it is possible to propose a gamma-distribution for modelling of pores size distribution in details made of MR material. Function of density of distribution in this case is

$$f(d) = \frac{\beta^{\alpha}}{\Gamma_{(\alpha)}} d^{\alpha-1} e^{-\alpha\beta} \qquad (12)$$

here Γ(α) is noncomplete gamma-function; α and β are parameters of distribution obtained from equations

$$\alpha = \beta d_c = d_c^2 / D_c = (1/K_B^2) \qquad (13)$$

$$\beta = d_c / D_c \qquad (14)$$

Here KB is coefficient of variation obtained by equation KB=σ/dc, σ is mean square deviation of value to be determined [16].

Function f(d) of density of pores size distribution (12) is presented on Figure 5 as continuous line 1 and determined by size of pores d and parameters α and β, obtained by experimental researches of MR material porous structure. However, qualification of obtained histograms on coinciding with gamma-distribution by criterion of Pearson didn't show 100% certainty of chosen function (12). However, it is possible to notice for this function that its region of definition is $\alpha \in [0-\infty]$, $\beta \in [0-\infty]$, $d \in [0-\infty]$. The proposed rule for pores size distribution isn't contradicted to reality of MR material porous structure for boundary values of pores diameter.

If parameter $\alpha \to 0$, d→0. It means that the porous structure has discontinuity.

If parameter $\alpha \to \infty$, the function f(d) tends to logarithmic normal distribution. For α=∞ D=0. The structure is ideal porous media, and its average diameter is dcs = d [6]. However, it is impossible to obtain α→∞ for MR material structure on technological reasons of stochastic porous structure as a result of MR material manufacturing. As it is presented in data of experimental research of porous structure for detail made of MR value of α is possible in a range from 0.6 to 3.6.

It is known that the shape of curve f(d) is determined by parameter α, and parameter β is a scale multiplier [16].

As generalization of presented analysis, it is possible to suppose that parameter α of function (12) determines adequately the pores size distribution for MR material.

This connection of parameter α and pores size distribution it is possible to consider, for example, as connection of distance between contours of elements of solid framework of porous space. It is possible to see it clearly for one-dimensional linear model presented on Figure 6.

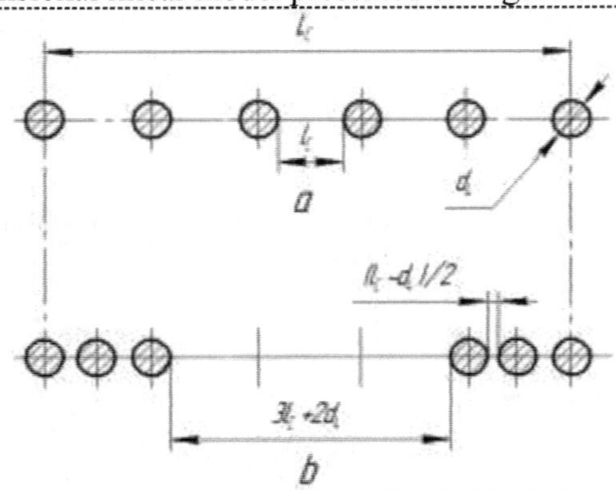

Figure 6. Linear models for analysis of discrete structures; a is porous media for α → ∞, D=0; B is porous media for ∞ > α > 0, D ≠ 0; Lc is basic length of model; lc is average distance between surfaces of little parts; dч is diameter of little part

Models of structure on Figures a and b have the same basic length Lc and average distance between surfaces of little parts lc, However, structure parameters α are different for these models. As it is seen from Figure 6b, distance between surfaces of little parts and placement of solid phase are different from these parameters of Figure 6a.

It is evidently that value of parameter of distribution α is connected with position of elements of solid phase relatively each other independently on rule of pores size distribution. Therefore, it is possible to use the non-dimensional parameter α as characteristics of inhomogeneity of macrostructure for MR material and other porous media with any other single-modal rule of pores size distribution.

To compare the porous media and to estimate an influence of their structures on processes of mass transport in it a scale characteristic of pore volume is necessary. It is proposed to use as this characteristic for porous media the average pore diameter dcs.

Let for two porous media the average diameters are equal dcs1 = dcs2, but α1 ≠ α2. However, the relative position of framework elements is different, and it is possible to suppose that influence of porous media on process of mass transport is not the same.

Therefore, comparison of porous media with different structure inhomogeneity by average diameter of pores are non-correct. Only for equality of parameter of distribution α and dcs these structures to be compared have equal geometric configuration of porous space.

Therefore, it is proposed to use a geometric model from for estimation of influence of variation of pores by size in a volume of porous detail on hydraulic properties of porous structure. This model is set of capillary with different diameter. A size of capillary in direction perpendicular to working media flow is changed by arbitrary rule. For direction of working media flow the size of pores doesn't change. All pores are connected with each other hydraulically.

It is proposed to use for this media as characteristic size of equation (1) its hydraulic diameter as

$$D_x = \frac{4F}{k} \tag{15}$$

here F is area of flow passage section in porous media; k is moistening perimeter.

It is possible to obtain the area of flow passage section for all pores of separate volume of porous media as

$$F = \sum_{i=1}^{N} \frac{\pi \cdot d_i^2}{4} p(d_i) .$$

A moistening perimeter of flow passage for all pores of separated volume of porous media equal to

$$k = \sum_{i=1}^{N} \pi \cdot d_i p(d_i)$$

By equation (15) and dependencies for F and κ it is possible to obtain equation for characteristic size of equation (1) as

$$D_x = d_{fh} = d_{cs}\left(1 + K_B^2\right) \tag{16}$$

here dcs is average diameter of pores obtained from equation (11); dfh is effective hydraulic diameter of porous media with arbitrary single-mode rule of pores size distribution.

By equation (13) it is possible to obtain equation for dfh for MR material as

$$d_{fh} = d_{cs}\left(1 + 1/\alpha\right) \tag{17}$$

It is seen from equation (17) that the characteristic size of MR material porous structure is determined by two parameters dcs and α.

Therefore, it is possible to assert that effective hydraulic diameter from equation (17) is more correct for determination of structure of porous detail made of MR material than this diameter from equation (7).

Problem 2. Obtaining of equations for calculation of hydraulic losses in MR material.

If to take into account equations (4), it is possible to obtain coefficient of resistance ξfh and Reynolds number Refh as

$$\xi_{fh} = \frac{2\Delta p \Pi^2 d_{cs}(1 + 1/\alpha)}{LV^2\rho}, \tag{18}$$

$$\mathrm{Re}_{fh} = \frac{V\rho d_{cs}(1 + 1/\alpha)}{\Pi\mu} \tag{19}$$

Dependency of ξfh and Refh is obtained by equation

$$\xi_{fh} = A_2 / \mathrm{Re}_{fh} \tag{20}$$

for laminar flow and

$$\xi_{fh} = A_3 / \mathrm{Re}_{fh} + B_1 \tag{21}$$

for transient flow.

It is possible to obtain values of constants A2, A3 and B1 from experimental researches of hydraulic losses in MR material by method presented in [3]. If to take into account these values and equations (18), it is possible to present equations (20) and (21) as

$$\frac{\Delta P}{L} = \frac{240}{2\Pi d_{cs}^2 (1 + 1/\alpha)^2} \mu V \tag{22}$$

for laminar flow and

$$\frac{\Delta P}{L}=\frac{220\mu V}{2\Pi d_{cs}^{2}\ (1+1/\alpha)^{2}}+\frac{2\rho V^{2}}{2\Pi^{2}d_{cs}(1+1/\alpha)} \quad (23)$$

for transient flow of liquid in MR material.

Critical value of Reynolds number is Refhk=10.

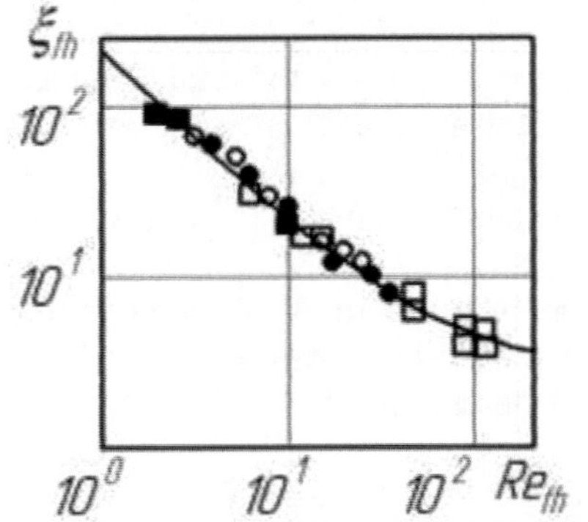

Figure 7. Dependency of hydraulic loss coefficient ξfh on Reynolds number Refh (dП= 0,09 mm): □ - П =0,741, α = 1,56; ■ - П =0,752, α = 1,48; ○ - П =0,608, α = 1,32; ● - П =0,614, α = 1,30

Graphic interpretation of equations (20) and (21) is presented on Figure 7. Experimental data in a limit of error (15….20%) coincides with calculation results by equations (19). As characteristic linear velocity Vx an average velocity of flow in pores is taken. It is obtained by average volume velocity V and porousness П. The effective hydraulic diameter is taken as characteristic size of porous structure. This diameter takes into account both the average diameter of pores dc, and relative variation of its size 1/α. As it is follow from equations (22) and (23), α in the range of its changing has influence on hydraulic losses in porous details made of MR material.

It is possible to use equations (22) and (23) for calculation of hydraulic losses in details made of MR material by well-known and new developed technologies.

2. Conclusion

1. The existed presentation of MR material by its hydraulic (average) diameter for hydrodynamic equations is not completely enough for understanding of influence of material structure on its hydrodynamic characteristics.
2. Researches of MR material structure by direct observation methods show, that for condition $\delta_f / D_c < 1$ (differently of $\delta_f / D_c > 1$) the relative minimal geometric size of porous detail has large influence on average diameter of porous media. Therefore, it is useful to separate this type of detail as a type of thin-wall structures made of MR material.
3. The relative placement of solid phase elements connects with value of parameters of density of pores size distribution α. Therefore, it is possible to use the dimensionless parameter α as characteristic of inhomogeneity of macrostructure of MR material.

The scale characteristics of porous volume as average diameter of pores dcs and parameter of inhomogeneity α are necessary for comparison of porous media and estimation of influence of its structure in processes of mass transport.

4. It is possible to determine the structure of porous detail made of MR material for hydraulic equations by effective hydraulic diameter dfh obtained by equation (17).

5. More precise dependencies for calculation of hydraulic losses in details of MR material are obtained. These dependencies take into account peculiarities of inner structure of detail determined by average diameter of pores dcs and parameter of structure inhomogeneity α.

This work was supported by the Ministry of Education and Science of the Russian Federation in the framework of the implementation of the Program ''Research and development on priority directions of scientific-technological complex of Russia for 2014– 2020.

3. References

[1] Levitan M.M., Rasin O.G. Efficiency of porous material application for intensification of thermal exchange in channels // Heat-mass exchange for systems with porous elements. Minsk, 1981. – PP. 91 – 99. [in Russian]

[2] Belov S.V. Porous materials for machine design. M.: Mashinostroenie, 1981, - 247 p. [in Russian]

[3] Izzheurov E.A. Formation of elements for structures of hydrodynamic channel made of elastic porous MR material for power plants. M.: Mashinostroenie, 2001.– 286 pp. [in Russian]

[4] Lykov A. V. Heat-mass exchange: Reference book. M.: Energia, 1978. 480 pp. [in Russian]

[5] Vasil'ev L.A., Tanaeva S.A. Thermophysical properties of porous materials. Minsk.: Nauka i tekhnika, 1971, 268 pp. [in Russian]

[6] Mints D.E., Shubert S.A. Hydraulic of granular materials. M.: Minkommunkhoz RSFSR, 1955, - 112 pp. [in Russian]

[7] Ponomarev Ju.K., Ermakov A.I., Simakov O.B, Mikhalkin I.K. Metallic counterpart of rubber: a material for vibration and shock protection / Physical metallurgy and heat treatment of metals, 2013, 1, pp. 8 - 13.

[8] Hong, J., Zhu, B., Ma, Y., Chen, L. Experiments on static characteristics of metal rubber particles /Journal of Beijing University of Aeronautics and Astronautics, 2012, 38 (2), pp. 218-221.

[9] Zhang, D., Scarpa, F., Ma, Y., Hong, J., Mahadik, Y. System identification methods for metal rubber devices / Mechanical Systems and Signal Processing, 2013, 39 (1-2), pp. 207-226.

[10] Zhang, D., Scarpa, F., Ma, Y., Boba, K., Hong, J., Lu, H. Compression mechanics of nickel-based superalloy metal rubber / Materials Science and Engineering A, 2013, 580, pp. 305-312.

[11] Belousov A. I, Izzheurov E.A., Setin A.D. Research of hydrodynamic and filter characteristics of porous material MR / Vibration strength and reliability of aircraft engines and systems. Kyibyshev, 1975, 2. - PP. 70-80. [in Russian]

[12] Ban A., Bogomolova A.F., Maksimov V.A., Nikolaevskiy V.P., Orandzhanyants V.G., Ryzhik V.M. Influence of earth material properties on a movement of liquid in it. M.: Gostekhizdat, 1962. – 151 pp. [in Russian]

[13] Zhavoronkov N.M., Aerov M.E., Umnik N.M. Hydraulic resistance and density of package of granular layer / Fisicheskaya himiya, 1946, Vol. 23, Issue 3, PP. 137–141. [in Russian]

[14] E. Glatt, S. Rief, A. Wiegmann, M. Knefel, E. Wegenke. Structure and pressure drop of real and Virtual metal wire meshes. Berichte Fraunhofer–Institut fűr Techno – und Wirtschaftmathematik JTWM, 2009, 157.

[15] Bogomolova A.F., Orlova N.A. Quantitative characteristics of porous space structure. PMTF, 1961,4.–PP.77-81. [in Russian]

[16] Hudson D. Statistics for physics. M.: Mir. 1970. – 296 pp. [in Russian]

IOP Conf. Series: Materials Science and Engineering **302** (2018) 012075 doi:10.1088/1757-899X/302/1/012075

Copper Nanoparticles: Synthesis and Biological Activity

A S Satyvaldiev[1], Z K Zhasnakunov[1], E Omurzak[2], T D Doolotkeldieva[3], S T Bobusheva[3], G T Orozmatova[1] and Z Kelgenbaeva[4]

[1]Faculty of Biology and Chemistry, Arabaev Kyrgyz State University, Bishkek, Kyrgyzstan
[2]Department of Chemical Engineering, Faculty of Engineering, Kyrgyz-Turkish Manas University, Chyngyz Aitmatov campus, Jal, 720038, Bishkek, Kyrgyzstan
[3]Department of Plant Protection, Faculty of Veterinary, Kyrgyz-Turkish Manas University, Chyngyz Aitmatov campus, Jal, 720038, Bishkek, Kyrgyzstan
[4]Institute of Chemistry, Kyrgyz National Academy of Sciences, Bishkek, Kyrgyzstan

emil.omurzak@manas.edu.kg;emilom@gmail.com

Abstract. By means of XRD and FESEM analysis, it is established that copper nanoparticles with sizes less than 10 nm are formed during the chemical reduction, which form aggregates mainly with spherical shape. Presence of gelatin during the chemical reduction of copper induced formation of smaller size distribution nanoparticles than that of nanoparticles synthesized without gelatin and it can be related to formation of protective layer. Synthesized Cu nano-powders have sufficiently high activity against the *Erwinia amylovora* bacterium, and the bacterial growth inhibition depends on the Cu nanoparticles concentration. At a concentration of 5 mg / ml of Cu nanoparticles, the exciter growth inhibition zone reaches a maximum value within 72 hours and the lysis zone is 20 mm, and at a concentration of 1 mg / ml this value is 16 mm, which also indicates the significant antibacterial activity of this sample.

1. Introduction

Synthesis and prediction of the properties of metal nanodisperse systems open wide opportunities for the creation of highly effective new drugs with biological activity for use in medicine and agriculture [1]. Metal nanoparticles show a pronounced biological activity including bacteriostatic and bactericidal effects [2]. Therefore, the study of the regularities in the manifestation of the biological activity of nano-disperse metals, depending on the method of their production, shape, size and physicochemical properties, is an urgent task.

The purpose of this study is to prepare Cu nanoparticles by chemical reduction method and to study their biological activity.

2. Experimental

2.1. Synthesis and characterization of Cu nanoparticles

Aqueous solutions containing Cu ions were prepared from copper sulphate crystal hydrate ($CuSO_4 \cdot 5H_2O$), 64% solution of hydrazine hydrate ($N_2H_4 \cdot H_2O$) was served as a reducing agent. As a stabilizer for Cu nanoparticles food gelatin was used, which was purified by washing three times in distilled water.

Nanosized Cu- powder synthesis is based on the chemical reduction of Cu (II) ions by hydrazine. Reducing activity of hydrazine is enhanced at pH> 10, therefore Cu nanoparticles were synthesized in alkaline and ammonia medium at pH of 11.

To synthesize Cu nanopowder in an alkaline or ammonia medium concentrated sodium hydroxide solution or a 10% ammonia solution was added into a 50 ml of an aqueous solution of copper ($C_{Cu}{}^{2+}$ = 0.05 M) a until the solution pH reached to 11. The solution is then heated to 50-60°C in a water bath and a hydrazine solution is added into the heated solution with an amount exceeding (in moles) 10 times the excess of copper ions. The solution with continuous stirring with a magnetic stirrer was held at this temperature for 30 minutes. The reduction product is then separated from the liquid using a centrifuge, washed with water until neutral (pH=) and ethanol. Obtained product was dried in drying oven at 55-60°C. During the synthesis, the gelatin solution was added to the copper salt solution in such amount that its content in the resulting mixture is 0.2%.

The phase composition of the products obtained from chemical reduction was established by X-ray diffraction (XRD) analysis. The diffractograms were taken on a RINT-2500 HV X-ray diffractometer (Kumamoto University, Japan) using copper-filtered radiation. Field emission scanning electron microscopy (FE-SEM) was used to determine the dispersion and morphology of the nanopowders. Micrographs of finely dispersed copper particles were taken on FE-SEM, JOEL JSM-7600F (Kumamoto University, Japan).

2.2. Biological activity of Cu nanopowders

A causative agent of fruit tree diseases – the gram-negative bacterium *Erwinia amylovora* (strain BK) was used as a test-object for the biological activity of copper nanoparticles. The bacterial strains were taken from laboratory collection bacteria from phytopathology laboratory of the Kyrgyz-Turkish "Manas" University.

The metal nanoparticles antibacterial action evaluation was carried out by the agar block method; for this, the culture of the test bacterium was grown on the surface of a solid nutrient medium, then holes with diameter of 5 mm were made using a sterile metal drill. 1 ml of an aqueous solution of copper nanoparticles was added to the wells at various concentrations and the Petri dishes with the cultures were further incubated at a constant temperature (27-28° C). The lysis zone of bacterial colonies around the agar blocks was observed and measured within 72 hours.

The biological activity of all Cu nanopowder samples (synthesized in alkaline and ammonia media, with and without gelatin) was investigated. Cu nanopowder solutions were prepared with a concentration of 5, 1, 0.5 and 0.1 mg per 1 ml.

3. Results and Discussion

3.1. Crystal structure, morphology and size of the nanoparticles

The reduction of Cu ions by hydrazine was carried out in an alkaline and ammonia medium with no use of stabilizer, and in second experiment with gelatin as a stabilizer. XRD diffractions of the obtained products are presented in figures 1 and 2.

The results show that the content of oxide phases is high in an alkaline medium. Along with metallic Cu, the product contains oxides of Cu_2O and CuO (figure 1a). In an ammoniacal medium, Cu^{2+} ions are mainly reduced to a metal, and univalent copper oxide is present in a small amount (figure 1b).

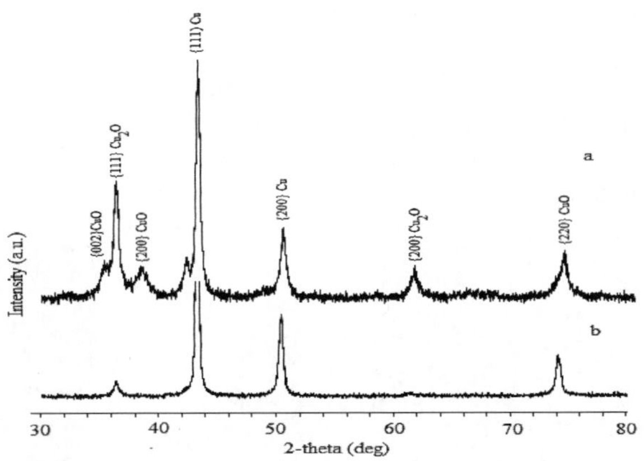

Figure 1. XRD patterns of Cu reduction with hydrazine products, in alkaline (a) and ammonia (b) media.

The size of the coherent scattering regions (CSR) of Cu is estimated from the broadening of the diffraction profiles by the Scherrer-Selyakov equation [3]. The sizes of Cu particles obtained in alkaline medium ranged from 20 to 43 nm, in ammonia medium they have sizes from 26 to 30 nm. Cu nanoparticles obtained in an ammonia medium have a narrower size distribution, compared to particles obtained in an alkaline medium.

The reaction medium composition effect can be explained by the fact that the reduction of Cu occurs from various forms of Cu ions. In alkaline medium at pH 10-11, Cu^{2+} ions are transferred to sparingly soluble hydroxide $Cu(OH)_2$, and in the ammonia medium a complex ion $[Cu\ (NH_3)\ _4]^+$ is formed. These forms of Cu are different in the dissociation constant and in the electrode potentials value. This quantities influence the recovery process and, accordingly, on formed products [4].

To improve the Cu nanoparticles aggregative stability, their synthesis was carried out in the presence of gelatin.

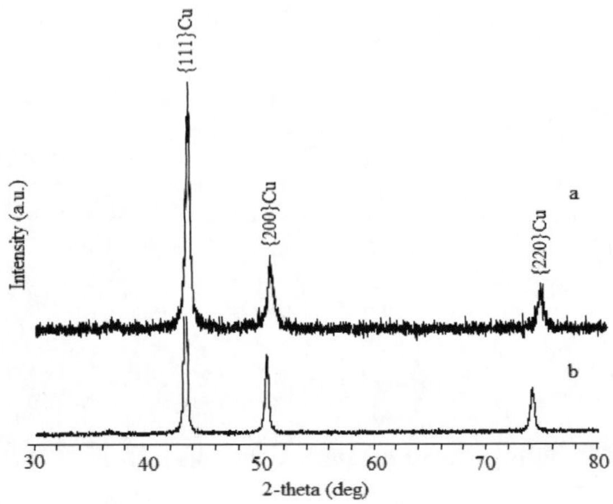

Figure 2. XRD patterns of Cu reduction with hydrazine products in the presence of gelatin in alkaline (a) and ammonia (b) media.

In the presence of gelatin in both alkaline and ammonia media, Cu^{2+} ions are almost completely reduced to metal (figure 2).

By the broadening of the diffraction profiles, the dimensions of the Cu nano-crystallites obtained in the alkaline medium in the presence of gelatin are ranged from 20 to 25 nm, and for Cu particles obtained from the ammonia with gelatin the size ranged from 22 to 33 nm. Such a narrow distribution of Cu particles obtained in the presence of gelatin in comparison with particles obtained without stabilizers is due to the fact that gelatin, forming a protective film on the surface of Cu nanoparticles, prevents aggregation and reduces an average size of the nanoparticles.

The nanopowders dispersion was studied by electron microscopy (figures 3 and 4).

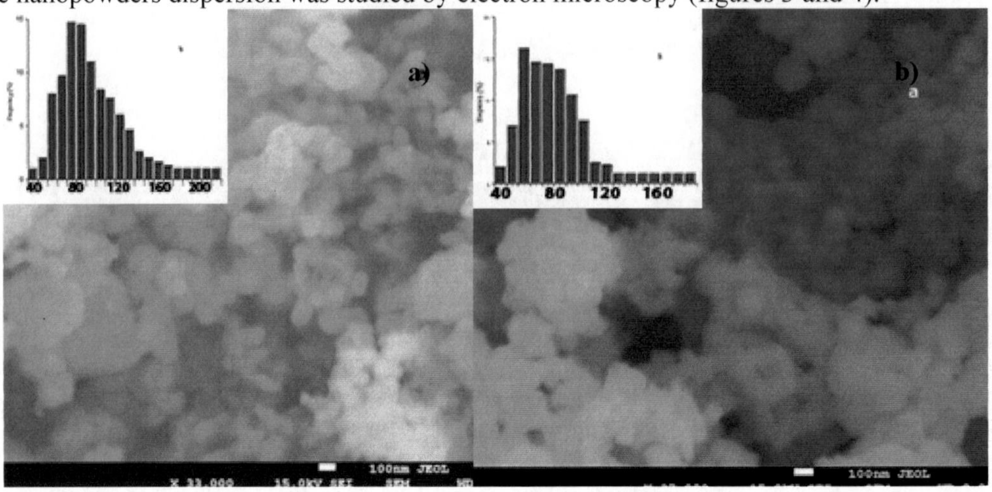

Figure 3. SEM images of Cu nanopowders produced in an alkaline medium (a) and Cu nanopowders obtained in an alkaline medium in the presence of gelatin (b) with their particle size histograms in their insets.

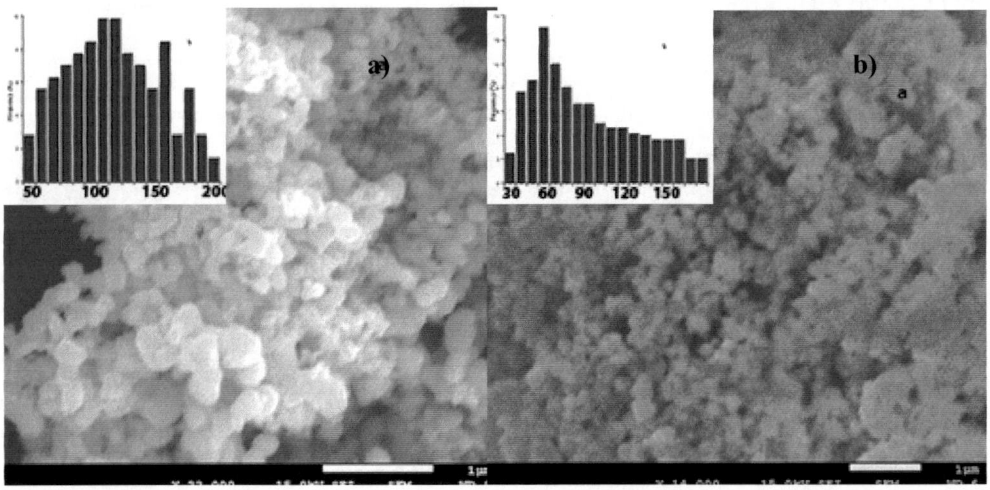

Figure 4. SEM images of Cu nanopowders obtained in an ammonia medium (a) and Cu nanoparticles obtained in the presence of gelatin in an ammonia medium (b) with their particle size histograms in their insets.

FESEM images of Cu nanoparticles synthesized in alkaline and ammonia media with and without gelatin show that in all four samples the particles are spherical in shape. Insets of the figures 3a and 4a show histograms of Cu nanoparticles synthesized in alkaline and ammonium media, respectively. As shown from the figures, Cu nanoparticles synthesized in an alkali medium have an average size of 70-100 nm, and in an ammonia medium 100-140 nm. When comparing the photomicrographs of Cu nanoparticles obtained in an alkaline medium without a stabilizer and with the addition of gelatin (insets of the figures 3a and 3b), one can see that when gelatin stabilizes, more fine particles are formed and the particle size distribution interval is somewhat narrower than in the case of a sample obtained without a stabilizer in an alkaline medium.

The average size of Cu nanoparticles obtained in an ammonia medium in the presence of gelatin (figure 4a) lies in the range of 50-70 nm. Such a small diameter and narrow size distribution of Cu nanoparticles obtained in the presence of a stabilizer may be associated with an increase of nanoparticles stability in the presence of gelatin [5].

A comparative analysis showed that the average sizes of the nanoparticles determined by the FESEM are substantially larger than the dimensions of the CSR. This is because the FESEM captures the total particle size, including the polymer stabilizer shell. In addition, Cu nanoparticles are prone to some aggregation, even in the presence of a stabilizer. These facts allow us to assume that peaks that appear in the region of 160-200 nm in all four histograms belong to agglomerates consisting of smaller nanoparticles.

3.2. Biological activity of the Cu nanoparticles

The biological activity of Cu nanopowders synthesized during hydrazine reduction in alkali and ammonia media in the absence and presence of gelatin was tested against the causative agent of *Erwinia amylovora* crop diseases (the causative agent of bacterial apple and pear burn). Figure 5 shows the zones of inhibition of the growth of *Erwinia amylovora* (red circle marked) under the Cu nanoparticles action synthesized by chemical reduction.

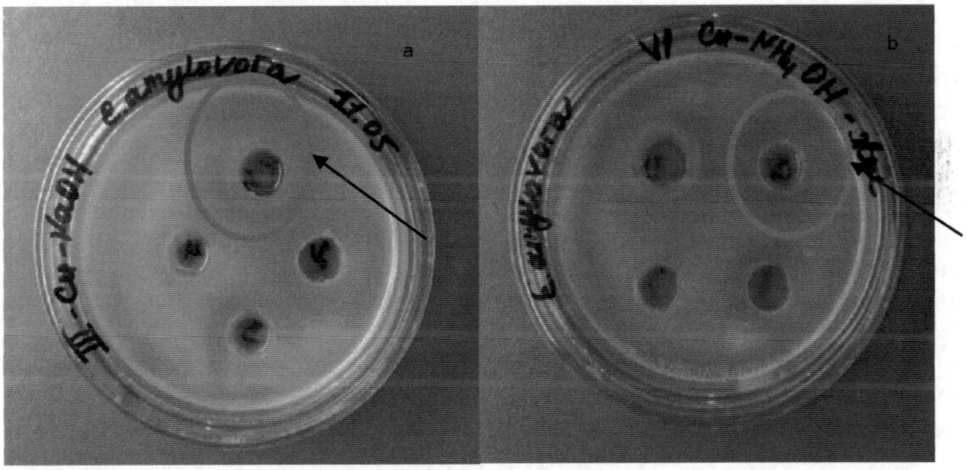

Figure 5. The zone of suppression of the growth of the *Erwinia amylovora* culture under the action of Cu nanoparticles obtained by chemical reduction with hydrazine in alkaline medium (a) and in ammonia medium (b).

The results of the study of the antibacterial activity of Cu nanopowders synthesized by the chemical reduction method show that all Cu nanopowders have a sufficiently high activity for the *Erwinia amylovora* [6] bacterial burn causative agent. The area of pathogen growth inhibition depends on the Cu nanoparticles concentration (table 1).

Table 1. Zone inhibition growth (mm) dependence of the bacterium *Erwinia amylovora* bacterium burner on the Cu nanoparticles concentration after 72 hours.

№	Copper nanopowders	Cu nanoparticle concentration			
		5 mg/ml	1 mg/ml	0,5 mg/ml	0,1 mg/ml
1	Cu-1	20,0	16,0	10,0	4,0
2	Cu-2	18,0	4,0	4,0	3,0
3	Cu-3	18,0	7,0	4,0	1,0
4	Cu-4	17,0	12,0	8,0	1,0

Cu-1 -copper nanopowder synthesized in alkaline medium.
Cu-2 copper nanopowder synthesized in ammonia medium;
Cu-3- copper nanopowder synthesized in alkaline medium in the presence of gelatin;
Cu-4 copper nanopowder synthesized in an ammonia medium in the presence of gelatin.

From the table 1 one can see that the Cu nanoparticles synthesized by the chemical reduction in an alkaline medium show the best results with respect to the causative agent of *Erwinia amylovora* bacterial burn. At a concentration of 5 mg / ml of Cu nanoparticles, the exciter growth inhibition zone reaches a maximum value within 72 hours and the lysis zone is 20 mm, and at a concentration of 1 mg / ml this value is 16 mm, which also indicates the significant antibacterial activity of this sample. Cu nanoparticles synthesized in alkaline medium at concentrations of 0.5 mg / ml and 0.1 mg / ml also show activity against this bacterium and the suppression zone is 10 and 4 mm, respectively. Biological activity of the Cu nanopowders obtained in alkaline medium in the presence of gelatin at concentrations of 1.0 mg / ml, 0.5 mg / ml and 0.1 mg / ml with respect to the causative agent of *Erwinia amylovora* bacterial burn is two or more times less than that of Cu nanopowders obtained in an alkaline medium in the absence of gelatin.

Comparison of the antibacterial activity of Cu nanopowders obtained in an ammonia medium in the absence and presence of gelatin shows that at a nanoparticle concentration of 5.0 mg /ml, the suppression zone of both samples practically coincides. At 1.0 mg / ml and 0.5 mg / ml concentrations, biological activity of Cu nanoparticles synthesized in the presence of gelatin is several times higher.

4. Conclusions

Copper nanoparticles with sizes less than 10 nm are formed during the chemical reduction, which form aggregates mainly with spherical shape and various sizes. The particle size distribution differs slightly depending the media and ranged from 20 to 40 nm. Presence of gelatin during the chemical reduction of copper can lead to smaller size distribution, which can be related to formation of protective layer.

Synthesized Cu nano-powders have a sufficiently high activity against the *Erwinia amylovora* bacterium, and the bacterial growth inhibition depends on the Cu nanoparticles concentration. Cu nanoparticles synthesized by the chemical reduction in an alkaline medium show the best results with respect to the causative agent of *Erwinia amylovora* bacterial burn. At a concentration of 5 mg / ml of Cu nanoparticles, the exciter growth inhibition zone reaches a maximum value within 72 hours and the lysis zone is 20 mm, and at a concentration of 1 mg / ml this value is 16 mm, which also indicates the significant antibacterial activity of this sample.

Acknowledgements

This research was supported by the Ministry of Education and Science of the Kyrgyz Republic.

References

[1] Valodkar M et al. **2011** Synthesis and anti-bacterial activity of Cu, Ag and nanoparticles: A green approach *Materials Research Bulletin*, **46** 384-389.
[2] Ramyadevi J et al **2012** Synthesis and antimicrobial activity of copper nanoparticles *Materials Letters* **71** 114-116.
[3] Bykkam S, Ahmadipour M et. al. **2015** Extensive studies on X-ray diffraction of green synthesized silver nanoparticles *Advances in nanoparticles*, **4** 1-10.

[4] Chekman I S, Gorchakova P V, Simonov P **2012** Nano-Cu: Synthesis, pharmacological and toxicological properties *Health of the Far East* **3** 48-54.

[5] Vegera A V, Zimon A D **2006** Synthesis and physico-chemical properties of gelatin stabilized Ag nanoparticles *Newsletter of Tomsk polytechnic University* **309** 60-64

[6] Babushkina I B, Borodulin V B, Korshunov G B, Puchinyan D M **2010** Comparative study of antibacterial action of iron and copper nanoparticles on clinical Staphylococcus aureus strains (in russian) *Saratov science-medicine journal* **6** 11-14.

IOP Conf. Series: Materials Science and Engineering **302** (2018) 012076 doi:10.1088/1757-899X/302/1/012076

Synthesis of Nanomaterials by the Pulsed Plasma in Liquid and their Bio-medical Applications

E Omurzak[1], Z Abdullaeva[2], A Satyvaldiev[3], Z Zhasnakunov[3], Z Kelgenbaeva[4], R Adil Akai Tegin[1], D Syrgakbek kyzy[3], T Doolotkeldieva[5], S Bobusheva[5] and T Mashimo[2]

[1]Faculty of Engineering, Kyrgyz-Turkish Manas University, Chyngyz Aitmatov campus, Jal, 720038, Bishkek, Kyrgyzstan
[2]Shock Wave and Condensed Matter Research Center, Kumamoto University, Kumamoto, Japan
[3]Faculty of Biology and Chemistry, Arabaev Kyrgyz State University, Bishkek, Kyrgyzstan
[4]Institute of Chemistry, Kyrgyz National Academy of Sciences, Bishkek, Kyrgyzstan
[5]Faculty of Veterinary, Kyrgyz-Turkish Manas University, Chyngyz Aitmatov campus, Jal, 720038, Bishkek, Kyrgyzstan

emil.omurzak@manas.edu.kg;emilom@gmail.com

Abstract. Pulsed plasma in liquid is a simple, ecologically friendly, cost-efficient method based on electrical discharge between two metal electrodes submerged into a dielectric liquid. We synthesized carbon-encapsulated Fe (Fe@C) magnetic nanoparticles with low cytotoxicity using pulsed plasma in a liquid. Body-centered cubic Fe core nanoparticles showed good crystalline structures with an average size between 20 and 30 nm were encapsulated in onion-like carbon coatings with a thickness of 2–10 nm. Thermal gravimetric analysis showed a high stability of the as-synthesized samples under thermal treatment and oxidation. Cytotoxicity measurements showed higher cancer cell viability than samples synthesized by different methods. Carbon coated ZnO nanorods with about 20 nm thickness and 150 nm length were synthesized by this method using different surfactant materials such as cetyl trimethylammonium bromide (CTAB) and sodium dodecyl sulphate (SDS). Cu and Ag nanoparticles of about 10 nm in size were also synthesized by the pulsed plasma in aquatic solution of 0.2 % gelatine as surfactant material. These nanoparticles showed high antibacterial activity for Erwinia amylovora and Escherichia coli.

1. Introduction

Nowadays, metal nanoparticles are gaining more interest as a new class of alternative antimicrobial agents because of the increasing of antibiotic resistance in microorganisms [1] presenting grave hazard for public healthcare [2]. But pure metal nanoparticles without any coating or treatment are toxic and cause health concerns in even trace amounts. Carbon has been used as biocompatible material and less affected in physiological conditions. So, carbon coated metal or semiconductor are therefore biocompatible materials. Iron nanoparticles have optimum magnetic response. Carbon coated iron nanoparticles are consequently best candidate to be used for hyperthermia, drug delivery, magnetic resonance imaging contrasts, biological labels etc.

Zinc oxide (ZnO), a semiconductor with unique electrical, optoelectronic, and luminescent properties. Because of these attractive properties, nanostructured ZnO have many important practical applications in catalysis, photoluminescence, and functional devices. Coating of ZnO with carbon can extend its above-mentioned applications even to more fields, particularly, bio-medical applications: as the luminescent probes of tissues for imaging by fluorescence microscope.

There are a number of synthesis methods for carbon coated Fe and ZnO nanoparticles have been reported such as chemical vapor deposition (CVD), arc discharge and so on. However, most of these methods lead to economic disadvantages including the need to maintain high temperatures, high pressures, and vacuum systems as well as use of expensive equipment. In addition, produced by these methods nanomaterials have quite high toxicity.

We have reported a simple, ecologically friendly, cost-efficient method based on the electrical discharge between two metal electrodes submerged into a dielectric liquid [3]. This is a versatile and easily configurable method for synthesis of metals, oxides, sulfides, carbides, etc by changing electrodes materials and the dielectric liquid. By choosing the electrical and experimental conditions, physical and chemical properties can be tuned [4]. In this paper, we report synthesis of biocompatible metal nanoparticles (Fe, Cu, Ag), carbon coated oxide nanostructure (ZnO) and their antibacterial activities.

2. Materials and Methods

All materials for electrodes and dielectric liquids were of high purity grade and obtained from Kojundo Co., Rare Metallic Co. Experiments were conducted using the pulsed plasma apparatus, which is schematically illustrated in figure 1. For synthesis of metal nanostructures, electrodes made of metal rods of about 5 mm in diameter and 12 mm in length and 200 ml dielectric liquid (toluene or ethanol) were used. As for zinc oxide nanostructures, electrodes were made of zinc and it was submerged into water. Surfactant materials with different polarities were added to water in order to achieve nanostructures resistance to aggregation and oxidation. Production yield of the samples differed according to the melting temperature of the being synthesized metals, for instance, low melting temperature metals such as Zn, can be produced up to few g/h.

X-Ray diffraction (XRD) pattern of the samples were taken using Cu Kα radiation, Rigaku RINT-2500VHF. Field Emission-Scanning Electron Microscope (FE-SEM) images of the products were taken by a JEM-2000FX.

Cytotoxicity of encapsulated metallic samples was evaluated by using the A549 cell line (Humanlung adenocarcinoma epithelial cells). Cell viability was measured by (4,5-dimethylthiazol-2-yl)-2,5-diphenyltetrazolium bromide) MTT and (2,3-bis-(2-methoxy-4-nitro-5-sulfophenyl)-2H-tetrazolium-5-carboxanilide) XTT assay kits purchased from Roche Diagnostics (Japan).

Antimicrobial activity of silver nanoparticle were conducted using the E-coli (ATCC 35218) bacteria in Plate Count Agar (Merck-M105463.0500) and Violet Red Bile Agar (Merck-M101406.0500) media by filter paper and hole methods.

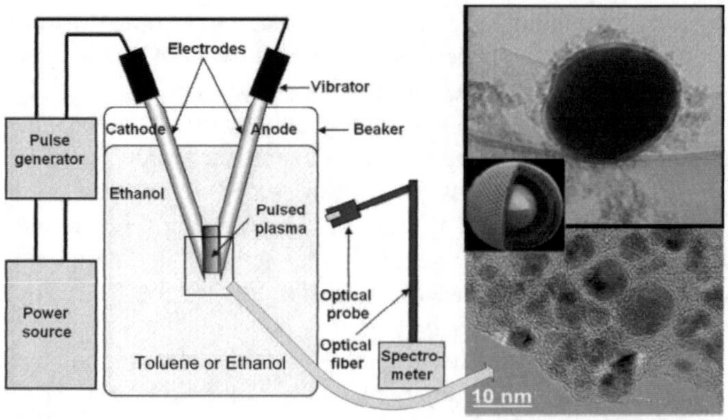

Figure 1. Schematic illustration of the pulsed plasma in liquid method and the Transmission Electron Microscopy images of the synthesized iron nanoparticles by this method.

3. Results and Discussion

Fe core nanoparticles with good crystalline structures of an average size between 20 and 30 nm were encapsulated in carbon coatings of 2-4 nm thickness, as shown in Fig.1. The synthesized samples showed good ferromagnetic properties at room temperature. The samples showed very high thermal stability (up to 700 °C) and high biocompatibility: more than 95 % of cells remained alive for carbon iron nanoparticles (see table 1). Such magnetic nanoparticles are perfect for using in hyperthermic cure of cancer cells, where external magnetic field fluctuates delivered to the target magnetic nanoparticles and by this generates heat. This heat is aimed to kill the cancer cells. Coating of metallic magnetic nanoparticles is critical for hyperthermia applications [5]. Table 1 represents a comparison of the cell viability of different samples. As we can see our samples showed highest cell viability among similar production methods and was equal to 95 %. In other words, only 5 % of the normal cells located in the vicinity of the cancer cell might be damaged due to targeted delivery by injection or other methods. We can conclude that magnetic nanoparticles by the pulsed plasma in liquid method are highly appropriate for hyperthermia treatments.

Table 1. Comparison of cell viability of the samples produced by different methods.

Method	Sample	Cell viability, %	Reference
Pulsed plasma in liquid	Fe@C	95	**Our work**
Catalytic CVD	Fe@C	2	Int. J. Nanomed. 5, 167 (2010)
Arc discharge	Fe@C	80	Cell Biol. Int. 32, 1001 (2008)
Aerosol particles	Fe	0.6	Int. J. Nanomed. 6, 167 (2011)

Carbon coated ZnO nanorods with about 20 nm thickness and 150 nm length were synthesized by this method. When different surfactant materials (CTAB, SDS) were used, ZnO nanorods with different morphologies were produced: CTAB induced formation of longer nanorods (average length of 118 nm and diameter of 28 nm) with smooth surfaces, while SDS resulted in longer nanorods (246 nm in length and 55 nm in diameter) with rougher surfaces (Figure 2).

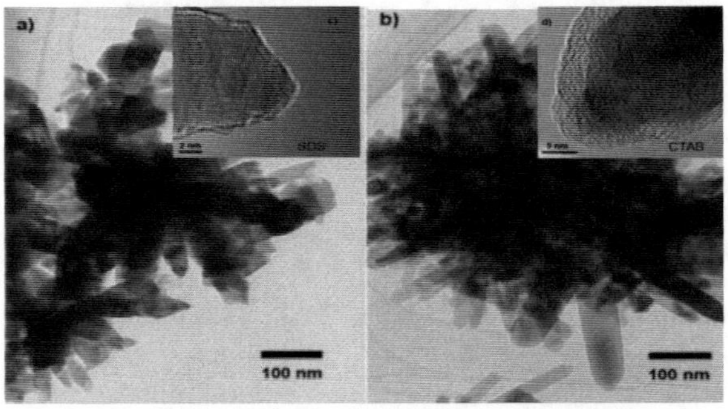

Figure 2. TEM images of the particles produced by the pulsed plasma in water solution of a) SDS and b) CTAB. Insets show HRTEM images of coated ZnO nanorods corresponding surfactants.

Closer look at the nanorod structure formed under SDS and CTAB conditions revealed that in both cases the surface of the nanorods were coated with carbon layers with different thickness: 2 nm in case of SDS and 5 nm for CTAB. Carbon is a best candidate for making materials biocompatible. Thus, we hope that our carbon-coated ZnO nanorods can be applied for biomedical applications, such as biomedical imaging (which includes fluorescence, magnetic resonance, positron emission tomography, as well as dual-modality imaging), drug delivery, gene delivery, and biosensing of a wide array of molecules of interest [6].

Silver nanoparticles were prepared by both chemical reduction and pulsed plasma (electrical discharge) in water methods. In addition, various surfactant materials were tested for keeping the nanoparticles' stability longer. Synthesis of Ag nanoparticles were based on chemical reduction of ions by hydrazine:

$$4Ag^{+} + N_2H_4 + 4OH^{-} = 4Ag + N_2 + 4H_2O \quad (1)$$

XRD analyses showed no effect of gelatin to Ag nanoparticles formation during Ag ions reduction to metallic Ag. Similar results were obtained in case of pulsed plasma synthesis of Ag. Effect of stabilizing agent SDS was also conducted and showed good result, where no reflections of metallic particles can be seen. However, small content of silver oxide was appeared. Particle size of the synthesized samples was small, about 10 nm in diameter.

Surfactant materials were used for increasing stability of silver nanoparticles. SDS surfactant was added to water in 0.4 % concentration and then the pulsed electrical discharge was conducted inside it. Obtained samples showed better stability and UV/visible absorption spectrometer confirmed that the nanoparticles aggregation over a time period of a week showed better stability than that of without surfactant. These samples were tested for antibacterial activity for some bacteria as described below.

Biological activity tests were conducted for the synthesized Ag nanoparticles produced by both chemical reduction and pulsed plasma method using *E-coli* bacteria.

Antibacterial activity of Ag nanoparticles prepared by chemical reduction method were determined for *E-coli* bacteria by filter paper method, where a round shaped special filter paper was submerged into a solution of Ag nanoparticles of certain concentration. Then this filter paper was placed to the bacteria cell culture in Petri dish. Figure 3 shows the E-coli bacteria culture with Ag nanoparticles lysis zones, which shows the area of the bacteria cell growth suppression zone. It is the part where bacteria growth was not possible due to the antibacterial effect of nanoparticles. Lysis zone for the Ag nanoparticles by chemical method was determined to be 5-7 mm, which is in the usual range value for such particles [7].

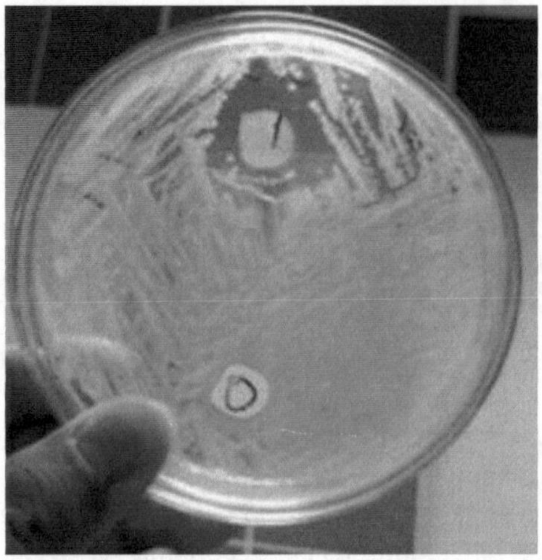

Figure 3. Effect of antibacterial properties of Ag nanoparticles produced by chemical reduction method to the *E-coli* bacteria cell culture.

Antibacterial activity of Ag nanoparticles prepared by electrical disharge method were also determined for *E-coli* bacteria by filter paper and hole method, which can be seen from the Figure 4. These are results for the Ag nanoparticles + SDS surfactant. Among surfactant materials SDS showed no antibacterial effect to *E-coli* bacteria, while CTAB had inhibited suppression zone when tested for antibacterial activity in pure state without any other particles. This urged us to use SDS surfactant material as stabilizing agent for nanoparticles and protects nanoparticles from aggregation, oxidation and other outer influences. Red oval line shows the effect of our samples. We can see that the bacteria around the nano Ag + SDS samples could not grow and growth suppression zones are clearly visible. In the hole method (Fig. 4b), where nano Ag + SDS sample was dropped to the holes had extraordinary effect to bacteria cell growth and almost totally killed them. Both filter paper method and hole method displayed that nano Ag + SDS sample exhibits excellent antibacterial property to the *E-coli* bacteria.

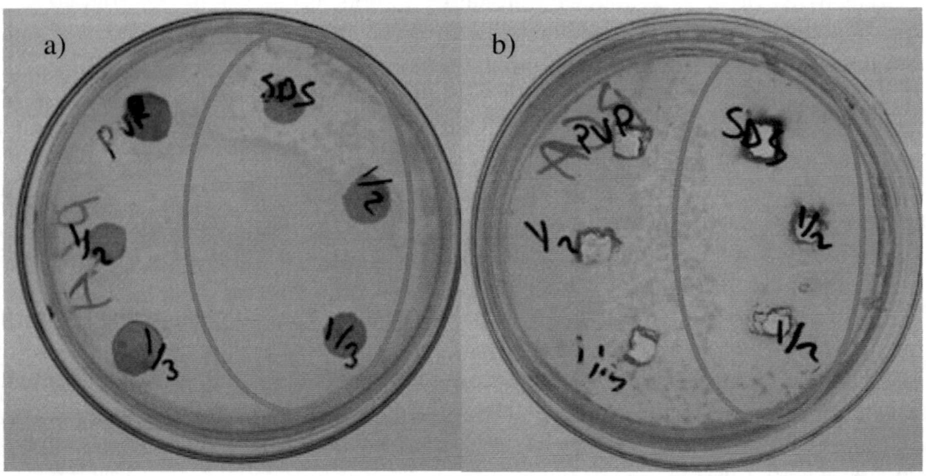

Figure 4. Effect of antibacterial properties of Ag nanoparticles produced by electrical discharge method in presense of SDS surfactant to the *E-coli* bacteria cell culture by the a) filter paper and b) hole methods

Table 2 shows results of the antibacterial test of our samples for the E-coli bacteria cells. As we mentioned above, CTAB surfactant material exhibits antibacterial property, thus, antibacterial activity of the Ag nanoparticles produced in presence of CTAB surfactant could not be clearly determined. However, SDS surfactant had no effect to bacteria. In addition, its stabilizing effect to Ag nanoparticles was excellent and nano Ag particles could be stable to external influence over time. Nano Ag + SDS sample produced by the electrical method showed better antibacterial property than the nano Ag produced by the chemical reduction method. This shows that the newly formed nanoparticles in plasma conditions can be quickly stabilized in-situ by SDS surfactant during formation process.

Table 2. Values of growth suppression zone of samples on *E-coli* bacteria culture.

Sample name	Lisys zone, mm	
	Filter paper method	Hole method
CTAB itself	1	7-10
SDS itself	0	0
Nano Ag by chemical method	5-7	-
Nano Ag + SDS by electrical discharge	7-9	15

4. Conclusions

We have described the synthesis of carbon encapsulated Fe nanoparticles and ZnO nanorods by simple and low-energy pulsed plasma in liquid method. Carbon coated Fe magnetic nanoparticles showed high biocompatibility: more than 95 % of cells remained alive for carbon iron nanoparticles. Carbon coating of ZnO nanorods can be achieved by using surfactant materials, which served as carbon source.

Nanopowders of Ag synthesized both by chemical and physical methods exhibited good antibacterial activity for *E-coli* and the suppression zone size was between 5 to 9 mm. Nano Ag + SDS sample produced by the electrical discharge method showed higher antibacterial activity than the nano Ag produced by the chemical reduction method.

Acknowledgements

This research was supported by the Ministry of Education and Science of the Kyrgyz Republic.

References

[1] Neu H C **1992** The crisis in antibiotic resistance. *Science*, **257** 1064-1066.qa

[2] Wise R, Hart T, Cars O. Streulens M, Helmuth R, Huovinen P **1998** Antimicrobial resistance is a major threat to public health *British Medical Journal*, **5 Sept.,** p. 609-612.

[3] Omurzak E et al **2007** Synthesis method of nanomaterials by pulsed plasma in liquid *J. Nanosci. Nanotechnol.*, **7** 3157-3159.

[4] Omurzak E, Abdullaeva Z, Iwamoto C, Ihara H, Sulaimankulova S, Mashimo T, **2015** Synthesis of hollow carbon nano-onions using the pulsed plasma in liquid *J. Nanosci. Nanotechnol.,* **15** 3703-3709.

[5] Giustini A J et al **2013** Magnetic nanoparticle hyperthermia incancer treatment. *Nano Life*, 1(01n02). doi: 10.1142/S1793984410000067

[6] Zhang Y, Nayak R T, Hong H, Cai W **2013** Biomedical Applications of Zinc Oxide Nanomaterials. *Curr Mol Med.* 13(10) 1633–1645.

[7] Das B, et al, **2015** Green synthesized silver nanoparticles destroy multidrug resistant bacteria via reactive oxygen species mediated membrane damage. Arabian Journal of Chemistry (2015), http://dx.doi.org/10.1016/j.arabjc.2015.08.008.

IOP Conf. Series: Materials Science and Engineering **302** (2018) 012077 doi:10.1088/1757-899X/302/1/012077

Some Calculated Research Results of the Working Process Parameters of the Low Thrust Rocket Engine Operating on Gaseous Oxygen-Hydrogen Fuel

V Ryzhkov[1] and I Morozov[1]

[1]Researcher, Samara National Research University, Samara, Russia

E-mail: ke_src@ssau.ru

Abstract. The paper presents the calculating results of the combustion products parameters in the tract of the low thrust rocket engine with thrust P ~ 100 N. The article contains the following data: streamlines, distribution of total temperature parameter in the longitudinal section of the engine chamber, static temperature distribution in the cross section of the engine chamber, velocity distribution of the combustion products in the outlet section of the engine nozzle, static temperature near the inner wall of the engine. The presented parameters allow to estimate the efficiency of the mixture formation processes, flow of combustion products in the engine chamber and to estimate the thermal state of the structure.

1. Introduction

The gas-dynamic parameters distribution in engine tracts is relevant for any low thrust rocket engines. The parameters determine the perfection of working processes in the combustion chamber and nozzle, which affect the thermal state of the structure and form the main (energy) parameters of rocket engines.

Until recently, most of the problems in the engines development were solved, mainly, by the experimental method, which was accompanied by the need to simulate operating conditions, high labor costs of manufacturing prototypes, experimental stands, fuel storage facilities, and thus the high cost of the experimental infrastructure. Engine development by this technology lasts a considerable period of time (up to 7 ... 10 years). The information received has clear limitations: the research is not carried out in the entire area of the change in the determining parameters. As a result, usually, only integral characteristics are determined, and at the last stage the thermal state of the structure is determined.

At present, numerical methods of investigation are becoming increasingly important for solving a number of problems for creating engines. It helps to reduce the development time, improve design quality and reduce the volume of experimental development.

Mathematical modeling is related to the object under study through the calculation domain, the initial and boundary conditions, but it is subject to the influence of specific factors that are absent in physical modeling, such as the replacement of a continuous domain by discrete one, replacement of differential equations describing physical processes by a system of algebraic equations, the using a number of limitations and assumptions.

Experimental methods are still indispensable at all stages of engine development: acceptance and approbation of technical solutions, assessment of the compliance of product parameters with the technical task requirements, engine development.

IOP Conf. Series: Materials Science and Engineering **302** (2018) 012077 doi:10.1088/1757-899X/302/1/012077

A promising direction is the integration of the results obtained in mathematical modeling to the computer design of low thrust rocket engines and further into the PLM system of the propulsion system and the spacecraft.

2. Calculations and result analyses

The computer environment ANSYS CFX is used to study the stationary spatial turbulent multicomponent chemically reacting flow of the working fluid in the sub-, trans- and supersonic regions.

Calculated studies of the low-thrust rocket engine working process (with thrust P ~ 100 N) were carried out using a mathematical model based on a software package [1], including the basic equations of conservation of mass, momentum, energy, concentration of fuel components in the form of Navier-Stokes. To simulate turbulence, a model based on Reynolds stresses (RSM) [2] was used, modernized for the "wall law" for the flow structure and heat-mass transfer in the near-wall region.

A diffusion combustion model (EDM) [3] with a simple chemical reaction has been chosen, since the mixing rate of fuel components is a limiting and not the rate of kinetic processes.

Effective Prandtl and Schmidt numbers are assumed to be equal «1».

Heat transfer on the wall of the engine chamber corresponds to the conditions of impermeability, the walls themselves are non-reactive and adiabatic.

The paper studies the following: the structure of the flow, the parameters of the low-thrust rocket engine working process, determining its efficiency and reliability, the main (energy) characteristics of the engine, and the thermal state of the engine structure.

Proceeding from this, it is reasonable to consider in the gas-dynamic distribution path of the engine (combustion chamber) the following three parameters: firstly, the total temperature of the combustion products as the main energy parameter, secondly, the axial velocity of the combustion products in the outlet section of the engine nozzle as a parameter that determines the specific impulse of the engine and, thirdly, the temperature of the combustion products in the engine region wall that determines the thermal state of the structure.

The realization of the described technology will be illustrated by the data obtained in the calculation of the low-thrust rocket engine (with thrust of 100 N) on gaseous oxygen-hydrogen fuel.

In order to ignite the fuel mixture in the combustion chamber of the rocket engine, an efficient, reliable and reliable electric spark ignition system was selected.

For the organization of the working process, one of the effective ways of mixing gaseous oxygen and hydrogen in the combustion chamber of the engine was selected, which based on the interaction of co-axial coaxial twisted gas flows of fuel components organized in several cascades [4].

The cooling of the engine chamber walls was solved with the help of gas curtains, the main parameters and design of which were selected using materials [5].

It is convenient to present the calculated results in the form of the parameters distribution patterns characterizing the quality of the working process in the longitudinal section of the engine chamber flowing part and in the characteristic cross sections, and in the distribution diagrams of the determining quantities.

Figure 1 shows the combustion products flow structure in the form of the streamlines distribution along the engine chamber flowing part.

The supply of gaseous oxygen and hydrogen is mainly carried out through tangential channels and effective mixing of the components takes place in the pre-chamber and the combustion chamber, as can be seen from the characteristic curvature of the flow lines. As the flow develops the flow lines of the combustion products become more ordered. The mechanism of turbulent mixing of gaseous oxygen and hydrogen flows is manifested in the area of the prechamber and the combustion chamber. It is characteristic for developed swirling currents; therefore, its use is an additional factor of influence on the completeness of the fuel combustion in the chamber without reducing the nozzle efficiency [6].

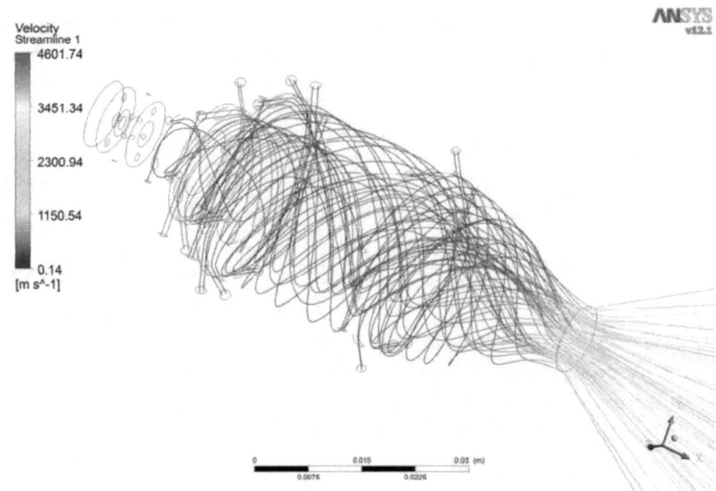

Figure 1. The streamlines distribution along the engine chamber flowing part

The total temperature distribution of the combustion products in the rocket engine chamber tract is an important factor of the working process efficiency evaluation in the low thrust rocket engine. The results are shown in figure 2.

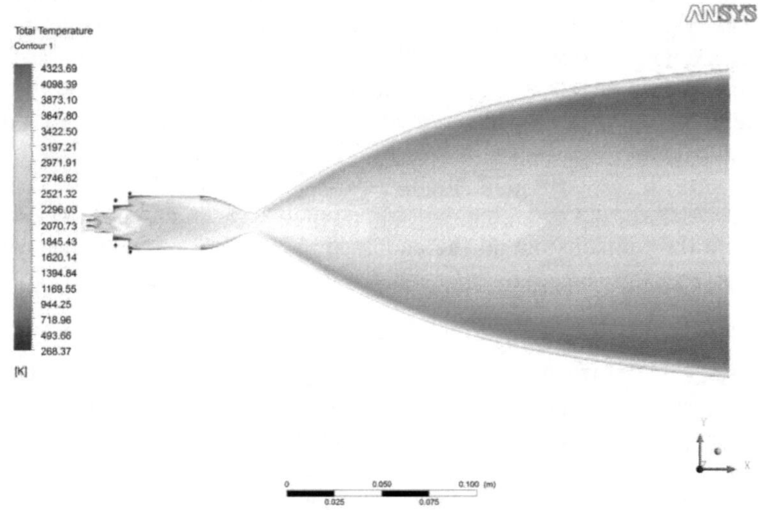

Figure 2. The total temperature distribution of the combustion products in the path of the engine

It should be noted the presence of extended zones of high temperature in the pre-chamber and the combustion chamber, starting from the section of supplying the main fuel components.

The figure shows the presence of specially organized wall zones with a reduced temperature of the combustion products along the engine chamber, which are generally lower than the permissible wall temperature; an admissible wall temperature is $T_{wall}^{available} \leq 2000$ K.

It can be seen that the intracameric processes are quite full, but the reserves for improving the mixing pattern are obvious. In support of what has been said, figure 3 shows the total temperature distribution of the combustion products in the final section of the combustion chamber

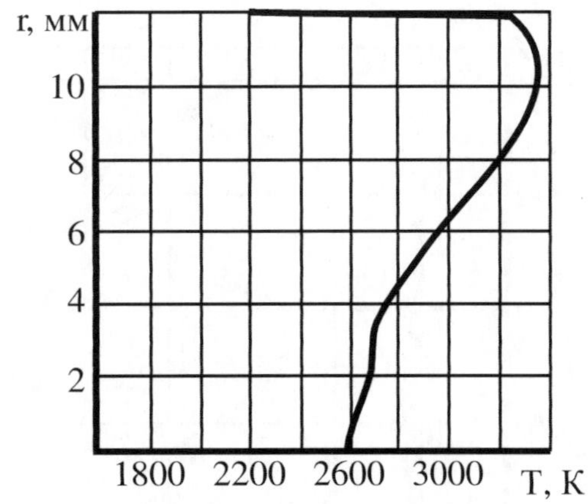

Figure 3. The total temperature distribution of combustion products in the final section of the engine combustion chamber

From the of the of the energy parameters formation view point of the low thrust rocket engine the numerical values and shape of the braking temperature diagram are interesting as a result of the processes of mixture formation and fuel conversion during combustion of fuel components. The requirement for the total temperature distribution is that in the central region of the chamber the temperature values should be approximated to the maximum for the accepted components mass ratio, and in the rocket engine wall area the temperature values do not exceed the wall temperature permissible for the selected structural material. Proceeding from this, the profile of the total temperature should ideally have a parabolic shape, facing its convex part in the direction of the combustion products flow.

The obtained combustion products total temperature distribution in the final section of the combustion chamber requires correction at least in part of the fuel redistribution between the stages of the mixture formation scheme.

The maximum specific thrust impulse in the rocket engine depends on the size and shape (completeness) of the axial velocity component distribution of the combustion products in the exit section of the nozzle. In this sense, the deformation of the velocity distribution from the minimum section of the nozzle to its cut is important. In the outlet section of the nozzle, the axial velocity component should tend to its maximum corresponding to the accepted degree of expansion of the combustion products in the nozzle.

Omitting the analysis of the reasons for the change in the total velocity and its components in the supersonic nozzle, let us consider the outlet section of the nozzle (Figure 4), where the diagrams of the flow velocity of the combustion products are presented. The axial velocity component of the combustion products has a characteristic distribution that directly affects the engine specific impulse formation. In the central region of the nozzle cross section, the oxygen-hydrogen fuel combustion products velocity reaches values U ~ 4300 m/s. In the region of the wall of the nozzle, a substantial decrease of the velocity is observed.

In general, the obtained velocity diagram meets the requirements for the desired distribution in the cross-section at the exit from the RDMT nozzle with a geometric expansion ratio $\overline{F}_{nozzle} = 250$.

The most important indicator of the low thrust rocket engine is the thermal state of the structure. At the calculation stage, this is largely facilitated by the temperature (total temperature) localization of the combustion products in the engine chamber wall region. These results it is convenient to present on the wireframe engine model. Figure 5 shows the combustion products temperature distribution in the area of the rocket engine wall as the basis for evaluating the low thrust engine thermal state with a single-layer chamber wall construction.

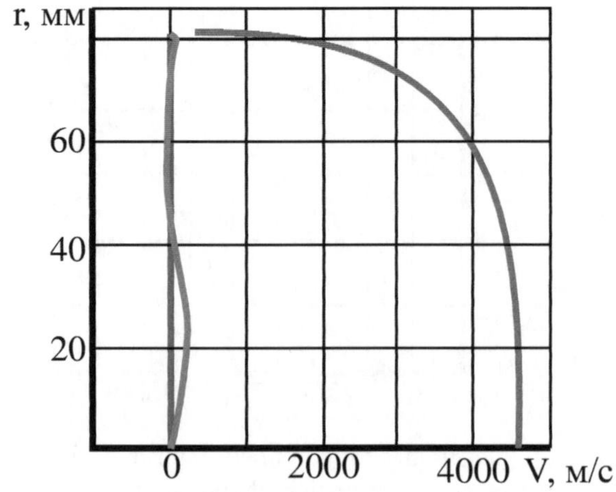

Figure 4. The total (axial) velocity distribution of the combustion products in the engine output section

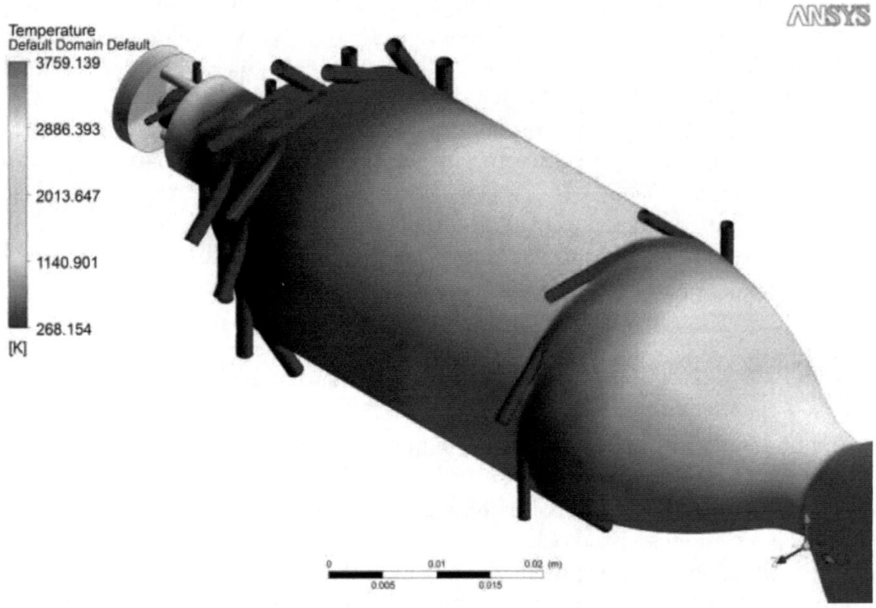

Figure 5. The combustion products temperature distribution in the engine chamber wall region

Comparing the values of the combustion products temperature with the permissible wall temperature of the selected structural material, it is necessary to find a compromise solution: either to replace the construction material (with higher values of $T_{wall}^{available}$) or to provide a lower temperature near the wall of the engine chamber, introducing changes in the mixture mixing scheme.

The obtained calculation information allows not only to determine the parameters and characteristics of the object under study, but also to outline ways to improve the working process and engine design.

High potential opportunities for the organization of an effective work process were laid down in the adopted scheme of RDMT on the basis of coaxial, swirling gas streams of rocket fuel components providing sufficiently high values of turbulence energy both local and integral. For the engine under investigation, the values of the specific thrust pulse at the level $I_{specific}^{vacuum}$ ~ 4200 м/с ($\varphi_{spec.impulse}$ ~ 0, 85) were obtained by calculation.

The methods for the engine improving are based on providing a more uniform and energy-consuming the total temperature distribution in the combustion chamber output section due to the working process intensification proper in the engine combustion chamber, and by increasing the prechamber flow rate and by using traditional methods of increasing the combustion chamber relative length. The thermal state problems of the engine structure should be solved by optimizing the gas curtains parameters.

3. Conclusion

The work shows that using a high-level mathematical model, taking into account real processes in low-thrust rocket engines, makes possible to determine a parameters set that allow to determine the low thrust rocket engine work process efficiency, to evaluate the integral parameters and the engine characteristics, depending on the determining factors. Analysis of calculated information makes it possible to outline ways to improve the engine working process both in the direction of improving the mixture formation quality and in the direction of ensuring the engine structure thermal state by bringing the combustion products temperature at the wall layers in line with the structural materials permissible temperature. It should be noted that the problem of the low thrust rocket engine design shape by models similar to that used in this work is solved by choosing the investigated variants by the efficiency criteria of the working process and by ensuring the engine structure thermal state.

Acknowledgments

This work was supported by the Ministry of education and science of the Russian Federation in the framework of the implementation of the Program of increasing the competitiveness of Samara National Research University among the world's leading scientific and educational centres for 2013-2020 years.

References

[1] ANSYS® *ANSYS CFX – Solver Modeling buide* (Release 11.0, ANSYS, Inc.) p. 566
[2] Wilcox D 1998 Turbulence Modelling for CFD (DCW Industries, Inc.)
[3] Varnatc Ju., Maas U., Dibbl R. Combustion. Physical and chemical aspects, modeling, experiments, formation of pollutants. Moscow : Fizmatlit, 2006, 352 p. (Rus.)
[4] Patent 2615883, Russian Federation, Small-draft rocket engine with multi-stage combustion chamber on hydrogen gas and oxygen (Raketnyj dvigatel' maloj tjagi (RDMT) s mnogokaskadnoj kameroj sgoranija na gazoobraznyh vodorode i kislorode) / Gradov V and Ryzhkov V // The applicant and patent owner "Samara State Aerospace University named after academician S.P. Korolev (National Research University)" (SSAU). – 2015157319. (Rus.)
[5] Volchkov, Je., 1983, Wall hung gas curtains (Pristennye gazovye zavesy) // Novosibirsk: Nauka, p. 239. (Rus.)
[6] Ryzhkov, V., 2009, Mathematical modelling and parametric research of flow of the twisted turbulent single-component stream of the propulsive mass in the trance - and supersonic areas of Laval nozzles // Vestnik of Saara State Aerospace University, pp. 382-391.

Measurements and Experimental Database Review for Laminar Flame Speed Premixed Ch4/Air Flames

I A Zubrilin[1], S S Matveev[1], S G Matveev[1] and D V Idrisov[1]

[1]Scientific and Educational Centre of Fluid Dynamics Research, Samara National Research University, Samara, Russia

Email: zubrilin416@mail.ru

Abstract. Laminar flame speed (S_L) of CH4 was determined at atmospheric pressure and initial gas temperatures in range from 298 to 358 K. The heat flux method was employed to measure the flame speed in non-stretched flames. The kinetic mechanism GRI 3.0 [1] were used to simulate S_L. The measurements were compared with available literature results. The data determined with the heat flux method agree with some previous burner measurements and disagree with the data from some vessel closed method and counterflow method. The GRI 3.0 mechanism was able to reproduce the present experiments. Laminar flame speed was determined at pressures range from of 1 to 20 atmospheres through mechanism GRI 3.0. Based on experimental data and calculations was obtained S_L dependence on pressure and temperature. The resulting of dependence recommended use during the numerical simulation of methane combustion.

1. Introduction

Development of new gas-turbine engines for power plants is very challenging task. The combustion chamber is one of the main components of any gas turbine engine. Design and working of the combustion chamber process of determining the essential characteristics of the engine. Currently in the design of the combustion chambers using Computational Fluid Dynamics (CFD) methods. CFD can quickly and accurately determine the basic characteristics of the designed (or modified) combustion chamber, such as the complete loss of pressure, air distribution laws, as well as emission characteristics and others. However, a qualitative of prediction of pollutant formation processes is impossible without the use of detailed chemical kinetics, as well as a qualitative description of combustion processes. One of the main parameters influencing the physics of combustion of fuel-air mixture is S_L [2]:

$$S_L = S_{L0}\left(\frac{T_u}{T_0}\right)^{\alpha}\left(\frac{P_u}{P_0}\right)^{\beta}, \qquad (1)$$

where T_u – the initial temperature of the unburned mixture,

P_u – the initial pressure of the unburned mixture

$S_{L_0} = C_1 + C_2(\varphi - C_3)^2$

where φ - the equivalence ratio

C_1, C_2 and C_3 - constants depending on the type of fuel

The exponents α and β are calculated as:

$\alpha = 2.18 - 0.8(\varphi - 1)$,

$\beta = -0.16 + 0.22(\varphi - 1)$.

As can be seen from the equation (1) S_L depends on the coefficient α and β. The CFD software mainly uses one of two equations to determine S_L. One of them is (1) and another one is from [3]:

$$S_L = FY_{F,u}^m \cdot \exp(-G / T^0) \frac{T_u}{T^0} \left(\frac{T_b - T^0}{T_b - T_u} \right)^n, \quad (2)$$

where $Y_{F,u}$ – the mass fraction of the fuel in the fuel-air mixture,
T^0 – the temperature of the inner layer of the laminar flame,
T_b – adiabatic equilibrium temperature of combustion products,
B, E, F, G, m, n – coefficients (presented in [3]).

As shown below, the calculation of the results of these two equations do not accurately superimposed on the experimental results, taken from open sources. Therefore, as proposed a process for preparing solutions according to the refined $S_L = f(\varphi, T_u, P_u)$.

Aims of this paper are to investigate different data on methane+air laminar flame speed and to determine the dependence of S_L on pressure and temperature.

2. Details of experiment

The measurements of CH4 and air mixtures flame speed were performed by experimental facility to determine the speed of the flame by the heat flux method. The heat flux method allows stabilization of flat adiabatic flames on a perforated burner due to possibility of balancing heat transfer between the flame and the burner plate. The edge of the burner plate is kept preheated to 368 K by a water circuit in the burner head, controlled by a water bath. The temperature difference between the burner plate and the unburned mixture forces negative heat flux from the plate to the gas, which is balanced by the positive heat flux from the flame back to the plate. The unburned gas temperature (T_u) was set in range from 298 to 358 K. The unburned gas pressure (P_u) was set to values of 1 atm. The range of equivalence ratios was limited to φ = 0.6-1.6. The experimental facility shown in the figure 1.

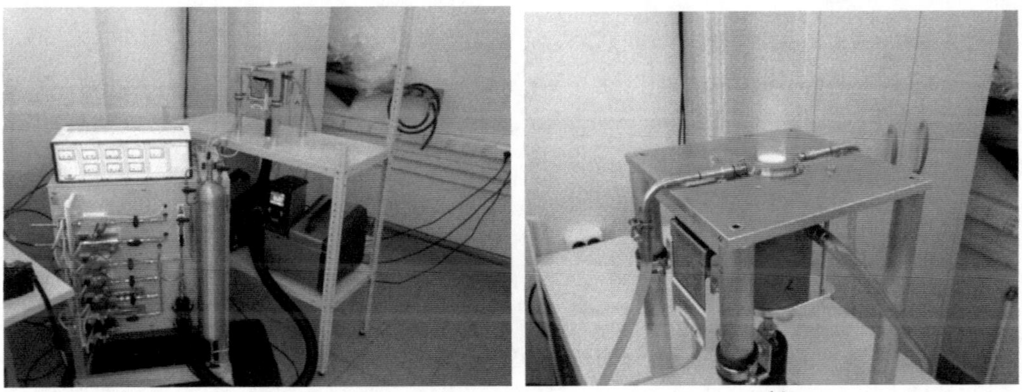

a) b)

Figure 1 - Test rig to measure S_L: (a) general view, (b) burner device.

3. Simulation

Simulation of adiabatic premixed one-dimensional flames was performed using Chemkin 4 code [4]. Grid-independent solutions were obtained for high-temperature mechanism GRI 3.0, which contains 325 reactions, 56 chemical species and were made by M. Frenklach et al [1]. The calculation was performed under the following initial conditions: $P_u = 1 \ldots 20\ bar$, $T_u = 298 \ldots 800\ K$, φ=0.6…1.6.

4. Approximation equation

Based on the simulation results several three-dimensional graphics were built. Two of them you can see in the figure 2: $S_L = f(\varphi, P_u)$ and $S_L = f(\varphi, T_u)$.

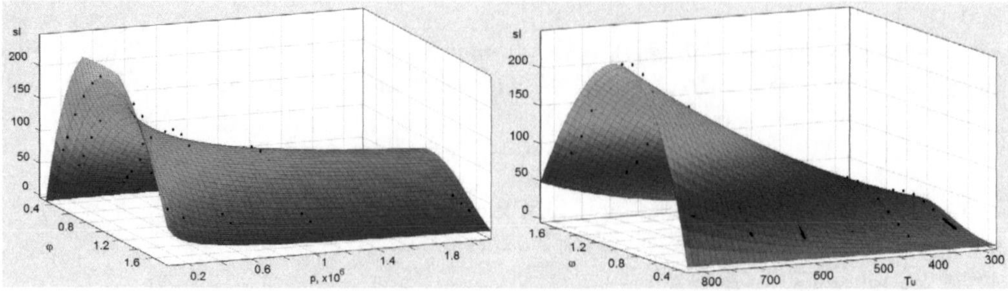

a) b)

Figure 2 – Calculated S_L dependence on φ and P_u with T_u =800 K (a) and dependence on φ and T_u with $P_u = 1\ bar$ (b)

The data obtained from the calculation can be described by the equation presented below:

$$S_L = (145\varphi^3 - 850\varphi^2 + 1265\varphi - 325)\left(\frac{T_u}{800}\right)^{5.1\varphi^2 - 10.8\varphi + 7.6}\left(\frac{p_u}{101325}\right)^{-0.4715\varphi^2 + 1.1\varphi - 0.966} \quad (3)$$

It should be noted that the use of this dependence is limited by the equivalence ratio range as φ=0,33…1,9.

5. Results and discussion

5.1 Temperature dependence

Figures 3 - 4 show values of S_L experimentally obtained, calculated by (1), (2), (3) and the mechanism GRI 3.0 in comparison with data of other authors [5-36].

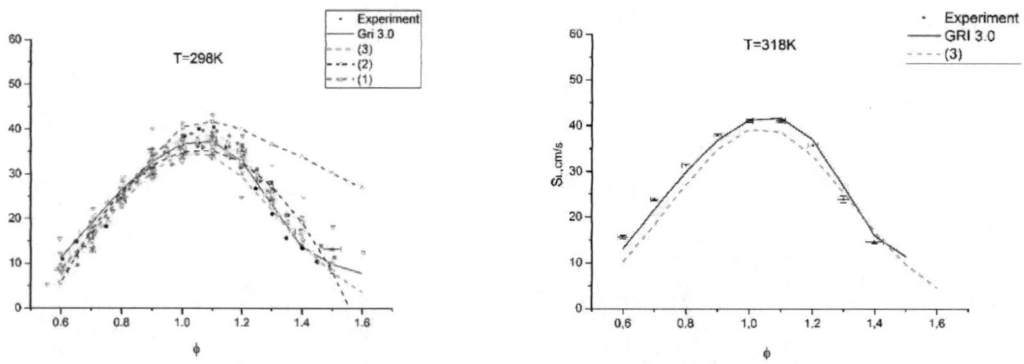

a) b)

Figure 3 – Laminar flame speed S_L (a) T_u = 298 K and (b) T_u = 318 K

(more details about the chart legend can be seen in Appendix).

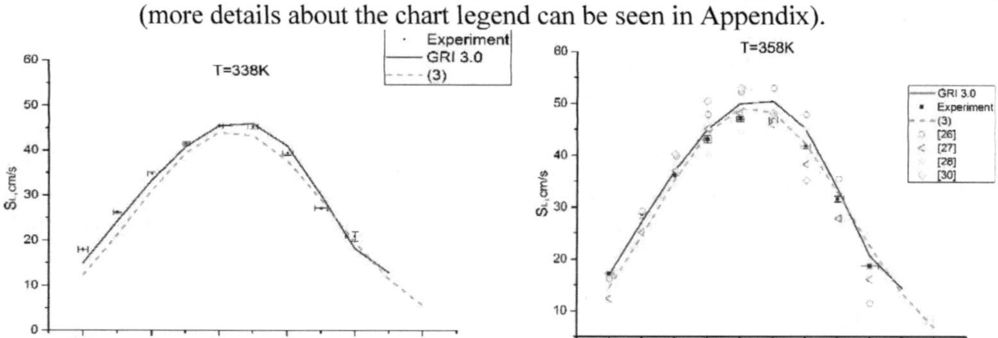

a) b)

Figure 4 – Laminar flame speed S_L (a) $T_u = 338$ K, (b) $T_u = 358$ K

(more details about the chart legend can be seen in Appendix).

Figures 3-4 show for this temperature range all data are in a good agreement, except equation (1) which shows higher values for rich equivalence ratio. Equation (2) shows a good prediction for all provided initial conditions. The experimental data of this work are in a good agreement with data of other authors [5-35] and with the data obtained in the works [1-4]. This indicates that used method is valid. Based on these graphs, it is clear that the kinetic mechanisms of chemical reactions GRI 3.0 is in good agreement with experiments and is suitable for further use in order to obtain the calculated values. This is necessary due to the lack of available experimental data for large pressures and temperatures. Equation (3) shows a bit of lower values according to results of GRI 3.0 simulations.

5.2 Pressure dependence

Figures 5 - 6 show values of S_L calculated by the equations (1), (2), (3) and the mechanism GRI 3.0 for initially temperature equal to 500 and 700K, and pressure equal to 10 and 20 bar.

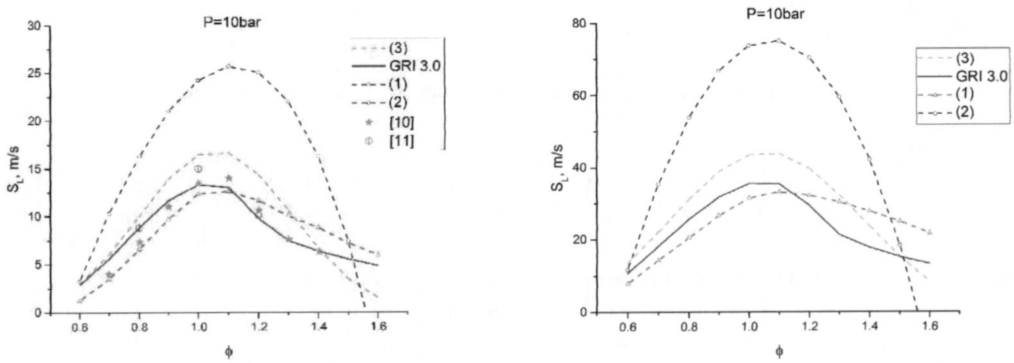

a) b)

Figure 5 - Laminar flame speed at P_u =10 bar, (a) T_u =300 K and (b) T_u =500 K.

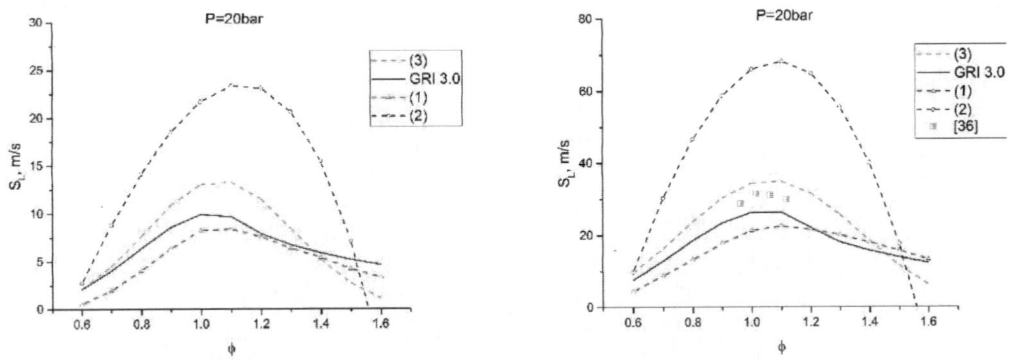

a) b)

Figure 6 - Laminar flame speed at P_u =20 bar, (a) T_u =300 K and (b) T_u =500 K.

As the presented graphs show, equation (2) significantly overestimates the value of S_L at high values of P_u in comparison with the experimental data. In turn, the calculation from equation (1) shows slightly understated S_L values with respect to the experimental data. Calculation by formula (3) allows obtaining S_L values, which agree satisfactorily with both experimental and kinetic calculations of the GRI 3.0 mechanism.

Figure 7 presents a comparison of the α and β values obtained in this paper with the data of other authors [37-39]. As can be seen from the presented graphs, the value of the degrees used in equation (2) is linear and does not correspond to the experimental data. Whereas the proposed values of α and β have extrema in the range 1 < ϕ <1.2, which corresponds to the experimental data and the calculations of other authors.

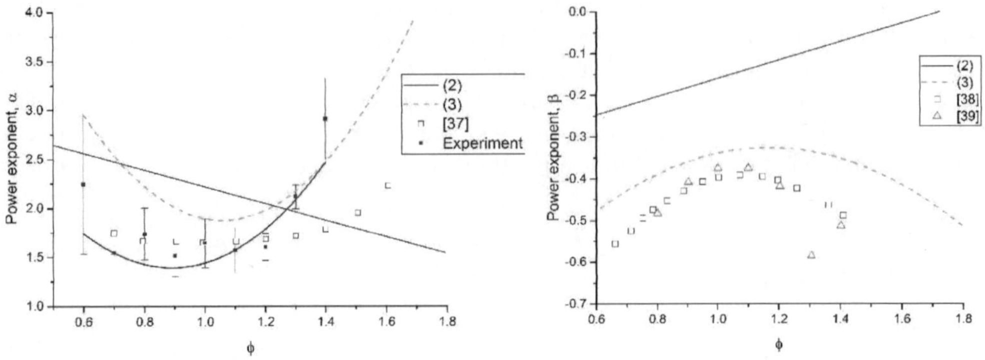

a) b)

Figure 7 – Power exponents α (a) and β (b) as a function of φ

6. Conclusions

During this work S_L was determined experimentally at atmospheric pressure and initial gas temperature in the range 298-358 K by heat flux method. S_L was also was determined by simulation with GRI 3.0 kinetic mechanism for initial temperature up to 800 K and pressure up to 20 bar. Based on results of simulations was developed equation (3). The presented experimental data and calculation using equation (3) are in good agreement with results of other authors and calculations using the kinetic mechanism GRI 3.0. The equation (3) can be used for modeling in the software packages.

Acknowledgments

This work was supported by the Ministry of education and science of the Russian Federation in the framework of the implementation of the Program "Research and development on priority directions of scientific-technological complex of Russia for 2014-2020" (RFMEFI58716X0033).

Appendix

Reference	Method	Symbol	Reference	Method	Symbol
5	Counterflow		20	Counterflow	
6	Closed vessel		21	Counterflow	
5	Counterflow		22	Bunsen flame	
7	Flat flame, heat flux		23	Closed vessel	
8	Closed vessel		24	Closed vessel	
5	Counterflow		25	Closed vessel	
9	Closed vessel		26	Closed vessel	
10	Closed vessel		27	Closed vessel	
11	Closed vessel		28	-	
12	Flat flame, heat flux		29	-	
13	Closed vessel		30	-	
14	Closed vessel		31	-	
15	Bunsen flame		32	-	
16	Flat flame, heat flux		33	-	
17	Flat flame, heat flux		34	-	

18	Closed vessel		35	-	
19	Closed vessel				

References

[1] GRI-Mech 3.0, The Gas Research Institute, http://www.me.berkeley.edu/gri_mech/
[2] M. Metghalchi, J. C. Keck Burning velocities of mixtures of air with methanol, isooctane, and indolene at high pressure and temperature, combustion and flame 48, 191-210, 1982 (Laminar Burning Velocity of Propane-Air Mixtures at High Temperature and Pressure M. METGHALCHI* and J. C. KECK combustion and flame 38, 143-154, 1980).
[3] Gottgens, J. Analytic approximations of burning velocities and flame thicknesses of lean hydrogen, methane, ethylene, ethane, acetylene and propane flames / J. Gottgens, F. Mauss, N. Peters // The combustion institute, 1992. – P. 7.
[4] CHEMKIN 10112, Reaction Design. San Diego; 2011.
[5] Vagelopoulos C. M. and Egolfopoulos F. N., 1998 Direct Experimental Determination of Laminar Flame Speeds, Proc. of the Combustion Institute 27, pp 513–9.
[6] S.C. Taylor PhD. Thesis, Vol. University of Leeds, 1991 F.N. Egolfopoulos, D.L. Zhu, C.K. Law, Symp. (Int.) on Combustion 23, pp 471-8.
[7] A. Vanmaaren, D.S. Thung, L.P.H. Degoey, 1996 Comb. Sci. Technol, 327-44.
[8] Aung K. T., Tseng L. K., Ismail M. A., and Faeth G. M., 1995 Response to Comment by S. C. Taylor and D. B. Smith on "Laminar Burning Velocities and Markstein Numbers of Hydrocarbon/Air Flames, Combustion and Flame 102, pp 526–30.
[9] Hassan M. I., Aung K. T., and Faeth G. M., 1998 Measured and Predicted Properties of Laminar Premixed Methane/Air Flames at Various Pressures, Combustion and Flame 115, pp 539–50.
[10] Rozenchan G., Zhu D. L., Law C. K., and Tse S. D., 2002 Outward Propagation, Burning Velocities, and Chemical Effects of Methane Flames up to 60 atm, Proc. of the Combustion Institute 29, pp 1461–69.
[11] Gu X. J., Haq M. Z., Lawes M., and Woolley R., 2000 Laminar Burning Velocity and Markstein Lengths of Methane-Air Mixtures, Combustion and Flame 121, pp 41–58.
[12] Bosschaart K. J., de Goey L. P. H., and Burgers J. M., 2004 The Laminar Burning Velocity of Flames Propagating in Mixtures of Hydrocarbons and Air Measured with the Heat Flux Method, Combustion and Flame 136, pp 261–9.
[13] Tahtouh T., Halter F., and Mounaim-Rousselle C., 2009 Measurement of Laminar Burning Speeds and Markstein Lengths Using a Novel Methodology, Combustion and Flame 156, pp 1735–43.
[14] Lowry, W., de Vries, J., Krejci, M., Petersen, E.,Serinyel, Z., Metcalfe, W., Curran, H., and Bourque,G., 2011 Laminar Flame Speed Measurements and Modeling o Pure Alkanes and Alkane Blends at Elevated Pressures, Journal of Engineering for GasTurbines and Power, in press.
[15] Kochar Y., Vaden S., Lieuwen T., Seitzman J., 2010 Laminar Flame Speed of Hydrocarbon Fuels with Preheat and Low Oxygen Content, AIAA-778.
[16] Selle, L., Poinsot, T. and Ferret, B., 2010 Experimental and numerical study of the accuracy of flame-speed measurements for methane/air combustion in a slot burner, Combustion and Flame 158, pp 146–154.
[17] Coppens, F.H.V., De Ruyck, J. Konnov, A.A., 2007 Combustion and Flame 149, pp 409-17.
[18] Halter, F., Chauveau, C., Djebaïli-Chaumeix, N., Gökalp, I. 2005 Proc. of the Combustion Institute, 30, p 201.
[19] Andrews, G. E. and Bradley, D., 1972 Combustion and Flame 18, p 133.
[20] Law, C.K. 1993 A Compilation of Experimental Data on Laminar Burning Rates, in Reduced Kinetic Mechanisms for Application in Combustion System, Peters, N. and Rogg, B., eds. Springer-Verlag, (Berlin).

[21] Peyman Zahedi and Kianoosh Yousefi, 2010 The 5thFuel & Combustion Conference of Iran, (Iran University of Science and Technology).
[22] Lowry, William Baugh, 2010 Effect of Blending on High-Pressure Laminar Flame Speed Measurements, Markstein Lengths, and Flame Stability of Hydrocarbons, Master's thesis, (Texas A&M University).
[23] Chen, X.F.; Sun, J.H.; Liu, Y. 2007 Microstructure of premixed propane/air flame in the transition from laminar to turbulent combustion, Chin. Sci. Bull. 52, pp 685–91.
[24] Huang Z., Zhang Y., Zeng K., Liu B., Wang Q., Jiang D. 2006 Measurements of laminar burning velocities for natural gas-hydrogen-air mixtures, Combustion and Flame 146, pp 302-11.
[25] Qin X., Ju Y. 2005 Measurements of burning velocities of dimethylether and air premixed flamesat elevated pressures, Proc. of the Combustion Institute 30, pp 233-40.
[26] Liao S.Y., Jiang D.M., Gao J., Huang Z.H. 2004 Measurements of Markstein numbers and laminar burning velocities for natural gas-air mixtures, Energy and Fuels 18, pp 316-26.
[27] Stone R., Clarke A., Beckwith P. 1998 Correlations for the LaminarBurning Velocity of Methane/Diluent/Air Mixtures Obtained in Free-Fall Experiments, Combustion and Flame 114, pp 546-55.
[28] Sharma S.P., Gupta C.P., 1981 Burning velocities and explosion pressures of acetylene-freon12-air mixtures in spherical bombs, Combustion and Flame 16, pp 213-22.
[29] Christopher Mumby 2010 Predictions of Explosions and Fires of Natural Gas/Hydrogen Mixtures for Hazard Assessment, A doctoral thesis submitted in partial fulfilment of the requirements for the award of degree of doctor of philosophy of Loughborough University.
[30] Elia M., Ulinski M., Metghalchi M., 2001 Laminar burning velocity of methane-air-diluent mixtures, Int of the ASME, vol. 123, pp 190-6.
[31] M. Fairweather, Hargrave, G. K., Ibrahim, S. S. and Walker, D. G., 2006 Studies of Premixed Flame Propagation in Explosion, Journal of Hazardous Materials 135(1-3), pp 58-65.
[32] Goswami, M., Derks, S.C.R., Coumans, K., Slikker, W.J., de Andrade Oliveira, M.H., Bastiaans, R.J.M., Luijten, C.C.M., de Goey, L.P.H., and Konnov, A.A. 2013. Combustion and Flame 160, p 1627.
[33] Lee, J.H., Kim, J.H., Park, J.H. and Kwon, O.C. 2010 Studies on properties of laminar premixed hydrogen-added ammonia/air flames for hydrogen production, Int. Journal of Hydrogen Energy, Vol. 35, No. 3, pp 1054-64.
[34] C. Heghes, 2006 C1-C4 Hydrocarbon Oxidation Mechanism, Ph.D. Dissertation, (Heidelberg University).
[35] Farrell, J. T., Johnston, R. J., and Androulakis, I. P., 2004 Molecular structure effects on laminar burning velocities at elevated temperature and pressure, SAE Technical Paper No. 2004-01-2936.
[36] Just, Th., 1994 http://combustion.berkeley.edu/gri-mech/version30/targets30/f6.html
[37] M. Christensen, E.J.K. Nilsson, Konnov, A.A.. The Temperature Dependence of the Laminar Burning Velocities of Methyl Formate + Air Flames. Fuel, submitted.
[38] Goswami, M.S. Derks, K. Coumans, W.J. Slikker, M.H. de Andrade Oliveira, R.J.M. Bastiaans, C.C.M. Luijten, L.P.H. de Goey, A.A. Konnov, 2013 The effect of elevated pressures on the laminar burning velocity of methane + air mixtures, Combustion and Flame, 160, pp 1627-35.
[39] Liao, S.Y. 2004 Determination of laminar burning velocities for natural gas, S.Y. Liao, D.M. Jiang, Q. Cheng, Fuel, Vol.83, pp 1247–50.

IOP Conf. Series: Materials Science and Engineering **302** (2018) 012079 doi:10.1088/1757-899X/302/1/012079

A Comprehensive Solution of the Problems of Ensuring the Strength of Gas Turbine Engine Compressor at the Design Stage

V V Vedeneev[1], M E Kolotnikov[1], P A Mossakovskii[1], L A Kostyreva[1], F A Abdukhakimov[1], P V Makarov[2], A A Pyhalov[3] and M A Dudaev[3]

[1] Lomonosov Moscow State University, 1 Michurinskii prospect, Moscow 119192, Russia
[2] "Salut" Gas Turbine Engineering Research and Production Center, 16 Budionny Ave, Moscow 105118, Russia
[3] Irkutsk State Transport University, 15 Chernyshevskogo str., Irkutsk 664074, Russia

vasily@vedeneev.ru

Abstract. In this paper we present a complex numerical workflow for analysis of blade flutter and high-amplitude resonant oscillations, impenetrability of casing if the blade is broken off, and the rotor reaction to the blade detachment and following misbalance, with the assessment of a safe flight possibility at the auto-rotation regime. All the methods used are carefully verified by numerical convergence study and correlations with experiments. The use of the workflow developed significantly improves the efficiency of the design process of modern jet engine compressors. It ensures a significant reduction of time and cost of the compressor design with the required level of strength and durability.

1. Introduction

Design of efficient low and high pressure compressors for modern jet engines is one of the most difficult problems in the design of jet engines. Requirements for the reduction of number of stages with the increase of the pressure ratio yields the increase of aerodynamic and structural loading of blades. This needs more refined analysis of possible operating regimes at the engine design stage. First, detailed analysis of blade flutter and forced vibrations in engine wheels, which can result in fast fatigue damage of blades, is needed. However, high-amplitude blade vibrations are not always successfully avoided, and an important problem of consequences of blade detachment arises. Even if a blade is broken off due to fatigue damage caused by high-amplitude vibrations, the engine damage should be localized, and a long safe flight at the auto-rotation regime should be possible.

In this paper we present a complex numerical workflow for analysis of blade flutter and high-amplitude resonant oscillations, impenetrability of casing if the blade is broken off, and the rotor reaction to the blade detachment and following misbalance, with the assessment of a safe flight possibility at the auto-rotation regime. The use of the workflow developed significantly improves the efficiency of the design process of modern jet engine compressors.

2. Blade Flutter Analysis

2.1. Method of Flutter Prediction

The energy method is used for flutter analysis [1]. We assume that the influence of the flow on natural blade modes and frequencies is negligible. Therefore, the airflow can result only in a small additional damping (for stability case) or additional energy inflow (for flutter case) without the change of natural modes and frequencies. Integrating the energy equation for a blade over an oscillation cycle, we obtain the change of the total energy over the cycle:

$$\Delta E = W = \int_{t_o}^{t_0+T} \int_S p(x,y,z,t)\vec{n}(x,y,z,t)\vec{v}(x,y,z,t)dsdt \quad (1)$$

where $E(t)$ is the total energy, W is the work done by unsteady pressure, T is the blade oscillation period, S is the blade surface, p is the flow pressure, $\vec{n}$ is the blade surface normal, and $\vec{v}$ is the velocity of the blade points. The following inequality is a criterion of flutter: $W > 0$.

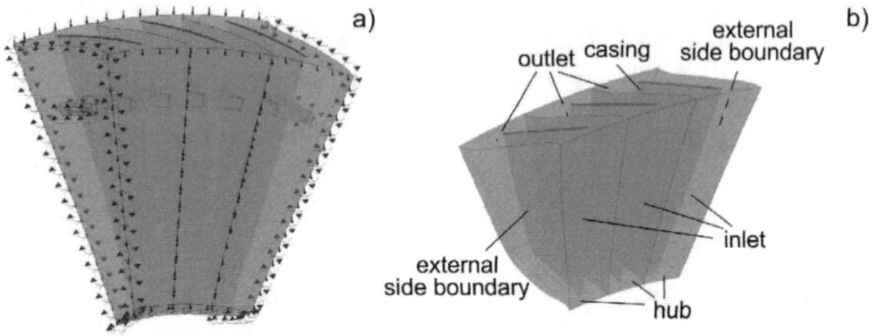

Figure 1. Model of three consecutive blade passages (a), boundary conditions (b).

Finite-volume model of the flow consists of three consecutive blade passages of one stage (Figure 1a). For unsteady fluid flow analysis, initial and boundary conditions (Figure 1b) are extracted from the steady-state flow calculated for the full compressor (where all stages are modelled), verified by full-scale compressor tests.

Mesh displacement in the form travelling wave corresponding to the wheel natural mode with a specified number of nodal diameters is applied to each blade surface:

$$\vec{u}(x,y,z,t) = A\cdot[\sin\omega t\cdot L_1(x,z) - \cos\omega t\cdot L_2(x,z)] \quad (2)$$

where A and ω are the blade oscillation amplitude and circular frequency, and $L_1(x,z)$, $L_2(x,z)$ are functions interpolating blade mode shapes corresponding to the node and antinodes of the wheel. The aerodynamic mesh deforms in accordance with the specified displacement of the boundary.

For modelling a forward (or backward) traveling wave, a phase lag $\omega t-\varphi$ and lead $\omega t+\varphi$ with respect to the middle blade are specified for neighbouring blades, where the phase shift $\varphi=2\pi m/N$ corresponds to the number of nodal diameters m (N is the number of blades).

2.2. Examples

As an example, we analysed the influence of various design parameters on flutter boundary, namely, radial and axial clearance, closure and opening of the inlet guide vane, radial flow irregularity, inter-blade tension in the mid-span shroud. The work done by pressure is calculated for the second and third natural modes of first-stage shrouded blades of a low-pressure compressor. A series of computational models, representing the change of one of the design parameters, was considered. The following configurations were studied:

1. Increased radial clearance by 0.5 mm
2. Increased radial clearance by 0.5 mm and shut by 1.5° inlet guide vane
3. Increased radial clearance by 0.5 mm and opened by 2° inlet guide vane
4. Increased radial clearance by 1 mm
5. Specified radial non-uniformity of the flow at the inlet

6. Decreased tension force in the mid-span shroud (simulation of the wear in shroud contact pairs)

Each mode was analysed in a full range of possible nodal diameters. Results for one flow regimes are presented here.

For the baseline mode, as well as for cases 1-5, calculation results predict the occurrence of flutter for the second mode at m=5 (Figure 2). It is seen that the change of the design parameters considered in cases 1-5, does not result in significant change of the work done by pressure over the oscillation cycle. In the case 6 the results show that the wear of the mid-span shroud contact pairs stabilises the blade, as the work for both modes becomes negative for all nodal diameters.

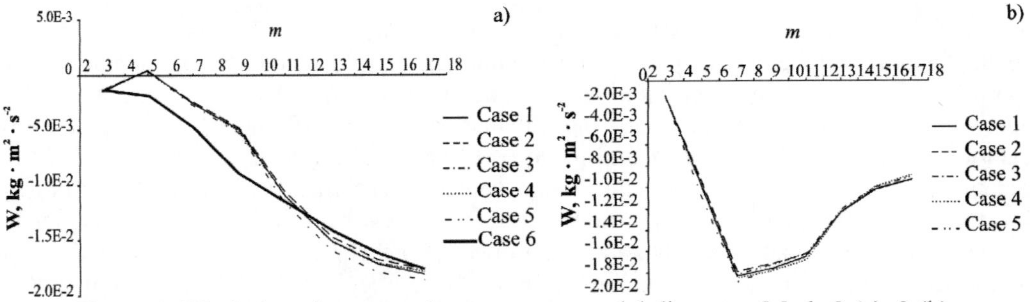

Figure 2. Work done by unsteady pressure vs nodal diameter. Mode 2 (a), 3 (b).

Comparison of calculation results with flutter test data shows a good agreement both in overall prediction (stability or flutter), and, for the flutter case, in the fastest growing mode.

3. Forced vibrations

When the engine is operating, the compressor blades are subject to periodically changing forces caused by the circumferential non-uniformity of the flow due to wakes of guide vanes and blades of preceding stages. When the frequency of these forces coincides with the natural frequency of the blade, resonance occurs. Possible resonance regimes can be found through Campbell diagram.

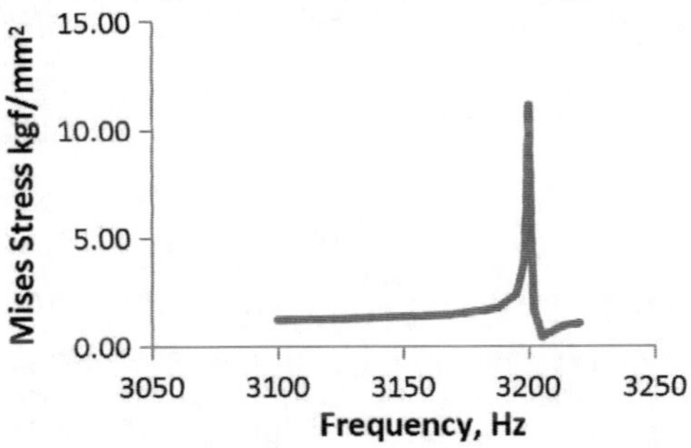

Figure 3. Frequency response at resonance regime.

The algorithm for calculating the amplitude of resonant oscillations consists in the following:

1. Calculation of the steady flow of air in the compressor with averaging of the parameters in the axial stator-rotor clearance;
2. Extraction of the model of the rotating wheel and guide vanes under consideration;
3. Calculation of the unsteady air flow in the selected stage without averaging of the parameters in the axial clearance, to obtain the unsteady pressure over the blade.
4. Transferring of the unsteady pressure field over the blade from CFD and CSD code.
5. Calculation of the amplitude of the resonant oscillations in the frequency domain (Figure 3).

4. Investigation of the impenetrability of engine cases due to blade breaking off.

Reliable computational methodology of impenetrability of the GTE cases after rotor blade breakage is a significant tool at the stage of design of a new product. Over the past several decades, the computational methods for assessing the impenetrability have undergone a significant evolution: from relatively simple semi-empirical techniques based on conservative energy criteria up to modern approaches to the analysis of supercritical behavior of structures using direct computer modeling in combination with a set of full-scale and virtual experiments needed for identification parameters of the calculation scheme. The main disadvantage of most of such calculation methods based on computer simulation is the low level of reliability of the results obtained. We present an original computational technique for assessing the impenetrability of GTE cases due to a blade breaking off, which has a higher predictive accuracy than existing ones [2-3].

4.1. The computational method for estimation the impenetrability of the GTE cases

The method is based on the experimental and computational procedure for solving the initial-boundary value problem by the finite element method in combination with the system of full-scale and virtual verification experiments used to adjust the calculation scheme (Figure 4). Verification of the calculation scheme consists in the classification of thermomechanical processes realized in the computational domain based on the characteristic values of deformations, strain rates, temperatures and parameters of the type of stress-strain state with subsequent reproduction of representative thermomechanical processes in verification experiments. The correction of the calculation scheme, which consists in the refinement of the parameters of the mathematical model (the parameters of the constitutive relations and the failure criteria for the materials used, the parameters of the contact interaction, etc.) is performed by reconciling the results of full-scale and virtual verification tests.

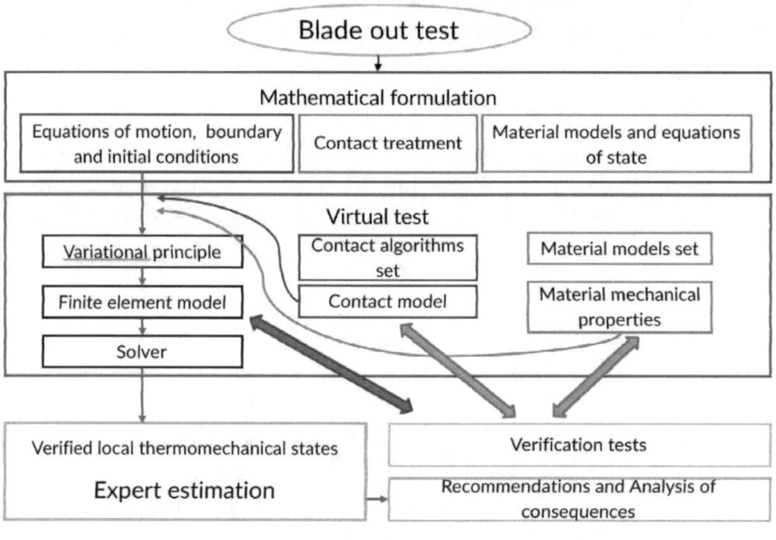

Figure 4. General scheme for the solution of the problem.

4.2. System of verification tests.

In this calculation technique, a set of verification tests consists of static and dynamic experiments (at normal and elevated temperatures) realizing thermomechanical states from a limited number of classes obtained as a result of solving the initial boundary value problem. A minimum set of verification tests for each material should include the following experiments: static uniaxial compression, tension and torsion (till failure) tests, dynamic uniaxial compression, tension and shear tests by the Kolsky method and its modifications, and ballistic impact tests.

4.3. Estimation of the quality of the methodology for assessing the impenetrability of the GTE cases

As an illustration of the efficiency of the calculation scheme used, the results of verification experiments using the Kolsky method for uniaxial compression and tension of metal samples, ballistic impact tests for metal and woven barriers (Figure 5), and bench tests for assessing the impenetrability of the metal casing GTE are shown (Figure 6). Quantitative and qualitative agreement of the obtained results shows the efficiency and reliability of the presented calculation technique at all stages of design and subsequent operation of the gas turbine engine (Figure 7).

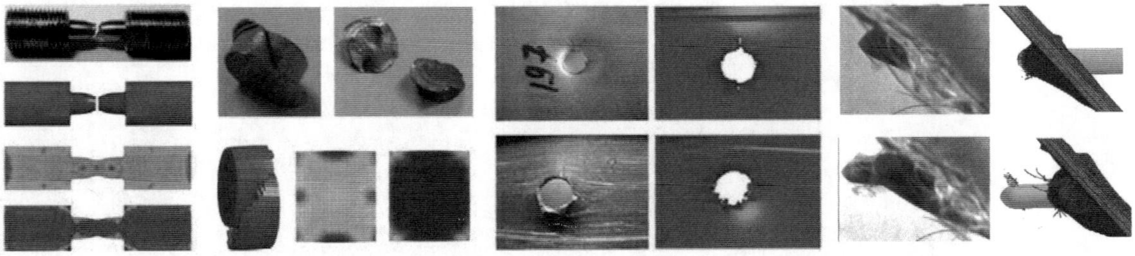

Figure 5. Comparison of results of full-scale and virtual verification tests .

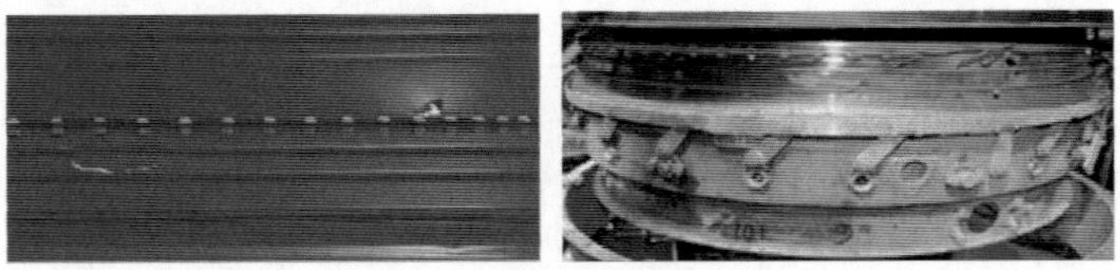

Figure 6. Comparison of the results obtained using the presented methodology with the data of bench tests.

Figure 7. Investigation of the composite case impenetrability.

5. Analysis of the rotor misbalance and bearing loading due to blade breaking off

In this section we consider the modeling of the rotor dynamics (Figure 8) in the case of a blade breakage [4-5]. Figure 9 shows the program of rotation frequency change of low-pressure rotor, starting from the standard spinning mode, first stage blade breakage and the autorotation mode.

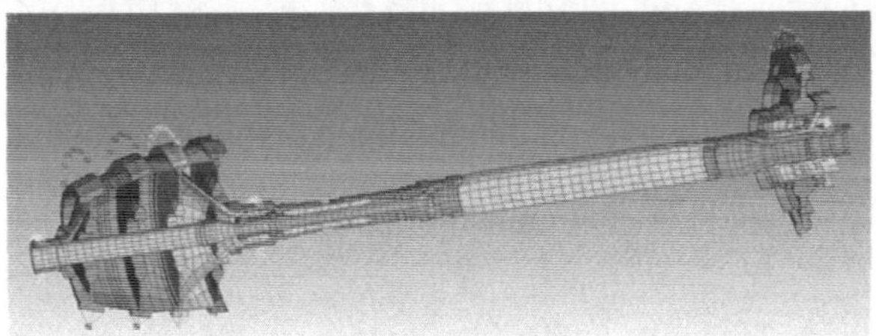

Figure 8. Finite element model of a low pressure rotor.

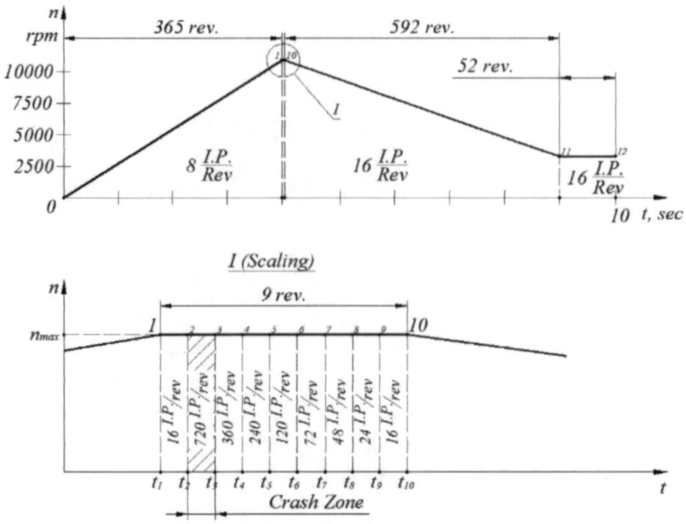

Figure 9. Rotation frequency change during the simulation of the blade breakage.

The total running time consists of four periods (Figure 9): starting from zero to the maximum speed; operation at maximum speed, at which the blade breakage is simulated; deceleration of the rotation speed; operation at autorotation mode (30% of the maximum speed).

To simulate the process of the blade breakage, the data obtained in the full-scale experiment are used in the form of measured location of the 1st stage wheel gravity center after the blade breakage. The results of the calculation (Figure 10) show a significant drop in the oscillations amplitude when switching to the autorotation mode, comparing to the blade breakage moment.

With calculated displacements of the low-pressure rotor supports (Figure 10) and rotor stress at the blade breakage and autorotation regimes, we estimate bending strength of the rotor shaft, fatigue strength of the rotor, bearing loads and durability, and finally, the safety of the engine operating at the autorotation regime.

The results of the rotor dynamics analysis during the blade breakage show the effectiveness of the experimental-theoretical approach described.

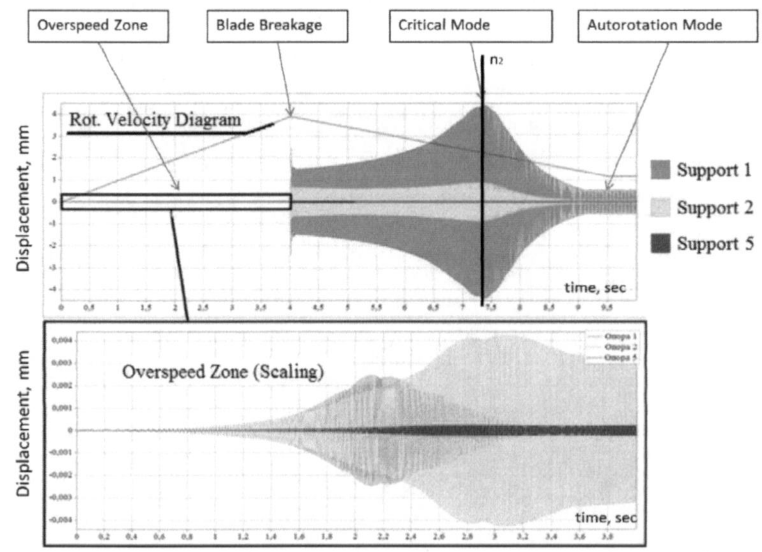

Figure 10. Displacement of low-pressure rotor supports during the simulation.

6. Conclusions

IOP Conf. Series: Materials Science and Engineering **302** (2018) 012079 doi:10.1088/1757-899X/302/1/012079

The numerical approach developed is an efficient tool for modeling the most dangerous regimes of compressor blades, rotor shaft and bearings. It ensures a significant reduction of time and cost of the compressor design for modern jet engines with the required level of strength and durability.

Bibliography

[1] Vedeneev V V, Kolotnikov M E and Makarov P V 2015 Experimental validation of numeric blade flutter prediction. J. Prop. Power. Vol. 31 (5), pp. 1281–1291.

[2] Mossakovsky P A, Dubyanskaya O G, Kolotnikov M E 2008 A new approach to assessing of GTE cases impenetrability with rotor elements. New technological processes and reliability of GTE. CIAM, Vol. 8, pp. 38-62.

[3] Mossakovsky P A, Antonov F K, Kolotnikov M E, Kostyreva L A 2012 On fan blade out event analysis of lightweight composite structures. Herald of propulsion engineering, Vol. 2, pp. 180-184.

[4] Pykhalov A A, Milov A E 2007 Contact problem of the turbomachine assembly rotors static and dynamic analysis. Irkutsk State Technical University Press. 192 p.

[5] Dudaev M A, Pykhalov A A 2015 Contact problem in the analysis of the turbomachine assembly rotors dynamic behavior. Sc. Bull. Novosibirsk State Tech. Univ. Vol. 3, pp. 113-129.

Development of Row of Vibration Insulators and its Mathematical Models on a Base of Common Multi-parameter Scheme of Element Axial Line

Yury K. Ponomarev

Samara National Research University, Russia, 443086, Samara, Moskovskoe shosse, 34.

E-mail: ponomarev-ssau@yandex.ru

Abstract: The mathematical model of deformation of a cable (rope) vibration insulator consisting of two identical clips connected by means of elastic elements of a complex axial line is developed in detail. The axial line of the element is symmetric relatively to the horizontal axis of the shape and is made up of five rectilinear sections of arbitrary length a, b, c, conjugated to four radius sections with parameters R_1 and R_2 with angular extent 90°. On the basis of linear representations of the theory of bending and torsion of mechanics of materials, applied mechanics and linear algebra, a mathematical model of loading of an element and a vibration insulator as a whole in the direction of the vertical Y axis has been developed. Generalized characteristics of the friction and elastic forces for an elastic element with a complete set of the listed sections are obtained. Further, with the help of nullification in the generalized model of the characteristics of certain parameters, special cases of friction and elastic forces are obtained without taking into account the nullified parameters. Simultaneously, on the basis of the 3D computer-aided design system, volumetric models of simplified structures were created, given in the work. It is shown that, with the help of a variation of the five parameters of the axial scheme of the element, in combination with the variation of the moment of inertia of the rope section and the number of elements entering the ensemble, the load characteristics and stiffness of the vibration insulators can be changed tens and hundreds of times. This opens up unlimited possibilities for the optimal design of vibration protection systems in terms of weight characteristics, in cost, in terms of vibration intensity, in overall dimensions in different directions, which is very important for aerospace and transport engineering.

1. Introduction

Development and application of rope vibration insulators began in a second half of XX century. In the beginning ropes of straight [1] and radius [2] shape was used. It was fixed in special supports as regular groups. Calculation models didn't exist in that time, stiffness characteristics was obtained by experiment. It was not so many variants for choice of stiffness: by diameter of rope, number of wires in cross-section, length and number of straight parts in element, radius of element curvature and number of elements. It was obtained [3] that it is impossible to obtain an isotropic vibration insulator with radius elastic elements to avoid additional resonances in a working frequency range: stiffness of vibration insulator is significantly different for three perpendicular directions for any parameters of rope elements. Different authors began to develop rope vibration insulators with different length of arc of element in a group, with different angles of inclination for planes of curve elements [4] and with two different shapes of elements in a group [5]. Because mathematical models of load characteristics were absent, this work was very difficult.

The aim of the present paper is development of mathematic models for row of rope vibration insulators with ten different parts of axial line for elastic element. Analysis of these models is a basement for development of vibration insulators with different shape and optimization of its structure by calculation for minimal size or weight, optimal strength characteristics with required load ability, optimal damping etc.

2. Materials and methods

As a basement a structure developed in Samara State Aerospace University was taken [6]. Researches of these structure characteristics were beginning in [7]. Main idea of this work is obtaining of simple mathematic models of different types of vibration insulator by changing some parameters in different compositions and nullification of some other parameters. The basement of this work is multi-parametric mathematical model of vibration insulator [6] and contemporary mathematic software (Mathcad, Maple).

As a common the rope element has straight parts *a*, *b*, *c* and radius parts R_1 and R_2 (Figure 1); elastic system of vibration insulator is described by number of wires in a cross-section of rope *n*, number of elements in vibration insulator *N* and diameter of rope element D_{mp}.

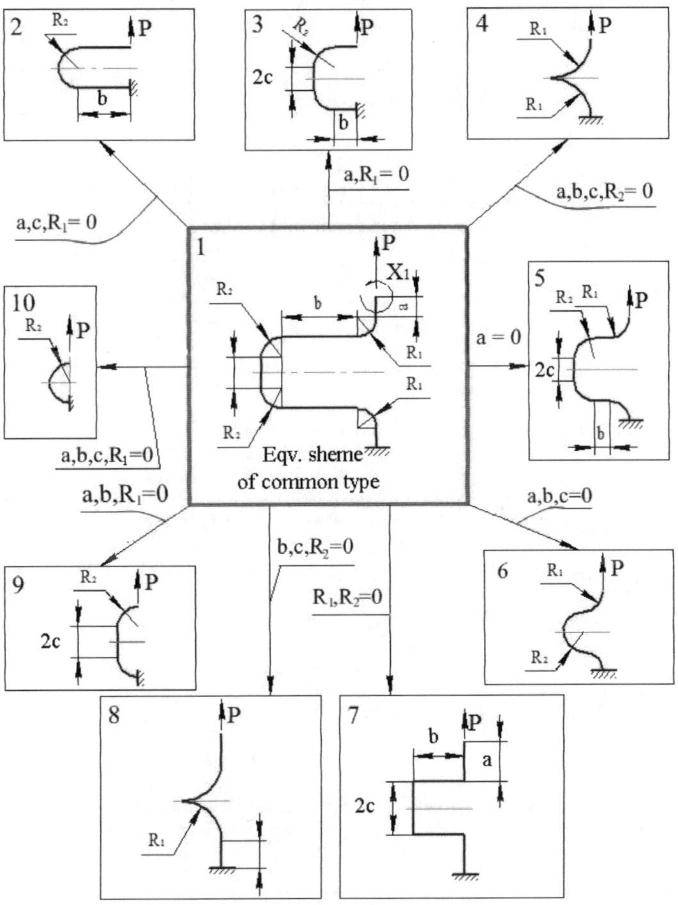

Figure 1. Scheme of elastic element of common multi-parametric type and possible particular cases of shapes for nullification of any parameters.

After nullification of same parameters of common scheme of elastic element, it is possible to obtain nine variants of elements developed from scheme 1. These variants and changing of n, N and D_{mp} allow changing of stiffness C_y and load ability P_y of rope vibration insulators in a wide range.

Table 1. Main equations for calculation of load characteristics P_y and stiffness C_y for rope vibration insulators presented in Table 2

Element number	Equations for P_y and C_y
1	$P_y = \frac{12EJ(2b+\pi R_1+\pi R_2)Ny}{\left\{\begin{array}{l}2(3\pi^2R_1^4 3\pi^2R_2^4-24R_1^4-24R_2^4+2b^4+48R_1^2R_2^2+24R_1^2b^2\\+24R_2^2b^2+4\pi R_1b^3+6\pi R_1^3b+4\pi R_2b^3+6\pi R_2^3b+3\pi^2R_1R_2^3\\+3\pi^2R_1^3R_2+24\pi R_1R_2^2b+24\pi R_1^2R_2b+6\pi^2R_1R_2b)\end{array}\right\}}$; $C_y = \frac{12EJ(2b+\pi R_1+\pi R_2)N}{\left\{\begin{array}{l}2(3\pi^2R_1^4 3\pi^2R_2^4-24R_1^4-24R_2^4+2b^4+48R_1^2R_2^2+24R_1^2b^2\\+24R_2^2b^2+4\pi R_1b^3+6\pi R_1^3b+4\pi R_2b^3+6\pi R_2^3b+3\pi^2R_1R_2^3\\+3\pi^2R_1^3R_2+24\pi R_1R_2^2b+24\pi R_1^2R_2b+6\pi^2R_1R_2b)\end{array}\right\}}$.
2, 3	$P_y = \frac{6EJN(2b+\pi R_2)y}{(3\pi^2R_2^4-24R_2^4+2b^4+24R_2^2b^2+4\pi R_2b^3+6\pi R_2^3b)}$; $C_y = \frac{6EJN(2\cdot b+\pi\cdot R_2)}{(3\pi^2R_2^4-24R_2^4+2b^4+24R_2^2b^2+4\pi R_2b^3+6\pi R_2^3b)}$.
4	$P_y = 2\pi EJNy/(R_1^2(\pi^2-8))$; $C_y = 2\pi EJN/(R_1^2(\pi^2-8))$.
5	$P_y = \frac{6EJN(2b+\pi R_1+\pi R_2)y}{\left\{\begin{array}{l}24R_1^2b^2+24R_2^2b^2+4\pi R_1b^3+6\pi R_1^3b+4\pi R_2b^3+6\pi R_2^3b+\\+3\pi^2R_1^4+3\pi^2R_2^4-24R_1^4-24R_2^4+2b^4+48R_1^2R_2^2+3\pi^2R_1R_2^3+\\+24\pi R_1R_2^2b+24\pi R_1^2R_2b+6\pi^2R_1R_2b\end{array}\right\}}$; $C_y = \frac{6EJN(2b+\pi R_1+\pi R_2)}{\left\{\begin{array}{l}24R_1^2b^2+24R_2^2b^2+4\pi R_1b^3+6\pi R_1^3b+4\pi R_2b^3+6\pi R_2^3b+\\+3\pi^2R_1^4+3\pi^2R_2^4-24R_1^4-24R_2^4+2b^4+48R_1^2R_2^2+3\pi^2R_1R_2^3+\\+24\pi R_1R_2^2b+24\pi R_1^2R_2b+6\pi^2R_1R_2b\end{array}\right\}}$.
6	$P_y = \frac{12EJN(\pi R_1+\pi R_2)y}{(6\pi^2R_1^4+6\pi^2R_2^4-48R_1^4-48R_2^4+96R_1^2R_2^2+6R_1^2R_2^3+6\pi^2R_1^3R_2)}$; $C_y = \frac{12EJN(\pi R_1+\pi R_2)}{(6\pi^2R_1^4+6\pi^2R_2^4-48R_1^4-48R_2^4+96R_1^2R_2^2+6R_1^2R_2^3+6\pi^2R_1^3R_2)}$.
7	$P_y = 6EJNy/b^3$; $C_y = 6EJN/b^3$.
8, 9, 10	$P_y = \frac{2\pi EJNy}{R_2^3(\pi^2-8)}$; $C_y = \frac{2\pi EJNy}{R_2^3(\pi^2-8)}$.

Table 2. Structures of vibration insulators developed on a base of elements from Figure 1 and in accordance to mathematical models of stiffness and load characteristics developed in the present paper (Table 1)

1	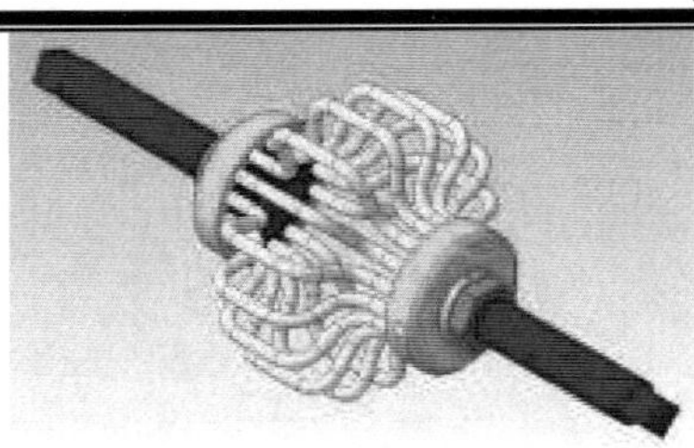	2	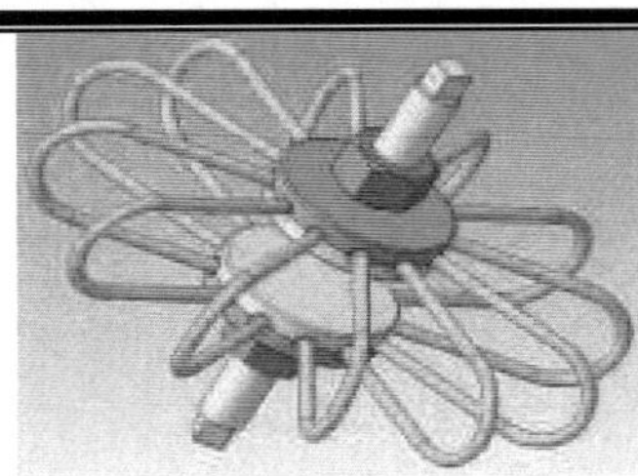
3	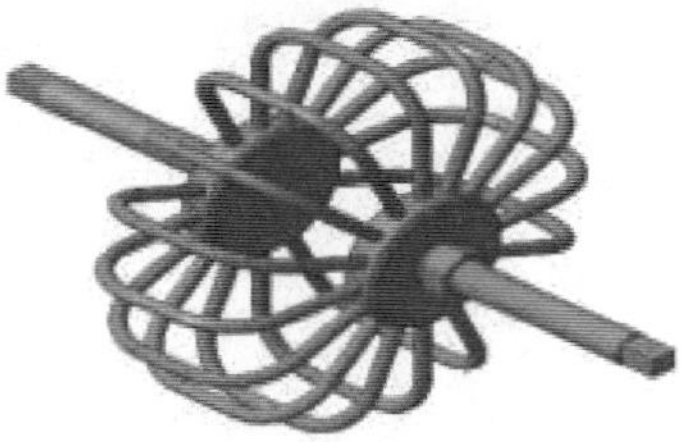	4	
5	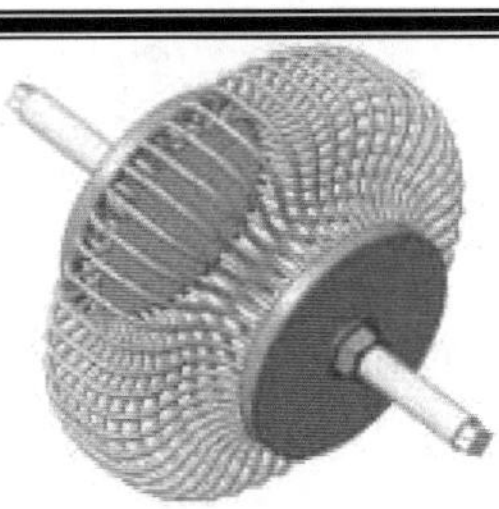	6	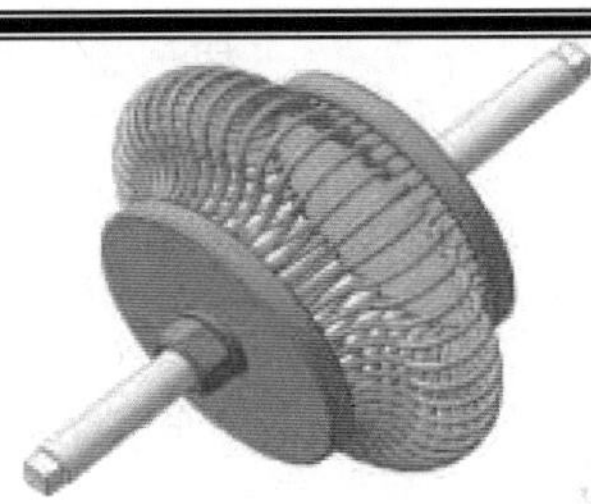
7		8	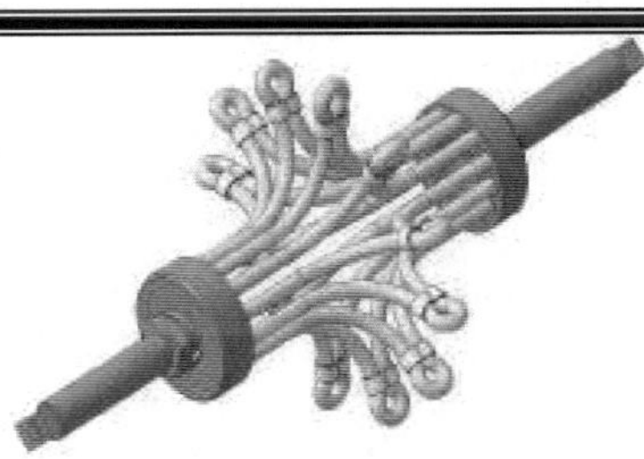
9		10	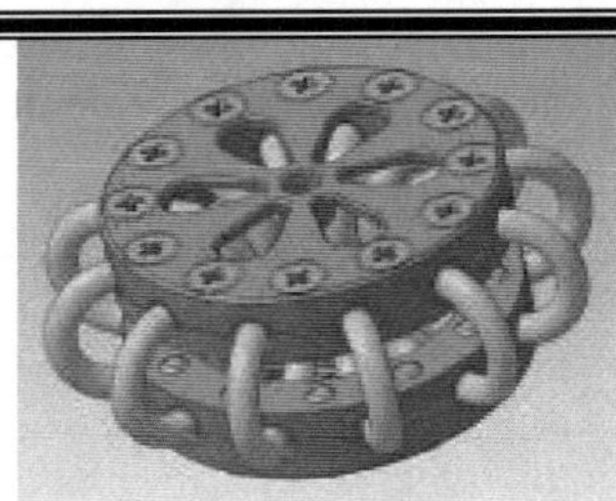

It is shown in the present paper that nullification of different parameters of equivalent scheme of elastic element allows obtaining of vibration insulator characteristics in a wide range (Figure 2).

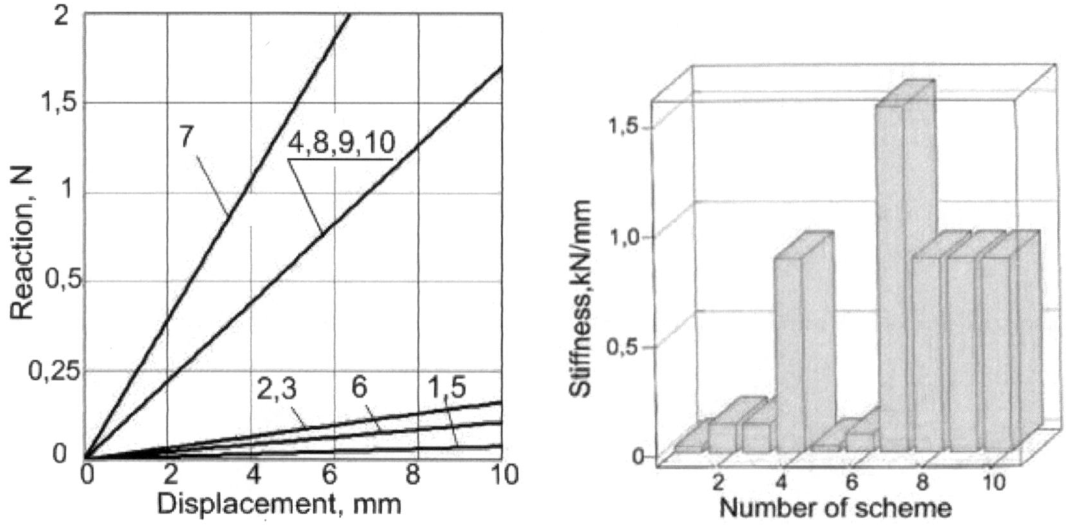

a b

Figure 2. Load characteristics (a) and stiffness (b) of vibration insulators from Table 2 (for all schemes J_x =1 mm^4, a = b = c = R_1 = R_2 = 20 mm).

Acknowledgment

This work was supported by the Ministry of Education and Science of the Russian Federation in the framework of the implementation of the Program ''Research and development on priority directions of scientific-technological complex of Russia for 2014– 2020".

References

[1] United States Patent 3 360 225. Shock and Vibration Mounts of the Cable Support Type. Carlo Camossi, 1967.

[2] USSR patent №236132, MCI F16F. Vibration insulator/ D.A. Volkov, V.D. Kirillov, B.V. Bolshakov (USSR). №1154296/25-28. Applied 06.05.67. Published 24.01.69. Bulletin №6. [in Russian]

[3] Ponomarev Yu.K., Kalakutsky V.I. Multi-layer vibration insulators with elastic elements of regular structure. Samara State Aerospace University, Samara, 2003, 256 pp. [in Russian]

[4] United States Patent №4586689. Rope vibration insulator with independent reaction on vertical and horizontal load / P. Lantero, 1984.

[5] Europe patent № 0059143, MCI F16F 7/14. Rope vibration insulator/ K. Prit. 1982.

[6] RF patent № 83113 U1, MCI F16F 7/14/ Vibration insulator/ Boyarov A.V., Vasyukov E.S., Gvozdev A.S., Kotov A.S., Melent'ev V.S., Ponomarev D.Yu., Ponomarev Yu.K. Applied 26.05.08, Published 20.05.09, Bulletin №14. [in Russian]

[7] Gvozdev, A.S. Development of parametric row of rope vibration insulators with П-shape elastic elements / A.S. Gvozdev, Yu.K. Ponomarev // Science and education for transport: proceedings of III all-Russian science-practical conference, 11-12 November 2010. – Penza: PTZhT. – 2010. [in Russian]

Development of Structure and Characteristics Calculation Method for Г- shape Rope Vubration Insulator

Yury K. Ponomarev

Samara National Research University, Russia, 443086, Samara, Moskovskoe shosse, 34.

E-mail: ponomarev-ssau@yandex.ru

Abstract: The paper gives an overview of the design of rope vibration insulators with elastic elements of the center line in the form of two rectilinear and one curved section. In the Russian-language scientific literature this type of rope vibration insulators received a stable name "Г-shaped vibration insulators" by analogy with the shape of the letter "gamma-Г" of the Greek alphabet and a similar letter of the Cyrillic alphabet. Despite the wide using of vibration insulators designed on this shape, its mathematical calculation model has not yet been developed. In this connection, in this article, for the first time on the basis of the "Method of Forces" and the "Mohr Method", an analytical technique has been developed for calculating the characteristics of a vibration insulator in the directions of three mutually perpendicular axes. In addition, the article proposes a new structure of a vibration insulator consisting of several tiers of elements of this type, based on a new patented technology for manufacturing quasi-continuous woven rings, proposed by the author of this article in co-authorship with several employees of the Samara National Research University. Simple formulas are obtained for calculating the load characteristics in three mutually perpendicular directions. This makes it possible to calculate the corresponding stiffness and natural frequencies of mechanical vibration protection systems. It is established that the stiffness of the vibration insulator in the direction of the Z axis is greater than the stiffness in the X and Y axis directions, however, if a vibration insulator with equal, or close to equal characteristics, along three axes has to be designed according to the technical specification, this can be done by selecting the parameters included in the equations given in article for load characteristics.

1. Introduction

One of first designers of Г-shape rope vibration insulators is Carlo Camossi [1]. A simple structure and manufacturing technology of vibration insulator presented on Figure 1 was proposed by him. This vibration insulator has three metal or plastic lathes with holes; sections of rope are passed through these holes. The lathes in the places of rope are pressed in some points to hold this rope. The middle lath is deformed at 90-degree angle after vibration insulator assembling. Two extreme lathes have holes for connection with protected object and a basement which is under a load of vibration.

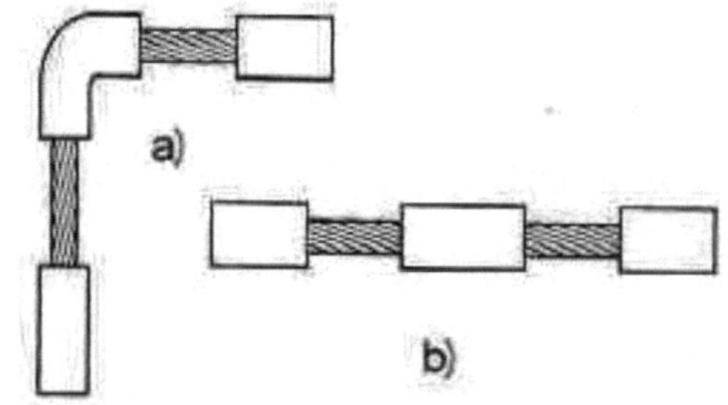

Figure 1. Vibration insulator of Camossi [1].

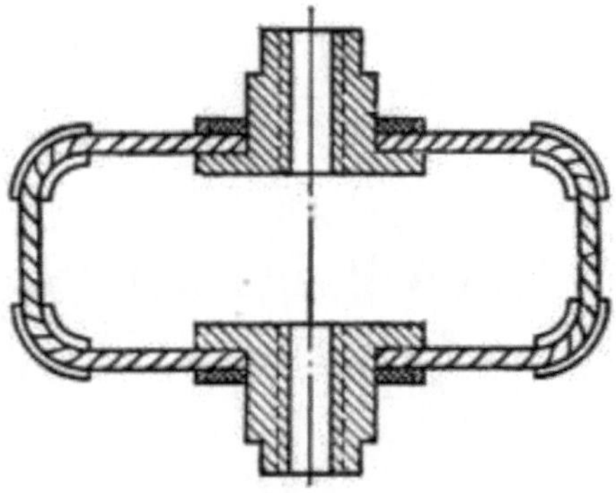

Figure 2. Vibration insulator of Tomilin [2].

L.S. Tomilin developed a spatial rope vibration insulator [2] as a group of Γ-shape or П-shape elements uniformly distributed on a circle and fixed in two round supports (figure 2).

Structure of Γ-shape vibration insulator as a group of ribbons with elliptic cross-section is presented on figure 3. The ribbons are winded by wire under stress to obtain large press load and damping properties [3].

In [4] a structure of vibration insulator as comb of Γ-shape elements with central guiding element made of deformed profile is proposed by V.S. Il'insky. Empirical equations for calculation of stiff characteristics of this vibration insulator are presented in [4] too.

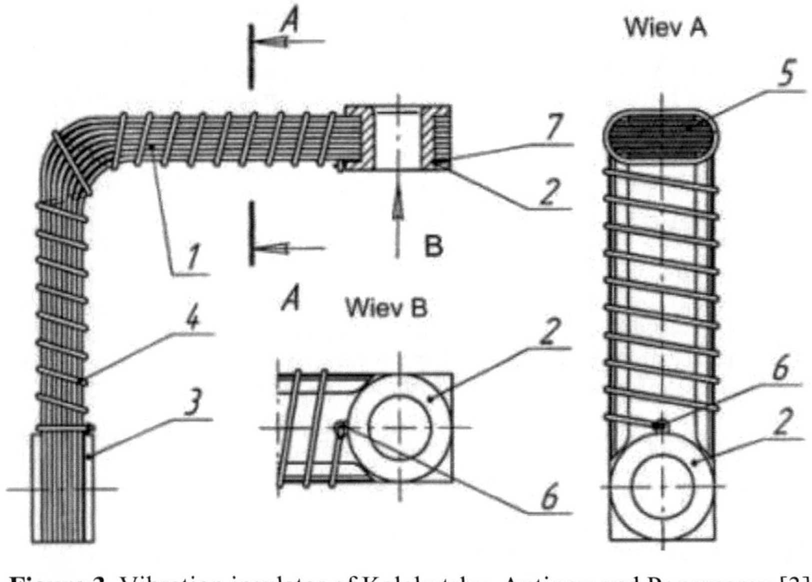

Figure 3. Vibration insulator of Kalakutsky, Antipov and Ponomarev [3].

As it is known for author of a present paper, there is no published analytic research of stiff characteristics for Γ-shape vibration insulators. The only research is [5]. Calculation of load characteristics of vibration insulator for three perpendicular directions obtained by finite element method in this paper. Results are presented in special criterial coordinates. It allows recalculation of these load characteristics in dimensional parameters if size of cross-section, length of straight parts and elastic modulus of rope material are known.

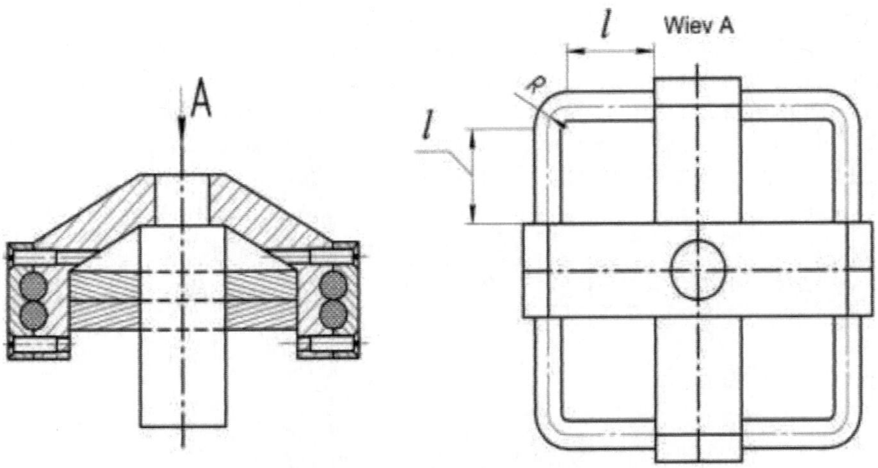

Figure 4. Scheme of two-layer rope vibration insulator with a group of Γ- shape radius-straight parts of axial line.

The aim of the present paper is development of work-able structure of rope vibration insulator with Γ-shape elastic elements and development of analytic model of deformation for this vibration insulator. The developed structure is presented on Figure 4.

A volume model is presented on figure 5. The shape of elastic elements is obtained by thermal fixation [6] in a special device.

The problem of development of mathematic model of load characteristic of vibration insulator for three perpendicular directions was solved by methods of strength of materials [5]. The used admissions are:

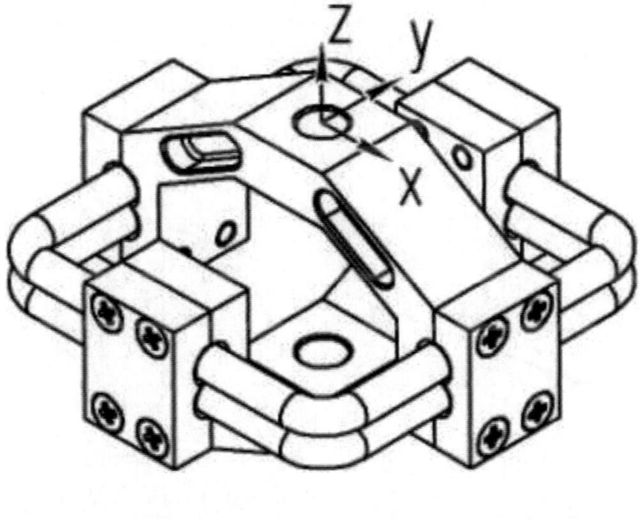

Figure 5. Volume model of vibration insulator with a group of eight Γ- shape rope elements.

1. Deformations of wires of elastic elements are little;
2. Spiral shape of wires in ropes elements didn't take into account;
3. Deformation of wire material is in accordance to Hook law;
4. Layers of rope elements have no interaction during vibration insulator deformation.

Because elastic elements of vibration insulator are fixed in supports, but one support has linear displacement relatively another one, elastic elements are non-determined statically. This problem was

IOP Conf. Series: Materials Science and Engineering **302** (2018) 012081 doi:10.1088/1757-899X/302/1/012081

solved by force method [7]. The main system of Γ-shape element was loaded for it consequently by linear loads P_x, P_y, P_z, unit moments and unit forces in accordance with method in [7]. For solution of static indeterminable the canonical equations of "force method" were obtained. "Superfluous" variables and equivalent bending moments were obtained from these equations, and load characteristics of element and vibration insulator were obtained for *x, y, z* axis directions as dependencies on geometrical and physical characteristics of vibration insulator:

$$P_x(x,l,R) = \frac{12NEJ_x(4l+\pi R)x}{(24Rl^3 - 48R^3l + 3\pi^2R^4 - 24R^4 + 10l^4 + 48R^2l^2 + 4\pi Rl^3 + 24\pi R^3 l)}; \quad (1)$$

$$P_y(y,l,R) = P_x(x,l,R); \quad (2)$$

$$P_z(z,l,R) = \frac{144EJ_x(\mu+2)(4l+\pi R)zN}{\begin{gathered}(1728Rl^3 - 1152R^3l - 768R^4\mu + 384l^4\mu + 72\pi^2R^4 - 576R^4 + 624l^4 + \\ + 2304R^2l^2 - 256R^4\mu^2 + 27\pi^2R^4\mu^2 + 192\pi Rl^3 + 864\pi R^3l + 1152Rl^3\mu - \\ - 1920R^3l\mu + 288\pi R^2l^2 + 90\pi^2R^4\mu + 1536R^2l^2\mu - 768R^3l\mu^2 + 96\pi Rl^3\mu + \\ + 936\pi R^3l\mu + 144\pi R^2l^2\mu + 252\pi R^3l\mu^2)\end{gathered}}. \quad (3)$$

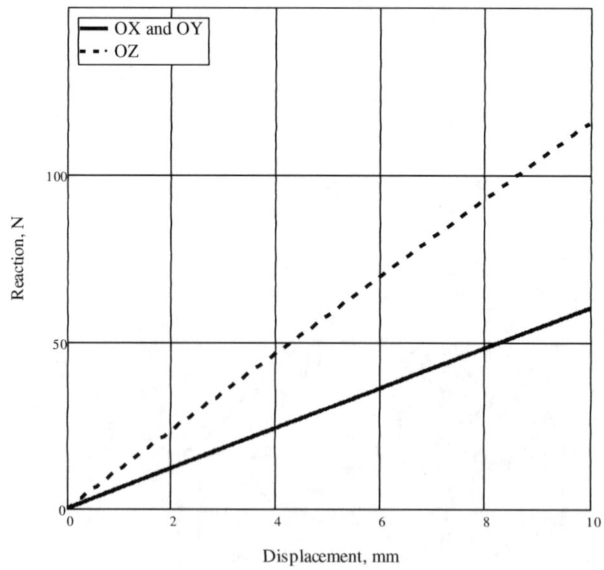

Figure 6. Load characteristics of vibration insulator for *OX, OY* and *OZ* axis direction: $l = 25$ mm; $R = 5$ mm; rope diameter $D = 5$ mm; number of wire in section $n = 49$; $N = 8$; $E = 2*10^5$ N/mm^2

Here E is elastic modulus for material of wire of rope, $\mu = 0{,}3$ is Poisson ratio, l, R are length of straight part and radius of axial line for rope elastic element (Figure. 4), $J_x = n(\pi d^4/64)$ is axial inertia moment for cross-section of rope, z is deformation of vibration insulator under static load, N is number of Γ-shape elements in vibration insulator.

Load characteristics of vibration insulator for three perpendicular directions for specific geometrical and physical parameters are presented on Figure 6 as an example. As it is seen from figure 6, stiff characteristics of vibration insulator for vertical and two horizontal directions are different. Beside it, stiffness of vibration insulator as first differential from equations (1–3) of respective displacement (static deformation) depends on geometrical parameters l, R. It allows obtaining of required stiffness for any mechanical system.

Acknowledgment
An experimental specimen of developed vibration insulator is tested on an aerospace unit.

This work was supported by the Ministry of Education and Science of the Russian Federation in the framework of the implementation of the Program «Research and development on priority directions of scientific-technological complex of Russia for 2014– 2020».

References

[1] United States Patent 3 360 225. Shock and Vibration Mounts of the Cable Support Type. Carlo Camossi, 1967.
[2] USSR Patent 241156, MPC F16F. Vibration insulator/ L.S. Tomilin (USSR). No 1169642/25-28. Applied 04.07.67. Published 01.04.69. Bulletin No 13.1. [in Russian]
[3] USSR Patent 2179664. MPC F16F 7/08. Antipov V.A., Kalakutsky V.I., Ponomarev Yu.K. et al. Published in Bulletin No 5 (2002 year). [in Russian]
[4] Il'insky V.S. Protection of radio-electronic and precise equipment against dynamic load //M.: «Radio i svyaz», 1982, 296 pp. [in Russian]
[5] Ponomarev Yu.K., Kalakutsky V.I. Multi-layer all-metal vibration insulators with regular structure of elastic elements. Samara: Samara science center of Russian Academy of science, 2003. – 168 pp. [in Russian]
[6] Lakhtin Yu.M., Rakhshtadt A.G. Thermal Treatment in Machine Manufacturing. M.: 1980. – 775 pp. [in Russian]
[7] Makarov E.G. Strength of materials on a base of Mathcad. «BHV - Peterburg», SPb:, 2004, - 512 pp. [in Russian]

IOP Conf. Series: Materials Science and Engineering **302** (2018) 012082 doi:10.1088/1757-899X/302/1/012082

The Problem of Ensuring Reliability of Gas Turbine Engines

Yu A Nozhnitsky[1]

[1]Central Institute of Aviation Motors (CIAM), 2 Aviamotornaya Street, Moscow, 111116, Russia

nozhnitsky@ciam.ru

Abstract. Requirements to advanced engines for civil aviation are discussing. Some significant problems of ensuring reliability of advanced gas turbine engines are mentioned. Special attention is paid to successful utilization of new materials and critical technologies. Also the problem of excluding failure of engine part due to low cycle or high cycle fatigue is discussing.

1. Introduction

International (in EU) and national (in US and Russia) requirements to advanced gas turbine engines (GTE) for civil aviation are considering cardinal improvement of ecological characteristics and increasing fuel efficiency [1, 2]. It may have created impression that a problem of provision of flight safety and dependability is not actual now. Indeed, modern aviation engines work without hard life time. Particular engines work continuously on aircraft without change up to 50 thousand hours. Life of critical (main) parts of engines for small-medium distance aircrafts is up to 20-40 thousand cycles (for "hot" and "cold" parts of engines) respectively. Engines have running time for in flight shoot down more than 50 thousand hours from the entrance of engine type in service (which is enough to get permission for ETOPS-flights at the 180-min approach from the nearest suitable airdrome), and more than 300 thousand hours in further operation [2].

At the same time, development of engines which meets formulated requirements is impossible without utilization of new engine architectures (engine with opened fan rotor, engine with geared fan, intellectual engine, etc.), without utilization of new materials (first of all different composite materials, intermetallic alloys, gradient materials, materials received using additive technologies, smart materials), without significant toughening of working conditions of engine parts (in terms of mechanical loads, thermal conditions, running time, including work at extremely high gasdynamic parameters – gas temperature before turbine wheel $T_3^* \geq 1950$ K, air compression $\pi_C^* > 60$) [1]. Besides, requirements for engine certification for confirmation of possible provision of safe service, continuously become stricter. In particular, during last year's new requirements (checking engine work under conditions of large ice crystals, mixed (ice-water) phase, under conditions of volcanic ash, confirmation of critical engine parts life taking into account possible defects, confirmation of engine work during single and flock birds ingestion, etc.) arose. At the same time requirements for increasing duration of ETOPS-flights and provision of possibility of engine utilization using reliability centered maintenance (RCM) have appeared.

Development of new advanced gas turbine engines with unique characteristics is impossible without successful utilization of new critical technologies. Critical engine technologies are technologies which provide quality improvement of engine competitiveness based on integration of new materials, new technological processes, design solutions, methods of designing and tests.

IOP Conf. Series: Materials Science and Engineering **302** (2018) 012082 doi:10.1088/1757-899X/302/1/012082

2. Critical gas turbine engine technologies.

Complexity of the task of ensuring strength reliability of gas turbine engine components was well illustrated by Rolls Royce in case of a high-pressure turbine (HPT) blade.

The blade operates at following conditions: pressure corresponding to the pressure at a depth of 500 m under water; temperature close to ½ of the surface temperature of the Sun, which is more than 200°C above the melting point of the alloy (try not to melt a piece of ice in the oven at 200°C); high rotation frequency (above 12000 rpm), flow velocity at the blade tips significantly exceeding the speed of sound, and high load from centrifugal forces, which corresponds to the gravity of a large bus); influence of non-stationary forces, causing intense fluctuations; influence of an aggressive environment. Even so, each blade develops power equivalent to the engine of Formula 1 car. Before the replacement, the blade should pass 15 million miles. To create such a blade, it is necessary to integrate ~15 technologies, while ensuring economic efficiency [2].

A turbine blade has a complex external shape and an extremely complex cooling system, which are determined based on 3D-gasdynamic, heat transfer and strength methods, while taking into account non-stationary effects. A modern blade of a high-temperature turbine is made by casting of single-crystal nickel superalloys of the IV generation, containing up to ~10% of rhenium and ruthenium with optimal crystallographic orientation. To ensure an acceptable thermal state, the blade has an extensive perforation, usually obtained by laser or electroerosion methods. The blade has a protection against high temperature corrosion of inner cavities and a maintainable thermal barrier coating of the external surface, providing both protection against oxidation and reducing the temperature of material of the blade. High-pressure turbines of advanced engines will operate at an even higher gas temperature than modern turbines. It can be expected that to manufacture blades for these turbines, silicides of refractory metals will be used, which retain their operability at higher temperatures than modern nickel superalloys without using expensive and scarce alloying elements and also have lower density. To successfully create blades for advanced high-temperature turbines, new heat-resistant coatings will be required, which retain their efficiency at higher temperatures than modern coatings based on partially stabilized zirconium oxide.

One of the most important tasks for successful creation of advanced engines is the task of ensuring high cooling efficiency of turbine blades. The use of double-walled blades with transpiration cooling is considered promising. Complexification of the blade design is impossible without improving the methods of its manufacture. Of great interest is the use of additive technology (AT) for the manufacture of blades with complex cooling layouts.

The next possible stage of development is the creation of turbine blisks using dissimilar materials [4]. It is possible to exclude in blisk joints between blades and disk. As a result, it is possible to reduce weight of a structure, increase efficiency, increase reliability (in particular, to eliminate the problems associated with fretting in joints). Blisks are widely using in compressor rotors. At the same time, requirements for materials of turbine disk and blades are significantly different. Casting single-crystal nickel superalloys are using for manufacturing of turbine blades, while powder alloys are mainly using for turbine disks. Blisks with component parts made of different materials may now be produced by powder metallurgy (by compacting separately fabricated blades and disks by hot isostatic pressing (Figure 1)), and in the future - using AT. Development of repair technologies is of great importance for widespread introduction of turbine blisks made from dissimilar materials.

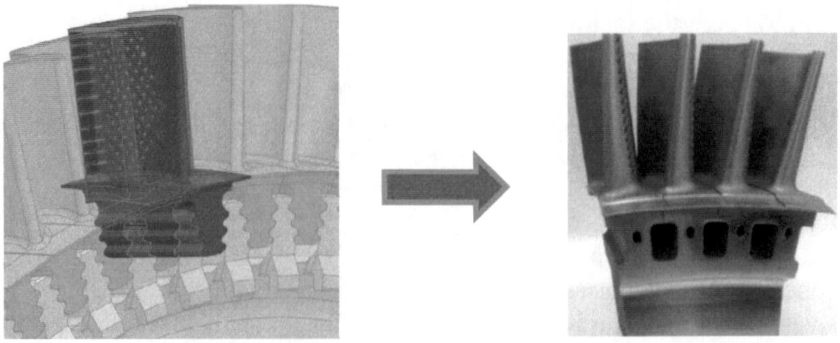

Figure 1. Section of bimetallic blisk of a turbine with cooled blades, developed in CIAM.

Without going into detail of all technologies needed to create a high-temperature turbine blade [2] with the required strength and reliability, we only note that all these technologies must be mutually linked and cost-effective. To develop these technologies, a multidisciplinary approach is required, and the extremely complicated process of creating a turbine blade depends on the structural and technological state of other engine parts, primarily other rotor and turbine stator parts.

The need for multidisciplinary approach is associated with a large number of often conflicting requirements for parts and components of a gas turbine engine. Some examples of recent activites in CIAM on multidisciplinary optimization of gas turbine engine components are given in [5, 6]. The role of robust design methods is also increasing. In addition, many of the requirements for the engine are in themselves of a complex nature. For example, in case of ingestion of bird flocks, hail, rain in the engine, there should be not only a hazardous failure (injection of fragments with large kinetic energy from engine case, non-localized fire, destruction of engine attachment points), but also unacceptable reduction of thrust. Therefore, in order to confirm compliance of an engine with this requirement, it is necessary both to develop methods for calculating blades damage when interacting with foreign objects entering a flow path, and methods for estimating engine thrust loss in the presence of blade damage. To increase an engine's resistance to the ingress of various foreign objects, it is necessary to develop and implement a large range of operational and design-technological measures. Of great importance, in particular, are development and implementation of methods for surface hardening of materials of fan and compressor blades.

New materials and design-technological solutions can be used in the engine only after reaching the VI level of technological readiness (on NASA scale) and the corresponding level of production readiness. Typically, the process of research of new structural material takes 10-15 years. All materials used in the engine must be carefully studied, which implies carrying out their special qualification - research of structural (realized in the design in the expected operating conditions) strength [7].

In advanced engines, we can expect usage of intermetallic alloys, alloys with a gradient of physical and mechanical properties; materials obtained using additive technology; smart materials. Various composite materials will find wide application.

Polymeric composite materials (PCM) are widely used for light fan blades (particularly on GE90, GENX, LEAP engines) and "cold" stator parts (including fan casings on GENX, LEAP). The use of these materials can significantly reduce engine weight, improve environmental performance (reduce noise level), improve reliability. Engine layout with an open fan rotor is one of the promising solutions that allow to significantly increase economic efficiency of an engine. However, there is no fan casing in this layout, the presence of which allows to reduce the criticality of a fan blade failure. Apparently, the only way to ensure operational safety is to confirm practical impossibility of destruction of a fan blade, and the most realistic way to implement this approach is to use a carbonplastic fan blade. Figure 2 shows a carbonplastic model of a fan blade prepared for testing on a spin rig.

Figure 2. The carbonplastic fan blade installed on a spin rig.

The use of metallic or intermetallic composite materials (primarily the use of materials with a titanium or titanium aluminide mould) for the manufacture of shafts, compressor blings, parts of engine attachments to an aircraft will also significantly reduce weight and improve reliability of engines. Over the years, activities were carried out on the use of monolithic ceramic materials for GTE parts [8]. These materials have a relatively low density and retain a sufficiently high strength to higher temperatures than nickel superalloys. However, the area of successful application of monolithic ceramic materials in gas turbine aircraft engines is limited mainly by elements of hybrid and ceramic bearings [9]. The use of hybrid rolling bearings (with ceramic rolling elements and metal rings) ensures an increase in life of bearings operating at high values of the parameter Dn (D – diameter between axes of rolling elements, n – angular velocity). The use of ceramic rolling bearings (with ceramic rolling elements, rings and a cage of, for example, carbon-carbon) can significantly reduce the requirements for lubrication and cooling of bearings, reduce weight. In this case, installation of ceramic rings on a metal shaft or in a metal housing should be carried out taking into account difference in coefficients of linear temperature expansion of the materials. To use materials based on ceramics for other engine parts, it was necessary to switch to ceramic composite materials (primarily silicon carbide reinforced with SiC fibers) with higher mechanical properties. For a long time, activities were also conducted on the use of carbon-carbon materials in GTE, which have high absolute and specific values of strength and rigidity at extremely high temperatures, and also a low coefficient of linear thermal expansion. However, to ensure reliable protection of parts, which were manufactured using these materials, from high-temperature oxidation, it turned out to be worthwhile to move from the carbon matrix to the ceramic matrix [8]. Currently, in the engines of the LEAP and GE9X families, ceramic composite materials, the use of which allows to reduce weight, increase fuel efficiency, and reduce the emission of harmful substances, have already found application. Figure 3 shows parts of a small-sized engine made of ceramic composite materials, which are under investigation in the CIAM.

a b

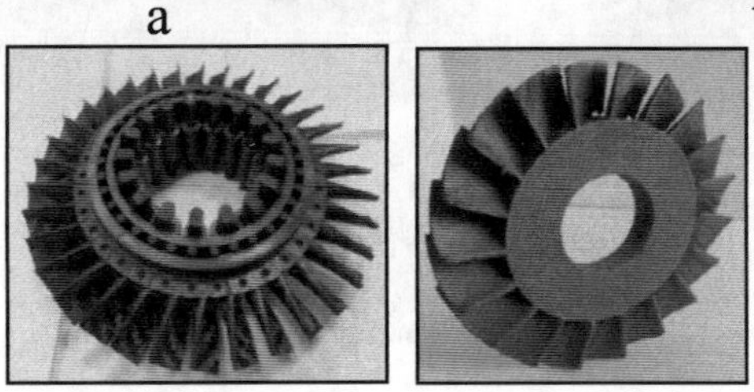

c

Figure 3. Parts made of ceramic composite materials.
a – turbine stator, b – flame tube, c – turbine wheel

Thorough activities on the creation of materials operable at extremely-high (over 1600 C) temperatures continue.

At the same time, special qualification of composite materials is much more complicated than in case of metallic alloys [2]. The anisotropy of physical and mechanical properties and structural irregularity of these materials necessitate a significant increase in the nomenclature of characteristics required for the design of components made from these materials and the number of test articles. Since the composite material is usually created for a particular component, it is not enough to determine characteristic values of a material realized in a component using simple shape samples, which is due to differences in structures of the material of the component and the sample, technologies for their production, and influence of scale effect. In addition, the characterization of composite materials is difficult due to specific properties of particular materials. For example, in the study of polymeric composite materials (PCM), great attention must be paid to obtaining data on the influence of process-dependent parameters and possible operational influences on the strength properties of materials. According to Boeing, when the PCM was introduced, the volume of tests had to be increased by a factor of 20 compared to using traditional alloys [10]. Advanced high-temperature materials must be tested at extremely high temperatures. This means that the test sample must be heated to the required temperature; clamped and loaded, providing centering; necessary measurements must be carried out. In this case, all the test equipment (heaters, screens, grippers, extensometers) can be made, for example, from refractory alloys (for testing in vacuum), carbon-carbon (for testing in a protective environment), ceramics or refractory alloys with coatings (for testing in air) [11]. It is also possible to carry out tests in air with samples located outside the heating zone, however, the thermal stresses arising in the test sample must be taken into account. It should also be noted that there is an insufficient number of standards that can be used for testing of samples made from composite materials.

Activities on the use of AT for manufacture and repair of aircraft engine parts are rapidly developing. AT allows you to obtain various parts that cannot be obtained by other methods, including parts with complex shapes, parts made from dissimilar materials, parts from materials with a gradient of properties. The use of AT in the manufacture and/or repair of aircraft engine parts can reduce the time and money spent on manufacturing/repair, reduce weight, reduce the number of parts and simplify the assembly, increase efficiency, durability and reliability.

CIAM (Magerramova L.A., et.al.) together with other enterprises (NIAT, Saturn, IL & ST, NPO System, Diaprom, SOVTEST ATE, etc.) generates scientific and technological know-how on the application of AT in design (including using topological optimization), manufacturing or repair of turbine blades with advanced cooling layouts (including double-wall blades), hollow turbine discs, hollow fan blades, compressors blisks, equipment for casting of cooled turbine blades (including ceramic rods), flame tubes sections of combustion chambers, front modules of combustion chambers, different stator parts (including cellular structures), ceramic parts of turbines. Some models of parts designed, manufactured with AT from metal alloys and undergoing studies in the CIAM are shown in Figure 4 [12,13].

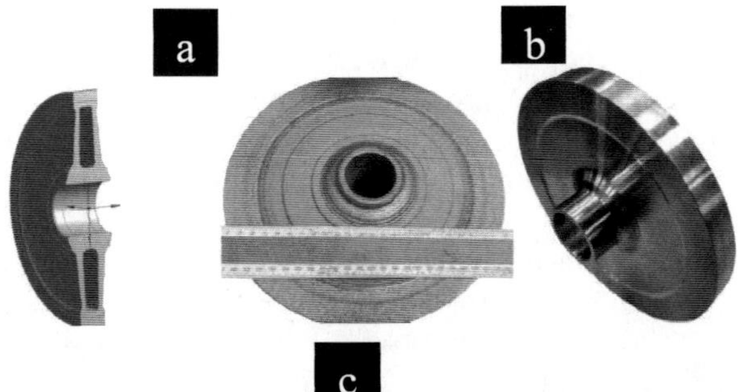

Figure 4. Some models of parts made from metal alloys, obtained with the help of AT
a – hollow fan blade, b – turbine blade, c – hollow turbine disk.

Reliable prediction of strength and longevity of aviation GTE parts does not lose its relevance. It is necessary to take into account complex mode of deformation, joint influence of various damaging factors, effect of a stress history on the strength. Widely applied principles of independent influence of damaging factors and a linear damage accumulation rule can give a significant error, and for the use of more accurate models a large amount of experimental data and application of probabilistic approaches are needed.

The task of prediction of longevity is simplified in the presence of one prevailing mechanism of damageability.

3. Prevention of fatigue failure.

For critical (main) engine parts of multi-engine aircraft, generally, the main damaging factor is low-cycle fatigue. In accordance with modern certification requirements, the life of these parts should be determined taking into account possible defects [11]. In this regard, currently, life of these parts is confirmed with the parallel use of two approaches - without taking into account initial defects (before formation of so-called engineering fatigue crack having a depth of ~0.4 and a length of ~0.8 mm), with an assessment of all hazardous areas of stress concentration and taking into account defects (based on a deterministic or probabilistic approach). Currently, certain metallurgical defects are necessarily taken into account (the presence of a solid α -phase in titanium alloys and ceramic inclusions in disks of powdered nickel alloys, porosity during casting or welding); technological defects (arising from production of holes in parts) and operational defects (from operational experience of similar builds). For the operation, the lesser of the obtained life values (with the reserves necessary to ensure operational safety) is allowed. The presence of large defects, dimensions of which are difficult to determine (unfavorable texture in heat-resistant titanium alloys, oxides and carbides along the boundaries of primary grains in powdered nickel alloys) should be eliminated by technological methods. The main methods of increasing longevity of parts, including taking into account possible presence of initial defects, are well known. However, in all these areas, research can and must continue. Much attention should also be paid to registration of the accumulated damageability for each part with subsequent evaluation of the residual life [16].

The main cause of the defect high appearing during operation of aircraft engines remains the high-cycle fatigue. This is due to the significant difficulty of predicting both the vibrational state of parts and the resistance of materials to high-cycle fatigue. When creating advanced engines, the problem of preventing destruction of parts from the high-cycle fatigue due to increased loading of turbomachines; the use of wide-chord blades with a dense frequency spectrum of natural frequencies; the use of builds with low damping; strengthening of coupling of vibrations; the growth of static stresses in parts and increase of temperature in these parts; increase in engine running time before removal for repair; the use of materials with reduced plasticity.

In recent years, a large amount of activities have been performed and continues to be carried out in the USA (as part of the special Air Force program, [17]), European countries and Russia, to prevent destruction from high-cycle fatigue, including prediction of vibratory stresses under forced oscillations, flutter analysis, non-contact methods of analysis of vibrations, investigation of permissible damages in the material, damping of vibrations, improvement of methods for determining resistance to high-cycle fatigue (study of the influence of stress cycle asymmetry on fatigue, the influence of other damaging factors, study of fatigue resistance of titanium alloys in the gigacyclic region, fatigue mechanisms of single-crystals, etc.), development of strengthening methods, development of robust design methods, etc. As an example of activities aimed at preventing destruction from high-cycle fatigue, Figure 5 shows a diagram of a system for diagnosing vibrations of turbomachine blades developed at CIAM and protected by 7 patents of Russian Federation. Sensors include gas pressure pulsation sensors, vibration sensors for stator parts, sensors for alteration of radial clearance in turbomachines. Signal processing is carried out using fast Fourier transformation, wavelet transformation, Prony's method. Currently, most studies of the vibrational state of turbomachine blades and resonant tests are carried out in Russia using these methods (see, for example, [18]). Systems for the non-contact vibration diagnostics of the blades (tip timing type) also find wide application in laboratory testing of engines, including dynamic spin rigs [19].

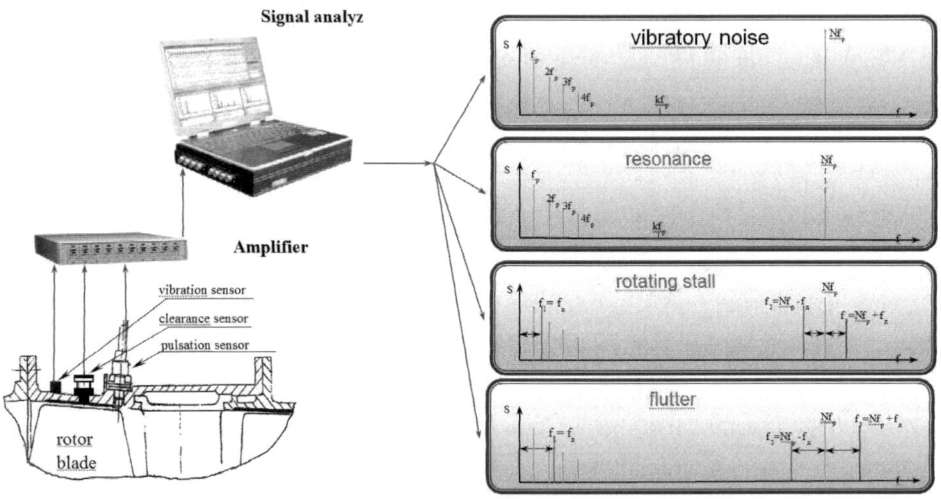

Figure 5. The non-contact system of vibration diagnostics of turbomachine blades in real time. $F_d = f_b \pm mf_r$, where N – the number of blades; m – number of deformation waves; f_r – rotor frequency; f_b – blade frequency

To ensure the reliability of aircraft engines, it is of great importance to develop methods for complex multi-parameter diagnostics of technical condition utilizing remote diagnostic centers. In recent years, there have been significant achievements in this direction. As an example, methods and means of diagnosing the technical condition of aircraft drives developed by CIAM [20] can be cited. Especially great opportunities are revealed by using methods and means of determining the kinematic error of gearing.

The task of ensuring strength reliability remains actual at the development of advanced aircraft engines, and the complexity of its solution increases for each new generation of engines.

Acknowledgments
I'd like to acknowledge the CIAM for the help in work organization and permission for publication.

References

[1] BabkinV.I., Lanshin A.I., Polev A.S., Fedyakin F.N. *Creation of Competitive Aircraft Engines of the Years 2025-2030*. Interindustry almanac, 49, 2015, pp.25-29.

[2] Nozhnitsky Yu. A., Baluev B.A., Fedina Yu.A. Experimental Investigation of Strength Reliability of Modern Gas Turbine Engines. Vestnik of the Ufa State Aerospace University, V.19. №3(69), 2015, pp.4-14.

[3] Value – the Sum of All Parts. The Magazine, Rolls-Royce, Issue 128, March 2011, pp.4-6.

[4] Magerramova L.A., Nigmatullin R., Vasilyev B., Kinzburskiy V. Design of a Bimetallic Blisk Turbine for a Gas Turbine Engine and Its Production Using Power Metallurgy Methods; ASME Paper GT2017-63560, 10p.

[5] Salnikov A., Danilov M. A Centrifugal Compressor Impeller: A Multidisciplinary Optimization to Improve Its Mass, Strength and Gas-Dynamic Characteristics. ASME Paper GT2017-64123, 10 p.

[6] Buyukli T., Salnikov A., Fedorchenko Yu. High-Loaded Compressor Blisk-type Impeller Multidisciplinary Optimization, ASME Paper GT2017-64116, 12p.

[7] Nozhnitsky Yu.A., Kachanov E.B., Golubovsky E.R., Kouevda V.K. Requirements for the Order and the Procedure of Estimating Structural Strength Characteristics of Metal Alloys for the Main and Critical Parts of Aviation Gas Turbine Engines. Vestnik of the Samara State Aerospace University, V.14, №3, part 1, 2015, pp.37-48.

[8] Nozhnitsky Yu.,A., Soudarev A.V. Evolution of Programs for Ceramic Gas Turbine Engine Design in the States of the Former Soviet Union. In "Ceramic Gas Turbine Design and Test Experience", Ed.M.Van Roode, M.K. Ferber, D.W. Richerson, ASME Press, 2002, pp.661-668.

[9] Petrov N.I., Lavrentyev Yu.L. Ways of Increasing the Reliability and Service Life of Bearing Assemblies of Modern Gas-Turbine Engines and Gears. Vestnik of the Samara State Aerospace University, V.14, №3, part 1, 2015, pp.228-237.

[10] Karimbaev T.D., Mezentsev M.A., Ezhov A.J. Engineering Design and Experimental Research of Non-metallic Parts and Components of the Hot Section of the Advanced Gas Turbine Engine. Vestnik of the Samara State Aerospace University, V.14, №3, part 1, 2015, pp.128-138.

[11] Warwick G. NASA-Led Consortium Will Bring Science to Art of Composites. Aviation Week and Space Technology, 2015-04-27, pp.1-4.

[12] Nozhnitsky Yu.A., Smirnov L.E. Ceramic, Carbon-carbon and other Composite Materials Tests High Temperature. In "Ultra High Temperature Mechanical Testing", Ed. By R.D. Lohr and M.Steen, Woodhead Publishing Limited, Cambridge, 1995, pp.184-192.

[13] Magerramova L.A., Nozhnitsky Yu.A., Vasilyev B.E., Kinzbursky V.S. The Use of Additive Technologies for Production of Advanced Gas-Turbine Engine Components. Technology of Light Alloys. 2015, №4, pp.7-13.

[14] Magerramova L.A., Vasilyev B.E., Kinzburskiy V.S. Novel Designs of Turbine Blades for Additive Manufacturing. ASME Paper GT2016-56084, 7p.

[15] Nozhnitsky Yu. A., Lokshtanov E.A., Dolgopolov I.N., Shashurin V.G., Volkov M.E., Ganelin I.I. Probabilistic Prediction of Aviation Engine Critical Parts Lifetime; ASME Paper GT2006-91350, 10p.

[16] Tumanov N.V., Lavrenyeva M.A., Vorobyeva N.A., Kalashnikova A.I. Application of the Cyclic Life Additivity Concept for the Determination of Service Life of Aeroengine Discs Made from Powder Alloys. Vestnik of the Samara State Aerospace University, V.14, №3, part 1, 2015, pp.49-59.

[17] Proceeding of the 10-th National Turbine Engine HCF Conference, March 8-11, 2005, New Orlean, USA.

[18] Khorikov A.A., Danilkin S.Yu., Mazikina T.I. Early Detection of Vibration Behaviour of Aircraft Engine Multistage Axial Compressor in Pre-Surge States During Rig Testing; Vestnik of the Samara State Aerospace University, V.14, №3, part 1, 2015, pp.148-159.

[19] Nozhnitsky Yu.A., Fedina Yu.A., Shadrin D.V., Servetnik A.N., Baluev B.A., Kanachkin A.V., Lepeshkin A.r., Tomashev A.A., Chernyshev S.A. New Possibilities of Using Spin Rigs to Provide Gas Turbine Engine Strength Reliability. Vestnik of the Samara State Aerospace University, V.14, №3, part 1, 2015, pp.71-87.

[20] Golovanov V.V., Vasilenko V.G., Zemskov A.A., Panov S.S., Emelianova A.A. *Diagnostic Methods and Tools for Condition-Based Maintenance of Aircraft Drives*. Vestnik of the Samara State Aerospace University, V.14, №3, part 1, 2015, pp.213-221.

Institute of Physics
Dirac House, Temple Back
Bristol BS1 6BE UK

ISSN: 1757-8981
ISBN 978-1-5108-5839-8

CURRAN ASSOCIATES INC.
proceedings.com

International Conference on Applied Sciences (ICAS2017)

IOP Conference Series: Materials Science and Engineering
Volume 294

Hunedoara, Romania
10 – 12 May 2017

IOP
Institute of Physics